D1257708

UNIVERSITY OF W
DISCARDED
ortage Avenue
Win eg, Manitoba R3B 2E9

Biology of the House Mouse

QL
1
.Z733
no.47

SYMPOSIA OF THE ZOOLOGICAL SOCIETY OF LONDON
NUMBER 47

Biology of the House Mouse

*(The Proceedings of a Symposium held at
The Zoological Society of London
on 22 and 23 November 1979)*

Edited by

R. J. BERRY

*Department of Zoology, University College London,
London, U.K.*

Published for

THE ZOOLOGICAL SOCIETY OF LONDON

BY

ACADEMIC PRESS

1981

ACADEMIC PRESS INC. (LONDON) LTD
24/28 Oval Road, London NW1 7DX

United States Edition published by
ACADEMIC PRESS INC.
111 Fifth Avenue, New York, New York, 10003

Copyright © 1981 by
THE ZOOLOGICAL SOCIETY OF LONDON
All rights reserved. No part of this book may be reproduced in any form by photostat, microfilm, or any other means, without written permission from the publishers

British Library Cataloguing in Publication Data

Biology of the house mouse. – (Symposia of the Zoological Society of London, ISSN 0084-5612; 47)
1. Mice – Congresses
I. Berry, R.J. II. Series
599.32'34 QL737.R666 74-5683
ISBN 0-12-613347-6

Printed in Great Britain at the Alden Press
Oxford London and Northampton

Contributors

BELLAMY, D., *Department of Zoology, University College, Cardiff, P.O. Box 78, Cardiff CF1 1XL, UK* (p. 267)

BERRY, R. J., *Department of Zoology, University College, Gower Street, London WC1E 6BT, UK* (p. 395)

BLACKWELL, J. M., *The Ross Institute, London School of Hygiene and Tropical Medicine, Keppel Street, London WC1E 7HT, UK* (p. 591)

BONHOMME, F., *Université de Montpellier 2, Laboratoire d'Evolution des Vertébrés, Place Eugène-Bataillon, F-34060 Montpellier Cedex, France* (p. 27)

BRITTON-DAVIDIAN, J., *Université de Montpellier 2, Laboratoire d'Evolution des Vertébrés, Place Eugène-Bataillon, F-34060 Montpellier Cedex, France* (p. 27)

BROTHWELL, D. R., *Institute of Archaeology, 31–34 Gordon Square, London WC1H 0PY, UK* (p. 1)

BULFIELD, G., *Department of Genetics, University of Leicester, University Road, Leicester LE1 7RH, UK* (p. 643)

DEOL, M. S., *Department of Genetics & Biometry, University College, Gower Street, London WC1E 6BT, UK* (p. 617)

EVANS, E. P., *The William Dunn School of Pathology, University of Oxford, South Parks Road, Oxford OX1 3RE, UK* (p. 127)

FESTING, M. F. W., *MRC Laboratory Animals Centre, Woodmansterne Road, Carshalton, Surrey SM5 4EF, UK* (p. 43)

GÖTZE, D., *Abteilung Immungenetik, Max-Planck Institut für Biologie, Corrensstrasse 42, 7400 Tübingen 1, West Germany* (p. 439)

GROPP, A., *Abteilung für Pathologie, Klinikum der Medizinischen Hochschule, Ratzeburger Allee 160, 24 Lübeck, West Germany* (p. 141)

JAKOBSON, M. E., *Department of Biology, North East London Polytechnic, Romford Road, London E15 4LZ, UK* (p. 301)

KLEIN, J., *Abteilung Immungenetik, Max-Planck Institut für Biologie, Corrensstrasse 42, 7400 Tübingen 1, West Germany* (p. 439)

LONGSTRETH, J. D., *Laboratory of Microbial Immunity, National Institutes of Allergy and Infectious Diseases, National Institutes of Health, Bethesda, Maryland 20205, USA* (p. 627)

LOVELL, D. P., *MRC Laboratory Animals Centre, Woodmansterne Road, Carshalton, Surrey SM5 4EF, UK* (p. 43)

LUSH, I. E., *Department of Genetics and Biometry, University College, Gower Street, London WC1E 6BT, UK* (p. 517)

LYON, M. F., FRS, *MRC Radiobiology Unit, Harwell, Oxon OX11 0RD, UK* (p. 455)

MACKINTOSH, J. H., *Sub-Department of Ethology, Birmingham University Medical School, Birmingham B15 2TJ, UK* (p. 337)

MARSHALL, J. T., *National Fish and Wildlife Laboratory, National Museum of Natural History, Washington, D.C. 20560, USA* (p. 15)

MORSE, H. C., III. *Laboratory of Microbial Immunity, National Institutes of Allergy and Infectious Diseases, National Institutes of Health, Bethesda, Maryland 20205, USA* (p. 627)

NADEAU, J. H., *Abteilung Immungenetik, Max-Planck Institut für Biologie, Corrensstrasse 42, 7400 Tübingen 1, West Germany* (p. 439)

PELIKÁN, J., *Institute of Vertebrate Zoology, Czechoslovak Academy of Sciences, Kvetna 8, Brno, Czechoslovakia* (p. 205)

PETERS, J., *MRC Radiobiology Unit, Harwell, Oxon OX11 0RD, UK* (p. 479)

ROBERTS, R. C., *Institute of Animal Genetics, West Mains Road, Edinburgh EH9 3JN, UK* (p. 231)

ROWE, F. P., *MAFF, Rodent Research Branch, Pest Infestation Control Tolworth Laboratory, Hook Rise South, Tolworth, Surbiton, Surrey KT6 7NF, UK* (p. 575)

SAGE, R. D., *Museum of Vertebrate Zoology, 2593 Life Science Building, Berkeley, California 94720, USA* (p. 15)

SEARLE, A. G., *MRC Radiobiology Unit, Harwell, Oxon OX1 0RD, UK* (p. 63)

SHIRE, J. G. M., *Department of Biology, University of Essex, Wivenhoe Park, Colchester CO4 3SQ, UK* (p. 547)

SMITH, J. C., *MAFF, Rodent Research Branch, Pest Infestation Control Tolworth Laboratory, Hook Rise South, Tolworth, Surbiton, Surrey KT6 7NF, UK* (p. 367)

THALER, L., *Université de Montpellier 2, Laboratoire d'Evolution des Vertébrés, Place Eugène-Bataillon, F-34060 Montpellier Cedex, France* (p. 27)

THORPE, R. S., *Department of Zoology, University of Aberdeen, Tillydrone Avenue, Aberdeen AB9 2TN, UK* (p. 85)

WAKELAND, E. K., *Abteilung Immungenetik, Max-Planck Institut für Biologie, Corrensstrasse 42, 7400 Tübingen 1, West Germany* (p. 439)

WALKOWA, W., *Institute of Ecology, Polish Academy of Sciences, 05-150 Dziekanow Lesny, Komianki, Near Warsaw, Poland* (p.427)

WALLACE, M. E., *Department of Genetics, University of Cambridge, Downing Street, Cambridge CB2 3EH, UK* (p. 183)

WARD, R. J., *Toxicology Department, May & Baker Limited, Rainham Road South, Dagenham, Essex RM10 7XS, UK* (p. 255)

WINKING, H., *Abteilung für Pathologie, Klinikum der Medizinischen Hochschule, Ratzeburger Allee 160, 24 Lübeck, West Germany* (p. 141)

Organizer and Chairmen of Sessions

ORGANIZER

R. J. BERRY, on behalf of the Zoological Society of London

CHAIRMEN OF SESSIONS

R. J. BERRY, *Department of Zoology, University College, Gower Street, London WC1E 6BT, UK*
H. GRÜNEBERG, FRS, *Department of Genetics and Biometry, University College, Gower Street, London WC1E 6BT, UK*
A. McLAREN, FRS, *MRC Mammalian Development Unit, Wolfson House, 4 Stephenson Way, London NW1 2HE, UK*
N. A. MITCHISON, FRS, *Department of Zoology, University College, Gower Street, London WC1E 6BT, UK*

Linkage Map of the Mouse

Compiled by

MURIEL DAVISSON AND T. H. RODERICK

Jackson Laboratory, Bar Harbor, Maine, USA

This map represents a cumulative summary of linkage information in the mouse since the first linkage group was established in 1915. Solid black lines depict the relative lengths of the individual chromosomes based on their proportional cytological lengths and an estimated total genetical length of 1600 cM. Hatched lines extend those chromosomes whose current mapped length exceeds the relative estimated length. Chromosome numbers are shown above the centromeres (represented by knobs). Gene symbols are given at the right of each chromosome and recombination distances at the left. These distances are based on classical genetical crosses, and genes which have been mapped by parasexual methods are given at the bottom of their respective chromosomes. Distances between the first marker and the centromere are given in parentheses when they have been determined using a Robertsonian chromosome, because they are often underestimated. When this distance is unknown the linkage group is centred on the chromosome. The relative certainty of the positioning of loci is indicated by the lines which show their location on the chromosome. Skeletal markers, those loci whose position is best established, are indicated by the longest lines; and loci with no lines are the least well positioned. Non-italicized loci are those whose order is uncertain and loci within brackets are ordered with respect to other genes but not each other. A complete listing of linkage data used in compiling the map is available in Davisson & Roderick (*In* Green, M. C. (*Ed.*) *Genetic variants and strains of the inbred mouse.* Stuttgart: Gustav Fischer Verlag (1981)). When publishing linkage information please include the following: type of cross, phase of the loci (coupling or repulsion) when appropriate, sex of the F_1 parent in backcrosses, number of recombinants, total number observed, and whether the data or parts of it have appeared before, including in *Mouse News Letter.*

Preparation of the linkage map of the mouse is supported by grants 75-03397 from the National Science Foundation and VC 17 from the American Cancer Society.

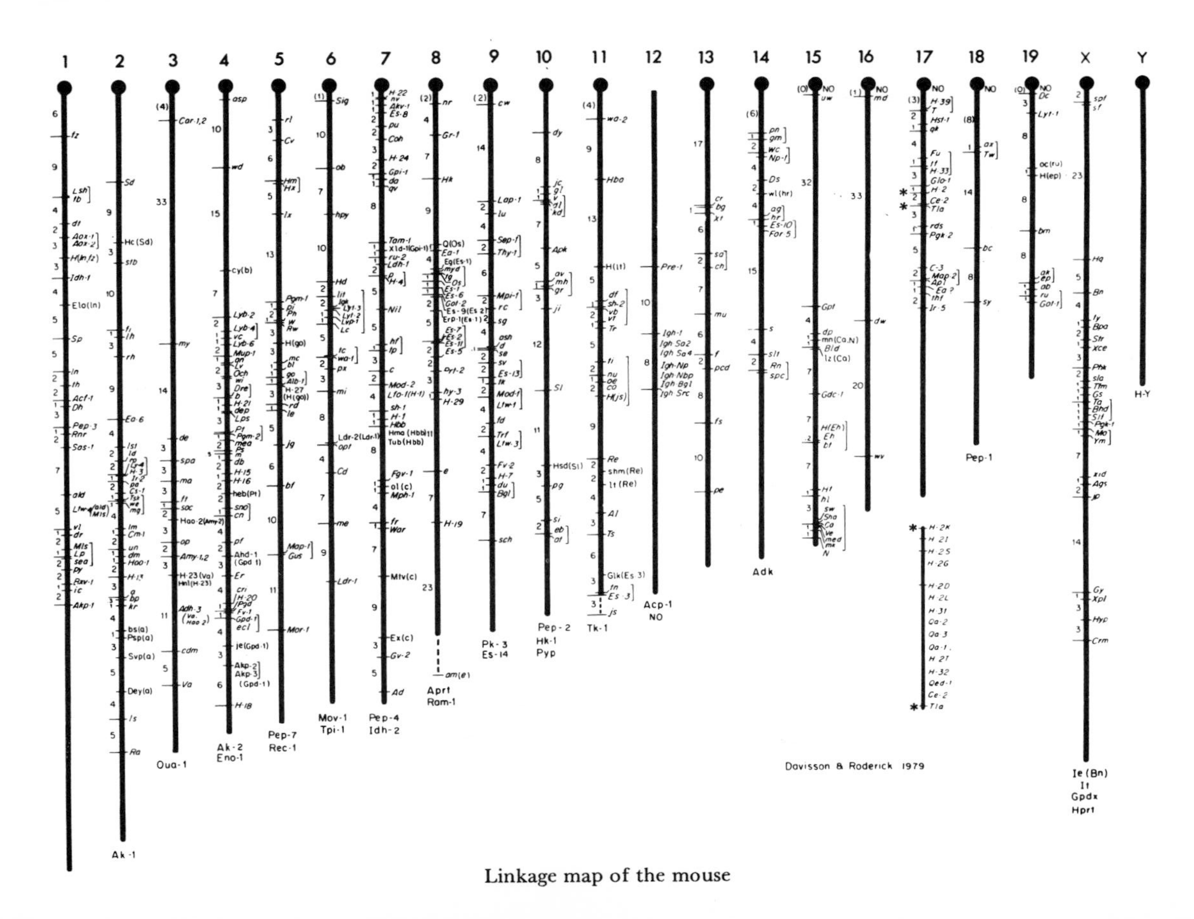

Linkage map of the mouse

Chromosome Map of the Mouse

Compiled by

A. G. SEARLE AND C. V. BEECHEY

MRC Radiobiology Unit, Harwell, Oxon, UK

The map shows approximate positions of breakpoints in inversions (In) and translocations (T), obtained from data given in *Mouse News Letter* or published elsewhere. The symbols for these are on the right of the chromosomes concerned, while symbols for some key marker loci (especially end markers) are given on the left. Positions given are often a compromise between those obtained by genetical and by cytological methods. Loci etc. at the distal end are not always the correct map distance from their neighbours because of scaling problems, shown by a distal break in the chromosome line. A proximal break indicates that the distance from centromere to nearest locus is not known yet. Robertsonian translocations include all laboratory-derived but not all wild-derived ones (q.v. chapter by Gropp & Winking, this volume pp. 141–181).

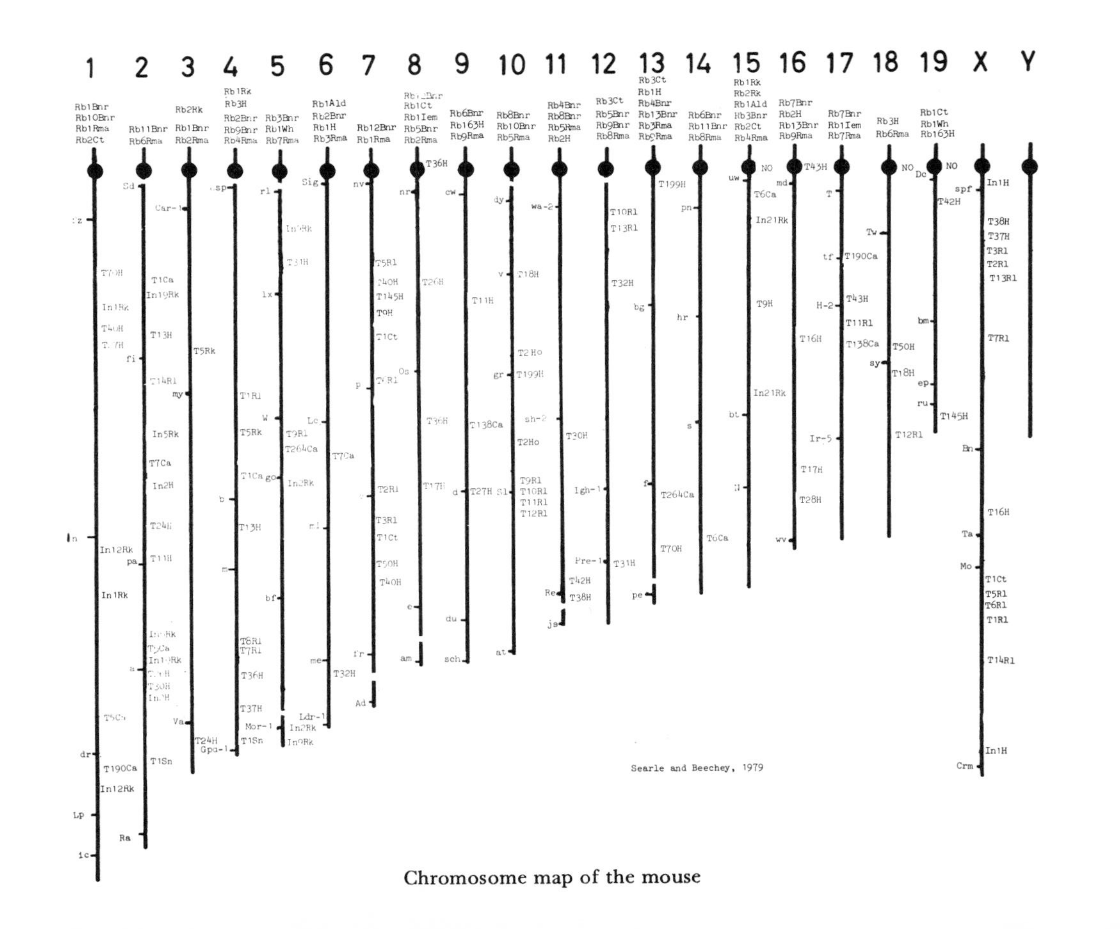

Chromosome map of the mouse

Preface

Probably more is known about house mice than any other mammal except possibly man. They have lived with and been kept by us for many centuries; and have become such a necessary research tool for many biologists that they are bought "off the shelf" from mouse breederies as if they were chemicals. Yet mice are more than commensal nuisances or laboratory material: they are real animals with social, physiological, and pathological stresses like any other mammal, including man. This means that wild mice are more relevant to biomedical research than many laboratory scientists like to think. After all, humans are not yet completely confined to sterile little boxes and fed on artificial food. Wild house mice provide a logical step in interpreting results obtained in the artificiality of the laboratory environment to the confusing heterogeneity of human life. Bronson (1979) has commented that "as a generality, we may be spending considerable time in the laboratory studying artifacts of utopian and constant conditions".

But as studies on wild mice are relevant to laboratory workers, so laboratory studies are relevant to field workers. If the information about gene action, behaviour and communications, physiological reactions, nutrition, genetical responses, and the like, which has been won in the laboratory could be used by ecologists, we would have an unparalleled understanding of a wild-living mammal.

This volume brings together contributions from field and laboratory scientists with the intention that each should complement the other. There have been many notable and comprehensive accounts of laboratory mice, beginning with Keeler's *The laboratory mouse* (1931), through the Jackson Laboratory's *Biology of the laboratory mouse* (Snell, 1941; Green, 1966) and Grüneberg's *Genetics of the mouse* (1943, 1952), to recent compilations such as *Origins of inbred mice* (1978, edited by Morse) and the Federation of American Societies for Experimental Biology's *Handbook on the mouse and rat* (1979), as well as books on specific aspects of mouse biology: Simmons & Brick on *The laboratory mouse: selection and management* (1970), Theiler on *The house mouse* (1972, almost entirely reproductive biology), Klein on the *Biology of the mouse histocompatibility-2 complex* (1975), and Silvers on *The coat colors of mice* (1979). In addition a twice yearly *Mouse News Letter* has been published since 1949; the mouse fancy has its own literature; and a number of semi-popular books have been written, such as Lauber's story of the Jackson Laboratory *Of man and mouse* (1971).

In contrast, the only book entirely devoted to non-laboratory mice is Crowcroft's delightful *Mice all over* (1966) (see also Berry, 1970), although general books on mammals inevitably draw extensively on house mouse work (e.g. Berry & Southern, 1970; Petrusewicz & Ryszkowski, 1970), and much information on mouse biology is contained in studies of rodent control (Hinton, 1918; Southern, 1954). Perhaps the only attempt before the present to recognize the value of linking laboratory with field interpretations has been *Contributions to behavior – genetic analysis. The mouse as a prototype* (Lindzey & Thiessen, 1970).

Most of the papers in this volume were given at a meeting held at the Zoological Society of London on 22 and 23 November, 1979 (Berry, 1980). A two-day meeting was far too short to review adequately house mouse biology, and several papers not delivered during the meeting have been added to this published version, including two from the East European ecological tradition (Drs Pelikán and Walkowa). There are differences of emphasis between authors; for example Festing & Lovell argue the danger of concentrating too much on wild mice, while other authors see great possibilities in supplementing inbred strains in the laboratory with mice caught in the wild. The main omissions are in molecular biology (which is only just beginning, as far as the mouse is concerned) and reproductive physiology (which has been excellently reviewed by Bronson, 1979). My thanks go to all who made the meeting a success and this book possible, particularly Dr H. G. Vevers and Miss Unity McDonnell of the Zoological Society of London, and Miss Janet Anderson for secretarial assistance. I am grateful to Drs Davisson, Roderick and Searle and Mr Beechey for permission to reproduce their gene and chromosome maps, which precede this Preface. The index has been compiled by Dr Miriam Harris.

This book covers (a) the relevance of studies on laboratory mice maintained under strictly controlled conditions of both genotype and environment to the conditions of less restrained life (in both mouse and man); (b) the potential contributions of natural populations of house mice to understanding inherited variation and congenital disease in man; and (c) the worthwhileness of collecting "new" inherited variants from wild mice as models and/or homologues of human conditions. My hope is that it will complement the Jackson Laboratory's *Biology of the laboratory mouse,* and be a complement to such works as the American Society of Mammalogists' *Biology of Peromyscus* (King, 1968).

Time will tell whether this will help towards a growing together of laboratory and field workers, and produce on the one hand a diminution in the absurd reductionism that too often characterizes laboratory-based biology, and on the other a transfer of critical thinking, concepts and techniques to singe some of the woolliness from ecological investigation.

REFERENCES

Berry, R. J. (1970). The natural history of the house mouse. *Fld Stud.* **3**: 219–262.

Berry, R. J. (1980). The great mouse festival. *Nature, Lond.* **283**: 15–16.

Berry, R. J. & Southern, H. N. (Eds) (1970). *Variation in mammalian populations.* London & New York: Academic Press.

Bronson, F. H. (1979). The reproductive ecology of the house mouse. *Q. Rev. Biol.* **54**: 265–299.

Crowcroft, P. (1966). *Mice all over.* London: Foulis.

Federation of American Societies for Experimental Biology. (1979). *Biological handbook. 3. Inbred and genetically defined strains of laboratory animals.* Bethesda: FASEB.

Green, E. L. (Ed.) (1966). *Biology of the laboratory mouse,* 2nd edn. New York & London: McGraw-Hill.

Grüneberg, H. (1943). *The genetics of the mouse.* Cambridge: Cambridge University Press.

Grüneberg, H. (1952). *The genetics of the mouse,* 2nd edn. The Hague: Nijhoff. (also *Bibliogr. genet.* **15**: 1–650.)

Hinton, M. A. C. (1918). *Rats and mice as enemies of mankind.* London: British Museum (Natural History).

Keeler, C. E. (1931). *The laboratory mouse.* Cambridge, Mass: Harvard U.P.

King, J. A. (Ed.) (1968). *Biology of* Peromyscus *(Rodentia).* Oklahoma: American Society of Mammalogists.

Klein, J. (1975). *Biology of the mouse histocompatibility-2 complex.* Berlin & New York: Springer-Verlag.

Lauber, P. (1971). *Of man and mouse.* New York: Viking.

Lindzey, G. & Thiessen, D. D. (Eds). (1970). *Contributions to behavior–genetic analysis. The mouse as a prototype.* New York: Appleton-Century-Crofts.

Morse, H. C. (Ed.) (1978). *Origins of inbred mice.* New York & London: Academic Press.

Petrusewicz, K. & Ryszkowski, L. (Eds) (1970). *Energy flow through small mammal populations.* Warsaw: Polish Scientific Publishers.

Silvers, W. K. (1979). *The coat colors of mice.* New York & Berlin: Springer-Verlag.

Simmons, M. L. & Brick, J. O. (1970). *The laboratory mouse: selection and management.* Englewood Cliffs: Prentice-Hall.

Snell, G. D. (Ed.) (1941). *Biology of the laboratory mouse.* New York: McGraw-Hill.

Southern, H. N. (Ed.) (1954). *Control of rats and mice* 3. *House mice.* Oxford: Oxford University Press.

Theiler, K. (1972). *The house mouse.* Berlin & New York: Springer-Verlag.

London, February 1981 R. J. BERRY

Contents

The Breeding, Inbreeding and Management of Wild Mice

MARGARET E. WALLACE

Patterns of Reproduction in the House Mouse

J. PELIKÁN

Genetical Influences on Growth and Fertility

R. C. ROBERTS

Diet and Nutrition

R. J. WARD

Ageing: with Particular Reference to the Use of the House Mouse as a Mammalian Model

D. BELLAMY

Physiological Adaptability: the Response of the House Mouse to Variations in the Environment

M. E. JAKOBSON

Behaviour of the House Mouse

J. H. MACKINTOSH

Senses and Communication

JANE C. SMITH

Population Dynamics of the House Mouse

R. J. BERRY

Structure, Dynamics and Productivity of Mouse Populations: A Review of Studies Conducted at the Institute of Ecology, Polish Academy of Sciences

WIERA WALKOWA

Population Immunogenetics of Murine *H-2* and *t* Systems

J. KLEIN, D. GÖTZE, J. H. NADEAU and E. K. WAKELAND

The *t*-Complex and the Genetical Control of Development

MARY F. LYON

Enzyme and Protein Polymorphism

JOSEPHINE PETERS

Mouse Pharmacogenetics

I. E. LUSH

Genes and Hormones in Mice

J. G. M. SHIRE

Wild House Mouse Biology and Control

F. P. ROWE

The Role of the House Mouse in Disease and Zoonoses

JENEFER M. BLACKWELL

Mutant Mice as Models for Human Genetical Deafness

M. S. DEOL

Expression of Murine Leukemia Viruses in Inbred Strains of Mice

JANICE D. LONGSTRETH and H. C. MORSE III

Inborn Errors of Metabolism in the Mouse

G. BULFIELD

Symp. zool. Soc. Lond. (1981) No. 47, 1–13

The Pleistocene and Holocene Archaeology of the House Mouse and Related Species

D. BROTHWELL

Institute of Archaeology, Gordon Square, London, UK

SYNOPSIS

Following the evolutionary divergence of murid lines by the Miocene, a number of species of *Mus* are clearly in evidence by Pleistocene times. Varieties of *Mus*, usually based on fragmentary evidence, are known from Europe, Africa and Asia at that time. The earliest association of the house mouse with an early urban community is at neolithic Çatal Hüyük in Turkey (*c.* 6500–5650 BC). Later, evidence of mouse infestation of towns is found at the sites of Kahun and Buhen in Egypt. Possibly the development of large urban complexes in the Classical world encouraged the spread of these small mammals into the western Mediterranean. House mice seem to have been established by the Iberian Bronze Age, with some evidence of their occurrence north in later Bronze Age levels in Holland (1220–820 BC). Movement into Britain may not have been until Iron Age times, but perhaps with no wide distribution before Saxon-Mediaeval phases.

Further distribution throughout the world may well have been initiated by the early European voyagers of the 15th–16th centuries.

ORIGINS

Although a number of major groups of mammals such as the carnivores, artiodactyls and primates showed considerable adaptive initiative by the Eocene, the rodents may best be regarded as creatures of the Oligocene and beyond. Since the Oligocene, they have become the most successful of all phylogenetic lines with maximum speciation occurring from the Pliocene onwards, and population numbers generally becoming greater than for any other mammal. Smallness of size, rapid rate of breeding and adaptiveness to changing ecological conditions have clearly assured their evolutionary success over the past 30 million years. In a recent review by Chaline (1977) the cricetine and murid lines are shown as diverging from each other by the Miocene. The emergence of different species of *Mus* could have occurred some time in the Pliocene. Compared with the voles, the Muridae have so far been of little interest in Quaternary environmental studies, but there is now a growing amount of archaeological material.

PLEISTOCENE MICE

By the Lower Pleistocene, an African variety of mouse, *Mus petteri*, with a size similar to *M. musculus*, occurs at the hominid locality of Olduvai. This appears to be a somewhat distinctive form, not closely associated with recent mice. Jaeger (1976) states that the Oldovan murids in general appear to have undergone few changes, perhaps as a result of the lack of pronounced environmental changes in this area during the Pleistocene. There is some similarity between this Lower Pleistocene variety of *Mus* at Olduvai and specimens from northern Africa, but Jaeger considers the species of southern Europe to be somewhat different.

Of Middle Pleistocene date, and also associated with a hominid site, are skeletal remains ascribed to the species *Mus musculus* from Choukoutien in China (Pei, 1936, 1939). Further Middle Pleistocene evidence of this species is also seen in western Asia and Europe, at Binagady in the Caucasus (Vereshchagin, 1967), in travertine deposits near Budapest (Jánossy, 1962) and at Chios in Greece (Storch, 1975). It may also occur (certainly a species of the genus *Mus*) at Ubeidiya in Israel (Haas, 1966). In view of the very fragmentary nature of fossil material considered to belong to this genus, species names can only be regarded as tentative symbols of possible evolutionary relationships, and I would not wish to imply other than this in the present brief review.

Upper Pleistocene evidence is a little more common, but there are in some instances questions of dating and of nomenclature. For instance, although Kurten (1968) accepts the possible Eemian dating of *M. musculus* from Kirkdale Cave in Yorkshire, Sutcliffe & Kowalski (1976) exclude the genus from small mammal faunas of the British Isles until Flandrian times. Problems of identification and nomenclature are exemplified by the transfer of the so-called *Mus piletus* remains from Pleistocene deposits near Palermo in Italy to that of the Maltese dormouse, *Leithia melitensis* (Bate, 1942a). Similarly, although Bate (1942b) separates the Palestinian *M. camini* from *M. musculus* on size, relative lengths of the molars and proportions of the mandible, Tchernov (1975) appears to accept all Upper Pleistocene material as *M. musculus*. He does, however, note some slight intraspecific variation during this time period, in the tuberculation of the molars and in the proportions of the skull.

Other claims of *Mus* species in the Upper Pleistocene include *M. minotaurus*, a large variety from Crete with a relatively short, robust snout and short, widely-curving upper incisors — with only minor notching on the outer cutting edge (Bate, 1942a). Fossil murid mice with size variation overlapping that of *M. musculus* are also known from Pleistocene deposits in Cyprus (Boekschoten & Sondaar, 1972). Conditions for the preservation of small mammals are of course by no means ideal in some areas, so that late Pleistocene *Mus* in India has so far been found only in the southern cave areas of Andhra Pradesh (Murty, 1975).

ARCHAEOLOGICAL MATERIAL

From the Holocene, far more evidence of sub-fossil varieties of *Mus* are known. Archaeological material has been found in such divergent localities as the Caucasus (Vereshchagin, 1967), Malta (Storch, 1970, 1972), Mallorca (Uerpmann, 1970), the Iberian site of Zambujal (Storch & Uerpmann, 1976), northern Germany (Reichstein, 1974) and Britain (Corbet, 1974). It is not my intention here to try and review all Holocene sites producing the house mouse or related species, but rather to concentrate on certain material. Beyond the Pleistocene, one of the most interesting questions is of course the antiquity of the association between mouse and man. As Haim & Tchernov (1974) point out in their study of rodents in the Sinai Peninsula, *M. musculus* has evolved a great versatility for occupying different biotopes from hot halophytic zones to oases, farms and cities. In this respect, the development of agriculture on the part of various human communities, and the concentration of some proportion of these populations into towns and cities by 6000 BC, provided an entirely new kind of environment for rodents to move into, and some have clearly accepted the adaptive challenge. It may well be, of course, that human habitations provided greater protection from predators than alternative habitats, provided that behavioural characteristics allowed a certain tolerance of humans!

In the Old World, an area of special importance both from the point of view of urban development and agriculture has been southwest Asia. Of special interest as regards the appearance of the house mouse on urban sites is the early Turkish town of Çatal Hüyük, a 32 acre neolithic site with successive building levels dated to between *c.* 6500 and 5650 BC (Mellaart, 1978). At one of the earliest levels (VIII) in a building complex which includes a shrine, a well-sealed red ochre burial under a platform had associated with it the bones of mice. In trying to explain the rodent bones, Mellaart (personal communication) questions whether they might indicate some special item of personal adornment, noting that not all the bones were represented. Cole (1972) offers the alternative possibilities that

TABLE I

Occurrence of bones at Çatal Hüyük, compared with a Pleistocene "Rodent Earth" assemblage at Westbury-sub-Mendip and a barn owl pellet sample. (Comparative data courtesy of P. Andrews, Q. Mould, and C. French.)

	Barn owl		Westbury "Rodent Earth"		Çatal Hüyük	
	No.	%	No.	%	No.	%
Mandible	94	2.0	144	3.5	129	15.4
Maxilla	46	1.0	99	2.4	77	9.2
Cranium fragments	241	5.2	139	3.4	47	5.6
Incisors	4	0.1	439	10.7	152	18.2
Molars	24	0.5	844	20.6	27	3.2
Scapula	81	1.7	13	0.3	0	0.0
Humerus	93	2.0	67	1.6	17	2.0
Radius	88	1.9	85	2.1	15	1.8
Ulna	92	2.0	68	1.7	11	1.3
Ribs	1016	21.8	39	0.9	3	0.4
Vertebrae	1210	25.9	167	4.1	56*	6.7
Pelvis	84	1.8	2	0.0	6	0.7
Femur	94	2.0	29	0.7	25	3.0
Tibia	90	1.9	97	2.4	71	8.5
Metacarpals/tarsals	332	7.1	784	19.1	99	11.8
Tarsals/carpals	79	1.7	71	1.7	35	4.2
Phalanges	999	21.4	1016	24.8	67	8.0
Totals	4667		4103		837	

* Caudals mainly.

they might have been a ritual offering of pet mice or a special food offering. If these seem exotic possibilities, one might recall that down to the time of Isaiah, it is recorded that groups of Jews met secretly in gardens to eat the flesh of pigs and mice as part of a very ancient religious rite (Epstein, 1971).

Fortunately, the Çatal Hüyük bones have been carefully saved and can be investigated in relation to these hypotheses. Through the kindness of my colleague, James Mellaart, these fragile bones have been handed to me for study and I can at present make the following comments prior to a full report later.

Firstly, it can be seen from Table I that the frequencies of different bones of the mouse skeleton are not those expected if the collection represents the decomposed bodies of about 75 corpses, which is the minimum number of animals represented. Skull fragments are well in evidence and tibias occur very commonly in comparison with both an owl pellet sample and a Pleistocene "Rodent Earth" assemblage from Somerset. In contrast, bones of the main

TABLE II

Distribution of age classes in the molars of the Çatal Hüyük mandibles. Wear grades adapted from Lidicker (1966) for use on the lower dentition.

Wear grades	3	4	5	6	7
Çatal Hüyük specimens	29	16	13	7	4

body area (ribs, scapulae and vertebrae) are noticeably uncommon, and indeed, the vertebrae remaining are virtually all caudal segments. Clearly this is a taphonomist's nightmare, and we are left with the distinct possibility that these were indeed derived from skins, used ritually in relation to human burial. Either as a result of bone fragmentation or immaturity, only a small number of measurements could usefully be taken for comparison with modern samples. As regards the age composition, approximately 50% of the long bones were evaluated as immature, from size and epiphyseal evidence. An alternative check on age was attempted, using a modification of Lidicker's (1966) scheme of attrition, originally prepared for the upper dentition but applied by me to the lower molars. It will be seen in the tabulation that the molar wear grades of most individuals are in the first three categories (Table II), and again suggest a predominance of younger individuals in the sample.

In the case of the interorbital breadth, one of the few skull dimensions which could be taken, it will be seen (Fig. 1) that the Çatal Hüyük specimens (marked as black dots) fall in the lower part of the *musculus* range. In terms of the lengths of the femur, tibia and humerus, measurements are very similar to those given in Leamy's (1974) study of osteometric traits for his random-bred mice. Similarly, measurements of the molar teeth were found to be well within the ranges provided by recently published German data on *M. musculus* (Reichstein, 1978).

The most distinction, although again one should keep in mind sample smallness, is seen in three of the recordable discontinuous traits. Compared with Berry (1963) and Berry & Searle's (1963) frequencies, the double accessory maxillary foramen is 9% less common in the Turkish sample. Also, a double mental foramen is 26% less common and the perforation of the olecranon fossa 27% lower in frequency.

Although we are not able to demonstrate at Çatal Hüyük definite infestation of housing by mice, there appears to be good evidence of this from early Egypt, although here, sadly, the actual remains have rarely been saved for study. Two sites are worth mentioning.

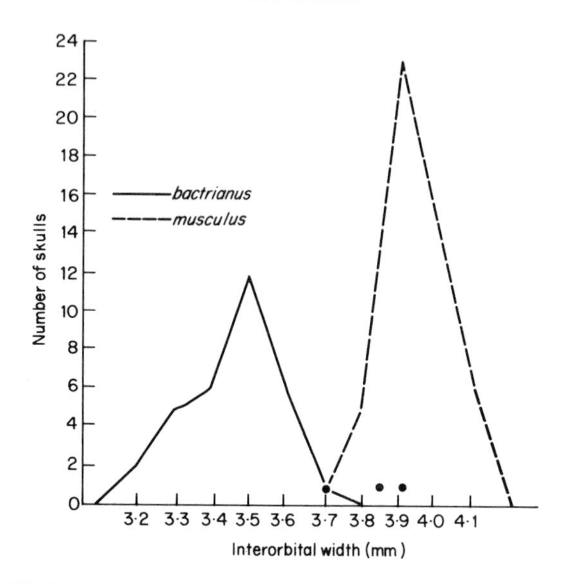

FIG. 1. Interorbital widths in two species of *Mus* (males only). Positions of three Çatal Hüyük specimens are given as black dots.

Petrie (in 1891) noted abundant rodent tunnelling in nearly every room of the houses of the Twelfth Dynasty workers' town at Kahun (*c.* 1700 BC). Further evidence of more than superficial links between rodents and human habitations occurs at the fortress town of Buhen (*c.* 1700 BC) near the Second Cataract (Dixon, 1972). Here, rodent remains were found in abundance in a part of the building complex, but unfortunately were not saved for study. In view of the fact that rats appear to have a more coastal distribution even today, and the Cairo spiny mouse (*Acomys cahirinus*), though sometimes found in houses, is more solitary in behaviour (Hufnagl & Craig-Bennett, 1972), the house mouse is surely likely to have been the major colonizer of these sites. Confirmation of this may be present at El Khattara, Egypt, although the possible Predynastic date is questionable, as they may be intrusive from later deposits (Gautier, personal communication). According to the evidence of two papyri, the mouse may have been used as food at times and also for medicinal purposes.

Owing to problems both of preservation and excavation, the occurrence of house mouse bones seems likely to remain uncommon on many sites. At the Urartu fortress, Karmir-Blur, in the Caucasus (mid-1st millennium BC) only two individuals were identified. At Pompeii, a site with unique preservation in some respects, the sum total of possible house mouse bones is one, and I suspect from the size and morphology of this tibia that it belongs to another species.

MICE AND HUMAN MOVEMENT

Although it will be some decades yet before the movement of house mouse populations into northern Europe can be plotted by reference to Holocene archaeological finds, we might perhaps briefly look at some of the evidence from Europe and western Asia.

However one views the present taxonomy of *Mus*, it seems likely that the southern European–Asian corridor with its varying environments and human population densities has played an important part in the microevolution of the house mouse. Besides its appearance at Çatal Hüyük and sites in the Caucasus, it has been identified at Apamea in northern Syria, though not until the Byzantine–Islamic period (Gautier, personal communication). Its distribution in the Mediterranean area must have been encouraged by the development of the advanced cultures of the Classical world. From a pre-Roman deposit in a burial chamber in Mallorca, Uerpmann (1970) found house mice with osteometric measurements within the modern size range, although the S'Illot site material could indicate smaller body size than usually seen in the western Mediterranean today. Something of the problems of zooarchaeological interpretation is seen in the fact that although the murid bones were associated with human burials, they seem likely to be derived from owl pellets deposited in the chamber above the grave.

In the case of the Iberian Bronze Age house mice from Cabezo Redondo, there is further evidence of size variation, compared with recent material, though samples are small (Storch & Uerpmann, 1969; Uerpmann, 1971). Uerpmann suggests that the reason for these possible size differences may well be environmental fluctuations, especially changes in climate. Further evidence of house mice has now been noted in the Iberian Peninsula (Storch & Uerpmann, 1976; von den Driesch, 1976). In some of this Iberian material there appear to be two different forms of house mouse, and Uerpmann (personal communication) believes that these may be indicative of the "domestic" and "wild" forms.

Although there is clearly growing evidence of the existence of house mice in southern Europe, a distribution which might have been assisted by human movements, there is surprisingly little late prehistoric evidence from more northern parts of Europe. Reichstein (1974) mentions early mouse evidence from northern Germany (in a paper mainly concerned with the spread of the rat). From the site of Bovenkarspel in north Holland, 59 fragments, mostly mandibles, were found in Bronze Age levels (^{14}C date between 3200 and 2700 years ago), and are considered to be contemporary (Ijzereef, personal communication). In contrast to this early date, the earliest evidence so far from France appears to be of Mediaeval date from the Abbey of Saint-Avit-Sénieur, Dordogne (Gautier & Ballmann, 1972).

Through the careful work of Dr Johannes Lepiksaar and his archaeological colleagues, *Mus musculus* has been identified tentatively at three Scandinavian sites. At the Mediaeval monastery of Gudhem, Västergotland (Lepiksaar, 1975) both cranial and post-cranial material occurs. Other post-cranial material, considered to be of the house mouse, occurs on the island of Oland in the Baltic, probably of post-Iron Age date (Lepiksaar, in press); also at the late Viking–early Mediaeval site of Oxiegarden Väst, prov. Scania (Lepiksaar, personal communication).

Finally, if we consider the evidence from Britain, indicative perhaps of one of the final European areas of conquest of the house mouse, it would appear to have been introduced to these islands by the final phase of prehistory. Corbet (1974) and Harcourt (1979) note its occurrence at the Dorset Iron Age settlement of Gussage All Saints in apparently sealed contexts, and it may well be that the Iron Age marked the beginning of a major intrusion. Its occurrence (Fig. 2) in Ossoms Eyrie Cave, Staffordshire, in pre-Roman levels (Yalden, 1977) might indicate a much greater antiquity, although I doubt it. The maxillary fragment from a late Iron Age ditch at Winterton, Lincolnshire (Thomas and R. Goodburn, personal communications) might not be of that date as the specimen occurred in a relatively shallow pre-Roman enclosure ditch. Also of questionable date is evidence of mice from the bottom of an Iron Age pit at Little Somborne, Hampshire (Locker, personal communication).

Possible Roman evidence occurs from the south coast to Yorkshire. At Portchester Castle, mice occur in more than one sample (Grant, 1975, 1977). In the Roman sewer at York, a complex in which the house mouse occurs with other small mammals, there is the slight chance of later intrusion into the deposits (Rackham, in Buckland, 1976). Roman levels of the Winnal Down (M3 Motorway) excavation have also produced further evidence (Maltby, personal communication). In the case of the Roman site of Caerleon in Gwent, a variety of small mammal bones including 5.9% of *Mus* occurs only in the town ruins, and possibly represents the later decomposed scatters of owl pellets (O'Connor, personal communication). Although good skeletal evidence is needed, the introduction of the house mouse to Ireland is thought to have occurred by Roman times (Wijngaarden-Bakker, 1974). There is no doubt that by Saxon-Mediaeval times the house mouse was well distributed. Pits in the Saxon town of Hamwih (Southampton) produced evidence of

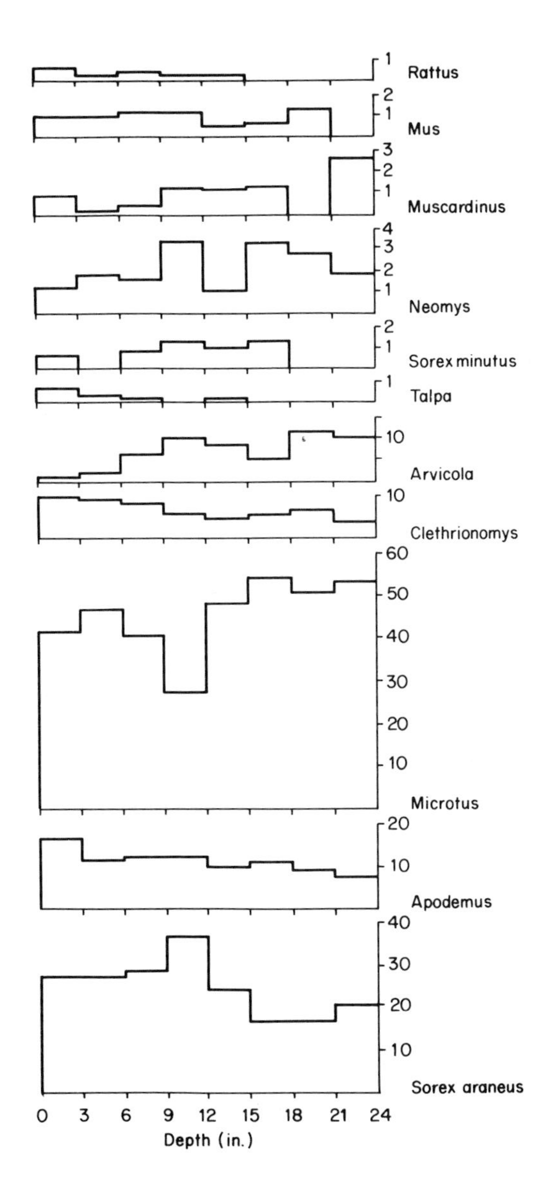

FIG. 2. Percentage of small mammal faunas including house mice, at varying depths in Ossoms Eyrie Cave, Staffordshire (scales differ according to sample sizes). Lowest levels are probably pre-Roman (after Yalden, 1977).

infestation by mice (Coy, personal communication) and similarly Middle Saxon levels in Vernon Street, Ipswich, provided additional material (Jones, personal communication). At more than one site in mediaeval Winchester, fragments of mouse have been identified (P. Shephard, personal communication). From mediaeval Lincoln (Brothwell, unpublished), Ludgershall Castle in Wiltshire (Jones, personal communication), late mediaeval Nonsuch Palace in Surrey (Locker, personal communication) and Barnard Castle, County Durham (Rackham, personal communication) have come further evidence.

CONCLUSIONS

If we take the limited Pleistocene and recent archaeological evidence together, we see the emergence of possibly a number of species of the genus *Mus* by early Pleistocene times, with geographically distinct forms in Eurasia and eastern Africa, and possibly also one or two island communities. Changes in human population size and the adaptive strategies of agriculture and urban development provided a protected environment which the house mouse colonized more effectively than any other small mammal in the vicinity of these changing human groups. We do not know whether mice may have been kept for ritual or other reasons, but this cannot be altogether ruled out. At least its occurrence in the building complex at Çatal Hüyük may have been turned to good use. The relatively late appearance of the house mouse in Britain and perhaps other parts of northern Europe is puzzling, and one asks why it did not appear with the early farmers. This position may be an accident of preservation, but is more likely to be related to the transport, and more importantly the nature of the packing, of trade objects. Clearly, the house mouse was well distributed by the time the early voyagers went in search of the spice lands of the East and westwards to the New World, and one can speculate about the extent to which these early ships may have carried the house mouse on these long journeys.

ACKNOWLEDGEMENTS

I have received information and comments from a variety of colleagues in zooarchaeology and I am extremely grateful to them all. In particular, I wish to mention Professor E. Tchernov, Dr J.

Lepiksaar, Professor J. Boessneck, Dr Hans-Peter Uerpmann, Dr A. Gautier, Dr T. Holland, Mr T. O'Connor, Mr J. Rackham, Dr A. Azzaroli, Miss B. Noddle, Professor R. J. Berry, Mr A. Currant, Dr A. van de Weerd, Mr G. F. Ijzereef, Dr L. H. van Wijngaarden-Bakker, Mr R. Goodburn, Miss Jenny Mann, Mrs Pauline Shephard and colleagues in the Winchester Unit, Miss Jennie Coy, Mr Mark Maltby, Mr R. Jones, Miss Alison Locker, Dr P. Andrews, Miss Q. Mould, and Mr C. French. Last but by no means least, my thanks to my archaeological colleague, James Mellaart, for making available for study the Çatal Hüyük house mouse bones.

REFERENCES

Bate, D. M. A. (1942a). New Pleistocene Murinae from Crete. *Ann. Mag. nat. Hist.* (11) **9**: 41–49.

Bate, D. M. A. (1942b). Pleistocene Murinae from Palestine. *Ann. Mag. nat. Hist.* (11) **9**: 465–486.

Berry, R. J. (1963). Epigenetic polymorphism in wild mouse populations. *Genet. Res.* **4**: 193–220.

Berry, R. J. & Searle, A. G. (1963). Epigenetic polymorphism of the rodent skeleton. *Proc. zool. Soc. Lond.* **140**: 577–615.

Boekschoten, G. J. & Sondaar, P. Y. (1972). On the fossil mammalia of Cyprus, I. *Proc. K. ned. Akad. Wet.* **75** (B): 306–325.

Buckland, P. C. (1976). The environmental evidence from the Church Street Roman sewer. In *The archaeology of York* **14/1**. London: York Archaeol. Trust.

Chaline, J. (1977). Rodents, evolution and prehistory. *Endeavour* **1**: 44–51.

Cole, S. (1972). Animals of the New Stone Age. In *Animals in archaeology*: 15–41. Brodrick, A. A. (ed.). London: Barrie & Jenkins.

Corbet, G. B. (1974). The distribution of mammals in historic times. In *The changing flora and fauna of Britain*: 179–202. Hawksworth, D. L. (Ed.). London: Academic Press.

Dixon, D. M. (1972). Population, pollution and health in Ancient Egypt. In *Population and pollution*: 29–36. Cox, P. R. & Peel, J. (Eds). London: Academic Press.

von den Driesch, A. (1976). Die tierischen Beigaben in den Gräbern der Siedlung "Cuesta del Negro" bei Rurullena/Granada. Appendix to: Lauk, H. D. Tierknochenfunde aus bronzezeitlichen Siedlungen bei Monachil und Purullena (Prov. Granada). *Studien über frühe Tierknochenfunde von der Iberischen Halbinsel* **6**: 112–117.

Epstein, H. (1971). *The origin of the domestic animals of Africa.* New York: Africana Publishing Co.

Gautier, A. & Ballmann, P. (1972). La faune d'un puits de l'abbaye de Saint-Avit Sénieur (XIe à XIIIe siècle, Dordogne). *Archéol. Médiev.* **11**: 355–379.

Grant, A. (1975). The animal bones. In *Excavations at Portchester Castle, I. Roman*: 378–408. Cunliffe, B. (Ed.). London: Rep. Res. Comm. Soc. Antiq. 33.

Grant, A. (1977). The animal bones. Mammals. In *Excavations at Portchester Castle, III. Mediaeval*: 213–239. Cunliffe, B. (Ed.). London: Rep. Res. Comm. Soc. Antiq. **34**.

Haas, G. (1966). *On the vertebrate fauna of the Lower Pleistocene site "Ubeidiya"*. Jerusalem: The Israel Academy of Science and Humanities.

Haim, A. & Tchernov, E. (1974). The distribution of myomorph rodents in the Sinai Peninsula. *Mammalia* **38**: 201–223.

Harcourt, R. (1979). The animal bones. In *Gussage All Saints. An Iron Age settlement in Dorset*: 150–160. Wainwright, G. J. (Ed.). (Department of the Environment Archaeological Report No. 10.) London: H.M.S.O.

Hufnagl, E. & Craig-Bennett, A. (1972). *Libyan mammals*. Harrow: Oleander Press.

Jaeger, J. J. (1976). Les rongeurs (Mammalia, Rodentia) du Pléistocène inférieur d'Olduvai Bed I (Tanzanie). Ière Partie: Les Muridés. In *Fossil vertebrates of Africa* **4**: 57–120. Savage, R. J. G. & Coryndon, S. (Eds). London: Academic Press.

Jánossy, D. (1962). Vorläufige Mitteilung über die Mittelpleistozäne Vertebratenfauna der Tarkö-Felsnische (NO-Ungarn, Bükk-Gebirge). *Annls hist. nat. Mus. natn. hung. Min. Paleont.* **54**: 155–174.

Kurten, B. (1968). *Pleistocene mammals of Europe*. London: Weidenfeld & Nicholson.

Leamy, L. (1974). Heritability of osteometric traits in a random-bred population of mice. *J. Hered.* **65**: 109–120.

Lepiksaar, J. (1975). Uber die Tierknochenfunde aus den mittelalterlichen Siedlungen Südschwedens. *Archaeozoological studies*: 230–239. Clason, A. I. (Ed.). Amsterdam: North Holland.

Lepiksaar, J. (In press). Animal remains found in Tornrör, Oland, Sweden. A study of a thanatocoenosis (late Iron Age to Recent times). *Striae* **10**.

Lidicker, W. Z. (1966). Ecological observations on a feral house mouse population declining to extinction. *Ecol. Monogr.* **36**: 27–50.

Mellaart, J. (1978). *Earliest civilisations of the Near East*. London: Thames & Hudson.

Murty, M. L. K. (1975). Late Pleistocene fauna from Kurnool Caves, South India. In *Archaeozoological studies*: 132–138. Clason, A. I. (Ed.). Amsterdam: North Holland.

Pei, W. C. (1936). On the mammalian remains from Locality 3 at Choukoutien. *Palaeont. Sin.* (C) **7**: 1–108.

Pei, W. C. (1939). A preliminary study on a new Palaeolithic station known as Locality 15 within the Choukoutien region. *Bull. geol. Soc. China* **19**: 147–187.

Petrie, W. M. F. (1891). *Illahun, Kahun and Gurob*. London: Nutt.

Reichstein, H. (1974). Bemerkungen zur Verbreitungsgeschichte der Hausratte (*Rattus rattus*, Linné, 1785) an Hand jüngerer Knochenfunde aus Haithabu (Ausgrabung 1966–69). *Die Heimat* **81**: 113–114.

Reichstein, H. (1978). *Mus musculus* Linnaeus, 1758 – Hausmaus. In *Handbuch der Säugetiere Europas*: 421–451. Niethammer, J. & Krapp, F. (Eds). Weisbaden: Akad. Verlags.

Storch, G. (1970). Holozäne Kleinsäugerfunde aus der Ghar-Dalam-Höhle, Malta. *Senckenberg. biol.* **51**: 135–145.

Storch, G. (1972). In Besenecker, H., Spitzenberger, F. & Storch, G. Eine holozäne Kleinsäuger Fauna von der Insel Chios, Agäis. *Senckenberg. biol.* **53**: 154–177.

Storch, G. (1975). Eine mittelpleistozäne Nager-Fauna von der Insel Chios, Agäis. *Senckenberg. biol.* **56**: 165–189.

Storch, G. & Uerpmann, H-P. (1969). Kleinsäugerfunde aus dem bronzezeitlichen Siedlungshügel "Cabezo Redondo" bei Villena in SO-Spanien. *Senckenberg. biol.* **50**: 15–22.

Storch, G. & Uerpmann, H. P. (1976). Die Kleinsäugerknochen vom Castro do Zambujal. *Studien über frühe Tierknochenfunde von der Iberischen Halbinsel* **5**: 130–138.

Sutcliffe, A. J. & Kowalski, K. (1976). Pleistocene rodents of the British Isles. *Bull. Br. Mus. nat. Hist. (Geol.)* **27**: 31–147.

Tchernov, E. (1975). Rodent faunas and environmental changes in the Pleistocene of Israel. In *Rodents in desert environments*: 331–362. Prakash, I. & Ghosh, P. K. (Eds). The Hague: Junk.

Uerpmann, H. P. (1970). Die Tierknockenfunde aus der Talayot-Siedlung von S'Illot (San Lorenzo/Mallorca). *Studien über frühe Tierknockenfunde von der Iberischen Halbinsel* **2**: 1–111.

Uerpmann, H. P. (1971). Osteologische Untersuchungen zur Klimageschichte von Zambujal. *Madrider Mitt.* **12**: 46–50.

Vereshchagin, N. K. (1967). [*The mammals of the Caucasus. A history of the evolution of the fauna*]. ((1959). Izdak. Nauk. [in Russian]). Jerusalem: Israel Programme for Scientific Translations.

Wijngaarden-Bakker, L. H. van (1974). The animal remains from the Beaker settlement at Newgrange, Co. Meath: First Report. *Proc. R. Irish Acad.* **74**: 313–383.

Yalden, D. (1977). Small mammals and the archaeologist. *Bull. Peakland Archaeol. Soc.* No. 30: 18–25.

Symp. zool. Soc. Lond. (1981) No. 47, 15–25

Taxonomy of the House Mouse

J. T. MARSHALL

National Fish and Wildlife Laboratory, National Museum of Natural History, Washington, D.C., USA

R. D. SAGE

Museum of Vertebrate Zoology, University of California, Berkeley, California, USA

SYNOPSIS

Current taxonomy of the house mouse, which unites all populations into one species, *Mus musculus,* needs amending because several forms overlap in distribution and differ from each other in skull morphology, ecology, and biochemistry. For instance, two wild-living forms with slender skulls and short tails, *Mus spretus* of the western Mediterranean rim and *M. abbotti* of the eastern Mediterranean, overlap the commensal, long-tailed *M. domesticus* in Spain and Yugoslavian Macedonia respectively. *Mus domesticus* and *M. musculus* should be considered as different species because of the narrow, permanent hybrid zone between them in Jutland. *Mus musculus* lives in farmhouses within Austrian fields where the steppe mouse, *M. hortulanus*, builds its granaries. European house mice can thus be considered as comprising five species: *Mus spretus, M. abbotti, M. hortulanus, M. musculus,* and *M. domesticus.*

Skulls of Asian house mice are alike in resembling those of European *Mus musculus,* from which they differ in being of smaller size and having more rounded contours. The dark-bellied, long-tailed, obligate house-dwelling *Mus castaneus* is apparently a species distinct from the white-bellied, short-tailed *M. molossinus.* The relationship between Asian and European house mice might be better understood if further field work or museum studies could reveal (1) the total distribution of the centrally located *Mus musculus* (?) *wagneri*; (2) areas of intergradation it might have with *musculus* to the west or *molossinus* to the east, and (3) whether there is contact between *M. castaneus* and *M. domesticus* in Nepal or eastern India.

INTRODUCTION

The following tentative nomenclature of house mice is adapted from that of Schwarz & Schwarz (1943) as amplified by Ellerman & Morrison-Scott (1951). Instead of treating all forms of house mice as one species as those authors did, we divide them into several species based on our ecological, biochemical and morphological findings in samples from some of the places where two forms overlap in distribution. This chapter presents the qualitative skull differences revealed when Sage's European and Moroccan voucher specimens were compared by Marshall with 23 taxa in the U.S. National Museum of Natural History (Appendix). Age differences were minimized by the selection of specimens whose cusp pattern on only the third upper molars is worn off.

Although we are fairly certain of the qualitative characters (Fig. 1) and appropriate scientific names of European populations, the Asian forms are in doubt because their skulls all look alike. Nevertheless,

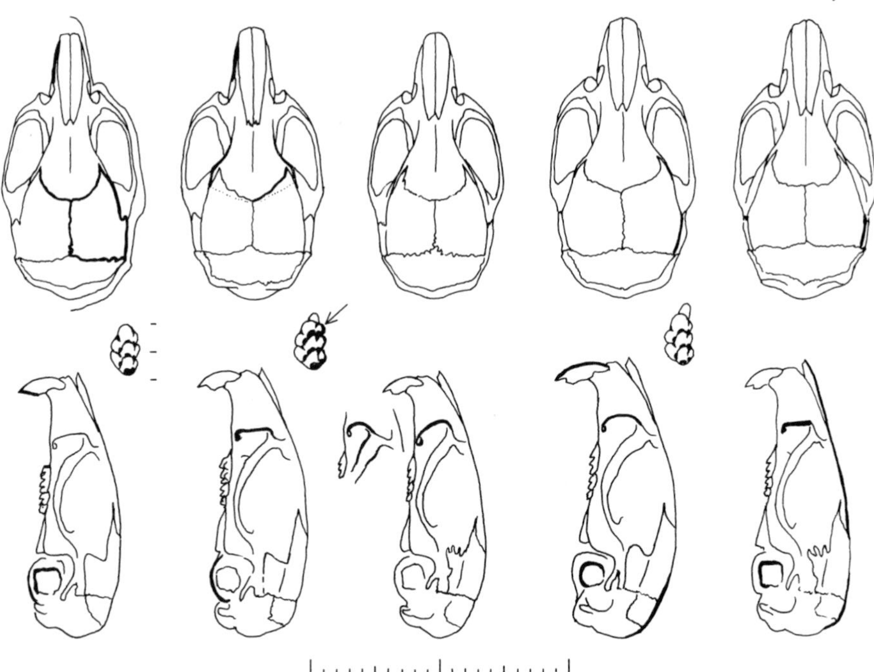

FIG. 1. Some qualitative differences seen between skulls of house mice, from left to right: *Mus spretus* inside a partial outline of *M. abbotti; M. hortulanus; M. castaneus* with inset of zygomatic plate of *M. molossinus*; *M. musculus*; and *M. domesticus.* Left upper anterior molar is shown for *Mus musculus,* characteristic of most taxa of house mice except for those of *M. abbotti* (left) and *M. hortulanus* (second from left). Scales are in millimeters.

east Asian samples probably represent two species related to *Mus musculus.* Field experience in southeast Asia (Marshall, 1977) added nothing to our understanding of the overlap of town and country mice for the reason that all the outdoor forms proved to be unrelated, native species. (These are the taxa listed by Schwarz & Schwarz,

1943, as *caroli, ouwensi,* and *formosanus,* which belong in *Mus caroli,* and the population from the Nilgiri Hills, which belongs in *M. nagarum*; see Marshall, 1977.)

EUROPEAN HOUSE MICE

Mus spretus, Lataste's Mouse

The bicoloured tail of Lataste's mouse is short, the fur of the ventral surface is white or buff with slate gray bases, the feet are white and the dorsum is ochraceous brown. The slender skull has a long narrow rostrum, smooth sutures, and oval cross-section through the cranium. The parietals are long at the expense of the lunate interparietal. The upper incisors are smoothly bevelled, without a notch. The zygomatic plate is inclined forward; the bullae are set at an angle to the transverse axis. There are only four intermolar palatal rugae (Lataste, 1883), whereas all other house mice have five. *Mus spretus* lives in natural vegatation of woodland and chaparral around the western rim of the Mediterranean from southern France (Britton & Thaler, 1978) to Spain ("*spicilegus,*" Sage, 1978), and from Morocco to Tunisia. An isolated population is at Cyrenaica, Libya. Two specimens with dark brown backs from Simbillawein, Daqahlia Province, Egypt, may represent this species.

Mus abbotti, Eastern Mediterranean Short-tailed Mouse

The skull is similar to that of *Mus spretus,* but is larger. The upper incisors usually show an angled or serrated posterior bevel, if they are not actually notched. (All the remaining house mice, mentioned below, have notched incisors.) The ventral wing of the parietal bone is deep and of triangular outline; the zygomatic plate is slightly wider than most of *Mus spretus* and the anterior root of the first upper molar is vertical (whereas other house mice have a slanting root). The short, bicoloured tail is a trifle longer than that of *Mus spretus.* There are 40 telocentric chromosomes. *Mus abbotti* occurs in natural vegetation and fields at least from Macedonia through Turkey and north-western Iran to the south coast of the Caspian Sea. Samples from the northern, humid portion of this area are brownish sooty on the back, white or buff with grey bases on the belly; whereas those from central and southern Turkey are a paler brown and whitish. Another name that probably belongs in synonomy with this species is *caudatus.* "*Mus musculus spicilegus*" has heretofore been used, but we deem this synonymous with *M. hortulanus.*

Mus hortulanus, Mound-builder, Hillock Mouse, Ährenmaus, Spike Mouse, Kurganchikovaya Mysh, Steppe Mouse

The family or small colony piles several kilogrammes of grain spikes around the base of a stalk, then paves this winter store with earth (as photographed in Festetics, 1961). The mound-builder has the same coloration as does Macedonian *Mus abbotti.* Its short tail averages about 65 mm for a head plus body length of 82 mm average among our 11 voucher specimens. There are the usual 40 telocentric chromosomes (in Sage's specimen, 10097). The skull is short and broad yet the rostrum is narrow. The anterior spine of the parietal is often reduced or wanting; the deep, ventrally-projecting wing of this bone has the same triangular shape as that of the previous two species. Unique attributes are an angular lateral outline opposite the anterior corners of the parietal, narrow zygomatic plate with concave anterior border, and forward position of the anterolabial cusp of the first upper molar (Fig. 1, arrow). The distribution of *Mus hortulanus,* an important member of the steppe fauna, is coextensive with that of the natural steppe vegetation and grain fields replacing it in eastern Europe.

Mus musculus, Linné's House Mouse

In dorsal view the cranium of *Mus musculus* is globular or tear-shaped, with short, broad rostrum. The skull is deep and convexly arched dorsally as seen in lateral view. The smoothly rounded, deep, ventral wing of the parietal is bent down at an angle and the anterior margin of the broad zygomatic plate is convexly arched in a semicircle. Upper incisors are less curved than those of other house mice. Coloration and proportions are similar to those of *Mus abbotti* and *M. hortulanus* except that *M. musculus* is less dusky and more brownish above. The tail, averaging about 73 mm, is shorter than the head-plus-body (about 85 mm). This white- or buff-bellied (with grey bases) house mouse is the one originally described from Uppsala, Sweden (Linnaeus, 1766). It extends from there to the north end of Jutland, and from the River Elbe eastward at least to Serbia, Yugoslavia (laboratory colonies and voucher specimens collected by Sage). This mouse lives in man-made habitats and in the north is dependent on buildings, at least in winter. It *encroached,* along with civilization, into the natural steppe vegetation that was the rightful home of the mound-builder, *M. hortulanus.*

An identical skull, only slightly smaller, characterizes the desert-dwelling *Mus musculus* (?) *wagneri.* This mouse, of northern interior

Asia, is pale, sandy brown on the back and pure white on the ventral surface, feet, and underside of the short tail. Its fur is long and silky; even the tail is furry. Unfortunately we cannot discern the distributional limits of *wagneri* from the literature because of the chaotic multiplicity of subspecific names used.

Mus domesticus, European House Mouse and *Mus poschiavinus,* Tobacco Mouse

The above two species, though biologically and chromosomally distinct (Gropp, Winking, Zech & Müller, 1972) are combined in this paragraph in order to highlight our findings that they are morphologically and biochemically identical. The European house mouse is confined to buildings, agricultural fields, and vegetation altered by man. Its tail is long – always as long as the head plus body, reaching or exceeding 100 mm in some samples from Iran. The skull is the most angular and least rounded of all the members of the house mouse group of species. The ventral wing of the parietal is invariably shallow and its suture with the squamosal follows a tortuous course. The straight ventral outline together with a nearly straight dorsal profile gives the skull a wedge shape in lateral view, with a high occiput. These traits reach an extreme development, with thickened angles of the cranial walls and a straight, vertical margin of the zygomatic plate, in the black, Alpine *Mus poschiavinus,* and the dark greyish brown, unicoloured population of *M. domesticus* from the River Elbe westward. The latter is also distributed eastward along the foot of the Alps to the shores of the Adriatic Sea, and is doubtless the progenitor of the "outbred Swiss albino laboratory mouse," which shares these cranial attributes exactly. A less extreme skull, with zygomatic plate like that of *Mus musculus,* is found in the paler-bellied, brown-backed mice of southern France, Spain, and Sicily (heretofore called *brevirostris* or *azoricus*). This is the type of skull seen in immigrants to England, America, and Australia. Finally, a slightly more slender skull characterizes all the long-tailed mice from northern Africa eastward across the deserts (*praetextus*) through Iran to Pakistan (*bactrianus*) and up the Himalayas to Kashmir and Nepal (*homourus*). Their supraorbital ridges are contracted medially at the anterior level of the parietals so that the sloping squamosal is visible in dorsal view. *Mus domesticus* adapts to local settings by extremes of coloration. In northern Africa this varies from a geographical mosaic of dark and of white-bellied colonies in Morocco to polymorphism in Egypt, Israel, and Syria (Atallah, 1978). Farther east, the sandy brown, desert style of coloration prevails (Arabia to Pakistan), clothing the mouse in a coat resembling the gerbils with which it is associated in desert habitats (Harrison, 1972). In north Africa the dark and white-bellied forms are often, but not invariably, segregated respectively in farms and oases. The Himalayan form (*homourus*) is uniformly dark on the back and white underneath with grey bases to the ventral fur. It requires buildings in the winter (S. Frantz, personal communication).

In view of the biochemical similarity of all these populations, their mosaic pattern of distribution with respect to colour phases, and the way they are carried everywhere by man, we feel that the use of subspecific names in *Mus domesticus* is confusing and unjustified. Genetical evidence to back our claim that *Mus domesticus* and *M. musculus* are different species comes from the discovery of a permanent, narrow hybrid zone between them (Ursin, 1952; Hunt & Selander, 1973). If these mice were conspecific their area of intergradation would be gradual and should be broader than the mere 20 km width of the hybrid zone across northern Jutland.

ASIAN HOUSE MICE

The skulls of the many named taxa of Asian mice, apparently related to *Mus musculus,* appear to be similar to each other. They have the flattened, globular skull of *Mus musculus* from which they differ in smaller size, more rounded contours, and less convex dorsal outline.

Mus molossinus, Japanese Wild Mouse

Living both indoors and in fields of Japan, *Mus molossinus* is a dark, brownish-grey mouse with white feet and a white underside to its short tail. The fur of the ventral surface is white with grey bases in Japan and Manchuria (a presumed wild-living subspecies, *manchu,* with short tail of about 50 mm) but more purely white in Korea (*yamashinai*). Distinctive biochemical attributes (Minezawa, Moriwaki & Kondo, 1979; Moriwaki, Shiroishi, Minezawa, Aotsuka & Kondo, 1979) as well as the small size and smooth skull seem to qualify *molossinus* as a full species, separate from *Mus musculus.* In many specimens the zygomatic plate appears short in lateral view (due partly to its lateral inclination), extending only half-way to the top of the rostrum.

Mus castaneus, Asian House Mouse

Mus castaneus joins the tobacco mouse in being man's closest indoor associate among undomesticated mammals. Like *Mus poschiavinus* it has no known feral range. (This does not apply to outdoor-living immigrants in Micronesia, presumably derived from *Mus castaneus*; see Marshall, 1962.) In Asia, *Mus castaneus* is limited to cities and towns mostly along the coast, from Bombay to Fukien and around Taiwan, Indonesia, and the Philippines. It is rare in Thailand, where *Rattus exulans* is the prevailing "house mouse." An Asian counterpart of north-west European *Mus domesticus, M. castaneus* is similar in its long tail, dark ventral colour, and the wiggly outline of the parietal-squamosal suture. *Mus castaneus* is much the smaller of the two and is usually tinted with ochraceous. Marshall (1977) overemphasized the importance of the peculiar anterior outline of the zygomatic plate (Fig. 1) as a diagnostic character of *castaneus* (and, incorrectly, of all Asian house mice). It is a frequent but not universal attribute. The subspecies *tytleri* of north central India is the same as *castaneus* except for its golden buff colour. According to K. Tsuchiya (personal communication) differences in C-bands and biochemical markers are sufficient to cause one to consider *castaneus* as a different species from *molossinus,* although they are closer to each other than either is to *Mus musculus,* and they do interbreed freely in the laboratory (M. Potter, personal communication). The important fact is, however, that they nowhere interbreed outside the laboratory (K. Tsuchiya, personal communication).

AREAS OF OVERLAP

Sage found *Mus domesticus* and *M. spretus* together on the same farms at Barcelona and Cadiz, Spain, and Azrou, Morocco (Sage, 1978). At Halbturn, Austria, he dug *Mus hortulanus* from mounds 500 m from houses in which he caught *M. musculus.* From Belgrade, Yugoslavia, home of *Mus musculus,* he went about 20 km to *M. hortulanus* mounds in the adjacent countryside. In Macedonia, beyond the range of *Mus musculus,* he collected live *Mus abbotti* in an area whence the comparatively rare *M. domesticus* was reported by B. M. Petrov (personal communication).

Among samples in the National Museum of Natural History, the specimens reported by Osborn (1965) include two species. From Cehennem Dere, in south central Turkey, are six *Mus domesticus* with white belly and long tail from a wooden house and one short-tailed (61 mm) *M. abbotti* from a "grassy area near spring." We identified the rest of Osborn's Turkish specimens as *Mus abbotti.* The only other overlap discerned so far among the specimens is around the western Mediterranean, where samples of *Mus spretus* are associated with collections of *M. domesticus* in all its varied colorations.

Turning to the literature we find *Mus domesticus* and *M. abbotti* coexisting in Grecian Macedonia at Thessaloniki (Bonhomme, Britton-Davidian, Thaler & Triantaphyllidis, 1978), and *M. domesticus* altitudinally separated from *M. castaneus* at Katmandu, Nepal (Marshall, 1977). We cannot identify the taxa involved in numerous other citations to variously mixed, interdigitated, or intergrading populations except those pertaining to Jutland, mentioned above in the account of *Mus domesticus.*

A WORD ON METHODS

The foregoing preliminary classification does not constitute an orgy of splitting nor an activity limited to house mice. Rather, it is the same, orderly approach being used by mammalogists upon a variety of complex genera such as *Apodemus, Peromyscus* and *Rattus.* Museum specimens have accumulated over the past 100 years and samples of different named taxa have been lumped together both in the trays and in the literature, awaiting study. Such study works on the premise that different species have skulls of different appearance, whereas races of the same species usually have similar skulls and do not occur together. In the *Rattus* complex, for instance, adequately studied species are removed one at a time from the conglomeration and are found to fit into previously known patterns of distribution (Musser & Chiu, 1979).

With house mice, a search for qualitative differences in the skulls in order to see what dimensions should be measured revealed the conspicuous traits of the littoral *Mus spretus* and the steppe-dwelling *M. hortulanus.* The remaining difficult taxa were tentatively grouped and their distributions plotted to reveal geographical gaps in our knowledge – the stage reached in this report. Some of the gaps can be filled from other museums; if not, further field work is necessary. Then detailed study of populations and analysis of measurements can proceed, as already started for skins and skulls of east Asian house mice by Jones & Johnson (1965, including an indictment of the work of Schwarz & Schwarz, 1943). No amount of biochemical, cytogenetic and breeding experiments, helpful as they are, can be a substitute for the skins and skulls with accurate locality and ecological data on their labels, in answering the crucial questions: Does *wagneri* meet and intergrade with *musculus* in the west, with *molossinus* in the east, and do *castaneus* and *domesticus* remain distinct in the vicinity of eastern India and Nepal?

CONCLUSIONS

The short-tailed mice appear to be geographically complementary. The entirely wild-living *Mus spretus* and *M. hortulanus* seem too far removed structurally from semicommensal mice to be considered representative of ancestral types, except that *spretus* and *abbotti* are close to each other morphologically, if not biochemically (Bonhomme *et al.*, 1978). Within the short-tailed group, *Mus musculus* is the most dependent upon man. Its origins will doubtless remain hidden until the distribution and ecology of *wagneri* and *molossinus* (*sensu lato,* see Appendix) are better known.

A long tail, in house mice, is associated with domesticity, as the literature has emphasized. Between them, the two house-dwelling, long-tailed species, *Mus domesticus* and *M. castaneus,* extend across the Eurasian continent (and well beyond it on both sides owing to transport by man). *Mus castaneus* has no known feral distribution; its cranial characteristics could easily be derived from a type represented by *M. molossinus.* It is hard to choose a point of origin from among the many outdoor, man-made habitats enjoyed by *Mus domesticus;* its skull bears no obvious similarity to those of other European forms except possibly *Mus abbotti.* Speculations about the origins and migrations accompanying human commerce of *Mus domesticus* and *M. castaneus* had best await determination of their relationship to each other.

Through much of their continent-wide ranges the two domestic mice broadly overlap the distributions of field mice, from west to east known as *Mus spretus* (northern Africa), *M. abbotti* (Greece, Turkey, Iran), *M. nagarum* (India), *M. cervicolor* (Burma), and *M. caroli* (south-east Asia). On the other hand, *Mus domesticus* is apparently the only medium-sized *Mus* found in the higher Himalayas and from Pakistan to Arabia. Does that mean it originated in the Middle East and that everywhere else it is an immigrant just as in the Americas and Australia?

APPENDIX

Specimens examined of house mice in the U.S. National Museum of Natural History, thought to be on their "natural" range. An asterisk identifies the sources of Sage's colonies. Traditional subspecies names are listed for ease in locating the samples and identifying them with descriptions in the literature.

Mus abbotti
  abbotti (dark) (*Yugoslavia), Turkey 6, northern Iran 103
  subspecies (pale) Turkey 15
Mus castaneus
  castaneus (dark, includes *urbanus*) India 17, Burma 1, coastal China 17, Taiwan 98, *south-east Asia to the Philippines 57
  tytleri (golden buff) north central India 4
Mus domesticus
  azoricus (pale belly) France 9, *Spain 6, Sardinia 1, Sicily 66, Italy 34
  bactrianus (pure white underparts) Pakistan 197
  brevirostris (same as *azoricus*) *Morocco 7
  domesticus (dark belly) France 13, Germany 19, Switzerland 3, *Yugoslavia 5
  gentilulus (desert coloration, small) Kuwait 3
  homourus (dark grey back, whitish belly) Kashmir 109, Hazara district, Pakistan 86
  praetextus (polymorphic mixture or mosaic of individuals resembling *bactrianus, brevirostris* and *domesticus*) *Morocco 5, Libya 109, Egypt 61, (*Israel), Iran 317
(*Mus hortulanus* *Austria, *Yugoslavia)
Mus molossinus
  manchu (very short tail) Manchuria 9
  molossinus Japan 17, Okinawa 3, Quelparte Island off Korea 3
  tantillus Szechuan, China 17
  yamashinai (whiter underparts) Korea 52
Mus musculus
  gansuensis (like *wagneri*) Kansu, China 18
  mongolium (like *wagneri*) Mongolia 17
  musculus (dark) Uppsala, Sweden 15, Kaliningrad, Russia 17, (*Czechoslovakia, *Yugoslavia)
  wagneri (pale brown with pure white underparts) Turkestan 19
Mus poschiavinus *Poschiavo, Switzerland 2, (*northern Italy)
Mus spretus
  hispanicus *Spain 28, France 1
  spretus *Morocco 428, Libya 34
  subspecies (dark brown back – species?) Egypt 2

REFERENCES

Atallah, S. I. (1978). Mammals of the eastern Mediterranean region; their ecology, systematics and zoogeographical relationships. *Säugetierk. Mitt.* **26**: 1–50.

Bonhomme, F., Britton-Davidian, J., Thaler, L. & Triantaphyllidis, C. (1978). Sur l'existence en Europe de quatre groupes de Souris (genre *Mus* L.) du rang espèce et semi-espèce, démontrée par la génétique biochimique. *C. r. hebd. Séanc. Acad. Sci., Paris* **287D**: 631–633.

Britton, J. & Thaler, L. (1978). Evidence for the presence of two sympatric species of mice (Genus *Mus* L.) in southern France based on biochemical genetics. *Biochem. Genet.* **16**: 213–225.

Ellerman, J. R. & Morrison-Scott, T. C. S. (1951). *Checklist of Palaearctic and Indian mammals.* London: British Museum (Nat. Hist.).

Festetics, A. (1961). Ährenmaushügel in Österreich. *Z. Säugetierk.* **26**: 112–125.

Gropp, A., Winking, H., Zech, L. & Müller, H. J. (1972). Robertsonian chromosomal variation and identification of metacentric chromosomes in feral mice. *Chromosoma* **39**: 265–288.

Harrison, D. L. (1972). *The mammals of Arabia* **3**. London: Benn.

Hunt, W. G. & Selander, R. K. (1973). Biochemical genetics of hybridization in European house mice. *Heredity* **31**: 11–33.

Jones, J. K., Jr & Johnson, D. H. (1965). Synopsis of the lagomorphs and rodents of Korea. *Univ. Kansas Publs Mus. nat. Hist.* **16**: 357–407.

Lataste, F. (1883). Note sur les souris d'Algérie. *Actes Soc. linn. Bordeaux* **37**: 13–33.

Linnaeus, C. (1766). *Systema Naturae.* 12th Ed. Holmiae.

Marshall, J. T. (1962). House mouse. In *Pacific islands rat ecology:* 241–246. Storer, T. I. (Ed.). Honolulu: Bishop Museum.

Marshall, J. T. (1977). A synopsis of Asian species of *Mus* (Rodentia, Muridae). *Bull. Am. Mus. nat. Hist.* **158**: 175–220.

Minezawa, M., Moriwaki, K. & Kondo, K. (1979). Geographical distribution of Hbb^p allele in the Japanese wild mouse, *Mus musculus molossinus. Jap. J. Genet.* **54**: 165–173.

Moriwaki, K., Shiroishi, T., Minezawa, M., Aotsuka, T. & Kondo, K. (1979). Frequency distribution of histocompatibility-2 antigenic specificities in the Japanese wild mouse genetically remote from the European subspecies. *J. Immunogenet.* **6**: 99–113.

Musser, G. G. & Chiu, S. (1979). Notes on taxonomy of *Rattus andersoni* and *R. excelsior*, murids endemic to western China. *J. Mammal.* **60**: 581–592.

Osborn, D. J. (1965). Rodents of the subfamilies Murinae, Gerbillinae, and Cricetinae from Turkey. *J. Egypt. public Hlth Ass.* **40**: 401–424.

Sage, R. D. (1978). Genetic heterogeneity of Spanish house mice (*Mus musculus* complex). In *Origins of inbred mice:* 519–553. Morse, H. C. III (Ed.). New York: Academic Press.

Schwarz, E. & Schwarz, H. (1943). The wild and commensal stocks of the house mouse, *Mus musculus* Linnaeus. *J. Mammal.* **24**: 59–72.

Ursin, E. (1952). Occurrence of voles, mice, and rats (Muridae) in Denmark, with a special note on a zone of intergradation between two subspecies of the house mouse (*Mus musculus* L.) *Vidensk. Meddr. dansk Naturh. Foren.* **114**: 217–244.

Symp. zool. Soc. Lond. (1981) No. 47, 27–41

Processes of Speciation and Semi-speciation in the House Mouse

L. THALER, F. BONHOMME and J. BRITTON-DAVIDIAN

Laboratoire d'Evolution des Vertébrés, Paléontologie et Génétique, Faculté des Sciences, Montpellier, France

SYNOPSIS

The application of the concepts and procedures of population genetics has dramatically modified the zoological picture of the house mouse in a very few years. The first shock came when Selander, Hunt & Yang (1969) showed that house mice from southern and northern Jutland (Denmark) are quite distinct biochemically. The second shock was due to Gropp, Tettenborn & Lehmann (1970), who discovered a mouse population in the Swiss Alps with $2n = 26$ chromosomes instead of the normal $2n = 40$. Starting with these pioneer works, a true systematic biology of the house mouse has been built up over the past ten years. From electrophoretic studies, it is now established that three biological species occur in Europe, one of them divided into two semispecies; many chromosomal semispecies are also recognized. Thus the so-called house mouse has been broken down into several distinct units. The question is how such a differentiation has occurred.

BIOCHEMICAL DIFFERENTIATION

Four biochemical groups have been defined in European mice, on the basis of genetical distances computed from electrophoretic data. Inter-group distances are about ten times greater than the mean distance between neighbouring populations of the same group (Table I). Although it is easy to apply biological criteria for species and semispecies to each group (Table II), serious problems remain about their proper Linnaean naming, and an *ad hoc* nomenclature (*Mus* 1, 2, 3, 4) has been proposed by Bonhomme, Britton-Davidian *et al.* (1978). This has been translated into formal taxonomic units in this paper as follows:

Mus 1 = *Mus musculus domesticus* + *M. m. brevirostris.*
Mus 2 = *Mus musculus musculus* (+ *M. m. wagneri* ?)
Mus 3 = *Mus spretus*
Mus 4 = *Mus spicilegus* (= *M. abbotti* = *M. hortulanus* ?)

(for the nomenclatorial problems involved see Marshall & Sage, this volume pp. 15–25.)

TABLE I

Biochemical differentiation of the four house mouse groups (from Bonhomme, Britton-Davidian et al., *1978).*

	Mus 4 Greece	*Mus* 1 Greece	*Mus* 3 France	*Mus* 3 Spain	*Mus* 1 France	*Mus* 2 Hungary
Mus 4 Greece	—	5	8	8	5	8
Mus 1 Greece	0.36	—	5	6	0	3
Mus 3 France	0.58	0.37	—	0	6	8
Mus 3 Spain	0.53	0.41	0.04	—	6	8
Mus 1 France	0.42	0.03	0.52	0.47	—	3
Mus 2 Hungary	0.53	0.37	0.68	0.62	0.43	—

Lower triangle: Nei's genetic distance.
Upper triangle: number of loci out of 20 with no allele in common.

TABLE II

Systematic relationships of the four biochemical groups.

Mus 2	Parapatric semispecies (narrow hybrid zone)		
Mus 3	Sympatric species	Allopatric species	
Mus 4	Sympatric species	Sympatric species	Allopatric species
	Mus 1	*Mus 2*	*Mus 3*

Biochemical Groups *Mus* 1 and *Mus* 2

Zoologists have long recognized two geographical races of mice in northern Europe. To the west mice are long-tailed and have a dark-coloured belly (*Mus musculus domesticus*); while to the east mice are not so long-tailed and have a light-coloured belly (*Mus musculus musculus*). The transition from one race to the other occurs along a boundary running north-south through Germany and then east–west across the Jutland peninsula of Denmark. Selander *et al.* (1969) showed that Danish populations south of the boundary line (*M. m. domesticus*) are much more different biochemically from those north of the boundary (*M. m.*

musculus) than they are from populations studied previously in the United States.

Choosing south Jutland populations to define the biochemical group *Mus* 1 we have identified it not only in other places where *M. m. domesticus* occurs (Atlantic Europe), but also in Spain, the south of France, Italy, Yugoslavia, Greece, Israel, Tunisia and Algeria. We conclude that *Mus* 1 occurs all around the Mediterranean. Most of the Mediterranean populations we have analysed belong to the *Mus musculus brevirostris* morphotype (tail a little shorter than *M. m. domesticus*, light-coloured belly). One sample (from Tatahouine, Tunisia) exhibits a "desert morphotype" frequently referred to as *M. m. praetextus*. We consider *Mus* 1 to be the typical inhabitant of Old World areas exposed to Mediterranean and Atlantic climates. *Mus* 1 has been introduced by man to the Americas and to many oceanic islands.

In the same way, we have used north Jutland populations to define the biochemical group *Mus* 2. We have been able to trace it to Hungary (Britton-Davidian, Ruiz Bustos, Thaler & Topal, 1978) and to Roumania (Thaler, Bonhomme & Britton-Davidian, 1978). European populations of *Mus* 2 should be referred to as *M. m. musculus*. However they closely resemble *Mus* 4 morphologically (see below), and this name has certainly been frequently misapplied; conversely many specimens of *Mus* 2 are currently misidentified. Despite this confused situation we believe it is safe to follow Schwarz & Schwarz (1943) in regarding *M. m. musculus* as extending through eastern Europe deep into the Asiatic part of the USSR.

The boundary line between *M. m. domesticus* and *M. m. musculus* has been studied in detail by Ursin (1952) using the methods of classical taxonomy. He noted that it was in fact a zone some 20 to 30 km wide in which the populations were morphologically intermediate between the two subspecies. Ursin concluded it was a hybrid zone. This stimulated a major electrophoretic investigation of Danish populations (Selander *et al.*, 1969; Hunt & Selander, 1973), which had far-reaching consequences:

(1) It confirmed Ursin's thesis. There is a hybrid zone between biochemical groups *Mus* 1 and *Mus* 2 matching the morphological boundary between the subspecies *domesticus* and *musculus*. On the continental scale this zone is not wide and deserves the now established biogeographical designation as a *narrow hybrid zone*.

(2) It showed that reproductive and genetical isolation are not equivalent. There is no reproductive isolation at all between *domesticus* and *musculus* in Denmark, but introgression between them occurs to a very limited extent, suggesting that the hybrids and

their progeny are less fit than their parents. Whatever limits the width of the hybrid zone, the absence of reproductive isolation between *domesticus* and *musculus* in nature supports the combining of them in a single biological species, but their genetical isolation contradicts it. The semispecies concept has proved useful in such cases. Following Selander and co-workers, we shall consider that *domesticus* (by extension the biochemical group *Mus* 1) and *musculus* (equivalent to *Mus* 2) interact as two parapatric semispecies belonging to the species *Mus musculus*.

(3) It revealed that the limited introgression observed involves the two semispecies differently. Penetration of *Mus* 1 alleles into the territory of *Mus* 2 proved to be on average more pronounced in terms of distance from the boundary line than the penetration of *Mus* 2 alleles into the territory of *Mus* 1. Moreover different alleles exhibit different degrees of penetration. These data indicate simultaneously that:

(a) Gene flow in *Mus musculus* occurs despite a discontinuous distribution of colonies.
(b) There may be differential selective control of the alleles introgressing into a semispecies.
(c) There is less restriction to introgression in *Mus* 2 than in *Mus* 1. This may be due either to a difference of gene flow or to a difference in selective control.
(d) A correlation exists between the *Mus* 1–*Mus* 2 boundary position and a steep climatic change: *Mus* 1 occupies a moister part of Jutland than *Mus* 2.

What about the interaction of *Mus* 1 and *Mus* 2 outside Denmark? A morphological boundary between *M. m. domesticus* and *M. m. musculus* was traced by Zimmermann (1949) from the Baltic Sea towards southern Europe. He did not identify the situation at the southern edge. We suggest tentatively that when reaching the Mediterranean region, *Mus* 1 changes from the *domesticus* to the *brevirostris* morphotype and becomes more difficult to distinguish morphologically from *Mus* 2, while the biochemical group *Mus* 4 (which is morphologically close to *Mus* 2) occurs on both sides of the *Mus* 1–*Mus* 2 boundary, thus adding to the difficulty.

Having identified *Mus* 1 in Greece, close to the Yugoslavian, Bulgarian and Turkish borders and having found *Mus* 2 in Hungary and south-eastern Roumania, we suggest that beyond the known *Mus* 1–*Mus* 2 boundary, the hybrid zone bends to the east and eventually reaches the Black Sea in Bulgaria. This is a very rough delineation of the boundary, and does not allow a definite identification at the moment. Nevertheless we feel it pertinent to note that the *Mus* 1–*Mus* 2 boundary line across Europe seems to fit generally with climatic factors, such as the mean January temperature (Fig. 1).

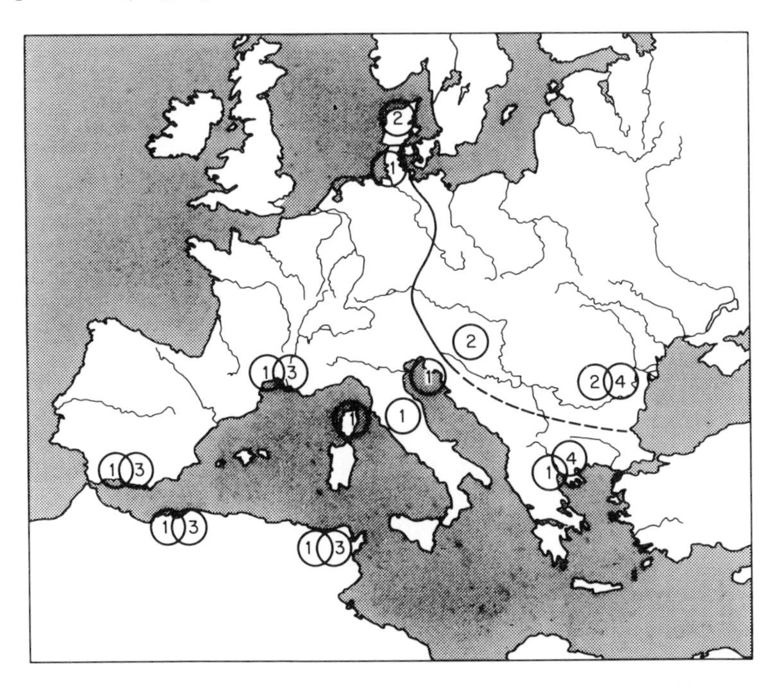

FIG. 1. Distribution pattern of the genus *Mus* in Europe. Circled numbers indicate trapping localities and biochemical group. The solid line represents the recognized (and the broken line the presumed) narrow hybrid zone between *Mus* 1 and *Mus* 2.

Biochemical Group *Mus* 3

In Mediterranean countries the so-called house mouse is commonly found outdoors as well as indoors. In many places outdoor populations exhibit morphological traits such as a very short tail. There has been much debate about these short-tailed outdoor Mediterranean mice, opinions varying from considering them as one or several species distinct from *Mus musculus* to denying them any separate taxonomic status. The accepted view until recently has been that of Schwarz & Schwarz (1943) who refer to them as two subspecies of *M. musculus* (*M. m. spretus* around the western Mediterranean and *M. m. spicilegus* in the east) occurring in sympatry with another subspecies of the same species (*M. m. brevirostris* in most places). Besides being contrary to the Rules of Nomenclature, this acceptance of two subspecies in the same locality merely avoids a biological problem: are short-tailed mice reproductively isolated or not?

The problem has been approached by electrophoretic methods in southern France (Britton, Pasteur & Thaler, 1976). There we were able to show that mice fall into two biochemical groups: the already well-known *Mus* 1 and others differing from both *Mus* 1 and *Mus* 2. We have called this second group *Mus* 3. In southern France, *Mus* 3 has almost never been found indoors whereas *Mus* 1 occurs both indoors and outdoors. In some outdoor habitats *Mus* 1 and *Mus* 3 breed side by side, but they do not interbreed in nature. First generation hybrids (F_1) would be very easy to detect since *Mus* 1 and *Mus* 3 have different fixed alleles at many loci. We have found no hybrids. Both field and morphological evidence support the conclusion that *Mus* 1 and *Mus* 3 are completely (or at least very highly) isolated reproductively: all *Mus* 1 are long-tailed (*brevirostris* morphotype) while all *Mus* 3 are short-tailed (*spretus* morphotype).

Mus 3 has been identified in Spain (Britton-Davidian *et al.*, 1978; Sage, 1978) and Algeria. In Spain it is not exceptional to trap it indoors. Where it occurs sympatrically (and sometimes syntopically) with *Mus* 1, it is reproductively isolated from it and exhibits the *spretus* morphotype. Thus *Mus* 1 and *Mus* 3 behave as sympatric species throughout the range of *Mus* 3. We may call the latter *Mus spretus* which we consider from morphological data to be a western Mediterranean species occurring in France, Spain (including the major Balearic islands), Morocco, Algeria and Tunisia. No short-tailed mouse has been recorded from Italy and Libya, nor does the species seem to extend east of France and Tunisia.

It is possible to hybridize *M. musculus* and *M. spretus* under laboratory conditions using either species as the male or the female parent (Bonhomme, Martin & Thaler, 1978). However it is much easier when the male is *M. spretus* since the females of this species are difficult to breed in captivity. Fully viable hybrids of both sexes have been obtained using various inbred laboratory strains as well as wild *M. musculus* (including specimens of *M. m. brevirostris* trapped in the territory of *M. spretus*). Female F_1 hybrids are fertile and have been back-crossed successfully to *M. musculus*, but male F_1 hybrids are sterile, as are most males from back-crosses. This sterility seems to be under multifactorial genetic control.

Biochemical Group *Mus* 4

In northern Greece we have shown that the short-tailed outdoor mice known there as "*Mus musculus spicilegus*" (Ondrias, 1965) which occur sympatrically with *Mus* 1 (*brevirostris* morphotype) are reproductively isolated from it and represent another biochemical group. This group differs not only from *Mus* 1 and *Mus* 2, but also from *Mus* 3, and has been named *Mus* 4 (Bonhomme, Britton-Davidian *et al.*, 1978).

In Roumania we acquired specimens of mound-building mice known there also as "*M. m. spicilegus*" (Hamar, 1960). These are well grain for winter. Despite this very peculiar behaviour, these mice proved to belong to the same biochemical group *Mus* 4 as Greek short-tailed mice (Thaler *et al.,* 1978). At the same time and place specimens of *Mus* 2 and *Mus* 4 were trapped. Electrophoresis indicated complete genetical isolation between these syntopic populations. Notwithstanding, they were so morphologically similar (both had short tails) that only electrophoretic data were capable of distinguishing the two groups. Clearly *Mus* 4 (*spicilegus*) and *Mus* 1 (*brevirostris*) behave as sympatric species in Greece.

It is equally certain that *Mus* 4 (*spicilegus*) and *Mus* 2 (*musculus)* behave also as sympatric species in Roumania. This contradicts a long-standing hypothesis that the indoor *M. m. musculus* of northern Europe (German, Poland. Scandinavia) have evolved from the outdoor short-tailed mice group of south-eastern Europe to which the mound-builder belongs (whatever its name: *spicilegus, hortulanus,* etc).

The distributions of *Mus* 3 and *Mus* 4 are separate, so it is impossible to observe their interaction in nature and to deduce from it their mutual taxonomic status. Many authors have combined them under the name of *spicilegus* following Miller (1909, 1912), for the sole reason that both are outdoor short-tailed mice. But the genetical distance between *Mus* 3 and *Mus* 4 is of the same order as that between them and any of the populations of *Mus musculus* with which they are sympatric, giving no grounds to refer to them as a single species (*Mus* 3 + *Mus* 4). We consider *Mus* 3 and *Mus* 4 as two allopatric species and name them respectively *Mus spretus* and *Mus spicilegus* (for nomenclatorial problems see Marshall & Sage, this volume pp. 15–25).

We have bred two fully viable female hybrids between a male *Mus spicilegus* from Roumania and a female from a laboratory strain of *Mus musculus.* These F_1 female hybrids have not reproduced when mated back to *M. musculus.*

Origin of the Present Distribution of the Four Main Biochemical Groups in Europe

No fossil *Mus* are known in any of the numerous post-glacial Palaeolithic deposits in Europe (see Brothwell, this volume pp. 1–13), which suggests that the ancestors of the living populations colonized Europe in Neolithic and later times. Yet the genetical distances between the four biochemical groups indicate divergence much earlier than the Neolithic; in other words, the four groups seem to have been well differentiated before spreading to Europe. On this assumption, we may interpret the hybrid zone between *Mus* 1 and *Mus* 2 as a secondary intergradation zone, implying that the parapatric *Mus* 1 and *Mus* 2 previously had allopatric distributions. The simplest suggestion is that *Mus* 2 (= *M. m. musculus*) migrated directly from Asia to Europe across Russia, while *M. m. domesticus* (which belongs to *Mus* 1) evolved from the Mediterranean *M. m. brevirostris* which was descended from north African *M. m. praetextus* via migrations through Spain and Italy (Schwarz & Schwarz, 1943). However *Mus* 1 is present all around the Mediterranean, so a migration route from the Middle East to western Europe across Turkey and the Balkan Peninsula cannot be excluded.

The present disrupted pattern of the *Mus* 1 and *Mus* 2 boundary, with a German tract and a Danish tract separated by the Baltic Sea, suggests a retreat of *Mus* 2 after its encounter with *Mus* 1. Following Hunt & Selander (1973), we may suppose that the boundary line position expresses a dynamic equilibrium controlled by climatic factors.

The western Mediterranean distribution of *Mus* 3 (= *M. spretus*) indicates that this shortest-tailed mouse probably differentiated in north-west Africa whence it migrated to south-west Europe (Schwarz & Schwarz, 1943). In Europe, and especially in southern France, *Mus spretus* is a very good indicator of the Mediterranean climatic belt (Fayard & Erome, 1976), but strangely enough it does not extend west of the Var river (Orsini, in press). Its absence from Italy is not yet understood.

From taxonomic descriptions in the literature, we may assume that *Mus* 4 is present from Yugoslavia to Iran through Turkey. This indicates a probable migration route from Asia to Europe south of the Black Sea. But *Mus* 4 is also present in Roumania and the Ukraine where the mound-builder type is very abundant on the wheat plains. So a migration route north of the Black Sea is also possible. While Greek and Iranian *Mus* 4 are not biochemically distinct (Darviche *et al.*, 1979), there is an appreciable genetical distance between Greek and Roumanian *Mus* 4 which have evolved different alleles at three loci. These suggest the existence of a southern biochemical subgroup *Mus* 4A and a northern 4B. The first subgroup could have used the southern migration route and the second subgroup the northern one.

We know almost nothing about the distribution of the biochemical groups *Mus* 1, 2 and 4 in Asia, and the zoological literature is difficult to evaluate. This absence of evidence does not rule out a provisional hypothesis that the three groups differentiated by classical allopatric speciation in separate areas of Asia. In the same way *Mus* 3 probably differentiated in Maghrebian Africa, an area long isolated by seas and deserts.

After this hypothetical differentiation from each other in isolation, the species and semispecies would have become sympatric and parapatric as they extended their distribution during the "Neolithic Revolution". Neolithic and historic man must have played a determining role in transporting mice across sea barriers, but perhaps more importantly by providing favourable habitats through farming and building practices.

CHROMOSOMAL DIFFERENTIATION

Cytogenetics proper are treated elsewhere in this symposium volume and we will restrict ourselves to the facts necessary for discussing the evolution of chromosomal races, drawing heavily on the stimulating contribution of White (1978).

Chromosomal races are not mere strange cytogenetical artefacts. Capanna and co-workers (Capanna, Civitelli & Cristaldi, 1977; Capanna & Riscassi, 1978), in a survey of Italian populations, have shown that such races are not rare, being distributed over significant fractions of the country (Fig. 2). The normal chromosomal complement of the house mouse is $2n = 40$ chromosomes, all acrocentrics. Each of the chromosomal races is characterized by being monomorphic for a number of Robertsonian translocations (centric fusions) between pairs of nonhomologous acrocentrics, resulting in the same number of metacentrics. The maximum number of fusions is ten, but for some reason the sex chromosomes and chromosome 19 never appear to be involved in a fusion, and so the maximum number of fusions observed so far is nine, resulting in $2n = 22$. Other observed numbers in monomorphic populations are 8 fusions ($2n = 24$), 7 fusions ($2n = 26$), and 6 fusions ($2n = 28$).

Chromosomal races may differ also by the fusions they harbour, that is the particular pair of acrocentrics associated in each fusion. There are races with $2n = 22$ which differ by as many as eight fusions out of nine. Capanna points out that a fusion must have originated from a unique event (chromosomal mutation); he considers that different fusions appeared sequentially and not

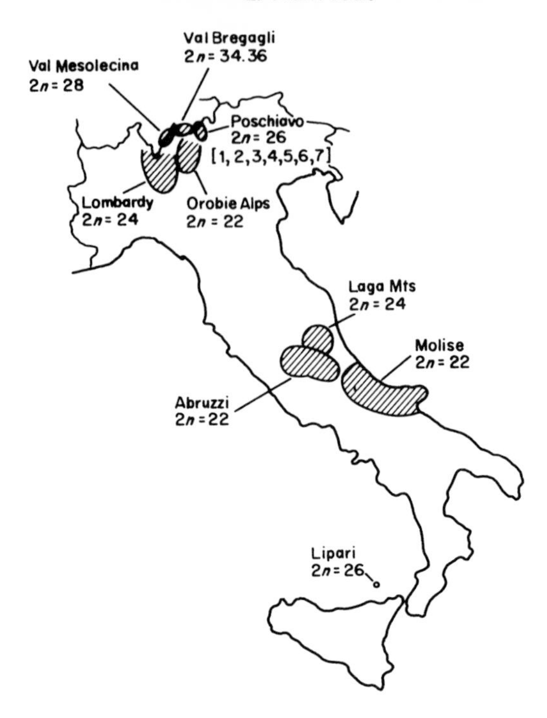

FIG. 2 Map of the Italian peninsula showing the approximate location of populations of mice with chromosomal fusions. The area not cross-hatched is presumably occupied by 40-chromosome mouse populations, but has only been sampled at a few points (from White, 1978).

simultaneously. There seems to be selection for an increasing number of fusions in chromosomal races although the contributors to each fusion do not seem to be significant. On the Italian mainland two geographical groups of chromosomal races have been defined, known as the Alpine and the Apennine systems. All races belonging to a system may be regarded as deriving from a single ancestor population in which the first fusion occurred. As new fusions appeared, they were fixed in a part only of the progeny, resulting in phylogenetic branching.

The interaction between a race with metacentric chromosomes and normal mice (as studied between the 22-chromosome Abruzzi race and the 40-chromosome mice around Rome) is of the semispecies type like that between *Mus* 1 and *Mus* 2. There is a narrow hybrid zone some 20 km wide in which populations are polymorphic, the same chromosome arms being present as acrocentrics and as parts of metacentrics. Experimental hybrids between a chromosomal race and 40-chromosome laboratory mice have a much lower fertility than their parents, due to frequent abnormalities at meiosis. This may account for the narrowness of the hybrid zone. The restraints to introgression seem to be even stronger than between *Mus* 1 and *Mus* 2. There seems also to be an asymmetry, metacentrics introgressing more easily into 40-chromosome populations than acrocentrics into the chromosomal race.

Experimental hybrids between two different chromosomal races are usually completely sterile owing to gross abnormalities at meiosis. This means that no polymorphic population is expected between contiguous chromosomal races, and none has been observed. In such cases distribution is parapatric without a hybrid zone. While no first generation hybrids have been trapped yet, such specimens probably occur but are not able to initiate any introgression. We may consider two such contiguous chromosomal races as semispecies since they are probably not reproductively isolated, but are completely genetically isolated and hence are approaching full speciation.

According to White (1978), the Italian mainland chromosomal races of the house mouse offer the best example known of stasipatric chromosomal speciation. White's model is two-fold. "Stasipatric" speciation refers to the spatial relationship between a new race and its ancestral species. The new race is seen as arising from a local population at a point surrounded by normal populations. This is a challenging alternative to the classical model of allopatric speciation which begins with the geographical disruption of a species, thus allowing for the differentiation of geographical races which eventually develop reproductive barriers as they evolve in isolation. "Chromosomal" speciation refers to the biological mechanism (i.e. lowered fertility of chromosomal heterozygotes) which produces some genetical isolation between neighbouring populations. The isolating chromosomal transformations (fusions) are interpreted by White as a protective device in a locally adapted population of mice against introgression from foreign populations.

We have scored several chromosomal races of the Alpine and Apennine systems for genetical differences by electrophoresis (Britton-Davidian, Bonhomme *et al.*, 1980). All have been shown to belong to the biochemical group *Mus* 1, as have the neighbouring 40-chromosome populations. In particular the Alpine chromosomal races share with the 40-chromosome populations of the same region a particular allele ($Es\text{-}10^d$) not known in the other parts of the range of *Mus* 1. All this fits well the stasipatric speciation model.

More surprisingly it has not been possible to show any degree of biochemical differentiation of a chromosomal race. Not only are all the alleles in a chromosomal race found also in the neighbouring 40-chromosome populations, but the amount of heterozygosity is about the same. This seems to indicate either that the chromosomal races are of a very recent origin, or that there remains enough gene flow between the populations to prevent divergence. These two explanations are not exclusive: the recent origin is obvious since the chromosomal races must have differentiated after the house mouse migrated to Italy a few thousand years BC; while there is indirect evidence of the persistence of some gene flow at least during the process of fixation of the centric fusions, since structural genes display the same polymorphism in chromosomal races as in 40-chromosome populations, implying recombination between homologous chromosome arms in heterozygotes for arm condition (i.e. separated as an acrocentric and fused in a metacentric). White (1978) argues that a chromosomal race will differentiate genetically. This is not contradicted by the electrophoretic evidence since most of the differentiation is likely to be controlled by regulatory genes, quite apart from the possibility that other structural genes than the ones scored by us may be involved.

White has suggested that natural selection operates at two levels in the process of stasipatric chromosomal speciation. Firstly there will be individual selection giving rise to a gene complex adapted to local conditions. Owing to gene flow, allele frequencies will be suboptimal. Then a centric fusion becomes fixed by chance in a small area, bringing a degree of protection against gene flow and allowing frequencies to move closer to the optimum. The next stage is when the centric fusion expands spatially by the differential extinction and initiation rates of colonies. In the now larger area where the first fusion was fixed, a second fusion may arise and be fixed in the same way, and so on. White calls this a chain process in chromosomal speciation.

It will probably be difficult to discover what precisely are the biological features giving an advantage to a chromosomal race. But we suggest that a clue may be in the already noticed asymmetry of the narrow hybrid zone between a chromosomal race and a 40-chromosome population: metacentrics introgress more easily into the 40-chromosome population than acrocentrics into the Robertsonian race. A number of hypotheses could explain this, but there is one which could be tested fairly easily — that the migration rate between localities is reduced in a chromosomal race as compared to the neighbouring 40-chromosome population. The basis of this suggestion is that *Mus* 1 was first established in Italy as an opportunist pioneer, well adapted to long distance propagation and facultative outdoor dwelling, while more recent populations,

UNIVERSITY OF WINNIPEG
LIBRARY
515 Portage Avenue
Winnipeg, Manitoba R3B 2E9

particularly those specializing in indoor commensalism, are maintained on more sedentary habits.

The Italian chromosomal races of the house mouse satisfy the biological criteria for semi-specific rank, and are more sharply isolated from 40-chromosome mice and from each other than are most recognized subspecies (for instance *M. m. domesticus* and *M. m. brevirostris*). So one could give them trinomial subspecific names without offending the Rules of Nomenclature. But there seems little point.

CONCLUDING REMARKS

The chromosomal semispecies of Italian mice support White's stasipatric chromosomal model. But is it a good *speciation* model in the form advocated by its author? We may question it at least in the house mouse and related species. Speciation is fully complete only when sister-groups are able to interact as sympatric species, like *M. musculus* and *M. spretus*. But these two (and indeed all other species of the subgenus *Mus*) exhibit the same 40-chromosome karyotype. Obviously these species did not evolve through chromosomal speciation. So, White's model appears to be only a *semispeciation* model in mice; it is by no way certain that any of the resulting semispecies will give rise to a new species.

The three species *Mus* 1 + 2, *Mus* 3, *Mus* 4 probably evolved through allopatric speciation. So did the semispecies *Mus* 1 and *Mus* 2. If the postulated geographical isolation had lasted longer, they would presumably have become full species. As it is they meet along a narrow hybrid zone – and this points to a state of dynamic equilibrium more likely to lead to the extinction of one of the semispecies than to progress towards speciation.

So most semispecies, whether they have arisen stasipatrically or allopatrically, do not seem to be on their way towards full speciation. Conversely, there is no evidence that fully developed sympatric species ever evolved from a state of narrow hybrid zone interaction. There is clearly much potential interest in a continuing study of house mouse speciation.

REFERENCES

Bonhomme, F., Britton-Davidian, J., Thaler, L. & Triantaphyllidis, C. (1978). Sur l'existence en Europe de quatre groupes de souris (genre *Mus* L.) du rang espèce et semi-espèce, démontrée par la génétique biochimique. *C. r. hebd. Séanc. Acad. Sci., Paris* **287** (D): 631–633.

Bonhomme, F., Martin, S. & Thaler, L. (1978). Hybridation en laboratoire de *Mus musculus* L. et *M. spretus* Lataste. *Experientia, Basel* **34**: 1140–1141.

Britton, J., Pasteur, N. & Thaler., L. (1976). Les souris du midi de la France: caractérisation génétique de deux groupes de populations sympatriques. *C. r. hebd. Séanc. Acad. Sci., Paris* **283** (D): 515–518.

Britton-Davidian, J., Bonhomme, F., Croset, H., Capanna, E. & Thaler, L. (1980). Variabilité génétique chez les populations de souris (genre *Mus* L.) à nombre chromosomique réduit. *C. r. hebd. Séanc. Acad. Sci., Paris* **290** (D): 195–198.

Britton-Davidian, J., Ruiz Bustos, A., Thaler, L. & Topal, M. (1978). Lactate dehydrogenase polymorphism in *Mus musculus* L. and *Mus spretus* Lataste. *Experientia, Basel* **34**: 1144–1145.

Capanna, E., Civitelli, M. V. & Cristaldi, M. (1977). Chromosomal rearrangement, reproductive isolation and speciation in mammals. The case of *Mus musculus*. *Boll. Zool.* **44**: 213–246.

Capanna, E. & Riscassi, E. (1978). Robertsonian karyotype variability in natural *Mus musculus* populations in the Lombardy area of the Po Valley. *Boll. Zool.* **45**: 63–71.

Darviche, D., Benmehdi, F., Britton-Davidian, J. & Thaler, L. (1979). Note: Données préliminaires sur la systématique biochimique des genres *Mus* et *Apodemus* en Iran. *Mammalia, Paris* **43**: 427–430.

Fayard, A. & Erome, G. (1976). *Mus musculus spretus* en Ardèche. *Mammalia, Paris* **40**: 689–690.

Gropp, A., Tettenborn, U. & Lehmann, E. (1970). Chromosomenvariation vom Robertsonian Typus bei der Tabakmaus, *M. poschiavinus*, und ihren Hybriden. *Cytogenetics* **9**: 9–23.

Hamar, M. (1960). O Systematike, rasproctranenü u ecologii courganchicovoi miski (*Mus musculus spicilegus* Petenyi, 1882) b rouminscoï narognoï respoublike. *Revue roum. Biol.* (Zool.) **5**: 207–219.

Hunt, W. G. & Selander, R. K. (1973). Biochemical genetics of hybridisation in European house mice. *Heredity* **31**: 11–33.

Miller, G. S. (1909). Twelve new European mammals. *Ann. mag. Nat. Hist.* (8) **3**: 415–422.

Miller, G. S. (1912). *Catalogue of the mammals of Western Europe.* London: British Museum (Natural History).

Ondrias, J. C. (1965). The taxonomy and geographical distribution of the rodents of Greece. *Säugetierk. Mitt.* **14**: 1–136.

Orsini, Ph. (In press). Notes sur les souris de Provence. *Annls Soc. Sci. nat. Toulon.*

Sage, R. D. (1978). Genetic heterogeneity of Spanish house mice (*Mus musculus* complex). In *Origins of inbred mice:* 519–553. Morse, H. C. (Ed.). New York: Academic Press.

Schwarz, E. & Schwarz, H. K. (1943). The wild and commensal stocks of the house mouse, *Mus musculus* Linnaeus. *J. Mammal.* **24**: 59–72.

Selander, R. K., Hunt, W. G. & Yang, S. Y. (1969). Protein polymorphism and genetic heterozygosity in two European subspecies of the house mouse. *Evolution* **23**: 379–390.

Thaler, L., Bonhomme, F. & Britton-Davidian, J. (1978). Molecular population genetics of mice. Significance of the Mediterranean Basin. *Hereditas* **89**: 149.

Ursin, E. (1952). Occurrence of voles, mice, and rats (Muridae) in Denmark, with a special note on a zone of intergradation between two subspecies of the house mouse (*Mus musculus* L.). *Vidensk. Meddr danske naturh. Foren.* **114**: 217–244.

White, M. J. D. (1978). Chain processes in chromosomal speciation. *Syst. Zool.* 27: 285–298.

Zimmerman, K. (1949). Zur Kenntnis der mitteleuropäischen Hausmäuse. *Zool. Jb.* (Syst.) 78: 301–322.

Symp. zool. Soc. Lond. (1981) No. 47, 43–62

Domestication and Development of the Mouse as a Laboratory Animal

MICHAEL F. W. FESTING and D. P. LOVELL

Medical Research Council, Laboratory Animals Centre, Carshalton, Surrey, UK

SYNOPSIS

Mice are the most widely used experimental mammals, having made important contributions in most areas of biomedical research. Currently there is a trend towards the increasing use of inbred and mutant strains in all disciplines. These are almost essential in modern immunological and cancer research. Mice have been domesticated for several thousand years, and have been used scientifically at least since 1664. Modern strains were developed in the early part of the 20th century from fancy, pet and, to a small extent, wild mice. The history of many strains has been recorded, though the accuracy and value of such a record must not be assumed.

The relationship between laboratory and wild mice has been investigated using various types of genetical markers. Studies of biochemical polymorphisms have revealed some new alleles in wild mice which were not known in laboratory stocks, but on the whole such markers do not suggest any fundamental differences between the two. Immunological markers on the other hand seem to show that wild and laboratory mice are very different. Such differences could have arisen either as a founder effect, or as a result of natural selection for survival in laboratory conditions.

Outbred Swiss-Webster mice appear to be rather similar genetically to island populations of wild mice. Studies of the relationships between inbred strains have been conducted using biochemical markers and mandible shape. These studies suggest that the C57 family of strains is not typical of other inbred strains of mice.

Studies of wild mice are of obvious scientific interest *per se*, and can provide new sources of inherited variation. However, suggestions that wild mice being more "natural" are a better model of man, and should be more widely used in general biomedical research are probably misguided. Modern research needs genetically defined inbred and mutant mice selected to fit each research project.

INTRODUCTION

The mouse is easily the most widely used experimental animal in biomedical research, accounting for over 60% of all animals used in the UK in 1978 (Home Office, 1978). Its popularity depends partly on its small size, good reproductive performance and the ease with which it can be handled, and partly on the wide range of genetically defined strains and mutants which are available.

Recently, geneticists have turned their attention to wild mice both as a source of new inherited variants, and in order to study the evolution of the species. This has revived interest in the relationship between wild and laboratory mice. The development of new techniques of biochemical genetics, immunogenetics, cytogenetics, numerical taxonomy, and the discovery of new genetical markers has now made it possible to attempt such studies. The aim of this chapter is to review briefly the contribution of the mouse to biomedical research and some of the available evidence on the origins of laboratory mice. The relationships between wild and laboratory mice, and between different strains of laboratory mice will also be considered, with a discussion of the implications of these findings for general biomedical research.

THE CONTRIBUTION OF GENETICALLY DEFINED MICE TO BIOMEDICAL RESEARCH

The contribution of the mouse to biomedical research has obviously been of immense importance. In many cases the role of the mouse has been simple, but still very important. Those white mice which were used to test and develop the first antibiotics have probably saved literally millions of human lives. However, the mouse can be developed to play a much more active role in research, and it is through the development of inbred strains, and later of congenic lines, mutants and other genetically defined types that the true potential of the mouse is now begining to be realized (Festing, in press).

The first inbred strain of mice, DBA, was developed by C. C. Little in 1909, but it was not until the 1920s and 1930s that geneticists and cancer research workers really began to understand the value of such animals and develop new strains on a large scale. By 1942 Dr L. C. Strong was suggesting that "... within the near future all research on mice should be carried out on inbred animals or on hybrid mice of known (genetically controlled) origin, where the degree of biological variability has been carefully controlled" (Strong, 1942). By that time work on inbred mice was beginning to be really fruitful. An early discovery was that tumours could frequently be transplanted within an inbred strain, whereas they would usually be rejected in genetically variable stocks. By 1916 the genetical laws governing tumour rejection via what would today be called an allograft reaction were formulated and tested experimentally by Little and Tyzzer (Klein, 1975). Later, an attempt by P. A. Gorer to study the nature of the genes that were responsible for this rejection led to the discovery of a blood group antigen, antigen 2, which was associated with tumour rejection. Antigen 2 is now known as the *H-2* complex, and research on this complex, largely made possible through the development of congenic strains of mice by G. D. Snell starting in 1947, has opened up whole new areas of immunological research.

In the meantime, inbred strains continued to play an important part in cancer research, largely as a vehicle for transplantable tumours which could be studied much more conveniently than their spontaneous counterparts. Eventually, the inbred or hybrid mouse bearing a transplantable tumour became the standard tool for evaluating anti-cancer drugs, and has provided most of the (pitifully) few effective drugs that are available today.

The Jackson Laboratory was founded in 1929 by C. C. Little. This institution has made a remarkable contribution to the development of the mouse as a laboratory animal, having been in the forefront of the development and use of inbred strains, congenic strains, and more recently recombinant inbred strains (Bailey, 1971), as well as conserving and investigating many mouse mutants. An early discovery was that mouse mammary tumours in susceptible strains were caused by an agent which could be transmitted through the mother's milk. This eventually led to many studies on viruses as a cause of cancer. The significance of this work to man is still not clear today, but it has undoubtedly resulted in the accumulation of a considerable amount of fundamental knowledge of the biology of murine tumour viruses and their relationships with their host.

The steep upward trend in the use of genetically defined mice in cancer research and immunology during the past 40 years (Festing, 1979a) has not always been paralleled in other disciplines. In toxicology and pharmacology, which are rapidly growing disciplines at present, few inbred animals are used, and even in cases where such animals are used the investigator is apparently often not fully aware of the way in which such strains could be used most effectively. The blame for this situation must lie with mammalian geneticists, who have often not bothered to inform their colleagues of the properties of these valuable animals (W. L. Russell, 1941). However, it looks as though this will change in the future. Not only will inbred and congenic strains be used in a wider range of different disciplines, but the value of mutants appears also to be becoming more widely recognized. The Staff of the Jackson Laboratory have often stated (Eva Eicher, personal communication) that the laboratory animal of the future is likely to be a mutant. The recognition of the value of the nude mouse in cancer research and immunology has created considerable interest in other mutants. In some fields such as studies of obesity (Festing, 1979b), the use of mutants is already well established.

These intellectual and practical developments of the laboratory mouse, together with a recent surge of interest in wild mice among mammalian geneticists and students of evolution, make it imperative that the origin and biology of the laboratory mouse should be more fully understood.

HISTORY OF DOMESTIC MICE

Keeler (1931) has made a thorough study of the "antiquity of the fancy mouse", and this section is based largely on his book. He found that the presence of mice was recorded in many ancient cultures. In ancient Egypt mice were probably considered to be a serious pest, and it may have been for this reason that the cat, a good killer of mice, was deified in about 2800 BC. As the cat was considered to embody all godly virtue, the mouse came to symbolize evil. It was also considered to be unclean in the ancient Hebrew culture in Palestine, as the Book of Leviticus states:

> These also shall be an abomination to you among the creeping things that creep upon the earth; the weasel, and the mouse, and the tortoise after his kind. . . . These are unclean to you among all that creep: whosoever doth touch them, when they be dead, shall be unclean until even.

However, mice were not evil in all ancient cultures. Homeric legend mentions Apollo Smintheus (God of mice) in about 1200 BC. Mice were worshipped by Cretan Teucri invaders who landed on the shore of Asia Minor to found colonies, following a decisive fight with the local Pontians in which victory for the Cretans was attributed to their Apollo who caused mice to gnaw the leather straps from the shields of their enemies. Temples were built in which mice were maintained at public expense. There is evidence that some of these mice were white.

This mouse cult spread to a number of other cities, including even Athens and Thespia on the Greek mainland. Aristotle refers to the white mice of Pontis, and Strabo (*c.* 25 BC) mentions the white mouse cultures in many Sminthian temples. Pliny mentions the use of white mice for auguries, noting "... how the lobes and filaments of their livers and bowels do increase or decrease in number according to the days of the Moon's age... ".

In Christian Europe the mouse fell into disrepute, becoming associated with witches and sorcerers. It was also considered to be "voluptuous and libidinous", and mice were often raised by curious churchmen in order to observe their wicked and lustful actions.

In the Orient, the mouse has always had a much higher social rating. Albino mice were also used for auguries, and between AD 307 and 1641 there were Government records of the trapping of mice in the wild, with a note of some 30 cases in which albino mice were caught. It is not clear how long different varieties of mice have been recognized in China, but the oldest Chinese lexicon, written in 1100 BC, has a word for "spotted mouse", and waltzing mice have been known since about 80 BC. These and other varieties of mice were established in Japan as pets and "fancy mice".

Wild mice in Japan are the subspecies *Mus musculus molossinus*, so it seems apparent that these fancy mice, which are not *Mus musculus molossinus*, were all derived from China. Keeler (1931) notes that the Chinese fancy mouse may be partly derived from *Mus bactrianus* (*wagneri*) of Tibet. It is not known whether these mice were crossed with *Mus musculus molossinus*, though there is some suspicion that this subspecies has made some contribution to the modern laboratory mouse. Kondo (1973) notes that a book *Breeding methods of striking mice* was published in 1787, and this has illustrations of black, white and spotted mice. Whether the term "striking" implies that the mice were used for fighting is not clear.

According to Keeler, "something over a hundred years ago several of these fancy varieties of the house mouse were taken from Japan to Europe by British traders... " Later these mice were used by scientists in investigations of the inheritance of the varietal characteristics, though until the re-discovery of Mendel's laws in 1900, the meaning of the data that they accumulated remained obscure.

In the meantime, in Europe the mouse had been used in scientific experiments as early as 1664, when Robert Hooke used a mouse in studies of the properties of air, and particularly oxygen. According to Birch's *History of the Royal Society*, published in 1756, in Vol. I page 428 of the proceedings of the Society there is a description of an experiment using a mouse to study the effects of an increase of pressure up to four times normal air pressure, in which the mouse survived (Masson, 1940). Unfortunately, it is not stated whether or not the mouse was wild. However, regular breeding of mice in the laboratory apparently did not start until about 1850. Presumably these mice were the ones imported from Japan, though it would be difficult to rule out the possibility that some European albino mice were crossed with them.

HISTORY OF LABORATORY MICE

The possible relationships between wild, fancy and laboratory mice are depicted in Fig. 1. The degree of relationship can be deduced to

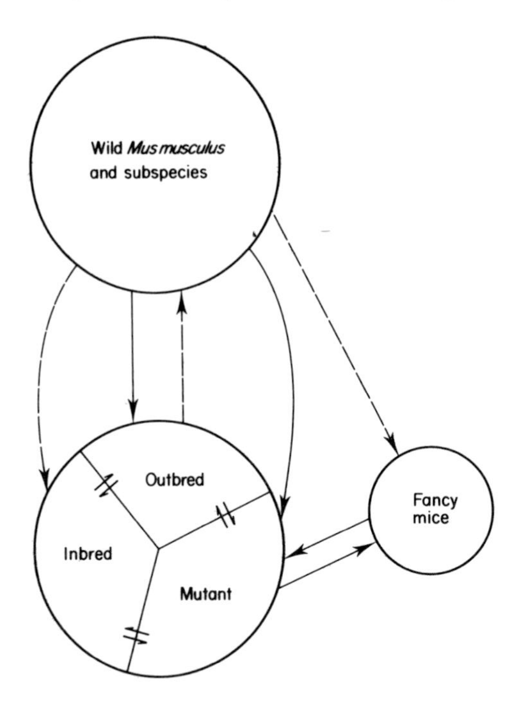

FIG. 1. Possible relationships between the three main mouse gene pools. Most laboratory mice are thought to have been derived from fancy mice in the late 19th century. These were in turn derived from wild mice (which are not themselves a homogenous entity) at various times in history dating back to at least 1100 BC. Direct inbreeding of wild mice has so far only been done on a very small scale. The exact relationships between these three groups will never be known, but biological data are probably more reliable than historical data.

some extent from historical records. As a start, it is clear that many of the mutants known in the laboratory were derived originally from fancy stock. Table I lists the origin of all the mutants described by M. C. Green (1966). A few mutants such as hairless (*hr*), pallid (*pa*) and extreme dilution (c^e) were found in wild caught mice. It is true that these represent only a very small fraction of the mutants listed in Table I, but it does imply that there has been some gene

TABLE I

Origin of mutants listed by M. C. Green (1966)

	No.	%
Laboratory stocks		
Inbred	112	35
Outbred	57	18
Unspecified	76	24
Fancy and "old"	23	7
Wild	6	2
Not known (mostly polymorphisms)	43	14
Totals	317	100

flow first from fancy and secondly from wild mice to laboratory mice during the last 70 years. Three inbred strains SF, SK and PERU inbred directly from wild mice by Dr Margaret Wallace should also be mentioned (Wallace, 1970).

The history of the development of most of the common inbred strains has been well documented. Charts developed initially by Staats (1966) and extended considerably by Potter & Lieberman (1967) and Klein (1975) have been published in a number of places. A dominant influence in the development of many of the American colonies was the colony established in Granby, Massachussets by Miss Abbie Lathrop. The origins of these mice are not known in detail, but they included not only fancy mice imported from various European countries, including Germany and England, but also wild mice from Vermont and Michigan, which were bred to the laboratory mice. Potter (1978) describes the Lathrop colony as ". . the American mouse melting pot", and suggests that the wild mice may have included mice of various subspecies, including *Mus musculus molossinus*, though this is based on biological rather than on historical data. The Lathrop stocks were supplied to C. C. Little who subsequently developed the C57 family of inbred strains, which have had an extremely important impact on biomedical research because they are so widely used. Little also developed the DBA strain from mice supplied by William Castle, which also came directly from the Lathrop colony. E. S. Russell (1978) notes that early work at the Jackson Laboratory included crosses of inbred strains of laboratory mice with *Mus bactrianus* and *Mus wagneri*, though it is not clear whether the crosses subsequently contributed to laboratory mouse stocks. Many of the most important strains can be traced back to the Lathrop colony.

The Lathrop stocks were not, however, the only stocks which contributed to the modern laboratory mouse. Other important origins included the "Bagg albino" stock which originally came from a mouse dealer in Ohio and were incorporated into the development of the BALB/c inbred strain; and the Swiss stock, imported into the USA by Dr Clara Lynch in 1926 (Lynch, 1969), which gave rise to a very large proportion of the outbred mice used in the world today, as well as to a number of inbred strains such as SWR, SJL and the four inbred strains BSVS, BSVR, BRVS and BRVR developed by Webster for resistance and susceptibility to St Louis encephalitis virus and *Salmonella* bacteria. These four inbred strains are not widely used today, but they may have made an important contribution to many of the outbred Swiss-Webster stocks which are very widely used. There are also several independently derived European strains.

Unfortunately, although this historical record is extensive, it is incomplete in many respects. Thus, although it can show that the Lathrop colony included wild mice, it does not show whether the contribution of such mice was quantitatively extensive, and whether the "wild genes" were spread throughout the colonies, or were maintained as separate stocks, which may subsequently have been discarded. Nor can the historical record show up the many cases of illegitimate matings, poor record keeping and distribution of inbred strains which were not fully inbred but which are known from biological data to have occurred.

BIOLOGICAL RELATIONSHIP BETWEEN WILD AND LABORATORY MICE

Information on the biological relationship between wild and laboratory mice can be obtained from quantitative characters such as behaviour, reproduction and growth. This aspect is considered in detail by Wallace (this volume, pp. 183–204). Data on biochemical, immunological, and cytogenetic polymorphisms are also available, as are data on retroviruses.

Quantitative Characters

Although this aspect is covered in detail by Wallace (this volume), it is obvious that the behaviour of wild mice is very different from laboratory mice, presumably as a result of countless generations of deliberate selection of the latter for tameness by man, combined with natural selection for ability to reproduce and survive under laboratory conditions.

Biochemical Polymorphisms

There have been many studies of biochemical polymorphisms in wild mice, and a review of this work is not appropriate here. It is clear from these studies that wild mice are highly heterogeneous, and that geographically distinct populations also differ in the degree and extent of polymorphism. These differences presumably arise as a result both of founder effects and as a result of selective forces favouring different genotypes under different conditions (Berry, 1970).

Roderick *et al.* (1971) studied six feral mouse populations from North Carolina, Vermont and Alberta, and 39 inbred strains at 17–18 biochemical loci. They concluded that among these 39 strains "variability is at least as great as in any single feral population, but probably less than that found among all feral populations of the species". Rather similar results can be obtained from examination of the 98 biochemical loci listed by Chapman *et al.* (1979). Of these loci, there are only six alleles listed which so far have been found only in wild *Mus musculus domesticus*, *Mus musculus musculus* or undesignated "wild" mice. However there are a further eight alleles found only in *Mus musculus molossinus* and seven in *Mus musculus castaneus* (Table II). This implies that although wild mice carry a number of alleles not found in laboratory stocks, and are therefore

TABLE II

Alleles at 98 "biochemical" loci that have been found in wild mice, but not yet in laboratory stocks (data from Chapman et al., *1979)*

Species and number of alleles found in wild mice but not yet found in laboratory stocks		Alleles in wild mice
M. m. molossinus	8	AgS^m, Apk^m, $Es\text{-}9^b$, $Gpt\text{-}1^c$, $Mor\text{-}1^b$, $Np\text{-}1^b$, $Prt\text{-}2^b$, $Prt\text{-}3^b$
M. m. castaneus	7	$Es\text{-}7^a$, $Es\text{-}7^c$, $Es\text{-}8^b$, $Gdc\text{-}1^d$, $Got\text{-}1^b$, $Gpt\text{-}1^c$, $Np\text{-}1^b$
M. m. musculus or *M. m. domesticus* or "wild"	6	$Alb\text{-}1^c$, $Erp\text{-}1^b$, $Es\text{-}6^b$, $Es\text{-}11^b$, Pgd^a, $Pgk\text{-}1^a$

a useful new source of such variation, wild and laboratory mice are not too dissimilar when biochemically polymorphic loci are considered. No doubt additional polymorphisms in wild mice will be detected as a wider range of populations are studied, and as the biochemical techniques themselves evolve. On the other hand, the range of variation in laboratory stocks will also increase as studies extend from the common inbred strains developed largely in North America to the other independently inbred strains developed in Europe, Japan and other parts of the world.

Immunological Polymorphisms

Interest in the immunogenetics of wild mice has also increased considerably in recent times, but it will take several years to accumulate data comparable in magnitude to the biochemical data. However it is already apparent that the major histocompatibility complex (*H-2*) is very different in wild and laboratory mice. This is evident from the data of Klein *et al.* (1978) (Table III) who developed a set

TABLE III

H-2 *Complex in wild and laboratory mice (Klein* et al., *1978)*

B10.W	Lines started	80
New	Lines established	35
New	*H-2* haplotypes[a]	20 (48)[b]
New	*H-2K* alleles	14 (23)[b]
New	*H-2D* alleles	19 (13)[b]

[a]Considering only alleles at *H-2K* and *H-2D* loci.
[b]Number known in laboratory stocks.

of B10.W congenic lines by backcrossing wild *H-2* haplotypes to an inbred C57BL/10 genetic background. Thirty-five lines were successfully established, and these had 20 new haplotypes not previously recorded in laboratory mice. As only 48 haplotypes differing at both *H-2D* and *H-2K* were known in laboratory mice, this implies a very significant difference between wild and laboratory stocks. This sample of mice also had 14 new *H-2K* and 19 new *H-2D* alleles compared with only 23 and 13 alleles known in laboratory stocks at the two loci, respectively. Such findings could imply that laboratory stocks are genetically very different from wild mice, and that the difference is due to a strong "founder effect". Such a conclusion appears to be in partial contradiction to the biochemical data discussed above. On the other hand, it is well established that the

H-2 complex is not selectively neutral (Klein, 1975 and this volume pp. 439–453), and that it is associated with a wide range of characteristics including growth, reproduction and resistance to certain viruses. Thus, it is possible that some of the wild haplotypes are not suited to survival in the laboratory environment, and have been eliminated by natural selection.

Further studies of non-*H-2* immunological polymorphisms may help to resolve this question.

Cytogenetics

With the rapid advances made in cytogenetics as a result of the development of chromosome banding techniques it is possible that such techniques could be used to study the relationship between wild and laboratory mice. However, according to Miller & Miller (1978), "All species of *Mus* examined so far have the same general chromosome banding patterns. This is not unexpected for the closely related subspecies *Mus musculus musculus* and *Mus musculus molossinus*, but it is surprising that the chromosomes of *Mus cervicolor*, *Mus caroli*, *Mus cookii*, *Mus booduga fulvidentris* and *Mus dunni* show no signs of inversions, reciprocal translocations or centric fusion translocations". On the other hand C-banding, which stains only the centromeric heterochromatin, shows polymorphisms which may be used to differentiate between inbred strains, and may show some differences between wild and laboratory mice. For example Forejt (1973) in a small scale study of seven inbred strains and six wild mice trapped in Czechoslovakia found a C-band polymorphism of the X chromosome in some of the wild mice which was not found in the laboratory mice. Much larger scale studies will be necessary before such data can be of much value in studying the evolution of the laboratory mouse. Silver staining of the nucleolus organizer regions is another potentially valuable technique (Miller & Miller, 1978). This can be used to distinguish between inbred strains, and the method is unrelated to C-banding.

Retroviruses

It has been demonstrated that certain endogenous viral gene sequences are transmitted through the germ line, and are present in normal cellular DNA in multiple copies (Callahan & Todaro, 1978). In *Mus* the total complement of these sequences amounts to about 0.04% of the cellular genome. These retroviruses have been used to study the relationships among different strains. This is possible because the viral genomes are not at allelic sites in different strains, but are strain dependent (Rowe & Hartley, 1978). However, studies of these retroviruses have not so far been used to show the relationship between wild and laboratory stocks of mice. In any case, more data will be required before the stability of these genetical markers can be determined.

BIOLOGICAL RELATIONSHIP AMONG LABORATORY MICE

The historical data suggest that a large proportion of the outbred mice in use today are of "Swiss" origin, and that many of the inbred strains of mice are closely related. However, historical records may be inaccurate as they rarely record accidental genetical contamination, or even deliberate outcrosses to fancy or wild mice.

Fortunately, many of the genetical markers already mentioned may be used to study the relationships between outbred stocks, inbred strains, and between substrains of the same inbred strain. In fact, there have been many studies of subline divergence using biochemical, immunological and morphological markers, but only a few attempts have been made to study the biological relationships between outbred stocks and between independent inbred strains.

The Genetics of Outbred Swiss-Webster Mice

There is some controversy on the degree of genetical variability present in so called "outbred" stocks of laboratory mice. Do these resemble wild populations, or has their previous history of inbreeding been such that they are genetically very uniform? Lynch (1969) stated that "Swiss" mice had a history of at least 12 generations of brother × sister inbreeding. Festing (1976) attempted to answer this question from studies of the phenotypic variability of inbred, F_1 hybrid, outbred and F_2 hybrid mice. When data on a number of different outbred stocks was pooled, it was shown that the outbred mice were less variable than F_2 hybrids, but significantly more variable than inbred strains or F_1 hybrids, implying some degree of inbreeding. In contrast O'Brien & Rice (1979) studied three different colonies of outbred Swiss-Webster mice using 20 biochemical markers representing at least 36 structural genetical loci. On average, 17.5% of loci tested were polymorphic, compared with similar estimates in wild mice of about 22–30%. The average degree of heterozygosity in each mouse was not significantly different from that of wild mice. They also found that the three different colonies were strikingly similar, with the same loci polymorphic in all three stocks in spite of decades of separation. Moreover, each colony was fixed for the same allele at the 29 monomorphic loci thus showing no evidence for recent random fixation of these alleles, although the colonies were not quite as variable genetically as wild mice. Obviously the degree of inbreeding in different outbred stocks will vary, but some of them at least are similar in this respect to wild island populations.

Relationships between Inbred Strains of Mice

The relationships between different inbred strains which are implied in the historical data can be investigated using a number of marker loci. Taylor (1972) carried out such an analysis of 27 inbred strains based on 16 loci. A 27 × 27 similarity matrix was calculated and subjected to an eigenvector analysis, similar in principle to a principal component analysis. This was used to map the strains in a two-dimensional array. He found that the strains fell into three groups. Strains of the C57 group, including C57BL/6, C57BL/Ks, C57L and C57BR were quite distinct from the main bulk of the strains, represented by, for example, CE, DBA/1, DBA/2 and BALB/c. Strains AU/Ss and SWR fell into a third subgroup.

In some ways these data are not in full agreement with the historical record, which fails to show that the C57 group of strains is in any way unusual. Moreover, some strains which might reasonably be expected to be biologically similar such as C58 and C57BL (which supposedly have a male parent in common), or CBA and AU (the latter supposedly having 50% CBA ancestry) do not appear to be biologically very closely related. This may be partly due to inaccuracies in the historical record, and partly due to the small number of genetical loci that are available for such an analysis. Data on the inherited markers in many inbred strains and substrains are rapidly being accumulated, and it is possible that Taylor's very useful analysis could now be extended to more strains and more loci.

Highly inherited polygenic characters may also be used in a similar analysis. One-dimensional data may simply be ranked to give a measure of the similarity between different strains. Such data can often be useful in planning experiments, as strains differing widely for a character of interest may be suitable for further study. Multidimensional data may be analysed by a variety of multivariate techniques to give a map of strain phenotypes analogous to Taylor's map of strain genotypes. One such phenotype, which has been found to be highly strain dependent, is the shape of the mandible (Festing, 1972, 1973, 1974a,b). Data on mandible shape in 62 colonies of inbred mice of most of the common inbred strains have been accumulated over a number of years. A total of 11 measurements are made on each mandible, and data on a representative sample of strains have now been subjected to a disciminant function analysis. Subsequent samples have been analysed using the pre-determined discriminant function constants (Festing, 1979a). The results are shown in Fig. 2. The two axes may be regarded as numerical measures of mandible shape, corrected (possibly rather crudely) for overall size. Each point represents the best estimate of the mandible shape of a particular colony, but the precision of the estimates varies widely, as sample sizes ranged from a single sample of about 12 mice to as many as 20 samples each of 10–20 mice. Measurements of mandible shape phenotype are also subject to experimental errors not found with Taylor's (1972) data.

Unfortunately, data of this sort cannot really be represented in a two-dimensional array without some distortion. Thus, strains which are related should map close to one another, but the converse is not necessarily true because strains that map close to one another in two dimensions may not do so in three or four dimensions. However, bearing in mind these limitations, a map of this sort can still be reasonably informative. Firstly, it is clear that closely related colonies do map in similar regions. The 16 colonies with a C57 ancestry including several "B10" congenic strains all map reasonably close to each other, in spite of the possibility of considerable genetical drift in mandible shape since the colonies were separated (Festing, 1973). These data support Taylor's data in showing the C57 group of strains to be rather unlike other strains of mice. Potter (1978) has suggested that C57BL may have had some *Mus musculus molossinus* ancestry as they both carry the same immunoglobulin allotypic markers, which are rare in other laboratory mice. Strain JU, which according to historical data has 50% C57BL ancestry (Falconer, 1960) appears to be genetically related to C57BL, as expected. DBA/2 and C3H/He also stand out as being genetically very distinct from the other strains. It is particularly interesting that although the eight colonies of C3H/He are closely related, the two colonies of C3H/Bi are quite distinct, and do not appear to be closely related to C3H/He. This observation is consistent with data on other skeletal characters (McLaren & Michie, 1954), and biochemical markers (McLaren & Tait, 1969; Krog, 1976) which suggests that strain C3H may have become genetically contaminated at some time in the past.

In contrast with Taylor's analysis, strain AU maps closely to CBA, as would be expected from historical data as it has 50% CBA ancestry. Finally, the congenic strain A.SW maps reasonably close to

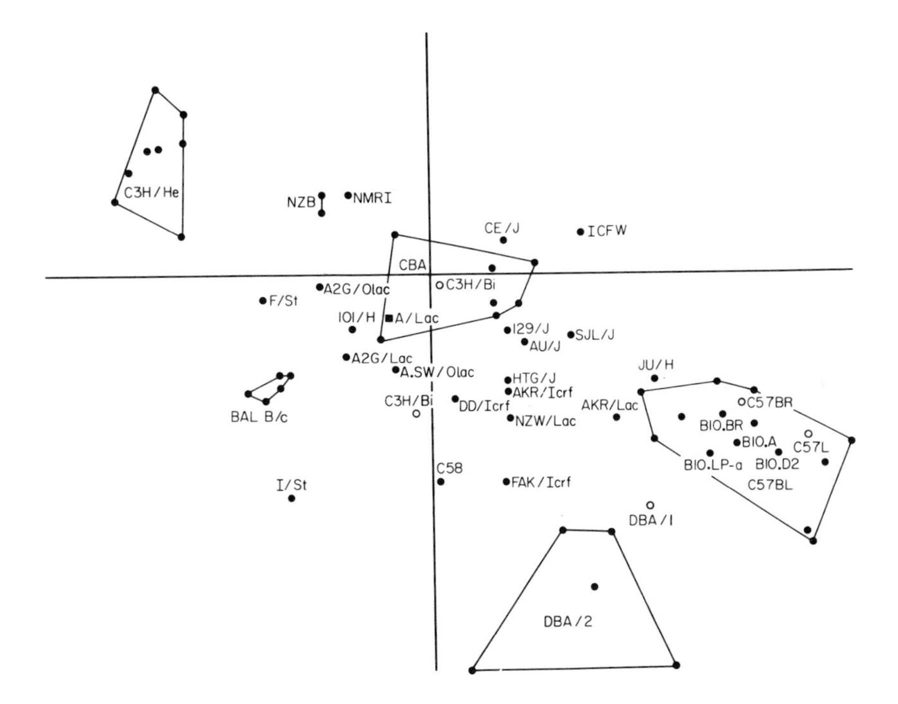

FIG. 2. Discriminant function analysis of 62 mouse colonies based on mandible shape. With a map of this sort, related colonies should be close to one another, though the converse may not be true. See text for further details and discussion. Convex hulls have been drawn round all colonies of C3H/He, CBA/Ca, C57BL (various sublines), DBA/2 and BALB/c. In these strains the points are not individually labelled in order to avoid congestion. Sublines of C3H/He include (from top in the figure): C3H/He-*mg*Lac, C3H/He-A^{vy}Icrf, C3H/HeLac, C3H/He-*mg*Olac, C3H/He-*mg*Bk, C3H/H, C3H/HeIci and C3H/HeIcrf. The CBA/Ca sublines included (top to bottom) CBA/H, CBA/CaOlac, CBA/CaLac, CBA/CaBk, CBA/CaIci, CBA/H-T6/Lac and CBA/H-T6/H. Within the CBA hull there is also a colony of C3H/BiIcrfHz (open circles) and A/Lac (square). The DBA/2 sublines included DBA/2Lac (top left), DBA/2/Olac (top right) DBA/2Ici (middle) DBA/2Icrf (bottom left) and DBA/2Bk. The BALB/c colonies were all very close, and included Icrf, Lac, Bk, Olac and Ici sublines. In the case of the C57BL hull, C57BR and C57L are shown as open circles, and a number of congenic strains on a C57BL/10 background are labelled. The colonies on the periphery, starting with the colony furthest to the left (near JU/H) and going clockwise are: C57BL/H-a^t, C57BL/10ScSnLac, C57BL/GoLac, C57BL/6By, C57BL/IcrfG, and C57BL/10JIci. The colony due west of B10.BR is C57BL/10ScSnOlac, and the colony to the east of B10.BR is C57BL/6J. The colony due south of the L of C57L is C57BL/Icrf-a^t (which is an independent mutation from that in C57BL/H-a^t). The colony very close to the bottom peripheral colony is C57BL/IcrfB.

We are indebted to Dr Mary Lyon for supplying mandible samples of mice from Harwell, and Mr A. Sebesteny for supplying samples of mice from the Imperial Cancer Research Fund colonies.

strain A/Lac and to strain A2G, which was derived from strain A following a genetical contamination.

DISCUSSION

Mice have made an undoubted contribution to biomedical research, sometimes in a rather humble capacity as "white mice" and sometimes in a much more sophisticated role as an inbred or mutant stock. The general trend, however is towards a great increase in the use of genetically defined mice, as the problems posed in research become increasingly difficult to solve, and as mice of an even wider variety become available.

Recently, wild mice have been investigated as a source of new inherited material, and it has become increasingly clear that wild and laboratory mice differ in many ways. Some of these differences may be due to a founder effect, and some may be due to selection. The degree of difference between the two depends to a large extent on the character studied. Behaviourally, it would normally be easy to distinguish the two, but biochemical marker genes may not show up such clear-cut differences. In any case, the implications for biomedical research of the difference between wild and laboratory stocks need to be discussed. While some research workers continue to develop genetically defined mice, others have suggested that laboratory mice are now so abnormal and have such a restricted degree of genetical variation that they can no longer be considered as a suitable model of a "normal" human population. These people suggest that research workers should seriously consider using wild mice in a much wider range of biomedical research than they are used at present.

Few geneticists would doubt that sampling wild mouse populations can provide new mutants, variants and chromosomal polymorphisms which can be of immense value in biomedical research, as well as being of interest to students of evolution. On the other hand, those who have bred wild mice in the laboratory can put forward some strong arguments for avoiding the use of such mice unless it is absolutely essential (see M. E. Wallace, this volume pp. 183–204). Wild mice are extremely difficult to handle, and can bite! Thus, unless there is a very compelling reason for their use wild mice should normally be avoided.

The argument that the degree of genetical variation present in laboratory mice is much lower than in wild mice is certainly open to question. True, inherited variants have been found in wild mice which are not present in laboratory populations, but conversely there are many hundreds of mutants held in laboratories which are not found in the wild. Man has favoured the survival of many types that would stand little chance in the wild. This is true also for a range of quantitative characters, where selection and inbreeding have substantially increased the range of phenotypes beyond what would normally be found in any normal wild population. Thus, it could well be argued that in fact laboratory mice are substantially more variable than their wild ancestors, in much the same way as the domesticated dog is now much more variable than its wild ancestors were.

Does the argument that laboratory mice are too genetically abnormal to be of value as a model for man have any vailidity? The answer again must be that there is no evidence that this is so. In the first place, outbred laboratory stock such as the colonies of Swiss-Webster mice studied by O'Brien & Rice (1979) appear to be genetically rather similar to island populations of wild mice. Such populations arise as a result of an initial sampling, giving a founder effect, followed by natural selection for fitness in a particular environment, which in this case happens to be the laboratory. True, such populations do not have the same social structure as wild populations, but nor would wild mice brought into the laboratory. However, it could be argued that these laboratory outbred stocks would provide a reasonable genetical model of wild mice should such a model be needed.

The trend in research is in any case away from outbred laboratory stocks towards inbred and congenic strains and mutants. This is because most workers regard genetics as a variable to be controlled and used rather than studied *per se*. It is a misunderstanding of the role of the animal as a model of man to assume that the model should be made to resemble man in every possible respect. On the contrary, it is precisely because models can be manipulated in ways not possible with humans that they are such powerful research tools. Many mutants such as nude, obese and muscular dystrophy strains would survive only a short time in the wild but are of obvious value in research. Similarly, inbred strains have been found to be remarkably powerful tools in immunology and cancer research, in spite of the fact that their breeding structure is highly unnatural. To insist on the wider use of "natural" stocks would put biomedical research back many decades. In any case, if a drug is found to be carcinogenic in a mouse, it would make very little difference in interpreting this finding in human terms to know whether the mouse happened to be inbred, outbred, or wild. The problem of extrapolating from mouse to man would still remain. The advantage of the inbred mouse in such cases is that if a drug is a carcinogen, there is more chance that this will be detected in an inbred mouse than in an outbred or wild mouse (Festing, 1979c; Haseman & Hoel, 1979).

The value of an inbred strain increases with the amount of background information on its characteristics. When data are available on a character of direct interest, such as susceptibility to a drug, it should be possible to choose susceptible and resistant strains for further study. A comparison of such strains may give considerable insight into the mode of action of the drug. On other occasions, there may be no obvious way of selecting a suitable pair of strains, and in this case the type of analysis carried out by Taylor (1972) on genetical polymorphisms, or presented here for mandible shape, in which some sort of estimate of the genetical similarity between strains is obtained, will be useful as a guide to choosing a set of strains which differ as much as possible. Thus, a study of the relationships among inbred strains is of more than academic interest.

In conclusion, it is clear that there is now a strong trend towards the exploitation of genetical variation as a tool in studying a wide range of problems in biomedical research. Wild mice offer a useful source of new genetical variation, but the future lies more in the full exploitation of inbred strains, congenic strains, recombinant inbred strains, selected strains, mutants and other types that may be developed rather than in the direct use of wild mice in the laboratory.

REFERENCES

Bailey, D. W. (1971). Recombinant-inbred strains: An aid to finding identity, linkage and functions of histocompatibility and other genes. *Transplantation* **11**: 325–327.

Berry, R. J. (1970). Covert and overt variation as exemplified by British mouse populations. *Symp. zool. Soc. Lond.* No. 26: 3–26.

Callahan, R. & Todaro, G. J. (1978). Four major endogenous retrovirus classes each genetically transmitted in various species of *Mus*. In *Origins of inbred mice*: 689–713. Morse, H. C. III (Ed.) New York, San Francisco & London: Academic Press.

Chapman, V. M., Paigen, K., Siracusa, L. & Womack, J. E. (1979). Biochemical variation: mouse. In *Inbred and genetically defined strains of laboratory animals. Part 1 Mouse and rat*: 77–95. Altman, P. L. & Katz, D. D. (Eds). Bethesda, Maryland: Fedn Am. Socs exp. Biol. Med.

Falconer, D. S. (1960). The genetics of litter size in mice. *J. cell. comp. Physiol.* **56** (Suppl. 1): 153–168.

Festing, M. F. W. (1972). Mouse strain identification. *Nature, Lond.* **238**: 351–352.

Festing, M. F. W. (1973). A multivariate analysis of subline divergence in the shape of the mandible in C57BL/Gr mice. *Genet. Res.* **21**: 121–132.

Festing, M. F. W. (1974a). Genetic reliability of commercially-bred laboratory mice. *Lab. Anim.* **8**: 265–270.

Festing, M. F. W. (1974b). Genetic monitoring of laboratory mouse colonies in the Medical Research Council Accreditation Scheme for the suppliers of laboratory animals. *Lab. Anim.* **8**: 291–299.

Festing, M. F. W. (1976). Phenotypic variability of inbred and outbred mice. *Nature, Lond.* **263**: 230–232.

Festing, M. F. W. (1979a). *Inbred strains in biomedical research.* London & Basingstoke: Macmillan.

Festing, M. F. W. (Ed.) (1979b). *Animal models of obesity.* London & New York: Macmillan.

Festing, M. F. W. (1979c). Properties of inbred strains and outbred stocks, with special reference to toxicity testing. *J. Toxicol. environ. Health* **5**: 53–68.

Festing, M. F. W. (In press). The contribution of inbred strains to biomedical research, past, present and future. In Proc. Symp. "*L'animal de laboratoire au service de l'homme*" Lyon 19–22 Sept. 1978.

Forejt, J. (1973). Centromeric heterochromatin polymorphism in the house mouse. *Chromosomal (Berl.)* **43**: 187–201.

Green, M. C. (1966). Mutant genes and linkages. In *Biology of the laboratory mouse*: 87–150. 2nd edn. Green, E. L. (Ed.). New York, Toronto, Sydney, London: McGraw-Hill.

Haseman, J. K. & Hoel, D. G. (1979). Statistical design of toxicity assays: role of genetic structure of test animal population. *J. Toxicol. environ. Health* **5**: 89–102.

Home Office (1978). *Statistics of experiments on living animals in Great Britain 1978.* London: HMSO (Cmnd. 7628).

Keeler, C. E. (1931). *The laboratory mouse.* Cambridge: Harvard University Press.

Klein, J. (1975). *Biology of the mouse histocompatibility-2 complex.* Berlin, Heidelberg, New York: Springer-Verlag.

Klein, J., Duncan, W. R., Wakeland, E. K., Zaleska-Rutczynska, Z., Huang, H-J. & Hsu, E. (1978). Characterization of H-2 haplotypes in wild mice. In *Origins of inbred mice*: 667–687. Morse, H. C. III (Ed.). New York, San Francisco, London: Academic Press.

Kondo, K. (1973). Improvement and establishment of inbred mouse strains in Japan. *Expl Animals (Japan) 22*, Suppl.: 271–276. (Proc. ICLA Asian Pacific Meeting on Laboratory Animals).

Krog, H. H. (1976). Identification of inbred strains of mice, *Mus musculus* I. Genetic control of inbred strains of mice using starch gel-electrophoresis. *Biochem. Genet.* **14**: 319–326.

Lynch, C. J. (1969). The so-called Swiss mouse. *Lab. Anim. Care* **19**: 214–220.

McLaren, A. & Michie, D. (1954). Factors affecting vertebral variation in mice I. Variation within an inbred strain. *J. Embryol. exp. Morph.* **2**: 149–160.

McLaren, A. & Tait, A. (1969). Cytoplasmic isocitrate dehydrogenase variation within the C3H inbred strain. *Genet. Res.* **14**: 93–94.

Masson, J. H. (1940). The date of the first use of guinea-pigs and mice in biological research. *J. S. Afr. Vet. Med. Ass.* **11**: 22–25.

Miller, O. J. & Miller, D. A. (1978). Cytogenetics. In *Origins of inbred mice*: 519–611. Morse, H. C. III (Ed.). New York, San Francisco, London: Academic Press.

O'Brien, S. J. & Rice, M. C. (1979). Genetic aspects of carcinogenesis and carcinogen testing. *J. Toxicol. environ. Health* 5: 69–82.

Potter, M. (1978). Comments on the relationship of inbred strains to the genus *Mus*: In *Origins of inbred mice*: 497–509. Morse, H. C. III (Ed.). New York, San Francisco, London: Academic Press.

Potter, M. & Lieberman, R. (1967). Genetics of immunoglobulins in the mouse. *Adv. Immunol.* 7: 91–145.

Roderick, T. H., Ruddle, T. H., Chapman, V. M. & Shows, T. B. (1971). Biochemical polymorphisms in feral and inbred mice (*Mus musculus*). *Biochem. Genet.* 5: 457–466.

Rowe, W. P. & Hartley, J. W. (1978). Chromosomal location of C-type virus genomes in the mouse. In *Origins of inbred mice*: 289–295. Morse, H. C. III (Ed.). New York, San Francisco, London: Academic Press.

Russell, E. S. (1978). Origins and history of mouse inbred strains: contribution of Clarence Cook Little. In *Origins of inbred mice*: 33–43. Morse, H. C. III (Ed.) New York, San Francisco, London: Academic Press.

Russell, W. L. (1941). Inbred and hybrid animals and their value in research. In *Biology of the laboratory mouse*: 325–348. 1st edn. Snell, G. D. (Ed.). New York: Dover Publications.

Staats, J. (1966). The laboratory mouse. In *Biology of the laboratory mouse*: 49–50. 2nd edn. Green, E. L. (Ed.). New York: McGraw-Hill.

Strong, L. C. (1942). The origin of some inbred mice. *Cancer Res.* 2: 531–539.

Taylor, B. A. (1972). Genetic relationships between inbred strains of mice. *J. Hered.* 63: 83–86.

Wallace, M. E. (1970). An unprecedented number of mutants in a colony of wild mice. *Environ. Pollution* 1: 175–184.

Symp. zool. Soc. Lond. (1981) No. 47, 63–84

Comparative and Historical Aspects of the Mouse Genome

A. G. SEARLE

Medical Research Council, Radiobiology Unit, Harwell, Oxford, UK

SYNOPSIS

For many centuries mouse genetical variants have been bred as pets and for show. This, combined with the success of mice as laboratory mammals, meant that they became widely used for genetical studies soon after the rediscovery of Mendel's Laws in 1900. Since 1910, numbers of known genes have approximately doubled in each decade to a present total of over 700, with an even more rapid growth in knowledge of enzyme variants in recent years. One major task has now been accomplished: the individual identification of all chromosomes and the assignment of linkage groups to them, through the use of Q- and G-banding techniques and of reciprocal translocations. The use of recombinant-inbred strains is becoming increasingly important for detecting linkages of polymorphic loci while somatic cell hybridization permits even nonvariant loci to be assigned to the right chromosome. This last technique is already proving of considerable use for comparative studies.

Comparisons of closely linked loci within the mouse genome reveal the existence of a large number of pairs or trios with very similar functions (*Amy-1* and *Amy-2*, *Car-1* and *Car-2*, various esterase loci and so on). This confirms the importance of duplication during mammalian evolution. There is also plenty of evidence for the existence of similar multiple allelic series in a wide range of mammals and for extensive homologous regions in different rodents, especially on chromosomes 7 and 8. It is well established that genetical information on the X chromosome tends to be conserved within the mammals but recent work has shown that mouse and man share a number of homologous autosomal regions even though separated for the last hundred million years or so. For instance, a large part of mouse chromosome 17 (containing the *H-2* and *Glo-1* loci) closely resembles part of human chromosome 6, while the same is true of mouse chromosome 4 and human chromosome 1. Knowledge of human and mouse genomes has far outstripped that of all other mammals and is rapidly increasing our understanding of the genetical events underlying the evolution of this class.

INTRODUCTION

This paper is really about two sorts of evolution: organic and cultural. Thus it is concerned not only with what has happened to the mammalian genome during the last 180 million years or so since its inception (as judged by comparisons of the house mouse with other mammalian species), but also with the remarkably rapid growth in human knowledge of the mouse genome which has taken place in the last 80 years, and how it has come about. Nevertheless, most of the mouse genome still remains to be discovered. What is known now must represent a highly selected sample of the whole, favouring, for example, loci with fully penetrant mutant alleles which have easily visible phenotypic effects. Thus we have learned much about mouse loci which affect coat colour, tail morphology or the proprioceptive system, but nothing about those controlling olfaction or tactile sensations, although it is clear that these are very important.

The same sorts of loci tend to be studied in all laboratory mammals, a shared selectivity which is a bonus for the comparative approach. The human geneticist has, of course, probed more deeply into hereditary clinical conditions but this has been partly compensated for by the parallel search for mouse (and other mammal) models of human genetical diseases, first fully documented by Grüneberg (1947). Only by considering how new ideas and methodologies have led to the exploration of previously unknown aspects of the mouse genome and of its organization can we fully appreciate the nature of the diverse lines of evidence which now throw light on how the mammalian genome has evolved since man and mouse first diverged from their common ancestor many millions of years ago. Therefore it seems necessary to deal with historical aspects of the mouse genome before discussing what comparative studies can tell us.

HISTORY OF MOUSE GENETICS

Man and mouse have been associated since at least the early Pleistocene (Berry, 1970), which probably means for over a million years. It is not known when a symbiotic component was added to this commensal relationship (i.e. pets as well as pests) but from documentary evidence we can deduce that it was several thousand years ago. For a very long time, genetical variants, from albinos to Japanese waltzing mice, have been kept as pets and bred in captivity. Clearly breeders must have had a good idea of how these characters were inherited, which ones bred true, disappeared on out-crossing and so on. However, there are few indications that this practical experience was ever analysed scientifically before the beginning of this century, though Grüneberg (1957) has related how, early in the last century, M. Coladon, a pharmacist from Geneva, systematically bred together mice with different coat colours and obtained results which we now realize were in excellent agreement with Mendelian expectation. This was over 30 years before Mendel's own results on peas were published. This experimental use of mice increased considerably during the 19th century and, when combined with the experience and available variants of the mouse fancy, made it comparatively easy to confirm Mendel's laws for animals soon after they were rediscovered in 1900.

Mouse Genetics since 1900

Growth in our knowledge

If we date the start of mouse genetics from 1900, then a perusal of Grüneberg's (1952) list of gene symbols and his accounts of the origins of known mutants suggests that about 13 of these, at eight different loci, were "genes of the mouse fancy" and thus available for study at that time. Lyon (1978) plotted growth in numbers of known mouse loci since 1910 and concluded that this was exponential, with the numbers almost doubling in each decade. The present position (with respect to variants rather than loci) is shown in Fig. 1 and confirms the general trend described by Lyon. It also

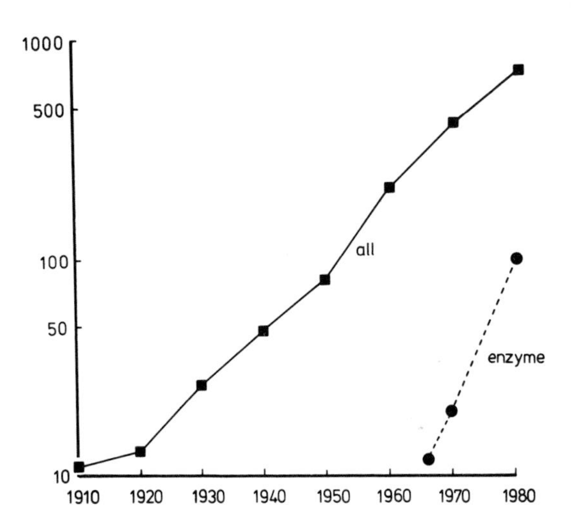

FIG. 1. Semi-logarithmic plots of numbers of genetical variants and of variant enzyme loci known in the mouse, against date. Numbers of 1950 and earlier decades derived from Grüneberg's (1952) listing of gene symbols; those for later decades based on the mutant gene list given in *Mouse News Letter* (February issues), with omission of *t*-haplotypes and provisional symbols.

shows that for one type of variant, in which particular enzymes are known to be affected, the doubling time has been markedly less than ten years. We can conclude that for some other types of variant (e.g. morphological ones) the doubling time must have been markedly more than a decade during the same period. Each broad component of the overall growth curve (comprising enzymic, immunological, morphological variation etc.) will presumably have different growth characteristics and will tend to plateau at different periods.

Knowledge of chromosome aberrations has shown the same sort of remarkable growth as that of enzyme variants in recent years, as Table I shows. These aberrations were unknown until they were deliberately induced with X-rays by Hertwig and Snell in the early 1930s (see Grüneberg, 1952). Only Snell's translocation, T(2;4)1Sn, remains from those days, but many more have been induced and conserved since, including a number of valuable X-autosomal ones. Robertsonian (whole-arm) translocations, on the other hand, have not been induced but have arisen spontaneously in laboratory stocks or has been recovered from wild mice. The latter group in particular have expanded very considerably in recent years, through the efforts of Gropp and co-workers (Capanna *et al.*, 1976; Gropp *et al.*, 1972) and have been used in a number of investigations.

Ramifications

Hand-in-hand with this very rapid rise in our knowledge of mouse genetics there has gone a continual increase in the number of different aspects of the subject which could be studied. "Nothing succeeds like success" and the concentration on genetics of the mouse, at the expense of other laboratory rodents, has given this discipline a built-in momentum. Table II gives a rough idea of the various stages in the development of mouse genetics which can now be distinguished, and their chronology. Just as the existence of the mouse fancy and of laboratory mice helped in the investigation of Mendelian segregation in animals, so the existence of large numbers of mice under genetical scrutiny led to the discovery of mutant alleles, like luxate (*lx*), brachyury (*T*), dominant spotting (*W*), and others (see Grüneberg, 1952). Many of these were of medical significance and therefore particularly worthy of study. As the numbers of mutants grew, so the opportunities for linkage studies did likewise. This growth also made the house mouse a more attractive possibility for the work on mammalian radiation genetics which followed World War II and the advent of the atomic age, although biomass was another important consideration. The development of a multitude of

TABLE I

Growth in the holdings of specific structural changes in mouse chromosomes from 1959–1979

Year	Reciprocal translocations		Robertsonian translocations		Insertions	Inversions
	X-autosomal	All	Wild-derived	All		
1959	1	9	0	0	0	0
1969	8	31	7	10	1	0
1979	12	52	55	65	2	8

Deficiencies and duplications have been omitted. Information obtained from various sources, especially *Mouse News Letter*.

TABLE II

Eleven stages in the historical development of mouse genetics, with approximate dates

Date	Stage
For millennia	Mutants bred as pets and for show
1900–1905	Application of Mendel's laws to animals
1905–	Studies on (mainly deleterious) mutants, especially those of medical interest
1915–	Linkage relationships
1933–	Induction of mutations and chromosome aberrations
1935–	Differences between inbred strains analysed (immunological, biochemical, behaviour etc.)
1956–	Studies on wild populations
1960–	Studies on X-chromosomal activity in females
1968–	Specific and subspecific differences explored
1970–1979	Assignment of linkage groups to chromosomes
1975–	Use of cell hybrids to locate homologous genes
1978–	Use of DNA probes

inbred strains for cancer research led to another boost, since it was clearly of great importance to discover the genetic basis of tumour rejection. This led to discovery of the *H-2* complex by Gorer and Snell (see Klein, 1975; Snell, Dausset & Nathenson, 1976), followed by a succession of other histocompatibility loci. This was really the beginning of the now very extensive study of hidden genetic polymorphisms in the mouse, using inbred strains at first and wild populations later. The Lyon hypothesis of X-inactivation in female mammals (Lyon, 1961) has led to extensive research on X-chromosomal activity in the mouse, man and other mammals.

Gene location

It is particularly instructive to consider how our present extensive knowledge of the arrangement and chromosomal position of about 350 mouse loci has been arrived at, since it shows the way in which many different lines of enquiry had to come together in order to achieve the desired result. As Table III shows, it has taken about 60 years to establish all the linkage groups (excluding the Y chromosome, any factors on which would only be separable by chromosome breakage). The first example of linkage (then described as reduplication) was reported by Haldane, Sprunt & Haldane (1915) and it is good to know that 64 years later the last-named author

TABLE III

Growth of the mouse linkage map

Linkage group	Chromosome	Original linkage	Author[a]
I	7	*p, c*	Haldane *et al.* (1915)
II	9	*d, se*	Gates (1927)
III	14	*hr, s*	Snell (1931)
IV	10	*r, si*	Keeler (1930)
		si, pg	Falconer & King (1953)
V	2	*pa, a*	Roberts & Quisenberry (1935)
VI	15	*Ca, N*	Cooper (1939)
VII	11	*wa-2, sh-2*	Snell & Law (1939)
VIII	4	*an, b*	Hertwig (1942)
IX	17	*T, Fu*	Reed (1937)
X	10	*v, ji*	Snell (1945)
XI	6	*wa-1*, Mi^{wh}	Bunker & Snell (1948)
XII	19	*ru, (je)*	Fisher & Snell (1948)
		ep, ru	Lane & Green (1967)
XIII	1	*fz, 1n*	Dickie & Woolley (1950)
XIV	13	*cr, f*	King (1956)
XV	18	*Tw, ax*	Lyon (1958)
XVI	3	*Va, de*	Curry (1959)
XVII	5	*pi*, W^{v}	Dickie & Woolley (1946)
		W^{v} not in LG III	Lane (1967)
XVIII	8	*Hk, Os*	Green *et al.* (1963)
XIX	16	*dw, wv*	Lane & Sweet (1974)
XX	X	*Ta, Mo*	Falconer (1953)
–	12	*Igh-1, Pre-1*	Taylor *et al.* (1975)

Mainly from Green (1970) with some later additions.
[a]References in Robinson (1972), except those of later date, given in this paper.

(Naomi Mitchison) is still alive and active, though in a different field. The loci concerned were albino (*c*) and pink-eyed dilute (*p*), and the epistasy of *c* over *p* made the analysis far from straightforward. Most of the subsequent linkages listed were obtained by conventional test procedures (Robinson, 1971), but the final one in the Table III listing (i.e. *Igh-1* and *Pre-1*) was first detected by the use of recombinant-inbred (RI) strains.

The very fruitful idea of developing RI strains and of using them for the direct determination of linkage, as well as for many other purposes, was conceived by Bailey (1971). Two unrelated inbred strains are crossed together and the offspring sib-mated, so as to give a parallel set of lines with different recombinations of the parental genes. If the parental strains have different alleles at a particular locus, then the derivative inbred lines will show a specific strain distribution pattern of fixation of one or other allele. If another segregating locus is closely linked to the first then it will tend to show the same strain distribution pattern, since the chances of recombination are low. If the two strains are well chosen, so that they differ genetically from each other at a large number of loci covering a substantial part of the genome, then there is a good chance of being able to fix the approximate position of a newly discovered polymorphic locus without having to carry out any of the usual rather tedious linkage test procedures. Use of the many sets of RI strains now available has led to the location of a large number of genes affecting behaviour, histocompatibility, biochemical traits and so on.

Once linkage groups had been established the next move was to assign them with the correct orientation to the right chromosomes. Both these procedures required the use of translocations, so were not possible until many translocation stocks had been assembled at Harwell as a result of studies in radiation cytogenetics (Carter, Lyon & Phillips, 1955, 1956; Lyon, Phillips & Searle, 1964; Searle, 1964; Batchelor, Phillips & Searle, 1966; Lyon & Meredith, 1966). For determination of the position of the centromere relative to the linkage group a metacentric marker chromosome can be used (Lyon, Butler & Kemp, 1968) or a reciprocal translocation with known breakpoint and good markers on each side (Searle, Ford & Beechey, 1971). If heterozygotes for such a translocation are intercrossed, complementation of gametes which have duplications and deficiencies of the chromosomal segments on the centromeric side of the breakpoint are expected to be much less common than those for the noncentromeric side, and appropriate markers can show this up.

Before linkage groups could be assigned to the right chromosomes, however, it was necessary to distinguish all of these individually, which was impossible on morphological criteria alone. This was successfully accomplished in 1971 by use of the quinacrine fluorescence and ASG banding techniques, and agreement was soon reached on how the chromosomes should be numbered (see Committee on Standardized Genetic Nomenclature for Mice, 1972) and on banding nomenclature (Nesbitt & Francke, 1973). The two chromosomes involved in reciprocal translocations could then be identified. If two such translocations were known to involve the same linkage group and therefore found to have a chromosome in common, then clearly that linkage group could be assigned to that chromosome. In this and other ways, most linkage groups had soon been allotted to their chromosomes (Miller & Miller, 1972, 1975). This process seems now to have come to an end, with the findings that LG XVI is really on Chr 3 (Eicher, 1979) rather than Chr 12 as originally thought, and that the linked *Igh-1* and *Pre-1* loci are towards the distal end of Chr 12 (Meo *et al.*, 1980). Thus all chromosomes (except the Y) now possess a linkage group.

Until very recently, gene location depended on the presence of two or more alleles being available in the species under study, so that recombination can be observed. With the techniques of somatic cell hybridization, however, it is now possible to place certain enzymic genes of which no variants are known in the mouse, on their correct chromosomes. Minna, Marshall & Shaffer-Berman (1975) have described the method, which usually involves the fusion of mouse spleen cells with cultured Chinese hamster cells and the isolation of a large number of hybrid clones. These tend to lose mouse chromosomes preferentially but to a very variable extent. If biochemical analysis of cell extracts from a number of clones shows that certain enzymes are always present together or absent together, and that their absence is associated with loss of a particular mouse chromosome, then the loci concerned can be regarded as syntenic and belonging to that chromosome. Further work with mouse reciprocal translocations may allow them to be located more precisely since these translocations effectively split a chromosome up into a number of smaller parts which can be studied separately. This cell hybridization method opens up immense possibilities for comparative work in different mammals on enzymic loci which can be expressed *in vitro;* some of the first results will be discussed later.

It should be mentioned here that it may be possible to locate some nonvariant mouse loci by another technique known as duplication-deficiency mapping (Eicher & Washburn, 1978) which makes use of the fact that translocation heterozygotes generate progeny with chromosome duplications and deficiencies. Some of these survive long enough for biochemical study. If the animal concerned is monosomic or trisomic for a particular enzymic or other locus, then this abnormal dosage may be detectable.

Another method for locating genes which may well gain in prominence is that of molecular rather than cell hybridization, with labelled DNA probes for specific genes or gene complexes which can then be located on particular chromosomes through the phenomenon of DNA/DNA annealing of homologous regions. This technique has been used by Valbuena *et al.* (1978) in an attempt to locate the mouse immunoglobulin heavy and light chain genes, but with limited success.

The study of polymorphisms

Of particular interest from the evolutionary point of view are those genetical and epigenetic polymorphisms which are found in wild populations of mice. The work on histocompatibility and other differences between inbred strains has, of course, revealed the sorts of variation which would be expected in wild populations also, but until recently not very much work was being carried out on these. Few of the early mutants, apart from pallid (see Grüneberg, 1952) seem to have been caught in the wild. Thus one of the first studies of polymorphisms in wild mice seems to be that started by Dunn & Suckling (1956) which was particularly concerned with the interesting properties of recessive members of the *T* complex, and which still continues. Another interesting polymorphism in wild populations which has been studied more recently is that for warfarin resistance (*War*) which is linked to the *Hbb* locus on Chr 7 (Wallace & MacSwiney, 1976), while Wallace (1971) has investigated a colony of wild mice in Peru which seems to contain an abnormally high number of mutant forms. However, most of the recent genetical interest in wild relations of the house mouse has been concerned either with (1) the investigation of enzyme and protein polymorphisms and the recovery of enzyme variants not found so far in laboratory mice (Chapman & Selander, 1979), or (2) the study of the many Robertsonian metacentrics found in natural populations of wild mice from Southern Europe, including the Poschiavo valley in Switzerland (*Mus poschiavinus*), the Rhaetian Alps, the Apennines, Sicily and so on. Investigations so far have concentrated on homologies, pairing behaviour and effects on fertility (Capanna *et al.*, 1976) but no doubt their scope will widen.

COMPARATIVE STUDIES

Comparisons of the nature and organization of genetic material in the house mouse and other mammals can throw light on a number of questions concerned with mammalian evolution, such as the relative importance of gene duplication, whether there is evidence for tetraploidization, whether particular chromosome segments remain intact for millions of years, what sorts of structural changes occur most frequently, and to what extent functionally related loci remain closely linked on one chromosome. In his very useful review of some comparative aspects of mammalian genetics, Lundin (1979) has discussed two separate types of homology: (1) that which exists within a species when a common ancestral gene becomes duplicated so as to give rise to a pair of "paralogous genes" (either through a regional event or through genomic doubling or tetraploidization) which may have equivalents in other species also, and (2) that which exists between different species when they possess "orthologous genes" with similar properties and clearly descended from a common ancestral gene. We shall deal first with some evidence for the latter type.

Homologies between Species

Allelomorphic series

Coat colour genes are particularly well known in mammals because of their value to the "fancy" and in distinguishing between different domesticated breeds. Table IV shows that a number of allelic series

TABLE IV

Allelic series affecting coat colour and found in the mouse and other mammals.

Name of series	Chief alleles	Apparent distribution
Albino	c^{ch}, c^{h}, c	Throughout mammals, including man
Agouti	A^{w}, a^{t}, a	Very widespread, but not known in man
Extension	E^{d}, e^{p}, e	Known in rodents, lagomorphs, carnivores, ungulates
Brown	b^{c}, b	As above
Dilute	d^{1}, d	Rodents, carnivores
Pink-eyed dilution	p^{d}, p	Rodents
Piebald (recessive)	s^{1}, s	Widespread but not known in man (dominant)
Dominant spotting with anaemia	W^{v}, W	Rodents, carnivores

From Searle (1968a, b).

of coat colour genes, all known in the mouse, are also found in many other groups of mammals (Searle, 1968a). The most widespread is the albino (c) locus, which is regarded as absolutely necessary for melanin pigmentation since it is the structural locus for the enzyme tyrosinase. The agouti locus, typically associated with banded hairs, is also very widespread, being present in monotremes and in many primates but not known in man. However, some other loci, such as that for pink-eyed dilution p, can only be clearly recognized in a much more restricted group of species.

Among the numerous coat colour genes which do not form allelic series a particularly interesting one is that known as beige (*bg*) in the mouse, which lightens coat colour by forming giant pigment granules. Although viable in the mouse there is good evidence that this gene is homologous with one in man associated with the lethal Chédiak–Higashi syndrome (Lutzner, Lowrie & Jordan, 1967; Padgett, 1968) and with similar conditions in mink and cattle. We can conclude that the gene concerned is an essential part of the mammalian genome, which is apparently connected with the proper functioning of lysosomes and other similar organelles, such as melanosomes (Paigen, 1978).

Homologous chromosome segments

As already mentioned, the first linkage to be reported in mice (and in vertebrates) was that between the *c* and *p* loci. Table V shows that this same linkage is found in other rodents and that the rat and mouse also share linked genes for the haemoglobin β-chain and for

TABLE V

Homologies between loci on mouse chromosome 7 and those in other rodent species

Species	Loci concerned with distances apart in cM
Mus musculus	*p* . . . 14 . . . *c* . . . 8 . . . *Hbb* . . . 15 . . . *War*
Rattus norvegicus	*p* . . . 19 . . . *c* . . . 7 . . . *Hbb* . . . 15 . . . *Rw*
Peromyscus maniculatus	*p* . . . 19 . . . *c*
Mesocricetus auratus	*b*[a] . . 36 . . . *c*

[a]Microscopical study has shown its similarity with *p*.
From Searle (1976).

warfarin resistance in similar positions, so that a large segment of mouse chromosome 7 is apparently found intact in the rat. It will be interesting to see whether the human albino gene (unlocated so far) is also linked to the *Hbb* locus, known to be on human chromosome 11. Since the *Ldh-1* locus is close to *p* on mouse chromosome 7 and the homologous human *LDH-A* locus is known to be linked to *Hbb* it seems possible that much of this mouse–rat segmental homology may extend also to man.

One other example of segmental homology in rodents seems of particular interest, namely that connected with mouse chromosome 8 (Table VI). Here there is very good evidence for gene duplication or further multiplication since seven esterase loci are known on this chromosome, including one cluster of two and another of three. A very similar arrangement is known in the rat, and Womack & Sharp (1976) have suggested homologies between three pairs of *Es* loci in

TABLE VI

Similarities between esterase and other loci on mouse chromosome 8 and on linkage groups in other rodents and a lagomorph.

Species	Loci concerned, with distances apart in cM
Mus musculus	*Ea-1** . . . . . . 8 . . . . . . *Es-1**. 2 . . *Es-6**, *Es-9* . . 7 . . *Es-2**, *Es-7*, *Es-11* . . 2 . . *Es-5* . . 17 . . *e**
Rattus norvegicus	*AgC**. 4 . . *Es-3* . . 2 . . *Es-2** . . . . *Es-4** . . . . 10 . . *Es-1**
Microtus ochrogaster	*Es-1* . . . . 17 . . *Es-2*
Oryctolagus cuniculus	*Es-1*, *Es-2* . . . 6 . . *Est-1*, *Est-2* . . . . 19 . . *e**

From Womack (1973), Searle (1976), Fox & van Zutphen (1979). Loci believed to be homologous are vertically aligned and given an asterisk.

the two species. Since the erythrocytic antigen (*Ea-1*) locus of the mouse seems homologous with *AgC* in the rat, while the extension (*e*) and esterase loci show similar linkage relationships between the mouse and the rather distantly related rabbit, it seems probable that rather a large segment of mouse chromosome 8 has remained more or less intact for quite a long stretch of mammalian evolution.

Other examples of probable homologous segments in mouse and rat have been found on mouse chromosomes 2 and 4 (Searle, 1976). Nesbitt (1974) has compared the G-banded chromosomes of these two species and estimated that homologous regions make up about 40% of the total genome.

Homologies within species

Genomic doubling

Ohno (1970) has shown the importance of duplication as an evolutionary process. As already mentioned, this could involve the whole genome (tetraploidization) or just individual loci or small segments. Lundin (1979) has looked for traces of ancestral tetraploidization in mouse and human chromosomes and has described possible paralogies between mouse chromosomes 1, 7 and 9 as well as between 2 and 17, in the form of loci with similar functions. Lundin has pointed out that two loci which were linked before tetraploidization might become unlinked in certain lineages thereafter because of random silencing of one or other of the duplicated loci, as well as through chromosomal rearrangements.

Gene duplication

As Table VII shows, a number of possible examples of gene duplication are now known in the mouse, in the form of very closely linked loci with very similar functions. The most convincing examples, listed here, concern biochemical loci, but other examples are known for genes with visible effects, for instance a triplet of genes giving dominant spotting and other pleiotropic effects (Searle & Truslove, 1970), and close linkage of naked, shaven, caracul, velvet (all affecting the hair) on chromosome 15 (Green, 1975). As far as the biochemical loci are concerned (Table VII), it looks as if similar close linkages are also found in primates wherever the same loci have been studied in them. The fact that one of the gene pairs (*Pgd* and *Gpd-1*) has even been found in *Neurospora crassa* (Jobbagy, Averca & Densya, 1975) suggests that there may be a functional need for certain duplicated loci to remain in close proximity. In any event, it seems probable that many of the

TABLE VII

Adjacent loci with closely related functions on mouse chromosomes and similar phenomena in other organisms.

Names	Symbols in mouse	Mouse chromosome	Known to be also closely linked in:
Aldehyde oxidase-1 and -2	*Aox-1, Aox-2*	1	
Salivary and pancreatic amylases	*Amy-1, Amy-2*	3	Man
Carbonic anhydrase-1 and -2	*Car-1, Car-2*	3	Guinea-pig, macaque
Alkaline phosphatase-2 and -3	*Akp-2, Akp-3*	4	
6-phosphogluconate dehydrogenase and glucose phosphate hydrogenase	*Pgd, Gpd-1*	4	Man, *Neurospora*
Esterases-1 and -6	*Es-1, Es-6*	8	Rat
Esterase-2, -7 and -11	*Es-2, Es-7, Es-11*	8	
Immunoglobulin heavy chain	*Igh-1* to *Igh-6*	12	Man, rabbit

From Womack & Sharp (1976), Searle (1976), Lalley, Minna & Francke (1978), Lundin (1979).

individual gene duplications (or any tetraploidization) must have occurred some hundreds of million years ago.

Mouse and Man

Clearly, biochemical loci in which the gene product can be characterized precisely are the best indicators of homology. Since *Homo sapiens* is the only other mammalian species beside the mouse in which extensive work on biochemical genetics has been carried out, a genetical comparison between man and mouse would seem especially desirable. Recent results of somatic cell hybridization experiments between mouse and Chinese hamster (Minna *et al.*, 1975; Lalley, Minna & Francke, 1978) have proved very informative. Table VIII shows the biochemical genes which have been located in both species, in order of their chromosome assignment in the mouse, but indicating also to which human chromosome they are now known to belong. There are ten instances involving eight chromosomes, in which two or more loci which are syntenic in the mouse are also syntenic in man, while there are only four

TABLE VIII

Syntenic loci in the mouse which have also been mapped in man, showing their synteny or otherwise in this species

Mouse chromosome number	Locus symbols	Human chromosome assignments
1	*Idh-1, Pep-3*	2 and 1
4	*Ak-2, Eno-1, Pgd, Pgm-2*	All 1p
5	*Pep-7, Pgm-1, Alb-1*	4
	Gus, Mor-1	7
	Map-1	15
7	*Hbb, Ldh-1*	11
	Idh-2	15
	Gpi-1, Pep-4	19
8	*Aprt, Got-2*	6 and 16
9	*Mod-1*	6q
	Mpi-1, Pk-3	15q
10	*Hk-1, Pyp*	10
	Pep-2	12
11	*Hba*	16
	Glk, Tk-1	17q
12	*Acp-1, Igh-1*	2 and 8
14	*Adk, Es-10, Np-1*	10, 13, 14
17	*H-2, Glo-1*	6p
X	*Ags, Gpdx, Hprt, Pgk-1*	All X

From Lalley *et al.* (1978), Davisson & Roderick (1979), Lundin (1979), McKusick (1978 and Personal communication).

chromosomes which have syntenic loci in mice, none of which are syntenic in man. This suggests that there has not only been extensive conservation of chromosome segments during mammalian evolution, but also a number of translocations of material since, for instance, mouse chromosomes 5 and 7 are each known to contain genetical material found in three different human chromosomes.

The finding that all four X-linked enzyme loci in the mouse are also on the human X-chromosome is in good agreement with Ohno's (1967) law of X-chromosome conservation in mammals. However, Francke & Taggart (1979) have deduced from somatic hybridization with the use of mouse translocation T(X;16)16H that the *Ags* locus is distal to *Hprt* in the mouse, which is the reverse of the arrangement in man.

Of particular interest is the finding that five syntenic loci on the short arm of human chromosome 1 are all located on mouse chromosome 4 (Lalley *et al.,* 1978). Figure 2 illustrates the banding karyotypes of these two chromosomes (not drawn to scale) and the most probable positions of loci. Those for mouse *Ak-2* and *Eno-1*

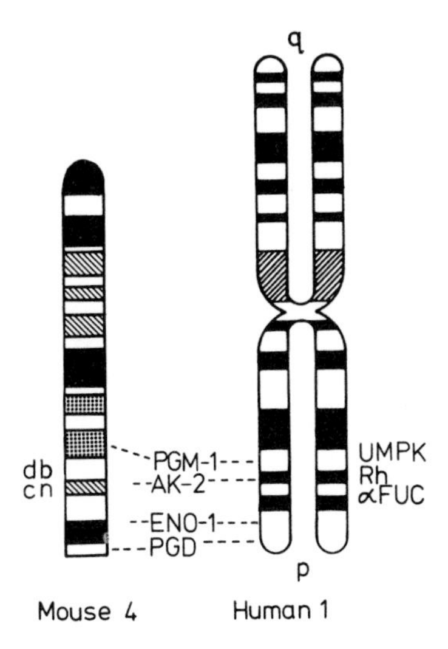

FIG. 2. Comparison of G-banded karyotypes of mouse chromosome 4 and human chromosome 1 to show some gene loci in an apparently homologous region. Information on human chromosome 1 loci from Cook & Burgenhout (1978), Hamerton & Cook (1979) and Donald, Wang, van der Hagen & Hamerton (1979). Information on mouse chromosome 4 from Nesbitt & Francke (1973), Lalley *et al.* (1978), Davisson & Roderick (1979) and Searle & Beechey (1981).

have not been placed on the Chr 4 linkage map yet, but are known to be on this chromosome by somatic cell hybridization. The *Gpd-1* locus is not shown, but is closely linked to *Pgd* in both species (Table VII). It can be seen that likely positions of the *Pgd*/PGD and *Pgm*-2/PGM-1 homologous loci on the murine and human chromosomes tally very well. If indeed this whole segment is homologous, then Fig. 2 shows that there should be homologies for the mouse diabetes (*db*) and achondroplasia (*cn*) loci on human chromosome 1 and for uridine monophosphate kinase (UMPK), rhesus blood group (Rh) and alpha-L-fucosidase (αFUC) on the distal part of mouse chromosome 4, unless they have been silenced (Lundin, 1979) or have arisen since rodent and primate ancestors diverged.

As already mentioned, the major histocompatibility complexes of mouse and man (*H-2* and *HLA*) show a number of striking similarities (Klein, 1975). It seems likely that the extent of the homologous regions on the chromosomes concerned may stretch well beyond *Glo-1* (Table VIII) since Mendell *et al.* (1979) have reported an HLA-associated gene on human chromosome 6p which causes spinal defects and may be homologous with part of the well-known *T* (brachyury) complex of the mouse, which is located about 12 units from *H-2* on its centromeric side. Further intensive studies on this unusually interesting chromosome region can be expected in both species.

CONCLUSIONS

This history of mouse genetics over the last 60 years has been one of continuous advance. Now that a major breakthrough has been successfully completed, namely the placing of all genes with known linkages on their correct chromosomes and with the right orientation, further progress will be much easier. The remarkable parallelism in the present state of our knowledge of the genetical content of human and mouse chromosomes (far outstripping that of any other mammal) means that a quasisymbiotic relationship between the two is tending to develop, each informational store being able to provide the other with many pointers to further fruitful research. Close contacts between human and mouse geneticists are therefore very desirable. From this point of view, the great distance of the relationship between man and mouse in terms of mammalian evolution is really an advantage, since it generates very useful clues to which parts of the mouse genome are an essential part of the mammalian genetical apparatus and which are concerned with evolutionary divergence within the class. As we have seen, a start has now been made on the delineation of these homologous regions between man and mouse, as well as on the more accurate positioning of particular loci within chromosomal G-bands or other regions (Nesbitt & Francke, 1973; Eicher & Washburn, 1978; Searle, Beechey *et al.*, 1979).

There is a very wide gulf between man-mouse comparisons which encompass nearly the whole of mammalian evolution and the attempt to find out what evolutionary changes are happening now in natural populations of mice. Nevertheless the extensive and rapidly expanding background knowledge of mouse genetics now available is of great value in this latter field also, allowing one, for instance, to relate structural changes to the loci which may be affected, to know what biochemical polymorphisms are likely to occur, and to look for such phenomena as linkage disequilibrium.

The other major breakthrough, namely somatic cell hybridization and the ability to locate genes on chromosomes by parasexual methods without prior knowledge of variants, should help to bridge the gap between the macroevolution of mammals and the micro-evolution of the house mouse. For it is likely to mean that, in the future, we can take close relatives of *Mus musculus* which have never been studied genetically, can hybridize their cells with those of Chinese hamsters or other species and locate the positions of a large number of loci in the same way as has been done for the laboratory mouse and is being done for a wide range of domesticated animals. By combining the technique with a careful search for changes in chromosomal patterns of G-banding it may be possible to trace the evolutionary history of the rodents (or at least the Muridae) in much the same way as Dutrillaux (1979) is mapping the chromosomal phylogeny of the primates.

All the signs suggest that in the next decade the outflow of genetical information on or relevant to the mouse will increase to something approaching a torrent. Under these circumstances it is more important than ever to have an efficient central source of mouse genetical information and advice, to pass results on quickly, to prevent confusion through ignorance or through conflicting systems of nomenclature and generally to ensure that scientific communication is facilitated in every possible way. The present system has worked well in the past but urgent thought is needed on whether it is the most suitable for the future.

ACKNOWLEDGEMENT

I am indebted to Mr Colin Beechey for preparing the text figures.

REFERENCES

Bailey, D. W. (1971). Recombinant-inbred strains. *Transplantation* **11**: 325–327.

Batchelor, A. L., Phillips, R. J. S. & Searle, A. G. (1966). A comparison of the mutagenic effectiveness of chronic neutron and γ-irradiation of mouse spermatogonia. *Mutation Res.* **3**: 218–229.

Berry, R. J. (1970). The natural history of the house mouse. *Fld Stud.* **3**: 219–262.

Capanna, E., Gropp, A., Winking, H., Noack, G. & Civitelli, M. V. (1976). Robertsonian metacentrics in the mouse. *Chromosoma (Berl.)* **58**: 341–353.

Carter, T. C., Lyon, M. F. & Phillips, R. J. S. (1955). Gene-tagged chromosome translocations in eleven stocks of mice. *J. Genet.* **53**: 154–166.

Carter, T. C., Lyon, M. F. & Phillips, R. J. S. (1956). Further genetic studies of eleven translocations in the mouse. *J. Genet.* **54**: 462–473.

Chapman, V. & Selander R. K. (1979). Polymorphisms: wild mouse. In *Inbred and genetically defined strains of laboratory animals. 1. Mouse and rat:*

220–227. Altman, P. L. & Katz, D. D. (Eds). Bethesda: Fedn Am. Soc. Exp. Biol.
Committee on Standardized Genetic Nomenclature for Mice (1972). Standard karyotype of the mouse, *Mus musculus*. *J. Hered.* **63**: 69–72.
Cook, P. J. L. & Burgenhout, W. G. (1978). Report of the committee on the genetic constitution of chromosome 1. *Cytogenet. Cell Genet.* **22**: 61–73.
Davisson, M. T. & Roderick, T. H. (1979). Linkage map of the mouse. *Mouse News Letter* **61**: 19.
Donald, L. J., Wang, H. S., van der Hagen, C. B. & Hamerton, J. L. (1979). Regional localization of PGD and ENO_1 on chromosome 1. *Abstr. Int. Workshop Human Gene Mapping* 5: 3. (Edinburgh).
Dunn, L. C. & Suckling, J. (1956). Studies of the genetic variability in wild populations of house mice I. Analysis of seven alleles at locus *T*. *Genetics* **41**: 344–352.
Dutrillaux, B. (1979). Comparison of the karyotypes of four Cercopithecoidea. *Cytogenet. Cell Genet.* **23**: 77–83.
Eicher, E. M. (1979). Personal communication. *Mouse News Letter* **60**: 50.
Eicher, E. M. & Washburn, L. L. (1978). Assignment of genes to regions of mouse chromosomes. *Proc. natl Acad. Sci., U.S.A.* **75**: 946–950.
Fox, R. R. & van Zutphen, L. F. M. (1979). Chromosomal homology of rabbit (*Oryctolagus cuniculus*) linkage group VI with rodent species. *Genetics* **93**: 183–188.
Francke, U. & Taggart, R. T. (1979). Regional mapping of *Sod-1* on mouse chromosome 16 and of HPRT and α-GAL on the mouse X, using Chinese hamster X mouse T(X;16)16H somatic cell hybrids. *Abstr. Int. Workshop Human Gene Mapping* 5: 159. (Edinburgh.)
French, E. A., Roberts, K. B. & Searle, A. G. (1971). Linkage between a hemoglobin locus and albinism in the Norway rat. *Biochem. Genet.* **5**: 397–404.
Green, M. C. (1970). Short history of the linkage map of the mouse. *Mouse News Letter* **50**: 8–9.
Green, M. C. (1975). The laboratory mouse, *Mus musculus*. In *Handbook of genetics*. **4**: 203–241. Kung, R. C. (Ed.). London: Plenum Press.
Gropp, A., Winking, H., Zech, L. & Miller, H. (1972). Robertsonian chromosomal variation and identification of metacentric chromosomes in feral mice. *Chromosoma (Berl.)* **36**: 265–288.
Grüneberg, H. (1947). *Animal genetics and medicine*. London: Hamish Hamilton.
Grüneberg, H. (1952). *Genetics of the mouse*. The Hague: Martinus Nijhoff.
Grüneberg, H. (1957). *Genes in mammalian development*. London: H. K. Lewis.
Haldane, J. B. S., Sprunt, A. D. & Haldane, N. M. (1915). Reduplication in mice. *J. Genet.* 5: 133–135.
Hamerton, J. L. & Cook, P. J. L. (1979). Committee on chromosome 1. *Abstr. Int. Workshop Human Gene Mapping* 5: 1 (Edinburgh).
Jobbagy, A. J., Averca, N. O. & Densya, C. D. (1975). Gene topography and function. I. Gene expression in germinating conidia of *Neurospora crassa*. *Biochem. Genet.* **13**: 813–831.
Klein, J. (1975). *Biology of the mouse histocompatibility-2 complex*. Berlin: Springer-Verlag.
Lalley, P. A., Minna, J. D. & Francke, U. (1978). Conservation of autosomal gene synteny groups in mouse and man. *Nature, Lond.* **274**: 160–163.

Lane, P. W. & Sweet, H. (1974). Personal communication. *Mouse News Letter* **50**: 44.

Lundin, L.-G. (1979). Evolutionary conservation of large chromosome segments reflected in mammalian gene maps. *Clin. Genet.* **16**: 72–81.

Lutzner, M. A., Lowrie, C. T. & Jordan H. W. (1967). Giant granules in leukocytes of the beige mouse. *J. Hered.* **58**: 299–300.

Lyon, M. F. (1961). Gene action in the X-chromosome of the mouse. *Nature, Lond.* **190**: 372–373.

Lyon, M. F. (1978). Standardized genetic nomenclature for mice: past, present and future. In *Origins of inbred mice:* 445–455. Morse, H. C. (Ed.). New York: Academic Press.

Lyon, M. F., Butler, J. M. & Kemp, R. (1968). The positions of the centromeres in linkage groups II and IX of the mouse. *Genet. Res.* **11**: 193–199.

Lyon, M. F. & Meredith, R. (1966). Autosomal translocations causing male sterility and viable aneuploidy in the mouse. *Cytogenetics* **5**: 335–354.

Lyon, M. F., Phillips, R. J. S. & Searle, A. G. (1964). The overall rates of dominant and recessive lethal and visible mutation induced by spermatogonial X-irradiation of mice. *Genet. Res.* **5**: 448–467.

McKusick, V. A. (1978). *Mendelian inheritance in man.* 5th edn. Baltimore: Johns Hopkins University Press.

Mendell, N. R., Ruderman, R. J., Demenais, F., Ruderman, J. G., Johnson, A. H., Amos, D. B. & Elston, R. C. (1979). The genetics of spinal development in man: Similarities with the mouse T locus. *Abstr. Int. Congr. Human Gene Mapping.* **5**: 35 (Edinburgh).

Meo, T., Johnson, J., Beechey, C. V., Andrews, S. A. & Searle, A. G. (1980). Linkage analysis of murine immunoglobin heavy chain and serum prealbumin genes establish their location on chromosome 12 proximal to the T31H breakpoint in band 12F1. *Proc. natl Acad. Sci. U.S.A.* **77**: 550–553.

Miller, D. A. & Miller, O. J. (1972). Chromosome mapping in the mouse. *Science, N. Y.* **178**: 949–955.

Miller, O. J. & Miller, D. A. (1975). Cytogenetics of the mouse. *A. Rev. Genet.* **9**: 285–303.

Minna, J. D., Marshall, T. H. & Shaffer-Berman, P. V. (1975). Chinese hamster X mouse hybrid cells segregating mouse chromosomes and isozymes. *Somatic Cell Genet.* **1**: 355–369.

Nesbitt, M. N. (1974). Evolutionary relationships between rat and mouse chromosomes. *Chromosoma (Berl.)* **46**: 217–224.

Nesbitt, M. N. & Francke, U. (1973). A system of nomenclature for band patterns of mouse chromosomes. *Chromosoma (Berl.)* **41**: 145–158.

Ohno, S. (1967). *Sex-chromosomes and sex-linked genes.* Berlin: Springer-Verlag.

Ohno, S. (1970). *Evolution by gene duplication.* London: Allen & Unwin.

Padgett, G. A. (1968). The Chediak-Higashi syndrome. *Adv. vet. Sci.* **12**: 239–284.

Paigen, K. (1978). Genetic control of enzyme activity. In *Origins of inbred mice:* 255–278. Morse, H. C. (Ed.). New York: Academic Press.

Robinson, R. (1971). *Gene mapping in laboratory mammals, Part A.* London: Plenum Press.

Robinson, R. (1972). *Gene mapping in laboratory mammals. Part B.* London: Plenum Press.

Searle, A. G. (1964). Genetic effects of spermatogonial X-irradiation on productivity of F_1 female mice. *Mutation Res.* **1**: 99–108.
Searle, A. G. (1968a). *Comparative genetics of coat colour in mammals.* London: Logos Press & Academic Press.
Searle, A. G. (1968b). An extension series in the mouse. *J. Hered.* **59**: 341–342.
Searle, A. G. (1976). Clues to homologous regions in mammalian autosomes. *Cytogenet. Cell Genet.* **16**: 430–435.
Searle, A. G. & Beechey, C. V. (1981). Maps of chromosomal variants: reciprocal translocations and insertions. In *Genetic variants and strains of the laboratory mouse.* Green, M.C. (Ed.). Stuttgart: G. Fischer.
Searle, A. G., Beechey, C. V., Eicher, E. M., Nesbitt, M. N. & Washburn, L. L. (1979). Colinearity in the mouse genome: a study of chromosome 2. *Cytogenet. Cell Genet.* **23**: 255–263.
Searle, A. G., Ford, C. E. & Beechey, C. V. (1971). Meiotic disjunction in mouse translocations and the determination of centromere position. *Genet. Res.* **18**: 215–235.
Searle, A. G. & Truslove, G. M. (1970). A gene triplet in the mouse. *Genet. Res.* **15**: 227–235.
Snell, G. D., Dausset, J. & Nathenson, S. (1976). *Histocompatibility.* New York: Academic Press.
Taylor, B. A., Bailey, D. W., Cherry, M., Riblet, R. & Weigert, M. (1975). Genes for immunoglobulin heavy chain and serum prealbumin protein are linked in mouse. *Nature, Lond.* **256**: 644–646.
Valbuena, O., Marcu, K. B., Croce, C. M., Huebner, K., Weigart, M. & Perry, R. P. (1978). Chromosomal locations of mouse immunoglobulin genes. *Proc. natn. Acad. Sci., U.S.A.* **75**: 2883–2887.
Wallace, M. E. (1971). An unpredicted number of mutants in a colony of wild mice. *Environ. Pollut.* **1**: 175–184.
Wallace, M. E. & MacSwiney, F. J. (1976). A major gene controlling warfarin resistance in the house mouse. *J. Hyg., Camb.* **76**: 173–181.
Womack, J. E. (1973). Biochemical genetics of rat esterase: polymorphism, tissue expression and linkage of four loci. *Biochem. Genet.* **9**: 13–24.
Womack, J. E. & Sharp, M. (1976). Comparative autosomal linkage in mammals: genetics of esterases in *Mus musculus* and *Rattus norvegicus. Genetics* **82**: 665–675.

Symp. zool. Soc. Lond. (1981) No. 47, 85–125

The Morphometrics of the Mouse: A Review

R. S. THORPE

Department of Zoology, University of Aberdeen, Aberdeen, UK

SYNOPSIS

Apart from being an important facet of evolutionary and systematic studies, the morphometrics of mice also impinge on a wide range of other disciplines including genetics, biometry, gerontology, developmental and experimental biology. Studies can involve identification, extent of variation, the interrelationships of characters and forms. Some of the methodological problems and sources of error are considered. A wide range of multivariate methods, including cluster, principal-component, factor, canonical and non-metric multidimensional scaling analysis, have been used with varying degrees of appropriateness. The application of some multivariate methods is critically considered, as is the combination of the methods of biometrical genetics and multivariate analysis. Environmental (including maternal) and genetical influences are involved in all morphometric characters: published estimates of heritability in mice are available for a wide range of characters. The various types of character may be very loosely ranked from low to high heritability as: non-metrical skeletal variants, body proportions/weight, skeletal dimensions, dental dimensions. There are a range of studies on the interrelationship of characters involving phenotypic, genetical and environmental correlations. There is insufficient information on the relationship of different types of character, but the interrelationships of characters within a type generally reflect anatomical proximity, except skull features which segregate into length and width groups. The multivariate differentiation within an isogenic strain with an allele difference at a single locus may be of relevance to evolutionary and taxonomic studies. Sexual dimorphism is widespread through character types but limited in extent and sometimes inconsistent in occurrence and direction. The discrimination between laboratory strains and sublines by canonical analysis is of considerable practical value. Multivariate estimates of variability within inbred, hybrid and outbred strains is relevant to experimental biology. Multivariate analysis of phenotypic traits can discriminate between sublines better than allozyme frequencies, and this is relevant to analysing population affinities. Studies of British mice, particularly on islands, reveal microgeographical variation and endocyclic rhythms in the extent of variance, and provide an example of rapid divergence due to the founder effect. There is considerable potential for multivariate morphometric studies of geographical and chromosomal races.

INTRODUCTION

Morphometrics, the quantitative analysis and characterization of form, is an important branch of evolutionary and systematic studies (Blackith & Reyment, 1971; Gould & Johnston, 1972; Thorpe, 1976; Oxnard, 1978) and impinges on a wide range of other disciplines such as developmental biology (Bookstein, 1977; Oxnard, 1978), experimental biology (Festing, 1976), gerontology (Bellamy *et al.*, 1973), genetics (Leamy & Sustarsic, 1978), biometrics (Thorpe, 1980), ecology (Berry, 1970b), and even behaviour (Thiessen, 1966; Johnston & Selander, 1973).

The morphometrics of the house mouse fall into four interrelated areas: identification, extent of variation, the interrelationships of characters, and the interrelationships of forms:

One can identify taxonomic categories or their hybrids and laboratory strains or sublines.

One can monitor the endocyclic rhythm in the extent of variation in natural populations or compare the variability of different classes of laboratory mice.

One can investigate the phenotypic, genetical and environmental correlations between characters for systematic, genetical or morphogenetic purposes.

One can investigate the relationships between strains and sublines of laboratory mice for practical purposes, between natural populations for evolutionary or taxonomic purposes and between sexes, age classes, or temporally segregated populations.

METHODOLOGY

Obviously the problems one encounters and the methods one employs in morphometric research are dependent on the aim of the particular study. Nevertheless, generalizations can be made. The problems and methods of analysing population affinities and geographical variation by morphometric methods are reviewed by Thorpe (1976) and a few pertinent points will be made below.

Accuracy and Repeatability of Character Measurement

Some characters, body proportions for example, are subject to recording error within and between researches because of the plasticity of the body, an imprecise definition of the character or a range of alternative recording procedures (Jewell & Fullagar, 1966). Novel methods that reduce this error and allow rapid character recording, such as Festing's (1972) procedure for measuring mandible dimensions, are to be welcomed.

Sources of Variation

There is a range of sources of variation that could introduce heterogeneity into the data and not only alter the mean character values, but more importantly influence the covariance between characters which could significantly influence matrix analyses. Some of the possible sources of heterogeneity include sexual dimorphism (p. 107), ontogenetic variations, e.g. growth and its allometric consequences, ageing (Bellamy *et al.*, 1973), temporal and seasonal changes in an individual (e.g. fat reserves), temporal changes in population composition (p. 111), habitat and microgeographical variation (p. 113) and bilateral asymmetry (p. 103).

Most of these sources of heterogeneity can be avoided by simple measures such as using one sex at a time, measuring one side of the specimen, using specimens collected at one time and of one age, and so on. Heterogeneity due to size variation generally needs statistical adjustment of the data.

Several studies have attempted this by expressing the character as a ratio of some "standard size" character such as body length (Schwarz & Schwarz, 1943; MacArthur & Chiasson, 1945; Ursin, 1952; Berry & Jakobson, 1975b; Berry & Peters, 1975; Leamy, 1981), body weight (Leamy, 1981), skull length (C. V. Green, 1933; Leamy, 1981) or even the sum of the size-dependent characters used (Festing, 1976).

There has recently been some debate on the value of ratios (Atchley, Gaskins & Anderson, 1976; Corruccini, 1977; Atchley & Anderson, 1978; Atchley, 1978; Albrecht, 1978; Dodson, 1978; Hills, 1978). Whilst there is no consensus of opinion on the suitability of ratios, they are best avoided as methods for adjusting for size independence unless one knows that there is a linear relationship between the variables and that the regression line passes through the origin. Regressing the character against a standard size variable (Gould, 1966; Thorpe, 1976; Atchley, 1978) is simple and preferable to other approximations such as those of Hills (1978) and Corruccini (1977).

Multivariate generalizations of size and shape may also be useful (Jolicouer, 1963) but the heterogeneity of the data and the inclusion of a "size" variable (Thorpe, 1976) are important if size and shape are not to be confused, as may be the case when raw data are used as by Dodson (1975, 1978).

Multivariate Methods

The simultaneous consideration of several characters by multivariate analysis is relevant to many areas of morphometric research (Blackith & Reyment, 1971; Gould & Johnston, 1972; Thorpe, 1976; Oxnard, 1978), although their limitations for studying growth and form are emphasized by Bookstein (1977). Multivariate techniques are described and discussed in Blackith & Reyment (1971), Gower (1972), Gould & Johnston (1972), Sneath & Sokal (1973), Clifford & Stephenson (1975), Thorpe (1976, 1980) and Harman (1976). Consequently they will not be treated in depth here, although the various methods used to investigate the morphometrics of mice will be briefly mentioned.

Most multivariate methods have two explicit or implicit stages. Firstly, the production of an association matrix (e.g. correlations between characters or "distances" between individuals or taxa) and secondly, the summarizing of the matrix into a more readily interpreted form.

The association matrix may be left unsummarized or in studies of geographic variation reduced to a simple network diagram between adjacent populations (Berry, 1967b). For the limitations of this see Thorpe (1976).

Cluster analysis, of which there are a multitude of methods (Sneath & Sokal, 1973), can be used to investigate the relationships between forms or characters (Leamy, 1977) by reducing the matrix to a hierarchical dendrogram. Mutually exclusive clusters of characters or "taxa" are portrayed.

Principal-component analysis is an ordination technique (Blackith & Reyment, 1971; Gower, 1972; Thorpe, 1980). The eigenvectors are extracted from a matrix of correlations or covariances between characters. The character loadings on the eigenvectors can be used to express relationships between characters (Bailey, 1956a; and see this volume p. 103). The principal component scores, which are the sum product of the eigenvector elements and character scores, can be used to express the relationship between forms. If one imagines the taxa as a cloud in hyperspace, with similar taxa close and dissimilar taxa far spart, then the first principal component is the best fit along the long axis of this cloud and expresses most variation between taxa. Subsequent principal components, at right angles to one another, express progressively less of the variation. This method requires the matrix to be positive semi-definite, which may not be the case if the phenotypic correlations have been partitioned into genetical and environmental correlations (Leamy, 1977).

Discriminant function or canonical analysis (Sneath & Sokal, 1973; Thorpe, 1976) ordinates two or more *a priori* defined groups so that there is minimum overlap and maximum separation between them. Properly used it can be a very effective method for identifying "unknown" individuals or groups, for investigating the relationships between forms and the interaction of characters (Festing, 1972, 1973, 1974a, b; Bellamy *et al.*, 1973; Leamy & Sustarsic, 1978) but it is demanding in its requirements of the data and can be misinterpreted (Sneath & Sokal, 1973; Thorpe, 1976, 1979). Caution is needed when this technique is used to compare the variability of groups as in Festing (1976), because the technique is based on the assumption of equal variability of the groups. Not all the variation is likely to be expressed in one canonical variate and the probability ellipsoids of groups may be in a different shape and orientated in different directions (i.e. inequality of the within-group covariance matrices). Nevertheless Festing's (1976) analysis of the variability of inbred, outbred, F_1 and F_2 laboratory mice compared such a large number of samples that one may have some confidence in the conclusion that variability in mandible shape tends to be greater in outbred than inbred mice. Consequently one is likely to need fewer inbred mice than outbred mice in an experimental study.

Factor analysis is essentially a psychometric technique (Harman, 1976) which has been used to investigate the relationships between characters in mice (Wallace & Bader, 1967; Leamy, 1975) and other animals (Brown, Barrett & Darrock, 1965). The factor loadings are analogous to the eigenvector loadings in principal-component analysis but are derived by a range of different recipes to ensure that the factors conform to the vague notion of simple structure (Harman, 1976). Studies that use factor analysis have, by the selection of an appropriate recipe, indicated (not mutually exclusive) groups of related characters. They have tended to be directed to the question "what are the groups of characters?" rather than the more fundamental question "are there groups of characters?" or more precisely "to what extent do the characters group?".

Non-metric multidimensional scaling is an ordination technique, which like factor analysis has its roots in psychometrics (Shephard, 1962a,b; Kruskal, 1964a,b); it offers solutions in one or more dimensions but attempts only to satisfy the rank order of the relationships expressed in a matrix. This technique is useful in giving an ordination when the matrix is not positive semidefinite as in Leamy's (1977) study of genetical and environmental correlations between characters. In some circumstances this technique has limitations, particularly if the data are very well structured or there are satellite groups (Thorpe, 1980).

Morphometry and Biometrical Genetics

Normally discussions of morphometry do not refer to biometrical genetics, but in studies of mice the two are often interlinked. Biometrical genetics may be involved in the choice of characters and the partition of the variance and correlation into genetical and environmental sources for subsequent morphometric analysis.

The phenotypic variation of a character (V_p) can be partitioned into genetical (V_g) and environmental (V_e) sources such that $V_p = V_g + V_e$ (Falconer, 1960). The genetical variance can be partitioned into various sources such as an additive (V_a), an interactive (V_i), a dominance (V_d) and a sex effect (V_s) such that $V_g = V_i + V_d + V_a + V_s$. The heritability ($V_a/V_p$) of a character determines its resemblance between parent and offspring and the expected response to selection.

The environmental variance can also be partitioned into various sources one of which is the maternal effect. This will be influenced by both the genotype and the environment of the mother and can be partitioned into pre- and post-natal effects, and so on (Bader, 1965a; Tenczar & Bader, 1966; El Oksh, Sutherland & Williams, 1967).

The phenotypic correlation can also be partitioned into environmental and genetical correlations (Bailey, 1956a; Leamy, 1977). As with the character variance this can be achieved by comparing appropriate classes of mice (Bailey, 1956a) or related animals, e.g. parent–offspring or sibs (Leamy, 1977).

Whilst no attempt will be made to review the procedures of biometrical genetics, a grasp of the reliability of published information may be useful. It is evident that in so far as partitioning correlations are concerned the procedure of Bailey (1956a) which relies on interstrain (inbred) correlations being genetical and intrastrain correlations being environmental, limits the study to laboratory mice and may depend on oversimplified assumptions. On the other hand the use of parent–offspring data (Leamy, 1977) can result in the occurrence of correlation coefficients which greatly exceed an absolute value of 1.0 (e.g. 1.5) and correlation matrices which are not positive and semidefinite (p. 100). Estimates of genetical correlations by selection experiments may be more reliable but are not available over such a comprehensive series of characters as above.

In so far as character variances are concerned Leamy's (1974) commendable study gives estimates for the heritability of 15 osteometric and three body proportion characters and weight in randombred (CV1) mice using various permutations of sons, daughters, dams and sires. The estimates for both sexes against sire are the most appropriate and the standard errors (which range from 0.12 to 0.29) indicate their reliability.

Self & Leamy (1978: table 4), estimated the heritability of 11 non-metrical skeletal variants, using regression, maximum likelihood correlation (MLC) and "Falconer's method". The heritability estimate for an individual character varies between methods but the average over the 11 characters is very similar for the various methods. The correlations between some series of estimates (e.g. Falconer's method V MLC) may be low, but if the range of "true" heritabilities is narrow this can be a meaningless statistic.

It is obvious that even in the most competent studies the estimates of heritability and partitioned correlations are subject to error, which in some cases can be quite large. This is relevant to studies that combine the methods of morphometry and biometrical genetics (Bailey, 1956a; Festing, 1976; Leamy, 1977) since both procedures are subject to error and variability due to the use of different methods. Studies that combine these measures (e.g. use a matrix of genetical correlations as an input for a morphometric analysis) may compound the errors and lability so that the results are of limited reliability. This has not been tested with mouse data by systematically comparing various combinations of genetical and morphometric methods to see if one obtains similar results, although in Leamy's (1977) study the initial clusters of genetic correlations can be clearly recognized in the two-dimensional non-metric multidimensional scaling analysis of the same data.

Information on the heritability of separate size and shape components derived from a principal-component analysis of phenotypic correlations would be of interest.

CHARACTERS

Types of Character

Characters may be loosely classified into (1) qualitative, which may be (a) two-state or (b) multistate (multinomial); or (2) quantitative, which may be (a) continuous variables, e.g. length; (b) meristic, e.g. number of vibrissae or (c) ordered into several arbitrary states, e.g. pale, medium and dark.

Studies of mouse morphometry use mainly characters that are expressed as qualitative two-state characters or quantitative continuous variables.

Non-metrical variants (also referred to as quasi-continuous variables, phenodeviants and epigenetic polymorphs) particularly of the skeleton but also of arterial systems (Froud, 1959), are well known in mice. They can be quantified in a variety of ways but are generally treated as qualitative two-state characters and expressed as a frequency (Berry, 1963).

These have been used in studies of laboratory mice since the 1950s (reviewed by Grüneberg, 1963) and in studies of natural populations (Weber, 1950; see references in Berry, 1977a,b). Berry & Searle (1963) and Berry (1963) describe and illustrate most known variants of the skeleton (49), review their occurrence and discuss their quantification. Skeletal variants involve foramen, fusion and dyssymphysis, etc. and are found in all parts of the skeleton. For example, of the 35 variants Berry (1964) used to investigate the affinities of the Skokholm population of the house mouse, 23 were from the skull, seven from the vertebral column, two from the girdles, with one each from the dentition, mandible and limbs.

Meristic characters are little studied in mice. Vibrissae numbers are fairly constant (Roderick & Schlager, 1966) although they vary more in tabby mice (Dun & Fraser, 1959). The number of caudal vertebrae is variable (C. V. Green, 1933; Barnett, 1965b) as is the number of tail rings (Fortuyn, 1912).

Most quantitative characters used in mouse morphometry are continuous variables, e.g. length and weights. They include linear dimensions of the skull, mandible, vertebrae, girdles, limbs, teeth and external body proportions (body, tail, foot, ear) as well as weight of the body and visceral organs (Table I). Most studies, with the exception of those by Berry and Leamy and their co-workers, are limited to one main category of characters.

The rationale for selecting the number and range of characters depends on the aim of the study. If one wishes to identify members of predefined groups then one needs only the minimum number of characters that will discriminate between these groups with the requisite degree of certainty. If, however, one wishes to assess the variability of a population, then different character sets may give different results. For example Berry (1970b) found that the endocyclic rhythm in the degree of variability in four osteometric and four odontometric characters was opposite to that in the non-metrical skeletal variants.

Similarly it could be argued that one needs to consider a range of characters when using morphometric methods to assess population affinities and geographical variation (Thorpe, 1976). What little information on mice there is would support this since Berry, Jakobson & Peters' (1978) study of the microdifferentiation of six populations of Faroe mice using five sets of characters (allozyme, body proportions and organ and body weights, scapula dimensions, mandible dimensions, non-metrical skeletal variants) showed that there was little similarity in patterns of interpopulation "distances" between the character sets. However, if one wishes to investigate the phyletic relationships, as would be pertinent to the above study of inter-island colonization, the methods of numerical phyletics may be robust, using different character sets even with poorly structured data (Mickevich & Johnson, 1976).

Character Control

Non-metrical skeletal variants

Both genetical and environmental factors are important in the control of non-metrical variants (Berry & Searle, 1963). Their control has been summarized as the polygenic (multifactorial) inheritance of the individual (and its mother) modified by tangible and intangible environmental factors and subject to the action of a threshold which is partly under genetical control. Berry's (1967a) discussion of third molar agenesis provides an example of the interaction of the various factors in controlling a non-metrical variant.

The maternal effect is important (Howe & Parsons, 1967) and includes such factors as the age, parity, weight and diet of the mother. Some factors, e.g. maternal diet, may have a clear influence on the incidence of variants in laboratory mice fed with extreme diets (Searle, 1954b; Deol & Truslove, 1957), but this may be less important in natural populations (Berry, 1963) due to the early elimination of extreme types. Moreover, Howe and Parsons' (1967) study of 25 variants in various laboratory strains suggests that, whilst environmental influences may be important in influencing the incidence of a particular variant within a strain, a multivariate estimate of the dissimilarity between strains is little influenced by this.

More recently Self & Leamy (1978) estimated the heritability and maternal effects of 11 variants of the skull and mandible in outbred laboratory mice. They estimate that on average the maternal effect is 0.08 (8%) with it ranging from zero (foramen ovale double) to 0.16 (maxillary foramen II) for individual variants.

TABLE I

The number and range of continuous variables used in mouse morphometry.

Author(s)	Dentition	Skull	Mandible	Vertebrae	Girdles	Limbs	"Soft anatomy" viscera	Body proportions	Weight
Bader, 1956		8							
Bader & Lehmann, 1965	3								
Bailey, 1956a			7	9	1				
Bailey, 1959		4			1	1			
Bellamy *et al.*, 1973							8	3	1
Berry, 1970b	4								
Berry & Peters, 1975							3	3	1
Berry & Jakobson, 1975a							7	3	1
Berry, Jakobson & Peters, 1978			11		9		6	3	1
Berry, Peters & Van Aarde, 1978							6	3	
Clarke, 1904		1						4	
Clarke, 1914								3	
Dynowski, 1963		8						4	1
Eisen & Legates, 1966									1
Elton, 1936								3	1
Evans & Vevers, 1938								4	
Festing, 1972			13						
Festing, 1973			13						
Festing, 1974a, b			11						
Festing, 1976			11						
C. V. Green, 1932	1	6							
C. V. Green, 1933	1	6							
Hanrahan & Eisen, 1973									1
Harrison, 1958								2	
Hill, 1959		1						3	
Hinton & Hony, 1916		1						4	

TABLE I (*continued*)

Author(s)	Dentition	Skull	Mandible	Vertebrae	Girdles	Limbs	"Soft anatomy" viscera	Body proportions	Weight
Jameson, 1898								4	
Laurie, 1946									1
Leamy, personal communication	6								1
Leamy, 1975		6	1		4	4		2	1
Leamy, 1977		6	1		4	4		2	1
Leamy & Hrubant, 1971	12								
Leamy & Sustarsic, 1978	6	3			1	3			1
MacArthur & Chiasson, 1945								4	
Mittwoch, 1979							1		
Moulthrop, 1942								4	
Nichols, 1944								2	
Schwarz & Schwarz, 1943		1						2	
Tenczar & Bader, 1966	3								
Thiessen, 1966							5		
Ursin, 1952		1							
Van Valen, 1965	1								
Wallace & Bader, 1967	24	3							

This representative but not comprehensive list of studies often excludes characters that do not fit into the above categories and characters that are not used throughout the study.

Their estimates of heritability, averaged over the 11 variants, give 0.13 using sires for Falconer's method and 0.20 for maximum likelihood correlation, respectively. Using mid-parent data the average heritability for the variants is even more similar (0.17 to 0.20) between methods.

Consequently the heritability of non-metrical variants is rather low and most variation is a consequence of intangible environmental factors (Searle, 1954a; Self & Leamy, 1978).

Skeletal dimensions

Leamy (1974) has estimated the maternal effects and heritability for a range of skeletal dimensions (mandible length, six skull, four girdle and four limb dimensions) in outbred laboratory mice. On average the maternal effects account for approximately a quarter of the variance (at 5 months) and were comparable in each set of characters, i.e. skull 0.21, mandible 0.18, girdle 0.26, and limb 0.29, although the estimates varied between individual characters.

The heritability for these 15 characters (taking the estimates of 5-month-old mice for both sexes on sire) show an average of 0.45 with the comparable estimates for each set, skull 0.44, girdle 0.46, limb 0.42, although the estimate for mandible length was higher at 0.57. This estimate of mandible length is in broad agreement with Festing's (1976) estimate using generalized shape as characterized by the first two variates of a canonical analysis.

Bailey's (1956b) estimate of the heritability of eight linear dimensions of the axis gives values of 0.32 to 0.54 with an average of 0.43. These are entirely commensurate with Leamy's estimates for other osteometric traits.

Leamy (1981) has discussed the heritability of the above osteometric traits when expressed as ratios based on skull length, body length and body weight. Some ratios, i.e. those based on skull length, tend to be more heritable than others although this does not address itself to the fundamental problem of ratios (p. 87).

Osteometric characters therefore tend to have moderately high heritability and are very strongly influenced by maternal effects.

Dentition

Studies of dentition generally involve the width of mandibular molars. Their control in laboratory mice has been investigated by Bader and Leamy and their co-workers. The maternal effect on molar width is important, and is responsible for up to a quarter or more of the total variation (Table IIa). It is not simply a function of maternal size, and there is unlikely to be any post-weaning compensation (unlike body weight) owing to the early termination of crown development (Tenczar & Bader, 1966).

The difference in response to the post-natal maternal environment between M_2 and M_3 is due to the different developmental histories of the two teeth as M_3 develops later (Tenczar & Bader, 1966).

The heritability estimates for molar widths are particularly high (Table IIb), and account for over half of the variance, at least in the anterior teeth.

TABLE II(a)

Maternal effect on the width of mandibular molars.

Author	M_1	M_2	M_3
Bader, 1965b	0.16	0.17	0.27
Leamy & Touchberry, 1974	—	0.29	0.13
Tenczar & Bader, 1966 (post-natal effect)	—	0.24	0.17

TABLE II(b)

Heritability of the width of mandibular molars.

Author	M_1	M_2	M_3
Bader, 1965b	0.66	0.67	0.47
Leamy & Touchberry, 1974	—	0.58	0.42
Bader, 1965a Bader & Lehmann, 1965 (genetic variance)	0.62	0.49	0.42

Viscera

I am not aware of a comprehensive study of the heritability of visceral characters although Thiessen's (1966) investigation of organ weights in relation to social position suggests that activity levels have a fairly high heritability and that in turn spleen size is significantly related to dominance. Barnett's (1965a, 1973) reviews of cold adaptation in mice indicate that there is an increase in weight of the stomach, intestine, liver and heart under certain conditions of cold adaptation. Also, under certain conditions of cold adaptation there is a greater increase during pregnancy of the number of corpora lutea and an increase in the weights of adrenals, kidney, heart and small intestine.

Body proportions

Body proportions such as head plus body, tail, foot and ear length are subject to environmental influences since mice raised under high temperatures have longer appendages (Harrison, 1958) whilst those adapted to cold conditions (Barnett, 1965a, b) tend to have shorter appendages. The tail in particular tends to be shorter and is thought to be a thermoregulatory organ.

Leamy's (1974) estimates of heritability for body, tail and total length generally show that tail length is more heritable than body length, while body proportions are generally less heritable (an average of 0.31 for both sexes on sire at 5 months) than skeletal dimensions. Maternal effects (an average of 0.28) are large and similar for all three proportions.

Body weight

The control of body weight and the heritability of sexual dimorphism in this character have been investigated over a period of years (for example, Falconer, 1953, 1960; Eisen & Legates, 1966; El Oksh *et al.*, 1967; Chai, 1971; Hanrahan & Eisen, 1973; Leamy, 1974) and no attempt is made to review this work here.

Suffice to say that body weight, as well as being under the polygenic influence of up to 100 genes (Falconer, 1960; Berry, 1970b) can also be strongly influenced by alleles at a single locus (Roderick & Schlager, 1966; Leamy & Sustarsic, 1978).

The influence of environmental factors such as cold (Barnett, 1961, 1965a, b, 1973) and overcrowding (Crowcroft & Rowe, 1961) is not always predictable. Body weight may or may not increase because of experimental overcrowding (Crowcroft & Rowe, 1961). Cold store mice tend to be heavy (Laurie, 1946), but adult laboratory mice transferred to cold lose weight while inbred mice reared in the cold are lighter (Barnett, 1965a), although they may return to normal weight after several generations.

The maternal effect on body weight is extremely important in young mice, although it reduces considerably with age owing to post-weaning compensation (Leamy, 1974; El Oksh *et al.*, 1967). Conversely, heritability tends (up to a point) to increase with age (El Oksh *et al.*, 1967; Leamy, 1974). Leamy's (1974) estimates of heritability for both sexes on sire at 5 months is 0.36, which is higher than that of El Oksh *et al.* (1967) but lower than those of Hanrahan & Eisen (1973) and Eisen & Legates (1966).

Conclusions

Bearing in mind the comments concerning relability of the estimates of heritability (p. 90), the fact that, strictly speaking, estimates pertain solely to the population used and that similar characters (e.g. M_2 and M_3) may have different heritabilities, several very tentative generalizations can be made.

Although most morphometric characters have an acceptable level of heritability (this is important for most evolutionary studies), there is no class of characters that is free from large environmental (particularly maternal) influences even though this may not be appreciated by some researchers.

Owing to the problems of reliability and extrapolation (above) with estimates of heritability, it is doubtful whether they are useful as criteria for choosing between character sets unless they are large and consistent. If one was to rank very loosely the types of character from low (*c.* 0.1/0.2) to high (*c.* 0.5/0.6) heritability, one might obtain the following series: non-metrical skeletal variants, body proportions/weight, skeletal dimensions, dental dimensions (molar widths). The various subsets of skeletal dimensions (except the mandible) have comparable "average" heritabilities.

Relationships between Characters

This section is concerned with the morphometric rather than genetical relationship between characters, although these approaches do impinge on one another.

Skeletal variants have been the subject of much genetical research (p. 95) although there has been little "biometric" analysis of their interrelationships. Truslove's (1961) investigation of the correlations between 29 skeletal variants showed that only a few (11) of the pairs of characters were significantly correlated, and then at a low level. Some of these correlations could be interpreted on the basis of anatomical proximity or mutual correlation with birth weight, but others were not readily explicable.

Quantitative characters, particularly the teeth and skeleton, have been subjected to much more morphometric analysis. These have included investigations of the relationships between various types of character, the relationship between characters of one type, the genetical, environmental and phenotypic correlations between characters and the morphometric effects of alleles at a single locus.

Relationships between types of character

Leamy (1975), using oblique factor analysis, investigated the relationship between six skull dimensions, four girdle dimensions, four limb lengths, two body proportions, weight, and mandible length of outbred laboratory mice at 1, 3 and 5 months of age.

Although the composition and order of factors differed between the 3- and 5-month-old mice and between sexes, they could generally be recognized as skull length, skull width, girdle, limb and tail.

This study was then followed up (Leamy, 1977) by separate analyses of the environmental and genetical correlations which had been partitioned using parent–offspring data. The resultant matrices were not positive semidefinite so cluster and non-metric multi-dimensional scaling (MDS) analysis were used to summarize them.

The cluster analysis (prior to MDS) of the genetical correlations showed four moderately distinct clusters which could be referred to as (1) body length + limb; (2) skull width + girdle; (3) skull length including mandible; and (4) tail length, weight and skull width. Leamy interprets the first two clusters as "appendicular" and the latter two as "axial". The MDS analysis does not reveal a clear structure (this may be a byproduct of the method, see p. 89 and Thorpe, 1980) although groupings similar to the "genetical" cluster analysis and the phenotypic correlations could be recognized. The environmental correlation matrix showed little clear structure by either cluster analysis or MDS.

Wallace & Bader (1967) also included different types of character (24 dental and three skull dimensions) in their analysis of character interrelationships in natural populations of the house mouse using orthogonal factor analysis. The five factors that were extracted showed that the skull length characters (one factor) were independent of the dental dimensions (four factors), including tooth lengths.

Berry & Jakobson (1975a) also used several types of character, i.e. body proportions, body weight, viscera weights and blood physiology, and give the correlation matrix between the characters for one natural population of mice. If one clusters this correlation matrix by single-linkage, it is evident that there is little clear structure to the matrix (Fig. 1) although the blood physiology characters form a distinct cluster. Body weight clusters with thymus weight, but generally there is a slight tendency for visceral weights to cluster independently of other characters. Consequently there is, with the exception of body proportions, a tendency for the characters to cluster by types.

In conclusion, then, it is apparent that different character types (with the exception of body proportions, Leamy, 1975, and above text) including different types of skeletal dimensions have a tendency to form separate factors for both genetic and phenotypic correlations. There is room for considerably more work on the interrelationships between different types of characters.

Relationships between characters of a single type

Skull dimensions

The skull length and skull width form distinct factors for both

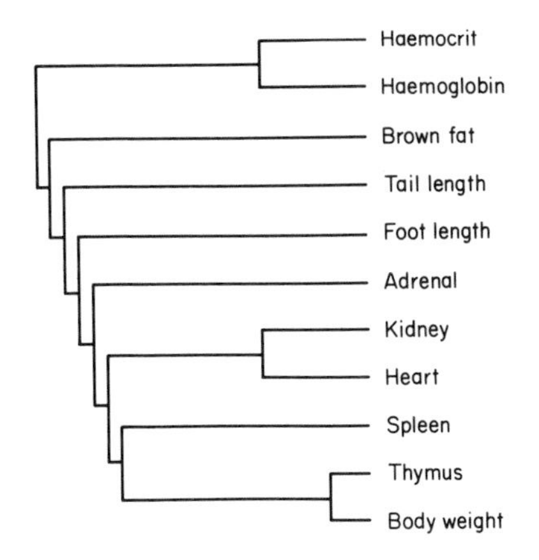

FIG. 1. Single-linkage cluster analysis of the correlations between 11 characters (adjusted for body length independence) measured on Skokholm mice. Data from Berry & Jakobson (1975a).

phenotypic (Leamy, 1975) and genetical (Leamy, 1977) correlations, and to some extent for environmental correlations (Leamy, 1977).

Mandible

Seven linear dimensions of the mandible have been investigated by Bailey (1956a) for inbred laboratory mice. Genetical and environmental correlations were obtained on the basis that interstrain correlations are genetical and intrastrain correlations are environmental. The principal-component analysis of the genetical correlations showed that the characters tended to be associated by anatomical proximity. Bailey interpreted "anterior" and "posterior" groups of characters which have functional significance, but these are not clearly defined. The environmental correlations showed a clearly defined grouping into anterior and posterior character groups. Unlike Leamy's (1977) genetical and environmental correlation matrices, Bailey's two matrices are highly correlated ($r = 0.72$, $p > 0.001$), which is why the two component analyses are comparable. Bailey's interpretation of the similarity of the environmental and genetical principal components is that genetical and environmental factors are influencing the same development pathways. However there is the possibility that it is a byproduct of the method of obtaining genetical and environmental correlations since neither Leamy's (1977) or Bader's (1965b) pairs of matrices show the same high positive correlation when tested.

Vertebral column

Bailey (1956a) also considered eight dimensions of the axis together with innominate length. As with the mandibular dimensions the "environmental" principal components revealed a clearer structure than the "genetical" principal components, although both indicated a generally similar picture and were interpreted as "arch" and "centrum" components.

Dentition

The interrelationship of dental variables has been more closely analysed than other sets of characters (Bader, 1965b; Bader & Lehmann, 1965; Wallace & Bader, 1967; Leamy & Hrubant, 1971).

Wallace & Bader (1967), using wild mice, investigated the widths and lengths of maxillary and mandibular molars on both the right- and left-hand-side giving 24 dimensions (plus three skull dimensions). Five orthogonal factors were extracted, four of which were odontometric and were interpreted as (1) width; (2) anterior length; (3) posterior length; and (4) M3. However – and this illustrates one of the limitations of this approach – alternative interpretations are possible for at least the first two factors which could be referred to as (1) general size (width bias); (2) anterior size (length bias). This does not emphasize length versus width although anterior versus posterior relationships are obvious in both interpretations.

Bader & Lehmann (1965) investigated the correlations between the widths of mandibular molars in wild, inbred, hybrid and outbred mice. In all four groups there is a tendency for the rank order of the correlations to be M_1/M_2, M_2/M_3, M_1/M_3. In other words, adjacent teeth are correlated and, as with Wallace & Bader (1967), the anterior two teeth are closely associated in contrast to the vestigial M_3. The correlations in the inbred and hybrid mice are lower than those in the outbred and wild mice presumably because in the former they are "environmental" rather than both environmental and genetical. Similar results were obtained by Bader's (1965b) analysis of genetical and environmental correlations of mandibular molar width and tooth row length in so far as M_1 and M_2 are more closely correlated than they are to M_3.

Leamy & Hrubant (1971) and Leamy & Sustarsic (1978) produced correlation matrices of maxillary and mandibular molar widths in their studies of six groups of inbred C57 BL mice that differed at only one locus. The pooled-within group correlation matrix from Leamy & Sustarsic's study (obtained by personal communication with L. Leamy) and the correlations from Leamy & Hrubant's study agree with the above studies of Bader and his co-workers in showing that the anterior–posterior position of the tooth is important, but disagree in indicating that M2 + M3 group together whilst M1 is less related.

If the characters are plotted against the first two eigenvectors extracted from Leamy & Sustarsic's pooled-within-group correlation matrix (Fig. 2), the importance of anatomical proximity can be seen.

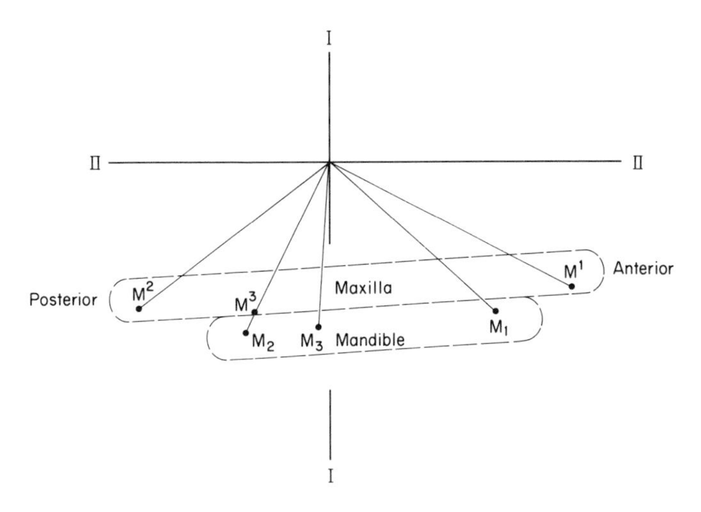

FIG. 2. Eigenvector analysis of the pooled-within-group correlations between molar widths. Data from Leamy & Sustarsic's (1978) study of C57BL mice (these are very similar to the matrix published in Leamy & Hrubant, 1971). The first and second eigenvectors account for 60% and 17% of the variation respectively. See text for explanation.

Although maxillary and mandibular teeth are segregated the dominant feature of the correlations is the separation of the first molars from the last two. The high correlation between occlusal partners (of obvious functional relevance) is portrayed as is the fact that the mandibular molars are more highly integrated than the maxillary molars.

Asymmetry

Bilateral asymmetry may be of two types, directional and fluctuating. If one side has consistently higher values than the other side it is known as directional asymmetry. If either side can have higher values it is known as fluctuating asymmetry.

Kidney weight in natural populations of mice is an example of directional asymmetry (Mittwoch, 1979), since the right kidney is consistently heavier than the left although the intensity of the asymmetry decreases with age. There is some suggestion of allomorphosis in this asymmetry since small mice from Pacific islands have a greater degree of kidney asymmetry than large mice from British and sub-Antarctic islands.

The molar widths of mice from natural populations are reported to exhibit fluctuating asymmetry (Bader, 1965a; Wallace, 1968). Leamy & Hrubant (1971) found that in the inbred laboratory mice they investigated the widths of M_1, M_3 and M^2 had significant directional asymmetry whilst the other molar widths showed fluctuating asymmetry.

With non-metrical skeletal variants that are capable of being expressed on both sides, one side may be more affected than the other. Searle (1954a) for example reports that in 60% of lateral variants there is significant asymmetry with the right side being the most frequently affected.

Morphometric differentiation due to a single locus

The pleiotropic effects of genes are well known but Leamy & Sustarsic (1978) provide an elegant example of how different alleles at a single locus can have widespread morphometric consequences in mice. Alleles at colour loci such as agouti (Leamy & Hrubant, 1971) and albino (L. Leamy, personal communication) can influence molar widths when on an isogenic background. Leamy & Sustarsic extended this approach to the multivariate case by investigating two sets of characters in C57BL/6 mice. The odontometric set of six characters was composed of the widths of the maxillary and mandibular molars whilst the osteometric set of eight characters was composed of body weight, three skull dimensions, three limb lengths and a girdle dimension.

Canonical analysis was carried out on each data set with each genotype comprising a group, i.e. *aa* (black), $a^t a$ (black and tan), $a^t a^t$ (black and tan), *Aa* (agouti), $A^{vy} a$ (viable yellow) and $A^y a$ (yellow). The canonical analyses (Fig. 3) and supplementary techniques show the multivariate divergence of these genotypes.

Leamy & Sustarsic (1978) point out the relevance of this to subline divergence and subspeciation. They suggest that the divergence of sublines of inbred strains may be due to the pleiotropic effect of alleles at a single locus and that the differences between subspecies* are commensurate with the level of divergence between the genotypes. The implication is that subspecies differences could

*It is true that some subspecies are erected on such small differences, but this ought not to be the case.

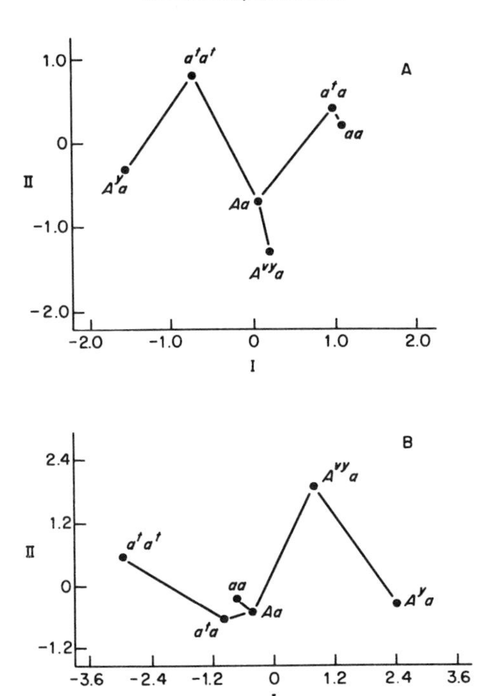

FIG. 3. Plots of the first two canonical variates for the six genotypes for both odontometric (A) and osteometric (B) data sets from Leamy & Sustarsic (1978). The group centroids are linked by the smallest Mahalanobis distances.

be due to the pleiotropic effects of a single locus. If this were true it would also be relevant to the computer simulation of raciation and speciation based on univariate and oligovariate systems (Endler, 1977) which would gain a greater degree of credibility if single genes had a widespread and intensive pleiotropic action in natural situations.

Apart from the problems of extrapolating from isogenic laboratory strains, which are largely free from selection pressures, to genetically diverse populations, which are subject to intense selection pressures and population turnover, there is the question of how many independent factors are involved in this morphometric divergence.

The divergence of the odontometric variables seems to be simply a function of size, since the general odontometric size of a group (computed as the sum product of the size (1st) eigenvector extracted from the pooled-within-group correlation matrix and the character scores) is highly correlated ($r = 0.89, p > 0.01$) with the first canonical variate of the odontometric analysis. However, the canonical analysis of osteometric divergence is not readily explained solely on general osteometric size (computed as above from the pooled-within-group correlations of just the seven osteometric characters) or weight, although these may be important. Moreover, the pattern of divergence is different for the two character sets (the osteometric and odontometric size of the genotypic groups is uncorrelated).

The confusing influence of size and its allomorphic consequences have long been known (Gould, 1966) and should not be allowed to unduly influence an assessment of racial affinities on which subspecies recognition or evolutionary deductions are based (Thorpe, 1976). Consequently, before one could extrapolate from Leamy & Sustarsic's interesting study (1978) one would wish to adjust for this size effect. Also, since one is dealing with rather fine differences, adjustment for other error sources and some sort of "control" or parallel study would be valuable to indicate the reliability of the findings. This is borne out by the fact that seasonal influences were revealed as being extremely important in a later unpublished study (L. Leamy, personal communication).

Conclusions

Studies of the interrelationships between different types of character generally reveal a tendency for the character to group according to type (with the exception of body proportions). Studies of the interrelationships between characters of a uniform type generally reveal the importance of anatomical proximity, although skull characters segregate into length and width groups. This, however, may be very dependent on the selection of characters. Analyses of the morphometric consequences of allelic differences at a single locus due to pleiotropy have far-reaching potential interest for evolutionary and systematic studies.

Relationships between Forms

This section discusses the relationship between temporal, sexual, laboratory and racial forms, but does not consider the relationship between ontogenetic forms (growth is considered by Roberts, this volume pp. 231–254, and ageing by Bellamy, this volume pp. 267–300).

All types of morphometric character, including non-metrical variants, show ontogenetic variation (Self & Leamy, 1978; see also Berry & Searle, 1963, for size dependence of variants). Some characters such as weight, body proportions, and skull dimensions grow throughout life (Dynowski, 1963), whilst others, such as molar widths, are constant in adult mice even though they are subject to wear. Differential growth and growth fields are apparent (C. V. Green, 1933; C. V. Green & Fekete, 1933; MacArthur & Chiasson, 1945) and ontogenesis can interact with factors such as sex (i.e. males grow faster, earlier). Ageing can influence a range of morphometric characters (Bellamy *et al.*, 1973).

Sexual dimorphism

Sexual dimorphism can result not only from sex-linkage but also from the effects of autosomal genes being expressed differently in the sexes owing to differences in the internal physiological environment (e.g. hormonal type) and external environment (due to different activities) between sexes (Eisen & Legates, 1966).

Sexual dimorphism in non-metrical skeletal variants is not thought to be extensive or intensive enough to be a problem when these variants are used to assess the similarity of natural populations (Berry, 1967a). Nevertheless, sexual dimorphism in variants is apparent in laboratory mice. For example, five vertebrae (instead of six) are more frequently found in males than females of the C3H strain (E. L. Green & Russel, 1951; McLaren & Michie, 1954) and 13 ribs (rather than 14) are more frequently found in males than females of the Bagg albino strain (E. L. Green, 1941).

Sexual dimorphism in variants differs between strains. Seven out of 21 variants showed significant sexual dimorphism in the C57 BL strain whilst four were sexually dimorphic in the A strain, yet only one of these variants was sexually dimorphic in both strains (Searle, 1954a). Similarly, the overall sexual dimorphism in C57 BL and BALB/c mice based on 25 variants is slightly greater in the latter strain (Howe & Parsons, 1967).

There is no evidence to suggest that sexual dimorphism in the frequency of variants is the result of the tendency for greater size in male laboratory mice (Searle, 1954a).

"Size" is one of the features which tends to show sexual dimorphism in mice. Body weight is generally greater in male than female laboratory mice, although the extent of the sexual dimorphism could differ between strains.

In natural conditions a more complicated pattern emerges. Berry (1970a) concludes that on average female wild mice are larger, although a collation of some of the patchy information on the subject (Evans & Vevers, 1938; Laurie, 1946; Crowcroft & Rowe, 1961; Dynowski, 1963; Berry & Tricker, 1969; Berry & Peters, 1975; Berry, Jakobson & Peters, 1978; Mittwoch, 1979) shows that in approximately a third of the natural populations males are heavier. Consequently in natural populations the extent and direction of sexual dimorphism in body weight is variable.

A greater body weight in males does not necessarily mean greater body proportions. Even though males are heavier in CV1 outbred mice, the females may or may not (depending on whether the sample is from parents or their adult offspring) have significantly longer bodies and tails (Leamy, 1974). Also in natural populations the situation is variable. Comparing mean values tabulated in Dynowski (1963) and Berry, Jakobson & Peters (1978) for natural populations, it is apparent that in about two-thirds of the populations females have higher mean body length and tail lengths, but in two-thirds of the population the males tend to have larger feet.

The extent of sexual dimorphism in body length also appears to vary geographically in so far as Schwarz & Schwarz (1943) use the extent and occurrence of sexual dimorphism in body length (and occasionally tail length) as diagnostic features for their subspecies.

As with body proportions, the direction of the sexual dimorphism in skeletal dimensions is not entirely dependent on body weight. Leamy's (1974) analysis of outbred CV1 mice, where males are heavier, shows that the pelvic girdle (as one would expect) is larger in females, whilst the pectoral girdle is larger in males. There is some tendency for this sexual dimorphism to be reflected in the associated limbs.

The skull dimensions of laboratory mice also show sexual dimorphism although the direction is not predictable. In Leamy's CV1 mice the skull dimensions (except zygomatic fenestral length) are generally significantly greater in males whilst in the inbred "little" strain (C. V. Green, 1932, 1933) the eight dimensions of the skull tended to be greater in females. In the natural populations that have been studied the skull dimensions tend to be greater in females. Hill's (1959) data on the mice of Tristan da Cunha and Gough indicate that females have large skulls for the seven dimensions recorded, while Dynowski (1963) also showed that in the Bialowieza population (females heavier) brain case length and skull height are greater in females although brain case depth is greater in males.

The dentition (molar widths), like the other character groups, can show slight sexual dimorphism which does not necessarily coincide with body weight. As sexual dimorphism in molar widths is slight in laboratory mice (Wallace, 1968) it has been dismissed as insignificant (Tenczar & Bader, 1966; Bader, 1965b), although with large enough samples it is detectable. Leamy & Hrubant's (1971) investigation using the C57 BL/6 strain showed that the widths of the maxillary (except M^3) and mandibular molars are greater in females although only significantly so in M_1 and M^3. Similarly in the inbred 129/J strain (L. Leamy, personal communication) the widths of M^2, M^3 and M_3 are significantly greater in females.

Little work has been done on odontometric sexual dimorphism in natural populations although Van Valen's (1965) data reveal that the direction of sexual dimorphism in M^1 widths fluctuates over time in the Skokholm population.

It is evident from Mittwoch's (1979) data on older mice from natural populations that irrespective of body weight, both the right- and left-hand kidneys tend to be heavier in males.

In conclusion, it is evident that sexual dimorphism is widespread throughout the various character systems although it is rather limited in extent and often inconsistent in direction and extent. Consequently, one cannot extrapolate with much confidence from one situation to another. There has been little work on the role of sexual dimorphism in natural populations of mice. Certainly there is nothing to compare with the elegant work on house sparrows, where multivariate morphometric analysis of sexual dimorphism in size has enabled the geographical and temporal variation in the extent of sexual dimorphism to be interpreted in terms of the behavioural differences between the sexes, their differential requirement for food, and their ability to obtain it (Johnston & Selander, 1973).

Laboratory mice

Although information on the morphometrics of some of the 200 strains and sublines of laboratory mice is available as a result of comparative studies of the non-metrical variants and mandible shape (e.g. Berry & Searle, 1963; Festing, 1972) and as a byproduct of other studies (e.g. weight, tail length and dentition in Searle, 1954a; Barnett, 1965b; Tenczar & Bader, 1966; Howe & Parsons, 1967), there is no comprehensive reference.

There have been numerous investigations of the non-metrical skeletal variants. Berry & Searle (1963) collate the information from some of the numerous investigations of non-metrical skeletal variants to compare the frequency of 35 variants in six inbred strains, i.e. C57BL/Gr, C57Br/cd, C3H/HeH, A/Gr, CBA/Gr and BALB/c. Differences between the strains were widespread and some strains had characteristic variants. It was such studies of laboratory mice that provided the basis for the study of microevolution in natural populations using these variants.

At a later date, Howe & Parsons (1967) investigated the mean divergence between three of these inbred strains (C57BL, C3H, BALB/c) and their hybrids based on the frequency of 25 variants. As with the above study they found large differences between strains, and by comparing the mean divergence of strains and their hybrids they deduced the "dominance" of each strain.

Whilst most non-metrical variants that are studied have been skeletal, the dentition (Grüneberg, 1965) and arterial system (Froud, 1959) also vary between inbred strains. In the latter case the differences were claimed to be commensurate with the differences between species.

Subline divergence within inbred strains has also been the subject of extensive study from the 1950s onwards (E. L. Green, 1953; Grüneberg, 1954) using non-metrical skeletal variants. From the systematic studies of this phenomenon (Deol *et al.*, 1957; Carpenter, Grüneberg & Russell, 1957; Grewal, 1962; Yong Hoi-Sen, 1972) it is apparent that subline divergence is a continuing process which cannot be explained by residual heterozygosity of the original stock.

Although the rate of divergence between sublines (as measured by skeletal variants) is related to the number of generations since they have been separated, it is about 1000 times greater (0.01 events per variant per generation) than the average spontaneous mutation rate for "major" genes of mice. It is thought that this is too great to be explained by polygenic control or pleiotropic interaction and that other self-replicating entities, such as viruses, may be influencing the frequency of these variants (Grüneberg, 1970; but see Beardmore, 1970).

Strain and subline divergence has also been investigated using linear dimensions of the skeleton. Using six skeletal measurements Bailey (1959) demonstrated, in agreement with the above studies, that sublines of inbred strains are divergent. The rates of divergence differed between strains (higher in BACB/An than C57BL/6) although this may be the byproduct of an inadequate sample of characters because these linear dimensions (unlike the non-metrical variants, Deol *et al.*, 1957; Carpenter *et al.*, 1957) show differing rates of divergence.

Festing's (1972, 1973, 1974a,b) investigations of strain and subline divergence using mandible dimensions have been directed more at the practical considerations of identification and quality control of commercial stocks than genetical theory. As such they provide an excellent example of "applied" morphometrics.

Mice belonging to eight strains (inbred, outbred and F_1) could be identified with considerable accuracy using canonical analysis (and supplementary techniques) based on 13 mandible dimensions

(Festing, 1972) even though several sources of error such as sexual dimorphism (see Table I in Leamy, 1974), age, size and environmental influences are likely to exist.

Festing (1973) extended this work to subline divergence, reducing the sources of error, as is commensurate with a "higher resolution" study, by using males of the same age. Nine sublines of C57BL/Gr and seven other strains were investigated as above which revealed that the extent of divergence between sublines is linearly related ($r = 0.60$) to the number of generations they have been separated. Consequently, subline divergence in mandible shape like the divergence in non-metrical variants, appears to be the result of the gradual accumulation of small changes over time. Sublines separated by 50–60 generations can be classified with only 5% overlap, and in general sublines can be classified with around 15% overlap. With more characters, which could very easily be recorded, one would expect a more complete discrimination of sublines.

Taylor (1972) investigated the divergence of several strains and sublines using 16 known genetical loci investigated by electrophoresis. The sublines were not well differentiated and in three strains (C57BL, C3H, A) there were identical pairs of sublines. Consequently the multivariate analysis of phenotypic traits appears, in this case, to be superior to allozyme comparisons in revealing genetical differences between "populations". This is of particular interest to the analysis of the affinities of natural populations since it contradicts the claim that the use of phenotypic traits to analyse affinities is somehow unsuitable (White, 1978) or relatively imprecise (see also Johnson, 1977, for some of the problems of using electrophoresis to assess "genetical" similarity).

Temporal variation

The phenotypic and genetical characteristics of a population can change over time. This can involve fossil and subfossil material or, with a rapid population turnover, it can be over a much shorter time-scale (such as years or seasons) and involve extant populations.

Temporal variation in the extant Skokholm population has been investigated by Berry and his co-workers over a decade or so using non-metrical skeletal variants and other characteristics (Berry, 1967a, 1977a, b; Berry & Jakobson, 1975b). Although the incidence of the variants does not fluctuate as much as their variance (Berry, 1967a) there is a measurable divergence over time (10 years) based on an analysis of 25 variants (Berry & Jakobson, 1975b) presumably as a response to selection pressures (Fig. 4). This is supported by biochemical evidence.

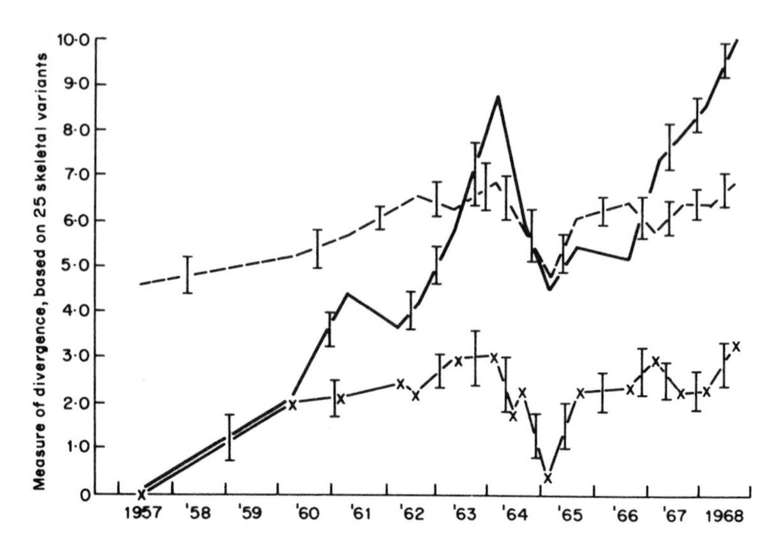

FIG. 4. Temporal variation in a natural population (Skokholm island). The continuous line shows the divergence from the previous sample, the line marked with crosses shows the divergence from the 1957 sample and the broken line measures the divergence from the neighbouring mainland population. (From Berry & Jakobson, 1975b.)

The incidence of both skeletal variants and dentition characters can be related to severe winters (Berry, 1967a). Van Valen (1965) suggests that a rare dental phenotype was lost to the Skokholm population owing to the drastic reduction in population size as a result of the severe winter of 1959/60.

There is a strong pattern of endocyclic (seasonal) variation in the variance of the non-metrical skeletal variants as well as skeletal and dental dimensions. About 50–90% of the mice on Skokholm die in the winter depending on the coldness of the early months of the year (Berry, 1968, 1970b). After this high winter mortality the variance of the four skeletal and four dental dimensions increases although the variance of the non-metrical variants decreases. This morphometric information, together with biochemical and physiological evidence, suggests that natural selection is acting in different ways at different times of the year, favouring one genotype in winter and another in summer.

Spatial variation

The study of the relationship between spatially segregated populations is the study of racial or geographical variation. The contribution of multivariate morphometric methods to this area has recently been reviewed by Gould & Johnston (1972) and Thorpe

(1976). Being an area of active research, new (or newly applied) methods such as numerical phyletics (Mickevich & Johnson, 1976) may be relevant to analysing the geographical variation of mice and other organisms.

Whilst numerical phyletics may not be the most appropriate method of analysing aspects of geographical variation such as the nature of transition ("hybrid") zones or clinal variation, they may be particularly suitable for analysing the colonization sequence in island groups. They could, for example, be used to investigate the Faroe mice (Berry, Jakobson & Peters, 1978).

The magnitude of spatial segregation can vary from microgeographical and habitat differences to that of an entire species or species complex involving consideration of population systematics.

Habitat and microgeographical variation

There has been little morphometric analysis of comparable mouse populations from different habitats (as opposed to populations from different localities which have different environmental conditions). Most information comes as a byproduct of studies with other aims.

Laurie (1946) investigated the reproduction of mice from four habitats (urban, flour buffers, corn ricks and cold stores) in southern England. From this study it is apparent that cold store mice tend to be heavier than those from the other three habitats with one female reaching a weight of 40.40 g, which is two to three times heavier than most mice. The average weights of fecund mice from the cold store, flour buffer, urban and corn rick habitats were *c.* 20, 16, 15 and 15 g respectively.

Mice also occur in other extreme commensal habitats such as coal mines. Elton (1936) reported mice from 550 m down an Ayrshire pit and Clegg (1965) comments on the mice from Yorkshire pits, but these studies do not reveal anything exceptional in their morphometry since they had "normal" body proportions and weight.

Bellamy *et al.*'s (1973) study of ageing in Skokholm mice revealed as a byproduct that mice from the centre of the island (bird observatory) differ phenotypically from mice from the sea cliffs. Multivariate discrimination of the age classes improved when habitat was taken into account.

One of the primary aspects of the habitat of a mouse population is whether it is commensal (indoor or outdoor) or wild. Although this concept has played an important role in hypotheses regarding the evolution or diagnosis of subspecies (Schwarz & Schwarz, 1943;

Zimmermann, 1949; Ursin, 1952; Berry, 1970a; Hunt & Selander, 1973; Marshall, 1977; Marshall & Sage, this volume pp. 15–25), there is little morphometric information regarding comparable commensal and wild populations. Nichols (1944) for example compares wild and commensal mice from Virginia, USA but goes no further than to imply a difference in skin thickness. Other studies involve geographical heterogeneity.

Several subspecies are regarded as being wild, indoor or outdoor commensals, and general morphometric characteristics have been attributed to mice from these habitats. Schwarz & Schwarz (1943) suggested that commensal forms tend to have longer tails, a reduction in the bulk of the face and size of the molars, a greater tendency to lose the third molars, and greater variability in size than wild mice. However, tail size has been shown to vary according to Allen's rule (Zimmermann, 1949), the morphometric differences may not be consistent with the habitat or subspecies (Ursin, 1952), habitat differences between some subspecies (*musculus* and *domesticus*) may not be clear cut (Hunt & Selander, 1973), and one cannot ascribe morphometric differences to habitat differences when so many other variables are involved.

Mice have the sort of population structure (rapid turnover and small demes), that enables microgeographical variation to develop quickly. Studies using non-metrical skeletal variants have shown that mice (indoor commensals) show microgeographical variation (Petras, 1967) and that populations can differ between rooms of the same building (Weber, 1950). Similarly, an investigation of corn rick mice (outdoor commensals) in Hampshire using 35 variants indicated differences between the populations in adjacent ricks (Berry, 1963). Although this microgeographical variation develops quickly, it is unlikely to have any temporal stability because of the seasonal fluctuation in numbers and dispersal of juveniles.

Macrogeographical variation

One or two generalizations can be made about the patterns of geographical variation in individual morphometric characters in mice, although it has not been the subject of much study.

The tail appears to be a thermoregulatory organ in mice and its length can vary with latitude according to Allen's rule (Zimmermann, 1949; Hunt & Selander, 1973; Marshall, 1977). This can be seen in mice from the British mainland (Berry, 1970a), where in southern England the tail is the same length as the body whilst in Scotland it is 20% shorter.

Other trends may also exist. For example, colour may be related to humidity (Marshall, 1977), and studies of island populations (Berry & Jakobson, 1975a) suggest that non-commensal populations approximate to Allen's and Bergmann's (size) rules.

Mice also tend to conform to the generalization that small mammals on islands tend to be comparatively large. The size of island mice is tabulated in Table III, but since the information comes from a range of sources (Hill, 1959; Berry, 1964; Berry & Jakobson,

TABLE III

The size of island mice. These average weights and lengths for island mice are from a range of studies and subject to many sources of error (see text). Where there are several estimates the highest and lowest are given.

Island	Weight (g)	Head and body length (mm)	Tail length (mm)	Hind foot length (mm)
Faroes				
1 Nolsøy	18.9	95.4	85.0	19.2
2 Sandøy	21.1	95.5	83.2	19.3
3 Thorshavn	16.2	85.8	76.5	17.4
4 Mykines	19.3	88.8	78.5	18.1
5 Hestur	14.3	83.6	75.3	17.7
6 Fugløy	12.4	86.5	74.4	17.7
Shetland				
1 Dunrossness	17.4	86.5	75.1	18.4
2 Scalloway	15.5	83.2	72.8	18.1
3 Bressay	16.7	86.5	78.5	18.7
4 Foula	23.9–26.3	97.1–98.3	81.2–83.7	18.1–18.3
Orkney				
1 S. Ronaldsay	17.2	87.2	72.6	18.1
2 Deerness	15.7	82.0	67.1	17.3
3 Harray	14.8	84.0	69.5	17.7
4 Sanday	17.9	87.6	73.1	17.7
5 N. Ronaldsay	17.5	85.5	69.2	18.0
6 N. Faray	26.2	99.0	84.0	19.0
Hebrides				
1 Bull of Lewis		93.5–98.0	81.5–85	18.5–18.7
2 Barvas (Lewis)		76.3–88.0	76.7–90.0	17.0–18.0
May	17.0–21.3	85.0–92.3	71.3–75.9	17.4–17.6
Skokholm	17.3–18.7	87.3–90.5	78.6–80.1	16.9–17.0
St Kilda		91.4–91.5	81.1–86.5	17.6–18.4
Macquarie	15.8	89.7	73.5	17.0
Marion	21.4	78.1	79.1	17.7
Medren	9.23–9.73			
Enewetak	8.9–9.5	75.2–76.6	66.5–68.2	15.1–15.5
Hawaii	12.93–13.18			
Tristan da Cunha		78	85	17.2
Gough		89	90	18.8

1975a, b; Berry, Jakobson & Peters, 1978; Berry, Peters & Van Aarde, 1978) and will involve a wide range of errors (sex, age, time, recording method, etc.), it is only meant as a very approximate guide. It is apparent that island mice tend to be large, although those from Enewetak and Hawaii are small. It may be that this tendency for increased size in isolates reflects an adaptive advantage in larger specimens for obtaining food in circumstances of not needing to escape down small holes from predators (which are infrequent on islands) (Corbet, 1961).

Except for the studies of British and other northern European island populations, there has been little work on the population affinities of mice using biometrical methods. The situation further east is confused by not knowing what species is the subject of study as there are several, previously unrecognized, sympatric species (J. T. Marshall, personal communication).

Berry (1963, 1967b) investigated the incidence of 35 non-metrical skeletal variants from 15 corn ricks in Hampshire, 11 widespread British populations and nine populations from across the world (18 variants). This and other work (Weber, 1950; Berry & Searle, 1963) established the use of these variants in characterizing genetically natural populations. The variation between rick populations suggests genetical drift and the uniformity of the British mainland populations suggests stabilizing selection.

Although the divergence (using 35 variants simultaneously) between the mainland populations was low, the islands of Skokholm (Wales) and May (Scotland) have divergent populations (Berry, 1964). The extinct St Kilda population, however, is not particularly divergent even though it had previously been considered as a subspecies owing to its large size. The rapid divergence of the Skokholm population from the neighbouring Welsh populations provides an example of rapid divergence due to the founder effect since the population had only existed for about 70 years at the time of study (Berry, 1964).

Berry, Jakobson & Peters' (1978) study of Faroe mice provides a further example of rapid divergence of island populations due to a few founders being introduced by man in the recent past. Populations from six of the Faroes were studied using four sets of morphometric characters (scapula dimensions, mandible dimensions, non-metrical skeletal variants and the weights of visceral organs plus body weight and proportions) and allozyme frequencies. A combination of the morphometric and genetical distances between these populations enabled a hypothesis to be constructed regarding the routes of inter-island colonization.

Other than the above work there has been little analytical morphometric work on geographical variation. Studies that describe mouse populations (Evans & Vevers, 1938; Elton, 1936; Moulthrop, 1942; Nichols, 1944; Harrison, 1958; Hill, 1959) tend to lack any quantitative comparison with other populations and tend to rely on size parameters such as weight and body proportions. These characters are strongly influenced by environmental factors (p. 97), sexual dimorphism, age (Dynowski, 1963) and method of recording (p. 86). Exceptions to this trend are given by Jones & Johnson (1965) and Dynowski (1963) who use a range of skull dimensions.

Taxonomic categories

Comprehensive morphometric analyses have had little part to play in describing or revising the taxonomic categories (species/subspecies) of mice. As with the description of populations (above) there is a tendency to rely on body proportions (Degerbøl, 1935; Moulthrop, 1942; Schwarz & Schwarz, 1943; Zimmermann, 1949; Ursin, 1952) although other characters such as condylo basal length may be used. The early studies involved some univariate quantification of the data and Ursin's (1952) study of the transition zone between *musculus* and *domesticus* in Denmark was the forerunner of Hunt & Selander's (1973) investigation of this zone uisng allozyme frequencies.

Recent taxonomic revisions (Marshall, 1977; Marshall & Sage, this volume pp. 15–25, personal communication) are even less dependent on morphometry, being based on qualitative descriptions of skull shape and tail length. Marshall (personal communication) concludes that species differences (he regards *musculus* and *domesticus* as separate species) involve skull shape whilst subspecific differences involve coloration as an adaptation to local conditions.

It is evident from theoretical considerations, from observations on other species (Gould & Johnston, 1972; Thorpe, 1976) and from studies of mice (Berry, Jakobson & Peters, 1978) that even comparatively local variation can involve a wide range of characters and that their multivariate morphometric analysis could have a good deal to offer evolutionary and taxonomic research.

The house mouse is of evolutionary interest because it not only provides an example of the founder effect (Berry, 1964) and allopatric "speciation" due to the Pleistocene effect which parallels other sympatric species (Thorpe, 1975, 1979; White, 1978), but also has a series of small, karyotypically divergent populations. These populations in the mountain valleys of Italy and Switzerland exhibit various Robertsonian fusions which may be either fixed or polymorphic (see White, 1978, for an overview of this). There has been no published account of the comparative morphometric divergence of these Robertsonian populations nor of any morphometric difference between individuals (within a single population) with different karyotypes, although the tobacco mouse (seven fixed metacentrics) is large and black (Fatio, 1869; Davatz, 1893; Marshall & Sage, this volume pp. 15–25) and has a wedge shape skull (J. T. Marshall, personal communication). The morphometric differences between these karyotypic forms may require multivariate analysis to elucidate them.

Where the morphometric consequences of Robertsonian fusions have not been investigated by multivariate methods, e.g. Gustavsson (1969) in cows and Hewitt & Schroeter (1968 and personal communication) in grasshoppers, morphometric divergence between karyotypes may not be revealed. However, multivariate methods such as canonical analysis have revealed significant morphological differences between different Robertsonian karyotypes within the same population in *Podisma pedestris* (Barton, 1979; G. M. Hewitt, personal communication).

Conclusions

The comparison of forms (sexual, temporal, racial, karyotypic, etc.) in natural populations is more a question of potential opportunity than achievement, although the analysis of temporal variation and island divergence (in conjunction with genetical and ecological studies) has contributed to our understanding of evolution.

DISCUSSION

Oxnard (1978), in his discussion of morphometrics, considered that it was too early to come to any firm conclusions regarding the contribution of morphometry to evolution and systematics. This is particularly true of mouse morphometrics.

It is apparent that morphometry is relevant to a wide range of disciplines and of both applied and theoretical value (see Introduction). Moreover there have been some comprehensive studies in which morphometry has played a part, for example character control and character relationships in laboratory mice (Leamy and co-workers), population biology of island mice (Berry and co-workers), and subline divergence.

Nevertheless, there are many more areas where morphometry could contribute far more than it has done, particularly in the analysis of raciation, speciation, "hybrid zones", the morphometric consequences of karyotypic divergence, and sexual dimorphism in natural populations. Morphometry has proved to be very useful in these areas in studies of other species (Gould & Johnston, 1972; Thorpe, 1976, 1979; Oxnard, 1978) and could be readily applied to a species like the house mouse which is of so much interest to microevolution.

ACKNOWLEDGEMENTS

I am grateful to the following researchers who corresponded with me whilst I was writing this review and who sent me papers and/or manuscripts: Dr J. T. Marshall, Professor E. Capanna, Professor A. Gropp, Professor E. Lehmann, Dr G. Hewitt, and Professor L. Leamy. I would also like to thank Professor L. Leamy for sending me a considerable number of matrices and tables which were so useful in writing this review and for his comments on the manuscript. I am also grateful to Professors F. W. Robertson and R. J. Berry for reading the manuscript. I would like to thank Academic Press for permission to publish Fig. 4 and The Society for Systematic Zoology for permission to publish Fig. 3.

REFERENCES

Albrecht, G. H. (1978). Some comments on the use of ratios. *Syst. Zool.* **27**: 67–71.

Atchley, W. R. (1978). Ratios, regression intercepts and the scaling of data. *Syst. Zool.* **27**: 78–83.

Atchley, W. R. & Anderson, D. (1978). Ratios and the statistical analysis of biological data. *Syst. Zool.* **27**: 71–78.

Atchley, W. R., Gaskins, C. T. & Anderson, D. (1976). Statistical properties of ratios. I. Empirical results. *Syst. Zool.* **25**: 137–148.

Bader, R. S. (1956). Variability in wild and inbred mammalian populations. *Q. Jl. Fla Acad. Sci.* **19**: 14–34.

Bader, R. S. (1965a). A partition of variance in dental traits of the house mouse. *J. Mammal.* **46**: 384–388.

Bader, R. S. (1965b). Heritability of dental characters in the house mouse. *Evolution* **19**: 378–384.

Bader, R. S. & Lehmann, W. H. (1965). Phenotypic and genotypic variation in odontometric traits of the house mouse. *Am. Midl. Nat.* **74**: 28–38.

Bailey, D. W. (1956a). A comparison of genetic and environmental principal components of morphogenesis in mice. *Growth* **20**: 63–74.

Bailey, D. W. (1956b). A comparison of genetic and environmental influences on the shape of the axis in mice. *Genetics* **41**: 207–222.

Bailey, D. W. (1959). Rates of subline divergence in highly inbred strains of mice. *J. Hered.* **50**: 26–30.

Barnett, S. A. (1961). Some effects of breeding mice for many generations in a cold environment. *Proc. R. Soc.* (B) **155**: 115–135.

Barnett, S. A. (1965a). Adaptation of mice to cold. *Biol. Rev.* **40**: 5–51.

Barnett, S. A. (1965b). Genotype and environment in tail length of mice. *Q. Jl. exp. Physiol.* **50**: 417–429.

Barnett, S. A. (1973). Maternal processes in the cold-adaptation of mice. *Biol. Rev.* **48**: 477–508.

Barton, N. (1979). *A narrow hybrid zone in the Alpine Grasshopper* Podisma pedestris. Ph.D. Thesis: University of East Anglia.

Beardmore, J. A. (1970). Viral components in the genetic background? *Nature, Lond.* **266**: 766–767.

Bellamy, D., Berry, R. J., Jakobson, M. E., Lidicker, W. Z., Morgan, J. & Murphy, H. M. (1973). Ageing in an island population of the house mouse. *Age and Ageing* **2**: 235–250.

Berry, R. J. (1963). Epigenetic polymorphisms in wild populations of *Mus musculus. Genet. Res.* **4**: 193–220.

Berry, R. J. (1964). The evolution of an island population of the house mouse. *Evolution* **18**: 468–483.

Berry, R. J. (1967a). The biology of non-metrical variation in mice and men. *Symp. Soc. Study hum. Biol.* **8**: 103–133.

Berry, R. J. (1967b). Genetical changes in mice and men. *Eug. Rev.* **59**: 78–96.

Berry, R. J. (1968). The ecology of an island population of the house mouse. *J. Anim. Ecol.* **37**: 445–470.

Berry, R. J. (1970a). The natural history of the house mouse. *Fld Stud.* **3**: 219–262.

Berry, R. J. (1970b). Covert and overt variation, as exemplified by British mouse populations. In *Variation in mammalian populations*: 3–26. Berry, R. J. & Southern, H. N. (Eds). London: Academic Press.

Berry, R. J. (1977a). The population genetics of the house mouse. *Sci. Progr., Oxford* **64**: 341–370.

Berry, R. J. (1977b). *Inheritance and natural history*. London: Collins New Naturalist.

Berry, R. J. & Jakobson, M. E. (1975a). Adaptation and adaptability in wild-living house mice. *J. Zool., Lond.* **176**: 391–402.

Berry, R. J. & Jakobson, M. E. (1975b). Ecological genetics of an island population of the house mouse. *J. Zool., Lond.* **175**: 523–540.

Berry, R. J., Jakobson, M. E. & Peters, J. (1978). The house mouse of the Faroe Islands: a study of microdifferentiation. *J. Zool., Lond.* **185**: 73–92.

Berry, R. J. & Peters, J. (1975). Macquarie Island house mice: genetical isolate on a sub-Antarctic island. *J. Zool., Lond.* **176**: 375–389.

Berry, R. J., Peters, J. & Van Aarde, R. J. (1978). Sub-Antarctic house mice: colonization, survival and selection. *J. Zool., Lond.* **184**: 127–141.

Berry, R. J. & Searle, A. G. (1963). Epigenetic polymorphism of the rodent skeleton. *Proc. zool. Soc. Lond.* **140**: 577–615.

Berry, R. J. & Tricker, B. J. K. (1969). Competition and extinction: the mice of Foula with notes on those of Fair Isle and St Kilda. *J. Zool., Lond.* **158**: 247–265.

Blackith, R. E. & Reyment, R. A. (1971). *Multivariate morphometrics.* New York: Academic Press.

Bookstein, F. L. (1977). The study of shape transformation after D'Arcy Thompson. *Math. Biosci.* **34**: 177–219.

Brown, J., Barrett, M. J. & Darrock, J. N. (1965). Factor analysis in cephalometric research. *Growth* **29**: 97–107.

Carpenter, J. R. H., Grüneberg, H. & Russell, E. S. (1957). Genetical differentiation involving morphological characters in an inbred strain of mice. II. American branches of the C57BL and C57BR strains. *J. Morph.* **100**: 377–388.

Chai, C. K. (1971). Analysis of the quantitative inheritance of body size in mice. VI. Diallel crosses between lines differing at a small number of loci. *Genetics* **68**: 299–311.

Clarke, W. E. (1904). On some forms of *Mus musculus, Linn.*, with description of a new subspecies from the Faroe Islands. *Proc. R. Phys. Soc. Edinb.* **15**: 160–167.

Clarke, W. E. (1914). Notes on the mice of St Kilda. *Scot. Nat.* 1914: 124–128.

Clegg, T. M. (1965). *The house mouse (*Mus musculus *Linn.) in some south Yorkshire coal mines.* Doncaster: Museum and Art Gallery.

Clifford, H. T. & Stephenson, W. (1975). *An introduction to numerical classification.* London and New York: Academic Press.

Corbet, G. B. (1961). Origin of the British insular races of small mammals and of the Lusitanian fauna. *Nature, Lond.* **191**: 1037–1040.

Corruccini, R. S. (1977). Correlation properties of morphometric ratios. *Syst. Zool.* **26**: 211–214.

Crowcroft, W. P. & Rowe, F. P. (1961). The weights of wild house mice (*Mus musculus*) living in confined colonies. *Proc. zool. Soc. Lond.* **136**: 177–185.

Davatz, F. (1893). *Mus poschiavinus* Fatio (Puschlaver-oder ouch Tabakmans gennant). *Jher. Nat. Forsch. ges. Grabund.* **36**: 95–103.

Degerbøl, M. (1935). On *Mus musculus specilegus* Pet. in Denmark. *Viddensk. Meddr dansk naturh. Foren.* **99**: 233–238.

Deol, M. S., Grüneberg, H., Searle, A. G. & Truslove, G. M. (1957). Genetical differentiation involving morphological characters in an inbred strain of mice. I. A British branch of the C57BL strain. *J. Morph.* **100**: 345–376.

Deol, M. S. & Truslove, G. M. (1957). Genetical studies on the skeleton of the mouse. Maternal physiology and variation in the skeleton of C57BL mice. *J. Genet.* **55**: 288–312.

Dodson, P. (1975). Functional and ecological significance of relative growth in *Alligator. J. Zool., Lond.* **175**: 315–355.

Dodson, P. (1978). On the use of ratios in growth studies. *Syst. Zool.* **27**: 62–67.

Dun, R. B. & Fraser, A. S. (1959). Selection for an invariant character, vibrissa number, in the house mouse. *Aust. J. biol. Sci.* **12**: 506–523.

Dynowski, J. (1963). Morphological variability in the Bialowieza population of *Mus musculus* Linnaeus, 1758. *Acta theriol.* **7**: 51–67.

Eisen, E. J. & Legates, J. E. (1966). Genotype-sex interaction and the genetic correlation between the sexes for body weight in *Mus musculus. Genetics* **54**: 611–623.

El Oksh, A. H., Sutherland, T. M. & Williams, J. W. (1967). Prenatal and postnatal maternal influence on growth in mice. *Genetics* **57**: 79–94.

Elton, C. (1936). House mice (*Mus musculus*) in a coal mine in Ayrshire. *Ann. Mag. nat. Hist.* (10) **17**: 553–558.

Endler, J. A. (1977). *Geographic variation, speciation and clines.* Princeton, New Jersey: Princeton University Press.

Evans, F. C. & Vevers, H. G. (1938). Notes on the biology of the Faroe mouse (*Mus musculus faeroensis*). *J. Anim. Ecol.* 7: 290–297.

Falconer, D. S. (1953). Selection for large and small size in mice. *J. Genet.* **51**: 470–501.

Falconer, D. S. (1960). *Introduction to quantitative genetics.* Edinburgh: Oliver & Boyd.

Fatio, V. (1869). *Faune des vertébrés de la Suisse* **1**. Geneva and Basle: H. George.

Festing, M. (1972). Mouse strain identification. *Nature, Lond.* **238**: 351–352.

Festing, M. (1973). A multivariate analysis of subline divergence in the shape of the mandible in C57BL/Gr mice. *Genet. Res.* **21**: 121–131.

Festing, M. (1974a). Genetic reliability of commercially-bred laboratory mice. *Lab. Anim.* **8**: 265–270.

Festing, M. (1974b). Genetic monitoring of laboratory mouse colonies in the Medical Research Council accreditation scheme for suppliers of laboratory animals. *Lab. Anim.* **8**: 291–299.

Festing, M. (1976). Phenotypic variability of inbred and outbred mice. *Nature, Lond.* **263**: 230–232.

Fortuyn, A. B. D. (1912). Über den systematischen Wert der japanischen Tanzmaus (*Mus wagneri* var. *rotans* nov. var.). *Zool. Anz.* **39**: 177–190.

Froud, M. D. (1959). Studies on the arterial system of three inbred strains of mice. *J. Morph.* **104**: 441–478.

Gould, S. J. (1966). Allometry and size in ontogeny and phylogeny. *Biol. Rev.* **41**: 587–640.

Gould, S. J. & Johnston, R. F. (1972). Geographic variation. *A. Rev. Ecol. Syst.* **3**: 457–498.

Gower, J. C. (1972). Measures of taxonomic distance and their analysis. In *The assessment of population affinities in man*: 1–24. Weiner, J. S. & Huizinga, J. Oxford: Clarendon Press.

Green, C. V. (1932). A genetic craniometric study of two species of mice and their hybrids. *J. exp. Zool.* **63**: 533–551.

Green, C. V. (1933). Differential growth in crania of mature mice. *J. Mammal.* **14**: 122–131.

Green, C. V. & Fekete, E. (1933). Differential growth in the mouse. *J. exp. Zool.* **66**: 351–370.

Green, E. L. (1941). Genetic and non genetic factors which influence the type of skeleton in an inbred strain of mice. *Genetics* **26**: 192–222.

Green, E. L. (1953). A skeletal difference between sublines of the C3H strain of mice. *Science, N.Y.* **117**: 81–82.

Green, E. L. & Russel, W. L. (1951). A difference in skeletal type between reciprocal hybrids of two inbred strains of mice (C57BLK and C3H). *Genetics* **36**: 641–651.

Grewal, M. S. (1962). The rate of genetic divergence of sublines in the C57BL strain of mice. *Genet. Res.* **3**: 226–237.

Grüneberg, H. (1954). Variation within inbred strains of mice. *Nature, Lond.* **173**: 674–676.

Grüneberg, H. (1963). *The pathology of development.* Oxford: Blackwell.

Grüneberg, H. (1965). Genes and genotype affecting the teeth of the mouse. *J. Embryol. exp. Morph.* **14**: 137–159.

Grüneberg, H. (1970). Is there a viral component in the genetic background? *Nature, Lond.* **225**: 39–41.

Gustavsson, I. (1969). Cytogenetics, distribution and phenotypic effects of a translocation in Swedish cattle. *Hereditas* **63**: 68–169.
Hanrahan, J. P. & Eisen, E. J. (1973). Sexual dimorphism and direct and maternal genetic effects on body weight in mice. *Theoret. appl. Genet.* **43**: 39–45.
Harman, H. H. (1976). *Modern factor analysis* (3rd edn). Chicago: University of Chicago Press.
Harrison, G. A. (1958). The adaptability of mice to high environmental temperatures. *J. exp. Biol.* **35**: 892–901.
Hewitt, G. & Schroeter, G. (1968). Population cytology of *Oedaleonotus*. I. The karyotypic facies of *Oedaleonotus enigma* (Scudder). *Chromosoma* **25**: 121–140.
Hill, J. E. (1959). Rats and mice from the islands of Tristan da Cunha and Gough, South Atlantic Ocean. *Results Norw. scient. Exped. Tristan da Cunha* No. 46: 1–5.
Hills, M. (1978). On ratios – a response to Atchley, Gaskins and Anderson. *Syst. Zool.* **27**: 61–62.
Hinton, M. A. C. & Hony, G. B. (1916). Notes on two collections of mice from Lewis, Outer Hebrides. *Scot. Nat.* **1916**: 221–227.
Howe, W. L. & Parsons, P. A. (1967). Genotype and environment in the determination of minor skeletal variants and body weight in mice. *J. Embryol. exp. Morph.* **17**: 283–292.
Hunt, W. G. & Selander, R. K. (1973). Biochemical genetics of hybridization in European house mice. *Heredity* **31**: 11–33.
Jameson, H. L. (1898). On a probable case of protective coloration in the house mouse (*Mus musculus*, Linn.). *J. Linn. Soc.* (Zool.) **26**: 465–473.
Jewell, P. A. & Fullagar, P. J. (1966). Body measurements of small mammals: sources of error and anatomical changes. *J. Zool., Lond.* **150**: 501–509.
Johnson, G. B. (1977). Assessing electrophoretic similarity: the problem of hidden heterogeneity. *A. Rev. Ecol. Syst.* **8**: 309–328.
Johnston, R. F. & Selander, R. K. (1973). Evolution in the house sparrow. III. Variation in size and sexual dimorphism in Europe and North and South America. *Am. Nat.* **107**: 373–390.
Jolicouer, P. (1963). The multivariate generalization of the allometry equation. *Biometrics* **19**: 497–499.
Jones, J. K. & Johnson, D. H. (1965). Synopsis of the lagomorphs and rodents of Korea. *Univ. Kans. Publs Mus. nat. Hist.* **16**: 357–407.
Kruskal, J. B. (1964a). Multidimensional scaling by optimizing goodness of fit to a non-metric hypothesis. *Psychometrika* **29**: 1–27.
Kruskal, J. B. (1964b). Non-metric multidimensional scaling: a numerical method. *Psychometrika* **29**: 115–129.
Laurie, E. M. O. (1946). The reproduction of the house mouse (*Mus musculus*) living in different environments. *Proc. R. Soc.* (B) **133**: 248–81.
Leamy, L. J. (1974). Heritability of osteometric traits in a random-bred population of house mice. *J. Hered.* **65**: 109–120.
Leamy, L. (1975). Component analysis of osteometric traits in random-bred house mice. *Syst. Zool.* **24**: 176–190.
Leamy, L. (1977). Genetic and environmental correlations of morphometric traits in random-bred house mice. *Evolution* **31**: 357–369.
Leamy, L. (1981). Heritability of morphometric ratios in random-bred house mice. In *Mammalian population genetics.* Smith, M. H. & Joule, J. (Eds). Athens: University of Georgia Press.

Leamy, L. & Hrubant, H. E. (1971). Effects of alleles at the agouti locus on odontometric traits in the C57BL/6 strain of house mice. *Genetics* **67**: 87–96.

Leamy, L. & Sustarsic, S. S. (1978). A morphometric discriminant analysis of agouti genotypes in C57BL/6 house mice. *Syst. Zool.* **27**: 49–60.

Leamy, L. & Touchberry, R. W. (1974). Additive and non-additive genetic variance in odontometric traits in crosses in seven inbred lines of house mice. *Genet. Res.* **23**: 207–217.

MacArthur, J. W. & Chiasson, L. P. (1945). Relative growth in races of mice produced by selection. *Growth* **9**: 303–315.

Marshall, J. T. (1977). A synopsis of Asian species of *Mus* (Rodentia: Muridae). *Bull. Am. Mus. nat. Hist.* **158**: 177–220.

McLaren, A. & Michie, D. (1954). Factors affecting vertebral variation in mice. I. Variation within an inbred strain. *J. Embryol. exp. Morph.* **2**: 149–160.

Mickevich, M. F. & Johnson, M. S. (1976). Congruence between morphological and allozyme data in evolutionary inference and character evolution. *Syst. Zool.* **25**: 260–270.

Mittwoch, U. (1979). Lateral asymmetry of kidney weights in different populations of wild mice. *Biol. J. Linn. Soc.* **11**: 295–300.

Moulthrop, P. N. (1942). Description of a new house mouse from Cuba. *Scient. Publs Cleveland Mus. nat. Hist.* **5** (5): 79–82.

Nichols, D. G. (1944). Further consideration of American house mice. *J. Mammal.* **25**: 82–84.

Oxnard, C. E. (1978). One biologist's view of morphometrics. *A. Rev. Ecol. Syst.* **9**: 219–241.

Petras, M. L. (1967). Studies of natural populations of *Mus*. IV. Skeletal variations. *Can. J. Genet. Cytol.* **9**: 575–588.

Roderick, T. H. & Schlager, G. (1966). Multiple factor inheritance. In *The biology of the laboratory mouse*: 151–164. Green, E. L. (Ed.). New York & London: McGraw-Hill.

Schwarz, E. & Schwarz, H. K. (1943). The wild and commensal stocks of the house mouse *Mus musculus* Linnaeus. *J. Mammal.* **24**: 59–72.

Searle, A. G. (1954a). Genetical studies on the skeleton of the mouse. XI. Causes of skeletal variation within pure lines. *J. Genet.* **52**: 68–102.

Searle, A. G. (1954b). Genetical studies on the skeleton of the mouse. XI. The influence of diet on variation within pure lines. *J. Genet.* **52**: 413–424.

Self, G. S. & Leamy, L. J. (1978). Heritability of quasi-continuous skeletal traits in a random-bred population of house mice. *Genetics* **88**: 109–120.

Shephard, R. N. (1962a). The analysis of proximities: multidimensional scaling with an unknown distance function. I. *Psychometrika* **27**: 125–140.

Shephard, R. N. (1962b). The analysis of proximities: multidimensional scaling with an unknown distance function. II. *Psychometrika* **27**: 119–140.

Sneath, P. H. A. & Sokal, R. R. (1973). *Numerical taxonomy*. San Francisco: Freeman & Co.

Taylor, B. A. (1972). Genetic relationships between inbred strains of mice. *J. Hered.* **63**: 83–86.

Tenczar, P. & Bader, R. S. (1966). Maternal effect in dental traits of the house mouse. *Science, N.Y.* **152**: 1298–1400.

Thiessen, D. D. (1966). The relation of social position and wounding to exploratory behaviour and organ weights in house mice. *J. Mammal.* **47**: 28–34.

Thorpe, R. S. (1975). Biometric analysis of incipient speciation in the ringed snake *Natrix natrix* (L). *Experientia* **31**: 180–182.

Thorpe, R. S. (1976). Biometric analysis of geographic variation and racial affinities. *Biol. Rev.* **51**: 407–452.

Thorpe, R. S. (1979). Multivariate analysis of the population systematics of the ringed snake *N. natrix* (L). *Proc. R. Soc. Edinb.* **78B**: 1–62.

Thorpe, R. S. (1980). A comparative study of ordination techniques in numerical taxonomy in relation to racial variation in the ringed snake *Natrix natrix* (L). *Biol. J. Linn. Soc.* **13**: 7–40.

Truslove, G. M. (1961). Genetical studies on skeletons of the house mouse. XXX. Search for correlation between minor variants. *Genet. Res.* **2**: 431–438.

Ursin, E. (1952). Occurrence of voles, mice and rats (*Muridae*) in Denmark, with a special note on a zone of intergradation between two subspecies of the house mouse (*Mus musculus* L.). *Vidensk. Meddr dansk naturh. Foren.* **114**: 217–244.

Van Valen, L. (1965). Selection in natural populations. IV. British house mice (*Mus musculus*). *Genetica* **36**: 119–134.

Wallace, J. T. (1968). Analysis of dental variation in wild-caught California house mice. *Am. Midl. Nat.* **80**: 360–380.

Wallace, J. T. & Bader, R. S. (1967). Factor analysis in morphometric traits of the house mouse. *Syst. Zool.* **16**: 144–148.

Weber, W. (1950). Genetical studies on the skeleton of the mouse. III. Skeletal variation in wild populations. *J. Genet.* **50**: 174–178.

White, M. J. D. (1978). *Modes of speciation*. San Francisco: W. H. Freeman & Co.

Yong, Hoi-Sen (1972). Is subline differentiation a continuing process in inbred strains of mice? *Genet. Res.* **19**: 53–59.

Zimmermann, K. (1949). Zur Kenntnis der mitteleuropäischen Hausmäuse. *Zool. Jb.* (Syst.) **78**: 301–322.

Symp. zool. Soc. Lond. (1981) No. 47, 127–139

Karyotype of the House Mouse

E. P. EVANS

Sir William Dunn School of Pathology, University of Oxford, Oxford, UK

SYNOPSIS

The mouse chromosomes have been known for 90 years. Many of those years passed with little reward in the search for markers which could be used to identify individual chromosomes with certainty. In the all acrocentric karyotype, a few autosomes could be recognized by their secondary constrictions and at times the sex chromosomes by their characteristic morphology. Over the last decade the development and use of banding techniques has brought about a dramatic change. Q- or G-banding has made it possible to recognize each chromosome and, with this innovation, all the known linkage groups have been assigned to chromosomes and correctly orientated with respect to the centromeres. In addition, the exact breakpoints of many rearranged chromosomes have been determined and in one chromosome, the genes accurately placed in relation to the G-band sequence. Banding has also confirmed existing information that, apart from Robertsonian translocations, few spontaneous rearrangements arise in the mouse. So far, comparisons of Q- and G-band patterns between inbred strains and in feral mice have indicated that there is no variation. Indeed, the band patterns in *Mus musculus* and its subspecies are also remarkably uniform. On the other hand, an array of variation has been encountered in the centomeric regions of both laboratory and feral mice. These include differences in C-bands, in secondary constrictions on nucleolar organizing chromosomes and in lateral asymmetry. Since these variants are constant for inbred strains, their observation may prove profitable in studies of mouse chromosome evolution.

In common with many of the other biological studies of the house mouse, those of the karyotype, both old and new, have largely been confined to that of the laboratory mouse.

HISTORY OF THE KARYOTYPE

1889–1970

Some aspects of the karyotype are old, but Crew & Koller (1932) commented that even up to that year, the many studies of the chromosomes of the mouse had been concerned mainly with their number or with sex chromosome constitution. They mention that in 1889 Tafani had given 20 as the haploid number, but Sobotta in

1895 made it 16 and Gerlach in 1906 reduced it even further to 12. With improved cytological techniques, the haploid number was re-established as 20 by Long (1908) and then further confirmed by Cox (1926) and by Painter (1926, 1927, 1928). Crew & Koller (1932) described all 40 of the mitotic chromosomes as being "rod-shaped" and on arranging them into arbitrary pairs, decided that the Y chromosome was probably the shortest and the X was one of the longest. Koller (1944) elaborated on this description and said that they were all composed of one long and one very short arm.

In the 1950s, the ever-increasing use of the mouse in genetical work, cancer research and in many other studies called for as thorough a knowledge of its chromosomes as was deemed possible. It was hoped that this knowledge would come from the concurrent technical advances made in the same period which had led to such improved presentations of mammalian chromosomes. These new preparations, however, although revealing a few additional characters, largely served to re-emphasize the rather featureless morphology of mouse chromosomes and the difficulty this presented in the positive identification of homologues. Apart from identification, much discussion ensued as to whether mouse chromosomes were telocentric, that is with strictly terminal centromeres and no short arms, or acrocentric with closely terminal centomeres and short, albeit invisible, arms (see Levan, Fredga & Sandberg, 1964). This discussion has continued to this day and, although it is commonly agreed that they are acrocentric, the term telocentric is still used in some laboratories. With regard to the improved preparations and chromosome recognition at mitosis, the X chromosome was identified by inference from its allocyclic condensation at cell division (Ohno, Kaplan & Kinosita, 1957). The Y chromosome was identified in some mice as being the odd one of the three distinctly smallest chromosomes (Stich & Hsu, 1960), which at times showed unique morphological features (Fig. 1) (Ford, 1966). An idiogram, derived from linear measurements (Levan, Hsu & Stich, 1962), indicated that the mitotic metaphase chromosomes ranged in size from two to five μm. The same authors also concluded that the curiously termed "rabbit-ear chromosomes" (see Fig. 1), originally described by Hsu, Billen & Levan (1961), were the combined result of a near centric secondary constriction and heteropycnotic area which could possible constitute a valuable landmark in the mouse karyotype. Further examples of these "landmarks" were recorded by Bennett (1965) who found up to eight chromosomes per cell with secondary constrictions, which she assumed must represent four pairs since they were usually present in even numbers. Their use as markers was, however, limited since they could only be related with certainty to the smallest pair

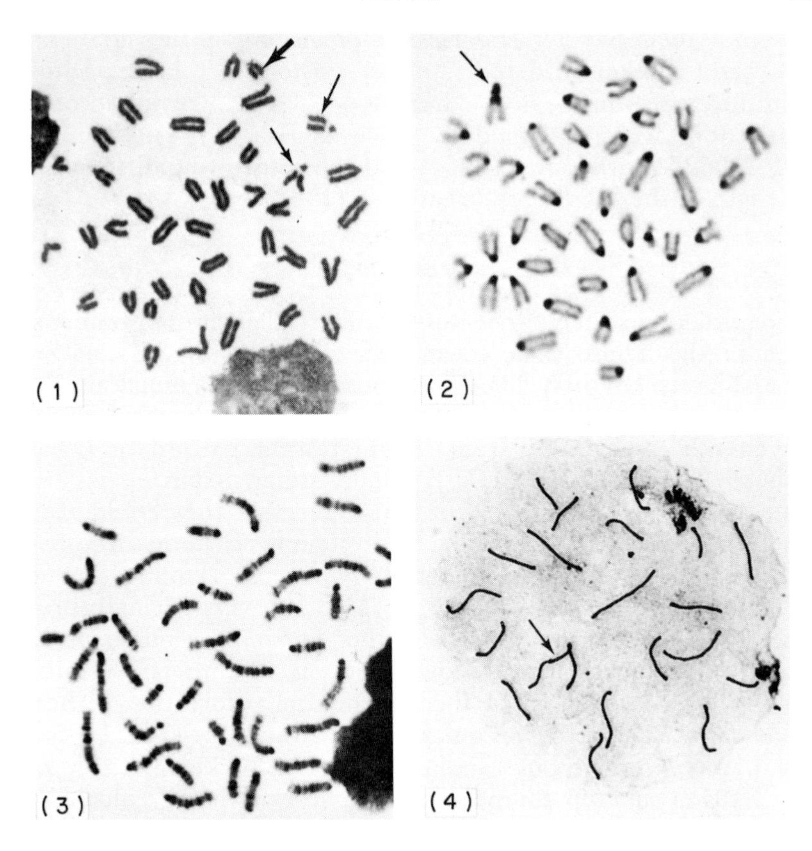

FIGS 1–4. (1) The karotype of the 1960s illustrating the few features for chromosome recognition: the "rabbit-ear" chromosomes (thin arrows) and the Y chromosome (thick arrow).

(2) C-banded karyotype illustrating an extreme variant of a double band on chromosome 18 (thin arrow).

(3) The high resolution G-banded karyotype of the 1970s.

(4) The 20 silver stained pachytene bivalents in a whole, sedimented spermatocyte. XY pair (thin arrow).

in the complement, a feature which was used by Evans, Lyon & Daglish (1967) to identify correctly one of the chromosomes involved in a Robertsonian translocation.

Further search for more precise markers in conventionally stained mitotic cells proved unrewarding, as did autoradiographic studies of differing chromosome replication patterns after ^{3}H-thymidine incorporation, which had proved useful in the identification of some human chromosomes (German, 1964). These only showed in the mouse that X and Y chromosomes could additionally be recognized by their late replication (Church, 1965; Galton & Holt, 1965).

Despite their paucity and lack of precision, subtle variations in these early markers did indicate the existence of intraspecific or strain differences in the mouse karyotype. Those involving secondary constrictions were described by Stich & Hsu (1960), Levan, Hsu *et al.* (1962) and Bennett (1965) and those involving differences in the length of the Y by Levan, Hsu *et al.* (1962).

Since 1970

The painstaking search for features of individual chromosome recognition in the mouse came to an end in 1970 with the emergence of banding techniques. The development of the quinacrine fluorescent technique (Caspersson *et al.*, 1968) enabled each one to be subsequently recognized from its fluorescent pattern or Q-bands (Francke & Nesbitt, 1971; O. J. Miller, Miller, Kouri *et al.*, 1971; Schnedl, 1971). Alternatively, it was found that they could also be recognized from their dark and light staining patterns or G-bands after partial denaturation followed by Giemsa staining (Schnedl, 1971; Buckland, Evans & Sumner, 1971). Some pairs originally proved difficult to distinguish from each other but improvements in these existing techniques and the introduction of others (Nesbitt & Francke, 1973) soon led to the unequivocal recognition of homologous pairs. Further, it was quickly realized that the Q- and G-bands were largely synonymous (Sumner, Evans & Buckland, 1971; Zech *et al.*, 1972) and up to the present time they have remained the mainstay for mouse karyotyping whereas other procedures such as reverse or R-banding (Dutrillaux & Lejeune, 1971) have found little favour.

During the development of these banding technqiues and later, it became increasingly apparent that the Q- and G-bands of normal mouse chromosomes were remarkably constant and it was the centromeric regions, akin to the secondary constrictions of the 1960s, which represented the most active sites of variation and change. Indications of this came, by design or by accident, from a number of procedures which, although they did not in themselves always provide a means of individual chromosome recognition, demonstrated an array of differences at the centromere. In a new analysis of secondary constrictions allied to the fluorescent karyotype, Dev, Grewal *et al.* (1971) showed strain differences in the presence or absence of prominent secondary constrictions on any of four pairs of different chromosomes. The C-bands, the intensely dark staining region around the centromere produced with Giemsa after chromosome denaturation with alkali (Pardue & Gall,

1970; Dev, Miller, Allderdice & Miller, 1972) or by hydrolysis and heat (McKenzie & Lubs, 1973), also showed differences between strains. C-bands can also be stained using Hoechst 33258 which produces bright fluorescence in this region (Natarajan & Gropp, 1972), a fluorescence which, although present in the chromosomes of *Mus musculus* and *M. m. molossinus,* is absent in other species of *Mus* (Hsu, Markvong & Marshall, 1978). The fact that some mouse chromosomes are nucleolar organizing and can be silver stained to show NORs (Goodpasture & Bloom, 1975), which locate sites of active 18S and 28S ribosomal RNA genes, provides another source of variation. Recently, Jonasson, Alves & Strom (1979), after BrdU incorporation followed by Giemsa staining of mouse chromosomes, have demonstrated interesting differences in the lateral asymmetry (Lin, Latt & Davidson, 1974) within the centromeric regions and have also shown the presence of some presumptive inversions.

THE MODERN KARYOTYPE, ITS ACHIEVEMENTS AND THE SEARCH FOR VARIATION

The discovery of the Q- and G-banding methods and their application to the mitotic chromosomes of the mouse resulted in the rapid publication of four separate versions of the karyotype (Buckland *et al.,* 1971; Dev, Grewal *et al.,* 1971; Francke & Nesbitt, 1971; Schnedl, 1971). Unfortunately, although the banded chromosomes had been numbered in the order of descending size in all four, variations between laboratories led to the production of four different numbering systems. This discrepancy was resolved by the Committee on Standardized Genetic Nomenclature for Mice (1972) and a standard karyotype was adopted but without band nomenclature. As a result of improved techniques which gave greater band resolution, the system of nomenclature proposed by Nesbitt & Francke (1973) was later adopted (Committee on Standardized Genetic Nomenclature for Mice, 1974). This system listed 312 distinct regions within the karyotype and had the advantage of being able to accommodate further band information in the future, either from higher resolution techniques or from more astute observations. This adopted system has, apart from some minor disagreements, stood the test of time and its continued use was recently endorsed (Committee on Standardized Genetic Nomenclature for Mice, 1979).

Undoubtedly, the most significant contribution made by the ability to identify each mouse chromosome has been in the placement of linkage groups. In the brief period between 1971 and 1973 most of the known linkage groups were assigned to chromosomes. Previously they had been known by Roman numerals in order of their discovery, but in 1972 it was proposed by the Committee that as soon as feasible the chromosome numbers in arabic, as represented in the standard karyotype, should be used in their place. The assignment of linkage groups to the chromosomes was achieved through the use of pairs of reciprocal translocations or pairs of reciprocals with Robertsonian translocations, each pair having a previously determined common linkage group. The subsequent identification by Q-banding of the common chromosome involved in the pair enabled the linkage group to be placed on that chromosome (see review by D. A. Miller & Miller, 1972). In addition, since the Q-banding also made it possible to identify the centromeric ends in relation to the breakpoint positions of the translocated chromosomes, a comparison of the genetical breakpoints permitted the linkage groups to be correctly orientated with respect to the centromere (D. A. Miller, Kouri *et al.*, 1971), a method which proved far less tedious than the existing methods of centromere location described by Searle, Ford & Beechey (1971). The further assignment of linkage groups has been reviewed by O. J. Miller & Miller (1975), and more recently the whole listing, with some amendments to past assignments, was published in *Mouse News Letter* (1979). The only unmapped chromosome which now remains is the Y, to which the possible genes responsible for male determination and H–Y antigen production have been assigned. Any proposed mapping of genes is confounded, in that recombination analysis is prohibited by the absence of chiasmata between X and Y. Furthermore, the prospect of obtaining some form of map from Y-autosomal structural rearrangements is eliminated by the fact that the few observed so far have all been male sterile (see O. J. Miller & Miller, 1975). The report of length differences in the Y chromosomes from different strains (Levan, Hsu *et al.*, 1962) has since been confirmed by a number of other workers, for example, Ford *et al.* (1975). An extreme form of variation, a submetacentric thought to have resulted by pericentric inversion, since there was no change in length compared to the normal Y of the strain, has been described by Winking (1978). Carriers were found to be fully fertile.

Q- and G-banding have also proved invaluable in the identification of mouse chromosomes involved in rearrangements. Unfortunately, the breakpoints of the many reciprocal translocations used in linkage group assignments (D. A. Miller & Miller, 1972) were determined by Q-bands which yield relatively imprecise linear information.

Furthermore, they were determined before a system of band nomenclature was adopted. Now that these translocations (and many more) have been re-studied using G-bands, they have yielded more accurate breakpoint location in terms of nomenclature. It is hoped that each chromosome will eventually be analysed in the manner described for chromosome 2 (Searle, Beechey *et al.*, 1979) where, in a combined genetical and chromosomal analysis of the linearly well-distributed breakpoints of ten reciprocal translocations, a number of gene loci have been positioned with some degree of certainty in relation to the G-bands. Chromosome 2 is, however, exceptional both for its involvement in so many reciprocal translocations and also for the useful spatial distribution of their breakpoints. For many of the other chromosomes to be analysed in this way, a considerable number of additional translocations would have to be induced.

All the reciprocal translocations used for linkage group assignments, and also many others, were induced by either irradiation or chemicals. Few have been reported to occur spontaneously in laboratory mice and probably none in feral mice. The same applies for paracentric inversions which can be induced, detected and then mapped by their inverted Q- or G-banded segments (Davisson & Roderick, 1973; Evans & Phillips, 1975). Since primary trisomy and monosomy act as embryonic lethals in the mouse (see Ford, 1975), there has been no call for their identification by banding in adults. On the other hand, a growing number of viable tertiary trisomics are being described and the more precise G-banding technique has proved useful in the interpretation of their origin and their use as a duplication-deficiency method for assigning genes to regions of mouse chromosomes (Eicher & Washburn, 1978).

Primary trisomy, amongst many other conditions involving whole or parts of chromosomes, is commonly found in malignancy. G-banding has been of value in determining the specific trisomies associated with the progression of spontaneous lymphoma (Dofuku *et al.*, 1975) and in leukaemias induced by radiation leukaemia virus (Wiener *et al.*, 1978). From the point of view of the detection of small rearrangements by Q- or G-banding, D. A. Miller, Dev *et al.* (1974) correlated a known genetical deletion with the deletion of a band on chromosome 7. Bennett (1975) reported that the brachyury allele T^{hp}, which involves a genetically determined deletion, was also cytologically detectable although Lyon *et al.* (1979) have failed to confirm the observation.

The vast number of laboratory and the few feral mice so far observed have been largely devoid of the spontaneous occurrence of abnormalities. However, many spontaneous Robertsonian translocations have been found. The origins, the relationships and banded chromosome constitution of these metacentric chromosomes have been summarized by Capanna *et al.* (1976) and recently catalogued in *Mouse News Letter* (1979).

The many Q- and G-banded normal karyotypes portrayed in the literature, although representing a variety of both closely and widely related strains together with hybrids of the laboratory mouse, show very similar banding patterns. The few disparities that arise can usually be resolved by reference to the greater or lesser resolving power of any one technique. Perhaps this apparent band stability is not surprising since in both the closely related and the more widely separated subspecies of *Mus,* the G-band patterns are remarkably uniform (Hsu, Markvong & Marshall, 1978; also see O. J. Miller & Miller, 1978).

From these and other observations it appears that detectable chromosomal variations in the mouse are confined to the centromeric regions. Forejt (1973) described C-band differences both between inbred laboratory strains and also within and between feral mice from central Bohemia. The chromosomes of the latter were not, however, identified as homologous pairs but tentatively paired on the basis of length and C-band size. Dev, Miller & Miller (1973) showed that in laboratory mice even closely related sublines could show up to five different pairs of chromosomes with C-band differences and in an extended survey of inbred strains, a further three pairs were added to this total (D. A. Miller, Tantravahi *et al.,* 1976). As Dev, Miller & Miller (1973) have shown, for example in a 129 × AKR F_1 hybrid mitosis, the striking C-band differences shown by some homologous pairs provide a powerful tool for the detection of heterozygosity and in some cases for establishing parental contributions. Recently, Burtenshaw, Brown & Evans (1979) found a chromosome 18 with a double C-band segregating in a laboratory stock. This probably represents the most extreme difference so far observed (Fig. 2).

To a lesser extent, since the detection of variation is not as precise as in the C-bands, the strain differences in secondary constrictions (Dev, Grewal *et al.,* 1971) can also be used to detect probable parental contributions. For example, this method was used by Ford *et al.* (1975) to confirm that an XXY mouse was the probable product of a functional XY oocyte.

The presence of secondary constrictions on some mouse chromosomes suggested that they were nucleolar organizing. This was confirmed by the use of *in situ* hybridization with radioactively labelled rRNA which showed that rDNA was present in some mice at the centromeric ends of pairs 15, 18 and 19 (Henderson *et al.*, 1974) and in others at those of 12, 16 and 18 (Elsevier & Ruddle, 1975). The introduction of silver staining to identify sites of active 18S and 28S ribosomal genes provided an alternative method for locating nucleolar organizing chromosomes. Using this method, O. J. Miller, Miller, Dev *et al.* (1976) and Dev, Tantravahi *et al.* (1977) confirmed that these were the five pairs involved and that strain differences did exist.

Yet another form of variation, or variations, in the centromeric regions of the mouse has recently been described by Jonasson, Alves & Strom (1979) during the development of a staining method for the detection of sister-chromatid exchanges in BrdU labelled chromosomes (Alves & Jonasson, 1978). In addition, the method resulted in an extremely clear presentation of the phenomenon of lateral asymmetry described by Lin, Latt & Davidson (1974). The high resolution of the centromeric region enabled Jonasson *et al.* (1979) to record differences in the amount of asymmetric material between the same pairs of chromosomes from four different inbred strains and also to observe presumptive "centromeric inversions" in some regions which were characteristic for each strain.

The failure to detect normal, as against induced, variation in regions other than centromeric has proved a disappointment. The high resolution G-banding methods now in use (Fig. 3) have proved rewarding in other species but not in the mouse. Either the variation is not present or still more refined banding methods are required for its detection. An alternative is to turn to meiotic observations but the conventional studies of the more condensed stages of this division have proved equally unrewarding from the point of view of detecting small differences (Evans, 1979). It is possible that pachytene analysis through observations of the synaptonemal complex (SC) may allow the identification of small rearrangements. A SC karyotype of the mouse has been prepared (Moses, 1977) but unfortunately, as a result of the age-old problem of the lack of "landmarks" (cf. Levan, Hsu *et al.*, 1962), the pachytene bivalents could only be arranged in decreasing order of length. Moses (1979) has reviewed the merits of the SC, as observed in whole spermatocytes under the electron microscope, as an indicator of chromosomal changes. Such work is, however, tedious and since it is now possible to view the SC under the light microscope (Fletcher, 1979; Forejt & Goetz, 1979; Dresser & Moses, 1979; Evans, Burtenshaw & Brown, in preparation) after spreading and silver staining (Fig. 4), the simplicity and rapidity of the method may encourage workers to look for signs of variation in these pachytene bivalents.

ACKNOWLEDGEMENTS

I am grateful to Mr M. D. Burtenshaw and Dr J. D. West for the mouse material used in Fig. 2.

REFERENCES

Alves, P. & Jonasson, J. (1978). New staining method for the detection of sister-chromatid exchanges in BrdU-labelled chromosomes. *J. Cell Sci.* **32**: 185–195.

Bennett, D. (1965). The karyotype of the mouse, with identification of a translocation. *Proc. natl Acad. Sci. U.S.A.* **53**: 730–737.

Bennett, D. (1975). The T-locus of the mouse. *Cell* **6**: 441–454.

Buckland, R. A., Evans, H. J. & Sumner, A. T. (1971). Identifying mouse chromosomes with the ASG technique. *Expl Cell Res.* **69**: 231–236.

Burtenshaw, M. D., Brown, B. B. & Evans, E. P. (1979). Personal communication. *Mouse News Letter* **61**: 56.

Capanna, E., Gropp, A., Winking, H., Noak, G. & Civitelli, M. V. (1976). Robertsonian metacentrics in the mouse. *Chromosoma* **58**: 341–353.

Caspersson, T., Farber, S., Foley, G. E., Kudynowski, J., Modest, E. J., Simonsson, E., Wagh, U. & Zech, L. (1968). Chemical differentiation along metaphase chromosomes, *Expl Cell Res.* **49**: 219–222.

Church, K. (1965). Replication of chromatin in mouse mammary epithelial cells grown *in vitro*. *Genetics* **52**: 843–849.

Committee on Standardized Genetic Nomenclature for Mice. (1972). Standard karyotype of the mouse, *Mus musculus*. *J. Hered.* **63**: 69–72.

Committee on Standardized Genetic Nomenclature for Mice. (1974). Personal communication. *Mouse News Letter* **50**: 2.

Committee on Standardized Genetic Nomenclature for Mice. (1979). Personal communication. *Mouse News Letter* **61**: 11.

Cox, E. (1926). The chromosomes of the house mouse. *J. Morph. Physiol.* **43**: 45–50.

Crew, F. A. E. & Koller, P. C. (1932). The sex incidence of chiasma frequency and genetical crossing over in the mouse. *J. Genet.* **26**: 359–383.

Davisson, M. T. & Roderick, T. H. (1973). Chromosomal banding patterns of two paracentric inversions in mice. *Cytogenet. Cell Genet.* **12**: 398–403.

Dev, V. G., Grewal, M. S., Miller, D. A., Kouri, R. E., Hutton, J. J. & Miller, O. J. (1971). The quinacrine fluorescence karyotype of *Mus musculus* and demonstration of strain differences in secondary constrictions. *Cytogenetics* **10**: 436–451.

Dev, V. G., Miller, D. A., Allderdice, P. W. & Miller, O. J. (1972). Method for locating the centromeres of mouse meiotic chromosomes and its application to T163H and T7OH translocations. *Expl Cell Res.* **73**: 259–262.

Dev, V. G., Miller, D. A. & Miller, O. J. (1973). Chromosome markers in *Mus musculus:* Strain differences in C-banding. *Genetics* **75**: 663–670.

Dev, V. G., Tantravahi, R., Miller, D. A. & Miller, O. J. (1977). Nucleolus organizers in *Mus musculus* subspecies and in the RAG mouse cell line. *Genetics* **86**: 389–398.

Dofuku, R., Biedler, J. L., Spengler, B. A. & Old, L. J. (1975). Trisomy of chromosome 15 in spontaneous leukemia of AKR mice. *Proc. natl Acad. Sci. U.S.A.* 72: 1515–1517.

Dresser, M. E. & Moses, M. J. (1979). Silver staining of synaptonemal complexes in surface spreads for light and electron microscopy. *Expl Cell Res.* 121: 416–419.

Dutrillaux, B. & Lejeune, J. (1971). Sur une nouvelle technique d' analyse du caryotype humain. *C. r. hebd. Séanc. Acad. Sci., Paris* 272: 2638–2640.

Eicher, E. M. & Washburn, L. L. (1978). Assignment of genes to regions of mouse chromosomes. *Proc. natl Acad. Sci. U.S.A.* 75: 946–950.

Elsevier, S. M. & Ruddle, F. H. (1975). Location of genes coding for 18s and 28s ribosomal RNA within the genome of *Mus musculus. Chromosoma* 52: 219–228.

Evans, E. P. (1979). Cytological methods for the study of meiotic properties in mice. *Genetics* 92: s97–s103.

Evans, E. P., Burtenshaw, M. D. & Brown, B. B. (In preparation). *Synaptonemal complexes observed by light microscopy in gravitationally sedimented mammalian spermatocytes and foetal oocytes.*

Evans, E. P., Lyon, M. F. & Daglish, M. (1967). A mouse translocation giving a metacentric chromosome. *Cytogenetics* 6: 105–119.

Evans, E. P. & Phillips, R. J. S. (1975). Inversion heterozygosity and the origin of XO daughters of BPA/+ female mice. *Nature, Lond.* 256: 40–41.

Fletcher, J. M. (1979). Light microscope analysis of meiotic prophase chromosomes by silver staining. *Chromosoma* 72: 241–248.

Ford, C. E. (1966). The murine Y chromosome as a marker. *Transplantation* 4: 333–335.

Ford, C. E. (1975). The time in development at which gross genome unbalance is expressed. In *The early development of mammals:* 285–304. Balls, M. & Wild, A. E. (Eds). Cambridge: University Press.

Ford, C. E., Evans, E. P., Burtenshaw, M. D., Clegg, H. C. Tuffrey, M. & Barnes, R. D. (1975). A functional "sex-reversed" oocyte in the mouse. *Proc. R. Soc. Lond.* (B.) 190: 187–197.

Forejt, J. (1973). Centomeric heterochromatin polymorphism in the house mouse. Evidence from inbred strains and natural populations. *Chromosoma* 43: 187–201.

Forejt, J. & Goetz, P. (1979). Synaptonemal complexes of mouse and human pachytene chromosomes visualised by silver staining in air-dried preparations. *Chromosoma* 73: 255–261.

Francke, U. & Nesbitt, M. (1971). Identification of the mouse chromosomes by quinacrine mustard staining. *Cytogenetics* 10: 111–116.

Galton, M. & Holt, S. F. (1965). Asynchronous replication of the mouse sex chromosomes. *Expl Cell Res.* 37: 111–116.

German, J. L. (1964). Identification and characterization of human chromosomes by DNA replication sequence. In *Cytogenetics of cells in culture:* 191–207. Harris, R. J. C. (Ed.). New York & London: Academic Press.

Goodpasture, C. & Bloom, S. E. (1975). Visualization of nucleolar organizer regions in mammalian chromosomes using silver staining. *Chromosoma* 53: 37–50.

Henderson, A. S., Eicher, E. M., Yu, M. T. & Atwood, K. C. (1974). The chromosomal location of ribosomal DNA in the mouse. *Chromosoma* **49**: 155–160.

Hsu, T. C., Billen, D. & Levan, A. (1961). Mammalian chromosomes *in vitro*. XV. Patterns of transformation. *J. natn. Cancer Inst.* 27: 515–541.

Hsu, T. C., Markvong, A. & Marshall, J. T. (1978). G-band patterns of six species of mice belonging to subgenus *Mus*. *Cytogenet. Cell Genet.* **20**: 304–307.

Jonasson, J., Alves, P. & Strom, A. (1979). Polymorphisms in the centromeric region of the chromosomes of the laboratory mouse. *Hereditas* **90**: 111–117.

Koller, P. C. (1944). Segmental interchange in mice. *Genetics* **29**: 247–263.

Levan, A., Fredga, K. & Sandberg, A. A. (1964). Nomenclature for centromeric position on chromosomes. *Hereditas* **52**: 201–220.

Levan, A., Hsu, T. C. & Stich, H. F. (1962). The idiogram of the mouse. *Hereditas* **48**: 677–687.

Lin, M. S., Latt, S. A. & Davidson, R. L. (1974). Microfluorometric detection of asymmetry in the centromeric region of mouse chromosomes. *Expl Cell Res.* **86**: 392–395.

Long, J. A. (1908). Chromosomes of the mouse. *Science, N.Y.* 27: 443.

Lyon, M. F., Evans, E. P., Jarvis, S. E. & Sayers, I. (1979). t-haplotypes of the mouse may involve a change in intercalary DNA. *Nature, Lond.* 279: 38–42.

McKenzie, W. H. & Lubs, H. A. (1973). An analysis of the technical variables in the production of C bands. *Chromosoma* **41**: 175–182.

Miller, D. A., Dev, V. G., Tantravahi, R., Miller, O. J., Schiffman, M. B., Yates, R. A. & Gluecksohn-Waelsch, S. (1974). Cytological detection of the c^{25H} deletion involving the albino (c) locus on chromosome 7 in the mouse. *Genetics* **78**: 905–910.

Miller, D. A., Kouri, R. E., Dev, V. G., Grewal, M. S., Hutton, J. J. & Miller, O. J. (1971). Assignment of four linkage groups to chromosomes in *Mus musculus* and a cytogenetic method for locating their centromeric ends. *Proc. natl Acad. Sci. U. S. A.* **68**: 2699–2702.

Miller, D. A. & Miller, O. J. (1972). Chromosome mapping in the mouse. *Science, N.Y.* 178: 949–955.

Miller, D. A., Tantravahi, R., Dev, V. G. & Miller, O. J. (1976). Q- and C-band chromosome markers in inbred strains of *Mus musculus*. *Genetics* **84**: 67–75.

Miller, O. J. & Miller, D. A. (1975). Cytogenetics of the mouse. *A. Rev. Genet.* 9: 285–303.

Miller, O. J. & Miller, D. A. (1978). Cytogenetics. In *Origin of inbred mice*: 591–611. Morse, H. (Ed.). New York & London: Academic Press.

Miller, O. J., Miller, D. A., Kouri, R. E., Allerdice, P. W., Dev, V. G., Grewal, M. S. & Hutton, J. J. (1971). Identification of the mouse karyotype by quinacrine fluorescence, and tentative assignment of seven linkage groups. *Proc. natl Acad. Sci. U.S.A.* **68**: 1520–1533.

Miller, O. J., Miller, D. A., Dev, V. G., Tantrahavi, R. & Croce, C. M. (1976). Expression of human and suppression of mouse nucleolus organizer activity in mouse-human somatic cell hybrids. *Proc. natl Acad. Sci. U.S.A.* 73: 4531–4535.

Moses, M. J. (1977). Microspreading and the synaptonemal complex in cytogenetic studies. In *Chromosomes today* **6**: 71–82. de la Chapell, A. & Sorsa, M. (Eds). Amsterdam: Elsevier/North Holland Biomedical Press.

Moses, M. J. (1979). The synaptonemal complex as an indicator of chromosome damage. *Genetics* **92**: s73–s82.

Natarajan, A. T. & Gropp, A. (1972). A fluorescence study of heterochromatin and nucleolar organization in the laboratory and tobacco mouse. *Expl Cell Res.* **74**: 245–250.

Nesbitt, M. N. & Francke, U. (1973). A system of nomenclature for band patterns of mouse chromosomes. *Chromosoma* **41**: 145–158.

Ohno, S., Kaplan, W. D. & Kinosita, R. (1957). Heterochromatic regions and nucleolus organizers in chromosomes of the mouse, *Mus musculus. Expl Cell Res.* **13**: 358–364.

Painter, T. S. (1926). A comparative study of chromosomes of the mammal. *Am. Nat.* **70**: 385–409.

Painter, T. S. (1927). The chromosome constitution of Gates's non-disjunction (V-o) mice. *Genetics* **12**: 379–392.

Painter, T. S. (1928). A comparison of the chromosomes of the rat and the mouse with reference to the question of a chromosome homology in mammals. *Genetics* **13**: 180–189.

Pardue, M. L. & Gall, J. G. (1970). Chromosomal localization of mouse satellite DNA. *Science N. Y.* **168**: 1356–1358.

Schnedl, W. (1971). The karyotype of the mouse. *Chromosoma* **35**: 111–116.

Searle, A. G., Beechey, C. V., Eicher, E. M., Nesbitt, M. N. & Washburn, L. L. (1979). Colinearity in the mouse genome: a study of chromosome 2. *Cytogenet. Cell Genet.* **23**: 255–263.

Searle, A. G., Ford, C. E. & Beechey, C. V. (1971). Meiotic disjunction in mouse translocations and the determination of centromere position. *Genet. Res.* **18**: 215–235.

Stich, H. F. & Hsu, T. C. (1960). Cytological identification of male and female somatic cells in the mouse. *Expl Cell Res.* **20**: 248–249.

Sumner, A. T., Evans, H. J. & Buckland, R. A. (1971). New techniques for distinguishing between human chromosomes. *Nature New Biol* **232**: 31–32.

Wiener, F., Ohno, S., Spira, J., Haran-Ghera, N. & Klein, G. (1978). Chromosome changes (trisomies no. 15 and 17) associated with tumor progression in leukemias induced by leukemia virus. *J. natn. Cancer Inst.* **61**: 227–237.

Winking, H. (1978). Personal communication. *Mouse News Letter* **58**: 53.

Zech, L., Evans, E. P., Ford, C. E. & Gropp, A. (1972). Banding patterns in mitotic chromosomes of the tobacco mouse. *Expl Cell Res.* **70**: 263–268.

Symp. zool. Soc. Lond. (1981) No. 47, 141–181

Robertsonian Translocations: Cytology, Meiosis, Segregation Patterns and Biological Consequences of Heterozygosity *

A. GROPP and H. WINKING

Institut für Pathologie der Medizinischen Hochschule Lübeck, West Germany

SYNOPSIS

Robertsonian (Rb) changes due to centric translocation of acrocentric chromosomes with the formation of metacentrics occur frequently in populations of *Mus musculus domesticus (brevirostris)* from Switzerland, Italy, and neighbouring areas. Chromosome identification by band staining has shown that free combinations between pairs of almost all the chromosome except no. 19 and the sex chromosomes, account for the great variety of Rb translocation metacentrics. The number of Rb chromosomes isolated from feral mice and introduced into laboratory strains is steadily increasing. The cytological properties of C-heterochromatin, the structure of the mitotic chromosomes, and particularly the pairing patterns in first meiotic prophase visible in electron microscope spreading preparations indicate a mechanism involving translocation with breakpoints within the pericentric heterochromatic segments of the acrocentrics.

Many of the consequences of Rb metacentric heterozygosity follow from first meiotic anaphase non-disjunction of chromosomes which had been involved in multivalent configuration in prophase. Rb metacentrics of feral origin usually produce higher rates of malsegregation than those arising in laboratory strains. However, great variations in rates occur, depending on the arm composition of the particular Rb metacentric and on the sex of the carrier (being considerably higher in females than in males). In the balanced products of first meiotic division of many (but not all) female Rb heterozygotes, an excess of the homozygous (acrocentric) type is found in both the oocyte and the balanced progeny. Heterozygosity for several independent Rb metacentrics follows the same rules, although cumulative rates of non-disjunction are lower than expected by mere statistical addition. Multiple heterozygosity with single or alternating homology of the chromosome arms showing formation of chains or rings with two to sixteen Rb metacentrics in meiosis I leads, in most cases, to high rates of non-disjunction, eventually causing postzygotic losses and a segregational type of subfertility. In addition, male-limited, hybrid-type sterility is often observed. It seems possible that recent electron microscope findings of non-homologous pairing and non-specific heterochromatin association in pachytene preparations may lead to an understanding of the factors determining hybrid sterility.

* Supported by a grant (Gr 71/45) of Deutsche Forschungsgemeinschaft (to A.G.) and in partial fulfilment of contract NICHD-NO1-HD-8-2858

Fertility barriers arising from such mechanisms may have played a role in recent evolution of the mouse, perhaps continuing at the present time. It can be argued that the occurrence of Rb-races in the wild is an expression of incipient or continuing speciation.

Finally, Rb metacentric heterozygosity can be used for the experimental induction of specific monosomy and trisomy, and thus represents a potent tool for studies in developmental genetics and pathology.

INTRODUCTION AND GENERAL FEATURES OF ROBERTSONIAN (Rb) WHOLE ARM CHANGES

The Robertsonian whole arm fusion or translocation type of chromosome rearrangement (Robertson & Rees, 1916; John & Freeman, 1975) has great interest due to its frequent occurrence among the chromosome changes which play a role in mammalian karyotype diversification. It is clear that in any discussion of the cytogenetics of the house mouse, *M. musculus,* an appropriate consideration of Robertsonian (Rb) variation is necessary, although acrocentrics represent the normal or, at least, the predominant chromosome type of the mouse (see E. P. Evans, this volume pp. 127–139). In fact, Rb arrangements occur at relatively high frequency in certain geographical areas, and Robertsonian heterozygosity may involve meiotic disturbances and impairment of fertility. It is possible that in natural populations such effects of heterozygosity have some significance as a vehicle of evolutionary processes in the species *M. musculus* (see Thaler *et al.,* this volume pp. 27–41).

The mechanism of Rb variation (Fig. 1) with the formation of a bi-armed (metacentric) chromosome from two acrocentrics

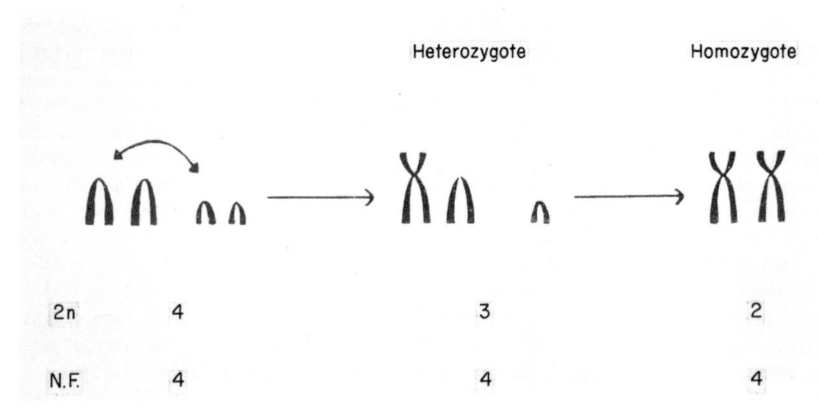

FIG. 1. Principle of Robertsonian (Rb) karyotype variation.

accounts for a change from a high to a low diploid number of chromosomes ($2n$). Evidence for the reverse process of separation of a Rb metacentric into two acrocentrics or fission is controversial and, at least, sparse (Hsu & Mead, 1969). Matthey (1945, 1966), in the course of his studies on several Rb systems in rodents, mainly in *M. (Leggada) minutoides* (recently more correctly assigned to the genus *Nannomys;* see Marshall & Sage, this volume, pp. 15–25), coined the term *Nombre Fondamental* (NF), defined as the total count of chromosome arms. The comparison of $2n$ and NF, together with the proof of homology between the metacentric arms and the original acrocentric chromosomes in meiotic preparations, represents a suitable description of a Rb system of whole arm rearrangement. This principle is demonstrated in Fig. 2 in which chromosomal sets of a laboratory mouse (*M. musculus,* strain NMRI, with only acrocentrics), of the "tobacco mouse" (*M. m.* (*poschiavinus*)) with seven pairs of Rb metacentric chromosomes, and of the F_1 hybrid of both are put together.

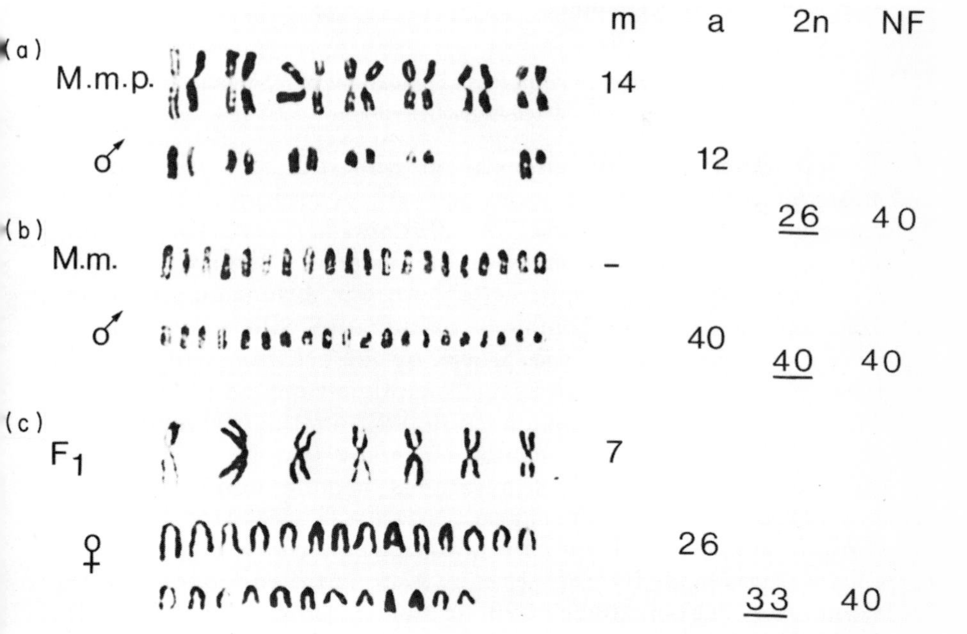

FIG. 2. Chromosome sets (conventional staining): (a) tobacco mouse (*M. m.* (*poschiavinus*)) male with 14 bi-armed (metacentric) chromosomes, (b) *Mus musculus* male (strain NMRI) with acrocentrics only, and (c) F_1 hybrid female from *M. musculus* and *M. m.* (*poschiavinus*) parents.

OCCURRENCE OF Rb IN LABORATORY BRED AND WILD LIVING HOUSE MICE

Rb Metacentrics in Laboratory Mouse Strains

Only four reports (Léonard & Deknudt, 1967; Evans, Lyon & Daglish, 1967; B. J. White & Tijo, 1968; Baranov & Dyban, 1971) describing bi-armed Rb chromosomes in laboratory mouse strains were published prior to 1976, while thereafter and up to the present time eight more examples of Rb metacentrics have been detected in laboratory bred animals. Curiously enough, these latter Rb chromosomes all appeared in lines derived from feral mice or carrying other metacentrics of feral origin (see Tables I & II, in the Appendix to this paper). The mouse lines mentioned in these Tables are, according to our knowledge, continuously bred in several laboratories (see Appendix, Table IIIa). Unfortunately, the unique observation of Arroyo Nombela & Rodriguez Murcia (1977) of a single female NMRI mouse with two Rb metacentrics, one of them involving an X chromosome (Rb2.3 and RbX.3), was only made after the death of the animal.

Rb Metacentrics found in Feral Populations of *M. musculus:* Geographical Origin

A first observation of an extreme Robertsonian variant of *M. musculus* with a series of seven pairs of metacentrics (Fig. 2) was made in the "tobacco mouse" or "*Mus poschiavinus*" (Fatio, 1869), a feral mouse population from the Valle di Poschiavo in South Eastern Switzerland (Gropp, Tettenborn & v. Lehmann, 1970). The more or less complete homology of the arms with acrocentrics of laboratory mouse origin was proven by the occurrence of regularly formed trivalents in diakinesis and meiotic metaphase I (Fig. 3d) of F_1 hybrids. However, it had been noted from the beginning of the chromosomal studies of the tobacco mouse that F_1 hybrids show reduced fertility due to segregational meiotic disorders and to postzygotic losses of unbalanced progeny. The assignment of the chromosome arms of the seven tobacco mouse Rb metacentrics (Fig. 3a) was made by Zech *et al.* (1972), based on proposals of the Committee on Standardized Genetic Nomenclature for Mice (1972).

Subsequently, more wild mouse specimens with serial sets of multiple, five to nine, pairs of Rb metacentrics were found in populations of unknown size in certain other areas of the Rhaetian Alps (Gropp, Winking *et al.,* 1972), in the Apennines (Capanna,

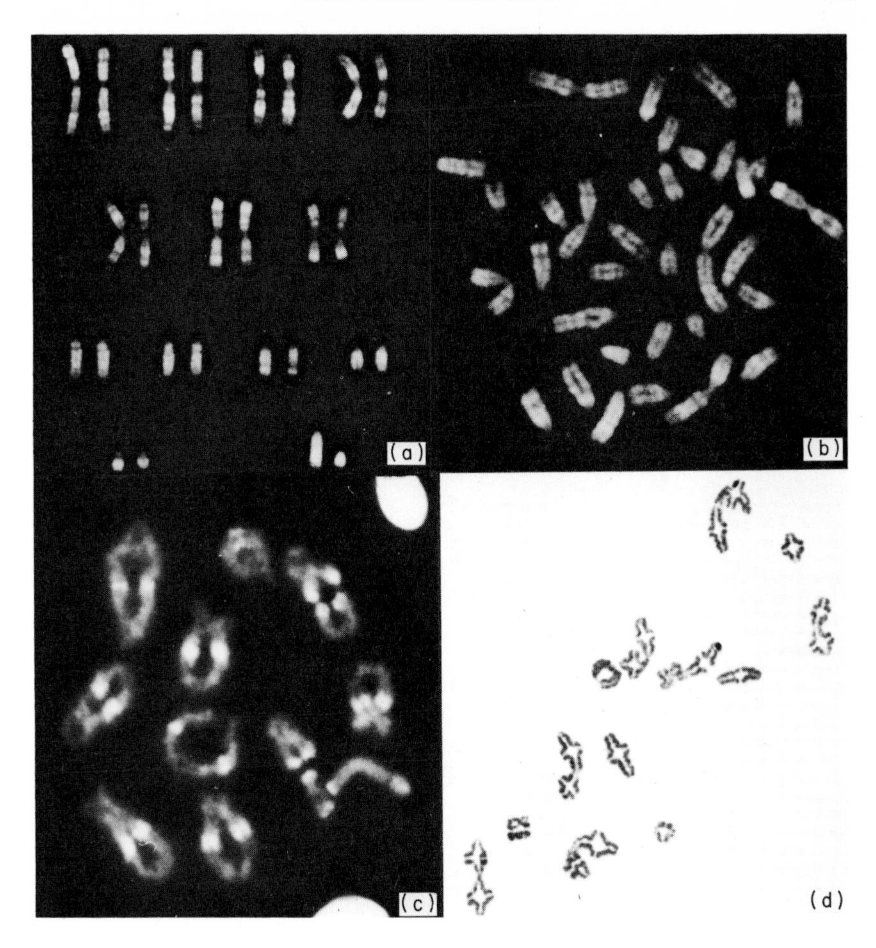

FIG. 3 (a) Karyotype of a male tobacco mouse (*M. m.* (*poschiavinus*)); seven pairs of Rb metacentrics Rb1–7 Bnr; Q-banding. (b) Mitotic metaphase of a tobacco mouse × laboratory mouse F_1-hybrid; heterozygosity for seven Rb metacentrics; Q-banding. (c) Meiotic metaphase I of a male tobacco mouse; seven ring bivalents; 33258 H-fluorescence with brightly fluorescing centromeres. (d) Meiotic metaphase I of female F_1-hybrid with seven trivalents; orcein staining.

Gropp *et al.*, 1976), in Sicily (v. Lehmann & Radbruch, 1977), in the Eolian Islands and in Dalmatia (H. Winking, A. Gropp, C. Redi, B. Dulic, E. Capanna, J. Uckert & G. Noack, in preparation).

Continued interest directed to the Rhaetian Alps and Lombardy, and the close collaboration of the authors with Professor E. Capanna (Rome) and Dr C. Redi (Pavia) allowed, during 1978 to 1980, the detection of more feral mouse populations with single or multiple Rb metacentrics of variable arm compositions in the Upper and Lower Valtellina, the Orobian Alps near Bergamo, in the area of Milano, and near Cremona (Winking *et al.*, in preparation; Table II). The same investigation led to the finding of all-acrocentric mice near Brescia and in Tortona south of the Po river, but heterozygous hybrid karyotypes were found in nearby localities of the zones bordering the aforementioned areas. Figure 4 shows the main sites where Rb metacentrics have been

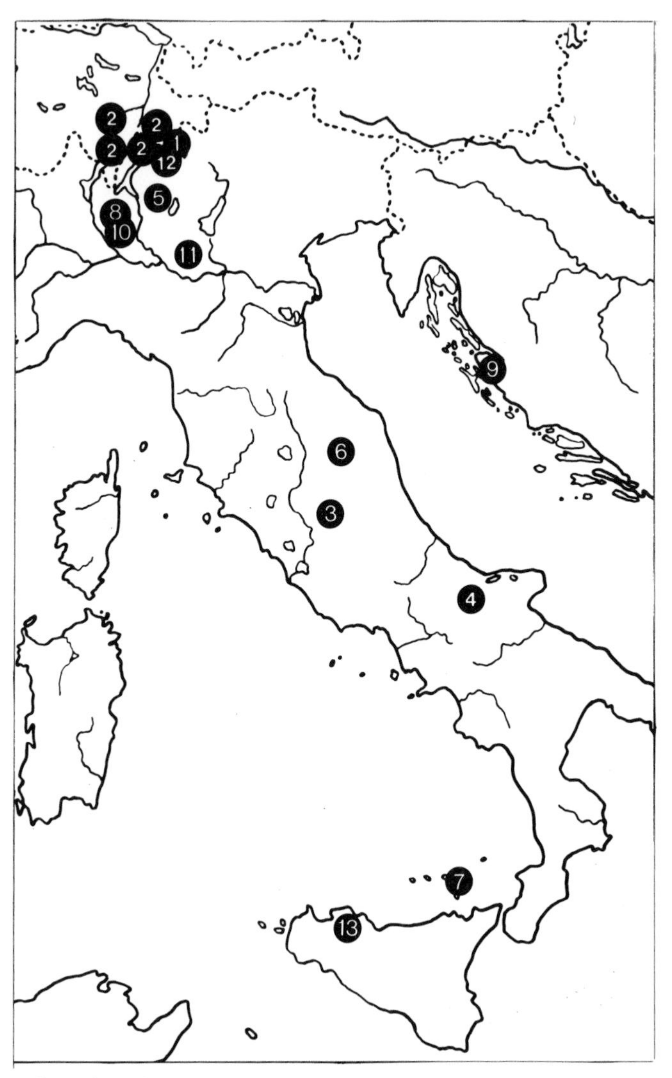

Fig. 4. Map of southern Europe with areas where mice with Rb metacentrics were found; the numbers refer to the numbering in Table I.I, Feral origin.

observed in the course of the survey conducted by our own group.

Outside the Rhaeto–Lombardian region, the Italian peninsula and islands, and Dalmatia, the occurrence of Rb metacentrics was recently reported from southern Germany (S. Adolph & J. Klein, personal communication) and, strangely enough, from Marion Island close to the Antarctic Convergence (Robinson, 1978). In this latter study, two females were found with one Rb metacentric ($2n = 39$). Another report of the occurrence of a spontaneous Rb metacentric mouse from India (Chakrabarti & Chakrabarti, 1977) needs confirmation.

Identification of Rb Metacentrics and Ascertainment of Homology of Chromosome Arms

The evaluation of shape, size, and especially Giemsa banding patterns of the arms of the mouse Rb metacentrics suggest complete or almost complete homology with the acrocentric chromosomes of the normal (laboratory) mouse (Nesbitt & Francke, 1973). At least, all evidence so far from comparative studies of the G-banding patterns of the arms of Rb metacentrics supports the assumption that there is identity of the bands with the respective acrocentric homologues. Therefore, it is reasonable that the arms of Rb metacentrics should be designated according to the numbering system of the acrocentrics in the basic karyotype of the mouse (Nesbitt & Francke, 1973), whereas the designation of the metacentric itself follows the rules of the Committee on Standardized Genetic Nomenclature for Mice (1972, 1974; see Gropp, Olert & Maurizio, 1971). The karyotype in Fig. 5 is an example of G-banding homology between the Rb arms and their acrocentric homologues in a hybrid male offspring of a male from Lipari (Eolian islands – see Table I.I) and of a NMRI female.

An evaluation of the arm composition of the 101 Rb metacentrics of feral origin identified so far (Table I.I) shows that each one of the acrocentric autosomal components of the mouse genome, except chromosome no. 19, occurs in many combinations – usually more than ten Rbs (Table IIIa). This includes the repeated occurrence of an Rb of same composition in separate populations or in different geographical areas between which a spread is easily possible (e.g., within the Rhaeto–Lombardian system) or unlikely (e.g., Lombardy and the Central Apennines). Chromosomes nos. 7 and 18 are only involved in seven and four different Rb metacentrics, respectively. In general, it appears that, with the exception of chromosome no. 19,

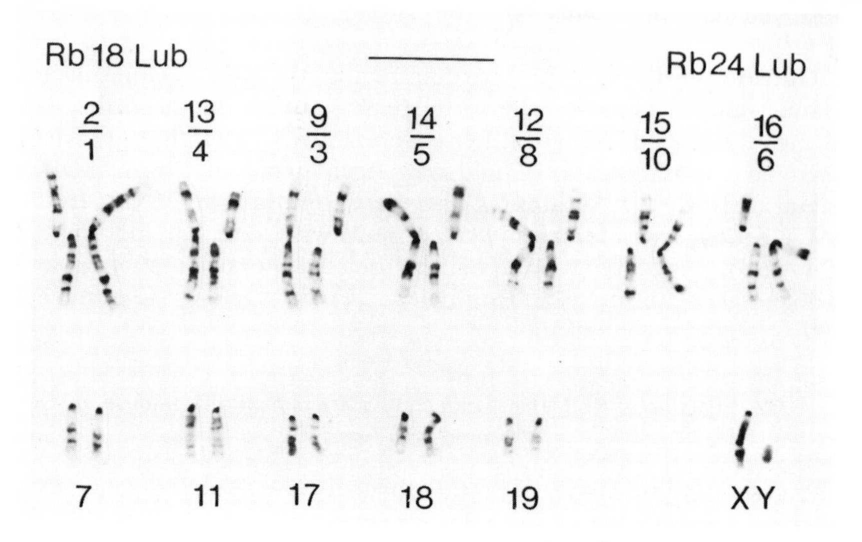

FIG. 5. Karyotype of a laboratory bred hybrid male offspring of a male from Lipari (Eolian Islands) and a NMRI female. Heterozygosity for seven Rb metacentrics (Rb18–24Lub); G-banding.

all autosomal acrocentric components homologous with nos. 1–18 contribute randomly to the formation of Rbs in feral populations. There is no convincing evidence that acrocentrics of similar size are preferentially associated in Rb translocations. On the other hand, the autosomal component no. 19 is represented only, albeit several times, in Rb metacentrics of laboratory origin (Table I.II). It is difficult to believe that its different involvement in Rbs of feral and laboratory origin is purely accidental. The X chromosome has never so far been found in Rb metacentrics of feral or laboratory origin, with the exception of the one and single example of the Rb(3.X) chromosome known from the report of Arroyo Nombela & Rodriguez Murcia (1977).

In a perfect Rb-system, the chromosome arm homology of an animal with an undetermined Rb metacentric can be determined in meiosis I of crosses with carriers of Rb of known and previously ascertained arm compositions. This method has been applied in all the investigations carried out in the Lübeck laboratory (Gropp, Winking *et al.*, 1972; Capanna, Gropp *et al.*, 1976; see Tables I and II). Similarly, the expectation that alternating arm homologies in hybrids derived from mice with different series of Rb metacentrics lead to the appearance of complex ring or chain multivalents (Fig. 6a and b) has been confimed (Capanna, Gropp *et al.*, 1976). However,

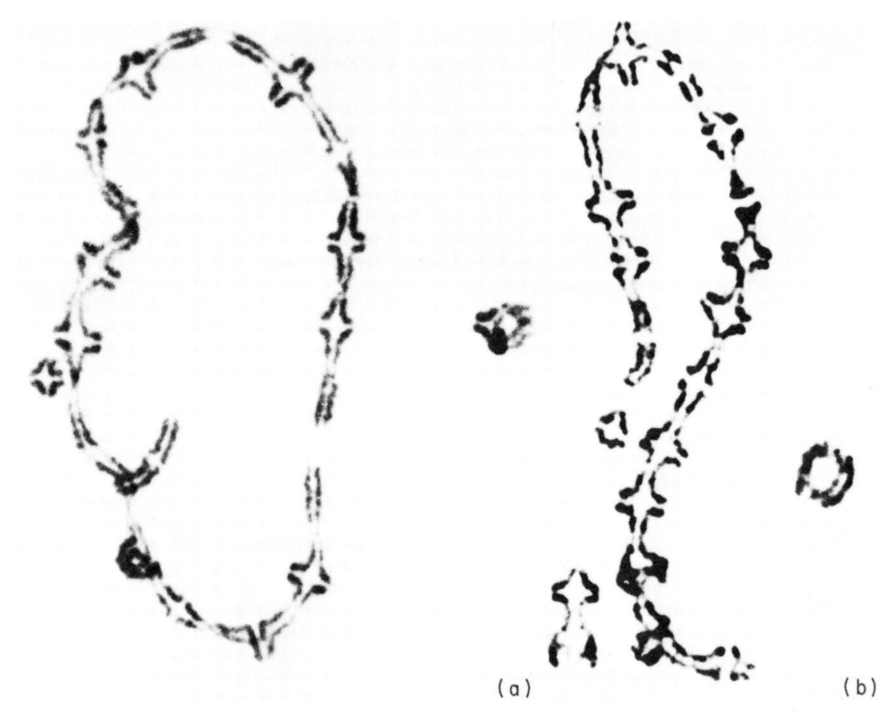

FIG. 6. Meiotic oocyte metaphase I figures from hybrids between polymetacentric mice. (a) Complete ring containing 16 Rb metacentrics: hybrid between populations 3 and 4 in Table I. (b) Chain containing 15 Rb metacentrics: hybrid between populations 1 and 3 in Table I.

it has to be recognized that the identity of G-banding patterns and the perfect formation of multivalents in the mouse, at least as shown on the level of the light microscope (see p. 154), does not exclude minor changes or rearrangements undetectable with the conventional methods of cytology (Cattanach & Moseley, 1973).

Isolation of Rb Metacentrics from Polymetacentric Carriers and Breeding of Lines with Single Rb Metacentrics in the Laboratory

Rb metacentric chromosomes have been isolated in several laboratories from feral mice carrying multiple metacentrics by back-crossing and selective breeding with laboratory strains, e.g. NMRI, (outbred strain) and C57BL/6. For the tobacco mouse metacentrics this work was largely done at Harwell (Cattanach & Moseley, 1973), as well as in Lübeck (Rb1 and Rb4Bnr) and by J. Klein (Rb7Bnr; see Klein, 1971).

In the meantime more Rb metacentrics have been isolated from other polymetacentric wild mouse stocks in the Lübeck laboratory (Rb8–10Bnr; Rina and Lub series). While Table I contains all so far identified Rb metacentric chromosomes, a list of available stocks with isolated single Rb metacentric chromosomes (hetero- or homozygous) held, according to our knowledge, at present in different laboratories is presented in Table IV.

Introgressive selective breeding into inbred mice has been done in the Lübeck laboratory over a minimum of five and usually more than 10 generations, for at present eight Rb metacentrics (Table V), using C57BL/6 background.

CYTOLOGICAL PROPERTIES OF MOUSE Rb METACENTRIC CHROMOSOMES

So far, only Robertsonian metacentric chromosomes occurring in live animals have been discussed in this report. However, Rb metacentrics are known to be common in established and permanent mouse cell cultures, e.g., in RAG which is originally derived from BALB/c cells (Klebe, Chen & Ruddle, 1974) and in strain L. The arms of RAG Rb metacentrics have been tentatively identified by Hashmi, Allerdice, Klein & Miller (1974), yet complete homology of these arms with normal mouse acrocentrics is difficult to ascertain, and often such homology is unlikely. On the other hand, isochromosomes with the shape of Rb chromosomes are not uncommon, e.g., in the RAG line (Nielsén, Marcus & Gropp, 1979).

In the following summary, reference is made to naturally occurring Rb metacentric chromosomes as well as to metacentrics found in established lines.

C-Banding; Centromeric Heterochromatin; Condensation Inhibition; Electron Microscopy of Mitotic Rb Metacentrics

In normal *M. musculus* chromosomes, the constitutive heterochromatin is located close to the centromere. It is known to contain a large amount of AT-rich satellite DNA (Pardue & Gall, 1970) which accounts for heavy C-banding and bright 33258-Hoechst fluorescence (Gropp, Hilwig & Seth, 1973). Laboratory strains and, even more, feral *M. musculus* subspecies show some size variation in the C-banded heterochromatic blobs (Miller *et al.*, 1976; see E. P. Evans, this volume pp. 127–139). A very distinct pattern of C-band variation has been described in *M. m. molossinus* chromosomes (Dev, Miller *et al.*, 1975). Rb metacentric chromosomes of "spontaneous" (laboratory) and feral origin show almost regularly duplex blocks of C-banded heterochromatin (Fig. 7). Sometimes they fuse into one large block, but usually the origin from the two blobs of both arms can be clearly recognized.

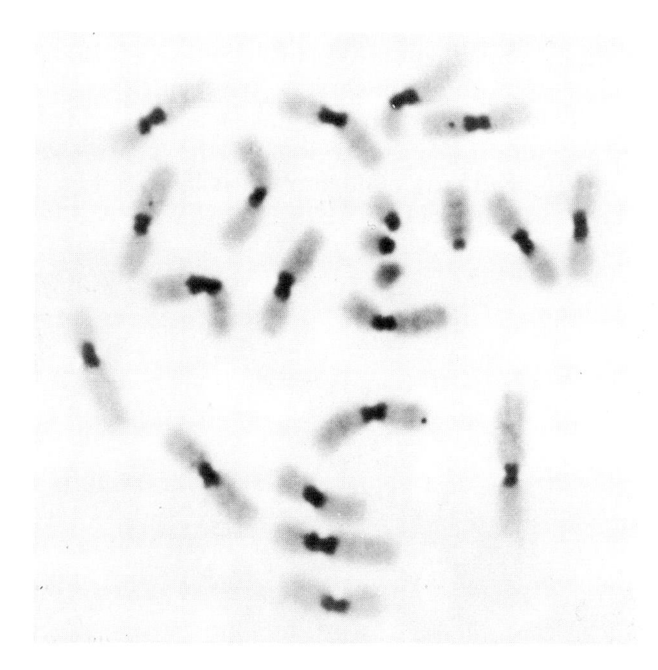

FIG. 7. C-banded metaphase of a male mouse with 18 Rb metacentrics; only chromosome No. 19, and X and Y are acrocentric. Duplicate appearance of darkly stained C-heterochromatin in the metacentrics.

From these observations it is unlikely that Rb variation leads to substantial losses of C-heterochromatin on the long arms of the fused acrocentrics (see, however, p. 156; also Comings & Avelino, 1972). Additional evidence for the conservation of C-heterochromatin as well as of pericentric silver-stained nucleolus organizer regions comes from studies on RAG- and L-chromosomes. The treatment of these cells with 33258 Hoechst causes condensation inhibition of the chromosomes which produces abnormal stretching, and provides an efficient means of analysis of the centromeric region (Nielsén *et al.*, 1979). With a similar method of 33258 Hoechst condensation inhibition, again on L-cells, Lau & Hsu (1977) demonstrated no apparent loss of Cd-bands (Eiberg, 1974). Although it may remain controversial whether Cd-banding stains centromeres, this observation shows that Rb variation does not involve losses of chromosomal segments, at least with regard to heterochromatin located distally to the centromere on the long arm of an acrocentic.

In electron micrographs of total preparations of metaphases in strain L cells (Schwarzacher, Gropp & Ruzicka, 1976), the chromosomes are made up of irregularly folded fibrils. These fibrils are about 20–40 nm thick, and they are tightly folded in the condensed parts of the chromosomes including the centric regions of Rb metacentrics. This composition and an apparent symmetry of the structural composition of the region of the centromere is even better shown (Fig. 8) if the condensation is inhibited by the action of 33258-Hoechst (Marcus *et al.*, 1979).

FIG. 8. Electron micrograph of 33258 H-treated Rb metacentric chromosome of strain L cell. (Preparation by H. G. Schwarzacher and F. Ruzicka, Vienna.)

Nucleolus Organizer Regions

In the laboratory mouse, silver stainable nucleolus organizer regions (NORs) have been observed near the centromeric C-bands of the shorter autosomes nos. 12, 15–19, though with variable involvement of these chromosomes among different inbred strains (Dev, Tantravahi *et al.*, 1977). However, a much wider variation of the localization of silver-NORs has been demonstrated in wild caught mice, particularly in some of the polymetacentric populations from Italy and Dalmatia (Winking, Nielsén & Gropp, 1980). Thus, silver-NORs were found in Rb metacentrics 13, 37 and 39 Lub in terminal sites of the arms corresponding to chromosomes nos 4 and 13 (Fig. 9). In the light of the occurrence of NORs on similar arm

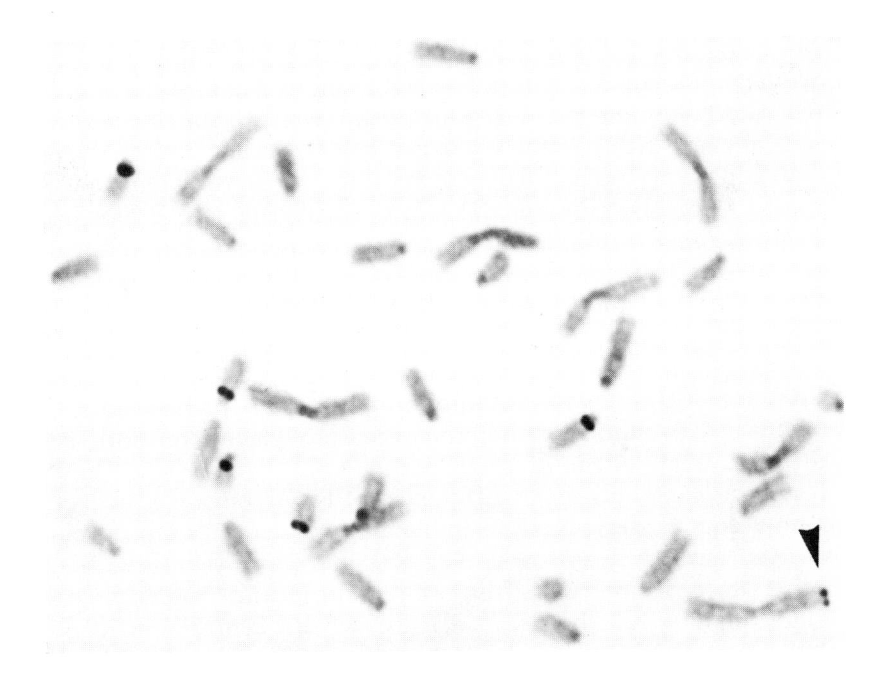

FIG. 9. Rb(3.4)39Lub with silver NOR in a terminal site on the arm corresponding to chromosome No. 4 (arrow). Other NORs in acrocentrics.

components in geographically separated Rbs, it may be suspected that the localization of NORs is a property acquired before the event of an Rb translocation happened.

Meiosis and Meiotic Pairing

In diakinesis and meiotic metaphase I of male and female Rb homozygotes regular ring-shaped bivalents of the metacentrics are often found (Fig. 3c), whereas hybrids with metacentric/acrocentric heterozygosity show trivalents (Fig. 3d) with perfect pairing (at the level of light microscopy). Similarly, quadrivalents and higher multivalents are formed if complex alternating arm homologies exist in the respective hybrids (p. 148; Fig. 6). There is no evidence that the formation and frequency of chiasmata in Rb metacentric homo- or heterozygotes show differences from normal due to the Rb condition (Polani, 1972).

Meiotic anaphase I malsegregation is one of the major features of Rb heterozygosity and is discussed below (pp. 156–160).

Electron microscopy of mid to late pachytene spreads from male metacentric heterozygotes of the mouse (Johannisson & Winking, 1979) using the method of Counce & Meyer (1973) revealed, for trivalents, a Y-shaped configuration with a short stump corresponding to paired non-homologous segments of the two acrocentrics, and two long branches with lateral elements corresponding to the paired homologous arms. Figure 10 demonstrates seven such figures (plus five autosomal bivalents and XY) in a pachytene spread of a male tobacco mouse F_1-hybrid. The apparent symmetry in the triangular stump of the Y-shaped configurations (Fig. 10: arrows indicate such regions in two of the trivalents, pointing to the paired centromeres of the acrocentrics) results from the adjustment of the non-homologous, probably entirely heterochromatic pericentric segments of the acrocentrics. Moreover, non-homologous pairing must be assumed also for the short heterochromatic segments on both sides of the centromere of the Rb metacentric.

It is not clear whether the differences of length between the paired non-homologous segments of individual trivalents (compare both trivalents marked by arrow in Fig. 10) have any significance, nor whether dislocations or shifting of structural units underlying the pseudosymmetry of the area adjacent to the centromeric region, reflect minor rearrangements with functional consequence. However, it seems reasonable to assume that crossover suppression, which was shown to be associated with heterozygosity of some of the tobacco mouse metacentrics by Cattanach (1978), is related to such structural phenomena.

EM preparations of pachytene spreads of quadrivalents and longer chain multivalents show that proximal chromosomal segments, though homologous, may remain unsynapsed on both ends of the chains (Winking & Johannisson, 1980). In addition, unsynapsed segments on the ends of chains are sometimes oddly thickened and occasionally one such segment is spatially associated with the XY-bivalent. Further investigations will show whether these peculiar observations are causally related to subfertility or sterility,

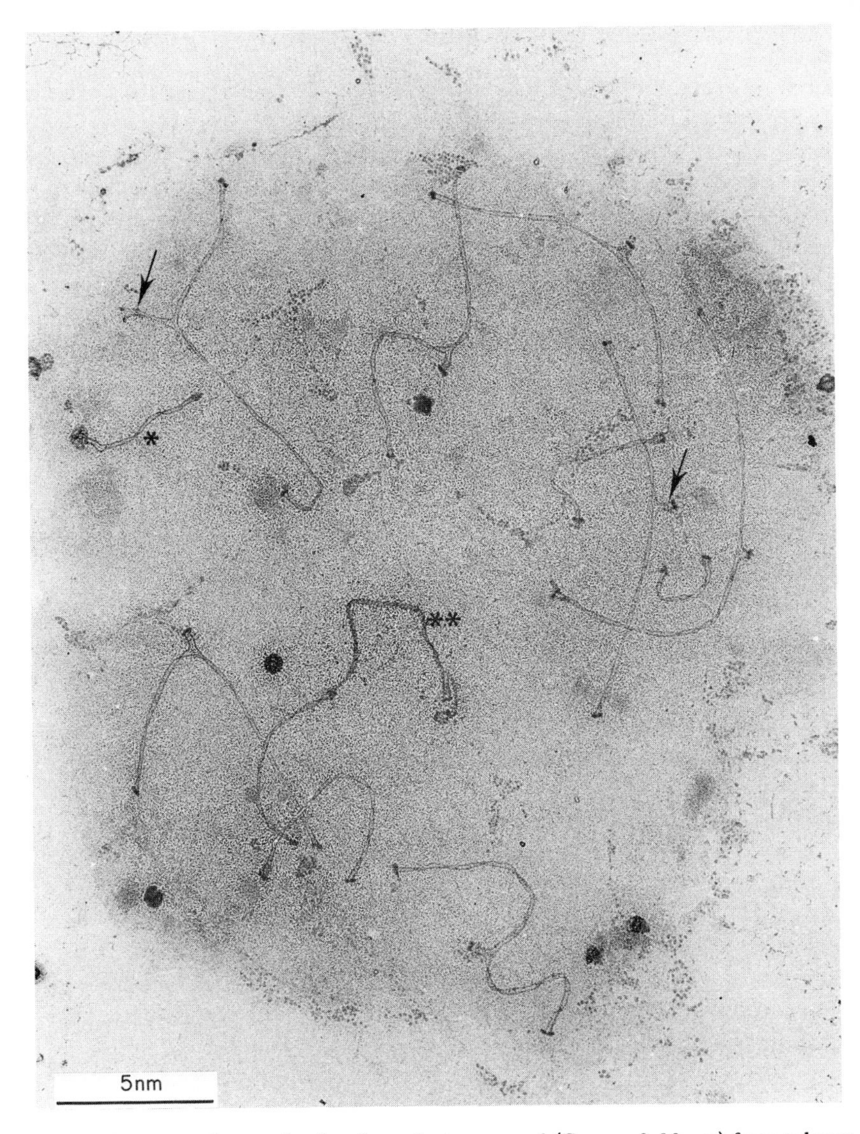

FIG. 10. Electron-micrograph of male pachytene spread (Counce & Meyer) from tobacco mouse × NMRI F_1-hybrid. Synaptonemal complexes of seven trivalents, five bivalents and XY. Arrows indicate two of the seven trivalents with different size of paired non-homologous segments of the acrocentrics. *Example of autosome bivalent. **XY-bivalent. (Preparation and photograph by R. Johannisson.)

which may occur in Rb metacentric hybrids with chains longer than trivalents.

Mode of Origin of Rb Whole Arm Rearrangement in the Mouse

John & Freeman (1975) discussed the possible modes of Rb exchange leading to whole arm rearrangement and came to the conclusion that no direct adhesion of chromosome ends is possible. They assumed that the Rb process depends on prior breakage, which may take place at different breakpoints. Some observations reported in the present study, e.g. the result of C-banding, etc., seem to rule out, at least for the Rb variation in the mouse, type 2, 5 and 6 (Fig. 11). The EM observations on trivalents discussed on p. 154 make type 3 unlikely, because it became clear from these studies that the elementary mouse chromosome is definitely acrocentric and possesses a short arm which gets lost when the Rb change occurs. The most likely mode then, seems to be type 4, although variants of this type including type 1 are similarly possible.

CONSEQUENCES OF Rb METACENTRIC HETEROZYGOSITY

Meiotic Anaphase I Malsegregation in Single Rb Metacentric Heterozygote

Noticeable meiotic anaphase I malsegregation with non-disjunction of chromosomes involved in metacentric/acrocentric trivalents was first shown to occur in mice heterozygous for mouse Rb metacentrics of feral origin (Tettenborn & Gropp, 1970; Cattanach & Moseley, 1973).

These studies were done in male meiosis, but meiotic anaphase I non-disjunction has been found to be very common in male *and* female Rb heterozygous mice. On the other hand, only low rates of malsegregation are reported from heterozygotes with Rbs of laboratory origin, e.g. Rb163H/+ (Evans *et al.*, 1967) or Rb1Iem (Baranov & Dyban, 1971), with the exception of Rb1Ct (B. M. Cattanach, personal communication).

Male meiosis

Table VI shows the rates of male meiosis I non-disjunction in a larger series of heterozygotes for Rb metacentrics in a mixed, partly maternally derived NMRI genome. They are calculated from chromosome arm counts in meiotic metaphase II plates (MII) which are expected to reveal malsegregation in anaphase I. It is evident that the rates of malsegregation in male heterozygotes show great variation (2–28%) between different Rb metacentrics of feral origin, whereas such rates are low (3–6%) in heterozygotes of the

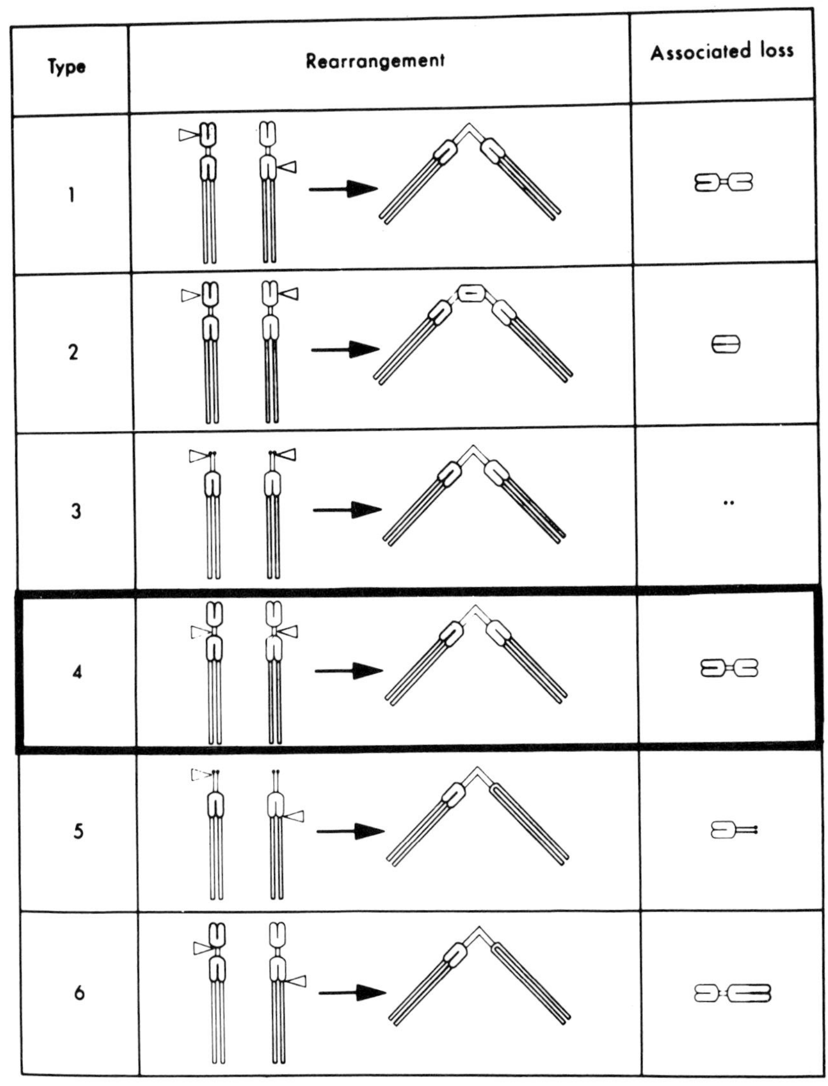

FIG. 11. Possible modes of Rb rearrangement according to John & Freeman (1975). Thick line surrounding "Type 4" inserted by the authors of this study. The blocks adjacent to the centromere represent C-heterochromatin.

Rbs of spontaneous laboratory origin of which data are available from our studies.

It cannot be ruled out that the introduction of a Rb metacentric of feral derivation into a laboratory strain genome (e.g., NMRI) is responsible in some cases for the relatively high non-disjunction rates. At the present time, this question is still open.

Nevertheless, systematic analyses currently carried out in Lübeck using wild trapped Rb hybrids with metacentric heterozygosity in the original wild type genome compared with Rb heterozygosity in mixed or laboratory genome support the assumption that the transfer of an Rb metacentric from an original or wild type to a laboratory mouse genome may cause an increase of the non-disjunction rates of the structural heterozygote (Gropp, Winking, Redi & Noack, unpublished). On the other hand, further modification of background, e.g. introduction from a mixed laboratory genome into inbred strains, does not seem to influence malsegregation rates (Nijhoff & de Boer, 1979: Rb4Bnr; Cattanach, 1979: Rb1Bnr).

It has been suggested that increased non-disjunction rates in the mouse depend, to a small extent, on morphologically undetectable minor structural variation between a feral Rb metacentric and the laboratory mouse genome (Cattanach & Moseley, 1973). It is certainly possible to explain the occurrence and the variability of non-disjunction events in Rb metacentric heterozygotes along these lines, including the finding of different rates of malsegregation in male heterozygotes of Rb metacentrics with similar arm composition but independent origin as shown in Table VII. The fact that in the (pseudo-?) homozygote of two such metacentrics the malsegregation rate is again low does not exclude the existence of minimal changes. Hypothetically, the pericentric structural differences of trivalents observed at the EM level (p. 154) may account, both in this case and also when Rbs of different arm composition are compared, for the variation of malsegregation rates encountered in independent Rb metacentrics (or trivalents, respectively).

Female meiosis

Previous studies on Rb(1.3)1Bnr/+ (Winking & Gropp, 1976), on Rb(8.17)1Iem/+ (Baranov & Dyban, 1971), on Rb(11.13)4Bnr/+ and Rb(16.17)7Bnr/+ (Gropp, 1973), showed that considerably higher meiotic non-disjunction rates occur in female than in male carriers of the same Rb metacentric. More extended current investigations involving meiotic metaphase II evaluations and the analysis of unbalanced versus balanced progeny of heterozygotes of 13 different Rb metacentrics of feral, and two Rbs (Rb1Ald/+ and Rb1Iem/+) of laboratory origin (Gropp, Winking, Redi, Noack, Kolbus & Louton, in press) confirm these earlier findings. In Rb7Bnr/+, Rb3Rma/+ and Rb13Lub/+ females, the difference amounts to a multiple of non-disjunction rates in the respective males, while the only cases with equal abnormality rates in both sexes, among the 15 tested Rb metacentrics, are those of Rb(11.13)-

4Bnr and Rb(8.12)5Bnr. The disposition of the oocyte to allow more non-disjunction than the spermatocyte (for the same Rb metacentric on identical strain background), is one of the intriguing facts emerging from the study of Rb metacentric heterozygosity. It is difficult to understand the reason and the mechanism of the greater tendency of the oocyte to produce malsegregation.

Another sex-dependent phenomenon observed in the first meiotic division of some but not all female Rb metacentric heterozygotes (see for Rb1Bnr/+: Winking & Gropp, 1976; for other Rbs: Gropp *et al.*, in press), is a distorted segregation of the rearranged chromosome. The result is a prevalence of offspring of homozygous type, i.e. without a metacentric in the balanced progeny apparently due to a preferential distribution of the Rb metacentric to the first polar body. Table VIII records data from a current systematic investigation of transmission of Rb chromosomes. In the balanced progeny of some female Rb metacentric heterozygotes the ratio of heterozygosity versus loss of the Rb metacentric in the oocyte (homozygosity) is 2:3, 1:2 or even 1:3 compared with an expected 1:1 ratio (Table VIII). However, in some other heterozygous conditions an almost equal distribution occurs. In the progeny of male heterozygotes there is no statistical deviation from 1:1, except in Rb(16.17)7Bnr (where mild *t*-mutation-like effects could account for segregation distortion). Evidence for segregation distortion, though with an opposite prevalence of the heterozygous type, has been found by Boué (1979) in the balanced progeny of human female Rb carriers. Data for this conclusion were collected in a European Collaborative Study on structural chromosomal anomalies in prenatal diagnosis.

Age

An increase of the non-disjunction rate (from 49% to 73%) together with a reduction of chiasma frequency has been noted with increasing age (from 7–10 weeks to 11–13 months) in female Rb(1.3)1Bnr/+ heterozygotes by H. Winking (unpublished). This increase in the progeny of a Rb metacentric heterozygote is definitely higher than that reported from ageing non-heterozygous mice (Yamamoto, Endo, & Watanabe, 1973; Speed, 1977). No measurable age effect on the non-disjunction rate has been noted in the progeny of Rb1Bnr/+ males (H. Winking, unpublished).

Meiosis I Segregation in Heterozygous Carriers of Multiple Rb Metacentrics

Mice heterozygous for several or many independent Rb metacentrics (polymetacentric individuals without arm homologies) form corresponding numbers of trivalents in meiosis I (see Figs 3d and 10). In such cases the overall non-disjunction rates (proportion of unbalanced segregation products) of heterozygotes with multiple Rbs can rise to very high values, such as 52% and 51% versus 68% and 77% in male compared to female tobacco mouse F_1 (Rb1–7Bnr/+) and CD–F_1 (Rb1–9Rma/+) (Winking & Gropp, 1976). It seems in these cases that the observed cumulative rates remain lower than the mere statistical addition of the non-disjunction rates of individual Rb metacentrics.

Animals with heterozygosity for two or more Rb metacentrics with partial homology of their arms form quadrivalents, pentavalents or longer chains or rings in meiosis I. Similar associations can be obtained in the laboratory by breeding animals derived from different feral populations. The occurrence of such hybrids in nature must be assumed, although complex Rb hybrids have so far not been observed in wild trapped mice. Yet, since mice with Rb(10.11)8Bnr and others with Rb(1.10)10Bnr (Gropp, Winking *et al.*, 1972) or mice of Milano I and II type (see Table II) may live in close vicinity, one can assume that progeny with quadrivalents (as formed by the 11.10:10.1 Rbs) or with a complex ring (Mil.I/Mil.II) may occasionally be produced. Hybrids bred in the laboratory from parents of the Upper Valtellina I and II (Table II) trapped in the same house showed a pentavalent 10+10.12/12.8/8.2+2 in meiosis I. More than half of their progeny was chromosomally unbalanced and lost owing to developmental disorders (unpublished observation). In all these cases, the meiotic malsegregation rates are high. However, each structural variation is unique, and different conditions show a different behaviour at meiosis.

Doubly metacentric heterozygous mice with monobrachial homology (see Fig. 12, top) show considerable rates of malsegregation, apparently with a prevalence of non-disjunction of both Rb metacentrics (Fig. 12, bottom right). In fact, it is characteristic of double Rb metacentric heterozygosity with arm homology that in hypermodal gametes the extra chromosome corresponds to the arm shared in the parent by both metacentric chromosomes (see pp. 164–165). The exact rates of non-disjunction are difficult to assess by usual techniques of M II counting, but approximative data can be obtained (Gropp, 1975). In such conditions, males may produce higher rates of unbalanced gametes than females, in contrast to the situation in single Rb heterozygotes (unpublished own observation; B. J. White *et al.*, 1974).

Fertility in Rb Metacentric Heterozygotes

A priori, two possible effects of gametogenesis impairment

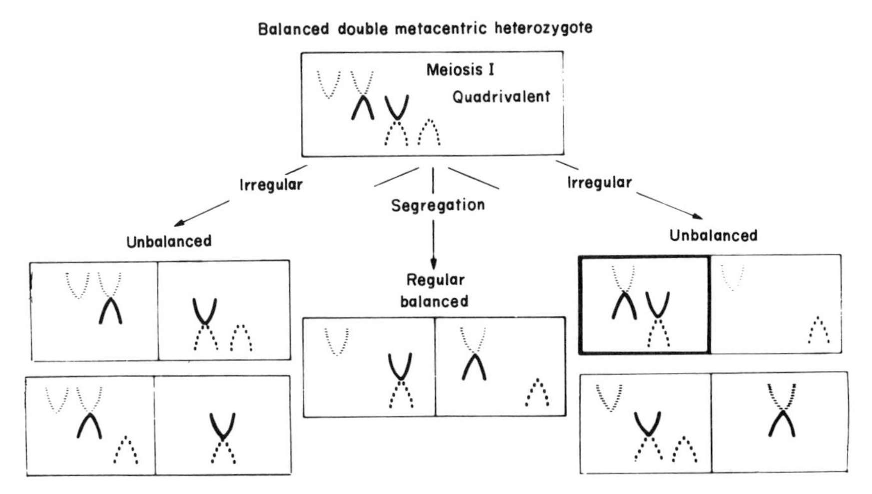

FIG. 12. Experimental design of specific induction of selective types of Rb metacentric non-disjunction using double Rb heterozygosity (see also Fig. 13). Unbalanced hyperhaploid (thick lined rectangle) and complementary hypohaploid (right of thick-lined rectangle) products of non-disjunction give rise, after fertilization, to trisomy and monosomy, respectively.

connected with Rb metacentric heterozygosity can be anticipated: one resulting from segregational disturbances of meiotic divisions; and the other connected with the hybrid constitution as such, corresponding to male-limited hybrid sterility (Haldane, 1922).

As discussed in the previous section, disorders of segregation of the chromosomes from multivalents are a prominent feature associated with single and multiple metacentric heterozygosity, in particular if Rbs of feral origin are involved. Anaphase I malsegregation is responsible for the occurrence of unbalanced, hypo- or hyperhaploid, gametogenic precursor cells, giving rise to chromosomally abnormal zygotes. In single Rb metacentric heterozygotes hypomodal monosomic and hypermodal trisomic embryos are expected, while in the cases of heterozygosity with several independent or with partially homologous Rbs, zygotes with multiple or combined monosomy and trisomy may be formed. Such unbalanced zygotes are subject to selective elimination and fertility can be markedly reduced as a consequence, although prezygotic selection does not play a role in this (Ford, 1972). In fact, there is convincing evidence from spermatogenesis that most, if not all, chromosomally unbalanced spermatogenic precursor cells develop into morphologically normal spermatozoa which are able to fertilize irrespective of their chromosomal abnormality (Stolla & Gropp, 1974). There is nothing against the assumption that the same is true for oogenesis, and that selection in the mouse system of meiosis I malsegregation acts exclusively in the postzygotic period of development (Gropp, 1975). Female carriers of single Rb metacentric heterozygosity, by virtue of their higher rates of non-disjunction, almost always produce considerably greater numbers of aneuploid zygotes than males.

In many cases of complex Rb metacentric heterozygosity in which quadrivalents and longer chains are formed in meiosis I, sex-dependent (i.e. male-limited) hybrid sterility is observed. It seems that this phenomenon is, in males, superimposed on segregational disturbances. Thus, several carriers of double metacentric heterozygosity with quadrivalent formation are male-sterile, but fertile in the respective females, e.g., Rb2Rma/Rb5Bnr, Rb2Rma/Rb1Iem, Rb5Bnr/Rb1Iem, Rb1Iem/Rb1Ct, and a few others. Semi-sterility observed in some other cases is also attributable to this condition. Histologically, a defect of spermatogenesis with the features of spermatogenic arrest after spermatocyte II stage is found (Döring, Winking & Gropp, 1975; Winking, 1980).

Heterozygosity with longer chains is associated with male sterility in most cases. Male hybrids with rings, even of larger size, do not show the spermatogenic failure characteristic of male hybrid sterility, but they may produce extremely high rates of unbalanced gametes. This is at least so in laboratory-bred complex Rb hybrids with metacentrics from different and independent mouse populations (Winking, 1980; further unpublished observations from current studies). It seems possible to explain the fact that longer chains have a greater tendency to induce male-dependent sterility than rings by noting the special behaviour of the ends of chains and the assumed spatial relationship of an unsynapsed region with the XY bivalent as discussed on p. 154. The importance of the association of multivalents with the XY as a cause of male sterility has previously been emphasized by Forejt (1979).

EVOLUTIONARY ASPECTS OF Rb-VARIATION IN THE MOUSE.

The karyotype, if considered as the structural mould of the genome on the cytological level, is constantly reshaped by forces acting upon the processess of evolution. It is clear that Rb variation provides the genome with a high grade of plasticity. As shown by some of the examples reported on p. 160, it is to a certain extent possible to simulate in laboratory breeding, events of hybridization which may have occurred in evolution. The permanent remodelling of the karyotype in the course of evolution allows repeated hybridization with ensuing meiotic Rb chromosomal heterozygosity. This can, in some cases and under certain conditions, represent a potent mechanism for building up reproductive barriers, thus promoting karyotype diversification and speciation as suggested in the stasipatric speciation model (M. J. D. White, 1968, 1978; Gropp, 1969).

It seems clear from recent studies that Rb heterozygosity is effective in building up such barriers only if (a) Rb metacentrics become introduced into the genetical background of a different deme (as compared to the similarly different laboratory mouse genome), or (b) complex meiotic pairing structures, in particular longer chains and rings, act with the consequences of malsegregation to produce a more efficient sterility barrier. Some of the special problems arising from the study of Rb metacentrics in feral mice from Alpine, Lombardian and Central Italian regions have been discussed by Capanna, Civitelli & Cristaldi (1977) and by M. J. D. White (1978). They come to the more general conclusion, as already suggested by Gropp, Tettenborn *et al.* (1970) for the tobacco *poschiavinus* mouse, that the Rb system in mice from these areas reflects processes of incipient speciation. It is the peculiarity of the tobacco *poschiavinus* mouse that this process seems to have reached a more advanced stage than in other populations.

However, it has to be recognized that recent studies by L. Thaler, J. Britton-Davidian & F. Bonhomme using enzyme markers (see Thaler, Bonhomme & Britton-Davidian, this volume pp. 27–41) seems to exclude a major or even noticeable impairment of gene flow between an Rb mouse population and the surrounding non-Rb population. Moreover, there is a need to settle the problem of taxonomic assignment of Rb feral mouse populations with regard to the respective distributions of *M. m. musculus, domesticus,* and *brevirostris* (see Marshall & Sage, this volume pp. 15–25). Many of the investigated Rb mouse populations, in particular those from the Apennines as well as the specimens from Dalmatia, belong to the *M. m. brevirostris* subspecies. But some others from the Rhaeto–Lombardian region, from the Eolian islands and from Sicily need a better taxonomic assignment to aid further discussion of the evolutionary role of Rb variation.

Independently of the wider implications of Rb karyotype evolution, certain facts are worth emphasizing:

(a) Rb association of arm components in feral populations is accidental, with the exception of chromosomes No. 19 and the sex chromosomes. It is unknown why chromosome no. 19 is exempt from taking part in Rb changes. The involvement of the sex chromosomes could have deleterious effects.

(b) It does not seem that there is a prevalence of particular Rb metacentrics in any population, nor is the distribution of any Rb so far known ubiquitous. However, it is interesting to note that the Rb3Bnr metacentric (Table IIIb) is often found and has a particularly wide distribution.

(c) It can be argued that silver-NORs (p. 153) on terminal sites had been acquired by certain chromosomes before they became involved in Rb translocations.

(d) Winking & Guénét (1978) reported the occurrence of a *t*-mutant ($t^{Lüb1}$) on the chromosome arm of the Rb(4.17)13Lub. This finding provides a tool for experimental research with a *t*-mutant on a cytological marker, but leaves open the question of an origin before or after the formation of the Rb metacentric. A *t*-mutational event before the formation of the Rb metacentric would not have allowed Rb homozygosity because of *t*-homozygote lethality. On the other hand, it certainly does not exclude the possibility of spread of a *t*-mutant within larger populations, and its incorporation into demes with already existing Rb heterozygosity by crossover between the centromere and *t*. Such facts, and others which might emerge from further work on the Rb system in the mouse, have to be reconciled with the knowledge obtained, e.g., from biochemical marker studies (see Thaler *et al.*, this volume p. 37) in order to understand fully the significance of Rb karyotype diversification.

MOUSE Rb TRANSLOCATIONS IN EXPERIMENTAL RESEARCH

If enough Rb metacentrics of different composition are available, they provide a useful tool for an experimental strategy (Fig. 12) allowing the induction of specific monosomies and trisomies in the mouse (Baranov & Dyban, 1972); Dyban & Baranov, 1978; Gropp & Kolbus, 1974; B. J. White *et al.*, 1974; Gropp, 1975). With their introduction into laboratory mouse strains (see Table IV), a sufficiently large reservoir of Rb metacentrics has been placed at the experimentalist's disposal to induce all theoretically possible whole arm monosomies and trisomies of the mouse. The most suitable experimental design is based on the use of a double Rb metacentric heterozygote with monobrachial homology (see Figs 12 and 13). Among different types of abnormal segregation (see p. 160), non-disjunction products of the two Rb metacentrics (a) either devoid of a Rb metacentric, or (b) containing both (Fig. 12, thick-lined rectangle; and Fig. 13, bottom right), are expected to occur.

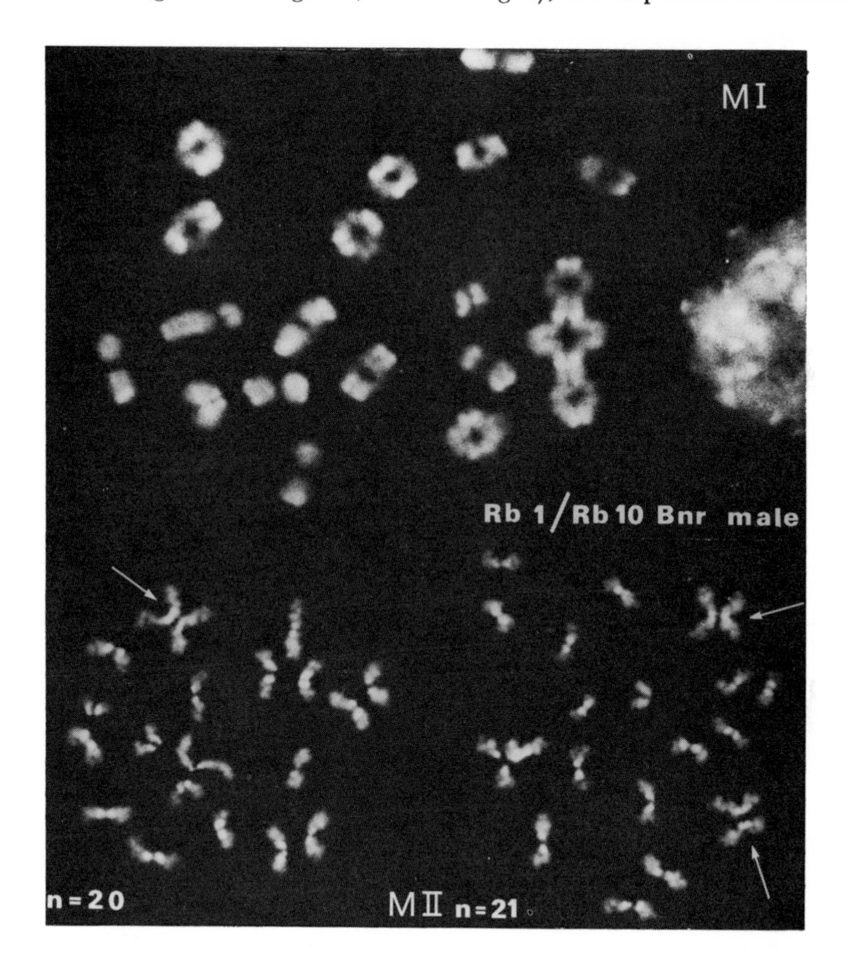

FIG. 13. Meiotic metaphase I (M I) of spermatocyte from a Rb(1.3)1/Rb(1.10)10Bnr male with quadrivalent (above). Below: Second meiotic metaphase plates (M II); left: modal haploid number ($n = \mathrm{NF}/2 = 20$) of chromosome arms, including one Rb metacentric; right: hypermodal haploid number of chromosome arms including both Rb metacentrics; the extra chromosome (arm) is a No. 1 from the (1.3)(1.10)Rbs. Fluorescence micrographs after 33258 Hoechst staining.

The chromosome corresponding to the arm common to both Rbs is then lacking in the resulting gamete in case (a), and present as an extra element in case (b). This means that monosomy is the consequence of fertilization in case (a), and trisomy in case (b). The type of monosomy or trisomy is determined, in this experiment, by the choice of the double Rb metacentric combination in a male and/or female progenitor.

Reviews of the developmental profiles of experimental trisomies are given by Gropp, Kolbus & Giers (1975), Gropp, Putz & Zimmermann (1976), Gropp (1978), Dyban & Baranov (1978) and Epstein (1980). Less, though increasing, effort has been directed to the study of monosomy (Dyban & Baranov, 1978; Epstein & Travis, 1979). It is obvious that the use of such experimental systems as an animal model of monosomy and trisomy can also contribute to problems of human teratology (e.g. Putz *et al.*, 1980), and other fields of research on developmental biology and pathology in animals and man.

ACKNOWLEDGEMENTS

The authors are grateful to Mrs G. Noack for technical help, to Mrs A. Sachon for the preparation of the figures, and to Mrs E. Hüttenhain for her great help in the preparation of the manuscript.

REFERENCES

Arroyo Nombela, J. J. & Rodriguez Murcia, C. (1977). Spontaneous double Robertsonian translocation Rb(2.3) and Rb(X.3) in the mouse. *Cytogenet. Cell Genet.* **19**: 227–230.

Baranov, V. S. & Dyban, A. P. (1971). Embryogenesis and peculiarities of karyotype in mouse embryos with centric fusion of chromosomes (Robertsonian translocation). *Ontogenez.* **2**: 164–176.

Baranov, V. S. & Dyban, A. P. (1972). Disturbance of embryogenesis in trisomy of autosomes arising in offsprings of mice with Robertsonian translocation (centric fusion) of T1IeM. *Arch. Anat. Leningrad* **113**: 67–77.

Boué, A. (1979). European collaborative study of structural chromosome anomalies on prenatal diagnosis. In *Prenatal diagnosis:* 34–46 *(Proc. 3rd European Conference on Prenatal Diagnosis of Genetic Disorders.)* Murken, J. D., Stengel-Rutkowski, S. & Schwinger, E. (Eds). Stuttgart: Enke.

Capanna, E., Civitelli, M.-V. & Cristaldi, M. (1977). Chromosomal rearrangement, reproductive isolation and speciation in mammals. The case of *Mus musculus. Boll. Zool.* **44**: 213–246.

Capanna, E., Gropp, A., Winking, H., Noack, G. & Civitelli, M.-V. (1976). Robertsonian metacentrics in the mouse. *Chromosoma* **58**: 341–353.

Cattanach, B. M. (1978). Crossover suppression in mice heterozygous for tobacco mouse metacentrics. *Cytogenet. Cell Genet.* **20**: 264–281.

Cattanach, B. M. (1979). *Mouse News Letter* **61**: 38.

Cattanach, B. M. & Crocker, A. J. M. (1979). *Mouse News Letter* **60**: 44.

Cattanach, B. M. & Moseley, M. (1973). Nondisjunction and reduced fertility caused by the tobacco mouse metacentric chromosomes. *Cytogenetics* **12**: 264–287.

Cattanach, B. M. & Savage, J. R. K. (1976). *Mouse News Letter* **54**: 38.

Cattanach, B. M., Murray, I. & Bigger, T. R. L. (1977). *Mouse News Letter* **56**: 37.

Chakrabarti, S. & Chakrabarti, A. (1977). Spontaneous Robertsonian fusion leading to karyotype variation in the mouse – first report from Asia. *Experientia* **33**: 175.

Comings, D. E. & Avelino, E. (1972). DNA loss during Robertsonian fusion in the studies of the tobacco mouse. *Nature New Biol.* **237**: 199.

Committee on Standardized Genetic Nomenclature for Mice (1972). Standard karyotype of the mouse *Mus musculus*. *J. Hered.* **63**: 69–72. (And *Mouse News Letter*, 1974.)

Counce, S. J. & Meyer, G. F. (1973). Differentiation of the synaptonemal complex and the kinetochore in locusta spermatocytes studied by whole mount electron microscopy. *Chromosoma* **44**: 231–253.

Dev, V. G., Miller, D. A., Tantravahi, R. R., Schreck, R. R., Roderick, T. H., Erlanger, B. F. & Miller, O. J. (1975). Chromosome markers in *Mus musculus*: differences in C-banding between the subspecies *M. m. musculus* and *M. m. molossinus*. *Chromosoma* **53**: 335–344.

Dev, V. G., Tantravahi, R., Miller, D. A. & Miller, O. J. (1977). Nucleolus organizers in *Mus musculus* subspecies and in the RAG mouse cell line. *Genetics* **86**: 389–398.

Döring, L., Winking, H. & Gropp, A. (1975). *Mouse News Letter* **52**: 42.

Dyban, A. P. & Baranov, V. S. (1978). *Cytogenetica Raswitija Miekopita juschich.* Moscow: Isdateldstwo Nauka.

Eiberg, H. (1974). New selective Giemsa technique for human chromosomes, Cd-staining. *Nature, Lond.* **248**: 55.

Epstein, Ch. J. (1980). Animal models for autosomal trisomy. In *Trisomy 21 (Down's Syndrome): Research perspectives.* Gerald, P. & de la Cruz, F. (Eds). Baltimore: University Park Press.

Epstein, Ch. J. & Travis, B. (1979). Preimplantation lethality of monosomy for mouse chromosome 19. *Nature, Lond.* **280**: 144–145.

Evans, E. P., Lyon, M. F. & Daglish, M. (1967). A mouse translocation giving a metacentric marker chromosome. *Cytogenetics* **6**: 105–119.

Fatio, V. (1869). *Faune des vertébrés de la Suisse.* **1**. Geneva & Basle: Georg.

Ford, C. E. (1972). Gross genome unbalance in mouse spermatozoa: does it influence the capacity to fertilize? In *The genetics of the spermatozoon*: 359–369. Beatty, R. A. & Gluecksohn-Waelsch, S. (Eds). Edinburgh & New York: University of Edinburgh.

Forejt, J. (1979). Meiotic studies of translocations causing male sterility in the mouse. II. Double heterozygotes for Robertsonian translocations. *Cytogenet. Cell Genet.* **23**: 163–170.

Gropp, A. (1969). Cytologic mechanisms of karyotype evolution in insectivores. In *Comparative mammalian cytogenetics*: 247–266. Benirschke, K. (Ed.). New York: Springer.

Gropp, A. (1973). Fetal mortality due to aneuploidy and irregular meiotic segregation in the mouse. *C. r. Colloque accidents chromosom. Réprod.* 255–269. Boué, A. & Thibault, Ch. (Eds). Paris: I.N.S.E.R.M.

Gropp, A. (1975). Chromosomal animal model of human disease. Fetal trisomy

and developmental failure. In *Teratology*: 17–33. Berry, C. L. & Poswillo, D. E. (Eds). Berlin, Heidelberg & New York: Springer.
Gropp, A. (1978). Relevance of phases of development for expression of abnormality. Perspectives drawn from experimentally induced chromosome aberrations. In *Abnormal fetal growth: Biological bases and consequences:* Naftolin, F. (Ed.). (Life Science Report 10) Dahlem Konferenzen.
Gropp, A., Hilwig, I. & Seth, P. K. (1973). Fluorescence chromosome banding patterns produced by a benzimidazole derivative. In *Chromosome identification–technique and applications in biology and medicine:* 300–306. Nobel Symp. No. 23 New York & London: Academic Press.
Gropp, A. & Kolbus, U. (1974). Exencephaly in the syndrome of trisomy No. 12 of the foetal mouse. *Nature, Lond.* **249**: 145–147.
Gropp, A., Kolbus, U. & Giers, D. (1975). Systematic approach to the study of trisomy in the mouse. II. *Cytogenet. Cell Genet.* **14**: 42–62.
Gropp, A., Olert, J. & Maurizio, R. (1971). Robertsonian chromosomal polymorphism in the mouse. *Experientia* **27**: 1226–1227.
Gropp, A., Putz, B. & Zimmerman, U. (1976). Autosomal monosomy and trisomy causing developmental failure. *Curr. Top. Pathol.* **62**: 177–192.
Gropp, A., Tettenborn, U. & v. Lehmann, E. (1970). Chromosomenvariation vom Robertson' schen Typus bei der Tabakmaus, *M. poschiavinus* und ihren Hybriden mit der Laboratoriumsmaus. *Cytogenetics* **9**: 9–23.
Gropp, A., Winking, H., Redi, C., Noack, G., Kolbus, U. & Louton, T. (in press). Consequences of Robertsonian heterozygosity in the mouse – Meiotic segregation and postzygotic abnormality in male versus female carriers of Rb chromosomes. *Heredity* (in press).
Gropp, A., Winking, H., Zech, L. & Müller, H. J. (1972). Robertsonian chromosomal variation and identification of metacentric chromosomes in feral mice. *Chromosoma* **30**: 265–288.
Haldane, J. B. S. (1922). Sex ratio and unisexual sterility in hybrid animals. *J. Genet.* **12**: 101–109.
Hashmi, S., Allerdice, P. W., Klein, G. & Miller, O. J. (1974). Chromosomal heterogeneity in the RAG and MSWBS mouse tumor cell lines. *Cancer Res.* **34**: 79–88.
Hsu, T. C. & Mead, R. M. (1969). Mechanisms of chromosomal changes in mammalian speciation. In *Comparative mammalian cytogenetics:* 8–17. Benirschke, K. (Ed.). New York: Springer.
Johannisson, R. & Winking, H. (1979). Synaptonemal complex of multivalents in mouse spermatocytes. *Europ. J. Cell Biol.* **20**: 122 (Abstract).
John, B. & Freeman, M. (1975). Causes and consequences of Robertsonian exchange. *Chromosoma* **52**: 123–136.
Klebe, R. J., Chen, T. R. & Ruddle, F. H. (1974). Controlled production of proliferating somatic cell hybrids. *J. Cell Biol.* **45**: 74–82.
Klein, J. (1971). Cytological identification of the chromosome carrying the IX linkage group (including H-2) in the house mouse. *Proc. natn. Acad. Sci. U.S.A.* **68**: 1594–1597.
Lau, Y.-F. Hsu, T. C. (1977). Variable modes of Robertsonian fusions. *Cytogenet. Cell Genet.* **19**: 231–235.
v. Lehmann, E. & Radbruch, A. (1977). Robertsonian translocations in *Mus musculus* from Siciliy. Experientia **33**: 1025–1026.
Léonard, A. & Deknudt, G. H. (1967). A new marker for chromosome studies in the mouse. *Nature, Lond.* **214**: 504–505.

Lyon, M. F. & Jarvis, S. (1979). *Mouse News Letter* **60**: 44.
Lyon, M. F., Mason, T. M. & Bigger, T. R. L. (1977). *Mouse News Letter* **56**: 37.
Marcus, M., Nielsén, K., Goitein, R. & Gropp, A. (1979). Pattern of condensation of mouse and Chinese hamster chromosomes in G2 and mitosis of 33258-Hoechst treated cells. *Expl Cell Res.* **122**: 191–201.
Marshall, J. T. & Sage, R. D. (1981). Taxonomy of the house mouse. In *Biology of the house mouse:* 15–25. Berry, R. J. (Ed.). London & New York: Academic Press.
Matthey, R. (1945). L'evolution de la formule chromosomiale chez les vertébrés. *Experientia* **1**: 50–78.
Matthey, R. (1966). Le polymorphisme chromosomique des *Mus* africains du sous-genre *Leggada.* Révision générale portant sur l'analyse de 213 individus. *Revue suisse Zool.* **73**: 585–607.
Miller, D. A., Tantravahi, R., Dev, V. G. & Miller, O. J. (1976). Q- and C-band chromosome markers in inbred strains of *Mus musculus. Genetics* **84**: 67–75.
Nesbitt, M. N. & Francke, U. (1973). A system of nomenclature for band patterns of mouse chromosomes. *Chromosoma* **41**: 145–158.
Nielsén, K., Marcus, M. & Gropp, A. (1979). Localization of NORs in chromosomes of mouse cell lines by a combined 33258-Hoechst and Ag-staining technique. *Hereditas* **90**: 31–37.
Nijhoff, J. H. & de Boer, P. (1979). A first exploration of a Robertsonian translocation heterozygote in the mouse for its usefulness in cytological evaluation of radiation-induced meiotic autosomal non-disjunction. *Mutat. Res.* **61**: 77–86.
Pardue, M. L. & Gall, J. G. (1970). Chromososomal localization of mouse satellite DNA. *Science, N.Y.* **170**: 1356–1358.
Phillips, R. J. & Savage, J. R. K. (1976). *Mouse News Letter* **55**: 14.
Polani, P. E. (1972). Centromere localization at meiosis and the position of chiasmata in the male and female mouse. *Chromosoma* **36**: 343–374.
Putz, B., Krause, G., Garde, T. & Gropp, A. (1980). Trisomy 12 versus Vitamin A induced exencephaly and associated malformations of the cranium, hypophysis and inner ear in the mouse embryo. *Virchows Arch. path. Anat. Histol.* No. 368: 65–80.
Robertson, W. M. & Rees, B. (1916). Chromosome studies. I. Taxonomy relationships shown in the chromosomes of *Lettegidae* and *Acrididae.* Chromosomes and variations. *J. Morph.* **27**: 179–331.
Robinson, T. J. (1978). Preliminary report of a Robertsonian translocation in an isolated feral *Mus musculus* population. *Mamm. Chrom. Newsl.* **19**: 84–85.
Schwarzacher, H. G., Gropp, A. & Ruzicka, F. (1976). Fine structure of 33258 H-treated chromosomes. *Hum. Genet.* **33**: 259–262.
Speed, R. M. (1977). The effects of ageing on the meiotic chromosomes of male and female mice. *Chromosoma* **64**: 241–254.
Stolla, R. & Gropp, A. (1974). Variation of the DNA content of morphologically normal and abnormal spermatozoa in mice susceptible to irregular meiotic segregation. *J. Reprod. Fert.* **38**: 335–346.
Tettenborn, U. & Gropp, A. (1970). Meiotic non-disjunction in mice and mouse hybrids. *Cytogenetics* **9**: 272–283.

Thaler, L., Bonhomme, F. & Britton-Davidian, J. (1981). Processes of speciation and semi-speciation in the house mouse. In *Biology of the house mouse:* 27–41. Berry, R. J. (Ed.). London & New York: Academic Press.
White, B. J. & Tjio, J. H. (1968). A mouse translocation with 38 and 39 chromosomes but normal N. F. *Hereditas* **58**: 284–296.
White, B. J., Tjio, J. H., van de Water, L. C. & Crandall, C. (1974). Trisomy 19 in the laboratory mouse. I. Frequency in different crosses at specific developmental stages and relationship of trisomy to cleft palate. II. Intrauterine growth and histological studies of trisomics and their normal littermates. *Cytogenet. Cell Genet.* **13**: 217–231.
White, M. J. D. (1968). Models of speciation. *Science, N. Y.* **159**: 1065–1070.
White, M. J. D. (1977). *Modes of speciation.* San Francisco: Freeman.
White, M. J. D. (1978). Chain processes in chromosomal speciation. *Syst. Zool.* **27**: 285–298.
Winking, H. (1980). Cytogenetic and histological observations in sterile males with Robertsonian translocations. *Cytogenet. Cell Genet.* **27**: 213.
Winking, H. & Gropp, A. (1976). Meiotic non-disjunction of metacentric heterozygotes in oocytes versus spermatocytes. In *Ovulation in the human:* 47–56. Crosignani, P. G. & Mitchell, D. R. (Eds). New York, San Francisco, London: Academic Press.
Winking, H. & Guénét, J. -L. (1978). *Mouse News Letter* **59**: 33.
Winking, H. & Johannisson, R. (1980). Pattern of pachytene pairing in mouse hybrids with chain and ring multivalents. *Clin. Genet.* **17**: 94.
Winking, H., Nielsén, K. & Gropp, A. (1980). Variable positions of NORs in *Mus musculus. Cytogenet. Cell Genet.* **26**: 158–164.
Yamamoto, M., Endo, A. & Watanabe, G. (1973). Maternal age dependence of chromosome anomalies. *Nature New Biol.* **241**: 141–142.
Zech, L., Evans, E. P., Ford, C. E. & Gropp, A. (1972). Banding patterns in mitotic chromosomes of tobacco mouse. *Expl Cell Res.* **70**: 263–268.

TABLE I

Rb metacentric chromosomes of the mouse (M. musculus)

Designation	Arm composition
I *Feral origin*	
1 S.E. Switzerland,[a] Poschiavo valley	
Rb1Bnr =	1.3
Rb2Bnr =	4.6
Rb3Bnr =	5.15
Rb4Bnr =	11.13
Rb5Bnr =	8.12
Rb6Bnr =	9.14
Rb7Bnr =	16.17
2 Several populations from[a] other Alpine valleys (incl. Mesolecina and Chiavenna)	
Rb8Bnr =	10.11
Rb9Bnr =	4.12
Rb10Bnr =	1.10
Rb11Bnr =	2.14
Rb12Bnr =	7.8
Rb13Bnr =	13.16
3 Central Italy, Apennine,[b] Abruzzi	
Rb1Rma =	1.7
Rb2Rma =	3.8
Rb3Rma =	6.13
Rb4Rma =	4.15
Rb5Rma =	10.11
Rb6Rma =	2.18
Rb7Rma =	5.17
Rb8Rma =	12.14
Rb9Rma =	9.16
4 Central Italy, Apennine,[b] Molise	
Rb10Rma =	1.18
Rb11Rma =	2.17
Rb12Rma =	4.11
Rb13Rma =	6.7
Rb14Rma =	3.13
Rb14Rma =	5.15
Rb16Rma =	8.14
Rb17Rma =	10.12
Rb18Rma =	9.16
5 Northern Italy (Alpie Orobie), near Bergamo	
Rb1Lub =	1.3
Rb2Lub =	2.8
Rb3Lub =	4.6
Rb4Lub =	5.15
Rb5Lub =	10.12
Rb6Lub =	11.13
Rb7Lub =	9.14
Rb8Lub =	16.17
Rb9Lub =	7.18
6 Central Italy (ACR), near Ancarano	
Rb18Lub =	1.2
Rb11Lub =	5.13
Rb12Lub =	3.9
Rb13Lub =	4.17
Rb14Lub =	6.16
Rb15Lub =	8.14
Rb16Lub =	10.12
Rb17Lub =	11.15
7 Southern Italy, Lipari, Isole Eolie	
Rb18Lub =	1.2
Rb19Lub =	4.13
Rb20Lub =	3.9
Rb21Lub =	5.14
Rb22Lub =	8.12
Rb23Lub =	10.15
Rb24Lub =	6.16
8 Northern Italy, near Milano (Mil. I)	
Rb25Lub =	2.4
Rb26Lub =	3.6
Rb27Lub =	5.15
Rb28Lub =	7.8
Rb29Lub =	10.12
Rb30Lub =	11.13
Rb31Lub =	9.14
Rb32Lub =	16.17

TABLE I. Continued.

Designation	Arm composition
9 Yugoslavia, near Zagreb (Zhadar)	
Rb33Lub	= 1.11
Rb34Lub	= 5.15
Rb35Lub	= 6.12
Rb36Lub	= 10.14
Rb37Lub	= 9.13
Rb38Lub	= 8.17
10 Northern Italy, near[c] Milano (Mil. II)	
Rb39Lub	= 3.4
Rb40Lub	= 2.8
Rb41Lub	= 6.7
Rb42Lub	= 5.15
Rb43Lub	= 10.12
Rb44Lub	= 11.13
Rb45Lub	= 9.14
Rb46Lub	= 16.17
11 Northern Italy (Cremona)[c]	
Rb47Lub	= 1.6
Rb48Lub	= 3.4
Rb49Lub	= 2.8
Rb50Lub	= 5.15
Rb51Lub	= 10.12
Rb52Lub	= 11.13
Rb53Lub	= 9.14
Rb54Lub	= 16.17
Rb55Lub	= 7.18
12 Northern Italy (Migiondo,[c] Upper Valtellina)	
Rb56Lub	= 1.3
Rb57Lub	= 2.8
Rb58Lub	= 4.6
Rb59Lub	= 5.15
Rb60Lub	= 10.12
Rb61Lub	= 11.13
Rb62Lub	= 9.14
Rb63Lub	= 16.17
13 Southern Italy, Sicily, Palermo[d]	
Rb1Sic	= 3.4
Rb2Sic	= 2.15
Rb3Sic	= 6.12
Rb4Sic	= 5.13
Rb5Sic	= 10.14
Rb6Sic	= 8.17
Rb7Sic	= 9.16

Designation	Arm composition	Reference
II *Laboratory origin*		
Rb1Ald	= 6.15	Léonard & Deknudt (1967)
Rb163H	= 9.19	Evans *et al.* (1967)
Rb1Iem	= 8.17	Baranov & Dyban (1971)
Rb1Wh	= 5.19	B. J. White & Tjio (1968)
Rb1Ct	= 8.19	Cattanach & Savage (1976)
Rb2Ct	= 1.15	Cattanach, Murray & Bigger (1977)
Rb3Ct	= 12.13	Cattanach & Crocker (1979)
Rb1H	= 6.13	Phillips & Savage (1976)
Rb2H	= 11.16	Lyon, Mason & Bigger (1977)
Rb3H	= 4.18	Lyon & Jarvis (1979)

[a] Gropp, Winking *et al.* (1972).
[b] Capanna, Gropp *et al.* (1976).
[c] Winking *et al.* (in prep).
[d] v. Lehmann & Radbruch (1977).

TABLE II

Rb metacentrics found in polymetacentric feral mice trapped in Rhaeto-Lombardian regions.

Region of origin	Arm composition																				Reference
	$\frac{1}{3}$	$\frac{4}{6}$	$\frac{5}{15}$	$\frac{11}{13}$	$\frac{8}{12}$	$\frac{9}{14}$	$\frac{16}{17}$	$\frac{2}{8}$	$\frac{10}{12}$	$\frac{7}{18}$	$\frac{2}{4}$	$\frac{3}{6}$	$\frac{7}{8}$	$\frac{3}{4}$	$\frac{6}{7}$	$\frac{4}{12}$	$\frac{10}{11}$	$\frac{2}{14}$	$\frac{13}{16}$	$\frac{1}{6}$	
Poschiavo	×	×	×	×	×	×	×														Gropp, Tettenborn *et al.* (1970)
Upper Valtellina I	×	×	×	×	×	×	×														Winking *et al.* (in prep.)
Upper Valtellina II	×	×	×	×		×	×	×	×												Winking *et al.* (in prep.)
Bergamo	×	×	×	×		×	×	×	×	×											Winking *et al.* (in prep.)
Cremona			×	×		×	×	×	×	×				×						×	Winking *et al.* (in prep.)
Milano I			×	×		×	×		×		×	×	×								Winking *et al.* (in prep.)
Milano II			×	×		×	×	×	×					×	×						Winking *et al.* (in prep.)
Roveredo	×												×			×	×	×	×		Gropp, Winking *et al.* (1972)
Chiavenna							×									×	×				Gropp, Winking *et al.* (1972)

TABLE IIIa

Frequency of involvement of autosomal acrocentric components in the formation of Rb metacentrics

Acrocentric homologue	No. of times occurring in	
	Rb of feral origin	Rb of laboratory origin
1	10	1
2	11	—
3	11	—
4	12	1
5	12	1
6	12	2
7	7	—
8	13	2
9	12	1
10	13	—
11	11	1
12	13	1
13	13	2
14	13	—
15	12	2
16	12	1
17	11	1
18	4	1
19	—	3

TABLE IIIb

Frequency of occurrence of metacentrics with identical arm composition in different populations of the house mouse.

Rb	No. of times occurring
1.2	2
1.3	3
1.6	1
1.7	1
1.10	1
1.11	1
1.18	1
2.4	1
2.8	4
2.14	1
2.15	1
2.17	1
2.18	1
3.4	3
3.6	1
3.8	1
3.9	2
3.13	1
4.6	3
4.11	1
4.12	1
4.13	1
4.15	1
4.17	1
5.13	2
5.14	1
5.15	8
5.17	1
6.7	2
6.12	2
6.13	1
6.16	2
7.8	2
7.18	2
8.12	2
8.14	2
8.17	2
9.13	1
9.14	6
9.16	3

TABLE IIIb. Continued

Rb	No. of times occurring
10.11	2
10.12	7
10.14	2
10.15	1
11.13	6
11.15	1
12.14	1
13.16	1
16.17	6

TABLE IV

Robertsonian metacentric chromosomes of feral and of laboratory origin.

I *Feral origin*	
Rb(1.3)1Bnr	HAR; JAX; LUB and others
Rb(4.6)2Bnr	HAR; JAX; LUB and others
Rb(5.15)3Bnr	HAR; JAX; LUB and others
Rb(11.13)4Bnr	HAR; JAX; LUB and others
Rb(8.12)5Bnr	JAX; LUB
Rb(9.14)6Bnr	HAR; JAX; LUB and others
Rb(16.17)7Bnr	HAR; JAX; LUB and others
Rb(10.11)8Bnr	HAR; JAX; LUB and others
Rb(4.12)9Bnr	LUB
Rb(1.10)10Bnr	HAR; JAX; LUB and others
Rb(3.8)2Rma	LUB
Rb(6.13)3Rma	HAR; LUB
Rb(4.15)4Rma	HAR; LUB
Rb(10.11)5Rma	LUB
Rb(2.18)6Rma	HAR; LUB
Rb(5.17)7Rma	LUB
Rb(9.16)9Rma	LUB
Rb(1.18)10Rma	LUB
Rb(2.17)11Rma	LUB
Rb(6.7)13Rma	LUB
Rb(8.14)16Rma	LUB
Rb(4.17)13Lub	LUB
Rb(7.8)28Lub	LUB
Rb(16.17)32Lub	LUB
Rb(6.12)3Sic	LUB
Rb(8.17)6Sic	LUB
II *Laboratory origin*	
Rb(6.15)1Ald	HAR; JAX; LUB and others
Rb(9.19)163H	HAR; JAX; LUB and others
Rb(8.17)1Iem	LEN; LUB
Rb(5.19)1Wh	HAR; JAX
Rb(8.19)1Ct	HAR; LUB
Rb(1.15)2Ct	HAR
Rb(11.16)3Ct	HAR
Rb(6.13)1H	HAR
Rb(11.16)2H	HAR
Rb(4.18)3H	HAR

HAR = MRC Radiobiological Research Unit, Harwell, Didcot, Oxfordshire OXII ORD, UK.

JAX = The Jackson Laboratory, Bar Harbor, Maine 04609, USA.

LUB = Medizinische Hochschule Lübeck, D-2400 Lübeck, West Germany.

LEN = Institute for Experimental Medicine, Acad. Med. Sci. USSR, Leningrad, USSR.

TABLE V

Rb metacentric chromosomes introduced by selective breeding through at least five, mostly more than ten generations into C57BL/6 (Work done by H. Winking, J. Uckert, and A. Gropp.)

Rb(1.3)1Bnr	Rb(16.17)7Bnr[a]	Rb(3.8)2Rma
Rb(8.12)5Bnr		Rb(4.15)4Rma
Rb(4.12)9Bnr		Rb(10.11)5Rma
Rb(1.10)10Bnr		Rb(9.16)9Rma

[a] Originally established by J. Klein.

TABLE VI

Malsegregation rates in male single Rb metacentric heterozygotes

Designation (arm composition)	Chromosome arm counts in male meiotic metaphase II (%)					Non-disjunction rates calculated on the base of > 20 × 2	No. of M II cells scored
	< 19	19	20	21	> 21		
I *Feral origin*							
Rb(1.3)1Bnr/+		8	85	7		14	400
Rb(4.6)2Bnr/+		10	79	11		22	300
Rb(5.15)3Bnr/+		13	73	14		28	300
Rb(11.13)4Bnr/+		12	74	14		28	400
Rb(8.12)5Bnr/+	2	3	93	2		4	300
Rb(9.14)6Bnr/+	1	5	89	5		10	300
Rb(16.17)7Bnr/+	1	2	95	2		4	300
Rb(10.11)8Bnr/+		4	95	1		2	600
Rb(4.12)9Bnr/+		5	91	4		8	400
Rb(1.10)10Bnr/+		2	96	2		4	400
Rb(3.8)2Rma/+	0.5	9	81.5	9		18	200
Rb(6.13)3Rma/+		4.3	92.3	3.3		6.6	300
Rb(4.15)4Rma/+		11	78	11		22	300
Rb(10.11)5Rma/+		6.6	86.2	7.2		14.4	600
Rb(2.18)6Rma/+		7	84.7	8.3		16.3	300
Rb(9.16)9Rma/+		11.7	80	8.3		16.6	300
II *Laboratory origin*							
Rb(9.19)163H/+		3	94	3		6	300
Rb(6.15)1Ald/+		8	89	2		4	600
Rb(8.17)1Iem/+		6	92.5	1.5		3	300

TABLE VII

Meiotic anaphase I malsegregation rates of male heterozygotes with metacentrics of identical arm compostion but different geographical origin.

Metacentric homo- or heterozygote, resp.	M II chromosome arms (%)					No. of M II cells scored
	< 19	19	20	21	> 21	
Rb(10.11)8Bnr/Rb(10.11)8Bnr		1,8	97,6	0,6		500
Rb(10.11)8Bnr/+		3,7	95,0	1,3		600
Rb(10.11)5Rma/+		6,6	86,2	7,2		600
Rb(10.11)8Bnr/Rb(10.11)5Rma		1,4	98,2	0,4		500

Origins: Rb(10.11)8Bnr = Alpine region (Val Bregaglia).
Rb(10.11)5Rma = Central Italy (Apennines, Abbruzzi).

TABLE VIII

Transmission of the metacentric in balanced (hetero- versus homozygous) progeny of single metacentric male and female heterozygotes.

Male				Female		
n	% heterozygous	% homozygous	Heterozygote	% heterozygous	% homozygous	*n*
271	48	52	Rb(1:3)1Bnr/+	36	64	261
132	52	48	Rb(11.13)4Bnr/+	47	53	87
154	52	48	Rb(8.12)5Bnr/+	51	49	160
151	59	41	Rb(16.17)7Bnr/+	38	62	89
110	54	46	Rb(10.11)8Bnr/+	40	60	120
125	49	51	Rb(4.12)9Bnr/+	46	54	107
			Rb(1.10)10Bnr/+	49	51	142
93	44	56	Rb(6.13)3Rma/+	26	74	62
124	48	52	Rb(4.15)4Rma/+	34	66	56
114	43	57	Rb(8.17)1Iem/+	47	53	98

Symp. zool. Soc. Lond. (1981) No. 47, 183–204

The Breeding, Inbreeding and Management of Wild Mice

MARGARET E. WALLACE

Department of Genetics, University of Cambridge, Cambridge, UK

SYNOPSIS

Management procedures, devised over 30 years' experience with wild mice from four widely separated geographical areas, are described for breeding, inbreeding and recording under normal laboratory conditions. A particularly useful breeding cage, together with minor adjunct equipment, is also described. Three small stocks, and one large one (70–80 pairs), have been studied, from initial matings between trapped mice to the fifth generation of sib-mating. Each stock has had an overall output of young which compared well with that of the better laboratory standard inbred strains, and each has achieved this in different ways. The percentage of females which fail to breed varies from 7–48% according to stock, but is consistent between generations; this is a response to restriction of activity since the occasional use of a wheel-cage establishes pregnancy, although a laboratory cage is necessary for parturition and litter-rearing. Three to 18% of litters, according to stock, are lost at or near birth. Only about 2% of young die between birth and 15 days old. These are very low rates compared with laboratory inbred strains. Mean litter size at birth is 5.5 ± 2.0. Variation in reproductive performance and other features qualifies them as a plentiful source of new genetical material. The lack of appreciable reduction in the fitness features measured, during inbreeding to a theoretical 60% loss of heterozygosity, suggests that wild mice segregate in more overdominant loci than do laboratory mice of all kinds at present.

INTRODUCTION

It is generally thought that, between them, laboratory stocks of mice offer an enormous pool of variation. There are over 300 genetically distinct standard sib-mated lines and several non-sib-mated stocks. However, most laboratory mice, and all laboratory inbred lines except the wild ones described here, are descended from the mouse fancy. Most inbred lines and their descendant lines were developed for research into only two fields, cancer and immunology (Festing, 1969), and the fancy itself (i.e. domestication) dates from 1400 years before Christ (Keeler, 1931; Cooke, 1978). These facts suggest that the variation so far obtained is far less than would be available were wild stocks to be tapped. New fields are developing all the time (the 1979 list of references added to Jackson Laboratory's *Subject strain bibliography of inbred strains of mice* lists 18 broad fields). It is encouraging that this variation is beginning to be systematically exploited, at least in *mus castaneus* and *M. molossinus* (Morse, 1978).

Wild genomes have been sampled, probably since about the beginning of the *t*-allele saga in 1956, but breeding of pure wild and between wild and laboratory stocks has not, until recently, been taken beyond the first generation (Connor, 1975; Connor & Bellucci, 1979). The agility of wild mice and of their F_1 with laboratory animals, and their tendency to bite their handlers, have deterred further breeding, and there is also a general view that wild mice breed poorly in laboratory conditions.

About 30 years ago a small number of wild mice (Harland, 1958) came into my hands and I began to breed them for a variety of purposes. Since then, our laboratory has always contained two or three colonies of wild *Mus musculus* (*Mus domesticus*, see Marshall & Sage, this volume pp. 15–25), and I have bred them over many generations (Table I).

This chapter has two purposes: to describe the laboratory conditions, equipment, and management procedures that have contributed to our success in breeding wild mice, and to give impressions and data on their variation in terms of animal size, juvenile viability and adult fertility and fecundity, in the early stages of sib-mating. Most laboratories planning to breed wild stocks in future will be forced through lack of space to inbreed; they also may wish to do so; and in general if inbreeding is successful, other forms of breeding will be. The requirement for quarantine under the new rabies law restricts the import of laboratory strains; local wild mice could be a source of variation.

> Verily, how basely men think of this kind of cattell and hold them no better than vermine, yet are they not without certain natural properties, and those not to be despised. (Pliny.)

HOUSING, EQUIPMENT AND MANAGEMENT

Housing and Management

Our housing and caging is the same for wild mice as for laboratory stocks, plus a deep box or chute. Room temperature is 20 ± 1 °C, lower than in many other laboratories; the "snuggable" design of the cages (below) makes this possible and the temperature is nearer

TABLE I

Wild stocks held in Department of Genetics, University of Cambridge, UK

Name and origin of stock[a]	No. of founders (Date trapped)	Main type of breeding	Standard inbred symbol (short name)	No. generations[b] 1979	Present whereabouts[c]	Author's published descriptions
Peru-Atteck	2♀ 2♂ (1961)[d]	non-sib then sib	—	26 +	Jackson	Wallace (1971a, 1979b)
Peru-Harland	1♀ (× CBA♂) (1968)	sib	PUH/Cam	22	Cambridge	Wallace (1979a, 1979b)
Peru-Emaus	1♀ 1♂ (1968)	sib	—	(14)	Extinct	Wallace (1979a)
Peru-Coppock	82♀ 66♂ (1976)	sib	—	5 + 1 random	Cambridge	Wallace (1979a, 1979b); Wallace & Berry (1978)
San Franciscan	2♀ 1♂ (1959)[e]	sib	SF/Cam	43 +	Jackson	Wallace (1971a, 1976)
Israeli	2♀ (× lab♂) (1961)	sib	IS/Cam	35 +	Jackson	Wallace (1971a)
Skokholm (Pembrokeshire)	2♀ 1♂ (1962)	sib	SK/Cam	29 +	Jackson	Wallace (1971a, 1979a); MacSwiney & Wallace (1978)
Plant Breeding Institute, Cambridge	4♀ 2♂ (1971)	sib + non-sib	(PBI)	9 non-sib	Cambridge	Wallace & MacSwiney (1976, 1979); MacSwiney & Wallace (1978)

[a] "Perus" mentioned by other authors are mainly from the Peru-Atteck stock

[b] Bracketed figure is for number at extinction; a number with + after it is for the time when exported

[c] Jackson = The Jackson Laboratory, Bar Harbor, Maine, USA. Cambridge = Dept. Genetics, Cambridge, UK. PBI is available to latter.

[d] Seven altogether left descendants early in breeding, but three did not contribute to main colony.

[e] Acquired when already inbred.

natural conditions for most wild mice. The ventilation system (15 changes per hour) makes a continuous low background noise; this deadens sudden loud noises, which would otherwise seriously disturb the day-time sleep and nursing. Each room has its individual smell, but the level in the wild mouse rooms is lower than in those used for laboratory mice: wild mice produce drier faeces and they lack the polyuria characterizing some of our strains of larger mice. The six-month quarantine imposed before the Peru-Coppock mice (see Table I) were obtained was beneficial because it allowed containment and eradication of a favus infection brought in by some of the trapped mice; most wild mice have a high level of internal parasitism, but none of the other wild stocks, which were not isolated, brought in any infection or contagion new to the usual (low) level tolerated under our "conventional" conditions.

All management procedures are the same as for laboratory mice, except for the use of a chute (below). Cage bottoms and bedding are changed, food topped up, and records made, in one operation weekly; water is topped up before and after the weekend, and bottles are cleaned monthly. Weekend work is required only for special procedures, like observing vaginal plugs. Woodwool is given to pregnant mice, and the nests (which are never soiled, see below) are transferred at the weekly cleaning until the young begin to leave them. Records are made according to a system popularly known as the "Cambridge System" (described in Wallace, 1971b, and elsewhere). Adults and young are handled at the weekly recordings, and no gloves are used for any procedures so far needed: ear-clipping, injections and the taking of saliva and urine samples. Handlers are bitten very rarely: hand-held mice attempt to bite less often than when forceps-held. Gloves were used during quarantine as a precaution, but were found to slow down the daily pipette feeding necessary at first to dose the favus-infected mice. Rooms and cages must be inspected before introducing wild mice, as some are very small compared with laboratory mice and can escape through very small holes and cracks.

Large Box or Chute

The agility and hand-biting tendencies of wild mice are maintained during inbreeding for at least 10 generations, despite inevitable selection against them even in our roomy cage and chute. When a cage is to be opened for inspection or changing, it is put into a smooth-sided box at least 45 cm high, with a floor area about $33\ cm^2$, and the cage-lid removed. The mice are encouraged to vacate the cage, which is then removed, as freedom to run and jump inside the box may be important during adaptation to the laboratory; each mouse is caught by hand, inspected while held by the tail with the forefeet grasping the cage lid or the handler's sleeve, and returned by the small chink available in our cage when the lid is slightly raised.

A chute was developed from the box when speedy cage-changing was essential (Wallace, 1968). This obviates catching the mice and is tailored to our cage design (below). In conjunction with vacuum-cleaning the cages, it brings the time for transfer of mice, cage-cleaning and replenishing food, from 1.5 min down to 1.0 min per cage, as short a time as is needed for laboratory mice without this apparatus (Wallace & Hudson, 1969).

Wheel-cage

The fairly high frequency of females failing to breed (below) is probably due to exercise restriction. During the first two generations of inbreeding, some pairs which had failed to breed within two months of mating were put in a cage with an exercise wheel; this usually resulted in pregnancy, but as it was not routinely done, the frequency of success is unknown. The pairs were then transferred to our ordinary cages, as the wheel cages were designed before our ethological studies (below) and were unsuitable for litter-rearing: either the females ate their young, or they carried them to the wheels with some nesting material and then whirled them all over the cage floor.

Breeding Cages

Our cages, popularly known as the Cambridge or Wallace cage, were designed for breeding and storing laboratory mice (Wallace, 1963, 1965b; and Fig. 1); they have unique features arising from our ethological studies. Our laboratory staff did a small comparative study with current designs and found a greater breeding potential for our cages. They are also pleasanter and easier to clean because the bedding is drier. The design ensures that faeces and urine are deposited only in the well-ventilated area, and that the mice do not build bedding up to the bottle spout so that it siphons the water out; the spout itself leaks less if jolted while being moved than is so with other designs. This dryness also reduces smell so that with weekly cleaning the level is no more than in laboratories with twice-weekly cleaning.

We consider our trouble-free experience of breeding wild mice to be mainly due to intelligent and gentle husbandry by technicians, and to the unique features of the cage design. Connor (1975) used a conventional "shoebox" cage in his study of wild mice and five out of ten sib-mated lines were lost by the seventh generation; Lynch (1977) lost seven out of ten by the fourth generation. The loss in Peru-Coppock, calculated similarly to Lynch's, works out at only 10% loss. Differences in genomes and management are unlikely to account for the whole of this contrast.

Wild mice build better nests than laboratory ones. Both wild and laboratory mice maintain a high nest temperature in our cage, which has a "snuggable" nest area, used also for sleeping. (See Jakobson, this volume pp. 301–335, for nesting and huddling as responses to cold.) This "snuggable" area is achieved by a close-fitting shelter over a bottom shallow enough for mice to build bedding easily up to the

Fig. 1. The "Cambridge" or "Wallace" cage. Internal dimensions: floor area 26.4 cm × 21.75 cm, height 7.75 cm. Note the nest under the solid shelter on the top, egress at the lower left side, sawdust-free area under the $\frac{1}{2}$ pint bottle, and exercise area under the open bars of the lid. This version is obtainable from Cope and Cope Ltd, 57 Vastern Road, Reading, Berks, UK. A similar version with polycarbonate as well as polypropylene base (27.5 cm × 20.5 cm × 10.0 cm.) is obtainable from Philip Harris Biological Ltd, Oldmixon, Weston-Super-Mare, Avon, UK.

lower edge of the shelter. This makes a long low draught-free tunnel open at only one end. The mice open up the nest as the young grow. The siting of the bottle and shelter ensures that water and food can be obtained only on the side opposite the nest, so that in turn faeces and urine are deposited only on that (well-ventilated) side. This siting also ensures that egress from the nest area is under the bottle so that the mice themselves keep the spout clear of loose bedding. Given a suitable design, mice exhibit hygienic behaviour.

The shallowness of the Cambridge cage allows it to have a larger floor area for the same volume as other cages. Wild mice use the large ventilated area for exercise, looping the loop or running up and down it. The repetition of particular patterns of exercise, the tenddency to chew the bars, and the habit of some wild mice of crumbling the pelleted diet without eating it, seem to be a response to restriction of activity, and underline the need for this to be allowed for in the cage design. Wheat is fed weekly to the wild mice, as it is to the laboratory ones. This does not lessen appreciably the diet-wasting, but it provides a harder surface to chew than pelleted diet, and may be beneficial in other ways: the mice eat it all within a day. Further design details are given in the references mentioned above.

VARIABILITY

General

Over the past few decades several measures of variability have been used in the study of inbred laboratory strains and of wild-caught populations to assess heterozygosity and population dynamics (see chapters in this volume by Berry, Peters and Lush). Only a few loci of the latest measure of these, biochemical polymorphism, were available for our earlier wild laboratory-bred stocks (Table I and Wallace, 1971a), but these stocks have since been used as sources of new alleles and new loci. Probably the largest number of loci used simultaneously to date is 68 (Berry, personal communication). Forty-nine loci were used in the study of 216 trapped Peru-Coppock mice (Berry & Peters, 1981). This large sample was caught at eight or nine different sites along 20 km of a river valley in Peru; a high level of heterozygosity per locus was found; and the allele frequencies showed that the sites are parts of a large interbreeding population. The frequency of rare variants compares well with that in other wild mice and in man (Wallace & Berry, 1978). For those of our wild stocks now held at the Jackson Laboratory, Bar Harbor, Maine,

USA (Table I), a list of references for variations found by different laboratories is maintained.

These studies are ample evidence that wild mice provide a good source of new variation. Impressions gained by their handlers during inbreeding in Cambridge follow.

Appearance

The stocks differ from each other enough for their handlers, given a specimen at random, to recognize from which stock it came. Colour varies (mainly in the ratio of black to yellow pigment along the hairs, dorsally and ventrally), size varies (e.g. Skokholm and PBI are about 30% bigger than the Peruvian mice), and conformation varies (tail/body length etc.; see Berry & Jakobson, 1975). Large size and agility are said to be physiologically correlated (Falconer, 1953) but all the "Perus" (Table I) are small and agile, while the larger PBI mice are also agile. Conformation and size are probably correlated with predation: the Skokholm mice, with no ground predators and few aerial ones, are larger than the Perus, and have fairly small pinnae and normally set eyes, whereas the Perus, under heavy predation by cats, are small with large thin pinnae and very prominent eyes.

Behaviour

Our wild mice are aggressive to each other. Males must be stored together before sexual maturity but thereafter Peru males must be thinned to twos and threes as they fight whenever the cages are cleaned, while sires may attack young of eight or more days in a clean cage. This is probably a response to the disruption of smell-mediated recognition of individuals and of territorial limits, following the change in bedding: transference of some soiled bedding can help. Peru females occasionally attack even their laboratory mate. Resident females in harems have sometimes killed a returned female if still lactating. PBI mice, more than the others, rattle their tails when confronting each other, and when being caught by their handler. Other behaviours vary also: the Perus are particularly jumpy when a thunderstorm is impending, and some stocks respond differently from laboratory ones to stress involved in transit (Wallace, 1976). Connor's (1975) study describes variation in six components of "wild" behaviour.

Size and Breeding Span

Birth weight in Perus is 1.2–1.5 g. Most Perus do not thrive if weaned at the usual 20–23 days; mice from sites where the ancestral wild-caught mice were small are left with their parents (and the new litter) until 5–9 weeks old (9–13 g), or where the trapped ancestors were large, until 5–6 weeks old (11–15 g); the average is 8 weeks and 12 g, in line with Pelikán's observations on trapped mice. Weights in Perus are maintained through the generations: adults which were small and slow-growing have young which are small as adults, and slow-growing. This is not necessarily hereditary, as we have been able to break a long run of small or poor-breeding San Franciscan mice (very inbred) by fostering one generation onto good laboratory mothers. Most female Perus are mated within two weeks after weaning, and first litters are born at an average of 3 months old; matings are usually judged infertile if no litter is due within 2 months of mating. Last sizeable litters are born at about 40 weeks, matings being terminated when reproductive performance falls. Despite later maturing, the fertility span compares quite well with that of our CBA/FaCam; it is about 6½ months for Perus (9½ months from birth) and 7½ months for CBA/FaCam (9½ months from birth). Our other (larger) wild mice mature earlier, but (when not killed earlier, see below), cease good breeding at about the same age, 9½ months. Bellamy's finding (this volume p. 287) that his Skokholm mice matured late is not unexpected in view of the small founder numbers of our two stocks, and underlines the heterogeneity in wild stocks.

All our wild-caught mice and their descendants breed evenly throughout the year, giving about one litter per month. This is probably a response to the even temperature, since our daylight length is not controlled (and see Berry & Jakobson, 1975; Pelikán, this volume pp. 205–229).

FERTILITY, FECUNDITY AND MORTALITY

Stocks and Breeding Policy

The stocks selected for this study were all sib-mated for four or more generations. Sixty per cent of the initial heterozygosity would in theory be lost during this time.

Three small stocks were chosen, since this would show how much could be achieved in a small space; these stocks happened to consist of some large mice (Skokholm and PBI, adult weight about 21 g and 19 g respectively) and some small (Peru-Emaus); no mutants were produced during inbreeding of any of them. One large stock (Peru-Coppock) was chosen since the larger amount of data would allow certain distinctions to be made with more confidence; this happened to consist only of small mice (adult weight about 17 g, the same weight as Peru-Emaus), and was unusual in producing a high number of mutants and containing mice with evidence of chromosome damage (Wallace & Berry, 1978; Wallace, in preparation).

All four stocks were bred for purposes other than the study of wild mice *per se,* so that the vital statistics given here happen to be those available within the limits of breeding policy. Mice in the small stocks were mated at sexual maturity in pairs (and trios in the early generations), and the mates kept together all the time. There were between 5 and 20 females breeding per generation. Young were killed at 15–18 days or left with the parents till weaning at about 22 days or more, and very little culling was done in the nest; matings were usually terminated soon after they produced about 20 young, in order to have several generations a year. There are few data therefore on breeding span. Sublines were split to replace inferior sublines.

The large stock, Peru-Coppock, started with slightly more than 10 pairs from each of eight sites and ended with only slightly fewer; only 10% were terminated through breeding problems. It was run in the same way as the small ones, except that culling was introduced after 15 days where mutants segregated, so as to reduce litter competition; also, although matings were never terminated before 20 progeny were born (some self-terminated earlier by death), the large majority of matings were allowed to breed until, as stated above, output began to fall off. There is thus a large amount of data on breeding span.

There are no data on life-span, but individual mice of all four stocks have been killed when at least a year old.

As the killing policy varied after 15 days, this is the age up to which in-litter mortality is measured. Most in-litter mortality occurs before 15 days when natural weaning onto solid food starts; mortality after this age probably has a different range of environmental causes from that before it.

Where a pair had not started to breed two months after mating, duplicate matings were made until either the original or the duplicate bred. Thus, only where the frequency of such matings was high, was there strong selection for fertility. No artificial selection was exerted in respect of whole-litter and in-litter mortality or litter size at birth. Owing to the breeding policy, there was a tendency to select for early sexual maturity in the three small stocks, but no special tendency in the large. The non-breeding females' nil outputs are excluded for calculations of per-female output, whole-litter and in-litter mortality, and litter size. The nil contribution of whole litters lost is also excluded from calculations of litter size at birth and in-litter mortality. The measures were kept separate so that if some features were environmentally rather than genetically controlled, this could be discerned from the data.

Preliminary Analysis

The data for fertility, fecundity and mortality for each stock, were collected into separate generations (and separate sites in Peru-Coppock). There was no heterogeneity between generations within each of the three small stocks, except for a slight but insignificant decrease in litter size in PBI mice. Matings between wild-caught mice of unknown age and relationship (generation 0), and the succeeding four sib-mated generations, are therefore pooled for the tables below. There was no heterogeneity between sites in the Peru-Coppock data, except for the small variation in percentage of non-breeding females mentioned above. Data for sites are pooled for the tables. Heterogeneity between generations 0, 1 and 2–4 was often important in the Peru-Coppock data but insignificant within generations 2–4, so the data are pooled in groupings 0, 1 and 2–4.

The Peru-Coppock data for generation 0 were then divided into litter sizes: there were no correlations between litter size and whole-litter loss or between litter size and in-litter mortality. This was also so for generation 1 and for generations 2–4. Generation 1 site 1 was then examined, and there was no correlation between age of mother and in-litter mortality, but there was a slight tendency for younger females to lose more whole litters than older ones. As this turned out not to be very marked in other sites and generations, and as generations 1–4 span the same ages, the data for different maternal ages are pooled in the tables.

Main trends

Fertility

The proportion of females which did not breed is a basic measure of fertility. There is probably only a very small component of male infertility, since males were occasionally mated polygamously or with a laboratory female, and very rarely was one infertile. Table II presents the percentage of infertile females per stock; the low rate for the Peru-Coppock mice and very significant heterogeneity

TABLE II

Fertility: number of ♀♀ which bred versus number of ♀♀ which did not breed

	Peru-Coppock (no. ♀♀ bred)					Other wild stocks (gens. 0–4) (no. ♀♀ bred)			
Gens.	Yes	No	Total	% ♀♀ infertile	Stock	Yes	No	Total	% ♀♀ infertile
0	120	16	136	11.76	Skokholm	34	31	65	47.7
1	98	3	101	2.97	Peru-Emaus	44	23	67	34.3
2–4	360	51	411	16.40	P.B.I.	108	8	116	7.0
Totals	578	70	648	10.80	Totals	186	62	248	25.0
Heterogeneity χ^2 (d.f.)			11.633(2)	($p < 0.01$)	Heterogeneity χ^2 (d.f.)			42.24(2)	($p \ll 0.001$)

between the other stocks indicate a characteristic rate for each stock. The Peru-Coppock stock also shows heterogeneity between generations 0, 1 and 2–4. The rate for 2–4 would be as low as in generation 0 if three exceptional very polygamous matings, where the male could have been infertile, were to be excluded. The rates for the two Peruvian stocks are markedly different.

Where individual progeny have low viability, their deaths are not likely to cause the mothers to lose the whole litter. Whole-litter losses, as opposed to part-litter losses, are therefore regarded as due to maternal failure of some kind: in initiating or maintaining lactation, for example, or defective maternal behavioural factors. Progeny inviability is likely to be confused with maternal failure only where a single mouse is born. Since young mothers tend to lose more whole litters than older ones, and more small litters are produced from younger mothers, the loss of litters of a single mouse is more likely to be due to maternal than to progeny features, so such losses are counted as whole-litter ones. Data on whole-litter loss are given in Table III. There is a smaller rate of litter loss than of female infertility throughout, and the rates for Perus are not clearly lower than those for the other two stocks. The significant heterogeneity between generations within the Peru-Coppock mice suggests an upwards trend: however, there is no significant difference between the loss rate in generation 1 and that in the later generations, whereas the difference between generations 0 and 1 is significant. There is clearly a discrete jump from generation 0 to generation 1.

Fecundity

Litter size at or shortly after birth is taken as a measure of fecundity, since this term does not discriminate between factors which cannot directly be distinguished from the data available. Such factors are ovulation rate, which is a maternal feature, and pre-natal mortality of progeny, which may be due to uterine defects or to viability features of individual progeny. The data for litter size are given in Table IV. These data do not take into account the trend for smaller litters at the beginning and end of the breeding span, but although the three small stocks were terminated partway through their span, their data are not thought to be biased as to size.

The standard deviations indicate homogeneity between generations in the Peru-Coppock mice, and homogeneity between the other three stocks. An overall average is 5.5 young per litter.

Mortality

The mortality rate of young between birth and 15 days derives from individual genotypes' intrinsic viability and their response to

TABLE III

Fertility: number of whole litters which lived 15 + days versus number of whole litters which died.

Peru-Coppock (no. whole litters)					Other wild stocks (gens. 0–4) (no. whole litters)				
Gens.	Lived	Died	Total	% died	Stock	Lived	Died	Total	% died
0	583	11	594	1.85	Skokholm	102	23	125	18.40
1	675	40	715	5.59	Peru-Emaus	128	4	132	3.03
2–4	2206	174	2379	7.89	P.B.I.	127	8	135	5.93
Totals	3463	225	3688	6.10	Totals	357	35	392	8.93
Heterogeneity χ^2 (d.f.)	22.9(2)	($p \ll 0.001$)			Heterogeneity χ^2 (d.f.)	20.61(2)	($p \ll 0.001$)		

TABLE IV

Fecundity: mean litter size at birth

Peru-Coppock			Other wild stocks (gens. 0–4)		
Gens.	Mean	s.d.	Stock	Mean	s.d.
0	5.47	2.02	Skokholm	5.82	2.29
1	5.48	2.02	Peru-Emaus	6.00	1.62
2–4	5.32	2.02	P.B.I.	4.89	2.00
Overall	5.38	2.02			
Range	1–14		Range	1–11	

competition with litter mates. Since no correlation was found between in-litter loss and litter size, the predominant factor is individual genotype. The data thus concern the genetics of the young: they are given in Table V.

There is very significant heterogeneity between generations in the Peru-Coppock mice and a just significant heterogeneity between the other wild stocks. Also (not shown), the loss in generations 2–4 in the former is subsignificantly higher than is the loss in the other stock with the high rate (Skokholm), and there is a significant difference between the average loss for Peru-Coppock and the average loss for the other three.

Interpretation and Comparison with Laboratory Stocks

Fertility

The fact that each of the four stocks in Table II has a characteristic proportion of non-breeding females argues a hereditary basis for fertility. The facts that there is no generation trend in the three small stocks, and that the heterogeneity between generations in the Peru-Coppock stock is not correlated with inbreeding, suggest that the loci concerned tend not to become homozygous. The markedly different rate between the two Peruvian stocks argues against the hereditary basis being determined principally by weight, for these mice are indistinguishable in body size. The heterogeneity between generations in the Peru-Coppock stock suggests a non-genetical explanation as follows: the less fertile mice tended to be the smaller, more retarded ones, and these were descended from the small (usually favus-infected) trapped mice. It seems that small retarded mice fail to breed, and those that do breed, provide a poor maternal environment which in turn produces small retarded mice.

This is not to deny an environmental component other than nurture interacting with the genotype: the rate of infertility under

TABLE V

Mortality of young: number of young which lived 15 + days versus number of young which died before 15 days

Peru-Coppock (no. young)					Other wild stocks (gens. 0–4) (no. young)				
Gens.	Lived	Died	Total	% died	Stock	Lived	Died	Total	% died
0	3178	31	3209	0.96	Skokholm	586	13	599	2.17
1	3660	95	3755	2.53	Peru-Emaus	892	8	900	0.89
2–4	11509	425	11934	3.56	P.B.I.	1847	31	1878	1.65
Totals	18347	551	18898	2.92	Totals	3325	52	3377	1.54
Heterogeneity χ^2 (d.f.)	62.11(2)	($p \ll 0.001$)			Heterogeneity χ^2 (d.f.)	~7.13(2)	($p < 0.05$)		

cage restriction is greater, perhaps a lot greater, than the rate in the wild, as the use of the wheel-cage shows. Possibly fertility in a cage is partly due to "placidity", in which case it could be an "unfit" character in a predated environment. Or the genotypic component may be related to buffering against stress; the Peru-Coppock mice, with high fertility, were preyed upon by cats – one may guess that they are buffered against the stress of this and so also against the stress of cage confinement. Runner's finding on implantation loss (quoted in Falconer, 1960: 61) suggests this idea. If true, Festing's (1968) figure of 19/360 (5.3%) infertile females for the commoner laboratory strains suggests that some well-domesticated mice retain this response to cage restriction.

There do not appear to be comparable data on infertility rates for individual laboratory inbred strains.

Turning to whole-litter losses in Table III, here again there seems to be a rate characteristic of each stock, arguing a hereditary basis; and again there is homogeneity between generations within the three small stocks, arguing either no initial heterozygosity or a delay in reaching it. The heterogeneity between generations in the Peru-Coppock mice does not necessarily argue genetical heterogeneity, although the rate increases with inbreeding. It has been stated that greater litter loss is correlated with younger mothers. As the trapped wild mice breeding in generation 0 were probably not all young when mated like those in later generations, the greater loss in youth has been relatively excluded from their data, so this could entirely explain the big difference between the small loss in generation 0 and later larger losses. Moreover, there cannot be much genetical heterogeneity since the females of generations 0 and 1 are similar in not being offspring of sib mated parents, unlike those of generations 2–4, yet they are dissimilar in rate of loss. Again, weight does not appear to be important since the heavier mice (Skokholm and PBI) have very different loss rates. An interaction between cage environment and genotype in terms of buffering against stress suggests itself, but it cannot be controlled in the same way because the Peru-Emaus stock has a low litter loss and a higher rate of infertility. The loss rates in all four stocks are not high compared with some of the commoner laboratory strains. These have been measured in terms of a "litter index" (Festing, 1968) and work out as a 3% loss in the best (A2G) and 43% in the worst (C57BR). One is tempted again to question the assumption that laboratory strains are universally adapted to a cage environment.

The hereditary differences *between* the wild stocks in whole-litter loss may or may not relate to their wild environment, and the explanation cannot be found without comparative studies in the field.

Fecundity

The litter sizes of about 5.5 young around birth in Table IV are smaller than the sizes for embryos seen in females near term in trapped mice (see Pelikán, this volume pp. 205–209). Mortality exactly at birth has not been measured.

The homogeneity of litter size between stocks seems to contradict the general finding that large mice have large litters (Falconer, 1960). The explanation may be that each stock has its optimum adult weight and correlated litter size, and that mice which are large for a given stock have large litters. Batten & Berry (1967) found that Falconer's correlation disappeared for large wild mice; optimum weight and litter size are probably controlled by the many factors which balance deaths before reproduction with the numbers needed for replacement. The homogeneity of litter size between stocks and between generations gives no clue as to how size is controlled in these stocks.

The homogeneity between generations is very surprising in view of the common experience with laboratory mice of decrease in litter size on inbreeding. It should be re-emphasized that the lack of any decrease in size on inbreeding applies to each of the eight sites in the Peru-Coppock mice. In the largest laboratory study on this feature (Falconer, 1960), 20 lines were carried on by sib-mating until they extinguished themselves by producing only one sex of offspring; three survived to 11 generations and one to 28 or more generations. All except the last three declined in litter size by about 0.5 young for every 10% increase of the inbreeding coefficient. On this basis, the initial Peru-Coppock litter size of 5.47, should, in the second to fourth generations, have dropped to an average of 3.02, whereas it only fell to 5.32.

The relative lack of inbreeding depression in the wild mice suggests yet again that progress towards homozygosity is retarded. It follows that fitness in wild mice is largely controlled by loci with over-dominance. The fact that three of the laboratory lines survived long inbreeding without depression shows that selection can favour maintenance of litter size, a conclusion amply borne out be a selection experiment (Falconer, 1960); this suggests that the four wild stocks have also survived because of intense selection.

The almost universal prevalence of inbreeding depression in laboratory mice may well be due to their having been divorced from the wild genome so long (thousands of generations) without much selection, or that they contained very few overdominant loci when inbreeding started. This has theoretical implications (Wallace, 1965a; Gilmour & Morton, 1971).

If this is correct, there is great improvement in litter size and other features of fitness to be obtained by introducing wild genomes and then selecting before inbreeding.

Lynch's small number of inbred mice showed a decline in litter size at weaning, but this may include certain nil contributions excluded in the present study; however her data support the idea of resistance to homozygosity (C. B. Lynch, personal communications). Connor & Bellucci (1979) found, during the first four generations of sibmating, less depression than expected on a no-overdominance model, both in litter size and in mortality to 30 days. This again supports the idea of overdominance retarding homozygosity.

Litter sizes vary in inbred laboratory mice in general. Taking as a representative sample the nine strains most commonly used for cancer and immunology (Festing, 1968) the range is from a mean of 7.6 young (C57BR) to 5.0 (NZW). These are slightly biased towards large size as compared with the four wild stocks, as the former included only the first four litters from animals which produced at least four, whereas the latter includes litters from all females who only had one, two and three litters (before dying or ceasing to breed). Again the four wild stocks, with 5.5 young per litter, compare well for practical and maintenance purposes.

The extremes in the range of sizes in all four stocks (Table IV) suggest that wild mice may occasionally produce litters almost as large as those recorded sporadically for laboratory stocks.

Mortality

The statistical heterogeneity between the three small stocks in Table V suggests a genotypic basis of some kind. The large increase in loss in the Peru-Coppock stock from generation 0 to generation 1 excludes a relation with the mother's genotype. The almost threefold increase in progeny mortality here, and the higher rate in generations 2–4, is as expected on the basis of a large number of lethals segregating in the wild population and becoming homozygous on inbreeding. The significantly higher mortality in Peru-Coppock than in the other stocks suggests there are more lethals segregating in Peru-Coppock than in the others. This is consistent with the finding, stated earlier, of a larger number of mutants here than in any other wild stock. The unusual nature of the Peru-Coppock stock precludes it from comparisons with laboratory stocks concerning inbreeding depression.

In the commonly used laboratory inbred strains (Festing, 1968), in-litter mortality is biased slightly in favour of lower mortality. It varies from 0.7 (DBA/2) to 3.9 (C57BR) per litter. This works out, for the measure used above in the wild stocks, as an overall mortality of 13% and 51% respectively. These are very high rates compared with the wild figures (about 2%). To reach the lowest laboratory figure, mortality in wild stocks would have to be at least five times as great for the period 15 days to weaning as it is for the period measured, 0–15 days. Possibly a high degree of progeny fitness due to over-dominance raises the threshold at which mice with poor viability die.

The relative freedom of wild stocks from in-litter mortality is surprising in view of the fact that they are not adapted to cage restriction. It encourages the view expressed above that the introduction of wild genomes to laboratory ones would make for improvement in reproductive performance.

Overall Production by Wild Mice

As stated under "Behaviour", above, the Peru-Coppock mice bred for about 40 weeks after mating at nine weeks (i.e. for 31 weeks). The average total production per female is about 40; this works out at a rate of 1.3 mice per female per week. For reasons given under "Stocks and Breeding Policy", the other wild stocks cannot be compared with laboratory inbred strains. In Festing's (1968) study, where breeders were replaced every 28 weeks from the mating date, the number of weaned young per female per week works out as 1.06 for CBA/CafCFW/Lac, which is a fairly representative strain. The Peru-Coppock females, which were killed only after breeding had already fallen off, do rather better.

It can be inferred from the statistics quoted above for laboratory strains that although all are acceptably productive overall, each achieves this by the combination of poor figures for some aspects of production and good figures for others, each strain differing from the others as to which aspects have good and poor figures. The four wild stocks have also been shown above to vary greatly in different aspects of productivity.

It can be concluded in general that there is no reason to expect wild stocks to be difficult to breed or inbreed, given a good environment, and that they vary greatly in different aspects of breeding. It is worth studying these aspects when setting up a particular stock so as to devise the best management procedures. For example, the Skokholm and Peru-Emaus stocks, with their high percentage of infertile females, are probably best mated in trios with subsequent killing of the less fertile female. Other practical procedures given here are those discovered by experience in the early stages of establishing the four wild stocks.

Incidentally, our experience of wild × laboratory crosses is that the hybrids are very large and prolific, and that "wildness" in terms of handling remains even after several back-crosses to the laboratory mice. This subject needs more study before generalizations can be made.

REFERENCES

Batten, C. A. & Berry, R. J. (1967). Pre-natal mortality in wild-caught house mice. *J. Anim. Ecol.* **36**: 453–463.

Berry, R. J. (Ed.) (1981). Population dynamics of the house mouse. In *Biology of the house mouse:* 395–425. London & New York: Academic Press.

Berry, R. J. & Jakobson, M. E. (1975). Adaptation and adaptability in wild-living house mice. *J. Zool., Lond.* **176**: 391–402.

Berry, R. J. & Peters, J. (1981). Allozymic variation in house mouse populations. In *Mammalian population genetics*. Smith, M. H. & Joule, J. (Eds). Athens: University of Georgia Press.

Connor, J. L. (1975). Genetic mechanisms controlling the domestication of a wild house mouse population (*Mus musculus* L.). *J. comp. Physiol. Psychol.* **89**: 118–130.

Connor, J. L. & Belluci, M. J. (1979). Natural selection resisting inbreeding depression in captive wild house mice (*Mus musculus*). *Evolution* **33**: 929–940.

Cooke, T. (1978). *Exhibition and pet mice*. Hindhead, Surrey: Saiga.

Falconer, D. S. (1953). Selection for large and small size in mice. *J. Genet.* **51**: 407–501.

Falconer, D. S. (1960). The genetics of litter size in mice. *J. cell. Physiol.* **56**: 153–167.

Festing, M. (1968). Some aspects of reproductive performance in inbred mice. *Lab. Anim.* **2**: 89–100.

Festing, M. (1969). Inbred mice in research. *Nature, Lond.* **221**: 716.

Gilmour, D. G. & Morton, J. R. (1971). Association of genetic polymorphisms with mortality in the chicken. *Theor. appl. Genet.* **41**: 57–66.

Harland, P.S.E.G. (1958). Skeletal variation in wild house mice from Peru. *Ann. Mag. nat. Hist.* (13) **1**: 193–195.

Keeler, C. E. (1931). *The laboratory mouse: its origin, heredity and culture.* Cambridge: Harvard University Press.

Lush, I. E. (1981). Mouse pharmacogenetics. In *Biology of the house mouse:* 517–546. Berry, R. J. (Ed.). London & New York: Academic Press.

Lynch, C. B. (1977). Inbreeding effects upon animals derived from a wild population of *Mus musculus. Evolution* **31**: 526–537.

MacSwiney, F. J. & Wallace, M. E. (1978). Genetics of warfarin resistance of house mice from three different localities. *J. Hyg., Camb.* **80**: 69–75.

Morse, H. C. (Ed.). (1978). *Origins of inbred mice.* New York: Academic Press.

Pelikán, J. (1981). Patterns of reproduction in the house mouse. In *Biology of the house mouse*: 205–229. Berry, R. J. (Ed.). London & New York: Academic Press.

Peters, J. (1981). Enzyme and protein polymorphism. In *Biology of the house mouse*: 479–516. Berry, R. J. (Ed.). London & New York: Academic Press.

Wallace, M. E. (1963). Laboratory animals: Cage design principles, practice and cost. *Lab. Pract.* **12**: 354–359.

Wallace, M. E. (1965a). The relative homozygosity of inbred lines and closed colonies. *J. theor. Biol.* **9**: 93–116.

Wallace, M. E. (1965b). The Cambridge mouse cage. *J. Anim. Techns* **16**: 48–52.

Wallace, M. E. (1968). A chute for the transference of hyperactive mice during cage-cleaning procedures. *Lab. Anim. Care* **18**: 200–205.

Wallace, M. E. (1971a). An unprecedented number of mutants in a colony of wild mice. *Environ. Pollut.* **1**: 175–184.

Wallace, M. E. (1971b). *Learning genetics with mice.* London: Heinemann.

Wallace, M. E. (1976). Effects of stress due to deprivation and transport in different genotypes of house mouse. *Lab. Anim.* **10**: 335–347

Wallace, M. E. (1979a). Continued high incidence of mutants in a Peruvian population of mice. *J. Hered.* **69**: 429–430.

Wallace, M. E. (1979b). Wild Peruvian mice. *Mouse News Letter* **61**: 28–30.

Wallace, M. E. (In preparation). *An inherited agent of mutation and chromosome damage in wild mice.*

Wallace, M. E. & Berry, R. J. (1978). Excessive mutational occurrences in wild Peruvian house mice. *Mutation Res.* **53**: 282–283.

Wallace, M. E. & Hudson, C. H. (1969). Breeding and handling small wild rodents: a method study. *Lab. Anim.* **3**: 107–117.

Wallace, M. E. & MacSwiney, F. J. (1976). A major gene controlling warfarin resistance in the house mouse. *J. Hyg., Camb.* **76**: 173–181.

Wallace, M. E. & MacSwiney, F. J. (1979). An inherited mild middle-aged adiposity in wild mice. *J. Hyg., Camb.* **82**: 309–317.

Symp. zool. Soc. Lond. (1981) No. 47, 205–229

Patterns of Reproduction in the House Mouse

J. PELIKÁN

Institute of Vertebrate Zoology, Czechoslovak Academy of Sciences, Brno, Czechoslovakia

SYNOPSIS

The house mouse has been Man's faithful follower since time immemorial. The commensal union of the two took place when Man the gatherer changed into Man the farmer and started to build permanent settlements. In these the house mouse found shelter, food and other necessary conditions for survival. This occurred not only because of the general ecological adaptability of mice towards human habitations but also because the species is able to reproduce in very marginal environments.

The available data suggest that with sufficient food and shelter, ambient temperature is the factor that limits the reproductive potential of the house mouse. In this respect, the virtually cosmopolitan *Mus musculus* breeds at temperatures up to $+32^{\circ}$C (e.g. in northern Africa, Australia, the West Indies and similar warm regions); and down to -20°C in human settlements in the arctic regions and in refrigerated rooms in the London docks. This range is considerably wider than that of any other mammalian species.

To synthesize and compare the reproductive capability of *Mus musculus* on a world-wide scale is impossible at the moment, and certainly beyond the scope of this chapter. Instead, this review has been consciously limited to pointing out reproduction indices, and variations in them to which our attention should be directed for future comparison and synthesis.

INTRODUCTION

The inbred house mouse is nowadays an indispensable laboratory animal for biological studies. It has also become a model animal for studies of reproductive processes, resulting in an exceptionally copious literature. Notwithstanding, it is relevant to consider the origin, ecological requirements, and reproduction of the wild species, *Mus musculus,* a rodent which has spread from the treeless steppes and grasslands of its extensive original range (from the shores of the Mediterranean Sea through central Asia up to China and Japan) (Serafiński, 1965).

Both in their original and in their acquired ranges, house mice live commensally in buildings and ferally in fields, where they temporarily populate corn or straw ricks.

The character and intensity of breeding are closely correlated with environmental conditions. In buildings, breeding is continuous under more or less constant conditions; the breeding of most outdoor mice has a distinctly seasonal character. These differences can be quantified in terms of the onset of sexual maturity, female oestrous cycles, length of the breeding season, litter size and its variation. Breeding intensity affects the growth rate and structure of the population.

This review focuses particularly on free-living house mouse populations. The data will serve for comparison between various geographical regions and provide criteria to judge reproductive output in confined laboratory stocks.

SEXUAL MATURITY IN RELATION TO BODY WEIGHT

In population samples, sexual maturity is most easily considered in relation to body weight. Determination of individual age by eye lens weight (Berry & Truslove, 1968), tooth wear (Breakey, 1963; Lidicker, 1966), or other criteria is more difficult and often as imprecise. In addition the size of the testes and glandulae vesiculares in males is usually taken as a sign of sexual maturity (Humiński, 1969), while in females the macroscopic signs of sexual activity include open vulva, swollen uterus, presence of a vaginal plug, presence of corpora lutea in the ovaries, macroscopic embryos, and active milk glands. Ideally observations should be accompanied by histological examinations.

These points can be illustrated by material from the lowland regions of Czechoslovakia (Table I), comprising mice trapped in buildings throughout the year as well as ones caught in fields and ricks between March and November. They are all weaned individuals taken in small break-back traps outside their nests.

It is evident that such traps catch individuals 6 g and more in body weight. The first sexually active individuals are found in the weight group of 8.1–10.0 g. This fits well with the observations of Hanák (1958) who studied the growth of young in a confined commensal population. He found that newborns weigh 0.84–1.11 g and 6.0 g is reached around the age of 20 days when the young start to leave the maternal burrow. House mice weighing 8–10 g are around 37 days old. According to Zaleska-Rutczyńska (1967), young females come into oestrus for the first time at the age of 30–40 days, which accords with this weight group.

TABLE I

Numbers of sexually mature individuals in relation to body weight in mice caught in lowland Czechoslovakia

	Body weight (g)									
	6.1–8.0	8.1–10.0	10.1–12.0	12.1–14.0	14.1–16.0	16.1–18.0	18.1–20.0	20.1–22.0	22.0 <	Total
Males										
Total no.	28	45	103	139	179	181	72	21	10	778
No. mature		1	13	69	139	162	70	21	10	485
% mature		22	12.6	49.6	77.6	89.5	97.2	100.0		62.3
Females										
Total no.	22	74	84	104	138	149	135	71	61	838
No. mature		1	20	61	110	141	135	71	61	600
% mature		1.4	23.8	58.6	79.7	94.6		100.0		71.6
No. females as % of total mice	56.8		44.9	42.8[a]	43.5[a]	45.2	65.2[b]	77.2[b]	85.9[b]	51.9
% immature	56.9		41.6[a]	38.0[b]	41.2	29.6[a]				44.8[a]
% mature			60.6	46.9	44.2[a]	46.5	65.8[b]	77.2[b]	85.9[b]	55.3[b]

[a]Significant at the 0.05 probability level.
[b]Significant at the 0.01 probability level.
Data from Pelikán (1974, with supplementary material).

The values given in Table I also indicate that sexual maturity occurs earlier and increases quicker in females than in males. This has also been found in confined populations (Crowcroft & Rowe, 1961), in ricks in England (Rowe, Taylor & Chudley, 1963, 1964), in feral populations in Bulgaria (Straka, 1966), and elsewhere. From 18.0 g onwards, all females are sexually mature whereas males are so from 20.0 g onwards.

The body weight at which 50% of individuals attain sexual maturity (computed by the method of Leslie, Perry & Watson, 1945), is 12.2 g for females and 13.1 g for males. Differences in sexual maturation between mice from fields, ricks and buildings are slight and insignificant: the weights at which 50% of females are sexually mature in these three habitats are 11.9, 12.2 and 12.4 g respectively; and in males, 12.9, 13.3 and 13.5 g respectively. From these data, and also the litter sizes in these habitats (see below), it appears that the house mouse may find more favourable environmental conditions in fields than in ricks or buildings.

Sexual maturity is also influenced by ambient temperature. Newborns weigh over 1 g at higher temperatures in summer months (Reading, 1966b) and their growth rate is affected (Knudsen, 1962). An additional complication is that newborns are heavier in small litters, and from mothers pregnant for the second or third time (Reading, 1966a).

In the Czechoslovak population described in this chapter, the mean body weight of males was 14.6 g; of females, 15.8 g (without embryos); and of males plus females, 15.2 g. Among immatures, the mean weight of males was 12.1 g; of females, 11.3 g; whilst in mature animals, the mean weight of males was 16.2 g and of females 17.5 g. It should be stressed that this is a central European population, i.e., *Mus musculus musculus* Linnaeus, 1758. Weights may be significantly different in other subspecies.

SEX RATIO

Sex ratio in a population is a very complex, dynamic phenomenon, influenced by sex-specific mortality of eggs and embryos, sexual maturity, changes in population structure which depend on breeding intensity, and population density. The last factor is of utmost importance for the control of the sex ratio: high population densities are accompanied by adverse effects on the behaviour of animals, leading to the elimination of sexually active males and other stress phenomena (Parkes, 1924; Schlager & Roderick, 1968; Kozakiewicz, 1976; etc.).

Changes in Sex Ratio during Growth

The sex ratio of *Mus musculus* at implantation and during embryonic development appears to be the same as among young in nests (MacDowell & Lord, 1925, 1926; Parkes, 1926). After leaving the nest and at the onset of sexual activity, competition and territorial behaviour change the sex ratios in different population components. Table I (bottom part) illustrates such changes. The percentages are derived for females, since they are of decisive importance for the reproduction and fate of the population.

An equal sex ratio is found among sexually immature individuals up to 10 g in body weight; the excess of females is statistically insignificant. In subsequent weight groups the proportion of immature females drops abruptly as females attain sexual maturity at a higher rate than males. Females comprise about 45% in the whole immature component, and the excess of males is statistically significant.

Among sexually mature individuals, the sex ratio is more or less equal in lower weight groups up to 18 g (except in the groups of 14.1–16.0 g in our material). Over 18 g animals are sexually mature, and the proportion of males declines sharply. Females constitute 86% of the highest weight group; and indeed predominate significantly in the whole mature component, making up 55% overall.

Despite these very marked changes at different stages of life, there is only a slight (and statistically non-significant) excess of females (51.9%) in the total population sample. It is clear that the overall sex ratio in a house mouse population depends upon whether sexually mature individuals (with an excess of females) or sexually immature ones (with an excess of males) are predominant.

Variation in Sex Ratio during the Year

As a result of different breeding intensities in indoor and outdoor populations, both the proportion of mature and immature individuals and the spatial activity of individuals (leading to a greater probability of catching mature males than females) may be different, thus affecting the sex ratio in the catch.

In buildings (Fig. 1B), the proportion of females in the catch is close to 50% throughout the year. In our material, the values varied roughly between 40% and 60%; a significant excess of females occurred only in March (59.1%) and April (59.5%). Males predominated significantly in November and December (only 41.3% females). These oscillations may be accidental or they may be due to

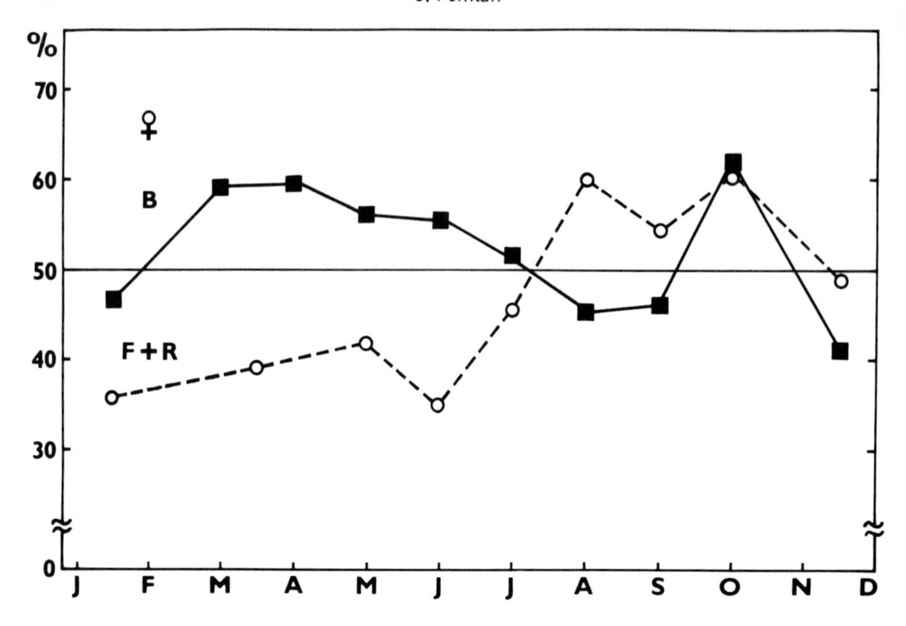

FIG. 1. Changes in sex ratio during the year in animals from buildings (B) and outdoor habitats (F + R).

spring migration of males into fields and their autumn return. In buildings, mice breed throughout the year and the sex ratio in successive monthly catches depends almost entirely on the fortuitous proportions of the immature and mature components of the population.

In the pooled data from fields and ricks, the proportion of females in the catch is low during the first part of the year (35.6% in January and February, and 34.4% in June). Between August and October, however, females predominate markedly. The pooled material from these three months shows a highly significant predominance of females ($n = 434$; female proportion 56.2%; $\chi^2 = 6.72; P < 0.01$). In November and December, the sex ratio is equal in the field and rick populations.

The activity of sexually mature males and their quicker migration from buildings to fields results in greater ease of trapping them at the beginning of the season. As the population increases during the breeding season, males are eliminated so that the proportion of females in the catch increases throughout the summer, particularly in ricks, which have a high density of animals. In late autumn, old females apparently die out so that the sex ratio tends towards equality again.

In data from elsewhere, females predominate in outdoor populations showing intensive breeding and the sex ratio becomes equal in winter (Evans, 1949; Crowcroft & Rowe, 1957; Rudishin, 1962; Rowe *et al.,* 1964; Lidicker, 1966; Berry & Jakobson, 1971; etc.). The mortality among mature males is greater than among females, and increases with population density, in both confined and outdoor populations (Brown, 1953; Petrusewicz, 1960; Reimer & Petras, 1967).

Sex ratio variation in the immature component is not correlated with population density (DeLong, 1967), as the aggressiveness of mature individuals does not directly affect the mortality of young in the nest (Lloyd & Christian, 1967, 1969).

At low and medium densities, males predominate; at high densities, encounters between adult males are more numerous and their mortality increases, as indicated by increased numbers of wounded males observed (Southwick, 1958). At peak densities and during the decline, females predominate (Petrusewicz & Andrzejewski, 1962; DeLong, 1967; Sutova, 1969; Naumov, Gibet & Shatalova, 1969, etc.). Moreover, the aggressiveness of adult males at high population densities increases their tendency to migrate; from ricks with a high density of animals, males left in much larger numbers than females (Rowe *et al.,* 1963). However, this increases their mortality rate, as the emigrants unambiguously lose in encounters with alien males in their territories (Crowcroft, 1955; Petrusewicz, 1959a, b; Rowe & Redfern, 1969).

Hence, the sex ratio of a catch will tell much about the situation in the population sampled.

BREEDING SEASON

We can determine both the total duration of breeding within the year and the breeding intensity, i.e. the percentage of sexually active individuals in a population at any one time. The important factor is whether the population sampled is commensal or feral. The ecological tolerance, adaptability and instantaneous response of *Mus musculus* to environmental conditions are exceptional in this respect (Barnett *et al.,* 1975; Berry & Jakobson, 1975; Bronson, 1979).

Of particular interest are populations showing a migratory character. In spring, individuals leave buildings and migrate to fields where they live through the growing season. After the cereal harvest, some of the feral individuals colonize straw ricks; a small fraction of their number even remain in them over the winter. However, as soon as the air temperature drops permanently below + 10°C, all field and most rick animals migrate back to buildings.

Duration of Breeding Season

Air temperature is of decisive importance here. For example, in the lowlands of South Moravia, Czechoslovakia, the annual mean air temperature is 9.5°C (125 days above 15°C, rainfall 565 mm) and mice stay in the fields from February till October, but in the cooler lowlands of North Moravia, 150 km to the north where the mean annual temperature is 8°C (91 warm days, rainfall 640 mm), they stay in the fields only from June till September, and their population densities are on average three times lower than in the more southern region (Zejda, 1975).

The nature of the climate in South Moravia is also shown by the maximum span of the breeding season, which extends from the first ten days of March (in which the first sexually active males are captured), till early November when the last litters are born (Pelikán, 1974) (Fig. 2). Commensal populations in buildings breed throughout the year, as reported from many parts of the world.

The situation of the rick populations is more complex. In South Moravia, straw ricks are built in July and August and are immediately invaded by rodents including the house mouse. This migration takes place halfway through the growing season so that a considerable number of individuals are sexually active. In the shelter, warmth and favourable trophic situation in the ricks, house mice breed throughout the winter, although the breeding intensity is low (see below). The ricks are taken apart by the following spring and the mice migrate from them to the surrounding fields (Gaisler, Zapletal & Holišová, 1967).

Breeding Intensity

In Moravia, breeding intensity is rather high in indoor house mouse populations. The percentage of sexually active males varies between 40% and 67% in successive months, with an annual mean of 52.2%; while the percentage of pregnant and lactating females varies between 28% and 52%, with an annual mean of 38.2%. In autumn, breeding intensity decreases in both sexes. This may be due to a real decline in sexual activity at the end of the growing season, or to a greater percentage of immature individuals in the population, or perhaps even to the immigration of immatures from the surrounding outdoor habitats at that time (Schwarz & Schwarz, 1943; Crowcroft & Rowe, 1958; Newsome, 1969).

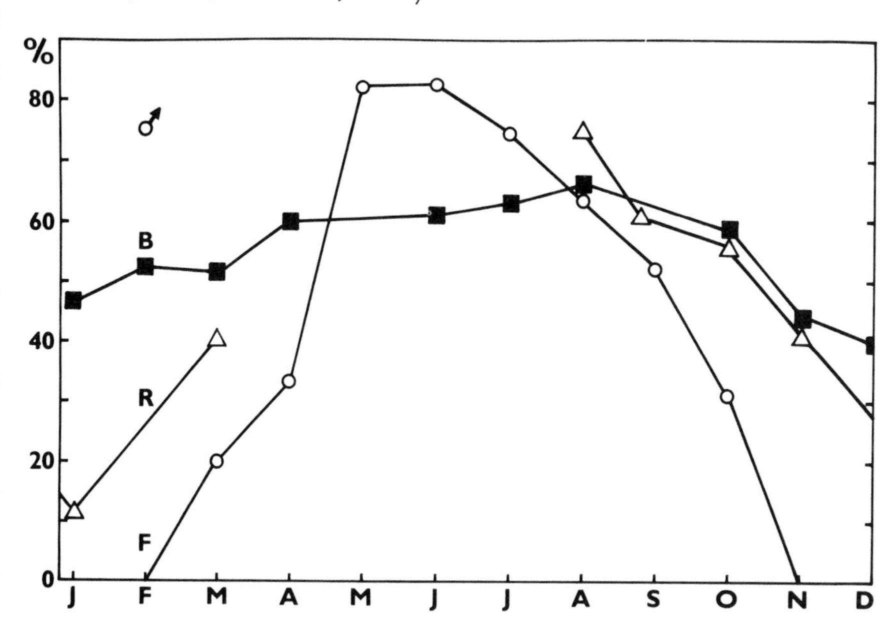

FIG. 2. Proportion of sexually active males throughout the year in mice from buildings (B), ricks (R) and fields (F).

In feral populations, breeding intensity has a markedly seasonal character. In Moravia, the maximum percentage of sexually active males in fields was 82.1% in May and 82.5% in June; in females it was 60.6% in July. In rick populations, breeding intensity drops in autumn later than in the field ones. Breeding may continue with a low intensity throughout the winter. Our estimate of the percentages of sexually active males are about 25% in December and 10% in January. The percentage of pregnant or lactating females is about 10% in December and negligible in January. The spring increase in these values is earlier in the ricks than in the fields.

Under favourable trophic conditions (which holds for corn ricks but not for straw ricks) house mice breed in ricks throughout the winter. The breeding intensity is invariably low, however, as evidenced by data from ricks in England (Venables & Leslie, 1942; Southern & Laurie, 1946; Rowe *et al.*, 1963), the USA (Linduska, 1942) and the Soviet Union (Fenjuk, 1950; Kozlov, 1962).

Although breeding intensity observed in buildings shows

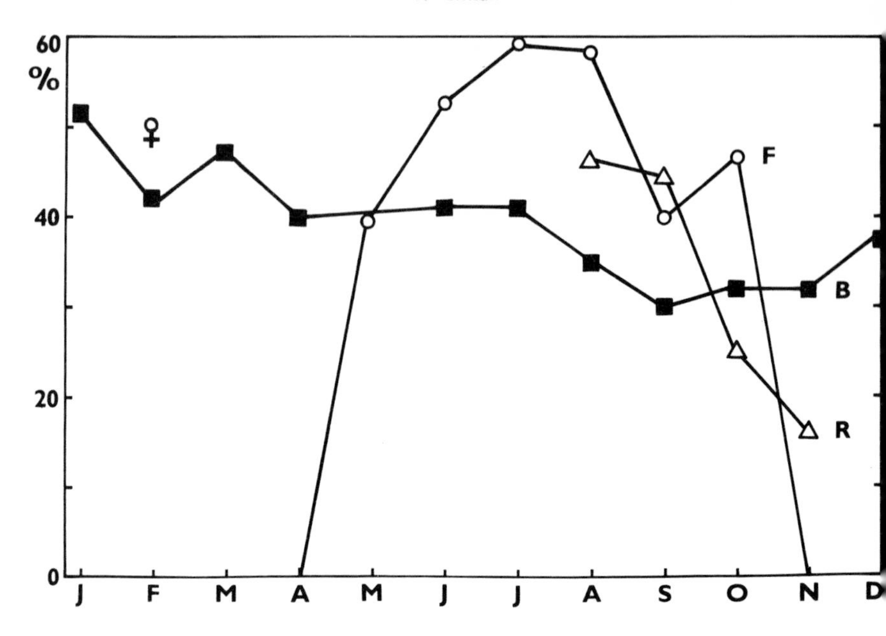

FIG. 3. Proportion of lactating and pregnant females throughout the year. Explanations as in Fig. 2.

irregular variation it apparently decreases in winter. It is worth noting that this occurs even in the warm regions of the original range of *Mus musculus,* e.g. the steppe regions of Astrakhan (Fenjuk, 1950) or the lowlands of the Amu Darya River (Soldatkin, Asenov, & Rubencik, 1959).

On the other hand, conditions in human settlements enable house mice to breed in extremely cold regions at the edges of their range, such as the Yamal Peninsula (Tupikova, 1947), the north-east of the USSR (Romanova, 1970), and elsewhere. The abundance of high calorie food enables house mice to breed even in deep freezes where they build nests inside frozen halves of carcases, as reported from London (Southern, 1954) and other towns (Pelikán, unpublished).

In populations in warmer regions which permanently inhabit fields, steppes or other grassland habitats, breeding may exceptionally last throughout the year (Richter, 1957; Hamar, 1960; Stohl, 1975). Even there, however, the intensity decreases in winter, as reported e.g. from California (Pearson, 1963), Mississippi (Smith, 1954), or the steppes and semi-deserts of the USSR (Lisicyn, 1953; Marin & Leonov, 1962; Fadeev, 1970), and is usually interrupted (Breakey, 1963; Lidicker, 1966; DeLong, 1967). In the arid regions of Australia, breeding is limited not by temperature but by the hardness of the soil (Newsome, Stendell & Myers, 1976).

The duration and intensity of breeding in feral populations are different in different years and are controlled by climatic conditions. For example, variation in breeding intensity and litter size (see below) in the permanently feral mice of Skokholm Island disappears immediately if they are kept in the laboratory (Berry, 1968, 1970a, b). High population density is an important limiting factor. Peak densities produce such strong stress situations as to block the oestrous cycle (Weir & De Friese, 1963; Chipman & Fox, 1966) and to stop breeding prematurely even under favourable conditions (Christian, 1956; Anderson, 1961; Lidicker, 1976; DeLong, 1978; etc.).

LITTER SIZE

Litter size is normally determined by the number of viable embryos or number of placental scars found in dissected females. The latter method gives higher values; it is often not possible to distinguish scars left by resorbed embryos, or to separate sets following two or more litters. The method of palpating embryos in live pregnant females is unreliable (Adamczyk & Walkowa, 1974).

Litter Size Range

The most usual litter sizes are four to eight embryos per set, six being the most frequent. The maximum values commonly reported are 12–13, rarely even 14 embryos per set (Tupikova, 1947; Zaleska-Rutczyńska, 1967); quite exceptionally, Freye & Freye (1960) found 19 young in the litter of an albino female.

Annual Mean Litter Size

In the literature, one often comes across values indicating the annual mean litter size. This single value serves to characterize the mean litter size that occurs in a given region, albeit with certain constraints (Pelikán, 1979). How strongly the annual mean litter size may be biased is set out in Table II, showing the quantitative changes in litter sizes of females from three different habitats in one region in the course of the year (Pelikán, 1974 and additional material).

It is evident that there are distinct differences between the mean litter sizes produced by females in buildings, ricks and fields: the

TABLE II

Variation in mean litter size in three different habitats in Czechoslovakia

	Buildings			Ricks			Fields		
Month	n	$\bar{x}$	$s_{\bar{x}}$	n	$\bar{x}$	$s_{\bar{x}}$	n	$\bar{x}$	$s_{\bar{x}}$
Jan.	15	4.87	0.215	—			—		
Feb.	11	5.27	0.469	1			—		
March	31	5.16	0.347	—			—		
April	52	6.10	0.206	—			—		
May	7	5.29	0.606	—			7	7.29	0.359
June	39	5.41	0.229	—			11	7.91	0.667
July	17	5.59	0.393	1	7.60		20	7.80	0.374
Aug.	29	6.45	0.411	4			17	8.23	0.379
Sept.	12	6.75	0.552	44	6.22	0.208	9	8.11	0.389
Oct.	2	5.42	0.753	3	5.88	0.639	8	7.25	0.250
Nov.	10			5			—		
Dec.	—			—			—		
Total	225	5.71	0.119	58	6.28	0.189	72	7.85	0.182
July–Nov.	70	6.11	0.255	57	6.30	0.190	52	8.06	0.199

Pooled material: $n = 355, \bar{x} = 6.24, s_{\bar{x}} = 0.099$.

annual mean litter size increases from 5.71 embryos in buildings to 6.28 in ricks and 7.85 in fields.

Since litter size varies during the year, these values are not strictly comparable. For that reason, we have separated out our data for the July to November period, but the mean values do not change significantly (see Table II, last line). Between July and November, the mean litter size was 6.11 in buildings, 6.30 in ricks, and 8.06 embryos per set in fields. Overall the mean from fields is significantly different from that in ricks ($t = 6.40$; $P < 0.01$) and in buildings ($t = 6.03$; $P < 0.01$). The difference between ricks and buildings is insignificant, however ($t = 0.60$; $P = 0.50$). In general, females in buildings produce litters about 25% smaller than those in fields.

During the growing season, house mice apparently find better environmental conditions (more light and heat, more food of better quality, mainly seeds) in fields, steppes and other grassland ecosystems than in buildings, and this is clearly reflected in the litter size (c.f. Tupikova, 1947; Rudishin, 1958; Kozlov, 1962).

The different litter sizes in outdoor and indoor house mouse populations may affect annual mean litter size if data from different habitats are pooled. In our material ($n = 355$) the annual mean litter size is 6.24 embryos, but this is an underestimate, as 72% of the females come from buildings in which the litter sizes are small, and only 20% come from fields where the litters are large.

Variation in Litter Size in the Course of the Year

Litter size ranges from small litters in the early growing season to maximum ones roughly in June, at the time of maximum day length. Thereafter litter size decreases again, the autumn values being usually higher than the spring ones. This pattern of changes is found in populations of all polyoestrous small rodents in the temperate zone of the northern hemisphere. However, variation in litter size in feral and commensal populations of *Mus musculus* may differ from the general pattern, depending on temperature conditions, food supply, population structure and density.

The available evidence show that litter size varies not only in free-living feral but even in commensal building populations, although the variation may be less here. The drop in June and increase in September (Table II) may be due to the immigration of pregnant females from outdoor habitats in which litter sizes are larger. The changes are statistically significant.

Our series of pregnant females from ricks and fields are too small for reliable testing; nevertheless, variation in the litter size is apparent even here. The maximum of 8.23 embryos per set in August is the highest monthly mean observed in Czechoslovakia; it may be due to good trophic conditions in fields with ripe cereal crops. The maximum mean litter size in our material (9.29 embryos) was found in seven pregnant females captured during the first ten days of August.

Variation in litter size during the year obviously affects the annual mean value since this depends on the time when most pregnant females are caught.

Litter Size and Female Body Size

There is a direct correlation between these values (Rahnefeld *et al.*, 1966; Elliott, Legates & Ulberg, 1968). In our material, regardless of the locality of capture, females up to 15.0 g of body weight have 5.83 ± 0.23 embryos per set; in the weight group 15.1–18.0 g, 6.38 ± 0.19 embryos; in the group 18.1–21.0 g, 6.56 ± 0.24 embryos; and in females weighing 21.1 g and over, 7.27 ± 0.31 embryos. This does not necessarily imply that populations with high mean body weights always have larger litters. No correlation was found between litter size and body size in populations of large body weight from Skokholm Island, the Isle of May, and Foula (Batten & Berry, 1967).

Geographical Variation in Litter Size

The annual mean litter size in feral populations of *Mus musculus* varies in different regions of its range, as does that in commensal indoor populations. Our mean of 5.71 embryos per set for indoor females from Czechoslovakia is very similar to a mean of 5.61 in laboratory stocks of albino mice in Poland (Czarnomska & Wezykowa, 1967). In Moscow, the mean is higher, 5.9, and in buildings in the south of the Yamal Peninsula it reaches 7.6 embryos per set (Tupikova, 1947). A similar tendency is found in populations of fields and steppes. The lowest mean has been recorded by Straka (1966) from Bulgaria, at 6.53 embryos per set. The mean increases towards the north: it is 7.85 in Czechoslovakia (Pelikán, 1974), 7.3 and 7.5 in the Ukraine (Naumov, 1940; Rudishin, 1962), and 9.3 in the environs of Moscow (Tupikova, 1947). In other words, there is a tendency for larger litter size with shorter breeding season in free-living rodents in northern regions (Spencer & Steinhoff, 1968). Apparently, the same trend holds even for indoor populations of *Mus musculus,* although more evidence is needed. Comparisons of the mean litter sizes in different geographical regions are hampered by the fact that the litter size is indirectly correlated with population density (Southwick, 1955, 1958; Bruce, 1962; Rowe *et al.,* 1964, etc.).

The growth and survival of a population are controlled not only by the litter size, but also by the number of young weaned from each litter, since this decreases in large litters (although the mean number weaned per litter is highest in litters just above the average: Mountford, 1968). Obviously survival of young is of greater importance for population growth than simple litter size.

EMBRYONIC RESORPTION

Numerous data are available on the mortality of eggs and embryos in *Mus musculus* (Hollander & Strong, 1950; Leziak, 1959; Lüning *et al.,* 1966, etc.). There are differences in this respect between pregnant females from different habitats. In Britain, Batten & Berry (1967) found that of the number of eggs shed, 2.5% died after implantation in females from fields; 10.4% in those from ricks; and 17.1% in those from buildings. In visibly pregnant females from fields in Czechoslovakia, 13.5% of sets and 1.9% of embryos were resorbed; from ricks, 6.9% sets and 1.4% embryos; from buildings, only 4.3% sets and 1.1% embryos (Pelikán, 1974).

Macroscopic examinations of embryonic mortality are not as accurate as histological methods but are nevertheless used. The uterus becomes swollen at the implantation site five days after fertilization, but it requires three to four days more for the embryos to attain the size of 4–5 mm when viable and resorbing embryos can be distinguished macroscopically (Snell, 1956; Wessel, 1967; Theiler, 1972; etc.). Hence, postimplantation mortality can be determined macroscopically from roughly the eighth or ninth day until the end of pregnancy (most frequently 19–21 days) (Enzmann, 1935; Dewar, 1968). In 355 visibly pregnant females from Czechoslovakia, regardless of habitat (Pelikán, 1974 and additional material), 7.04% of sets and 1.51% of embryos showed signs of resorption. The latter value corresponds with the values of 2.7% and 0.94%, found respectively by Laurie (1946) in England and by Straka (1966) in outdoor populations in Bulgaria.

Intra-uterine mortality increases with population density, attaining very high values in ricks, where there is limited space available for individual animals. Thus, Southwick (1958) compared low and high density populations in English ricks and found an increase of the proportion of sets affected by resorption from 14.3% to 24.4%, and of embryos resorbed from 1.6% to 2.0%; Rowe *et al.* (1964) recorded even higher values of 22.0% to 37.5% sets affected and 6.2% to 13.4% embryos resorbed.

High resorption rates appear to occur during winter months, as shown by our values for October (15.38% of sets, 2.06% of embryos) and January (13.33% and 3.95%, respectively). This may be due to unfavourable light and temperature factors in winter as well as to the high population density resulting from the concentration of the animals in buildings since autumn. However, the maximum resorption rate in our data occurred in May (28.57% of sets, 6.38% of embryos). No explanation of this fact is available.

Embryo resorption rate has been found to increase with increasing body weight of the female (i.e. net weight without embryos). In our females divided into groups by weight, the percentage of sets affected by resorption increased from 5.6 to 12.7%. This, however, may have been due to the increase in litter size (see below).

Resorption Rate and Litter Size

This relation is very important, as it suggests there is a physiological limit to the number of embryos that can be nourished by a female (Lüning *et al.*, 1966). As shown in Table III, embryonic resorptions were determined in sets of two to nine embryos. Maximum

TABLE III

Resorption rate in relation to numbers of implanted embryos

	Number of implanted embryos											
	1	2	3	4	5	6	7	8	9	10	11	12
Sets examined	3	4	9	38	61	76	71	56	22	12	2	1
% affected	—	25.0	22.2	7.9	8.2	5.3	5.6	7.1	9.1	—	—	—
Embryos examined	3	8	27	152	305	456	497	448	198	120	22	12
% resorbed	—	12.5	11.1	2.0	2.6	1.1	0.8	1.6	1.5	—	—	—

resorption rates occur in small-sized litters, but decrease with increasing litter size and are lowest in sets of six or seven embryos. In large litters, the resorption rate is slightly increased again.

This variation in resorption rate suggests that females with very small litters are physiologically affected individuals, as indicated also by the small numbers of embryos implanted, while females with a large number of implants are apparently incapable of nourishing such a large set.

Our evaluation also indicates that the left and right uterine horns have a different effect on postimplantation resorptions: mortality in the right horn was almost double that in the left (5.19% and 2.60% for sets, and 1.71% and 0.95% for embryos respectively). The differences approach the level of statistical significance. The average numbers of normal, viable embryos were 3.24 in the right and 3.11 in the left ($t = 1.18; P = 0.25$).

The above observations agree with the results of two English studies (Laurie, 1946; Rowe *et al.*, 1964) which report larger numbers of both implantations and resorptions in the right uterine horns. However, Lidicker (1966) and Quadagno (1967) found higher egg production in the left uterine horn but equal numbers of implantations in both horns in North American populations, which suggests higher mortality in the left horn. Obviously the problem requires further study using histological methods.

Cases of superfoetation in *Mus musculus* are well-known and thoroughly investigated (Rollhäuser, 1949; Bloch, 1952; Bilewicz, 1957). Such cases are rare, however, as in other species of rodents.

Resorption Rate during Embryonic Development

As mentioned above, resorbing embryos up to about 4 mm in length cannot be distinguished from viable ones macroscopically. In 6 mm long embryos (Table IV), 5.71% sets were affected, with 0.86% embryos resorbing. Maximum values are found among embryos 9–10 mm long (31.03% sets, 5.14% embryos); the resorption rate decreased above this with increasing age of the embryos, although there appeared to be an increase in resorption rate again in the oldest embryos.

Intra-uterine mortality manifests itself also in decreased mean litter size if only normal, viable embryos are counted (Table IV, last two columns). Up to a 6 mm embryo size, the mean is 6.45 per set; 7–12 mm, 6.14; and in 13 mm and over, 5.83 embryos per set. The difference between the two extreme means is statistically significant

TABLE IV

Resorption rate in relation to embryo length

Maximum length of embryo (mm)	Sets			Embryos			
	No. examined		% affected	No. examined		% resorbed	Mean litter size[a]
2	58			367			
4	97			629			6.45 ± 0.13
6	35	2	5.71	232	2	0.86	
8	26	4	15.38	179	7	3.91	
10	29	9	31.03	175	9	5.14	6.14 ± 0.19
12	29	4	13.79	187	9	4.81	
14	23	3	13.04	145	4	2.76	
16	32	1	3.12	206	1	0.48	5.83 ± 0.22
18	14	2	14.28	75	2	2.67	
20	12			53			
Total	355	25	7.04	2248	34	1.51	6.24 ± 0.10

[a] Calculated only from viable embryos.

($t = 2.39$; $P = 0.02$). Whole embryonic sets are not infrequently lost but it is difficult to estimate how often.

POPULATION STRUCTURE

In population analyses of *Mus musculus,* age structure can be determined by marking individuals (e.g. Berry & Jakobson, 1971) or estimated by tooth wear (e.g. Breakey, 1963; Lidicker, 1966) or eye lens weight (Berry & Truslove, 1968). For some purposes, body weight can be used, while sometimes it may be sufficient to recognize merely the four basic groups of the "structure cross", i.e. males, females, immatures, and adults.

Populations with a marked seasonal pattern of reproduction consist in spring mostly of adult "overwinterers" born in the preceding year, all of them medium weight individuals. Body weight increases when plant growth begins, and the animals attain sexual activity and start breeding (see Fig. 4, spring). By the autumn, a high proportion of immature individuals occurs in the catch, which shows a wide variation of weights and a predominance of adult females in the highest weight groups.

In peak years, a high proportion of immatures are caught as early as mid-summer, and breeding is terminated. In years of more normal numbers, the vast majority of adults die in late autumn and early winter. The survivors consist very largely of young of the year which are not sexually mature. They tend to lose weight slightly over the winter.

This pattern may vary in time in feral populations, depending on the amount and time of migration between buildings, fields and ricks as well as on the breeding intensity.

Commensal populations of *Mus musculus* breed throughout the year, and their age structure shows little variation during the year. Although the data summarized in Fig. 4 (right) come from animals caught in the winter months (December to February), the immature component of this sample is 38.5% of the total.

The most complicated situation is found in rick populations. These are affected by the temporary character of the habitat, by the rather high population density, by the breeding intensity, and by immigration. As a rule, the weight structure of such a population is rather like a feral one in summer, since breeding continues longer in ricks, unless very high densities are reached.

Although variation in weight can only approximately indicate the true age structure of a population, it is the only simple criterion in

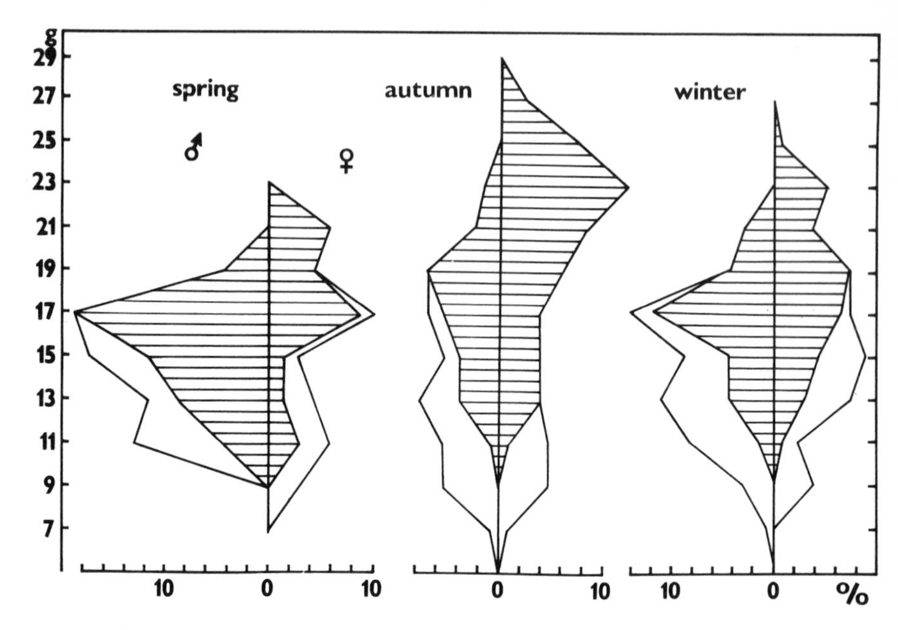

FIG. 4. Weight structure of lowland populations of *Mus musculus:* samples taken in fields in March to May (left) and in August to October (middle); in buildings taken in December to February (right). Open areas, sexually immature individuals; hatched areas, sexually mature individuals.

live-trapping studies (Breakey, 1963; Pearson, 1963; etc.). Determining the weight structure of a population is essential for bioenergetic investigations.

ACKNOWLEDGEMENTS

My thanks are due to Ing. V. Baruš, D.Sc, Corresponding Member to the Czechoslovak Academy of Sciences and Director of the Institute of Vertebrate Zoology in Brno, for consultations and valuable comments on the manuscript; to Dr R. Obrtel, C.Sc, for comments and emendations of the text as well as for translating the Czech version into English; and to Professor R. J. Berry, of University College London, for the invitation to write this contribution.

REFERENCES

Adamczyk, K. & Walkowa, W. (1974). Estimating the number of newborn animals in enclosed populations of laboratory mice. *Acta theriol.* **19**: 247–257.

Anderson, P. K. (1961). Density, social structure and nonsocial environment in house-mice populations and the implications for regulation of numbers. *Trans. N. Y. Acad. Sci.* (2) 23: 447–451.

Barnett, S. A., Munro, K. M. H., Smart, J. L. & Stoddart, R. C. (1975). House mice bred for many generations in two environments. *J. Zool., Lond.* 177: 153–170.

Batten, C. A. & Berry, R. J. (1967). Prenatal mortality in wild-caught house mice. *J. Anim. Ecol.* 36: 435–464.

Berry, R. J. (1968). The ecology of an island population of the house mouse. *J. Anim. Ecol.* 37: 445–470.

Berry, R. J. (1970a). The natural history of the house mouse. *Fld Stud.* 3: 219–262.

Berry, R. J. (1970b). Covert and overt variation, as exemplified by British mouse populations. *Symp. zool. Soc. Lond.* No. 26: 3–26.

Berry, R. J. & Jakobson, M. E. (1971). Life and death in an island population of the house mouse. *Expl Geront.* 6: 187–197.

Berry, R. J. & Jakobson, M. E. (1975). Adaptation and adaptability in wildliving House mice (*Mus musculus*). *J. Zool., Lond.* 176: 391–402.

Berry, R. J. & Truslove, G. M. (1968). Age and eye lens weight in the House mouse. *J. Zool., Lond.* 155: 247–252.

Bilewicz, S. (1957). [Observations on the genesis of superfoetation in white mice]. *Przegl. Zool.* 1: 292–305. [In Polish.]

Bloch, S. (1952). Untersuchungen über Superfetation an der Maus. *Schweiz. med. Wochenschr.* 82: 632–648.

Breakey, D. R. (1963). The breeding season and age structure of feral house mouse populations near San Francisco Bay, California. *J. Mammal.* 44: 153–167.

Bronson, F. H. (1979). The reproductive ecology of the house mouse. *Q. Rev. Biol.* 54: 265–299.

Brown, R. Z. (1953). Social behaviour, reproduction and population change in the house mouse (*Mus musculus*). *Ecol. Monogr.* 23: 217–240.

Bruce, H. (1962). The importance of the environment on the establishment of pregnancy in the mouse. *Anim. Behav.* 10: 389–390.

Chipman, R. K. & Fox, K. A. (1966). Oestrous synchronization and pregnancy blocking in wild house mice (*Mus musculus*). *J. Reprod. Fert.* 12: 233–236.

Christian, J. J. (1956). Adrenal and reproductive responses to population size in mice from freely growing populations. *Ecology* 37: 258–273.

Crowcroft, P. (1955). Territoriality in wild house mice, *Mus musculus* L. *J. Mammal.* 36: 299–301.

Crowcroft, P. & Rowe, F. P. (1957). The growth of confined colonies of the wild house mouse (*Mus musculus* L.). *Proc. zool. Soc. Lond.* 129: 359–370.

Crowcroft, P. & Rowe, F. P. (1958). The growth of confined colonies of the wild house mouse (*Mus musculus* L.): the effect of dispersal on female fecundity. *Proc. zool. Soc. Lond.* 131: 357–365.

Crowcroft P. & Rowe, F. P. (1961). The weights of wild house mice (*Mus musculus* L.) living in confined colonies. *Proc. zool. Soc. Lond.* 136: 177–186.

Czarnomska, A. & Wezykowa, J. (1967). [Numbers of mice born and surviving in litters in relation to strain, season of the year and number of preceding litters]. *Zwier. Labor.* 6: 7–13. [In Polish.]

DeLong, K. T. (1967). Population ecology of feral house mice. *Ecology* 48: 611–634.

DeLong, K. T. (1978). The effect of the manipulation of social structure on reproduction in house mice. *Ecology* 59: 922–933.

Dewar, A. D. (1968). Litter size and the duration of pregnancy in mice. *Q. J. exp. Physiol.* 53: 155–161.

Elliott, D. A., Legates, J. E. & Ulberg, L. C. (1968). Changes in the reproductive processes of mice selected for large and small body size. *J. Reprod. Fert.* 17: 9–18.

Enzmann, E. V. (1935). Intrauterine growth of albino mice in normal and in delayed pregnancy. *Anat. Rec.* 62: 31–45.

Evans, F. C. (1949). A population study of house mice (*Mus musculus* L.) following a period of local abundance. *J. Mammal.* 30: 351–363.

Fadeev, G. S. (1970). [Reproduction of the house mouse in the cultured landscape of the Iliisk depression]. *Mater. Pozn. Fauny Flory SSSR* NS (zool.) 9: 231–239. [In Russian.]

Fenjuk, B. K. (1950). [Quantitative estimation of mice and voles, problem of their prognosis in the South-East]. *Gryz. borba nimi* 3: 89–125. [In Russian.]

Freye, H. A. & Freye, H. (1960). Die Hausmaus. *Neue Brehm-Büch.* No. 268: 1–104.

Gaisler, J., Zapletal, M. & Holišová, V. (1967). Mammals of ricks in Czechoslovakia. *Přiodov. Pr. Česk. Akad. Věd.* 1: 299–348.

Hollander, W. F. & Strong, L. C. (1950). Intrauterine mortality and placental fusion in the mouse. *J. exp. Zool.* 115: 131–150.

Hamar, M. (1960). [On the systematics, distribution and ecology of the house mouse in Roumania]. *Revue Biol. Buc.* 5: 207–220. [In Russian.]

Hanák, V. (1958). Beitrag zur Kenntnis der postnatalen Entwicklung der Hausmaus, I. Körper- und Haarwachstum. *Věst. Čsl. Spol. Zool.* 22: 279–292.

Humiński, S. (1969). Biomorphological studies on testes and male accessory glands in some species of the families *Muridae* and *Microtidae* found in Poland. *Zool. pol.* 19: 213–225.

Knudsen, B. (1962). Growth and reproduction of house mice at three different temperatures. *Oikos* 13: 1–14.

Kozakiewicz, M. A. (1976). The effect of population density and sex ratio on the mortality of juvenile laboratory mice in experimental populations. *Acta theriol.* 21: 339–351.

Kozlov, N. P. (1962). [Peculiarities of reproduction of the house mouse in dry steppes of the Stavropol district]. *Biul. Mosk. Obs. Isp. Prir.* 47: 117–120. [In Russian.]

Laurie, E. M. O. (1946). The reproduction of the house-mouse (*Mus musculus* L.) living in different environments. *Proc. R. Soc.* (B) 133: 248–281.

Leslie, P. H., Perry, J. S. & Watson, J. S. (1945). The determination of the median body weight at which female rats reach maturity. *Proc. zool. Soc. Lond.* 115: 473–488.

Leziak, K. (1959). Studies on retarded pregnancy in mice from inbred matings (sib-mating). II. Resorption of foetuses. *Folia biol. Kraków* 7: 267–275.

Lidicker, W. Z. (1966). Ecological observations on a feral house mouse population declining to extinction. *Ecol. Monogr.* 36: 27–50.

Lidicker, W. Z. (1976). Social behaviour and density regulation in house mice living in large enclosures. *J. Anim. Ecol.* 45: 677–699.

Linduska, J. P. (1942). Winter rodent populations in field-shocked corn. *J. Wildl. Mgmt* **6**: 353–363.

Lisicyn, A. A. (1953). [Reproduction and mortality of the house mouse in conditions of steppes at Salsk]. *Sb. Nauch. Rabot Privolg. Protivoepid. Stancii* **1**: 81–109. [In Russian.]

Lloyd, J. A. & Christian, J. J. (1967). Relationship of activity and aggression to density in two confined populations of house mice (*Mus musculus* L.). *J. Mammal.* **48**: 262–269.

Lloyd, J. A. & Christian, J. J. (1969). Reproductive activity of individual females in three experimental freely growing populations of house mice (*Mus musculus* L.). *J. Mammal.* **50**: 49–59.

Lüning, K. G., Sheridan, W., Ytterborn, K. H. & Gullberg, U. (1966). The relationship between the number of implantations and the rate of intra-uterine death in mice. *Mutation Res.* **3**: 444–541.

MacDowell, E. C. & Lord, E. M. (1925). Data on the primary sex-ratio in the mouse. *Anat. Rec.* **31**: 143–148.

MacDowell, E. C. & Lord, E. M. (1926). The relative viability of male and female mouse embryos. *Am. J. Anat.* **37**: 127–140.

Marin, S. N. & Leonov, J. A. (1962). [Peculiarities of ecology of the house mouse in Karakum desert at the Aral sea]. *Vop. Ekol. Kiev* **6**: 95–96. [In Russian.]

Mountford, M. D. (1968). The significance of litter-size. *J. Anim. Ecol.* **37**: 363–367.

Naumov, N. P. (1940). [The ecology of the hillock mouse, *Mus musculus hortulanus* Nordm.]. *Zh. Inst. Evol. Morf.* **3**: 33–77. [In Russian.]

Naumov, S. P., Gibet, L. A. & Shatalova, S. P. (1969). [Dynamics of the sex ratio in density fluctuations of mammals]. *Zh. obshch. Biol.* **30**: 673–680. [In Russian.]

Newsome, A. E. (1969). A population study of house mice. *J. Anim. Ecol.* **38**: 341–377.

Newsome, A. E., Stendell, R. C. & Myers, J. H. (1976). Free-watering a wild population of house mice — a test of an Australian hypothesis in California. *J. Mammal.* **57**: 677–686.

Parkes, A. S. (1924). Studies on the sex ratio and related phenomena: V. The sex ratio in mice and its variations. *Br. J. exp. Biol.* **1**: 323–334.

Parkes, A. S. (1926). Studies on the sex-ratio and related phenomena: I. The foetal retrogression in mice. *Proc. R. Soc. Lond. (B)* **95**: 551–558.

Pearson, O. P. (1963). History of two local outbreaks of feral house mouse. *Ecology* **44**: 540–549.

Pelikán, J. (1974). On the reproduction of *Mus musculus* in Czechoslovakia. *Přírodov. Pr. Česk. Akad. Věd.* **8**: 1–42.

Pelikán, J. (1979). Sufficient sample for evaluating the litter size in rodents. *Zool. Listy* **28**: 289–298.

Petrusewicz, K. (1959a). Differences in male and female quantitative dynamics in confined populations of mice. *Int. congr. Zool.* **15** (10): 1–2.

Petrusewicz, K. (1959b). Further investigation of the influence exerted by the presence of their home cages and own populations on the results of fights between male mice. *Bull. Acad. pol. Sci.* **7**: 319–322.

Petrusewicz, K. (1960). Some regularities in male and female numerical dynamics in mice populations. *Acta theriol.* **4**: 103–137.

Petrusewicz, K. & Andrzejewski, R. (1962). Natural history of a free-living population of house mice (*Mus musculus* L.) with particular reference to groupings within the population. *Ecol. pol.* (A) **10**: 85–122.

Quadagno, D. M. (1967). Litter size and implantation sites in feral house mice. *J. Mammal.* **48**: 677.

Rahnefeld, G. W., Comstock, R. E., Singh Madho & NaPuket, S. R. (1966). Genetic correlation between growth rate and litter size in mice. *Genetics* **54**: 1423–1429.

Reading, A. J. (1966a). Effects of parity and litter size on the birth weight of inbred mice. *J. Mammal.* **47**: 111–114.

Reading, A. J. (1966b). Influence of room temperature on the growth of house mice. *J. Mammal.* **47**: 694–697.

Reimer, J. D. & Petras, M. L. (1967). Breeding structure of the house mouse, *Mus musculus*, in a population cage. *J. Mammal.* **48**: 88–99.

Richter, H. (1957). Zur Wintervermehrung der Ahrenmaus (*Mus m. musculus* L.) und der Feldmaus (*Microtus arvalis* Pallas) in Mittelmecklenburg. *Arch. Freunde NatGesch. Mecklenb.* 3: 133–140.

Rollhäuser, H. (1949). Superfoetation in mouse. *Anat. Rec.* **105**: 657–663.

Romanova, G. A. (1970). [Ecology of the house mouse at the north-eastern edge of its range]. *Ekologija* **1**: 98–99. [In Russian.]

Rowe, F. P. & Redfern, R. (1969). Aggressive behaviour in related and unrelated wild house mice (*Mus musculus* L.). *Ann. appl. Biol.* **64**: 425–431.

Rowe, F. P., Taylor, E. J. & Chudley, A. H. (1963). The numbers and movements of house mice (*Mus musculus* L.) in the vicinity of four corn ricks. *J. Anim. Ecol.* 32: 87–97.

Rowe, F. P., Taylor, E. J. & Chudley, A. H. (1964). The effect of crowding on the reproduction of the house mouse (*Mus musculus* L.) living in corn-ricks. *J. Anim. Ecol.* 33: 477–484.

Rudishin, M. P. (1958). [Distribution and population dynamics of small rodents in Western forest-steppe of Ukrainian RSR]. *Vidav. A.N. Ukr. RSR Kiev:* 3–27. [In Ukrainian.]

Rudishin, M. P. (1962). [Peculiarities of the reproduction of common rodents in Western parts of Ukrainian RSR]. *Vopr. Ekol. Kiev* **6**: 122–123. [In Ukrainian.]

Schlager, G. & Roderick, T. H. (1968). Secondary sex ratio in mice. *J. Hered.* **59**: 363–365.

Schwarz, E. & Schwarz, H. K. (1943). The wild and commensal stocks of the house mouse, *Mus musculus* L. *J. Mammal.* **24**: 59–72.

Serafiński, W. (1965). The subspecific differentiation of the Central European house mouse (*Mus musculus* L.) in the light of their ecology and morphology. *Ekol. Pol.* (A) **13**: 305–348.

Smith, W. W. (1954). Reproduction in the house-mouse (*Mus musculus* L.) in Mississippi. *J. Mammal.* **35**: 509–514.

Snell, G. D. (1956). *Biology of the laboratory mouse.* New York & London: McGraw-Hill.

Soldatkin, I. S., Asenov, G. A. & Rubencik, J. V. (1959). [Numbers, reproduction and mortality of the house mouse in the oasis of Amu-Darja basin]. *Gryz. borba nimi* **6**: 79–89. [In Russian.]

Southern, H. N. (Ed.) (1954). *Control of rats and mice.* 3. *House mice.* London: Oxford University Press.

Southern, H. N. & Laurie, E. M. O. (1946). The house-mouse (*Mus musculus*) in corn ricks. *J. Anim. Ecol.* **15**: 134–149.

Southwick, C. H. (1955). Regulatory mechanisms of house mouse populations: social behaviour affecting litter survival. *Ecology* **36**: 627–634.

Southwick, C. H. (1958). Population characteristics of house mice living in English corn ricks: density relationships. *Proc. zool. Soc. Lond.* **131**: 163–175.

Spencer, A. W. & Steinhoff, H. W. (1968). An explanation of geographic variation in litter size. *J. Mammal.* **49**: 281–286.

Stohl, G. (1975). Die Regelung der Fortpflanzungstätigkeit bei *Mus musculus spicilegus* Petényi. *Vertebr. hung.* **16**: 55–72.

Straka, F. (1966). Untersuchungen über die Biologie der Hausmaus (*Mus musculus* L.) unter Freilandbedingungen. *Plant Sci.* **3**: 87–95.

Sutova, M. (1969). [Importance of structural factors in processes of population dynamics of *M. m. spicilegus* Pet.]. *Studii Cerc. Biol.* (Zool.) **21**: 377–384. [In Roumanian.]

Theiler, K. (1972). *The house mouse.* Berlin & New York: Springer.

Tupikova, N. V. (1947). [Ecology of the house mouse in central part of the USSR]. *Fauna Ekol. Gryz.* **2**: 5–67. [In Russian.]

Venables, L. S. V. & Leslie, P. H. (1942). The rat and mouse populations of corn ricks. *J. Anim. Ecol.* **11**: 44–68.

Weir, M. W. & De Friese, Y. C. (1963). Blocking of pregnancy in mice as a function of stress. *Psychol. Rep.* **13**: 365–366.

Wessel, M. (1967). Untersuchungen zum Genitalzyklus der Hausmaus (*Mus musculus* L.) und Entwurf einer Normentafel der Embryonalentwicklung. *Zool. Jb.* (Anat.) **84**: 375–424.

Zaleska-Rutczyńska, Z. (1967). [Problems of fertility and fecundity in laboratory animals with special respect to the white mouse]. *Zwier. Lab.* **5**: 89–93. [In Polish.]

Zejda, J. (1975). Habitat selection in two feral house mouse lowland populations. *Zool. listy* **24**: 99–111.

Symp. zool. Soc. Lond. (1981) No. 47, 231–254

Genetical Influences on Growth and Fertility

R. C. ROBERTS

Institute of Animal Genetics, West Mains Road, Edinburgh, U.K.

SYNOPSIS

Body weight is a highly heritable trait and responds readily to selection. While the total response is fairly predictable, the pattern of the response varies among populations with different genetical parameters, and also because of accidents of gene sampling between generations. It is postulated that much of the genetical variance found among wild populations is a consequence of their chance genetic origin – the founder effect. While there may be a range of variation over which there is little natural selection on body weight, extreme deviants in any direction can be shown to be less fit. It is argued that there could be similar stabilizing selection operating on fertility, but that during evolution, there has been selection for the largest litters that can survive over an average range of conditions. This usually results in considerable mortality, but also allows an increase in fertility should conditions temporarily improve.

INTRODUCTION

There is by now abundant evidence that genes influence both body weight and fertility. Many mutant genes have been isolated which influence both of these traits, sometimes simultaneously. Such mutant genes, however, are to varying degrees pathological in their effects, and because they are seriously deleterious, occur at very low frequencies. They are not usually detectable in populations where they are not deliberately maintained, and as such, are of no immediate consequence in the context of natural populations. But it is equally clear that other genes, which for present purposes can be labelled "normal" alleles, also contribute to the natural variation both in body weight and in fertility. This kind of genetical variation can be studied in the laboratory, where pedigrees can be kept, by the standard techniques of quantitative genetics. The measure of the importance of genetic influence is frequently expressed as the *heritability*. This is a useful parameter, since it governs the extent to which relatives resemble one another (essentially what we mean when we say a trait is "genetic") and it also allows us to predict responses to selection. If one were to say that the heritability of body weight is in the region of 30% to 40%, and that of fertility traits not more than half of this amount, few people who know the subject would quibble about its generality. For sure, estimates can be produced that lie outside these narrow ranges, and there are reasons for that. But the generality is such that the *cognoscenti* might not even bother to enquire what species was being discussed – it could equally well be laboratory mice or broiler chickens (not fish, though). In short, body weight tends to be highly heritable – where almost half of the natural variation may be due to genes. Fertility traits are much less heritable, some of them being of very low heritability indeed, e.g. conception rates in cows. Whether the same would apply to conception rates in mice is uncertain, but it would be a good guess that it would.

Two points merit brief further consideration – the different heritabilities of different traits, and their applicability across species or populations. The thinking on the first of these goes back to Fisher's (1930) "fundamental theorem of natural selection". The argument, stripped to bare essentials, states that traits close to natural fitness will have been exposed to millennia of natural selection, and any available genetical variation in the trait will have become largely exhausted in the process. Thus, such traits will have little genetical variance (of this kind) left, and they will show low heritabilities as a result. This does not mean, however, that gene products do not influence such traits – of course they do. It is just that "superior" alleles will have been selected and thus tend to become fixed in any particular population; they will be largely the same for all individuals and they will therefore not contribute to variation between individuals. The considerations stemming from Fisher's theorem refer only to the *additive* genetical variance, on which selection acts. It is beyond the scope of this paper to pursue this further (see Roberts, 1967a, for further discussion), but in the fitness traits especially, there remains a substantial amount (usually) of *non-additive* genetical variance, arising from dominance and interactions between loci. This non-additive variance leads to the observed effects of inbreeding, and its complement – heterosis – to which the fitness traits are particularly susceptible. To the extent that fertility traits are closer to natural fitness than body weight is, the differences between them in general levels of heritability become amenable to genetical interpretation, and are a reflection of the evolutionary history of the traits.

If we accept that fertility and body weight each have similar evolutionary histories across different species, we can accept also that those species will be similar in the genetical architecture of analogous traits. The generality quoted earlier makes some sense. Having said that, a caveat is necessary. Strictly speaking, heritabilities as estimated refer only to the actual population supplying the measurements and under the conditions under which those measurements were taken. Thus, as Monteiro & Falconer (1966) found, the heritability of body weight in the mouse alters according to the age at measurement, and Falconer (1960a) showed that for a given age, the heritability observed could be influenced by the plane of nutrition. Any particular estimate of heritability should thus be considered as the product of a unique set of circumstances – indeed, the same point could be made of any experimental result. Nevertheless, if we took all estimates of heritabilities of body weight in the mouse at, say six to eight weeks of age, and similarly all heritabilities of the number born in first litters, the two sets of estimates would be virtually non-overlapping. The generality that body weight is more heritable than fertility holds – as a generality – and any exceptions should not cause undue concern. Chance and circumstances will see to it that there are exceptions.

In view of this, what might be the relevance of genetical work with laboratory mice to natural populations of wild mice? Speaking strictly, we cannot tell, because we cannot exercise laboratory procedures under non-laboratory conditions. We can, however, hazard a guess: any differences will be due more to a difference in the conditions, rather than to any gross difference in the genetical make-up of laboratory and wild mice. Some reviewers (Roberts, 1965a, b; Eisen, 1974) have discussed the applicability of genetical work with laboratory mice to domestic livestock, and found adequate parallels. In evolutionary terms this is a quantum jump compared to the applicability of mice to other mice, so perhaps we need do no more than exercise the customary care when extrapolating beyond the range of our data.

It is against this background that the genetical investigations of body weight and fertility will be considered.

GENETICAL VARIATION IN BODY WEIGHT

Differences in Body Weight between Strains and Populations

There is abundant evidence that inbred strains of mice, kept under the same laboratory conditions, differ in body weight at the same age. The evidence will not be reviewed in detail here – see Poiley (1972) for extensive data on the topic. This fact alone is sufficient proof of genetical variation in body weight, though it tells us nothing of the nature of such genetical variation, nor how easily it may be exploited by selection. It tells us nothing either of the adaptive significance of body weight (if any), for the origin of such genetical variation is presumably adventitious, or largely so, and arose from the historical separation of inbred strains during their formation. We may suppose, however, that what pertains to laboratory strains will apply, at least in part, to wild populations that are reproductively isolated. Direct comparisons between natural populations are of course complicated by possible – indeed likely – differences in the environment. Even so, Berry and his colleagues (Berry & Jakobson, 1975; Berry, Peters & Van Aarde, 1978; Berry, Jakobson & Peters, 1978) have argued convincingly for genetical variation among island populations of wild mice, including genetical variation in body weight. These studies ranged from the Faroe Islands to the Australian sub-Antarctic. The Faroe populations were compatible with the colonization of separate islands by small numbers of effective founders, and the analogy with the differentiation of inbred strains (above) is clear. This is not to say that Faroe mice are highly inbred by laboratory standards; it is just that similar forces are at work, and that genetical differentiation could not occur unless there had been initial genetical variation.

Because of this founder effect, we have to be cautious about inferring genetical differences between populations when wild-caught samples are brought into the laboratory for further study. The numbers caught are usually rather small, and of those caught, fewer breed. In addition, who is to say whether mice successfully trapped are a random sample of the population they are purported to represent? With this caveat, however, studies on wild mice in a standard laboratory environment confirm the existence of ample genetical variation, either between or within wild populations, with respect to body weight (Plomin & Manosevitz, 1974; Barnett *et al.*, 1975; Ebert & Hyde, 1976; Lynch, 1977). Of these, only Plomin & Manosevitz compared populations derived from different localities, two from Texas and one from Colorado. Body weights were clearly different at 200 days, though not at 22 days.

Vagaries of sampling apart, differences between populations tell us nothing about the adaptive significance of body weight. It would be equally plausible to argue that each population is carefully adapted to its own ecological niche as it would be to maintain that body weight is of little consequence, reflecting little more than the historical accidents of founder effect with subsequent non-directional drift. Common sense would dictate that perhaps both elements might have contributed to the current state, and only supplementary evidence can suggest their relative importance. Such supplementary evidence may derive from two distinct sources. The first is well illustrated by Berry, Jakobson & Peters (1978), who calculated genetical distances for several characters among different populations from the Faroes. In their analyses, they included body weight, organ weights, allozyme variants and indices of skeletal shape. What they found was that distances calculated from different characters were poorly correlated with each other. Even correction of the skeletal parameters for body size failed to improve the correlations of distances among the skeletal parameters, suggesting to the authors that size *per se* was unimportant. Had body size lent coherence to the remaining skeletal data, then the adaptive significance of body size would have been indicated. As it is, the conservative conclusion is that the adaptive significance of body weight could not be established, within the bounds of this data set.

The second approach is to examine the genetical properties of mouse populations by selection. The responses themselves offer some guide to the evolutionary history of the trait, as outlined earlier. In addition, the reproductive performance of lines selected for body weight allows the correlated effects on fitness to be examined.

Responses to Selection for Body Weight

Selection for body weight in the mouse has a long history, dating from the pioneering studies of Goodale (1938) and MacArthur (1944). Comprehensive reviews of the field, up to the dates of publication, have been provided by Roberts (1965a) and by Eisen (1974), and specialized aspects of the topic were further discussed by Roberts (1979), bringing the bibliography more or less up to date. To avoid repetition, if nothing else, no attempt will be made to be equally exhaustive here; references will be chosen purely to illustrate the main findings.

In the same way that Berry and his colleagues showed founder effects on island colonies, so do selection experiments reflect the gene content of their various base populations. More than that, however, even a given gene content does not necessarily yield a predictable outcome, because during selection – or even propagation without selection – genes can be lost through accidents of sampling. This may be true of genes even favourable to the direction of selection, particularly if they are at low frequency in small populations. This is the essence of Kimura's (1957) oft-quoted paper on chance fixation, extended by Robertson (1960) in a theoretical treatment of limits to selection. Thus, a selected line may become fixed for a particular allele even though a better one had originally been available, because the better allele was accidentally lost in the process. It will be intuitively obvious that this accidental loss will be less likely as the population size increases, as the frequency of the favourable allele increases, and as the magnitude of its effect increases. Chance fixation can be a major factor in the limits to selection ultimately reached; initial responses will be subjected to general drift – chance fixation being a special case.

The repeatability of selection responses was tested by Falconer (1973), with specific reference to body weight at six weeks of age in the mouse. He had six lines selected for high body weight, six for low and six unselected control lines. All derived from the same base population. The effects of drift were very clear in the control lines. Initially their mean weights varied from about 22 g to 23.5 g. After 20 generations of random propagation, the range was fully from 21 g to 25 g. The initial sampling, even from the same families for pairs of lines, had generated significant differences in body weight. Subsequently, the control lines diverged further because of drift. The effects of random drift were equally evident (or almost so) in the selected lines. After 20 generations, the large lines varied in mean weight from about 32 g to 35 g, the small lines from 13 g to 16 g. So, despite the differences between replicated lines due to drift, they nevertheless reached similar end points. The overall effect of drift must therefore be judged to be small, compared to the effect of the selection, even with the rather small population sizes. The effective number of parents (selection being within families) was never more than 32 in Falconer's lines, and frequently less because of some sterility.

Whether the responses are judged to be similar or dissimilar is very much a matter of outlook and emphasis. Having pointed out the similarities of the weights reached after 23 generations of selection, Falconer was still able to show "forcibly how dissimilar the replicates were over the first part of the selection", especially when the divergence between upwards and downwards selection was compared. Falconer said that his results led to a "clear warning" that deserves to be quoted in full: "single selection experiments on the scale of one of these replicates can be very misleading about the rate of response, and particularly about the asymmetry, if judged from the first 5 or even 10 generations". In conjunction with this warning. Falconer reported that the range of realized heritabilities among replicates was from 25% to 46% for the high lines, and from 16% to 50% for the low.

Falconer's experiment has been given some space here because it is the most comprehensive on record. Though the point did not figure among Falconer's declared objectives, we might pause to consider what kind of framework it provides for thinking about populations of mice in the wild. The main difficulty here is the uncertainty about effective population sizes in the wild. Even though Laurie (1946) was able to capture 2368 mice from one wheat rick, the original invasion may have been a small number of effective parents. If we use Falconer's effective number of 30 or so as a not unrealistic model for many populations, then we should not overestimate the effect of drift on body weight in the short run. And despite drift errors being cumulative (Hill, 1971), they need not amount to all that much in the long run either. As the population size increases, drift becomes less; but if the population size decreases, then inbreeding inevitably occurs, and though this may generate considerable genetical differentiation *at the time,* the concomitant loss of genetical variance will allow less scope for further genetical differentiation thereafter. After Falconer's (1973) report, the population size of all experimental lines in Edinburgh was doubled; after a further 40 generations of random mating, the control lines are no more divergent now than they were then. Whether this lack of further divergence is due to the increased population size or to the accumulated effects of inbreeding (i.e. to loss of genetical variance) is a moot point. Generally speaking, unless population sizes are relatively small and relatively stable, drift will not be detectable as a gradual accumulation that proceeds indefinitely. It will occur in fits and starts, corresponding to a bottleneck in population size through some crisis. The small sample may well generate a change in mean body weight, while the inevitable inbreeding will reduce the scope of further drift. Any migration between populations will of course have the opposite effect, and prevent genetic isolates being developed.

The amount of selection practised in Falconer's experiment was presumably far greater than any selection on body weight operative in the wild. We need not suppose, therefore, that any differentiation through drift among wild populations will be swamped by the effects of selection. It may be argued, however, that unlike laboratory populations, wild mice may be subject to spatial and temporal variation in the environment. To the extent that different populations may become adapted to specialized niches, selective forces are inescapably implied. The balance of the evidence is probably against this happening, as far as body weight is concerned, as discussed by Berry and his colleagues, quoted earlier.

Other reports testify to the high heritability of body weight, notably the extensive studies of Eisen and his colleagues in North

Carolina. Eisen (1978) found heritabilities of six-week weight to be from 42% to 55%, depending on the method of estimation. A point of particular interest in the North Carolina work is the simultaneous selection for body weight and traits negatively correlated with it. Large mice generally have larger litters and longer tails, and vice versa, among other correlated traits. Eisen (1978) selected for the antagonistic relationships involving litter size and Eisen & Bandy (1977) selected similarly in a replicated experiment involving body weight and tail length. In agreement with theory, the changes in body weight were less when selection involved also a second trait. The implications for what may happen in the wild are obvious. If body weight is part of an adaptive complex involving other traits (as will undoubtedly be the case), then the scope for changing body size – depsite its high heritability – may be severely restricted if there is concomitant selection for traits negatively correlated with it.

The growth curve of the mouse, as in other mammals, can be described in general terms as sigmoid, and a variety of mathematical functions have been employed to obtain the best statistical fit. The logistic equation usually proves to be as good as any. Eisen (1976) comprehensively reviews the effects of selection on growth curves, with the following general conclusion: although the constants in the equation can be shown to have been altered, the overall *shape* of the growth curve is generally unaffected, the differences being mostly due to the re-scaling of the two axes. A study by McCarthy & Doolittle (1977) set out to change the shape of the curve by a variety of procedures: to change five-week and 10-week weights in opposite directions, or else to change one while holding the other constant. While their attempts were not uniformly successful, and in some cases agreed rather poorly with prediction, they nevertheless showed that it was perfectly feasible to alter the shape of the curve. It is only fair to add, however, that they were most successful with the somewhat sophisticated procedure of restricted indices. As a matter of personal opinion, I should find it hard to imagine selective forces in the wild operating in this manner, as there must be more urgent matters demanding selective attention. But it would be foolhardy to dismiss the possibility, particularly if conditions were to favour early maturity (rapid early growth) combined with a small mature size. Other possible combinations would seem to me to be even less plausible.

To conclude this section, body weight has been shown to be highly heritable, and on the argument presented earlier, this is compatible with the idea (though not proof of it) that body weight is not a major component of natural fitness. It is reasonable to postulate that the variation in body weight found in the wild is therefore either a consequence of chance genetical origin (founder effect), or else the product of nutrition and other environmental variables. And as we saw, drift may also contribute to some differentiation between populations. It would be a worthwhile experiment, for those with an appetite for handling wild mice in captivity, to select them for body weight, to test whether the genetical parameters for wild mice resemble those of synthetic laboratory populations, and lead to similar responses. My guess is that they would, and if this were so, it would be grist for the mill of those who argue against the adaptive significance of body weight. We should be wary, though, of pushing that argument too far. If we think of the genetical situation within populations, there may be a range of body weights around the mean where all mice are more or less equally fit, though possibly for different reasons. For instance, the larger mice may be more successful in establishing territory, but the smaller mice may more easily meet their nutritional demands. But there must be bounds on permissible departures from a limited range, beyond which extreme size in either direction becomes a crippling handicap. Whether a trait is judged to be adaptive or not is nothing more than a statement of the amount of variation in that trait that we are prepared to accept, and it is merely a question of degree.

In the next section, the influence of body size on fertility – which is more obviously a component of fitness – will be examined, to identify some of the reasons why extreme deviants in body weight may be less fit.

Effects of Body Weight on Fertility

Selection for body weight leads to a well-documented correlated effect on litter size: larger mice have larger litters, while small size leads to small litters (see Eisen, 1974; Roberts, 1979). Other things being equal, large mice should therefore be fitter. Other things, however, are not equal, and large mice have their own reproductive difficulties. Indeed, Lerner (1954) argued that directional selection for any metric trait would lead to reduced fitness, and that part of his argument has never been seriously challenged.

The problem with large mice is their proneness to sterility. It is of no use to them to have potentially large litters if they have no litter at all. Eisen, Hanrahan & Legates (1973) and Falconer (1973) both illustrate the problem in sharp relief. Roberts (1967b) reports a large line that was lost through sterility. We need go no further to appreciate that there *has* to be some kind of stabilizing selection for body weight. Extreme deviants in either direction are not fit, and that is that.

Some of the reproductive difficulties of large mice seem to arise because they get too fat. An offshoot of the line that was lost (above) was saved by mating it at an earlier age (Roberts, 1974) before fat had accumulated. The fertility of another large line was helped considerably by relaxing selection (Roberts, 1966), even though there was no additive genetical variance in body weight remaining in that line. This example proves that at least some of the negative correlation between large size and fertility is environmental in terms of its immediate origin. However, even though the immediate physiological cause may be environmental, it does not remove the ultimate genetical involvement if large mice inexorably (though perhaps not unavoidably) get fat. Many workers have found that large mice get fat, though in some cases the increased fatness is not apparent until later ages (Robinson & Bradford, 1969; Timon, Eisen & Leatherwood, 1970; Bakker, 1974; Sutherland, Biondini & Ward, 1974; McPhee & Neill, 1976; Hayes & McCarthey, 1976; Eisen & Bandy, 1977; Eisen, Bakker & Nagai, 1977).

What might happen in the wild? If large size leads to fatness and fatness to sterility, one answer would be to advance sexual maturity, so that some breeding is done before the fat accumulates. We may imagine that there must be some selection anyway for early maturity and rapidity of reproduction. But in the case of large mice, this would not seem to be adequate compensation for the drastic shortening of the length of their reproductive life. Roberts (1961) compared the lifetime production of two large strains and two small strains. The two small strains each produced 11 litters over their lifetime, as against three and five, respectively, for the two large strains. The result was that the small mice weaned almost twice as many offspring as the large mice. More striking, however, was the lifetime production of a cross between a large and a small strain; the cross-bred mice weaned three times as many offspring, over their lifetime, as the better of the two parental strains.

It may be only partly relevant that the parental strains of this cross differed in body size. Roberts (1967b) reported improvements in fertility among crosses of large lines, though the lifetime production of those crosses was not examined. But we may speculate in this context, given the superiority of hybrids, whether there may not be some selection in the wild for any propensity to out-cross — to seek a mate from outside the population in which the mouse grew up. This could be regarded as an extension of the incest taboo, and we can at least specify one of the conditions which might favour it. It is easiest to express the notion verbally if we allow ourselves some "selfish gene" thinking. If a mouse (or any diploid organism) produces a hybrid offspring, then only half of the gametes of that offspring will transmit the gene that we have in mind. If, instead, the mouse mates within its own population, then there is some probability that the gene we are monitoring becomes homozygous with a replica of itself, and when such an offspring in turn breeds, all of its gametes will transmit the gene in question. We can therefore see how the balance might swing: out-crossing will become favourable when hybrid offspring, with single copies of the gene, can transmit it more frequently than the more inbred offspring with two copies. Following an out-cross, there will be less advantage to further out-crossing, and offspring with two copies of the gene may do better than hybrids, until inbreeding depression (see later) reduces fertility again. Yanai & McClearn (1972a, b) were able to show the preference of females for mating with unrelated males, both among inbred and random-bred mice. We therefore have a behavioural mechanism for promoting higher fertility by producing hybrid offspring preferentially. But the selective advantage of this depends on the reproductive superiority of these hybrids, as noted above.

The Mediation of Gene Effects on Growth

We shall now examine briefly by what mechanisms genes may influence body size, and consider the basis of some of the genetical variation we may observe.

Increased growth could be obtained in one of two ways. Either the mouse could eat more food, or else the same amount of food could be used more efficiently. Selection for increased growth in the laboratory has generally altered both: larger mice eat more and also convert it more efficiently. Falconer (1960a), Fowler (1962), Rahnefeld *et al.* (1965), Lang & Legates (1969), Timon & Eisen (1970), Eisen & Bandy (1977) and finally, Eisen, Bakker & Nagai (1977) all agree that both food intake and efficiency are altered by selection. In addition, Sutherland, Biondini, Haverland *et al.* (1970) showed that both food intake and the efficiency of conversion respond to separate selection, confirming that each trait is under some genetical control. There is a complex relationship between intake and efficiency which affects the interpretation of these results. It is inappropriate to seek out the complexities here, but the main feature of the relationship is the following. If an average-sized mouse needs, say, 15 g of mouse food to keep itself alive for a week, without additional growth, then the mouse that eats 16 g, and grows a bit, is clearly more efficient than the mouse that eats 15 g and whose weight stays still. Against that, a mouse that gets by on 14 g (still without growing) is more efficient than the mouse whose intake of 15 g just keeps it going. But at marginal intakes, there is usually a positive correlation between intake and efficiency. At the other end of the scale, excessive voracity can lead to inefficiency, if the high intake of food is not converted fully into a weight gain. It could be accompanied by a higher heat loss, or an energetically costly body composition, e.g. excess fat.

If laboratory results can be translated into the field (or barn), what kind of selection on appetite and efficiency might we find? It seems reasonable to suppose that high efficiency would always be at a premium, and though there is some genetical variation in the trait, its heritability is generally lower than that of body weight. It is likely, therefore, that over its evolutionary history, there has been considerable selection for efficiency. It is also plausible to argue that because of the complex relationship between appetite and efficiency, outlined above, there may have been some selection for an intermediate level of intake; mice eating less or more may have a lower efficiency. A low efficiency might not matter as long as food is plentiful and constantly available. But mice that let their efficiencies slip in times of plenty leave descendants who are poorly equipped to meet the next shortage, for survival means the surviving of crises. There may therefore be recurring cycles of weeding out of inefficient mice.

Coleman (1978), in a penetrating review, speculates on the kind of selection that occurs in the wild with specific reference to lipid metabolism. He notes that several species of desert rodents, when brought into the laboratory, can develop symptoms of clinical diabetes. Many animals become hyperphagic, obese, hyperinsulinaemic and show some glucose intolerance, while a few might develop the more extreme symptoms of hyperglycaemia, glucosuria and a ketonic form of diabetes. Coleman points out that in the feral state, these rodents have a limited food supply and develop normally. Further, the very features that cause problems in the plenitude of the laboratory are associated with a metabolically thrifty genotype that, Coleman speculates, is the product of natural selection. During periods of excess of food, the potential hyperphagia, coupled with an increased rate of lipogenesis, allows the rapid accumulation of fat stores for use in days of privation. It is pointed out that obese mice may survive up to 30 days of starvation, whereas normal mice would be dead in two or three days. Coleman suggests that hyperinsulinaemia, as the most consistent feature of the syndrome, may be the key to the improved lipid anabolism. Whatever the exact nature of the mechanisms, we have a model of the kind of selection that may operate on appetite control, with its consequences for improved efficiency.

Although he does not note it as a specific concern, there is a corollary of Coleman's review that we should note *en passant*. It is that genes beneficial in the wild may be highly deleterious in the laboratory, and find themselves selectively eliminated.

Any discussion of the mediation of gene effects on growth at the biochemical level is both beyond my scope and outwith my competence. The reader is referred to Shire's (1976) exhaustive review of genetical variation in endocrine systems as an excellent place to start. Shire documents abundant evidence of genetical variation, both in various aspects of hormone production and in the responses of target organs. As one example of the hormonal changes brought about by selection for body weight, Pidduck & Falconer (1978) found that increased growth in their strains was partly due to an increased amount, or activity, of circulating growth hormone, while reduced growth was due, again in part, to a reduced sensitivity of the target organs. Clearly, there is immense scope for selection at this level, though a caveat is necessary: just because we observe changes in mechanisms following a change in body size, we need not suppose that the changes in mechanisms were necessarily a direct cause of differential growth; they could just as well be the consequences of differential growth. As an example, we may consider cell number and cell size. It is obvious that if a mouse is to be bigger, then it must either have more cells or bigger cells, or some combination of the two. Robinson & Bradford (1969) suggested that selection for body weight alters cell number rather than cell size, a conclusion to which Priestley & Robertson (1973) somewhat cautiously lend support. In contrast, Falconer, Gauld & Roberts (1978a) found that both cellular components had been altered by selection, and that in some organs, changes in cell mass were as great as changes in cell number. More than that, when large and small mice were compared at the same weight (as distinct from the same age), then the organ sizes were the same and the number of cells was the same; from which it follows that cell size was also the same. Given the size of an organ, then its cellular components became predictable irrespective of strain or age. In other words, the effect of the selection had been to alter developmental age relative to chronological age. Large mice simply grew faster, and one of the results was to increase both cell number and cell size. It is therefore impossible to say that the changes in the cellular components were in any sense the cause of a change in body weight.

Before we dismiss the cellular basis of growth regulation, however, we should note the implications from aggregation chimaeras between large and small mice (Roberts *et al.*, 1976; Falconer, Gauld & Roberts, 1978b). The proportion of large cells in any individual mouse is a matter of chance, and can vary over the whole range. The body weight of the ensuing chimaera is linearly and directly proportional to the number of large cells, as if growth depended on the cellular genotype throughout the whole body. No particular organ acted as if it was controlling growth. The nine organs included in these studies, taken together, accounted for all of the variance in growth; indeed, any one of them on its own gave a reasonable prediction. If there is a growth-controlling organ, then its cellular composition must correlate highly with those included, which in turn correlated highly among themselves.

The purpose of this section has been two-fold. The first was to hint – no more – at the manifold nature of the raw material on which selection for body weight can act. The second was cautionary: to suggest that we ought not to be over-anxious to deduce causation from association. It may be tempting, at first, to link an animal's growth to the food that it ingests, and the efficiency with which it converts its food into animal product. This, however, ignores the problem of what controls the animal's appetite in the first place. Radcliffe & Webster (1976) postulate that food intake is closely related to the animal's impetus (rats, in their case) for laying down protein. Certainly, there is the well-known phenomenon of compensatory growth, following a period of inadequate feeding, whereby animals tend to revert to a normal weight for age. Indeed, were this not to occur under wild conditions, animals could never recover from temporary deprivations. Growth control must therefore ignore ephemeral perturbations; this may be one of the reasons why the growth control system has proved so intractable to experimental attack.

GENETICAL VARIATION IN FERTILITY

Responses to Selection

Selection for fertility traits has not enjoyed the attention expended upon various aspects of body size and growth rate. The first definitive account of a selection experiment for litter size was given by Falconer (1960b), though he had published a preliminary account earlier (Falconer, 1955). He produced a high line with a mean litter size of nine live young at birth, and a low line with a mean of six. The heritability of the divergence between the two lines was only 13%; the high line on its own gave a low estimate of 8%, though the low line showed a higher value of 23%. However, because the high line was more variable, the actual responses were not so asymmetrical as the heritability values.

Subsequent experiments have in general exceeded Falconer's somewhat modest responses. Bateman (1966) using what was essentially a form of mass selection, generated a two-fold difference (with a high line of 11) over 12 generations. Bateman does not quote heritabilities but he was, at least at times, selecting intensely. Bakker, Wallinga & Politiek (1978), selecting for large litter size only, increased it from eight to 14, over 29 generations, with a realized heritability of 11%. Eisen (1978) reports the impressive value of 16 young born in a line selected for litter size, though he had started from the high base level of 12. The heritability approached 20%, depending on method of estimation. All of these studies give uncomplicated results. However, Bradford (1968, 1971) presents some interesting variations. He increased litter size from nine to 12 over 11 generations, from a cross-bred derived from eight inbred lines. He was less successful when selecting from a four-line cross, his improvement being only about one offspring per litter. He further selected from the four-line cross after it had previously undergone seven generations of selection for weight gain, and this time failed to improve litter size at all. He was equally unsuccessful when selecting for increased litter size following superovulation. Furthermore, Bradford did not observe the usual correlated effect on litter size when selecting for weight gain. He noted that the genetical correlation between the two traits varies among populations. Batten & Berry (1967) had independently come to the conclusion that body size and litter size need not be correlated; indeed, they went a step further, and claimed that in the case of their island mice, natural selection had operated against such a correlation.

The results of laboratory experiments, on balance, indicate that additive variance for litter size is usually present, though in variable amounts, and that the trait usually responds to selection. On the genetical argument presented in the Introduction, this would suggest that litter size is less closely related to natural fitness than is sometimes supposed. Batten & Berry (1967) make the same point, and we shall return to it briefly in a later section.

Non-additive Genetical Variance in Litter Size

Non-additive variance stems from dominance and interaction effects, and is defined as that part of the genetical variance not amenable to selection. It may nevertheless contribute to the resemblance between certain classes of relatives, particularly full sibs. Crudely, non-additive variance refers to the special effects of combinations of genes, either at the same locus or at separate loci. When an animal breeds, these combinations are broken up and we observe only those effects that genes exercise singly, giving rise to the additive variance. Non-additive variance comes into play during inbreeding and crossing, where levels of heterozygosity (among other genetical effects) are altered. It is the basis of inbreeding depression and heterosis.

There is no room here to review the copious literature on inbreeding and crossing in the mouse. Eisen (1974) provides access to this literature. Suffice to say that litter size declines by about half a mouse per 10% increase in inbreeding coefficient, and that litter size is restored on crossing. Even standard laboratory inbred strains usually (though not inevitably) show heterosis in litter size, despite the fact that such strains are the peculiarities that have survived the inbreeding process, and thus represent nothing except themselves. The vast majority of lines fail to withstand inbreeding, and become extinct through infertility and inviability. The survivors are therefore not random representatives of the base population from which they were drawn. Falconer (1971) was able to capitalise on this fact by forming nine inbred lines from his strain selected for high litter size (see earlier). Inbreeding depression in litter size immediately set in, at about the expected rate noted above. However, by maintaining sublines and practising selection for litter size, Falconer was able to maintain four of the lines through 11 generations of sib-mating. At that point he crossed them and the derived cross-breds had a mean litter size of 1.5 mice above that of the original selected strain. Falconer postulated that rare recessive genes, perhaps as many as 30 or possibly more, had arrested the original response, and that these recessive genes had been exposed by inbreeding and eliminated by selection.

What do these general considerations lead us to expect with wild mice? An extension of the argument that traits close to fitness display little additive genetical variance is that such traits should also have considerable non-additive variance left. Traits close to natural fitness should therefore be particularly susceptible to the effects of inbreeding. As far as litter size is concerned, there has been little systematic work on the genetical parameters of wild-caught mice.

Two reports suggest that wild-caught mice are not particularly sensitive to inbreeding. Lynch (1977) reported that litter size in wild-caught mice declined with inbreeding at the standard laboratory rate, while the mean litter sizes of her surviving sublines altered little over six generations of sib mating. Connor & Bellucci (1979) similarly employed extensive subline replacement, but even so, five of their ten inbred lines, from wild-caught mice, failed to survive the inbreeding. Litter size declined under their conditions, but only slowly at first. They deduced that inbreeding was being counteracted by natural selection, possibly involving heterozygous advantage. They certainly found substantial heterozygosity in four of their five lines, even after 20 generations of sib-mating.

If litter size in wild mice was one of the major determinants of natural fitness, then inbreeding might be expected to have drastic effects on it. The evidence from the two studies just quoted is that this is not so, and that litter size in the wild has not been subjected to previous natural selection much different to that pertaining under laboratory conditions.

The Components of Fertility

Litter size is a complex trait determined sequentially by ovulation, fertilization, implantation and embryonic survival — even without the perinatal and postnatal hazards that determine the number of offspring that themselves survive to breed. We shall ignore here any genetical influences of the male or litter size, and female sterility will also be excluded. We shall concentrate on the normal range of variation found among fertile animals.

Ovulation rate responds to artificial selection. Land & Falconer (1969) selected both for natural and induced ovulation rates, with substantial responses. The lines selected for high and low natural ovulation, however, did not differ in the number of young born, despite a difference of seven ova shed (21 v. 14). Land & Falconer's induced ovulation lines, on the other hand, differed by about two young at birth, when allowed to ovulate naturally. The genetical correlation of 0.33 which they report between natural and induced ovulation shows that, despite some genetical overlap, the two traits are substantially different, as Bradford (1968) had found in his selection programme. Land (1970) was able to show genetical influences both on FSH activity and on ovarian sensitivity, and that both are positively correlated, genetically, with body weight. This explains why selection for body weight often (though not invariably) changes ovulation rate, and vice versa.

The most extensive study of selection for components is that reported by Bradford (1969). He selected separately for ovulation rate and embryo survival, both in the presence and absence of superovulation. Briefly, embryo survival responded to selection in both cases, increasing litter size by two mice or so in the untreated line. Litter size did not increase in the superovulated line, but embryo survival improved, in proportionate terms, because of a reduction in ovulation rate. Selection for high natural ovulation gave a response of about two ova over ten generations, but without any increase in number born, as Land & Falconer (1969) had also found. Selection for a high induced ovulation rate gave no response.

By and large, therefore, genetical manipulation of the components of litter size has little effect on the number born. Nevertheless, litter size itself *can* be manipulated genetically, so what happens to the components? Falconer (1960b, 1963) found that the response in his high line was entirely attributable to increased ovulation rate, while in his low line, ovulation rate had not been altered but embryonic deaths had increased markedly in the post-implantational stage. Bateman (1966), working with similar material, essentially confirmed Falconer's result, except that the embryonic mortality in his low line was distributed evenly before and after implantation. Bradford (1969) also found that his high litter size line had increased in ovulation (without affecting mortality), noting that the reciprocal effect when selecting for ovulation had not been found (see above). Though perhaps the evidence is too meagre to generalize, it is so far entirely consistent. Selection for litter size yields qualitatively different responses in the two directions: high litter size means more eggs, low litter size greater mortality.

Studies comparable to the laboratory ones on the components of litter size were conducted on wild mice by Batten & Berry (1967). Their material derived from several island and mainland populations. Ovulation rates were low by laboratory standards, seldom exceeding ten. Nevertheless, they found extensive embryonic mortality; a fairly constant fraction of one-third of all eggs were lost, more of the losses occurring before implantation than afterwards. The authors invoke deleterious genetical factors to explain these deaths, but this is open to question. Certainly, inbreeding studies on wild mice (Lynch, 1977; Connor & Bellucci, 1979) do not suggest that wild mice are particularly prone to the exposure of recessive lethals. Further, Southwick's data (quoted by Batten & Berry, 1967) show that embryonic death increased with population density. It seems likely that much of the embryonic death in wild mice is of environmental origin, and that the lethality has not been eliminated by natural selection since it affords a reservoir of higher fertility should environmental circumstances prove favourable.

So, what price litter size in wild populations? As Batten & Berry (1967) point out, optimum litter size is probably submaximal, since survival may be reduced in large litters. And as we saw earlier, mice with large litters may have a shorter length of reproductive life (Roberts, 1961). In that study there was the complication of large body size. But in another case, even a line selected for high litter size (14 as against eight for the control line) was overtaken by the control line, in terms of cumulative number born, by 28 weeks of age (Wallinga & Bakker, 1978). Under a system of continuous pairing, their high line was unable to sustain its high litter size over successive parities, unless the male was removed to prevent post-partum fertilization.

Few would dispute that an intermediate litter size would be favoured by natural selection, extremes in either direction being less fit. But this statement avoids a more critical question – what determines the level of intermediacy? A glib answer might be that the exact level will depend on the amount of environmental support. But if we are to invoke adaptation on that scale, for a character of low heritability, we need a lot of time in a very constant environment. And even were this so, it would leave the population potentially very vulnerable to environmental change. Extreme adaptation can be self-defeating, and the fittest mouse is probably the one with the largest number of surviving and fertile descendants that could – conceptually – cope with a fairly broad range of environments. Over its evolutionary history, the mouse will have been subjected to natural selection for increased litter size – litters about as large as can be sustained *on average* in a variety of conditions, with some spare capacity in fertility that can be cashed in when conditions are favourable. The genetical evidence is entirely compatible with this – the amount of non-additive variance in the trait reflecting past natural selection for larger litters, but with enough additive variance remaining to show that litter size is not the only component of fitness, and that it could, if need be, still respond to further selection.

CONCLUSIONS

It has been argued that both body weight and litter size, under natural conditions, have been subjected to at least a mild form of stabilizing selection whereby extreme deviants in either direction are selectively eliminated. But that statement, as it stands, is a trivial one; it is probably true of all traits at all times. A more meaningful question might be: is there a narrow range, or a wide one, over which the effects of natural selection are not easily detectable? The problem here is that we cannot compare different traits on the same scale; there is no logical way of deciding whether corpora lutea counts are more variable than tail length. We can, however, begin to make objective comparisons if we ask how much of the variation is attributable to genetical causes, and to what kind of genetical causes. The answer is clear: body size has a considerable amount of additive variance, giving high estimates for the heritability, whereas fertility traits have lower heritabilities and more non-additive variance. Body weight changes are readily brought about by selection; fertility is more subjected to the effects of inbreeding and crossing. Though most of the evidence comes from laboratory populations, studies on wild mice do not suggest that they are particularly different in their genetic architecture. If we accept the premises set out in the Introduction, then it is at least a reasonable speculation that body weight has not been under the strong influence of natural selection during its evolutionary history. Fertility, on the other hand, has been the subject of considerable natural selection; indeed, if we take fertility and viability together, they *are* natural selection – and there is nothing else that natural selection can be. The component traits of fertility (and litter size is only a component) can still show some additive variance. This is explained, at least in part, by the complex interactions among the components; changes in one component may be buffered by compensating changes in other components, with perhaps little effect on fertility as a global trait. In a sense, the main conclusion was stated at the beginning. What I have attempted in this review is to marshall some of the evidence to support it.

ACKNOWLEDGEMENTS

I am greatly indebted to Dr E. J. Eisen for his patient and friendly discussion of points that arose during the preparation of this manuscript. I am particularly grateful to him for his struggle to read the manuscript in a very rough form – and I apologise to him for dealing selectively with his valuable comments.

I wish to thank Professor R. J. Berry and Dr Carol B. Lynch for drawing my attention to some critical references. To them, and particularly to the authors concerned, I also apologise for any omissions.

REFERENCES

Bakker, H. (1974). Effect of selection for relative growth rate and bodyweight of mice on rate, consumption and efficicency of growth. *Meded. LandbHoogesch. Wageningen* **1974**: 74–78.

Bakker, H., Wallinga, J. H. & Politiek, R. D. (1978). Reproduction and body weight of mice after long-term selection for large litter size. *J. Anim. Sci.* **46**: 1572–1580.

Barnett, S. A., Munro, K. M. H., Smart, J. L. & Stoddart, R. C. (1975). House mice bred for many generations in two environments. *J. Zool., Lond.* **177**: 153–169.

Bateman, N. (1966). Ovulation and post-ovulational losses in strains of mice selected from large and small litters. *Genet. Res.* **8**: 229–241.

Batten, C. A. & Berry, R. J. (1967). Prenatal mortality in wild-caught house mice. *J. Anim. Ecol.* **36**: 453–463.

Berry, R. J. & Jakobson, M. E. (1975). Adaptation and adaptability in wild-living house mice (*Mus musculus*). *J. Zool., Lond.* **176**: 391–402.

Berry, R. J., Jakobson, M. E. & Peters, J. (1978). The house mice of the Faroe Islands: a study in microdifferentiation. *J. Zool., Lond.* **185**: 73–92.

Berry, R. J., Peters, J. & Van Aarde, R. J. (1978). Sub-Antarctic house mice: colonization, survival and selection. *J. Zool., Lond.* **184**: 127–141.

Bradford, G. E. (1968). Selection for litter size in mice in the presence and absence of gonadotrophin treatment. *Genetics* **58**: 283–295.

Bradford, G. E. (1969). Genetic control of ovulation rate and embryo survival in mice. I. Response to selection. *Genetics* **61**: 905–921.

Bradford, G. E. (1971). Growth and reproduction in mice selected for rapid body weight gain. *Genetics* **69**: 499–512.

Coleman, D.L. (1978). Diabetes and obesity: shifty mutants? *Nutr. Rev.* **36**: 129–132.

Connor, J. L. & Bellucci, M. J. (1979). Natural selection resisting inbreeding depression in captive wild house mice (*Mus musculus*). *Evolution* **33**: 929–940.

Ebert, P. D. & Hyde, J. (1976). Selection for agnostic behaviour in wild female *Mus musculus*. *Behav. Genet.* **6**: 291–304.

Eisen, E. J. (1974). The laboratory mouse as a mammalian model for the genetics of growth. *Proc. World congr. Genet. Appl. Livestock Prod.* **I**: 467–492.

Eisen, E. J. (1976). Results of growth curve analyses in mice and rats. *J. Anim. Sci.* **42**: 1008–1023.

Eisen, E. J. (1978). Single-trait and antagonistic index selection for litter size and body weight in mice. *Genetics* **88**: 781–811.

Eisen, E. J., Bakker, H. & Nagai, J. (1977). Body composition and energetic efficiency in two lines of mice selected for rapid growth rate and the F_1 crosses. *Theor. appl. Genet.* **49**: 21–34.

Eisen, E. J. & Bandy, T. (1977). Correlated responses in growth and body composition of replicated single-trait and index selected lines of mice. *Theor. appl. Genet.* **49**: 133–144.

Eisen, E. J., Hanrahan, J. P. & Legates, J. E. (1973). Effects of population size and selection intensity on correlated responses to selection for postweaning gain in mice. *Genetics* **74**: 157–170.

Falconer, D. S. (1955). Patterns of response in selection experiments with mice. *Cold Spring Harbor Symp. Quant. Biol.* No. 20: 178–196.

Falconer, D. S. (1960a). Selection of mice for growth on high and low planes of nutrition. *Genet. Res.* **1**: 91–113.

Falconer, D. S. (1960b). The genetics of litter size in mice. Symposium on mammalian genetics and reproduction. *J. cell. comp. Physiol.* **56** (Suppl. 1): 153–167.

Falconer, D. S. (1963). Qualitatively different responses to selection in opposite directions. In *Statistical genetics and plant breeding:* 487–490. Hanson, W. D. & Robinson, H. F. (Eds). (Publ. 982 Nat. Acad. Sci.) Washington D. C.: Nat. Res. Council.

Falconer, D. S. (1971). Improvement of litter size in a strain of mice at a selection limit. *Genet. Res.* **17**: 215–235.

Falconer, D. S. (1973). Replicated selection for body weight in mice. *Genet. Res.* **22**: 291–321.

Falconer, D. S., Gauld, I. K. & Roberts, R. C. (1978a). Cell numbers and cell sizes in organs of mice selected for large and small body size. *Genet. Res.* **31**: 287–301.

Falconer, D. S., Gauld, I. K. & Roberts, R. C. (1978b). Growth control in chimaeras. In *Genetic mosaics and chimaeras in mammals.* Russel, L. B. (Ed.). New York: Plenum Publishing Corporation.

Fisher, R. A. (1930). *The genetical theory of natural selection.* Oxford: Oxford University Press.

Fowler, R. E. (1962). The efficiency of food utilization, digestibility of foodstuffs and energy expenditure of mice selected for large or small body size. *Genet. Res.* **3**: 51–68.

Goodale, H. D. (1938). A study of the inheritance of body weight in the albino mouse by selection. *J. Hered.* **29**: 101–112.

Hayes, J. F. & McCarthy, J. C. (1976). The effects of selection at different ages for high and low body weight on the pattern of fat-deposition in mice. *Genet. Res.* **27**: 389–403.

Hill, W. G. (1971). Design and efficiency of selection experiments for estimating genetic parameters. *Biometrics* **27**: 293–311.

Kimura, M. (1957). Some problems of stochastic processes in genetics. *Ann. math. Statist.* **28**: 882–901.

Land, R. B. (1970). Genetic and phenotypic relationships between ovulation rate and body weight in the mouse. *Genet. Res.* **15**: 171–182.

Land, R. B. & Falconer, D. S. (1969). Genetic studies of ovulation rate in the mouse. *Genet. Res.* **13**: 25–46.

Lang, B. J. & Legates, J. E. (1969). Rate, composition and efficiency of growth in mice selected for large and small body weight. *Theor. appl. Genet.* **39**: 306–314.

Laurie, E. M. O. (1946). The reproduction of the house mouse (*Mus musculus*) living in different environments. *Proc. R. Soc.* (B) **133**: 248–281.

Lerner, I. M. (1954). *Genetic homeostasis.* Edinburgh: Oliver & Boyd.

Lynch, C. B. (1977). Inbreeding effects upon animals derived from a wild population of *Mus musculus. Evolution* **31**: 526–537.

MacArthur, J. W. (1944). Genetics of body size and related characters. I. Selecting small and large races of the laboratory mouse. *Am. Nat.* **78**: 142–157.

McCarthy, J. C. & Doolittle, D. P. (1977). Effects of selection for independent changes in two highly correlated body weight traits of mice. *Genet. Res.* **29**: 133–145.

McPhee, C. P. & Neill, A. R. (1976). Changes in body composition of mice selected for high and low eight week weight. *Theor. appl. Genet.* **47**: 21–26.

Monteiro, L. S. & Falconer, D. S. (1966). Compensatory growth and sexual maturity in mice. *Anim. Prod.* **8**: 179–192.

Pidduck, H. G. & Falconer, D. S. (1978). Growth hormone function in strains of mice selected for large and small size. *Genet. Res.* **32**: 195–206.

Plomin, R. J. & Manosevitz, M. (1974). Behavioural polytypism in wild *Mus musculus*. *Behav. Genet.* **4**: 145–157.

Poiley, S. M. (1972). Growth tables for 66 strains and stocks of laboratory animals. *Lab. Anim. Sci.* **22**: 757–779.

Priestley, G. C. & Robertson, M. S. M. (1973). Protein and nucleic acid metabolism in organs from mice selected for larger and smaller body size. *Genet. Res.* **22**: 255–278.

Radcliffe, J. D. & Webster, A. J. F. (1976). Regulation of food intake during growth in fatty and lean female Zucker rats given diets of different protein content. *J. Nutr.* **36**: 457–469.

Rahnefeld, G. W., Comstock, R. E., Boylan, W. J. & Singh, M. (1965). Genetic correlation between growth rate and feed per unit gain in mice. *J. Anim. Sci.* **24**: 1061–1066.

Roberts, R. C. (1961). The lifetime growth and reproduction of selected strains of mice. *Heredity* **16**: 369–381.

Roberts R. C. (1965a). Some contributions of the laboratory mouse to animal breeding research. Part I. *Anim. Breeding Abstr.* **33**: 339–353.

Roberts, R. C. (1965b). Some contributions of the laboratory mouse to animal breeding research. Part II. *Anim. Breeding Abstr.* **33**: 515–526.

Roberts, R. C. (1966). The limits to artificial selection for body weight in the mouse. II. The genetic nature of the limits. *Genet. Res.* **8**: 361–375.

Roberts, R. C. (1967a). Some evolutionary implications of behaviour. *Can. J. Genet. Cytol.* **9**: 419–435.

Roberts, R. C. (1967b). The limits to artificial selection for body weight in the mouse. III. Selection from crosses between previously selected lines. *Genet. Res.* **9**: 73–85.

Roberts, R. C. (1974). Selection limits in the mouse and their relevance to animal breeding. *Proc. World Congr. Genet. appl. Livestock Prod.* **1**: 493–509.

Roberts, R. C. (1979). Side-effects of selection for growth in laboratory animals. *Livestock Prod. Sci.* **6**: 93–104.

Roberts, R. C., Falconer, D. S., Bowman, P. & Gauld, I. K. (1976). Growth regulation in chimaeras between large and small mice. *Nature, Lond.* **260**: 244–245.

Robertson, A. (1960). A theory of limits in artificial selection. *Proc. R. Soc.* (B), **153**: 234–249.

Robinson, D. W. & Bradford, G. E. (1969). Cellular response to selection for rapid growth in mice. *Growth* **33**: 221–229.

Shire, J. G. M. (1976). The forms, uses and significance of genetic variation in endocrine systems. *Biol. Rev.* **51**: 105–141.

Sutherland, T. M., Biondini, P. E., Haverland, L. H., Pettus, D. & Owen, W. B. (1970). Selection for rate of gain, appetite and efficiency of feed utilization in mice. *J. Anim. Sci.* **31**: 1049–1057.

Sutherland, T. M., Biondini, P. E. & Ward, G. M. (1974). Selection for growth rate, feed efficiency and body composition in mice. *Genetics* **78**: 525–540.

Timon, V. M. & Eisen, E. J. (1970). Comparisons of *ad libitum* and restricted feeding of mice selected and unselected for postweaning gain. I. Growth, feed consumption and feed efficiency. *Genetics* **64**: 41–57.

Timon, V. M., Eisen, E. J. & Leatherwood, J. M. (1970). Comparisons of *ad libitum* and restricted feeding of mice selected and unselected for postweaning gain. II. Carcase composition and energetic efficiency. *Genetics* **65**: 145–155.

Wallinga, J. H. & Bakker, H. (1978). Effect of long-term selection for litter size in mice on lifetime reproduction rate. *J. Anim. Sci.* **46**: 1563–1571.

Yanai, J. & McClearn, G. E. (1972a). Assortative mating in mice and the incest taboo. *Nature, Lond.* **238**: 281–282.

Yanai, J. & McClearn, G. E. (1972b). Assortative mating in mice. I. Female mating preference. *Behav. Genet.* **2**: 173–183.

Symp. zool. Soc. Lond. (1981) No. 47, 255–266

Diet and Nutrition

R. J. WARD

Toxicology Department, May & Baker Limited, Dagenham, Essex, UK

SYNOPSIS

Little is known about the dietary requirement of the laboratory mouse and even less about that of the house mouse. Most estimates of the requirements are based on adequate diets or on the assumption that the mouse has similar requirements to the rat. A review of the known food habits of the house mouse and closely related species unfortunately does not shed any light on the requirements because the necessary analytical information is not yet available.

INTRODUCTION

Little is known about the nutrient requirements of the laboratory mouse, and even less about those of wild house mice, although it seems reasonable to suppose that they may be similar.

The house mouse and other closely related species depend on the food available in the immediate area of their habitat and consequently nutrient intake will vary, particularly from season to season, whereas that of the laboratory mouse is more or less constant throughout its life.

Because of the lack of information on the house mouse (*Mus musculus*) it has been necessary to include closely related species such as the wood mouse (*Apodemus sylvaticus*) and the deer mouse (*Peromyscus maniculatus bairdi*) in this review.

NUTRIENT REQUIREMENTS

Although mice have been used extensively as laboratory animals for many years (over 3×10^6 were used in the UK in 1977), research into the nutrient requirements for growth, reproduction, lactation or maintenance has received little attention. As a result, most estimates of the nutrient requirements for mice are based on:

(a) data accumulated many years ago involving mouse strains or diet ingredients that are no longer available or cannot be identified;

(b) experimental results derived from studies not directed towards the establishment of requirements;

(c) estimated nutrient consumption by mice fed diets producing "acceptable performance" or

(d) the assumption that mouse requirements are similar to those of the rat (National Academy of Sciences, 1978).

In the report, *Dietary Standards for Laboratory Animals* (1977), prepared by the Diets Advisory Committee of the Medical Research Council's Laboratory Animals Centre, rats and mice are considered together "as insufficient evidence of differences was found to justify separate nutrient recommendations."

When considering the nutrient requirement of a species, it is important to ask the question "requirement for what". Most diets are formulated to provide rapid post-weaning growth that will result in maximum body weight at maturity. If this type of diet is fed throughout the life of the animal it will become obese and will probably develop one or more spontaneous tumours later in life. There have been many papers over the last few years on the effect of restriction of food intake on the incidence of tumours and longevity but these have been almost exclusively concerned with the rat. Recently, studies have been made on the incidence of spontaneous tumours and longevity in mice (Roe & Tucker, 1973). The results of one experiment, given in Table I, show that, although in this case the length of life was unaffected, there was a dramatic decrease in the number of tumours in those receiving a restricted diet.

TABLE I

Effect of dietary intake on tumour incidence (Roe & Tucker, 1973)

Diet	Number of mice dead at 18 months	Number of tumours
5 g d^{-1}	23	4
ad libitum (1/cage)	15	32
ad libitum (5/cage)	22	23

In a more recent experiment (Tucker, 1979) male and female mice were given diet *ad libitum* or at a rate of 4 g/day. The results (Table II) show that a 20% decrease in food intake increased the survival rate at 24 months by 25% in the males and 29% in the females, and

TABLE II

Tumour incidence and survival in mice (Tucker, 1979)

	♂		♀	
Food intake	5 g[a]	4 g[b]	5 g[a]	4 g[b]
% survival (24 months)	72	90	68	88
% bearing tumours	80	52	82	46

[a] *Ad libitum.*
[b] *Restricted.*

the percentage of animals bearing tumours was decreased by 32% in the males and 31% in the females.

Restriction of food intake is not practical when conducting long-term studies with rodents, particularly in carcinogenicity tests when large numbers of animals are necessary. Over the past few years a diet has been developed with a low protein and a low energy content (Labsure RDM).* Feeding this type of diet *ad libitum* has a similar effect to restriction of nutrient intake. As far as I am aware, no studies have yet been carried out on mice but in the rat there is a considerable increase in the life span of the animal. After two years, only two of 15 controls receiving a normal rodent diet were still alive whereas all but four of 15 given the low protein/low energy diet were still alive. The average mature weight of these animals was about 70% of that of the controls (Ward & Clark, in preparation). Animals receiving this type of diet have to eat continuously in order to satisfy their energy appetite. This is in contrast to the eating habits of laboratory rodents and presumably wild rodents, which consume most of their food at night. But are laboratory rodents necessarily normal rodents? Some doubt was raised when it was found that the ALAT (alanine amino transferase) level in the plasma of the rats given the low protein diet was considerably higher than that of those given a normal diet. A comparison of the values of the ALAT levels in the plasma of wild rats caught in Wales with laboratory animals (Table III) suggests that these are similar to the laboratory rats given the normal diet. The reason for the raised ALAT values in the deprived rats is at present not known.

CHEMICAL COMPOSITION OF ADEQUATE DIETS

The calculated composition of two diets which have proved successful in the large-scale breeding of laboratory rodents is given in Table IV.

* No longer available, replaced by RDM 22.

TABLE III

Mean values of serum alanine amino transferase

Laboratory rats	CRM diet	31.92 ± 1.76 mu ml^{-1}
Laboratory rats	RDM diet	81.25 ± 6.86 mu ml^{-1}
Wild rats	Natural diet	49.5 ± 4.40 mu ml^{-1}

TABLE IV

Calculated analysis of diets for breeding mice

	ALGH	*CRM*
Crude oil	4.8%	2.35%
Crude protein	17.3%	17.45%
Crude fibre	3.9%	4.26%
Calcium	0.7%	0.86%
Phosphorus	0.5%	0.74%
Metabolic energy	2915 cal kg^{-1}	2870 cal kg^{-1}
Energy/protein ratio	168	164

Whether the mouse requires the level of the nutrients available in these two diets has not been fully established.

Many studies on the protein requirement of mice have been made with purified or semi-purified diets. Goettsch (1960) found that 13.6% casein (about 12.0% protein) was the minimal dietary protein concentration that supported acceptable growth, reproduction and lactation. Bruce & Parkes (1949) reported slight differences in the reproductive performance of mice fed natural ingredient diets that contained 13.6%, 15.0% or 18.4% digestible protein. In considering all the available data, the National Academy of Sciences publication (1978) recommended an 18% dietary protein content for reproduction and 12.5% for growth. This would suggest that both the commercial diets have adequate protein for reproduction but about 50% more than that required for growth and maintenance.

Both the ALGH and CRM diets were formulated following the pioneer work of Guzy & Rapp (1975) in which the importance of the energy to protein ratio of the diet was demonstrated. These workers showed that there was an optimum ratio for the number and weight of young in the litter (Table V). For rats this ratio was found to be 180; a value for mice would be of considerable interest. With natural ingredients and without the addition of large quantities of oil to enhance the energy content of the diet, the protein content of the diet must be limited in order to achieve the energy/protein ratio as suggested by Guzy & Rapp.

TABLE V

Effect of protein to energy ratio on reproduction in rats (Guzy & Rapp, 1975)

Ratio	210	195	180	170	160
No. weaned young/litter	9.32	9.12	9.61	8.95	9.05
Individual weight of weaned young/litter (g)	45.4	46.6	47.8	46.3	45.6
Litter weight (g)	423	425	429	414	412

The requirement for the majority of the minerals is unknown and has been estimated from the content of adequate diets. In contrast, that for fat-soluble vitamins is reasonably well documented although some values have been based on studies in which natural products, such as Cod Liver Oil, have been used as a source of the vitamins (National Academy of Sciences, 1978). The requirement for water-soluble vitamins has been assessed on adequate diets for germ-free mice (Luckey, Bengson & Kaplan, 1974) where there is no microbial synthesis of vitamins.

The breeding of inbred strains of mice presents other problems not only in the chemical composition of the diet but also in the hardness of the cube. At the Jackson Laboratory four different diets have been developed which have proved best in the strains for which they are used in terms of reproductive performance and health (Hoag & Dickie, 1966). The chemical composition and strains to which the diets are fed are given in Table VI. These diets differ from those described above and most other commercial diets produced in the UK in (a) the very high fat contents, and (b) the low fibre contents even in the "general" diet. Unfortunately it is not possible to

TABLE VI

Calculated analysis of diets used in Jackson Laboratory (after Hoag & Dickie, 1966)

	Diet 1	*Diet 2*	*Diet 3*	*Diet 4*
Components				
Protein	21.2	24.9	21.3	17.9
Fat	10.9	8.2	7.1	11.2
Fibre	1.7	2.1	1.1	1.1
Ash	4.7	4.7	4.8	4.3
Ca	1.4	0.9	1.0	0.9
P	0.9	1.1	0.9	0.8
Strains	C57BL/6J A/J CBA/J	DBA/2 A/HeJ C57L/J DBA/1J	DBA/2J AKR/J	General diet for all other strains

calculate the energy/protein ratios of these diets with the information available, but with the large amount of fat it would appear that the metabolisable energy would be high and the energy/protein ratio would also be high.

As mentioned above, cube hardness is a very important factor with some strains of mice. Ford (1977) reported that for several strains of mice there was an optimum hardness beyond which maximum food intake is prevented and reduced growth results. Mice fed a hard diet had difficulty in consuming enough to stay alive.

FOOD PREFERENCES AND AVAILABILITY

There have been a limited number of studies of the food available to the house mouse and its near relatives. Berry & Tricker (1969) reported that insects form a major part of the diet of the house mouse and the proportion of insects in the stomach increases during the summer even though the availability of seeds is increasing at the same time (Whitaker, 1966; Berry, 1968). A similar study by Watts (1968) of the foods eaten by wood mice showed that animal matter was eaten in appreciable quantities only during May, June and July (Fig. 1). This preference for food of animal origin, when

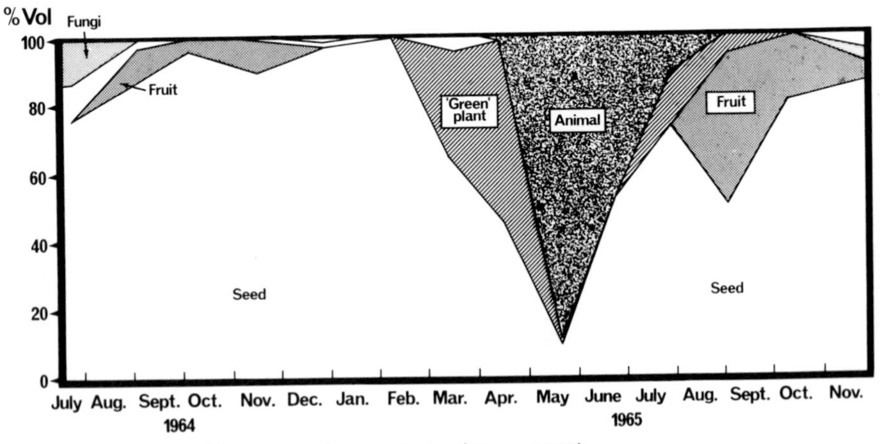

FIG. 1. Summary of foods eaten by wood mice (Watts, 1968)

available, is probably important for the survival of the species because several workers have shown that mice fail to thrive on a pure cereal diet. Studies involving the use of C57BL mice by Deol & Truslove (1957) showed that, of the common cereals, oats gave the best results in a reproductive trial extending over five litters, although it was inferior to the control diet (Table VII). The main difference observed was the interval between first and second litters, indicating that post-partum mating did not occur. No reason for this or the other observed differences between the control diet and the cereals is given by the authors. The most obvious difference in the nutrient intake is in the 18.6% protein content of the control diet and the 10% present in the cereals. The higher oil content of the oats may have resulted in an increased palatability over the other cereals with a corresponding increase in food eaten and hence a higher protein intake. Unfortunately no food intake values are reported, if indeed they were recorded. Another not so obvious difference is in the vitamin A intake which would be extremely low in an all-cereal diet. For over 50 years it has been known that a deficiency of vitamin A causes a partial or complete infertility (Sherman & McLeod, 1925) and an interruption in the oestrous cycle (Evans & Bishop, 1922). It seems possible that a deficiency of vitamin A may have been responsible for the observations of Deol & Truslove (1957).

Food preference studies by Southern (1954) also showed that mice had a preference for rolled oats over wheatmeal (Table VIII), and they had a similar preference for breadcrumbs. They also preferred rolled oats with added olive oil rather than with added sugar. Whether this was an effort to satisfy their energy requirement more easily or whether the diet was more palatable is not clear. Another experiment showed a slight preference for rolled oats mixed with groundnut oil over olive oil, with a complete rejection of linseed oil. However, the mice preferred sausage-rusk mash to rolled oats.

An extensive study of the foods eaten in the wild by different species of mice was made by Whitaker (1966) in Vigo County, Indiana, from 1962–65. Analysis of the stomach contents showed a difference in preference for the various available foods. Seeds of foxtail grass were the most important food of the house mouse, and about 20% of the food of the species consisted of these (Table IX). Seeds of non-cultivated and cultivated plants made up another 43.1%. Of the animal foods, lepidopteran larvae were a major food, contributing 14.6%, with other animal foods another 9.5%.

Various kinds of unidentified and cultivated seeds made up 42.7% of the food of the white-footed mouse (Table X). Over 35% of the stomach contents were of animal origin, over 11% being lepidopteran larvae.

Lepidopteran larvae were the most important food of the

TABLE VII

Studies on C57BL mice (adapted from Deol & Truslove, 1957)

Diet	Average litter size (5 litters)	Mean birth Weight (g)	% surviving to 60 days	Average time (days) between 1st & 2nd litters	% content of:	
					Protein	Oil
Control (Diet 86)	5.38	1.40	55.1	36	18.6	2.2
Oats	5.15	1.26	52.5	59	10.0	4.0
Wheat	4.65	1.26	37.4	63	11.3	1.7
Buckwheat	4.40	1.36	38.6	72	11.0	2.0
Barley	3.77	1.34	29.6	97	10.0	1.7

TABLE VIII

Food preferences of house mice (adapted from Southern, 1954)

Foodstuff	Amount eaten (g)
Rolled oats (R0)	17.25
Wheatmeal	8.75
Breadcrumbs	19.00
R0	1.50
R0 + 10% sugar	0.50
R0 + 5% olive oil	50.00
R0 + 20% olive oil	11.50
R0 + 20% linseed oil	0.00
R0 + 20% groundnut oil	16.00
Sausage – rusk mash	26.50
Rolled oats	16.50

TABLE IX

Food of Mus musculus *(house mouse) (Vigo County, Indiana 1962–65, adapted from Whitaker, 1966)*

Foodstuff	% volume
Seeds of *Setaria* (Foxtail Grass)	20.2
Seeds of non-cultivated plants and grasses	20.2
Lepidopteran larvae	14.6
Other animal foods (other larvae etc.)	9.5
Miscellaneous vegetation	10.0
Fungi (*Endogone*)	2.0
Seeds of cultivated plants	22.9

TABLE X

Food of Peromyscus leucopus *(white-footed mouse) (Vigo County, Indiana 1962–65, adapted from Whitaker, 1966)*

Foodstuff	% volume
Unidentified seeds	13.1
Miscellaneous vegetation	12.0
Lepidopteran larvae	11.4
Mast	8.1
Identified seeds (wheat, *Setaria*)	28.3
Other animal foods	24.3
Green vegetation	3.3

deer mouse (Table XI), forming 15.4% of the food of the species. Nearly 33% of the food consisted of cultivated crops with wheat and soya beans each forming 10% of the diet. Other animal food accounted for a further 17.5%, and green vegetation, 20%. The

TABLE XI.

Food of Peromyscus maniculatus bairdi *(deer mice) (Vigo County, Indiana 1962–65, adapted from Whitaker, 1966)*

Foodstuff	% volume
Lepidopteran larvae	15.4
Seeds of cultivated plants	33.0
Seeds of non-cultivated plants	14.6
Other animal foods	17.5
Miscellaneous vegatation	19.8

habitat of the animal would be expected to influence the food chosen by the animal. Whitaker (1966) compared the major foods of house and deer mice in cultivated habitats. Two of these (Table XII) show that the two species have definite preferences. In a field of

TABLE XII

Comparison of major foods of house and deer mice in cultivated area habitat (adapted from Whitaker, 1966)

	Soya bean		Wheat	
	House mouse	*Deer mouse*	*House mouse*	*Deer mouse*
Setaria seeds	48.9	0	4.0	0.4
Lepidopteran larvae	19.3	15.6	27.1	26.8
Sorghum halepense seeds	12.6	0	0	0
Soya beans	3.5	24.5	0.6	0
Miscellaneous vegetation	1.5	8.3	3.3	22.2
Wheat seeds	0	8.2	24.9	9.3

soya beans the house mouse had a strong preference for *Setaria* seeds whereas the deer mouse preferred soya beans. In a wheat field the house mouse preferred wheat seeds and the deer mouse miscellaneous vegetation. In both habitats lepidopteran larvae formed a large proportion of the food intake of both species.

Why the preference for lepidopteran and other larvae? Of the many possible reasons there are two which are most relevant. Rosen

& Levinger (1972) have reported that when rats were offered a free choice between a protein-rich and a protein-poor diet it was found that the animals grew faster and showed a very strong preference for the protein-rich diet. Alternatively, lepidopteran larvae contain β-carotene, the precursor of vitamin A (Goodwin, 1971). The intake from this source would be low, but Kuhn & Brockman (1933) showed that a daily dose of only 2.5 μg β-carotene would restore the oestrous cycle.

It would be of considerable interest to calculate the intake of the various nutrients by the house mouse under natural conditions. This would enable a comparison to be made between the nutrient intake of the wild mouse and the laboratory-bred mouse. Unfortunately the exact composition of the foods of the mouse is not well enough documented to allow an estimate of the intake of the various nutrients. It is also doubtful whether the chemical composition of some of the foods is known.

Perhaps as a result of this symposium this information will become available in the near future.

REFERENCES

Berry, R. J. (1968). The ecology of an island population of the house mouse. *J. Anim. Ecol.* 37: 445–470.

Berry, R. J. & Tricker, J. K. (1969). Competition and extinction: the mice of Foula with notes on those of Fair Isle and St Kilda. *J. Zool., Lond.* **158**: 247–265.

Bruce, H. M. & Parkes, A. S. (1949). Feeding and breeding of laboratory animals. IX. A complete cubed diet for mice and rats. *J. Hyg., Camb.* **47**: 202–208.

Deol, M. S. & Truslove, G. M. (1957). Genetical studies on the skeleton of the mouse. XX. Maternal physiology and variation in the skeleton of C57BL mice. *J. Genet.* **55**: 288–312.

Dietary Standards for Laboratory Animals (1977). Report of the Laboratory Animals Centre Diets Advisory Committee. *Lab. Anim.* **11**: 1–28.

Evans, H. M. & Bishop, K. S. (1922). On an invariable and characteristic disturbance of reproductive function in animals reared on a diet poor in fat soluble Vitamin A. *Anat. Rec.* **23**: 17–18.

Ford, D. (1977). Influence of diet pellet hardness and particle size on food utilization by mice, rats and hamsters. *Lab. Anim.* **11**: 241–246.

Goettsch, M. A. (1960). Comparative protein requirements of the rat and mouse for growth, reproduction and lactation using casein diets. *J. Nutr.* **70**: 307–312.

Goodwin, T. W. (1971). Pigments–Arthropods. In *Chemical zoology* **6**: 279–306. Florkin, M. & Scheer, M. T. (Eds). London: Academic Press.

Guzy, K. & Rapp, K. (1975). Zum Energie- und Rohproteinbedarf der Han Sprague Dawley Ratte wahrend der Aufzucht und Zucht. *Tierlaborat.* 2: 44–57.

Hoag, W. G. & Dickie, M. M. (1966). Nutrition. In *Biology of the laboratory mouse:* 39–43. Green, E. L. (Ed.). New York: McGraw Hill.

Kuhn, R. &. Brockman, H. (1933). Einfluss der Carotine auf Wachstum Xerophthalmie, Kolpokeratose und Brunstcyclus. *Klin. Wschr.* **12**: 972–973.

Luckey, T. D., Bengson, M. H. & Kaplan, H. (1974). Effect of bioisolation and the intestinal flora of mice upon evaluation of an Apollo diet. *Aerosp. Med.* **45**: 509–518.

National Academy of Sciences (1978). *Nutrient requirements of domestic animals* No. 10. Nutrient requirements of laboratory animals. Washington.

Roe, F. J. C. & Tucker, M. (1973). Recent developments in the design of carcinogenicity tests in laboratory animals. *Proc. Eur. Soc. Study Drug Tox.* **15**: 171–177.

Rosen, M. & Levinger, I. M. (1972). The influence of protein on food consumption in the rat. *Lab. Anim.* **6**: 287–293.

Sherman, H. C. & Macleod, F. C. (1925). The relation of vitamin A to growth, reproduction and longevity. *J. Am. Chem. Soc.* **47**: 1658–1662.

Southern, H. N. (1954). The foods that mice prefer and the way in which they eat them. In *Control of rats and mice* 3. *House mice:* 99–119. Chitty, D. & Southern, H. N. (Eds). Oxford: Clarendon Press.

Tucker, M. (1979). The effect of long-term food restriction on tumours in rodents. *Int. J. Cancer* **23**: 803–807.

Watts, C. H. S. (1968). The foods eaten by wood mice (*Apodemus sylvaticus*) and bank voles (*Clethrionomys glareolus*) in Wytham Woods, Berkshire. *J. Anim. Ecol.* **37**: 25–41.

Whitaker, J. O. (1966). Food in *Mus musculus, Peromyscus maniculatus bairdi* and *Peromyscus leucopus* in Vigo County, Indiana. *J. Mammal.* **47**: 473–486.

Symp. zool. Sol. Lond. (1981) No. 47, 267–300

Ageing: with Particular Reference to the Use of the House Mouse as a Mammalian Model

D. BELLAMY

Department of Zoology, University College Cardiff, Cardiff, UK

SYNOPSIS

The term ageing covers a very large interdisciplinary area of research heavily biased in the direction of care of the elderly. The approach of the gerontologist will be clarified in relation to the concept of biological development. The various ways of investigating ageing will be reviewed, wherever possible drawing examples from research involving the use of the laboratory mouse as an experimental model. Particular emphasis will be placed on the fact that the free living house mouse does not live long enough to age in the same way as the various protected laboratory strains.

TERMINOLOGY

Over the years many terms describing the various aspects of the life-cycle have accumulated, such as "development", "differentiation", "maturation" and "senescence", all of which have been established by specialists who are usually concerned with the study of only one temporal phase of life. The definitions are usually employed from a restricted viewpoint and have the effect of cutting the life-cycle into arbitrary segments. Some of the divisions may have no biological relevance when a broader perspective is taken and the terminology can actually impede research and communication.

From the outset it must be realised that ageing, as an inevitable progressive deterioration of function, is not universal. Some protozoa, unicellular algae, malignant cell lines and the vegetative parts of plants appear to be capable of unlimited cultivation as clones without a general decline in vigour. Lack of ageing also characterizes the germ cell line as it is transmitted in evolution. In this respect, ageing appears to be the inevitable outcome of a bodily organization where reproductive cells require the support of a relatively large volume of cellular resources not directly concerned with reproduction. The position of simple metazoan organisms such as *Hydra* and the planarian worms which appear to contain a reservoir of embryonic-type cells needs further investigation.

In organisms that do deteriorate with time, four landmarks in the life history are: the time of fertilization; the time of sexual maturation; the time when growth ceases; the time of death. This raises the question, "Is the period of growth related to the duration of life?" Put more specifically, "Is ageing a continuation of any of the processes of growth?" This is an interesting starting point in any discussion of the nature of ageing because from this aspect growth and ageing are viewed as phenomena which relate to the pattern of the total life history. It also makes it possible to ask subsidiary questions concerning the likelihood of control of ageing through the interplay of factors which are known to be important in the regulation of growth.

There is a close interplay between ageing and development and there are difficulties in defining both terms (Table 1). To many biologists the term "ageing" is used to describe any time-dependent change with respect to life history. However, it is necessary to place some restriction on the definition in order to arrive at a category of changes which is useful in formulating problems open to experiment. For example, a common feature of definitions of ageing is that it is distinguished from the time-dependent changes of development; development includes the events of differentiation, growth and maturation which aid survival until the individual is a reproductively competent adult, whereas ageing processes lead to a failure to adapt to the environment and ultimately result in death. Points of difficulty arise because in this context many developmental events are the obvious precursors of ageing phenomena, and some changes which begin before or shortly after birth continue unabated throughout life. Despite this, many workers would restrict the term "ageing" and eliminate from consideration any changes which do not render the individual more likely to die in a given time interval as it grows older. Good examples of human systems that deteriorate early and which do not show up as an increase in mortality are the regression of the human female reproductive system, loss of scalp hair and the loss of adaptability of the eye (Fig. 1). This particular difficulty of definition has been recognized and another category, "senescence", introduced to include only those events which contribute to the decreased resistance to death. However, it is generally accepted that of the many ageing phenomena, very few could be proved not to influence mortality. Also, the force of mortality depends very much on the environment. For example, it is of no obvious disadvantage

TABLE I

Some definitions of ageing

1	"An ageing process is any process occurring in an individual which renders that individual more likely to die in a given time interval as it grows older."
2	"Ageing can be defined as that universal attribute of all multicellular organisms which is characterised by time-dependent, reproducible alterations in structure, which may or may not be peculiar to each species or strain."
3	"Senescence is that part of the total ageing process which occurs during the last trimester of an adult existence and during which time related structural and functional changes of a degradative nature predominate in certain organs and tissues, such as to lead ultimately to the diminished capacity of the individual to survive the assaults of both the internal and external surrounding environments."
4	"Old age is a major involution of the living organism."
5	"All living matter changes with time in both structure and function, and the changes which follow a general trend constitute ageing. Ageing begins with conception and ends only with death."
6	Ageing is "a decrease of intracellular water, with maintenance of extracellular water and a varying fat mass."
7	An ageing process is "one that occurs in all members of a population, that is progressive and irreversible under usual conditions, and that begins or accelerates at maturity in those systems which undergo growth and development."
8	Ageing is "a gradual decline in an organism's adaptation to its normal environment following the onset of reproductive maturity."
9	Ageing is described as "a biological process which causes increased susceptibility of an organism to diseases."
10	"Ageing is merely the vector sum of a number of morbid processes, most of which take time to develop and often a long time to reach serious climax."
11	"Ageing can be defined as an increasing probability for the individual to contract one of the degenerative diseases."
12	"Ageing processes may be defined as those which render individuals more susceptible as they grow older to various factors, intrinsic or extrinsic which may cause death."
13	"Ageing stands for the mere increase in years without overtones of increasing deterioration and decay. Senescence means ageing accompanied by that decline of bodily faculties and sensibilities and energies which ageing colloquially entails."
14	"Ageing is the process of progressive entropy or the decrease in the information content of the organism until a minimum is reached which is incompatible with life."

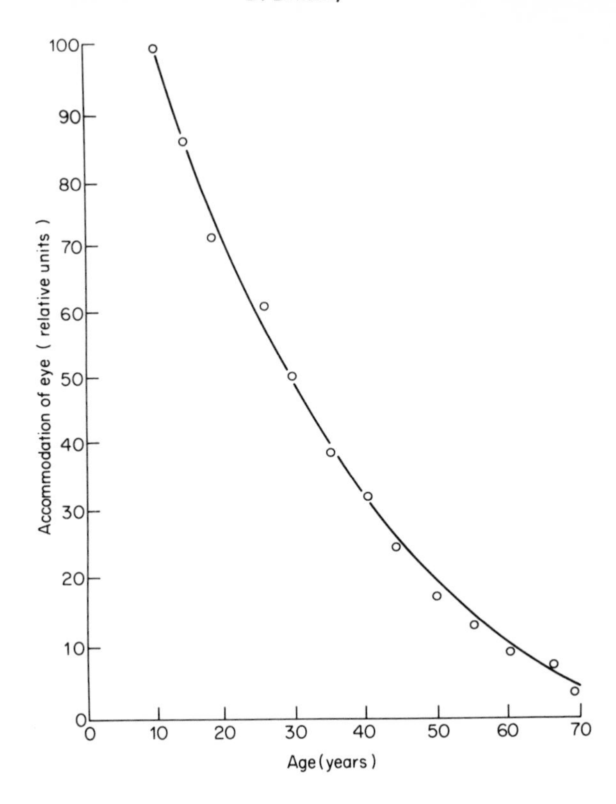

FIG. 1. Changes in the accommodation of the human eye with age (from Birren, 1964).

for civilized man not to be able to run a five-minute mile at the age of 50. Nevertheless, the well-documented decline in human athletic performance from the second or third decade could be included as an early ageing process in a less sophisticated society, where predators had the ability to run faster than human prey. There is also the possibility of delayed secondary responses to early primary changes, which may increase the chances of death after a considerable lag-period. For example, there is a marked decrease in bone density in human females that is partly related to the decline in the female reproductive system. The mean age for menopause is about 50 years whereas the mean age at which half of the bone loss has taken place occurs almost 20 years later (Fig. 2). Comparable data for men show that the loss of bone is considerably less and starts at least a decade later. This faster and earlier loss of bone in women is associated with a dramatic rise in fractures of the long bones which has no counterpart in men. In the later years complications, such as

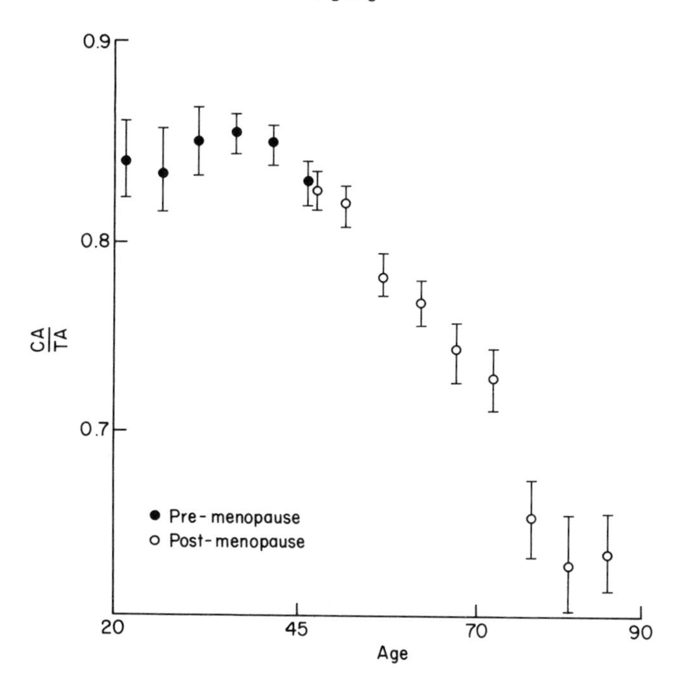

FIG. 2. Bone loss in women between the ages of 20 and 90 (from Gallagher & Nordin, 1973). Bone density was calculated from hand radiographs of normal pre- and post-menopausal women. $CA/TA = 1 - [1 - (C/T)]^2$ where C is cortical thickness and T total thickness.

pneumonia, arising from fractures are an important cause of death.

Perhaps our difficulties of definition arise because it is implicit in most current classifications that ageing phenomena have unique causes. It is by no means clear that there is such a compartmentation and the view may be advanced that development and ageing share the same basic cellular processes. Further, if the term development is used comprehensively to cover the whole life-cycle, ageing will then appear as a subordinate process. Differences between early and late events of development are that early events lead to perfection of function and late events result in the deterioration of function. From this aspect gerontology is the scientific study of the irreversible deterioration in those structures and functions which have a definite peak or plateau of development in all members of the species. This definition means that the study of ageing does not begin at an arbitrary age of the subject. Neither need the study be restricted to the period after the system in question has reached a maximum level of function because the causes of deterioration may be found in the earlier phase of life.

Largely because of its restrictive connotations the term senescence will not be used here and, in view of the importance of environment in the expression of failures in biological organization, ageing will be considered as a loss of precision which results from chemical deterioration, errors and variability of gene expression in the systems specifying form and function. Ageing is thus expressed as a loss of adaptability resulting from a decline in tissue and functional reserves. Any system that meets these criteria will be considered, even if there is no obvious link with the increased probability of death. The definition implies that a major feature of ageing is that the homeostatic systems of the body become less efficient in combating fluctuation in the external world. On this basis it is true to say that ageing begins before there is a marked increase in mortality. In the human, a steady decline in physiological performance is first noticeable during the third decade of life; the corresponding stage in the laboratory mouse (lifespan 800 days) occurs at about 100 days; in experimental *Drosophila* cultures (life-span 27 days) ageing of some systems begins at about 10 days. Therefore it appears that in this context phenomena of ageing may occupy well over half the maximum lifespan.

The wide temporal range that the gerontologist has to cover, together with the fact that ageing is most clearly manifest at the most complex level of many physiological systems yet can only be fully comprehended at the simplest level, means that the study of ageing has an exceptionally broad intellectual base. This interdisciplinary aspect is illustrated in Table II which, starting from the general definition already given, lists the objects of study and processes of interest to the gerontologist. This wide range also means that gerontology must proceed by the selection of appropriate models from the entire living world. It is therefore of interest to investigate the value of the house mouse in this extensive field.

In general the mouse, the rat, the guinea-pig and the rabbit age in the same ways and they are used interchangeably as models for human ageing. Rats and mice are the nearest the experimenter is likely to get in controlled studies relevant to human ageing because they are realistic in terms of housing space and costs. Whilst the mouse is more economical in space, the rat has the practical advantage of size, particularly where physiological and biochemical work requires repetitive blood taking, as well as the provision of large quantities of tissue or sizeable parts for easy sampling.

Another important point is that the study of ageing must

TABLE II

Scope of research on ageing

General definition: Gerontology is the scientific study of deterioration in those structures and functions that have a definite peak or plateau of development in all members of the species.

Objects of study	Processes of interest
Molecules	Chemical deterioration
	Enzyme synthesis
Enzymes	Differential expression of genes
Organs	Growth
	Physiological regulation
Organisms	Reproduction
	Behavioural adaptibility
Populations	Heredity
	Natural selection

proceed by modelling across the whole range of evolution. Therefore discussing work in the mouse alone would give a bizarre view of the ageing field. This article has been written to provide a broad perspective into which current and future work on the mouse may be fitted.

Until recent years this comprehensive viewpoint has not been accepted by most research workers and this has been partly responsible for an unbalanced research strategy and compartmentation of ideas which have tended to isolate ageing research from the mainstream of biological knowledge. Also, there has been much fruitless methodological argument as to whether one should first develop a general testable hypothesis, describe the ageing organism in detail, or enumerate possible causes of ageing which are then eliminated one by one. Whilst the decisive approach in science is inevitably an experimental one, all viewpoints are valid and it is apparent that the popular yet extremely biased position of model-making has militated against the more tedious collection of information on the natural history of ageing.

MACHINE ANALOGIES

Arising from problems of definition, many advances in science have come by posing complicated problems in terms of machine analogies. Ageing is no exception to this generalization, in that a number of mechanical analogies have been proposed in attempts to define the general picture. Because time is the most obvious element, it is not surprising that ageing has long been viewed as a clock-like process. This has led, in turn, to the view that a molecular clock deteriorates. Further ideas along this line raise the possibility that if the clock could be tampered with, particularly in order to make it run more slowly, the rate at which age deterioration occurs could be slowed down and life thereby prolonged.

As judged by the different times at which various systems reach their peak performance and the different rates at which they deteriorate, it is clear there must be several clocks operating, even in the simplest organisms. Also, since the onset of some characteristic ageing phenomena can be preferentially accelerated or delayed by experimental treatment, some or all of the clocks must function independently of each other. On the whole, the clock concept has not been very useful in defining ageing because the analogy does not match a process that once it has run down, cannot be re-started.

Another mechanical analogy, based on the continuous playing of a gramophone record, although it highlights the most fundamental characteristic of cellular function, namely the capacity to repeatedly transcribe coded information with decreased fidelity, suffers from the same disadvantage as the clock idea, in that the record of life can only be played once. If it is assumed that life is controlled by the repeated playing of the same record, from what we know about the biochemical transcription process, the music, i.e. the integrated function of the cell, is not quite the same each time the record is played. This is not only because a "wearing out" of the record of life may occur by repetitive use and it may show cracks due to chemical deterioration with the passage of time, but also because the music of life itself affects the score which is heard on subsequent playing; to begin with, playing improves the score, but later, playing causes it to deteriorate. Providing that the record player and the record are taken to have these unusual technical properties, the analogy represents development in the broadest sense. It takes account of feedback from the transcribed information governing cellular function, which is able to alter the "groove pattern" for the next playing. However, as with the clock analogy, it is necessary to postulate the existence in the body of several "records" with separate "players" to take account of the multifactorial basis of ageing.

A comparison that illustrates the important evolutionary aspect of ageing is the "space probe analogy". This likens an individual life to the voyage of a space probe that has been designed and programmed to pass close to a distant planet and transmit its findings back to base. Once the mission is completed the craft will continue to function but increasingly with the passage of time, various units will begin to break down owing to inevitable chemical deterioration of the key components, and, perhaps, frictional wearing. Eventually the probe will "die" when its main transmitter fails. This model may be taken to represent an organism that is programmed by evolution to last a certain length of time in relation to the stresses of its surroundings, but once this period has been exceeded, various bodily systems fail to respond appropriately because the shape and chemistry of the various components are no longer appropriate to function. The analogy is particularly useful in highlighting the fundamental importance of chemical deterioration in ageing. It is also apt in that it takes account of the fact that the lifespan evolved in the wild, i.e. a period comparable to that taken by the space probe to achieve its objective, is the most important biological feature of ageing. Against this may be set the prolonged lifespan of man and his domesticated animals in protected environments, where life continues beyond the lifespan set by the environment because of the absence of mortality factors that played a part in the past evolution of the species. On this view, death in old age occurs because an ordered programme of development is exceeded when the species is in a greatly more favourable environment than the one in which it evolved.

Any imperfection of function which results in a progressive deterioration in the organism could be classed as an ageing effect and from this viewpoint alone, it is likely that, as in the space probe, there are many causes and consequently many processes of ageing. This follows from considering that the length of the natural life-cycle as governed by disease and predation is the endpoint of evolution. The way in which this ecological lifespan is established by natural selection need not be identical or even similar in different species. A common metabolic cause of ageing in the various organs within an individual is also unlikely on the grounds that they differ so much in both structures and function. The available experimental evidence supports the idea of organ ageing being chemically self-contained and not dominated by systematic factors. This multifactorial theory of ageing is opposed to the less likely unitary hypothesis which states that there is only one process which is responsible for the general loss of adaptability in all organs.

THE CONCEPT OF FINITE LIFE-SPAN

When a population of laboratory mice is maintained under controlled laboratory conditions, all members have the same environmental history, and presumably have many of their genes in common, yet they do not die at the same instant in time. If the number of individuals remaining alive at set intervals out of a given number recorded at birth is plotted against time, instead of a rectangular graph, a curve is usually obtained (Fig. 3). There is a great spread in the age at death

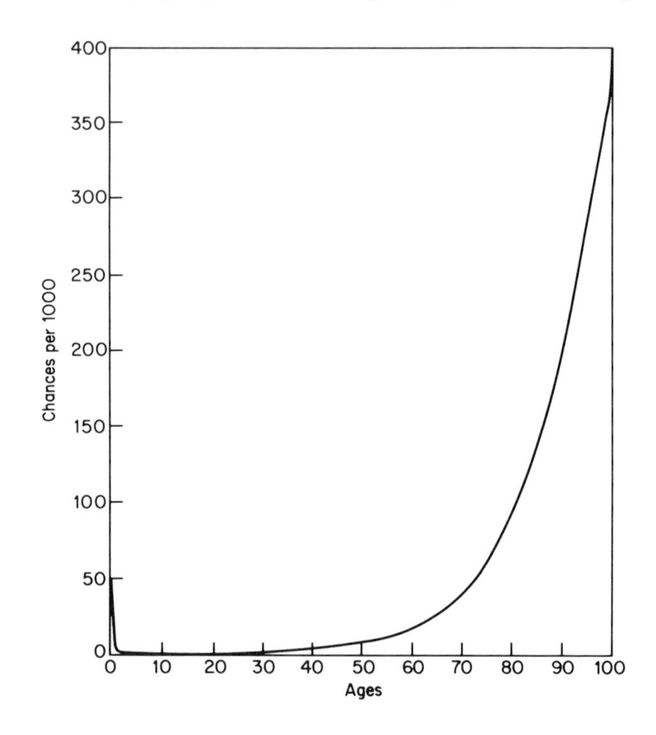

FIG 3. Probability of human deaths in relation to age (from Strehler, 1960).

and this is taken to mean that there is variability in resistance to death in the population. The actual shape of the graph, called a mortality curve, gives no information about the rate of ageing in individuals. It is merely an accurate age-frequency distribution at the moment of death; the shape of the curve is not fundamentally related to the rate or kind of ageing of a given population. Ruling out accidents, the spread in the timing of death indicates that either the population was not uniform in genotype, or that if it was genetically homogeneous, events occurred at random from birth which resulted in different rates of ageing.

Another way of expressing the data is to measure the rate of death at different ages (Fig. 4). These values are termed age-specific death rates and show for human populations in advanced countries that once the mean lifespan is exceeded, the

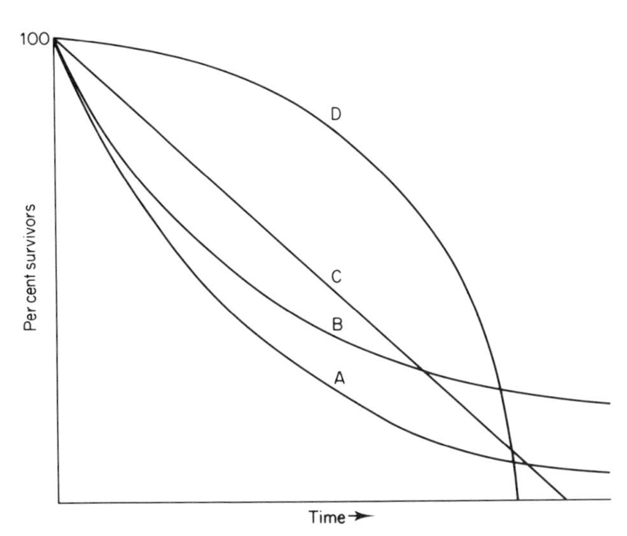

FIG. 4. Various types of mortality curve (from Bang, 1969). A Constant death rate. B Decreasing death rate. C Constant number of deaths per unit time. D Increasing death rate.

death rate – and therefore the probability of dying – doubles at regular intervals.

The curve of human mortality is at a minimum around the age of 12 and then rises slowly at first and latterly much faster. Age 12 is therefore the actuarial prime of life; at 12 one is more likely to survive one further year, month or minute than at any other age. A distinctive feature of the curve is the smoothness that defines the force of mortality in later life. Any complete theory of the origin and evolution of ageing must explain the smoothness and coherence of this curve of increasing mortality.

A particular difficulty in the use of mortality curves derived from long-lived species is that the terminal deaths of a particular population may refer to individuals that experienced environmental conditions greatly different from those that the younger members of the same species now experience in the same place. This can usually be overcome with the commonly used laboratory animals, particularly mice, which are bred in constant conditions with a well established genotype. For other longer-lived animals an approximation may be obtained by calculating generation life tables each based on the observed mortality of a single generation of births. However, the recorded mortality of most human populations does not permit the calculation of complete tables for generations separated by more than 10 to 20 years so that it can never be asserted that the mortality curve truly represents current age-related mortality.

Animals tend to have a lifespan which is characteristic of the species, implying that ageing is under genetical control. It is this genetical element of longevity which probably accounts for the thousand-fold variation in the maximum lifespan between species in captivity (Table III). The data in Table III refer to studies in which

TABLE III

Age distribution of maximum observed life-spans of animals

Maximum lifespan (yr)	Numbers of species with reliable records					
	Invertebrates	Vertebrates				
		Fish	Amphibians	Reptiles	Birds	Mammals
0–5	8	3	2	—	4	6
5–10	5	—	—	—	4	4
10–20	3	—	1	—	1	2
20–30	—	—	3	3	8	7
30–40	1	1	—	1	2	1
40–50	1	—	—	1	—	—
50–60	1	2	—	—	1	—
60–70	—	—	—	—	3	—
70–100	—	1	—	3	3	1
100 +	—	—	—	—	1	1
Totals	19	7	6	8	27	22

maximum lifespan was assessed from the survivors of small initial populations; most of the records came from zoos with good records of animal husbandry, and, in the case of long-lived fish, from maximum scale counts in a large catch. The species distribution between the various age-classes is probably indicative only of the ease with which the species may be studied. The overall range of lifespan from the smallest metazoans, with a maximum lifespan of a few days, to man with a lifespan of over 100 years is about 1:3500. This is about the same longevity range found in the higher plants. The existence of this well defined lifespan pattern indicates that the variation in the characteristic lifespans of species depends on the genetical constitution selected in the course of evolution.

The importance of genotype in determining longevity comes out more clearly from studies of inbred experimental animals where environment can be controlled very precisely. For the laboratory mouse, inherited variants can be obtained that differ by a factor of about three in maximum lifespans between strains (Table IV).

TABLE IV

Mean lifespans of several strains of laboratory mice and rats

Strain	Mean life-span (days)			
	Females		Males	
	Number	Longevity mean (S.E.)	Number	Longevity mean (S.E.)
AKR	79	312 (9.4)	79	350 (10.8)
NZB	111	441 (12.1)	110	459 (13.3)
A	68	558 (19.7)	65	512 (21.1)
BALB/c	33	561 (30.3)	35	509 (26.3)
C57BL	29	580 (35.8)	31	645 (34.2)
C57L	26	604 (27.6)	22	473 (30.9)
LACG	40	617 (26.2)	36	536 (38.9)
A2G	51	644 (19.4)	49	640 (21.8)
C57BR/cd	46	660 (22.7)	45	577 (29.8)
LACA	38	664 (29.9)	41	660 (38.5)
129/RrJ	36	666 (23.2)	35	699 (29.8)
C3H	193	676 (9.8)	147	590 (18.6)
DBA/1	39	686 (33.3)	35	487 (35.9)
CE	23	703 (37.3)	20	498 (48.5)
DBA/2	23	719 (35.4)	22	629 (42.1)
NZW	20	733 (42.8)	28	802 (34.0)
WA	22	749 (40.1)	25	645 (29.9)
P	10	782 (51.9)	14	729 (42.9)

Hybrids of the F_1 generation have been shown to have a mean lifespan that is greater than that of either of the parent inbred lines. Differences in the longevity of laboratory populations are associated with differences in the rate of decline in physiological vigour and incidence of pathologies. With respect to the latter point, the frequency distribution and extent of pathological changes in human subjects are clearly genetically conditioned. For example, at the histological level generalized atheroma is considered almost a normal finding in the aortic arch for a person over the age of 60, yet is rarely found in the pulmonary artery of the same person. With respect to genotype, twin studies indicate a greater concordance of cause of death from cancer and tuberculosis in one-egg twins compared with two-egg twins. Also cause of death in mouse strains is strain specific. Deaths from cancer in the CBA mouse result mainly from hepatomas; in the AKR/J mouse, lymphoid leukaemia is more common, while mammary tumours have a high frequency in CBA female mice. Removal of one major cause of death by selection can be expected to unmask another major cause with a high incidence at a greater chronological age.

Statistically, the genetical element in the human lifespan is illustrated by the extra four years which is added to the mean life-span by having all four grandparents surviving to 80 years of age. On the other hand, the environmental influence is indicated statistically by the average life expectancy in cities being five years shorter than that in a rural environment.

Even the laboratory lifespan of inbred rodents may be greatly affected by different laboratory environments. For example, two cohorts of C57/BL/6J American male mice, one maintained in New York City and the other in Maine, both derived from the same colony, differed greatly in average longevity, maximum longevity and age at which mortality rate began to increase (Fig. 3); 50% survival occurred at about 18 months in the Maine Laboratory and at about 30 months in New York. It must be concluded that diet and general laboratory conditions were responsible for these large differences.

This dependence of lifespan on environment is a common finding in other animals. Some of the early work which was carried out on simple aquatic animals such as rotifers and water fleas pointed to the importance of temperature, chemical composition of the surrounding environment and nutritional status. The nutritional status is of considerable importance, first because the two extremes of over- and under-nutrition lead to a decrease in longevity, but also because moderate reduction in calorie-intake which retards growth, extends lifespan of animals at the evolutionary extremes of bodily organization.

The influence of the environment may be explored using mortality survival curves obtained for natural populations. These vary considerably with regard to general shape (Fig. 5). Most differ considerably from the rectangular type already discussed, in that the plateau is either attenuated or absent altogether. It is possible that some of the extreme deviations from the rectangular type represent the operation of mortality factors unrelated to ageing.

A differential mortality rate appears to be a sex-linked feature of many animals. In humans, the male death rate is higher than that of the female over the entire lifespan. At birth, the ratio of males to females is 106:100. There is evidence that a differential death rate of males is a feature of intra-uterine life with estimated primary sex ratios of between 110–160 males to every 100 females. A longer lifespan is a common feature of females in laboratory rodents but there are examples of mouse strains where females have the shorter lifespan. With male groups that were castrated before puberty (Table V) the difference between male and female longevity disappears. In this respect it is still a valid viewpoint that the sex difference in longevity is in some way connected with different patterns

TABLE V

Survival curves of male human subjects castrated at different ages

Age castrated at (yr)	Median lifespan
8–14	76.3
15–19	72.9
20–29	69.6
30–39	68.9
Non-castrate	64.7

From Hamilton & Mestler (1969).

of development and is not due to the action of recessive alleles on the X chromosome.

The fact that species have a characteristic lifespan has prompted correlation analysis of various species characters in an effort to define those which play a dominant role in longevity.

On a comparative basis, there is a highly significant relation between lifespan and body weight in mammals. This regression accounts for about 60% of the variance of lifespan of 63 species of placental mammals (Fig. 6). A better fit of lifespan is obtained when regressions are calculated on brain weight (accounts for 76% of

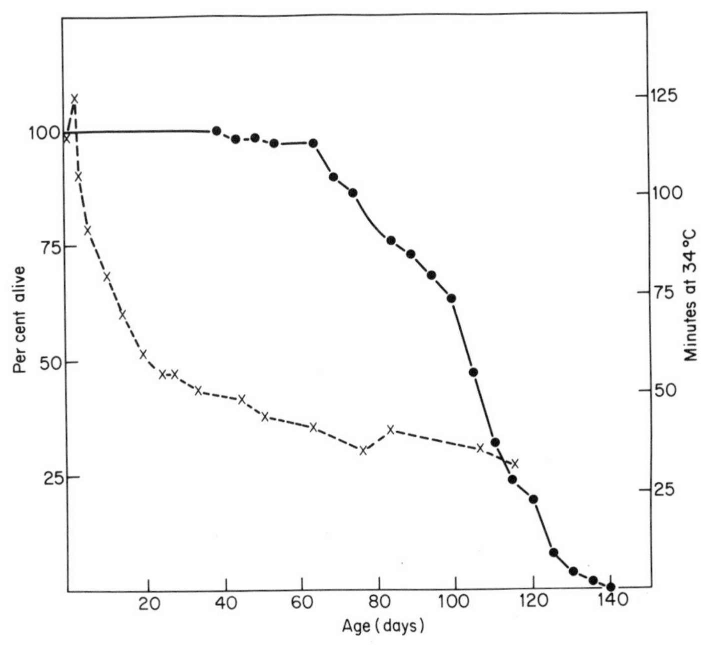

FIG. 5. Mortality of male *Drosophila subobscura* at two temperatures (from Maynard Smith, (1966).

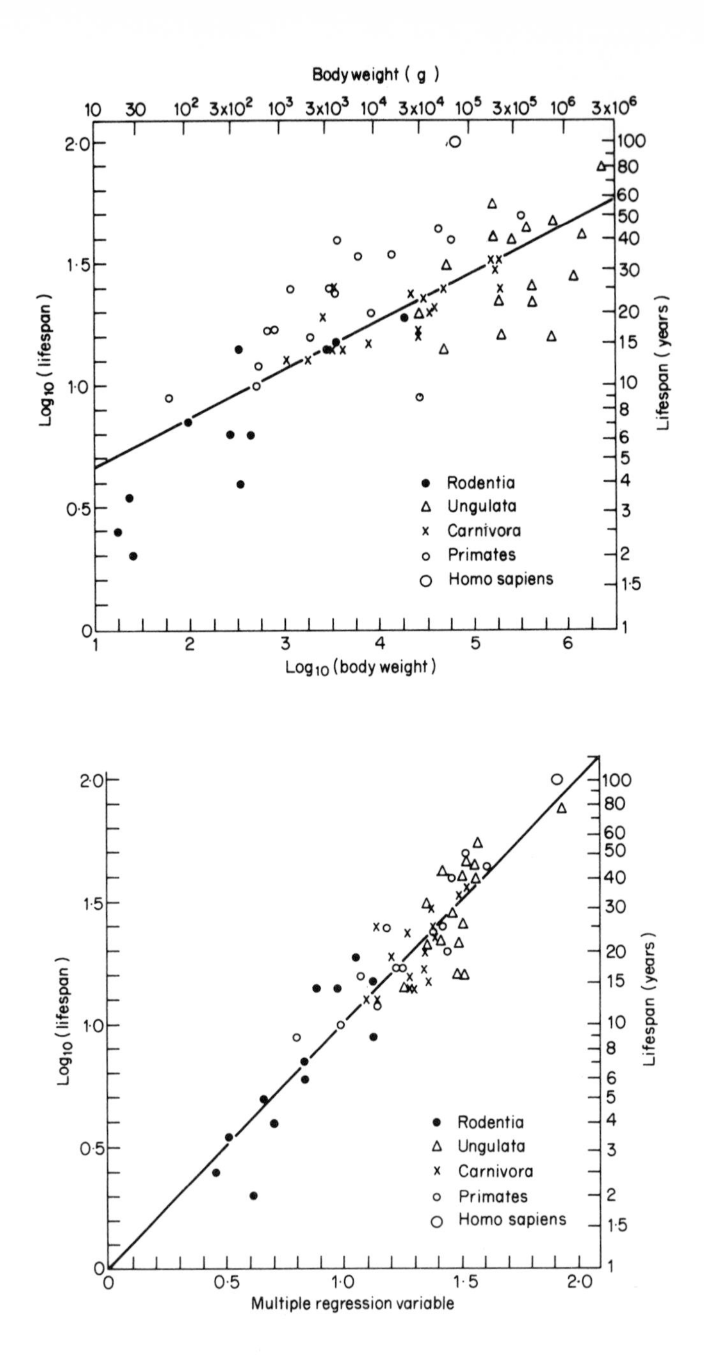

FIG. 6. Relation of mammalian lifespan to body and brain weight (from Sacher, 1959).

lifespan variance), although the wide scatter in this plot indicates that other factors are also involved in the lifespan variance.

Although generally in large species the brain is relatively smaller than that in related small species, the different parts of the brain show disproportionate evolutionary development. Evidence is beginning to emerge indicating that the "newer" parts are the ones that age first. Also there is a trend in numerous lines for a successive increase in relative brain size as well as a relatively larger brain. This implies that a larger brain was a selective advantage, possibly because there is greater central nervous control over bodily functions. The simplest interpretation of this relationship is that it is an advantage to have a brain and a disadvantage to have a body; which gives rise to the common theoretical view that the rate of individual ageing is related to the extent of total control of the body's physiological processes. This is also thought to be the basis for hybrids frequently having a greater longevity than inbred lines, i.e. it is postulated that hybrids by virtue of possessing two distinct control systems have better stabilizing mechanisms compared with inbred types. Because body weight correlates highly with brain weight the view has been taken that evolution of the brain has been a major factor in changing lifespan by allowing the maintenance of an improved physiological regulation nearer to some kind of biochemical ideal. On this idea, brain development has tended to reduce the magnitude of physiological variations and so reduce the probability of irreversible deleterious changes that contribute to ageing. Although this idea is attractive, because mortality seems to be associated proximally with failures in physiological regulation, there are no comparative data on the efficiency of homeostasis within the mammals to substantiate it. There is no correlation between body size or brain size and longevity within species. There is also a need to consider that there are many advantages in having a large body and that the bigger the body the longer it takes to collect the necessary resources to build it.

AGEING IN THE NATURAL ENVIRONMENT

Consideration of ageing in relation to development takes the discussion of longevity firmly into the realm of evolution, with questions arising on the nature of the wild lifespan compared with the maximum observed lifespan and extended longevity of laboratory animals.

Most life tables and mortality curves have been constructed using laboratory species or domesticated populations. For many

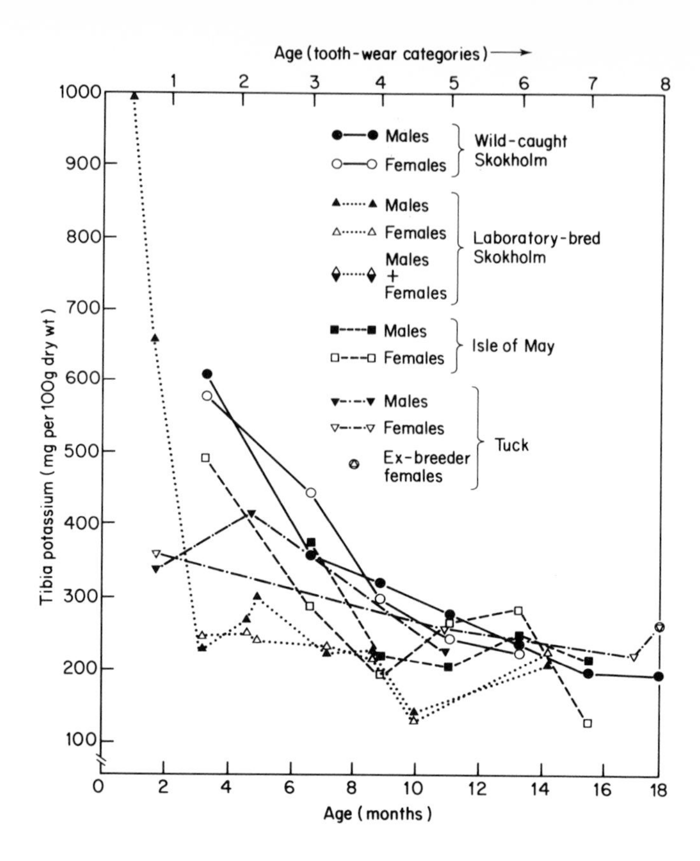

FIG. 7. Chemical composition of wild-caught and laboratory-bred mice (from Morgan & Bellamy, 1975).

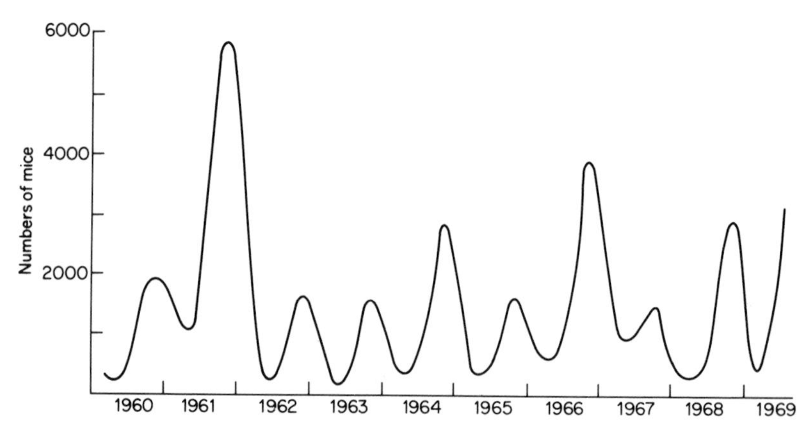

FIG. 8. Fluctuations in balance between birth and death in a wild population of the house mouse (from Berry & Jakobson, 1971).

species it is known that these tables frequently show maxima of life expectancy which are much higher than the greatest lifespans recorded in the wild. There is a strong possibility that this discrepancy may throw light on the nature of ageing, but very few accurate comparisons of lifespans have been made for species in different environments. This is because of the great practical difficulties in obtaining reliable life tables from the natural environment. Under experimental conditions lifespan is affected by environmental variables such as nutrition and temperature (Fig. 7).

As already discussed, life tables are simply constructed by counting the numbers of individuals of different ages in a population. This tends to give a static picture, whereas the wild population is governed by at least five basic numerical changes. An input of individuals occurs through births and immigration; an output occurs through deaths (natural and by predation) and emigration; also, over several lifespans, the total population size may change drastically (Fig. 8).

Information about mortality in the wild is collected by three methods according to the habits of the species in question. These methods involve knowing respectively: age at death; numbers of survivors out of an initial definite number; and age structure of the population at one point in time with mortality rate inferred from the fall in the size of the older age classes.

Mortality in wild populations is simple to measure when reproduction and migration can be ignored and the population examined repeatedly as it decreases in size. However, this is only possible with animals which have an annual life-cycle, a short reproduction period and a small range of movement. More often than not, the death rate in a population is difficult to measure and one must resort to estimating birth rate in steady-state populations.

The two most commonly used practical methods of assessing age of wild animals are the mark-recapture technique and the evaluation of some feature in captive animals that is known to be affected in a well-defined way by the passage of time. Use of the latter method often requires that the structure in question be calibrated by the mark recapture method. This method has been used to measure the life-span of mice living wild on the Welsh island of Skokholm (Table VI). The least satisfactory field parameters are those based on continuous variables such as growth or wear and tear.

Broadly speaking, ecologists attempt to describe how organisms behave in nature and try to explain such fundamental questions as why certain organisms live in a particular place, and what regulates their numbers and maintains particular species patterns. More specifically, many ecologists are concerned with obtaining information on

TABLE VI

Life tables for wild house mice on Skokholm Island

	Date of birth											
	March–April			May–June			July–August			September–October		
Month	Obs. nos.	No. of[a] survivors	Life expectancy (weeks)	Obs. nos.	No. of[a] survivors	Life expectancy (weeks)	Obs. nos.	No. of[a] survivors	Life expectancy (weeks)	Obs. nos.	No. of[a] survivors	Life expectancy (weeks)
May–June	276	1000	13.9									
July–August	—	530.0	13.3	307	1000	15.5						
Sept.–Oct.	93	283.0	12.5	139	600.0	14.3	1383	1000	14.3			
Nov.–Dec.	30	152.3	10.9	114	360.0	12.3	798	600.0	12.3	518	1000	11.0
Jan.–Feb.	—	68.5	10.2	72	162.0	13.3	—	270.0	13.4	—	400.0	12.4
March–April	12	30.8	15.8	28	72.9	15.7	183	121.5	15.8	71	160.0	15.8
May–June	3	18.5	14.7	11	43.7	14.7	82	72.9	14.7	54	96.0	14.7
July–August	0	11.1	13.0	0	26.2	12.9	—	43.7	13.0	—	57.6	13.0
Sept.–Oct.		6.7	10.0		15.7	9.9	25	26.2	10.1		34.6	10.1
Nov.–Dec.		4.0	5.2		9.4	5.1	—	15.7	5.3		20.7	5.3
Jan.–Feb.		0.4	4.3		0.9	4.3	0	1.6	4.9		2.1	4.9
March–April		0	0		0	0		0.2	4.3		0.2	4.3
May–June								0	0		0	0

[a] Calculated from mortality rates.
From Berry & Jakobson (1971).

the sizes of populations in successive developmental stages, so that a life table may be constructed and an attempt made to determine factors that regulate population size in relation to chronological age. The census of populations and the definition of stages at which mortality factors operate are necessary first steps in estimating ecological productivity. However, generally, neither animal nor plant ecologists pay attention to the determination of specific mortality factors operating in natural populations on an age basis.

If, with further study, it turns out that the maximum laboratory lifespan of organisms is always greater than that in the wild, it may be that part of the increase in longevity in captivity is due to the experimenter selecting genotypes for good adaptation to laboratory conditions. A more likely explanation however, is that the major causes of death in wild populations are accidents, predation and disease, which are largely age-independent. There are many natural populations where deaths appear to follow an accident-rate curve. It is the natural lifespan in the wild that must be under selection pressure. Furthermore, in asking which processes render the individual more likely to die in a given time interval as it grows older, we are likely to obtain different answers from laboratory strains compared with wild types. This is not to say that age changes leading up to death in wild-type populations have a different fundamental basis.

It is interesting to ask whether evolution can ever favour a deterioration of function. Advantageous regression probably occurs in the early age involution of organs such as the thymus (Table VII) which perform their role early in life and where there may be physiological disadvantages in retaining function. It is arguable that senile changes never occur in truly wild populations of animals living under the conditions in which they evolved. However, this may not apply to the early decline in the female reproductive system which is a feature of laboratory populations.

In the wild mouse on Skokholm Island off the coast of Wales, the population is carried through the winter by sexually inactive females which begin reproducing in the spring. Laboratory stock derived from this island race lose reproductive efficiency quickly and are infertile at six months of age. Here we may have an example of accelerated ageing under laboratory conditions. The only relevant experimental data is that the age-related fall in fertility of female rats can be reduced by food restriction. This points to the need for further co-operative work on wild mice and their laboratory counterparts. So far comparative work has been confined to studies of the inorganic chemical patterns of bones and blood vessels (Fig. 9).

TABLE VII

Effect of age and cortisol treatment on the cellular composition of the cortex of mouse thymus

	Number of cells[a] (mean ± S.E.)					
Cell type	Young mice (1)	v[b]	Old mice (2)	v[b]	Cortisol-treated mice (3)	v[b]
Lymphocytes	295.86 ± 12.68	0.20	137.17 ± 11.27 1–2 P < 0.001	0.40	82.21 ± 9.34 2–3 P < 0.001	0.70
Epithelial	19.00 ± 2.41	0.59	13.08 ± 2.59 1–2 P < 0.2	0.97	12.47 ± 1.08 1–3 P < 0.1	0.53
Macrophages	1.95 ± 0.30	0.72	17.54 ± 2.11 1–2 P < 0.001	0.59	22.26 ± 1.47 2–3 P < 0.1	0.41
Fibroblasts	0.41 ± 0.14	1.61	6.08 ± 1.19 1–2 P < 0.001	0.96	5.18 ± 0.59 1–3 P < 0.001	0.70
Plasma	0.53 ± 0.17	1.25	1.41 ± 0.42 1–2 P < 0.2	1.23	1.03 ± 0.24 1–3 P < 0.1	1.38
Mitotic index	1.46		0.89		1.27	
Total cell population	319.36 ± 13.41	0.20	186.04 ± 11.33 1–2 P < 0.001	0.30	124.0 ± 9.75 1–3 P < 0.001	1.28

[a]Grid area: 80 μm × 80 μm.
[b]Coefficient of variation.
$n = 22$.
From Bellamy & Alkufaishi (1972).

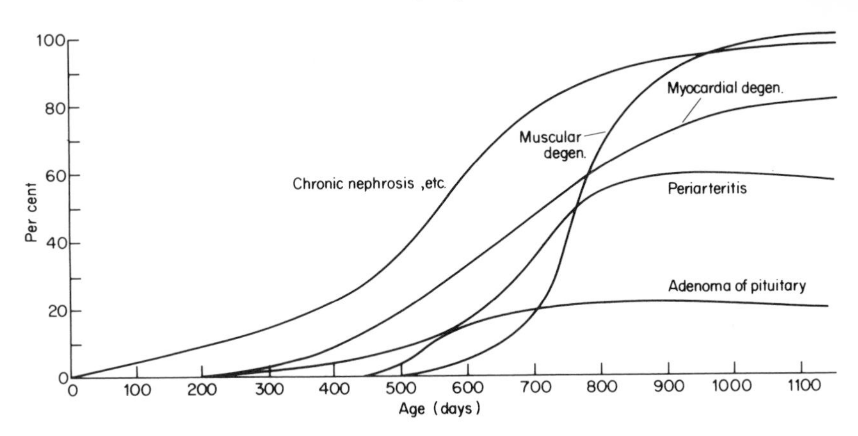

FIG. 9. Time of onset of various diseases in the rat (from Simms, Berg & Davies, 1959).

THE CHEMICAL NATURE OF AGEING

Ageing, to a chemist, describes those chemical reactions such as oxidation, cross-linking and thermal rearrangements that are slow, irreversible and eventually result in the complete conversion of the starting material into one or more products with new properties. It is implied that chemical ageing occurs because the starting material is unstable when considered over periods expressed in weeks, months, years and decades. Chemical stability depends very much on environment; metal structures in desert regions hardly rust at all but are rapidly oxidized in a temperate climate. Also, organic compounds that have survived millions of years embedded deep in fossil rocks may disappear in a few hours when exposed to air.

Since living organisms are essentially composed of chemical aggregations and interlocking reactions confined in space, ageing must be viewed as retrogressive three-dimensional chemical change. However, by convention, an ageing system, either living or inanimate, is not one which simply shows chemical change with time; change must be shown to underlie a deterioration in some arbitrary, desirable property of the system. For example, man-made rubber, plastics, paint, leather and paper may be said to age because they contain molecules with reactive groups which by intra- and intermolecular reactions, slowly change, resulting in a loss of elasticity, transparency, etc. Chemical changes in rubber bands on storage are described as ageing because they eventually result in the bands snapping under low tension. In this respect ageing of elastic bands follows a human-type mortality curve (Fig. 10) when "snapping" is

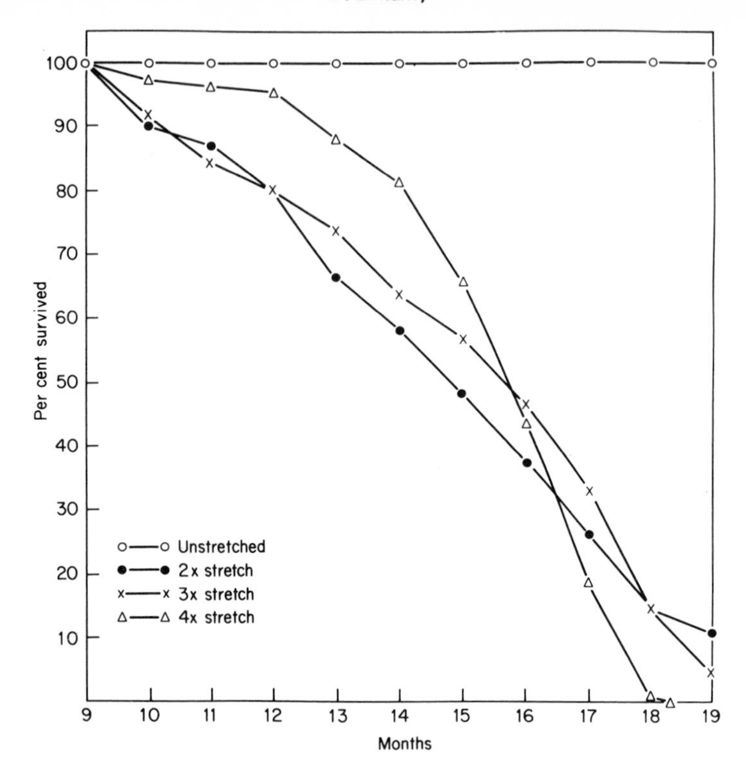

FIG. 10. Survival of stretched rubber bands (from Fels, 1969).

taken as the point of death although the underlying oxidation process may have built up steadily to this arbitrary snapping point from the time the bands were manufactured. In some other natural polymers similar chemical changes occur whilst the material is within the organism, although at the moment none of these obviously deleterious changes can be connected with an increased chance of mortality of the whole organism.

Chemical ageing results in several types of change with respect to the state of the product compared with that of the starting material. A random molecular arrangement may be converted into a regular array as in crystallization and polymerization. Highly ordered arrays may be converted into less ordered structures, as seen in the unfolding of tightly and complexly folded polypeptide chains during denaturation of proteins. Uniformity in composition may give way, by intermolecular reactions of increasing complexity, to a diversity of composition.

Two other points of definition need to be discussed which relate to the influence of time on the rate and nature of the chemical reaction. Some kinds of chemical ageing occur at random and the time that has elapsed from the starting point does not affect the chance or outcome of a particular molecule subsequently undergoing the reaction. Radioactive decay is such a process, where the event of atomic disintegration is random and the probability of its occurrence is independent of how long the atom has been in existence. At any one time the number of atoms disintegrating is proportional only to those present and this proportion is constant. This situation, in fact, refers to many chemical reactions which follow what is known as first-order kinetics. Other kinds of chemical ageing involve a more progressive type of change where the extent of the change that has already taken place affects the rate and nature of the immediate reaction. Rusting and corrosion of thick metal bars is of this kind where the build-up of products tends to protect the metal from further reaction.

Chemicals that make up living systems are not resistant to chemical ageing in the categories just described and there can be little doubt that all kinds of reactions occur. Two obvious ones that we know a little about are denaturation of proteins and the condensation of inert complexes, such as lipofuchsin pigment which accumulates in post-mitotic tissues such as the brain, and pseudo-elastin derived from ageing collagen. There are probably many others which cannot be detected because of the lack of specific tests.

Chemical ageing is only seriously detrimental in a living organism if the ageing material cannot be replaced, or if non-removable products that accumulate eventually reach a level that disturbs or displaces vital structures. A basic assumption of the concept of protein turnover is that denatured proteins are degraded and replaced. On the other hand, it does not seem as though lipofuchsin deposits can be eliminated naturally from brain cells, so that aged neurones, with pigment occupying a large proportion of the cellular volume, are likely to be inefficient. Similarly, connective tissue accumulation, which occurs widely and predictably in many organs, may eventually block off cells from nutrient supplies or interfere more fundamentally with the function of cells by the release of inhibitory factors.

Chemical ageing is likely to be very serious when it occurs in self-replicating molecules of DNA and long-lived RNA templates. It is known that DNA does deteriorate in seeds and germ cells, but at the same time evidence is available to indicate that this kind of chemical damage can be detected and repaired. This poses the question of possible shifts in balance between degradation and repair with age favouring the loss of parts.

Living systems can also age by mis-specification of enzymes which govern the fidelity of DNA translation and transcription. This is the basis of the error theory of ageing. Errors exist in macromolecules that are not assembled with complete accuracy, so that faulty molecules could be responsible for further errors and, if the errors accumulated in newly synthesized proteins or DNA, could cause cell death. Recent work with the liver of ageing laboratory mice has indicated levels of mis-specification of histones rising from about 5×10^{-6} per residue to 20×10^{-6} per residue in old animals.

The fact that organisms age and die implies that some imperfections at the chemical level cannot be corrected either during the individual life-span or in evolutionary time. Looked at in this way ageing is the logical outcome of life being a highly co-ordinated assembly of reactive chemicals which only exists through the natural selection of biochemical mechanisms that are able to detect, inhibit, reverse or replace molecules that either undergo slow chemical change or that accumulate errors through faulty synthesis. The fact that ageing eventually predominates in laboratory organisms may be taken to indicate that it is advantageous for natural selection to retard ageing in the natural environment for only a limited period of time. If we consider life in this way, as the evolution and maintenance of a biochemical strategy to retard and perhaps counteract ageing, gerontology is placed in a central position in biological thought and this makes it all the more surprising that little serious attention was paid to ageing in the historical development of biochemistry.

Much of the confusion concerning the chemical definition of ageing comes from the ill-considered use of analogies. For example, comparison of the body with a machine brings up inappropriate concepts of "wear and tear". Although ageing of machines occurs partly because of chemical ageing (such as rusting and perishing), which does have a counterpart in the living body, a very important part of machine ageing occurs through frictional forces which do not apply in living systems. There is no evidence that components of living tissue age in proportion to normal usage, notwithstanding that excessive loading of joints in human subjects may result in subsequent malfunction of the joint in old age. In this respect, it is a characteristic of life that the chemical efficiency of organs increases with the demands made upon them. Furthermore, the two relevant phenomena of work-hypertrophy and compensatory-hypertrophy are, to a great degree, independent of age. Regeneration of limbs and internal organs on the other hand, which has no counterpart in the non-living world, does decline with age but there is no evidence that this decline is related to the number of chemical demands that have been placed upon the mechanism.

PATHOLOGY AND AGEING

Early workers, in seeking to clarify a chemical change in terms of a fundamental ageing phenomenon, regarded it necessary to rule out pathological processes. Those who entered gerontology from medical fields took great pains to emphasize that every effort should be made to clarify the distinction between the pathological features of ageing from what were termed the fundamental physiological aspects; only the latter aspects should be considered in formulating theories of ageing. From this standpoint, definition of ageing in the human population is greatly complicated because it is likely that primary biochemical defects may give rise to pathologies in a random way (Table VIII). Experimental animals also show pathological ageing as evidenced from the onset of latent infectious diseases, some of which, such as pneumonia and ear disease, become clearly developed in advanced old age.

Apart from disease and pathologies such as pneumonia that are well defined as being abnormal from independent evidence, it is extremely difficult to make a distinction between normal and abnormal ageing. One approach is to take the lowest common denominators in the longest lived and shortest lived individuals as being probable primary causes. In this respect, it is worth stressing that, as yet, no case of human old age has been found in which all or several organs and functions show only slight or no changes with ageing. However, the existence of a few human individuals who die in extreme old age with a relatively small number of degenerative features, indicates that there is a fundamental base-line from which most individuals deviate to different extents.

This whole problem of defining the fundamental features of ageing was well summarized by Korenchevsky (1962), the founder of experimental gerontology. His view was that two differential criteria should be used:

(a) Changes which occur only in the majority of old organisms but not in all old individuals, should not be considered as features of pathological ageing, because they are not necessarily present and are apparently avoidable in old age.

TABLE VIII

The ten leading causes of death in the United States, 1900 and 1959

Rank	Cause of Death	Death rate per 100000 population	Per cent of deaths from all causes
1900			
1	Pneumonia and influenza	202	11.8
2	Tuberculosis	194	11.3
3	Diarrhoea and enteritis	143	8.3
4	Disease of the heart	137	8.0
5	Cerebral haemorrhage	107	6.2
6	Nephritis	89	5.2
7	Accidents	72	4.2
8	Cancer	64	3.7
9	Diphtheria	40	2.3
10	Meningitis	34	2.0
1959			
1	Diseases of the heart	364	38.6
2	Cancer and other malignancies	148	15.7
3	Cerebral haemorrhage	108	11.5
4	Accidents	50	5.4
5	Certain diseases of early infancy	39	4.1
6	Pneumonia and influenza (except of newborn)	33	3.5
7	General arteriosclerosis	20	2.1
8	Diabetes mellitus	16	1.7
9	Congenital malformations	12	1.3
10	Cirrhosis of liver	11	1.2

Note: Because of changes in classification procedures, the disease categories for the two years are not strictly comparable.
Source: Various reports from the National Office of Vital Statistics, United States Public Health Service.
From Strehler (1960).

(b) Those features in old individuals, which are the same or very close to the features in all adult or young persons, should be considered as associated with normal physiological ageing.

The assumptions are that there is one, or only a limited number of basic changes and that all individuals age in the same way. Unfortunately, neither of these two assumptions has any experimental basis. A common dilemma is exemplified by the classic autopsy of a male aged 102 which revealed generalized atherosclerosis, thrombosis of the left femoral, popliteal and tibial arteries and of the accompanying veins, terminal bronchopneumonia, a carcinoid tumour of the ilium, enlargement of the thyroid and chronic emphysema; the cause of death was gangrene of the left foot!

Whilst there can be little doubt that many pathologies in old age result from the decreased resistance to microbial disease, studies on germ-free mice have shown that infections are not an important fundamental aspect of ageing. Ageing, germ-free mice show the same pattern of non-infectious pathological conditions as do conventionally maintained animals. This is particularly true for kidney glomerulosclerosis – a thickening of the glomerular basement membrane associated with hyaline deposits in the wall of the glomerulus – which has been described as the most prevalent lesion of the vascular system of laboratory mice. The incidence and severity of this lesion is the same in germ-free and conventional mice. This is also the case for the incidence of muscular degeneration which occurs within a narrower age-range than glomerulosclerosis. There is also clear evidence that the introduction of germ-free conditions to a laboratory where rodent strains are already well managed by conventional methods does not increase lifespan; in some experiments germ-free conditions have actually reduced longevity!

The fact that certain diseases show a well-defined age incidence has led to several ideas in which the pathology is seen as the outcome of a developmental process. One such human disease is gout, which when it attacks men does so usually between the ages of 40 and 50. This disease is marked by painful inflammation of the joints caused by urate deposits and is the major manifestation of hyperuricemia. Hyperuricemia may appear in the third decade before there is any discomfort in the joints. This is a hereditary condition with a well-defined genetical mechanism, which suggests that other diseases may arise, with secondary complications that are late manifestations of inborn metabolic disorders, implying that genetical variation in susceptibility to common primary processes of ageing may account for non-uniform pathological change.

A developmental origin for some pathologies is evident from the

TABLE IX

Frequency distribution of major sites of arteritis in ageing RF mice

Artery	Frequency of involvement
Renal	32/64
Aorta	28/64
Coronary	10/64
Ovarian	9/64
Uterine	9/64
Splenic	6/64
Axillary	5/64
Adrenal	3/64
Hepatic	3/64
Gastric	2/64
Cervical/mediastinal	2/64
Pancreatic	1/64
Thymic	1/64
Spermatic	1/64
Superior mesenteric	1/64

fact that they appear to have a highly localized incidence within the body. For example, arteritis occurs frequently in the renal artery of old mice but is seldom seen in the arteries supplying other tissues (Table IX). It is difficult to escape the conclusion, when the basic structure and function of all arteries are very similar, that a subtle programmed aspect of phenotype expression is responsible for the pathology.

Degenerative conditions have a diverse and complex origin and one is immediately struck by the difficulties, not only in defining the cause of death in old organisms, but also the nature of disease. Taking the broad view that disease is a malfunction that affects people in a random fashion begs the question, are fatal degenerative conditions, individually described as diseases, in fact ageing processes? The answer to this question depends on whether disease is defined as being due to a random event, unrelated to age, or as a progressive deterioration that occurs in only a fraction of the population at a given age. In the latter situation, taking Korenchevsky's dictum, the only difference between an ageing process and a disease is that an ageing process occurs in all of the population. To a certain extent, this type of discussion points to the futility of arguing that only those clinical phenomena should be considered within the province of gerontology that are manifest in the bulk of the population, particularly when it is known that for all populations phenotypic variability is widespread. Stating this another way, if each disease could be cured no doubt we would all encounter many diseases. These would occur at different chronological ages dependent upon genotype and environment. They would all be expressions of a range of deteriorating developmental programmes proceeding at different rates; fundamental age changes of uniform occurrence in all organs are expressed as non-uniform expressions of diseased phenotypes depending on the interplay between genes and environment at the level of individuals.

HOMEOSTASIS AND DEVELOPMENT

When we come to examine the reasons for the increased force of mortality in old organisms, there are two possible levels of function, corresponding to the organ and the cell, at which the failure of the living system may be considered in meaningful terms. At the level of the organ, there appears to be little evidence that malfunctions, either through disproportionate growth or the failure of repair mechanisms, occur on a scale sufficient to increase the chances of the organisms dying. The gross chemical composition of tissues in laboratory stock also shows a remarkable constancy with age (Table X). On this plane, however, one is clearly dealing with the deterioration of complex interacting multicellular systems, comprising sense organs, nervous system, endocrine system and effectors. Taken together several deviant sub-systems may together result in a failure to produce the correct degree of response in relation to alterations in the external and internal environment. Ageing is clearly seen when physiological stress is imposed. Lack of ability of a sub-system to respond appropriately to a disturbed environment does not appear to be due to the failure of any single component. For example, the endocrine system probably fails to meet demand because of both a fall in the rate of secretion of hormones and the inability of the target tissue to make the appropriate response to the hormones. Often, structural changes in several components of the effector organs are alone sufficient to limit mechanical aspects of an endocrine response.

Death in the context of organ function involves a series of events which occur over a small fraction of the lifespan, possibly amounting to only a few hours. That is to say, organisms die as the result of internal fluctuations in body chemistry which can no longer be contained through homeostatic regulation; a small shift in metabolism which could be counteracted in youth becomes amplified to the point of preventing a vital function. Thus, death in old organisms may result from minor changes in the environment.

TABLE X

Electrolyte content[a] of tissues in young[c] and old[b] C57 mice

	Kidney				Muscle				Brain			
	Young		Old		Young		Old		Young		Old	
	M	F	M	F	M	F	M	F	M	F	M	F
N	10	10	10	10	10	10	10	10	9	9	9	9
Potassium	93.00	92.25	88.00	89.73	89.18	88.75	79.25	98.05	98.90	94.75	93.25	95.83
(m-equiv/g wet wt)	± 2.14	± 1.69	± 2.10	± 2.30	± 3.53	± 6.28	± 2.24	± 3.18	± 1.92	± 1.81	± 2.73	± 1.77
Sodium	77.75	83.50	88.00	89.18	21.67	23.00	24.50	25.55	49.73	47.25	48.25	50.00
(m-equiv/g wet wt)	± 2.25	± 3.25	± 7.53	± 3.20	± 1.32	± 0.73	± 0.82	± 2.35	± 1.84	± 1.12	± 1.75	± 1.72
Water (% wet wt)	75.33	74.47	75.75	74.98	71.78	75.79	73.17	72.67	79.28	79.43	78.49	76.97
	± 0.35	± 0.52	± 0.30	± 1.13	± 3.24	± 0.51	± 0.93	± 1.30	± 0.7	± 0.29	± 0.24	± 0.48

[a]Means ± S.E.M.
[b]Old = 18 months.
[c]Young = 2 months.

The normal low tolerance and limited capacity for self-regulation of metabolic systems imply that the individual processes concerned in regulation rarely operate under conditions which an engineer might consider optimal and that the resulting organs, even at the peak of development, are never quite perfect. Tolerance and imperfection appear to be the necessary prerequisites, as well as the consequences, of natural selection, with genetical polymorphism possibly arising as of a balance of defects. This is another way of saying that the population carries a genetical load and every indiviual has some structural defect resulting from a deleterious combination of genes which will eventually show up with time. At this point gerontology impinges upon the fundamental problems of contemporary biology which are concerned with the unfolding of the chemical programme of the fertilized ovum against the background of a fluctuating external environment. The house mouse in the wild can only take us a very small part of the way along this path but the many laboratory strains that have been established which vary in lifespan and causes of death are invaluable mammalian models.

REFERENCES

Bang, F. B. (1969). Senescence as a biological problem. In *Biology of populations*: 414–430. Sladen, B. K. & Bang, F. B. (Eds). Amsterdam: Elsevier.

Bellamy, D. & Alkufaishi, H. (1972). The cellular composition of thymus: A comparison between cortisol-treated and aged C57/BL mice. *Age & Ageing* **1**: 88–98.

Berry, R. J. & Jakobson, M. E. (1971). Life and death in an island population of the house mouse. *Expl Geront.* **6**: 187–197.

Birren, J. E. (1964). *The psychology of ageing*. London: Prentice Hall.

Fels, I. G. (1969). A model system for molecular ageing and senescence. *Gerontologia* **15**: 308–316.

Gallagher, J. C. & Nordin, B. E. C. (1973). Oestrogens and calcium metabolism. *Front. Horm. Res.* **2**: 98–117.

Hamilton, J. B. & Mestler, G. E. (1969). Mortality and survival: comparison of eunuchs with intact men and women in a mentally retarded population. *J. Geront.* **24**: 395–411.

Korenchersky, V. (1962). *Physiological and pathological ageing.* Basel: Karger.

Maynard Smith, J. (1966). Theories of ageing. In *Topics in the biology of ageing:* 1–27. Krohn, P. L. (Ed.). London: Wiley.

Morgan, A. J. & Bellamy, D. (1975). Age changes in some organic constituents of bone and aorta. A comparative study in wild-trapped and laboratory mice (*Mus*). *Age & Ageing* **4**: 209–223.

Sacher, G. A. (1959). Relation of lifespan to brain weight and body weight in mammals. *Ciba Colloquium on Ageing* **5**: 115–133. London: Churchill.

Simms, H. S., Berg, B. N. & Davies, D. F. (1959). Onset of disease and longevity of rat and man. *Ciba Colloquium on Ageing* **5**: 74. London: Churchill.

Strehler, B. L. (1960). *The biology of ageing*. Baltimore, USA: Waverley.

GENERAL REFERENCES AND BACKGROUND READING.

Bellamy, D. (1967). Hormonal effects in relation to ageing in mammals. *Symp. Soc. exp. Biol.* 21: 427–454.

Bellamy, D. (1970). Ageing and endocrine responses to environmental factors: with particular reference to mammals. *Mem. Soc. Endocr.* **18**: 303–339.

Birren, J. E. (1964). *The psychology of ageing.* London: Prentice Hall.

Krohn, P. L. (Ed.) (1960). *Topics in the biology of ageing.* London: Wiley.

Korenchevsky, V. (1962). *Physiological and pathological ageing.* Basel: Karger.

Wolstenholme, G. E. W. & O'Connor, M. (Eds). (1954-). *Ciba Foundation Colloquia on ageing.* London: Churchill.

Symp. zool. Soc. Lond. (1981) No. 47, 301–335

Physiological Adaptability: the Response of the House Mouse to Variations in the Environment

M. E. JAKOBSON

Department of Biology, North East London Polytechnic, Romford Road, London E15, UK

SYNOPSIS

House mice show varied physiological responses when exposed to climatic challenges. These responses reflect individual adaptability if associated with improved fitness.

This review concerns itself primarily with the response of house mice to cold. From numerous laboratory studies, the expected response of a non-hibernating small mammal when exposed to continuous cold is to maintain body temperature by increased heat production. This basic response involves both shivering and non-shivering mechanisms; currently emphasis is laid on the role of brown adipose tissue in increased non-shivering thermogenesis of cold-adapted animals.

Variations from the basic pattern of response (e.g. in brown fat) are discernible from studies in which the pattern of cold exposure is discontinuous, or associated with other stresses, or related to genetical variations. Such studies emphasize the difficulty (especially for field biologists) of assessing the adaptive value of mechanisms evaluated in purposefully reductionist laboratory studies. More complex designs may uncover aspects of adaptability previously hidden; for example only recently have unequivocal data been published on the ability of *Mus musculus* to undergo bouts of daily torpor in the laboratory and in the field.

Much variation is found in physiological features of mice within and between geographically distinct populations. Some features show trends similar to those found in laboratory studies: relative brown fat weight correlates inversely with meteorological temperature. Other features differ: winter mice do not have the high-metabolizing response expected of cold-acclimated stocks. The biological validation of particular responses of feral mice as adaptive in their natural habitat is elusive.

INTRODUCTION

Physiological adaptability is the capacity to adjust physiologically, usually rapidly, in response to environmental challenges and insults

(Berry & Jakobson, 1975). Adaptability is essentially a characteristic of individuals, being one feature of their inherited adaptation. Adjustment within the individual occurs when some physiological function is sensed as deficient (Adolph, 1972). The response may be judged adaptive if individuals showing it have increased survival (and increased reproductive fitness) compared to individuals not showing the response. It is these adaptive responses, rather than change as such, that constitute adaptability.

This review concentrates on studies on the adaptation of the house mouse to cold environments, partly derived from studies on the house mouse itself, and partly from studies on other small mammals. The literature abounds in descriptions of the mechanisms of physiological flexibility of small mammals in cold. These mechanisms are commonly asserted to indicate the adaptation, acclimation or acclimatization of the individual to its environment (Hart, 1961; Bligh & Johnson, 1973). The terms reflect differences in the type of cold exposure on the animals, as perceived by the investigator. Cold acclimation responses relate primarily to those observed in the laboratory in controlled, usually constant cold. Cold acclimatization studies observe changes in animals subjected to more natural environments, where the investigator assumes levels and changes in external climate influence response. Adaptation covers both responses, and also that reflecting genetical differences between individuals (Hart, 1961; Barnett, 1965).

THE HOUSE MOUSE AS A HOMEOTHERM

A basic fact of life for a house mouse is its high surface area to volume ratio (Dawson, 1967), which favours rapid heat exchange with the environment. Body temperature is usually reported to lie in the range 34–40°C at most environmental temperatures (e.g. Hart, 1951; Bernstein, 1966). The external temperature at which mammals respond to cold by increased heat production (the lower critical temperature) is usually estimated by plots of minimum metabolism against environmental temperature (Fig. 1). For mice this is in the region of 30–32°C (Herrington, 1940; Hart, 1950; Mount, 1971; Hudson & Scott, 1979). This is similar to the zone of preferred temperatures chosen by mice placed in thermal gradients. This zone ranges from 31–36°C, with young animals or animals reared in the cold opting for higher substrate temperatures (Ogilvie & Stinson, 1966; Lynch *et al.*, 1976). In most habitats (even commensal ones) environmental temperature is rarely this high, but nest and huddle

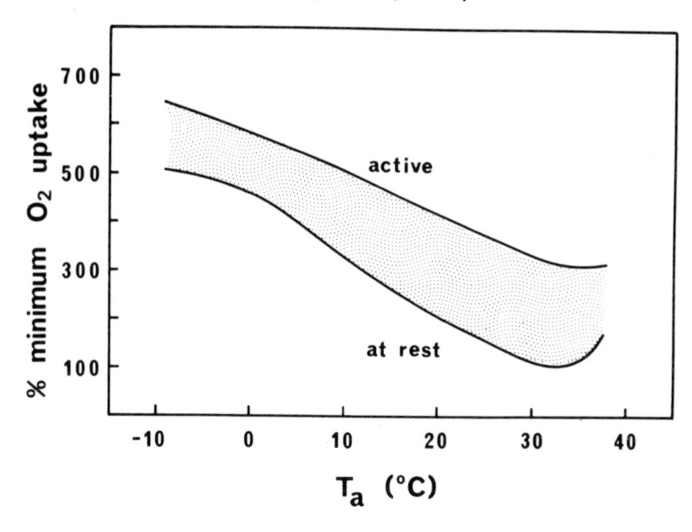

FIG. 1. Metabolism of white mice at rest and during maximal exercise at different ambient temperatures (T_a). After Hart (1950).

temperatures (Barnett, 1956; Barnett, Munro *et al.*, 1975) suggest that mice may spend significant periods of their rest phase at temperatures within or close to thermal neutrality. Bryant & Hails (1975) suggest that young mice huddling at 22°C may at times be under heat stress.

In its active phase, during foraging and social behaviour on or near the ground surface, a mouse will probably experience temperatures below thermal neutrality. Its thermal relations with its environment will more accurately be described by some line above the minimum level (Fig. 1). This is because thermogenesis due to activity is additive to that of cold thermogenesis (Hart, 1950, 1952). Only in extreme cold does activity substitute cold thermogenesis, when shivering is dominant (in rats, Hart & Jansky, 1963). Enforced exercise at 20°C and 30°C raised body temperature 1–2°C, caused a small drop at 10°C and caused hypothermia at 0°C (Hart, 1951). If a mouse normally experiences cold when active, then at times heat loss mechanisms may also be activated.

At rest in the cold, heat conservation depends on fur and circulatory adjustments. Mount (1971) found that core-ambient thermal insulation improved about 25% when mice were moved from 30°C to 22°C. At 22°C insulation of furred mice was $0.418°C\,m^2\,h\,kg^{-1}$, compared to $0.275°C\,m^2\,h\,kg^{-1}$ for genetically hairless mice. Despite differences in activity and shape of the two stocks, this suggests that fur contributes 30–40% of total insulation, which is similar to results obtained from excised pelts (Barnett, 1959; Dawson & Webster, 1967). The lower critical temperature for hairless mice was 32–34°C, and metabolism at 22°C was over 40% higher than for furred mice. Cold-adapted mice show a 20–80% improvement in fur insulation, as judged from smooth and ruffled pelts (Barnett, 1959; Hart, 1956). Remaining insulation reflects vasoconstriction and differential cooling of peripheral tissues: the average temperature of body tissues in mice at rest at 0°C may be up to 3.5°C lower than colonic temperature (Hart, 1951).

In hot environments, or during exercise, heat loss is increased by blood flow to the skin and tail. Loss of tail in mice reduces resistance to heat (Harrison, 1958). The small size of a mouse means that the reserve of water available for evaporative heat loss is limited, and the primary response to heat loads is likely to be behavioural (Taylor, 1977). Saliva-spreading may occur but is not well documented in mice (Hart, 1971; Hudson & Scott, 1979). Cutaneous evaporation markedly increases at air temperatures of 35°C (Edwards & Haines, 1978). When placed at 41°C, mice can limit their rise in body temperature to 41–42°C for about one hour, before succumbing to the lethal level of 44°C (Wright, 1976). Acclimation of adult mice to 36°C over two to four weeks can reduce basal metabolic rate 11–36%, depending on the strain (Pennycuik, 1967).

In summary, most mice experience high environmental temperatures only in their rest phase. During activity, when the nest is vacated, physiological mechanisms of heat production are switched on.

MECHANISMS OF ENHANCED HEAT PRODUCTION IN COLD

Acute exposure to cold results in an immediate metabolic response to maintain homeothermy. Continued exposure results in further adjustments (acclimation or acclimatization), the levels of which are related to the level of cold. These adjustments can usually be shown to be adaptive in terms of enhanced survival.

General models of response of small mammals to cold have emerged from numerous studies, regularly reviewed (Barnett & Mount, 1967; Smith & Horwitz, 1969; Chaffee & Roberts, 1971; Hart, 1971; Janksy, 1973; Gale, 1973; Himms-Hagen, 1976) or the subject of symposia (e.g. Girardier & Seydoux, 1978; Wang & Hudson, 1978). Reviews solely on the house mouse are few (Barnett, 1965, 1973).

A general survey of heat production mechanisms is presented here.

Wherever possible, data are presented on house mice. However, the model crosses both species boundaries (hibernators and non-hibernators) and methodology boundaries (cellular and subcellular *in vitro*; laboratory and field *in vivo*). Most of the studies are from small mammals, but these may still be ten or more times larger than house mice. Factors whose importance is influenced by body size (Chaffee & Roberts, 1971; Heldmaier, 1972; Schmidt-Nielsen, 1972) may need reassessment in the future.

Shivering and Non-shivering Thermogenesis

A primary division of thermogenic mechanisms is into shivering thermogenesis (ST), restricted to the somatic skeletal musculature (Hemingway, 1963), and non-shivering thermogenesis (NST), encompassing remaining mechanisms. Cold-induced non-shivering thermogenesis can occur in many tissues, including muscle, and is mediated by the sympathetic nervous system (Jansky, 1973; Himms-Hagen, 1976). This is shown, on transfer into cold, by increased catecholamine loss in urine (in rats, Leduc, 1961; Shum, Johnson & Flattery, 1969). Levels of adrenaline decline after about a week, but noradrenaline levels stay high.

These catecholamines may facilitate shivering through increased mobilization of fat and glycogen, but the main effect of noradrenaline in small mammals is to stimulate non-shivering thermogenesis. Levels of cold-induced non-shivering thermogenesis, and its estimates from injected noradrenaline, are inversely related to body size (Heldmaier, 1972). Adjustment to cold is characterized by a shift from shivering to non-shivering thermogenesis. Total metabolism in cold-acclimated stocks is typically higher than in warm-acclimated stocks at all temperatures (in rats, Heroux, 1963; Foster & Frydman, 1979).

In mice this shift in thermogenic pattern is neatly demonstrated by Mejsnar & Jansky (1971) (Fig. 2). Warm-acclimated (28°C) mice shivered in cold tests below 20°C, while cold-acclimated (9.5°C) mice did not shiver at the temperatures tested. Oxygen consumption of cold-acclimated mice was higher at all temperatures, with maximum metabolism 20% larger. Lynch *et al.* (1976) and Lacy, Lynch & Lynch (1978) also found that stocks kept at 5°C had higher basal metabolic rates than those at 22°C.

Injection of noradrenaline increases non-shivering thermogenesis up to three times basal levels. Cold-acclimated mice show enhanced sensitivity to noradrenaline, and up to 80% greater maximum response (Jansky *et al.*, 1969; Mejsnar & Jansky, 1971; Lacy *et al.*,

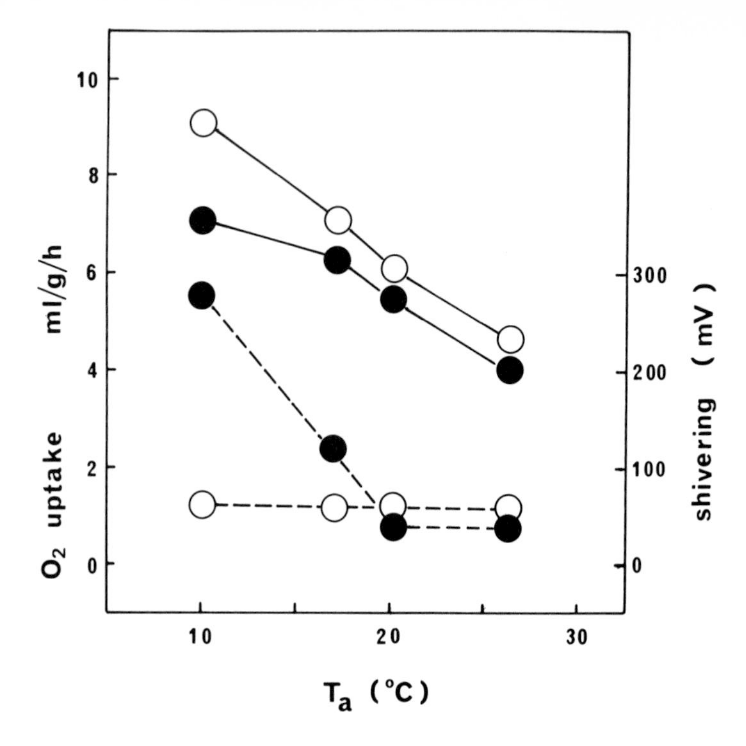

FIG. 2. Oxygen uptake (——) and shivering (– – – –) of mice acclimated at 28°C (●) and at 9.5°C (○). After Mejsnar & Jansky (1971).

1978). Response to injected adrenaline is also higher, but the thermogenic effect is not as great as to noradrenaline (Mejsnar & Jansky, 1971).

Shivering and non-shivering thermogenesis are initiated by different neural pathways. Shivering is largely a function of spinal cord and skin temperatures, while non-shivering thermogenesis relates to hypothalamic and skin temperatures. These pathways appear relatively independent in the guinea-pig (Brück & Wünnenberg, 1970), but in rats the intensity of both forms of thermogenesis can be influenced by both spinal and hypothalamic thermoreceptors (Banet, Hensel & Liebermann, 1978). In rats, shivering is probably initiated only when non-shivering thermogenesis has reached its maximum (or been inhibited by propranolol), causing a slight drop (1–2°C) in body temperature (Banet & Hensel, 1977). In *Peromyscus maniculatus*, in a similar test, body temperatures may drop considerably more before shivering begins (Lilly & Wunder, 1979). The details of control in the mouse are unknown. On cooling of isolated muscle from mice, discharge from cold receptors within the muscle increases; in muscles taken from cold-acclimatized mice the rate of increase and the maximum response is reduced (Banet & Seguin, 1967).

Cold acclimation increases survival time in extreme cold (Hart, 1953a). White mice acclimated to 30°C, 20°C, 10°C and 1°C were able to survive 200 minutes at test temperatures approximating −8.5°C, −18°C, −20°C and −21°C respectively. The greatest improvement occurred between mice kept at 30°C and 20°C. Changes in cold resistance associated with acclimation were largely acquired within the first two weeks after transfer to the new acclimation temperature, being complete within five weeks (Hart, 1953b). Differences in survival were abolished if the mice were fasted for 24 hours before the test (Hart, 1957).

Sites of Non-shivering Thermogenesis

The contribution of different tissues to non-shivering thermogenesis is currently undergoing revision. Brown adipose tissue (BAT) has long been recognized as one of the most thermogenic tissues (per unit mass) in adult non-hibernating small mammals (Smith, 1961; Smith & Horwitz, 1969), but its small mass (1–2% body weight) led to the belief that its contribution to total heat production must be small. In a comprehensive review of non-shivering thermogenesis, Jansky (1973) apportioned NST amongst tissues as follows: muscle 50%; liver 25%; gut 10%; brown fat less than 10%. However it was known that removal of areas of brown fat (the interscapular pads) could reduce noradrenaline-induced heat production up to 60%, leading Himms-Hagen (1976) to suggest that brown fat might release some humoral factor to stimulate NST elsewhere. Recently, new methods of blood flow analysis using labelled microspheres have enabled Foster & Frydman (1978, 1979) to propose more simply that brown fat itself can contribute up to 60–70% of cold-induced or noradrenaline-induced heat production (Table I). These estimates depend on many assumptions, such as that thermogenesis is proportional to blood flow, and the authors provide evidence to back their claims. Their data strongly suggest that brown fat can be the dominant site of thermogenesis in cold-acclimated rats, and codominant with muscle (shivering) in warm-acclimated rats. Heart and diaphragm increase their contribution in proportion to the extra respiratory and circulatory load (cardiac output increases up to 63% at −19°C in cold-acclimated rats).

The distribution of brown fat has been described in the rat and newborn mice (Cameron & Smith, 1964; Cameron, 1975) and in

TABLE I

Percentage contribution of tissues to cold-induced thermogenesis in rats, estimated from increases in blood flow

	Warm-acclimated at 28°C		Cold-acclimated at 6°C	
Brown fat (% body weight)	0.74		1.34	
Test temperature	6°C	– 19°C	6°C	– 19°C
Brown fat	37	29	72	61
Skeletal muscle	33	39	8	10
Liver and gut	7.2	1.4	0.9	2.4
Heart, ribcage & diaphragm	10.9	22	12.6	17.5
Other tissues	11.9	8.6	16.5	9.1

(After Foster & Frydman, 1979.)

adult *Peromyscus maniculatus* (Rauch & Hayward, 1969). Its location is more restricted than white fat. Heldmaier (1974) found 54% of brown fat occurred in the interscapular region of mice acclimated to 30°C; 30% occurred in subscapular and suprasternal regions, 20% in dorsal cervical areas and 5% as subvertebral deposits. Unlike rats, no perirenal brown fat was found.

Functionally the distinction between white fat and brown fat is that the former is a store of energy to be liberated for biological work in other tissues, while brown fat may release energy into the blood solely as heat. Thus in starvation in the warm, brown fat stores are not depleted in newborn rabbits, even at death (Heim & Kellermayer, 1967). In a cold environment, brown fat is rapidly depleted.

Anatomically brown fat is characterized by a "double" sympathetic innervation, one to the vascular supply, the other to the adipocytes themselves (Daniel & Derry, 1969). Over 150 000 β-adrenergic receptor sites have been reported per hamster brown fat cell (Cannon, Nedergaard *et al.*, 1978), over 100 times that in white fat, and nearly twice that in rat myoblasts (Atlas, Hanski & Levitski, 1977). Typically the fat is stored in multilocular form (many small droplets rather than the single large globule of white fat); mitochondria are numerous and have densely packed cristae. Enzyme patterns are distinctive, e.g. increased activity of carnitine acetyl and carnitine palmitoyl transferases (Hahn, Beatty & Bocek, 1976). However Cannon, Nedergaard *et al.* (1978) emphasize that these are not foolproof distinctions, especially on an interspecific level, and regard the features described as typical for brown and white fat as the extremes of a continuum.

In summary, total brown fat is often less than 1% of body weight in mice. Its weight may be estimated by interscapular brown fat, representing about 50% of the total. Wet weight is a convenient measure, but lipid-free weight correlates better with its activity (J. S. Hayward, in Chaffee & Roberts, 1971). The latter measure would have advantages in field studies, as it may buffer short-term variations of cold experience due to handling procedures prior to the death of the animal.

Changes in Brown Adipose Tissue in the Cold

Acclimation to cold over four weeks increases brown fat mass in laboratory mice in proportion to the cold stress (Heldmaier, 1974). Acclimation to 10°C increased total brown fat by a factor of 2.5 (Fig. 3). All regions of brown fat showed significant hypertrophy, except dorsal cervical brown fat (which might play a role in warming spinal receptors associated with shivering). Dorsal cervical brown fat

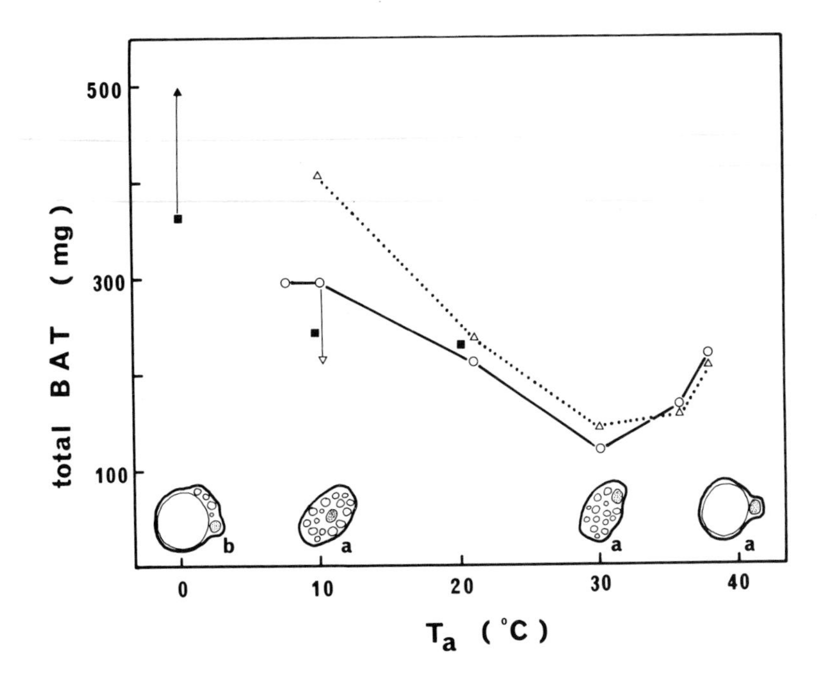

FIG. 3. Brown fat (BAT) weight in mice responding to acclimation or cold shock temperatures. (○—○) mice after 4 weeks acclimation; (△····△) hairless mice after 4 weeks acclimation; (■) effect after 11 × 2.5 h cold shocks to furred mice kept at 30°C; (——▶) effect of 11 × 5 h cold shock at 0°C; (——▷) effect of huddling in 5 mice at 10°C. Nature of fat storage within brown fat cells (a) in acclimated stocks; (b) after cold shocks at 0°C. After Heldmaier (1974, 1975a, b).

did increase significantly in hairless mice, which overall showed a 50% greater hypertrophy in relative weight ($mg\,g^{-1}$) than furred mice at 10°C.

Heldmaier (1974) found that brown fat mass also increased in mice acclimated to 37.5°C, but the fat content of the cells was unilocular, resembling (though smaller than) white fat (Fig. 3). Pospisilova & Jansky (1976) also report an increase in relative brown fat weight from 6.6 $mg\,g^{-1}$ to 10.2 $mg\,g^{-1}$ in mice acclimated from 32°C to 11°C, but with the small samples used, the trend was not significant. In both studies, mice showed a drop in average body weight at the lowest temperatures of acclimation (7°C and 11°C); they were unable to successfully acclimate mice to temperatures lower than this in the absence of nest material (cf. Hart, 1953a; Barnett, 1964).

Rapid changes can occur in brown fat on transfer of animals into cold (Cameron & Smith, 1964). A doubling in mass occurs within a week of transfer of rats from 28°C to 4°C (Desautels, Zaror-Behrens & Himms-Hagen, 1978). Ikemoto, Hiroshige & Itoh (1967) reported a 40% increase in interscapular brown fat wet weight (and a 70% increase in fat-free mass) in SM mice after five days at 5°C. Brown fat growth can be stimulated by daily injections of noradrenaline or thyroxine in the warm (Leblanc & Villemaire, 1970), suggesting hormonal mediation in the cold response. In hamsters, melatonin and day length also influence brown fat growth (Heldmaier & Hoffman, 1974).

Within brown fat cells of rats mitochondria proliferate (Suter, 1969), and show a greater than four-fold increase in mass within one to three weeks (Anderson *et al.*, 1970). The size of lipid droplets decreases and their number increases, thereby increasing surface area, in hamsters (Ahlabo & Barnard, 1974).

Mechanisms of Thermogenesis in Brown Adipose Tissue

Substrate mobilization and brown fat proliferation

On exposure to cold, the primary switch for non-shivering thermogenesis in brown fat is the arrival of noradrenaline at the cell surface (Himms-Hagen, 1976). In intact rats the response of brown fat is proportional to neural stimulation (Flaim, Horwitz & Horowitz, 1977), and involves stimulation of both α- and β-adrenoreceptors, though the latter predominate (Horwitz, 1978). Plasma levels of noradrenaline play a minor role in stimulating brown fat (Seydoux & Girardier, 1978) but may stimulate substrate mobilization or thermogenesis in other tissues (Horwitz, 1978).

Binding of noradrenaline to the cell surface results in depolarization, increase in conductance and a consequent increase in metabolism (Girardier, Seydoux & Clausen, 1968; Horowitz, 1972; Seydoux & Girardier, 1978). The immediate intracellular events are reviewed by Severson (1979) and authors in Wang & Hudson (1978). Concentrations of cyclic AMP rise activating protein kinase systems, finally activating the rate-limiting hormone-sensitive triglyceride lipase. Of the free fatty acids released, up to half may be released into the blood (Cannon, Nedergaard *et al.*, 1978), though in cold-acclimated animals all may be metabolized within the cell (Portet *et al.*, 1974). Within brown fat cells, the free fatty acids are broken down in the mitochondria via β-oxidation and the Krebs cycle.

ATP-dependent means of thermogenesis

The work of Edelman and colleagues (e.g. Edelman, 1976a, b) has highlighted the importance of thyroid hormones and the activity of membrane pump mechanisms in the understanding of the increased basal metabolism of homeotherms compared to poikilotherms.

Increased activity of sodium pumps in brown fat is well documented: blocking with ouabain results in a 60–70% drop in thermogenesis of isolated hamster brown fat cells (Horwitz, 1973), confirming that the cell membrane can be an important site of thermogenesis in the cold (Horwitz, 1978). Other ATP-utilizing "futile cycles" have been proposed, e.g. calcium cycling across mitochondrial membranes (Christiansen, 1971; Hittelman, Fairhurst & Smith, 1967), and a corticosteroid and insulin sensitive cycle in carbohydrate metabolism of infant rats (Hahn, Seccombe, Kirby & Skala, 1978).

The significant role of ATP-using mechanisms is consistent with the high levels of creatine phosphate and high activities of phosphocreatine kinase (20 times that of white fat) in rats and NMRI mice (Berlet, Bonsmann & Birringer, 1976). Cold acclimation increases creatine levels in hamsters (Cannon, Nedergaard *et al.*, 1978).

ATP-independent means of thermogenesis

During mitochondrial respiration, the free energy from substrate oxidation is believed to be converted into a proton electrochemical gradient across the inner membrane, and ATP production at ATP translocase sites is linked with the flow of protons back into the matrix (the chemiosmotic theories of Mitchell, 1976). Brown fat mitochondria differ from other tissues in that the electrical re-entry of protons into the matrix may at times occur along routes independent of the proton-translocating ATPase system, thereby dissipating their energy solely as heat (Nicholls, 1976a, b). The presence of this pathway appears to relate to the presence of a particular polypeptide in the membrane, characteristically with a mol. wt of about 32,000 daltons. This has been found in hamsters (Nicholls, Bernson & Heaton, 1978) and in rats (Desautels *et al.*, 1978).

This by-pass is normally closed, being inhibited by the presence of purine nucleotides, GDP, ADP, etc. (Nicholls *et al.*, 1978). The inhibition is abolished by increased levels of fatty acids (Heaton & Nicholls, 1976) or acyl Co-A esters (Cannon, Sundin & Romert, 1977), which are assumed to compete for the polypeptide site and facilitate proton flow when bound to the 32,000 Dalton protein. Characteristics of brown fat mitochondria are reviewed by Cannon & Lindberg (1979).

Using this model, Desautels *et al.* (1978) found that within the first hour of transfer of rats from 28°C to 4°C there was a significant increase in purine nucleotide binding in brown fat continuing to a maximum within a week, and settling to a level five times above pre-exposure levels. Increases in the amount of the 32,000 dalton protein were also detectable directly within a day, reaching a maximum in two to three weeks. Himms-Hagen & Desautels (1978) report a 45% increase in purine-binding in brown fat mitochondria taken from inbred mice (C57BL/6J: +/+ or +/ob) within three hours of exposure to 4°C. This confirms the generality of rapid response in mice as well as rats and hamsters.

This is a model of loose-coupling and not uncoupling: existing ATP production sites can still function and are merely competing for protons. The latter are likely to be abundant as brown fat mitochondria may have over five times as many respiratory chains as liver cells, while having 40% fewer ATP synthetase sites as heart muscle (Cannon & Vogel, 1977). However the precise balance of the competing thermogenic mechanisms in brown fat is unresolved.

Non-shivering Thermogenesis in Tissues other than Brown Fat

The work of Foster & Frydman (1979) suggests that brown fat may contribute 40–70% of cold-induced heat production. This leaves 30–60% to be accounted for. Of the remaining tissues muscle (skeletal and heart) is the major contributor (Table I). Non-shivering thermogenesis in muscle is similar to that in brown fat, but has no loose-coupling mechanisms (Himms-Hagen *et al.*, 1978). In rat, oxygen uptake increases on noradrenaline infusion (Grubb & Folk, 1976), and on increased blood flow (Grubb & Folk, 1978). As sympathetic innervation is not great, plasma catecholamines may be a significant stimulus. Both α- and β-adrenoreceptors are involved (Grubb & Folk, 1977), with depolarization occurring to a lesser extent (10%) than in brown fat (Teskey, Horwitz & Horowitz, 1975). Cold-acclimation is accompanied by increase in number of ATP-translocase sites (Himms-Hagen *et al.*, 1978). The interaction, if any, between shivering and non-shivering mechanisms in muscle is undocumented. Non-shivering thermogenesis increases in liver and muscle slices from cold-acclimated Swiss mice (at 5°C), and is largely ouabain sensitive, suggesting enhanced sodium transport mechanisms (Stevens & Kido, 1974).

Cold exposure results in numerous endocrine changes, e.g. in blood levels of insulin, adrenal corticoids, prolactin and thyroid hormones (see Thompson, 1977, for brief review). The immediate role of many of these in cold thermogenesis may be low (Jansky, 1978). The increase in basal metabolism of cold-acclimated stocks may reflect a direct effect of thyroid hormones, or a potentiation of catecholamines (Himms-Hagen, 1976). Ikemoto *et al.* (1967) report that the response to noradrenaline in hypothyroid SM mice occurred only in the presence of thyroxine, although the cold-induced increase in brown fat weight was similar in hypothyroid and normal mice. The role of adrenaline-induced thermogenesis in cold acclimation is receiving renewed attention (Mejsnar, Cervinka & Jansky, 1976; Jansky, 1978; Vybiral & Andrews, 1979).

In summary, it is the presence of conflicting evidence rather than consistent negative evidence that results in little emphasis being placed at present on cold-induced non-shivering thermogenesis mechanisms other than noradrenaline mediated ones.

VARIATIONS ON A THEME

The mechanisms described above rely heavily on the study of animals maintained in artificially simple constant cold environments. Animals may be precipitously transferred to the acclimation temperature, to which they respond without access to nesting material or companions, but with food and water *ad libitum*. The animals are usually laboratory stocks, whose genetical variability is unknown or limited. It is well recognized not only that this is a very simple model of the natural world, but also that it may have different properties. For example, study of rats exposed to natural climatic changes, or use of wild stocks in constant cold, can result in altered response patterns (e.g. Heroux, 1963; Barnett, 1965; Barnett, Munro *et al.*, 1975).

Heroux (1970, 1974) considers that certain responses in acclimation may represent failures, rather than successes, of adaptability.

A few examples follow to illustrate variations in response that can occur when the laboratory experimental designs are made a little more complex, arguably moving closer to the complexities of the natural world.

Early Experience of Cold

A house mouse may be born without effective physiological thermoregulation (Chew & Spencer, 1967). Shivering is not observed till day 3 (Arjamaa & Lagerspetz, 1979), and fur appears about day 8. Non-shivering thermogenesis as judged by response to adrenaline is not detectable till day 5, reaching a maximum at 17–19 days (Lagerspetz, Auvinen & Tirri, 1966). It is about this time that enhanced maternal nesting thermoregulatory behaviour declines (Lisk, 1971; Lynch & Possidente, 1978). Bryant & Hails (1975) report that mice aged 8–10 days could thermoregulate when isolated at 25°C: below this age mice had higher metabolic rates in huddles compared to isolation (when they cooled more quickly); above this age, huddling conserved energy in the expected manner.

Lynch *et al.* (1976) looked at the effect of early exposure to cold on later cold acclimation. Mice were reared from day 0–25 either at 5°C or 25°C, in litters kept constant at four to avoid effects of litter size on response or mortality (Barnett & Neil, 1972). Weaned mice from both stocks were then exposed singly to 5°C from days 25–50. At the end of this period mice reared in the cold had higher brown fat weights, chose warmer substrate temperatures (a predisposition to huddling?) and built smaller nests than the controls which responded to cold by building larger nests and exhibiting less brown fat hypertrophy. The balance between physiological and behavioural thermoregulation is intricate and variable. These data support the idea that maternal behaviour can influence the future pattern of adaptability in litters (Barnett, 1973).

Genetical Differences

Despite the wide range of genetically different stocks available, relatively little work has been published on genetical variations in thermoregulatory adaptability of mice. In some studies, the physiological effects of single genes can be investigated. The study of hairless mice (Mount, 1971; Heldmaier, 1974, 1975a) has proved a useful model to emphasize the relative value of fur, showing both increased metabolic and brown fat hypertrophy response (Fig. 3), while the behavioural mechanisms of huddling (Heldmaier, 1975a) or operant behaviour (Baldwin, 1968) can minimize physiological stress.

Mice homozygous for the obese (*ob*) gene show numerous thermoregulatory defects such as lowered metabolic rates, low body temperature, lowered survival ability in cold and noradrenaline sensitivity about one half that of controls (Trayhurn & James, 1978). Bray & York (1979) fully review the syndrome of this and other genetically obese strains. The poor thermoregulatory ability of the *ob/ob* mice stimulated Himms-Hagen & Desautels (1978) to investigate the "loose-coupling" ability of brown fat mitochondria of these mice. On cold exposure there was no increase in the rate of purine nucleotide binding (see above), suggesting a failure of the ATP-independent heat production mechanism. York, Bray & Yukimura (1978) could stimulate the Na/K-ATPase activity of liver and kidney slices with thyroid hormone (T3) in control mice, but not in *ob/ob* mice. Thus ATP-dependent mechanisms also appear defective. York *et al.* (1978) suggest that the many and varied physiological defects in obese mice may reflect some common defect of membrane architecture that affects all membrane-associated functions.

Genetical studies on strain differences usually deal with many gene differences, of unknown thermoregulatory importance. Pennycuik (1967) measured metabolism of mice at temperatures of 26–36°C. Among seven inbred strains, differences were observable only at 26°C, the extremes being DBA with metabolic rates 45% higher than 101 or C3H. Contrasts occurred at several test temperatures when comparing four random-bred stocks. In a further study on strain differences in acclimation patterns after transfer for two weeks to 33°C (from 21°C), Pennycuik (1972) showed that some outbred lines (e.g. R70) had an ability to decrease metabolism at high temperatures, while others (Wild) did not. The F_1 and F_2 progeny of these stocks showed similarity to the R70 stocks below the lower critical temperature, and to the wild stock above this temperature. In complex adaptability measures, F_1 hybrids do not show consistent relation to one parental strain or another, though some studies report hybrids with metabolism characteristics intermediate between parental stocks (Schlesinger & Mordkoff, 1963; Gorecki & Krazanowska, 1970).

The studies of Barnett and colleagues are now classic long-term studies on the effect of cold on morphological, reproductive and survival patterns over many generations of inbred, random-bred or wild-caught mice (Barnett & Widdowson, 1965, 1971; Barnett, 1965, 1973). Efficient nest-building behaviour was essential for survival in constant cold (– 3°C), and morphological changes (e.g. larger livers, longer intestines, less total body fat) may relate to changes in rates of food assimilation and thermogenesis. F_1 hybrids of inbred strains survived better than parental strains when placed alone without nesting material at – 3°C for three weeks (Barnett, 1964): all inbred C57BL died, 77% of strain A, 54% GFF, 33% A2G, while F_1 crosses averaged 21% mortality, compared to 30% in random-bred stock. C57BL mice had lower metabolic rates than A2G, but F_1 hybrids were even lower (Barnett, 1965). Given nest material, adult mortality was nil, except for 18% for C57BL (Barnett & Manly, 1956). Selection of random-bred mice for 12 generations in the cold resulted in increased growth capacity when compared at 21°C with controls – at 16 weeks they weighed 13% more (Barnett & Scott, 1963). Stocks of wild-caught mice, kept in the cold for nine generations and returned to 21°C, were superior to controls in reproductive performance and growth, having more fat, longer intestines, but lighter heart, stomach and kidneys (Barnett, Munro *et al.*, 1975).

Lacy *et al.* (1978) studied thermoregulatory characteristics of two lines of mice (outbred stock), subjected to 11 generations of bi-directional selection for high and low nesting ability. Nesting was measured by the weight of presented cotton taken, but the nests were removed after completion. Thus while being selected for nest-building ability, the mice did not benefit from their nests, both stocks being exposed to the same temperature (22°C). Any change in thermoregulatory ability was interpreted as pleiotropic or linked effects of the genes controlling nesting behaviour. The high-nesting line had higher basal metabolic rates, lowered food consumption and higher body temperatures than the low-nesting line. No differences were found in noradrenaline response, brown fat weight or thermal preference.

Pennycuik (1979) also studied changes in thermoregulatory features in mice that had been selected for reproductive performance at high temperature over 12 or more generations. While no differences were detectable at the temperature at which selection occurred, the stocks with high fertility showed higher metabolism below thermal neutrality.

The results of these studies are varied, but emphasize that the selection within one environment or on one characteristic, can produce differences in adaptability potential in other environments. Lacy & Lynch (1979) report on the heritabilities of thermoregulatory characteristics of 225 litters of heterogeneous stock (HS/Ibg). Heritabilities of basal metabolic rate, noradrenaline response (estimating nonshivering thermogenesis), body temperature and interscapular brown fat weight were all less than 0.1. Those for thermoregulatory behaviour (nesting, thermal preference and food consumption) were intermediate (0.3), and that for total body weight 0.4. They conclude from this that the physiological mechanisms have been pushed to their genetical limits by natural selection leaving minimal levels of additive genetical variance.

Short Cold Shocks

Intermittent exposure to varied levels of cold is the likely experience of mice in natural populations. Too few studies have modelled this in the laboratory. Hart (1953b) found that short cold exposures for 2–3 hours over eight days increased survival time in extreme cold. Mice exposed to frequent short cold shocks (18 times for 10 minutes over two days prior to survival tests) increased survival time in proportion to the temperature of the cold shock: e.g. given short exposures to −5°C, only 25% of treated mice had died by the time all controls had died when exposed to −20°C (Leblanc *et al.*, 1967). This short-term adaptation did not result in increased sensitivity to noradrenaline, though urine catecholamine levels were raised compared to controls under cold stress.

Heldmaier (1975b) studied the effect of daily cold shocks (2.5 h) on the trophic response of brown fat in mice. Cold shocks of 20°C for 11 days produced the same increase in brown fat weight as four weeks' continuous exposure. Cold shocks of 10°C produced little further increase, but cold shocks of 0°C for 11 days produced a response greater than the maximum obtainable in standard acclimation studies. This response increased when cold exposure was increased to five hours a day (Fig. 3). The histological character of brown fat after short cold shocks showed the fat concentrated in a large globule with only few small droplets – not the expected multilocular pattern. The relation of the histological appearance to thermogenic capacity is unknown. Cameron (1975) reports age-dependent shifts to unilocular fat storage in Swiss albino mice kept at 26°C.

These selected studies illustrate that adaptability response to intermittent cold may differ significantly in both extent and mechanism from that in response to continuously imposed cold.

Huddling

Huddling is a mechanism usually prevented in most laboratory studies on physiological mechanisms. However it is clearly a common mechanism in rodents for energy conservation (Hart, 1971). Huddles of five mice can reduce food consumption over 30% in the cold (Prychodko, 1958). Grouping reduced oxygen uptake in mice at 15°C and 8°C (especially during their rest phase), but not at 31°C (Mount & Willmott, 1967). Metabolism of 15–20-day-old mice decreased up to 50% in huddles of seven mice at 25°C, though further increase in huddle size had no effect (Bryant & Hails, 1975). Heldmaier (1975a) used the weight of brown fat as a measure of perceived temperature in huddles of furred and hairless mice exposed to 10°C. Groups of two to four furred mice had 15–20% lower brown fat weights than single mice, and groups of five had almost 40% less (Fig. 3). Hairless mice, while having greater brown fat mass than furred mice in all groupings, also showed up to 35% drop in brown fat weight in huddles. Heldmaier estimated that the perceived temperature in the largest huddles was up to 10°C warmer than air temperature, even in the absence of nests.

Food Shortage and Daily Torpor

Mice in most laboratory studies are given food and water *ad libitum*, a poor mimic of natural conditions where cold and food shortage are the more likely partners. The response of many other small rodents to these factors is daily torpor (Hudson, 1978), which has now been recorded in feral house mice (Morton, 1978), and the mechanism described in laboratory mice by Hudson & Scott (1979).

Hudson & Scott found that under normal laboratory conditions (food and water *ad libitum*) albino mice typically maintained body temperatures between 36 and 39°C irrespective of environmental temperature (2.5–38°C). Within 24 hours of removal of food supply, five out of 14 mice went into torpor, from which they could be roused independently of external heat. Restricted food intake (2 g millet per day) also elicited torpor (Fig. 4). Shivering was suppressed during entry into torpor, but was clear during arousal even at body temperatures of 18°C. Oxygen uptake was lower at a given body temperature during entry into torpor, than during arousal. Length of torpor ranged from 5–46 hours. Body temperatures would be maintained at 16–19°C, even if environmental temperature dropped below this, implying functional thermoregulation at the low level. The total pattern is quite consistent with that documented for numerous other species, for which torpidity is a well-accepted adaptation (Hudson, 1978). The occurrence of torpor in huddles of mice was not investigated. Torpor to similar

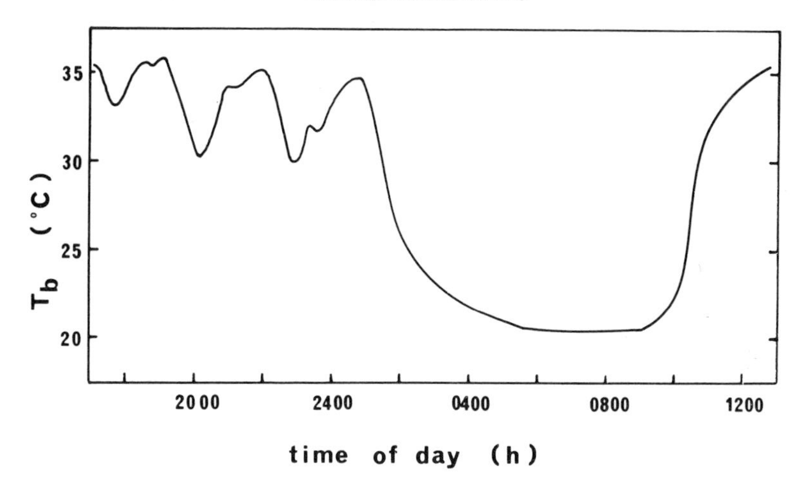

FIG. 4. Daily torpor in a laboratory mouse fed on restricted diet at 19.5°C. After Hudson & Scott (1979).

low body temperatures has also been elicited in house mice by severe water deprivation (Fertig & Edmonds, 1969).

PHYSIOLOGICAL ADAPTABILITY IN NATURAL POPULATIONS

This review so far has selected results obtained from controlled laboratory experiments, where new insights on adaptability mechanisms may come from improved techniques and moves to more complex experimental designs. There still remains a hiatus between the increased understanding of these mechanisms as such, and the understanding of their relative importance in particular instances in the natural world.

In the absence, at present, of easy monitoring techniques for free-living mouse-sized mammals, the adaptability of wild mice has largely to be estimated from responses of individuals removed (however temporarily) from their natural habitat. These responses are interpreted in the context of habitat characteristics that are usually assumed, rather than shown, to be relevant to the animal. Correlations may emerge, but mechanisms are not demonstrated. For example, the study of metabolism in wild animals using standard physiological techniques which prevent behavioural responses (e.g. Hart & Heroux, 1953, 1963; Jakobson, 1971) must be recognised as a limited mimic of the natural adaptability response in the wild (Heroux, 1970). Nonetheless, these studies provide some guidelines on the nature of adaptability mechanisms in these animals.

Geographical Variation

One approach in comparing adaptability in geographically distinct populations is to catch animals, and keep them in the laboratory until they have acclimated, before measuring response. This essentially investigates genetical differences in adaptability insofar as they manifest themselves in one particular environment. As seen in previous sections, this is likely to survey only part of the adaptability available to the species.

This approach has demonstrated several geographical trends in studies of closely related species of small mammals. Greater nest-building activity occurred in northern populations of *Peromyscus* spp. than in southern forms (King, Maas & Weisman, 1964). Latitude correlated with muscle glycogen content, and inversely with plasma glucose levels in a comparison of 13 species of rodent (Galster & Morrison, 1976). Hayward (1965a, b) found few differences in of liver glycogen and a hypertrophy of brown fat greater than in the meadow vole, which maintained higher glycogen levels (Galster & Morrison, 1976). Hayward (1965a, b) found few differences in metabolism or body composition relatable to geographical origins of six races of *Peromyscus*, when acclimated to laboratory conditions. Golden spiny mice (*Acomys russatus*) from Mount Sinai had enhanced non-shivering thermogenic capacity and greater resistance to cold than mice from Dead Sea populations (Borut, Haim & Castel, 1978). Ladygina (1952, quoted in Hart, 1971) found that mice from higher latitudes had metabolic rates higher than mice from lower latitudes.

As the ecological distribution of most of these species is limited, geographical studies on adaptability must usually cross species boundaries. This is not necessary in the case of the house mouse, whose wide distribution, like that of man, provides one of the few opportunities to study adaptability on a world-wide basis within a mammalian species.

Estimates of interpopulation variation of up to 13 measures (organ weights and blood characters) in house mice from islands of north-east Europe are given by Berry & Jakobson (1975) and Berry, Jakobson & Peters (1978). Animals were not standardized to laboratory conditions, and the variation largely reflects both genetic and acclimatization differences. These studies were followed up on island populations in more constant climates (Table II): Macquarie Island, Marion Island and South Georgia, close to the Antarctic convergence (Berry & Peters, 1975; Berry, Peters & Van Aarde, 1978; Berry, Bonner & Peters, 1979), and islands nearer the equator,

TABLE II

Characteristics of mice from geographically distinct populations

Location	Climate (°C)	Weight (g)	Interscapular brown fat ($mg\,g^{-1}$)	Blood haemoglobin ($g\,(100\,ml)^{-1}$)
South Georgia	2	18	11.0	18.3
Macquarie Is.	4	16	6.2	13.7
Marion Is.	5	21	3.7	18.1
Hawaii	21	12	4.5	16.5
Enewetak Atoll (2 islets)	28	9	1.7–2.2	16.6–17.4

(Mean values after Berry *et al.*, various sources, see text.)

Enewetak Atoll and Hawaii (Berry & Jackson, 1979). Mice from warm regions can be small, those from cold regions can be big: there is not strict adherence to Bergmann's Rule (Table II). Haemoglobin levels, known to increase in house mice in response to seasonal cold in temperature climates (Maclean & Lee, 1973; Berry & Jakobson, 1975), show no strict geographical trend.

Interscapular brown fat (expressed as $mg\,g^{-1}$ body weight) is six times heavier in mice from South Georgia than in mice from Enewetak Atoll. Taking all islands studied, brown fat weight correlates significantly with meteorological temperatures (Fig. 5: $r = -0.585$, $P < 0.001$, $n = 31$). The relation ($0.144\,mg\,g^{-1}$ rise in interscapular brown fat weight for every 1°C drop in meteorological temperature) is similar in slope and position to that found in the acclimation studies on laboratory mice by Heldmaier (1974, 1975b). This provides a degree of mutual validation. (N.B. Interscapular brown fat expressed as $mg\,g^{-1}$ is a convenient measure, but will not eliminate the size factor – small mice possessing proportionately more brown fat than larger mice).

The variations in Fig. 5 present a framework for further questions rather than answers. They may be due to genetical contrasts in physiological or behavioural thermoregulation – the populations certainly differ in their gene pools (see papers by Berry and by Peters in this volume). Or they may be due to habitat contrasts between the islands (e.g. soil depth, wind, rain) in so far as they impose microclimates on the mice that differ from the crude meteorological estimate of air temperature. Only further ecological study can determine the balance of the physiological and behavioural attributes, and their correlation with underlying genetical or environmental contrasts.

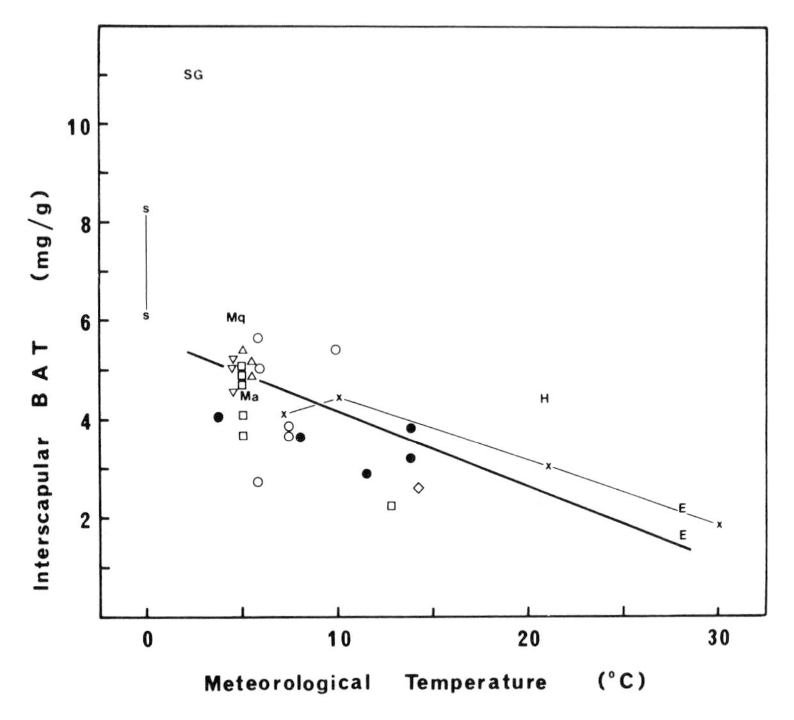

FIG. 5. Interscapular brown fat (BAT) weight and estimated climate in geographically distinct populations of the house mouse. South Georgia (SG), Macquarie Is. (Mq), Marion Is. (Ma), Hawaii (H), Enewetak Atoll (E), Isle of May (seasonal data, •), Skokholm (◇), Orkney islands (□), Shetland islands (▽), Faroe islands (○), mainland Britain (△). Laboratory mice acclimated to cold (x—x) or given cold shocks at 0°C (s——s), after Heldmaier (1974, 1975b, see Fig. 3).

Data taken from nearest meteorological station are either mean annual temperature or mean monthly temperature if sample collected within a month.

Seasonal Variation

Animals exposed to natural climatic variations, or sampled from wild populations in different seasons, characteristically show in winter a replacement of shivering with non-shivering thermogenesis, an increased sensitivity to noradrenaline, and many features similar to cold-acclimated animals (Hart & Heroux, 1963; Heroux, 1963; Lynch, 1973). However, winter animals differ from cold-acclimated ones in having lower, and not higher, metabolic rates at all test temperatures when compared to summer animals (i.e. the opposite to Fig. 2). Some populations studied do not show this, resembling the acclimation response (Wunder, Dobkin, & Gettinger, 1977, on the prairie vole), though ecophysiological studies often show different patterns of adaptability in different years (Wunder, 1978).

Winter increase in haemoglobin levels has been followed in free-living individuals in *Peromyscus* spp. (Sealander, 1962), and house mice trapped in winter have higher haemoglobin levels than those trapped in summer (Maclean & Lee, 1973; Berry & Jakobson, 1975). Seasonal cycles of basal metabolism and thyroid activity, with low levels in winter, are reported in several species of voles (Delost & Naudy, 1956; Rigaudière & Delost, 1966).

Acclimatization of free-living house mice has been studied over several years on two islands, Skokholm off west Wales, and the Isle of May in the Firth of Forth. Mean values for basal metabolism tend to decrease in winter on Skokholm (Berry, Jakobson & Moore, 1969; Jakobson & Moore, 1971), but may increase on the Isle of May (Triggs, 1977), where metabolic response to noradrenaline could also be shown to increase in winter (Jakobson & Triggs, in prep.), as also do brown fat weights (Berry & Jakobson, 1975).

The mean metabolic response to a cold test (225 minutes at 5°C) can vary greatly in successive weekly samples of mice (Jakobson, 1971), yet an overall pattern emerges if data are presented to define the normal "bounds" of response within which the response of individual mice is likely to fall. This is done in Fig. 6 where the 95%

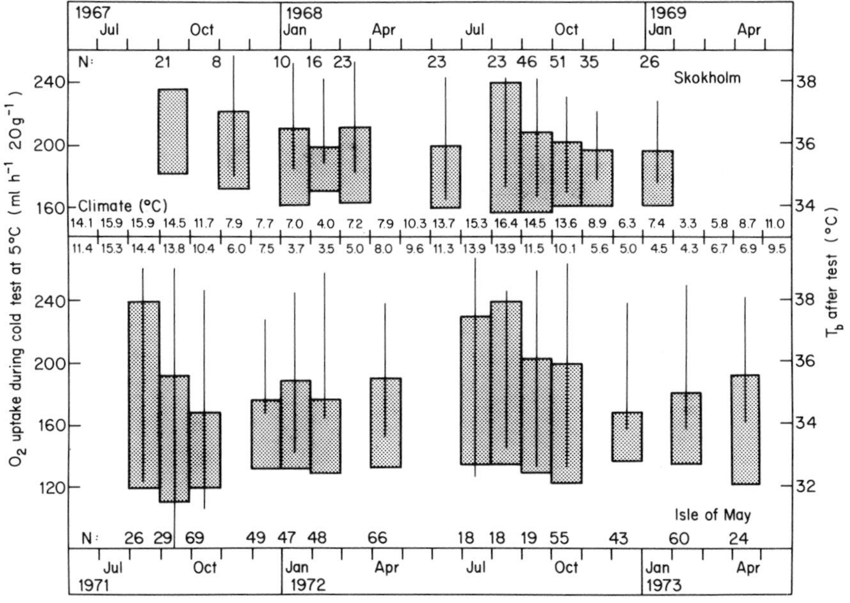

FIG. 6. Response to cold tests in male feral mice over two winters on two islands. 95% range for each month in oxygen uptake (hatched blocked) and body temperature (vertical bars) after 200 (Isle of May) or 225 minutes (Skokholm) at 5°C. Climate is mean monthly temperature on adjacent mainland. Oxygen uptake standardized within islands to the 20 g level after Jakobson (1971).

range (1.96 × s.d. about the mean) are plotted for a number of months across two winters on two islands. From each summer to winter there is a narrowing of the "bounds of normality" representing the lack of the high-metabolizing response in winter mice. Mice with a low-metabolizing response are found at all seasons. There is an association between mean monthly temperature and the position of the upper 95% limit ($r = 0.485$, $P < 0.10$ for Skokholm; $r = 0.809$, $P < 0.001$ for the Isle of May). The lower 95% limit does not correlate with temperature. Data on the normal range of body temperatures are also given in Fig. 6: here the lower 95% limit correlates inversely with climate ($r = -0.891$, $P < 0.001$ for Skokholm; $r = -0.506$, $P < 0.01$ for Isle of May), i.e. it was less likely to find mice leaving the cold test with low body temperatures in winter. The upper temperature limit is independent of climatic temperature.

In summary, winter mice in feral populations lacked a high metabolic response (unless commensal with man, Jakobson, 1971), and resisted body temperature decline in the cold more than summer mice. The lack of high-metabolizers partly reflects a drop in activity in the test (Triggs, 1977), and partly insulation (Jakobson, 1971).

The Isle of May mice had a much wider range of post-test body temperatures (30–39°C) than Skokholm mice (34–38°C). Triggs (1977) reported mice in a natural nest, and in artificial field nests, in huddles up to a dozen with body temperatures with as great a range as from the cold test. One mouse from a natural nest had a body temperature of 26.8°C 20 minutes after disturbance, and was able to heat up to "normal" temperature within half an hour at 6°C. It may be that house mice on the Isle of May do utilize torpor as an adaptability mechanism. Morton (1978) commonly found feral house mice huddling with the insectivorous marsupial *Sminthopsis crassicaudata*: both species went torpid in cold weather, with mouse body temperatures as low as 18–20°C at air temperatures of 11–12°C. Previously, reports of torpor in house mice have been circumstantial (e.g. Degerbøl, 1942, see Berry, 1970).

Only studies such as these, directly in the field on unrestrained animals, can validate the cold-test observations. The latter, while describing some mechanisms of physiological adaptability (e.g. Fig. 6) still leave the investigator largely to speculate on their biological role in the unhindered private life of the mouse.

Adaptability and Survival

The harsh definition of adaptability at the start of this article related it to increased fitness. Winter-caught wild deer mice and rats do indeed survive longer in extreme cold tests than summer-caught animals (Hart & Heroux, 1953, 1963). The studies of Barnett (Barnett, 1964, 1965) are exceptional in that they follow many generations in the cold, monitoring both survival and reproductive success. While adult mortality on transfer to cold (− 3°C), given nesting material, was nil for strains A and GFF, and 18% for C57BL, the percentage of young born that survived to weaning was 67% for strain A, 31% for C57BL, and 18% for GFF (Barnett & Manly, 1956). They were unable to maintain a breeding stock of GFF. Fitness thus involves both the energy demands of individual survival and the extra energy or behavioural demands of pregnancy and lactation (Barnett & Little, 1965). The characteristics of survivors may be quite variable, e.g. in liver glycogen or food consumption (Barnett, Coleman & Manly, 1960). Any group classified as "survivors" by the observer may still consist of a continuum from animals about to die to those well able to survive and breed (Barnett & Manly, 1956).

Samples of mice taken from wild populations will also contain such heterogeneity in the face of many and unknown mortality agents. Mortality in feral house mice, while highest in winter, is high at all times of year; life expectancy at birth is of the order of 100 days (Berry & Jakobson, 1971).

The analysis of survival characteristics in the wild through the capture and release of tested individuals, and the plotting of subsequent survival from field data, has repeatedly demonstrated that no simplistic physiological answer is available for survival in the natural environment (Berry, Jakobson & Triggs, 1973; Jakobson, 1978). No measure taken in isolation is a consistent indicator of survival through winter. Multivariate analysis within a sample can clearly detect patterns of potential survivors, e.g. using measures of body size, metabolism in warm and cold and haemoglobin levels (Jakobson, 1978). These patterns seemed to represent a "more stable" response to the imposed tests. But these within-sample descriptions give less than 35% success rate in predicting the survival of mice in another winter. Similarly, low predictive value was found even for mice tested in the same winter at a different time (Triggs, 1977, and personal communication). Such essential biological validation of these "survival" patterns is still elusive.

What determines the response of mice to these tests, over and above the stress or stimulation of the novel event? Differences will reflect the animal's past thermal experience (e.g. from its mother's nest with litter mates; to the nature of its adult nesting and huddling behaviour; to the variable cold shocks experienced during foraging expeditions voluntarily undertaken). At another level, its future survival depends on its behaviour in the context of unknown challenges of the future winter climate and food supply, and the interaction of these experiences (partly imposed, partly chosen) on the existing physiological state and potential adaptability of the animal.

Physiological and behavioural mechanisms may buffer each other (Lynch *et al.*, 1976), or exacerbate the cold stress, given that cold-acclimated mice are more likely to enter and explore cold environments (Barnett, Hocking & Wolfe, 1978). The study of physiological responses alone can only give a partial estimate of the biological responses of the animal, even to one factor (cold) assumed important in influencing its survival chances. Seasonal changes in genotype frequency at a phosphoglucomutase locus in natural populations of *Apodemus sylvaticus* have been correlated with differing abilities to mobilise liver glycogen during fasts (Leigh-Brown, 1977a, b). Seasonal cycles in genotype frequency at a haemoglobin β-chain locus occur in Skokholm house mice. Spring samples, compared to autumn samples, have fewer Hbb^s/Hbb^d heterozygotes (Berry & Murphy, 1970). Within autumn samples taken in later years, older mice, compared to younger ones, had fewer heterozygotes (Berry, 1978). This is interpreted as some advantage of heterozygosity at this locus to young mice in autumn, that disappears through winter, or with age. Over five autumns between 1971 and 1977 (Berry, Jakobson & Peters, unpublished), data have been collected on blood haemoglobin (g/100 ml), and these show that mice homozygous for Hbb^d had higher values (mean 17.2 ± 0.20 s.e.) than heterozygotes (16.4 ± 0.13, $P < 0.01$) or mice homozygous for Hbb^s (16.2 ± 0.18, $P < 0.001$). This physiological correlate with genotype does not by itself account for the apparent survival contrasts.

CONCLUSIONS

Physiological adaptability is but one aspect of the armoury of the house mouse in its adaptation to the environment. Laboratory studies succeed in isolating the mechanisms of physiological adaptability, but usually preclude any alternative avenue of adaptability to the animal. In nature, a mouse will exercise its complete battery of adaptability, physiological and behavioural, within the limits of its genotype. Its response will be the interaction of this overall adaptability with the heterogeneity and malleability of its habitat.

Field studies provide numerous descriptions of physiological change and variation. The precise mechanisms causing change usually remain speculation, and the interpretation is still largely an act of faith. It remains problematical, especially in as ecologically exploitive a species as the house mouse, to generalize on the relative importance of any mechanism studied in isolation to survival in the field.

Studies both in the laboratory and in the field must attempt to become more comprehensive. Only then can more confident models of the roles of different types of adaptability in the biology of the house mouse evolve.

REFERENCES

Adolph, E. F. (1972). Some general concepts of physiological adaptations. In *Physiological adaptations: desert and mountain:* 1--7. Yousef, M. K., Horvath, S. M. & Bullard, R. W. (Eds). New York and London: Academic Press.

Ahlabo, I. & Barnard, T. (1974). A quantitative analysis of triglyceride droplet structural changes in hamster brown adipose tissue during cold exposure and starvation. *J. Ultrastruct. Res.* **48**: 361–376.

Anderson, H. T., Christiansen, E. N., Grav, H. J. & Pedersen, J. I. (1970). Thermogenesis of brown adipose tissue reflected in mitochondrial respiratory control. *Acta physiol. Scand.* **80**: 1–10.

Arjamaa, O. & Lagerspetz, K. Y. H. (1979). Postnatal development of shivering in the mouse. *J. therm. Biol.* **4**: 35–39.

Atlas, D., Hanski, E. & Levitski, A. (1977). Eighty thousand β-receptors in a single cell. *Nature, Lond.* **268**: 144–146.

Baldwin, B. A. (1968). Behavioural thermoregulation in mice. *Physiol. Behav.* **3**: 401–407.

Banet, M. & Hensel, H. (1977). The control of shivering and non-shivering thermogenesis in the rat. *J. Physiol., Lond.* **269**: 669–676.

Banet, M., Hensel, H. & Liebermann, H. (1978). The central control of shivering and non-shivering thermogenesis in the rat. *J. Physiol., Lond.* **283**: 569–584.

Banet, M. & Seguin, J. J. (1967). Afferent discharge produced by muscle cooling in mice maintained at normal and low environmental temperatures. *Can. J. Physiol. Pharmacol.* **45**: 319–327.

Barnett, S. A. (1956). Endothermy and ectothermy in mice at − 3°C. *J. exp. Biol.* **33**: 124–133.

Barnett, S. A. (1959). The skin and hair of mice living at low environmental temperatures. *Q. Jl exp. Physiol.* **44**: 35–42.

Barnett, S. A. (1964). Heterozygosis and the survival of young mice in two temperatures. *Q. Jl exp. Physiol.* **49**: 290–296.

Barnett, S. A. (1965). Adaptation of mice to cold. *Biol. Rev.* **40**: 5–51.

Barnett, S. A. (1973). Maternal processes in the cold-adaptation of mice. *Biol. Rev.* **48**: 477–508.

Barnett, S. A., Coleman, E. M. & Manly, B. M. (1960). Mortality, growth and liver glycogen in young mice exposed to cold. *Q. Jl exp. Physiol.* **45**: 40–49.

Barnett, S. A., Hocking, W. E. & Wolfe, J. L. (1978). Effects of cold on activity and exploration by wild house mice in a residential maze. *J. comp. Physiol.* **123**: 91–95.
Barnett, S, A. & Little, M. J. (1965). Maternal performance in mice at -3°C: food consumption and fertility. *Proc. R. Soc.* (B) **162**: 492–501.
Barnett, S. A. & Manly, B. M. (1956). Reproduction and growth of mice of three strains, after transfer to -3°C. *J. exp. Biol.* **33**: 325–329.
Barnett, S. A. & Mount, L. (1967). Resistance to cold in mammals. In *Thermobiology*: 411–477. Rose, A. S. (Ed.). London: Academic Press.
Barnett, S. A., Munro, K. M. H., Smart, J. L. & Stoddart, R. C. (1975). House mice bred for many generations in two environments. *J. Zool., Lond.* **177**: 153–169.
Barnett, S. A. & Neil, A. C. (1972). The growth of infant mice at two temperatures. *J. Reprod. Fert.* **29**: 191–201.
Barnett, S. A. & Scott, S. G. (1963). Some effects of cold and of hybridity on the growth of mice. *J. Embryol. exp. Morph.* **11**: 35–51.
Barnett, S. A. & Widdowson, E. M. (1965). Organ weights and body composition in mice bred for many generations at -3°C. *Proc. R. Soc.* (B) **162**: 502–516.
Barnett, S. A. & Widdowson, E. M. (1971). Organ weights and body composition of parturient and lactating mice, and their young at 21°C and -3°C. *J. Reprod. Fert.* **26**: 39–57.
Berlet, H. H., Bonsmann, I & Birringer, H. (1976). Occurrence of free creatine, phosphocreatine and creatine phosphokinase in adipose tissue. *Biochim. Biophys. Acta* **437**: 166–174.
Bernstein, S. E. (1966). Physiological characteristics. In *Biology of the laboratory mouse*: 337–350. Green, E. L. (Ed.). New York: McGraw-Hill.
Berry, R. J. (1970). The natural history of the house mouse. *Fld Stud.* **3**: 219–262.
Berry, R. J. (1978). Genetic variation in wild house mice: where natural selection and history meet. *Am. Scient.* **66**: 52–60.
Berry, R. J., Bonner, W. N. & Peters, J. (1979). Natural selection in House mice (*Mus musculus*) from South Georgia (South Atlantic Ocean). *J. Zool., Lond.* **189**: 385–398.
Berry, R. J. & Jackson, W. B. (1979). House mice on Enewetak Atoll. *J. Mammal.* **60**: 222–225.
Berry, R. J. & Jakobson, M. E. (1971). Life and death in an island population of the house mouse. *Expl Geront.* **6**: 187–197.
Berry, R. J. & Jakobson, M. E. (1975). Adaptation and adaptability in wild-living house mice (*Mus musculus*). *J. Zool., Lond.* **176**: 391–402.
Berry, R. J., Jakobson, M. E. & Moore, R. E. (1969). Metabolic measurements on an island population of the house mouse during the period of winter mortality. *J. Physiol., Lond.* **201**: 101–102*P*.
Berry. R. J., Jakobson, M. E. & Peters, J. (1978). The house mice of the Faroe islands: a study in microdifferentiation. *J. Zool., Lond.* **185**: 73–92.
Berry, R. J., Jakobson, M. E. & Triggs, G. S. (1973). Survival in wild-living mice. *Mamm. Rev.* **3**: 46–57.
Berry, R. J. & Murphy, H. M. (1970). Biochemical genetics of an island population of the house mouse. *Proc. R. Soc.* (B) **175**: 225–227.
Berry, R. J. & Peters, J. (1975). Macquarie Island house mice: a genetical isolate on a sub-Antarctic island. *J. Zool., Lond.* **176**: 375–389.

Berry, R. J., Peters, J. & Van Aarde, R. J. (1978). Sub-Antarctic house mice: colonization, survival and selection. *J. Zool., Lond.* **184**: 127–141.

Bligh, J. & Johnson, K. G. (1973). Glossary of terms for thermal physiology. *J. appl. Physiol.* **35**: 941–961.

Borut, A., Haim, A. & Castel, M. (1978). Non-shivering thermogenesis and implication of thyroid in cold labile and cold resistant populations of Golden spiny mouse (*Acomys russatus*). *Experientia* Suppl. **32**: 219–227.

Bray, G. A. & York, D. A. (1979). Hypothalamic and genetic obesity in experimental animals: an autonomic and endocrine hypothesis. *Physiol. Rev.* **59**: 719–809.

Brück, K. & Wünnenberg, W. (1970). "Meshed" control of two efferent systems: non-shivering and shivering thermogenesis. In *Physiological and behavioral temperature regulation*: 562–580. Hardy, J. D., Gagge, A. P. & Stolwijk, J. A. J. (Eds). Springfield, I11.: Thomas.

Bryant, D. M. & Hails, C. J. (1975). Mechanisms of heat conservation in the litters of mice (*Mus musculus* L.). *Comp. Biochem. Physiol.* **50A**: 99–104.

Cameron, I. L. (1975). Age-dependent changes in the morphology of brown adipose tissue in mice. *Tex. Rep. Biol. Med.* **33**: 391–396.

Cameron, I. & Smith, R. E. (1964). Cytological responses of brown fat in cold exposed rats. *J. Cell Biol.* **23**: 89–100.

Cannon, B. & Lindberg, O. (1979). Mitochondria from brown adipose tissue – isolation and properties. *Methods in Enzymology* **55**: 65–78.

Cannon, B., Nedergaard, J., Romert, L., Sundin, S. & Svargarten, J. (1978). The biochemical mechanism of thermogenesis in brown adipose tissue. In *Strategies in cold: natural torpidity and thermogenesis:* 567–594. Wang, L. C. H. & Hudson, J. W. (Eds). New York: Academic Press.

Cannon, B., Sundin, U. & Romert, L. (1977). Palmitoyl coenzyme A: a possible physiological regulator of nucleotide binding to brown adipose tissue mitochondria. *FEBS Lett.* **74**: 43–46.

Cannon, B. & Vogel, G. (1977). The mitochondrial ATPase of brown adipose tissue. Purification and comparison with the mitochondrial ATPase from beef heart. *FEBS Lett.* **76**: 284–289.

Chaffee, R. R. J. & Roberts, J. C. (1971). Temperature acclimation in birds and mammals. *A. Rev. Physiol.* **33**: 155–202.

Chew, R. M. & Spencer, E. (1967). Development of metabolic response to cold in young mice of four species. *Comp. Biochem. Physiol.* **22**: 873–888.

Christiansen, E. N. (1971). Calcium uptake and its effect on respiration and phosphorylation in mitochondria from brown adipose tissue. *Eur. J. Biochem.* **19**: 276–282.

Daniel, H. & Derry, D. M. (1969). Criteria for differentiation of brown and white fat in the rat. *Can. J. Physiol. Pharmacol.* **47**: 941–945.

Dawson, N. J. (1967). The surface area/body weight relationship in mice. *Aust. J. biol. Sci.* **20**: 687–690.

Dawson, N. J. & Webster, M. E. D. (1967). The insulative value of mouse fur. *Q. Jl exp. Physiol.* **52**: 168–173.

Degerbøl, M. (1942). Mammalia. *Zoology Faeroes* Pt. 65: 1–133.

Delost, P. & Naudy, J. (1956). Cycle saisonnier de la thyroide chez les Rongeurs sauvages non-hibernants. *C.r. Séanc. Soc. Biol.* **150**: 906–909.

Desautels, M., Zaror-Behrens, G. & Himms-Hagen, J. (1978). Increased purine nucleotide binding, altered polypeptide composition and thermogenesis in brown adipose tissue mitochondria of cold-acclimated rats. *Can. J. Biochem.* **56**: 378–383.

Edelman, I. S. (1976a). Transition from the poikilotherm to the homeotherm: possible role of sodium transport and thyroid hormone. *Fedn. Proc. Fdn. Am. Socs. exp. Biol.* **35**: 2180–2184.

Edelman, I. S. (1976b). Thyroid hormone: thermogenesis and the bioenergetics of Na^+ pumps. In *Biogenesis and turnover of membrane macromolecules:* 169–177. Cook, J. S. (Ed.). New York: Raven Press.

Edwards, R. M. & Haines, H. (1978). Effects of ambient water vapor pressure and temperature on evaporative water loss in *Peromyscus maniculatus* and *Mus musculus*. *J. comp. Physiol.* **128**: 177–184.

Fertig, D. S. & Edmonds, V. W. (1969). The physiology of the house mouse. *Scient. Am.* **221**: 103–110.

Flaim, K. E., Horwitz, B. A. & Horowitz, J. M. (1977). Coupling of signals to brown fat: α- and β-adrenergic responses in intact rats. *Am. J. Physiol.* **232**: R101–R109.

Foster, D. O. & Frydman, M. L. (1978). Nonshivering thermogenesis in the rat. II. Measurements of blood flow with microspheres point to brown adipose tissue as the dominant site of calorigenesis induced by noradrenaline. *Can. J. Physiol. Pharmacol.* **56**: 110–122.

Foster, D. O. & Frydman, M. L. (1979). Tissue distribution of cold-induced thermogenesis in conscious warm- or cold-acclimated rats re-evaluated from changes in tissue blood flow: the dominant role of brown adipose tissue in the replacement of shivering by non-shivering thermogenesis. *Can. J. Physiol. Pharmacol.* **57**: 257–270.

Gale, C. C. (1973). Neuroendocrine aspects of thermregulation. *A. Rev. Physiol.* **35**: 391–430.

Galster, W. & Morrison, P. (1975). Carbohydrate reserves of wild rodents from different latitudes. *Comp. Biochem. Physiol.* **50A**: 153–157.

Galster, W. & Morrison, P. (1976). Contrasting patterns of energy substrates in tundra and meadow voles. *Physiol. Zool.* **49**: 445–451.

Girardier, L. & Seydoux, J. (Eds) (1978). Effectors of thermogenesis. *Experientia* Suppl. **32**: 1–345.

Girardier, L., Seydoux, J. & Clausen, T. (1968). Membrane potential of brown adipose tissue. A suggested mechanism for the regulation of thermogenesis. *J. gen. Physiol.* **52**: 925–940.

Gorecki, A. & Krazanowska, H. (1970). Oxygen consumption in two inbred mouse strains and F_1 hybrids. *Bull. Acad. pol. Sci.* (Sci. biol.) **18**: 115–119.

Grubb, B. & Folk, G. E. (1976). Effect of cold acclimation on norepinephrine stimulated oxygen consumption in muscle. *J. comp. Physiol.* **110**: 217–226.

Grubb, B. & Folk, G. E. (1977). The role of adrenoreceptors in norepinephrine-stimulated $\dot{V}O_2$ in muscle. *Eur. J. Pharmacol.* **43**: 217–223.

Grubb, B. & Folk, G. E. (1978). Skeletal muscle $\dot{V}O_2$ in rat and lemming: effect of blood flow rate. *J. comp. Physiol.* **128**: 185–188.

Hahn, P., Beatty, C. H. & Bocek, R. M. (1976). Carnitine transferases in brown fat of newborn and developing primates. *Biol. neonate* **30**: 30–34.

Hahn, P. Seccombe, D., Kirby, L. & Skala, J. (1978). Control of phosphoenolpyruvate carboxykinase in brown adipose tissue of infant rats. *Experientia* Suppl. **32**: 61–67.

Harrison, G. A. (1958). The adaptability of mice to high environmental temperatures. *J. exp. Biol.* **35**: 892–901.
Hart, J. S. (1950). Interrelations of daily metabolic cycle, activity and environmental temperature of mice. *Can. J. Res.* (D) **28**: 293–307.
Hart, J. S. (1951). Calorimetric determination of average body temperature of small mammals and its variation with environmental conditions. *Can. J. Zool.* **29**: 224–233.
Hart, J. S, (1952). Use of daily metabolic periodicities as a measure of the energy expended by voluntary activity of mice. *Can. J. Zool.* **30**: 83–89.
Hart, J. S. (1953a). The relation between thermal history and cold resistance in certain species of rodent. *Can. J. Zool.* **31**: 80–98.
Hart, J. S. (1953b). Rate of gain and loss of cold resistance in mice. *Can. J. Zool.* **31**: 112–116.
Hart, J. S. (1956). Seasonal changes in insulation of the fur. *Can. J. Zool.* **34**: 53–57.
Hart, J. S. (1957). Climate and temperature induced changes in the energetics of homeotherms. *Revue can. Biol.* **16**: 133–174.
Hart, J. S. (1961). Physiological effects of continuous cold on animals and man. *Br. med. Bull.* **17**: 19–24.
Hart, J. S. (1971). Rodents. In *Comparative physiology of thermoregulation.* **II**: 1–149. Whittow, G. C. (Ed.). New York: Academic Press.
Hart, J. S. & Heroux, O. (1953). A comparison of some seasonal and temperature-induced changes in *Peromyscus*: cold resistance, metabolism, and pelage insulation. *Can. J. Zool.* **31**: 528–534.
Hart, J. S. & Heroux, O. (1963). Seasonal acclimatization in wild rats (*Rattus norvegicus*). *Can. J. Zool.* **41**: 711–716.
Hart, J. S. & Jansky, L. (1963). Thermogenesis due to exercise and cold in warm and cold-acclimated rats. *Can. J. Biochem. Physiol.* **41**: 629–634.
Hayward, J. S. (1965a). The gross body composition of six geographic races of *Peromyscus. Can. J. Zool.* **43**: 297–308.
Hayward, J. S. (1965b). Metabolic rate and its temperature-adaptive significance in six geographic races of *Peromyscus. Can. J. Zool.* **43**: 309–323.
Heaton, G. M. & Nicholls, D. G. (1976). Hamster brown-adipose-tissue mitochondria. The role of fatty acids in the control of the proton conductance of the inner membrane. *Eur. J. Biochem.* **67**: 511–517.
Heim, T. & Kellermayer, M. (1967). The effect of environmental temperature on brown and white adipose tissue in the starving newborn rabbit. *Acta Physiol. Acad. Sci. Hung.* **31**: 339–346.
Heldmaier, G. (1972). Cold-adaptive changes of heat production in mammals. In *Proc. Int. Symp. Environ. Physiol (Bioenergetics)*: 79–81. Smith, R. E. (Ed.). Dublin: FASEB.
Heldmaier, G. (1974). Temperature adaptation and brown adipose tissue in hairless and albino mice. *J. comp. Physiol.* **92**: 281–292.
Heldmaier, G. (1975a). The influence of the social thermoregulation on the cold-adaptive growth of BAT in hairless and furred mice. *Pflügers Arch.* **355**: 261–266.
Heldmaier, G. (1975b). The effect of short daily cold exposures on development of brown adipose tissue in mice. *J. comp. Physiol.* **98**: 161–168.
Heldmaier, G. & Hoffman, K. (1974). Melatonin stimulates grown of brown adipose tissue. *Nature, Lond.* **247**: 224.
Hemingway, A. (1963). Shivering. *Physiol. Rev.* **43**: 397–422.

Heroux, O. (1963). Patterns of morphological, physiological and endocrinological adjustments under different environmental conditions of cold. *Fedn Proc. Fedn Am. Socs exp. Biol.* 22: 789–792.
Heroux, O. (1970). Pathological consequences of artificial cold acclimatization. *Nature, Lond.* 227: 88–89.
Heroux, O. (1974). Physiological adjustments responsible for metabolic cold adaptation and possible deleterious consequences. *Revue can. Biol.* 33: 209–222.
Herrington, L. P. (1940). The heat regulation of small laboratory animals at various environmental temperatures. *Am. J. Physiol.* **129**: 123–139.
Himms-Hagen, J. (1976). Cellular thermogenesis. *A. Rev. Physiol.* **38**: 315–351.
Himms-Hagen, J., Cerf, J., Desautels, M. & Zaror-Behrens, G. (1978). Thermogenesis and their control. *Experientia* Suppl. 32: 119–134.
Himms-Hagen, J. & Desautels, M. (1978). Mitochondrial defect in brown adipose tissue of obese (*ob/ob*) mouse – reduced binding of purine nucleotides and a failure to respond to cold by an increase in binding. *Biochem. Biophys. Res. Commun.* **83**: 628–634.
Hittelman, K. J., Fairhurst, A. S. & Smith, R. E. (1967). Calcium accumulation as a parameter of energy metabolism of brown adipose tissue. *Proc. natn. Acad. Sci. U.S.A.* **58**: 697–702.
Horowitz, J. M. (1972). Neural control of thermogenesis in brown adipose tissue. In *Proc. Int. Symp. Environ. Physiol* (*Bioenergetics*): 115–121. Smith, R. E. (Ed.). Dublin: FASEB.
Horwitz, B. A. (1973). Ouabain-sensitive component of brown fat thermogenesis. *Am. J. Physiol.* **224**: 352–355.
Horwitz, B. A. (1978). Plasma membrane involvement in brown fat thermogenesis. *Experientia* Suppl. 32: 19–23.
Hudson, J. W. (1978). Shallow, daily torpor: a thermoregulatory adaptation. In *Strategies in cold: natural torpidity and thermogenesis:* 67–108. Wang, L. C. H. & Hudson, J. W. (Eds). New York: Academic Press.
Hudson, J. W. & Scott, I. M. (1979). Daily torpor in the laboratory mouse, *Mus musculus*, var. albino. *Physiol. Zool.* 52: 205–218.
Ikemoto, H., Hiroshige, T. & Itoh, S. (1967). Oxygen consumption of brown adipose tissue in normal and hypothyroid mice. *Jap. J. Physiol.* **17**: 516–522.
Jakobson, M. E. (1971). Acclimatization to cold in house mice living on an island. *Int. J. Biometeor.* **15**: 330–336.
Jakobson, M. E. (1978). Winter acclimatization and survival of wild house mice. *J. Zool., Lond.* **185**: 93–104.
Jakobson, M. E. & Moore, R. E. (1971). Season and metabolic rate in house mice on an island. *J. Physiol., Paris* **63**: 296–299.
Jakobson, M. E. & Triggs, G. S. (In preparation). *Seasonal acclimatization of house mice on two islands.*
Jansky, L. (1973). Non-shivering thermogenesis and its thermoregulatory significance. *Biol. Rev.* **48**: 85–132.
Jansky, L. (1978). Hormonal thermogenesis of "non-norepinephrine" type. *Experientia* Suppl. 32: 169–175.
Jansky, L., Bartunkova, R., Kockova, J., Mejsnar, J. & Zeisberger, E. (1969). Interspecies differences in cold adaptation and non-shivering thermogenesis. *Fedn Proc. Fdn. am. Socs exp. Biol.* **28**: 1053–1058.
King, J. A., Maas, D. & Weisman, R. (1964). Geographical variation in nest size among species of *Peromyscus. Evolution, Lancaster, Pa.* **18**: 230–234.

Lacy, R. C. & Lynch, C. B. (1979). Quantitative genetic analysis of temperature regulation in *Mus musculus*. I. Partitioning of variance. *Genetics* **91**: 743–753.

Lacy, R. C., Lynch, C. B. & Lynch, G. R. (1978). Developmental and adult acclimation effects of ambient temperature and temperature regulation of mice selected for high and low levels of nest-building. *J. comp. Physiol.* **123**: 185–192.

Ladygina, K. M. (1952). *Zool. Zh.* **31**: 736 ff. (see Hart, J. S., 1971).

Lagerspetz, K. Y. H., Auvinen, L. & Tirri, R. (1966). The postnatal development of the metabolic response to adrenaline in mice. *Ann. Med. exp. fenn.* **44**: 67–70.

Leblanc, J., Robinson, D., Sharman, D. F. & Tousignant, P. (1967). Catecholamines and short-term adaptation to cold in mice. *Am. J. Physiol.* **213**: 1419–1423.

Leblanc, J. & Villemaire, A. (1970). Thyroxine and noradrenaline on norandrenaline sensitivity. *Am. J. Physiol.* **218**: 1742–1745.

Leduc, J. (1961). Catecholamine production and release in exposure and adaptation to cold. *Acta physiol. Scand.* 53 (Suppl. 183): 1–101.

Leigh-Brown, A. J. (1977a). Genetic changes in a population of field mice (*Apodemus sylvaticus*) during one winter. *J. Zool., Lond.* **182**: 281–289.

Leigh-Brown, A. J. (1977b). Physiological correlates of an enzyme polymorphism. *Nature, Lond.* **269**: 803–804.

Lilly, F. B. & Wunder, B. A. (1979). The interaction of shivering and nonshivering thermogenesis in deer mice (*Peromyscus maniculatus*). *Comp. Biochem. Physiol.* **63C**: 31–34.

Lisk, R. D. (1971). Oestrogen and progesterone synergism and elicitation of maternal nest-building in the mouse (*Mus musculus*). *Anim. Behav.* **19**: 606–610.

Lynch, G. R. (1973). Seasonal changes in thermogenesis, organ weights and body composition in the white-footed mouse, *Peromyscus leucopus*. *Oecologia* **13**: 363–376.

Lynch, G. R., Lynch, C. B., Dube, M. & Allen, C. (1976). Early cold exposure: effects on behavioural and physiological thermoregulation in the house mouse, *Mus musculus*. *Physiol. Zool.* **49**: 191–199.

Lynch, C. B. & Possidente, B. P. (1978). Relationships of maternal nesting thermoregulatory behaviour in house mice (*Mus musculus*) at warm and cold temperatures. *Anim. Behav.* **26**: 1136–1143.

Maclean, G. S. & Lee, A. K. (1973). Effects of season, temperature and activity on some blood parameters of feral house mice, *Mus musculus*. *J. Mammal.* **54**: 660–667.

Mejsnar, J., Cervinka, M. & Jansky, L. (1976). Substitution between calorigenic effects of noradrenaline and adrenaline and their different inhibition by propranolol. *Physiol. bohemoslov.* **25**: 201–206.

Mejsnar, J. & Jansky, L. (1971). Nonshivering thermogenesis and calorigenic action of catecholamines in the white mouse. *Physiol. bohemoslov.* **20**: 157–162.

Mitchell, P. (1976). Vectorial chemistry and the molecular mechanics of chemiosmotic coupling: power transmission by proticity. *Biochem. Soc. Trans.* **4**: 399–430.

Morton, S. R. (1978). Torpor and nest-sharing: their ecological significance in an insectivorous marsupial, *Sminthopsis crassicaudata*, and the house mouse, *Mus musculus*. *J. Mammal.* **59**: 569–575.

Mount, L. E. (1971). Metabolic rate and thermal insulation in albino and hairless mice. *J. Physiol., Lond.* **217**: 315–326.
Mount, L. E. & Willmott, J. V. (1967). The relation between spontaneous activity, metabolic rate and the 24-hour cycle in mice at different environmental temperatures. *J. Physiol., Lond.* **190**: 371–380.
Nicholls, D. G. (1976a). Bioenergetics of brown adipose tissue mitochondria. *FEBS Lett.* **61**: 103–110.
Nicholls, D. G. (1976b). Hamster brown-adipose-tissue mitochondria. *Eur. J. Biochem.* **62**: 223–228.
Nicholls, D. G., Bernson, S. M. & Heaton, G. M. (1978). The identification of the component in the inner membrane of brown-adipose-tissue mitochondria responsible for regulating energy dissipation. *Experientia* Suppl. **32**: 89–93.
Ogilvie, D. M. & Stinson, R. (1966). The effect of age on temperature selection by laboratory mice (*Mus musculus*). *Can. J. Zool.* **44**: 511–517.
Pennycuik, P. R. (1967). A comparison of the effects of a variety of factors on the metabolic rate of the mouse. *Aust. J. exp. Biol. med. Sci.* **45**: 331–346.
Pennycuik, P. R. (1972). Inheritance of patterns of oxygen consumption in mice. *Aust. J. biol. Sci.* **25**: 1031–1037.
Pennycuik, P. R. (1979). Selection of laboratory mice for improved reproductive performance at high environmental temperature. *Aust. J. biol. Sci.* **32**: 133–151.
Portet, R., Laury, M. C., Bertin, R., Senault, C., Hlusko, T. & Chevillard, L. (1974). Hormonal stimulation of substrate utilization in brown adipose tissue of cold-acclimated rats. *Proc. Soc. exp. Biol. Med.* **147**: 807–812.
Pospisilova, D. & Jansky, L. (1976). Effect of various adaptational temperatures on oxidative capacity of the brown adipose tissue. *Physiol. bohemoslov.* **25**: 519–527.
Prychodko, W. (1958). Effect of aggregation of laboratory mice (*Mus musculus*) on food intake at different temperatures. *Ecology* **39**: 500–503.
Rauch, J. C. & Hayward, J. S. (1969). Topography and vascularisation of brown fat in a small non-hibernator (deer mouse), *Peromyscus maniculatus. Can. J. Zool.* **47**: 1301–1323.
Rigaudière, N. & Delost, P. (1966). Variations saisonniéres du métabolisme de base chez les petits Rongeurs sauvages non-hibernants (Microtinés). *C.r. Séanc. Soc. Biol.* **160**: 1581–1586.
Schlesinger, K. & Mordkoff, A. M. (1963). Locomotor activity and oxygen consumption in two inbred strains of mice. *J. Hered.* **54**: 177–182.
Schmidt-Nielsen, K. (1972). *How animals work.* Cambridge: University Press.
Sealander, J. A. (1962). Seasonal changes in blood values of deer mice and other small mammals. *Ecology* **43**: 107–119.
Severson, D. L. (1979). Regulation of lipid metabolism in adipose tissue and heart. *Can. J. Physiol. Pharmacol.* **57**: 923–937.
Seydoux, J. & Girardier, L. (1978). Control of brown fat thermogenesis by sympathetic nervous system. *Experientia* Suppl. **32**: 153–167.
Shum, A., Johnson, G. E. & Flattery, K. V. (1969). Influence of ambient temperature on excretion of catecholamines and metabolites. *Am. J. Physiol.* **216**: 1164–1169.
Smith, R. E. (1961). The thermogenic activity of the hibernating gland in the cold acclimated rat. *Physiologist* **4**: 113.

Smith, R. E. & Horwitz, B. A. (1969). Brown fat and thermogenesis. *Physiol. Rev.* **49**: 330–425.

Stevens, E. D. & Kido, M. (1974). Active sodium transport: a source of metabolic heat during cold adaptation in mammals. *Comp. Biochem. Physiol.* **47A**: 395–397.

Suter, E. R. (1969). The fine structure of brown adipose tissue. 1. Cold induced changes in the rat. *J. Ultrastruct. Res.* **26**: 216–241.

Taylor, C. R. (1977). Exercise and environmental heat loads: different mechanisms for solving different problems? *Int. Rev. Physiol.* **15**: 119–146.

Teskey, N., Horwitz, B. & Horowitz, J. (1975). Norepinephrine-induced depolarisation of skeletal muscle cells. *Eur. J. Pharmacol.* **30**: 352–355.

Thompson, G. E. (1977). Physiological effects of cold exposure. *Int. Rev. Physiol.* **15**: 29–70.

Trayhurn, P. & James, W. P. T. (1978). Thermoregulation and nonshivering thermogenesis in the genetically obese (*ob/ob*) mouse. *Pflügers Arch.* **373**: 189–193.

Triggs, G. S. (1977). *Ecology, physiology and survival of an isolated mouse population.* Ph.D. Thesis: University of London.

Vybiral, S. & Andrews, J. F. (1979). The contribution of various organs to adrenaline-stimulated thermogenesis. *J. therm. Biol.* **4**: 1–4.

Wang, L. C. H. & Hudson, J. W. (Eds). (1978). *Strategies in cold: Natural torpidity and thermogenesis.* New York: Academic Press.

Wright, G. L. (1976). Critical thermal maximum in mice. *J. appl. Physiol.* **40**: 683–687.

Wunder, B. A. (1978). Yearly differences in seasonal thermogenic shifts of prairie voles (*Microtus ochrogaster*). *J. therm. Biol.* **3**: 98.

Wunder, B. A., Dobkin, D. S. & Gettinger, R. D. (1977). Shifts in thermogenesis in the prairie vole (*Microtus ochrogaster*). *Oecologia* **29**: 11–26.

York, D. A., Bray, G. A. & Yukimura, Y. (1978). Enzymatic defect in obese (*ob/ob*) mouse – loss of thyroid-induced sodium-dependent and potassium-dependent adenosine-triphosphate. *Proc. natn. Acad. Sci. U.S.A.* **75**: 477–481.

Symp. zool. Soc. Lond. (1981) No. 47, 337–365

Behaviour of the House Mouse

J. H. MACKINTOSH

Sub-Department of Ethology, Birmingham University Medical School, Birmingham, UK

SYNOPSIS

The house mouse, particularly in its domesticated laboratory form, has been the subject of an extremely large number of behavioural investigations. Studies have been continuous over the last 50 years and a large body of data has accumulated relating behavioural factors to physiology, genetics, population control and communication. It is therefore a paradox that there remain numbers of important areas where the information available is sketchy or merely inferential. This imbalance is a product of the success of the mouse as a laboratory animal, the particular interests of the laboratory investigators and also the difficulty of studying the animal under natural conditions.

Behaviour structure may be approached at a number of different levels, beginning with the description of the individual motor patterns of which the behaviour is composed and the relationship between them, and progressing to social organization and to interactions with other factors. Approximately 50 individual elements of behaviour have been described in the mouse, which by analysis can be shown to be organized into a number of distinct motivational groups. Strain, sex and situational differences in these elements are almost exclusively confined to variations in frequency of occurrence, but a possible example of a difference in morphology exists in one island population. The social organization of the house mouse, on the other hand, is strongly influenced by a variety of factors including available space, group composition and history, population density and the genetical and physiological state of the individuals involved. The organization tends to be a despotic dominance under confined or crowded conditions, and territorial under less restricted ones.

Population dynamics and social behaviour are clearly related, and it is suggested that behavioural flexibility is a major factor contributing to the success of the mouse.

INTRODUCTION

The house mouse, particularly in its domesticated laboratory form, has been the subject of a vast number of behavioural investigations. Indeed, it is probable that in the league table of most studied animals, its only rival is the white rat. It was examined in some detail by Uhrich (1938) and Retzlaff (1938), and has been the focus of much interest ever since; thus even if pharmacological studies, which are extremely numerous, are neglected, there is very considerable literature. It is therefore a paradox that there remain numbers of important areas where the information available is limited or merely inferential.

The popularity of the mouse is an outcome of the fact that it is a singularly convenient laboratory animal by reason of its size and ease of management. Availability is perhaps not the best reason for choosing an animal for behavioural studies, but intensive development, for example in cancer research, has led to the production and documentation of an imposing array of inbred strains. These variants facilitate genetical investigations and also the selection of a type with any desired trait, for example, aggressiveness. Small size, on the other hand, is not a desirable characteristic for field studies in a subject and when it is linked with largely nocturnal habit and timidity, it is not surprising that in spite of the economic importance of the mouse the ratio of laboratory studies to those undertaken in natural situations is of the order of several hundreds to one.

This imbalance, together with the very obvious differences between the laboratory cage and any field situation, has frequently led to doubts as to the validity of behavioural studies in confined conditions. The problems involved will be discussed later as they remain to an extent unresolved, but it must be pointed out that whatever value there is in inferences drawn from laboratory studies as to the likely nature of the behaviour of the wild house mouse, it is necessary to study the laboratory mouse as a system in itself. More mice are used in experiments than any other animal and therefore a detailed knowledge of their behaviour is of considerable importance.

The behaviour of the mouse may be approached at a number of different levels. Thus the first topic will be a description of the basic motor patterns which make up the behavioural repertoire of the mouse, and this will be followed by an examination of the way in which these elements are related to each other. These studies establish the fine structure of the behaviour and provide a basis for the investigation of strain and sex differences. They also allow comparisons to be made with other species and a few such examples will be included as they provide additional information as to the nature of mouse behaviour. The final level is that of social structure and discussion of this subject will inevitably raise questions as to the relationship between environment and behaviour and of the interaction between behaviour and physiological status.

ELEMENTS OF BEHAVIOUR

Many authors have described a few motor patterns in the mouse, but the number who have attempted an exhaustive check-list is limited. Noirot (1958) gave details particularly in the area of maternal behaviour and Van Abeelen (1963, 1966) published a detailed list covering both social and non-social behaviour. Clarke & Schein (1966) used 13 elements and Estep, Lanier & Dewsbury (1975) 12 for males and 12 for females. Van Oortmerssen (1971) produced a list of 45 items and that of Grant & Mackintosh (1963) described 50 elements with most emphasis on social interaction. This list has provided a basis for much subsequent work and it is given in outline in Table I. As may be seen, it includes elements covering a wide range

TABLE I

Brief description of the elements of behaviour in the mouse. In all cases where the description involves two animals A is the actor and B the mouse at which the element is directed.

Element	Brief description
Explore	Investigation of surroundings
Attend	A looks at B
Approach	*a*
Nose	A investigates facial region of B
Sniff	A investigates urogenital region of B
Investigate	Investigation other than Nose or Sniff
Follow	*a*
Attempt Mount	Incomplete male sexual behaviour
Mount	Full male sexual pattern
Genital Groom	Grooming own genital region
Push Under	A pushes under B
Crawl Over	A crawls over B
Push Past	A pushes past B
Groom	A grooms B
Leave	Undirected movement away from B
Stretched Attend	A extends body toward B while attending
Walk Round	A walks round B
Exploring Approach	Approach and Explore combined
Crouch	*a*
Straight Legs	Standing with legs extended
Evade	Head or forebody movement away from B
Retreat	*a*
Flee	Fast, almost uncontrolled movement away from B

TABLE I (*continued*)

Element	Brief description
On Back	Lying flat on back
Freeze	Complete cessation of movement
Sideways Posture	A oriented sideways to B
Offensive Sideways	As Sideways Posture with head toward B
Defensive Sideways	As Sideways Posture with head away from B
Upright Posture	A stands on hind legs facing B
Offensive Upright	As Upright Posture with head forward
Defensive Upright	As Upright Posture with head back
Oblique Posture	A stands facing B with forelegs extended
Threat	Sharp head or forebody movement toward B
Attack	A approaches B rapidly and bites on distal flank
Bite	[a]
Over	A positioned at right angles over B
Chase	[a]
Aggressive Groom	A vigorously grooms B's nape
Rattle	Sinusoidal tail movement
Circle	Movement in a circle at a distance from B
Zigzag	A moves from side to side of cage in front of B
Wash	Forepaws groom face
Self-Groom	[a]
Dig	Sawdust scraped back with forepaws and kicked back with hindpaws
Push Dig	Sawdust pushed forward with forepaws
Displacement Dig	Abbreviated digging movements
Eat	[a]
Drink	[a]
Displacement Groom	Abbreviated face-washing
On Bars	Climbing on bars of cage
Off Bars	Climbing off bars of cage

[a]Self-explanatory.

of complexity, some, such as "Threat", are simple motor units, whereas others, such as "Explore", are highly inclusive. Slater (1978) has reviewed the advantages and disadvantages in lumping and splitting in the construction of a behavioural inventory and has also enumerated the perils of the use of elements of unequal status in the same list. However, there are clear advantages in an extensive even if imperfect check-list as it avoids some of the problems inherent in the use of still more inclusive categories. The upright postures provide an example of this. Grant & Mackintosh (1963) recognized three variants of these elements, "Offensive Upright", "Defensive Upright" and "Upright Posture", which, as is reflected in the names, differ not only morphologically but also in their motivational background. Where uprights are lumped into a single unit, as has frequently been the case, the spread of causation has resulted in a degree of confusion, e.g. Scott (1945) and Tedeschi *et al.* (1959) regard uprights as aggressive, whereas Warne (1947) considered them to be submissive. The tendency to use inclusive categories persists and can at best lead to a loss of data and at worst to misinterpretation.

ELEMENT GROUPS

Grant & Mackintosh (1963) pointed out that although the elements in their list were classified under broad motivational headings this was done without the provision of supporting data. Subsequent analysis has confirmed the distinctions made and Mackintosh, Chance & Silverman (1977) published a classification based on sequence analysis. This is reproduced in modified form in Table II and it can be seen that this method defines three element groups which are respectively:

(1) non-social behaviour;

(2) social investigation and sexual behaviour; and

(3) agonistic behaviour which is divided into three closely linked sub-groups: (3a) aggressive behaviour; (3b) ambivalent behaviour; and (3c) flight behaviour.

The groups are defined on the basis that the elements of which they are composed show strong sequential links only with others within the group. That is, not only will all the elements tend to precede and succeed each other, but they will also show similarities in the responses they produce in other mice and in the types of behaviour that engender them. Group (3) is subdivided because, as would be expected, direct sequential links between (3a) and (3c) are uncommon. They do, however, occur as responses to each other and are connected via (3b). The elements in this subgroup show a gradation in their affinities and these are shown in Fig. 1. Those elements labelled offensive link strongly with aggression, and those

TABLE II

Element groups

Element groups				
(1)	(2)	(3a)	(3b)	(3c)
Explore	Investigate	Threat	Offensive Sideways	Evade
Wash	Nose	Aggressive Groom	Offensive Upright	Retreat
Self-Groom	Sniff	Attack	Sideways Posture	Flee
Dig	Follow	Bite	Upright Posture	On Back
Push Dig	Attempt Mount	Over	Defensive Sideways	Oblique Posture
Eat	Mount	Chase	Defensive Upright	Kick
Drink	Genital Groom	Rattle		Crouch
Displacement Groom	Push Under	Circle		Straight Legs
Displacement Dig	Crawl Over	Zigzag		On Bars
Leave	Push Past	Walk Round		Off Bars
Attend	Groom			Freeze
Stretched Attend (Approach)				

labelled defensive relate in the same way to flight, and those without qualification are intermediate in character.

What is not immediately clear from this is that the two major types of ambivalent element, "Upright Postures" and "Sideways Postures", differ in their balance. Sideways postures tend as a whole to be more closely related to aggression and uprights to flight. Fig. 2 shows a fan diagram derived from correlations of sequence analysis data and clearly shows this effect, each sideways posture being more closely correlated with "Attack" than is its corresponding upright posture. The data have also been subject to cluster analysis* and Fig. 3 is a dendrogram derived from this method. The element relationships that it reveals support the classification produced by the simpler techniques.

Detailed description of the links between elements is out of place in a general paper of this kind, but one or two points of particular

*Ward's method.

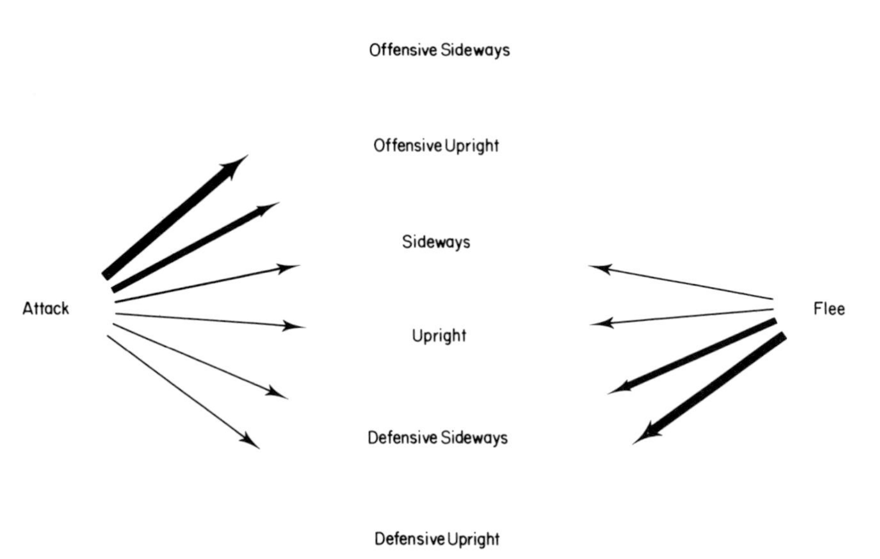

FIG. 1. Sequential links between representative elements of groups (3a) and (3c) (Attack and Flee) and the elements of (3b). The width of the arrows is approximately proportional to the transition probability.

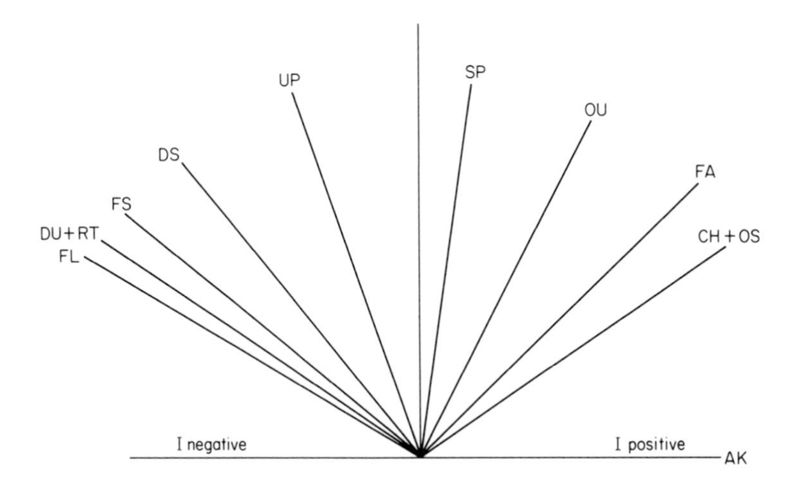

FIG. 2. Cosine model based on sequence analysis of mouse behaviour. FL = Flee, DU = Defensive Upright, RT = Retreat, FS = On Back, DS = Defensive Sideways, UP = Upright Posture, SP = Sideways Posture, OU = Offensive Upright, FA = Over, CH = Chase and OS = Offensive Sideways. Elements to the right of the vertical are positively correlated with AK (Attack), those to the left negatively.

interest should be made. Firstly, the names of two elements have been changed since previous publication. "Submissive posture" has

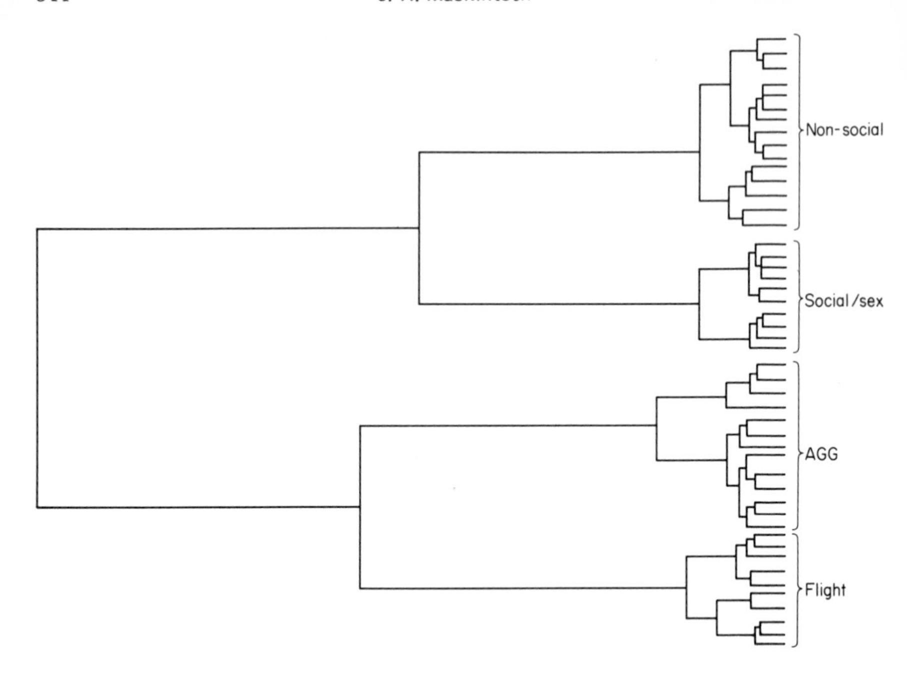

FIG. 3. Dendrogram based on cluster analysis of mouse behaviour. Each terminal branch represents one of the elements in Table I.

become "On Back", and "Aggressive Posture" has been changed to "Over". They were named originally in analogy with similar elements in the Norway rat where the performance of the Aggressive posture–Submissive posture pair can be seen to cause a diminution of the tendency for the more dominant animal to behave aggressively. Analysis has revealed that they do not have the same effect in the mouse.

Secondly, although in general there are few sequential links between the groups some elements take up an intermediate position. An example of this is "Aggressive Groom", which shows strong links both with sexual behaviour and with aggression. A possible explanation of this may lie in the postural similarities of Aggressive Groom, Over, Attack and Attempt-Mount. Dawkins & Dawkins (1976) have described a phenomenon which they called postural facilitation whereby a stage in a sequence is partially determined by the morphology of the preceding one. If the explanation proposed here is correct, then the concept may be extended to include a degree of motivational indeterminacy produced by the occurrence of similar motor patterns in a number of different categories. The method described so far defines element groups which correspond to the principal motivational systems and the section which follows examines each of these systems in turn with the object of giving a general outline of mouse behaviour.

Agonistic Behaviour

The principal focus of attention of behaviour studies in the mouse has been aggression. Many have been concerned with the effects of environmental, social and genetical factors on levels of aggressive behaviour and therefore discussion of these will be postponed to later sections. Some indications of the structure of agonistic behaviour have been given in previous paragraphs. The general point can be made, however, that the mouse is in many circumstances highly aggressive. The number of laboratory studies on aggression testify to the ease with which mice can be persuaded to fight under those conditions. For example, in enclosures Southwick reported a rate of one fight per mouse per hour (Southwick, 1955b), and the same author (Southwick, 1958) records levels of wounding of up to 45% in free living wild mice.

Sexual Behaviour

The mouse is typical of many rodents in that the female, apart from some investigatory activity, takes a relatively passive role in sexual behaviour, and courtship as such is absent. Males identify females and their reproductive condition by means of olfactory cues (Kalkowski, 1968) and females also show interest in male urine (Scott & Pfaff, 1970). A phenomenon that is probably related to this is that female olfactory threshold is lowest shortly before the sexually active phase, as has recently been demonstrated by Schmidt (1978).

The mating sequence is in general approach by the male followed by investigation, particularly genital sniffing and following if the female retreats, succeeded by "Attempt Mounts" or "Mounts" with palpation of the female and pelvic thrusts. Males when they dismount normally groom their own genitals and may also show a pattern of behaviour "Push Under".

Sexual behaviour has been described in detail by Van Abeelen (1963), Noirot (1958) and Estep *et al.* (1975), and the process of mating and sexual activity together with its physiological correlates has been studied by McGill (1962, 1963, 1967, 1969), Land & McGill (1967), and McGill, Cowen & Harrison (1968).

Parental Behaviour and Nest Building

Parental behaviour in the mouse consists of retrieving unweaned young, grooming them, nest building and lying with the pups in a lactating position. Noirot, who has undertaken extensive research into parental behaviour in mice (Noirot, 1964b, 1965, 1969a), reports that both males and females show all the components of parental behaviour including the adoption of the lactating position, although this is of course unproductive in the case of males. She has demonstrated the effect of stimuli from the pup on maternal behaviour and has shown that different stimuli affect different components of this behaviour, e.g. retrieving occurred in response to an ultrasonic call from the infant, whereas grooming and lactation did not show the same dependence (Noirot, 1964a). In spite of these differences, however, the sequence Retrieve – Lick – Nest Build – Lactating Position which occurs when pups are brought back to the nest, is relatively constant (Noirot, 1969b). Maternal behaviour builds up during pregnancy, but there is some conflicting evidence as to which components are affected, e.g. Noirot & Goyens (1971) state that retrieving is unaffected, whereas Fraser & Barnett (1975) show that pup carrying increases during the later stages.

Although nest building occurs in the absence of young, it is included with parental behaviour as not only does it occur within the retrieval sequence but also the efficiency with which nests are built exerts a strong influence on litter survival (Southwick, 1955a).

Nest building as such has been studied by Van Oortmerssen (1971) who developed a standardized quantitative test which measures the amount of "fraying" of paper strips. We have used this test to examine nest building behaviour in samples of wild mice from the Island of May, and found that although they had been trapped, moved from the island, tested for a week at the Psychology Department of St Andrew's University and sent by train to Birmingham, they showed indication of an annual fluctuation in nest building absent from laboratory stock.

Other behavioural traits showed equal persistence in these mice and they will be discussed in relation to the social structure of wild mice.

Non-Social Behaviour

Exploratory behaviour

Mice intensively examine new surroundings and re-examine familiar ones. They move about sniffing the substrate and from time to time rear up in a position "Scan", which appears to be concerned with sampling of airborne information. There has been no finely detailed analysis of exploration in the mouse but Barnett & Cowan (1976) provide a general review of exploratory behaviour which is, however, biased towards the rat.

The characteristics of exploration in the mouse that have been revealed include the following:

(1) as would be expected, it shows a diurnal rhythm (Poirel, 1968);

(2) the information gained appears to be highly specific. Jakob & Claret (1967) showed that mice could distinguish between two very similar activity machines, and the same authors also demonstrated that the information received by a mouse when exploring was well retained for at least two days; and

(3) Garcia-Segura (1977) showed that experience of different qualities of surroundings affects the pattern of exploration.

Feeding and maintenance activities

Food preferences in mice have been most studied in connection with the development of pest control techniques, (e.g. Rowe, Smith & Swinney, 1974) and this was also the rationale for Crowcroft's studies. He showed (Crowcroft, 1966) that mice tended to visit many feeding points in the course of a single foray and also that they exerted considerable selection as to which grains they ate out of the many available. More recently, a detailed analysis of the time properties of feeding behaviour was undertaken by Weipkema (1968, 1971). Hoarding food does not appear to be as organized an activity in mice as in many rodents, although some food is brought back to the nest (Manosevitz, 1967).

"Grooming" and "Digging" both show internal sequences of elements, "Face Groom" tending to precede "Body Groom" and "Genital Groom", and "Dig" – which has two sub-components, one the withdrawal of the substrate with forepaws and the other kicking it backwards with the hind legs – precedes "Turn" and "Push Dig".

Activity cycles

Activity tends to be a blanket term covering all forms of motor pattern and it therefore follows that if activity is cyclical, then so is all the behaviour. Measurement of mouse activity shows a diurnal rhythm (Poirel, 1968), which tends to have dawn and dusk peaks of activity, plus a series of minor short-term peaks, large during the dark phase and small during the light one. These short cycles do not appear to be synchronized in a mouse colony as a general measurement of activity in a mouse room revealed a smooth pattern of change over 24-hour periods.

Activity is affected by the social structure of a group of mice (Fujimoto, 1953), and this may well account for the variation in activity found when different numbers of mice are caged together (Dews, 1953). As has been reported for a number of species, activity in females varies with the oestrous cycle (Gutmann, Lieblich & Gross, 1975).

SEX DIFFERENCES IN BEHAVIOUR

Figure 4 demonstrates the results of an experiment in which the frequency of individual elements in males and females of the BALB/c strain was compared under the same conditions. There were considerable frequency differences, but long-term studies indicate that there are no qualitative differences; all the elements of behaviour described can occur in both sexes.

A striking feature of the results in Fig. 4 is that the aggression shown by females is extremely low. This is not untypical of caged mice and as St John & Corning (1973) point out, female mice have often been regarded as non-aggressive, a viewpoint that was reinforced by the fact that the earlier studies suggested that they do not become aggressive as a result of isolation or of testosterone administration, as do males (Beeman, 1947; Tolman & King, 1956). Reports of female aggression, however, have been in existence for a long time; Retzlaff (1938) for example, showed that some females attacked and killed castrates, and also other females which were to some extent unfamiliar to them.

In addition workers who have examined the mouse in naturalistic conditions have always been aware of the existence of female aggression (Lloyd & Christian, 1967; Southwick, 1955b). More recently the characteristics of female aggression have been defined by a number of different authors. Although there is some fluctuation during the oestrus cycle, with a low level at oestrus (Hyde & Sawyer, 1977) females tend to be most aggressive when pregnant (Noirot, Goyens & Buhot, 1976), or lactating (Haug, 1972; Gandelmann, 1972). Aggression peaks after parturition and is high for 13 days and then declines (Gandelmann, 1972). Its intensity varies between females and between lactating periods. It has also been shown recently that aggression can be induced in females by the administration of androgens (Evans & Brain, 1978). There is some dispute as to the nature of the motor patterns involved when female mice

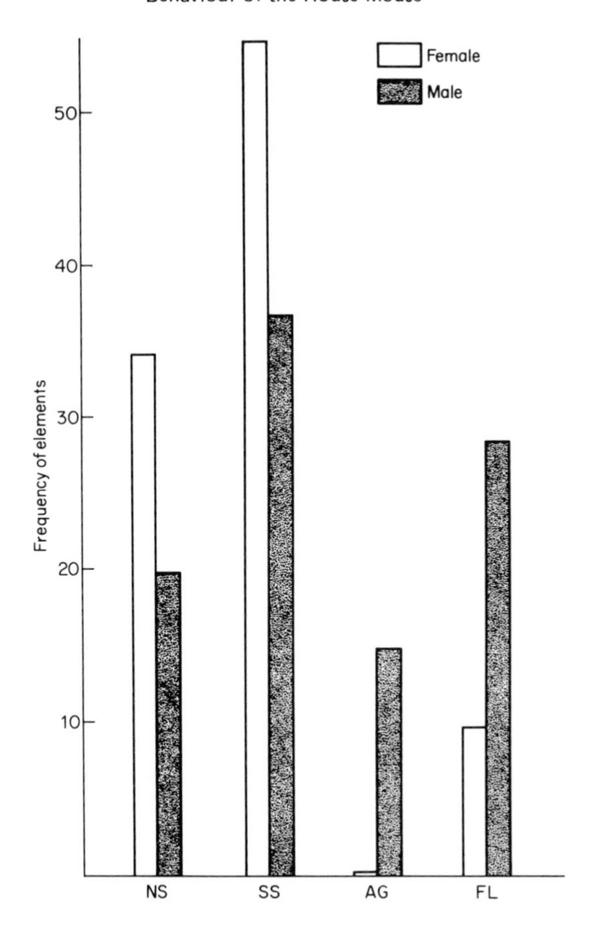

FIG. 4. Comparison of behaviour of male and female mice giving numbers of elements in each category/mouse/observational period. NS = Non social; SS = Social Investigation and Sexual Behaviour; AG = Aggression; FL = Flight.

attack. Ebert (1976) reported that the behaviour was indistinguishable from that of males, but Noirot (1965) noted that females tended to be more precipitate and that they were less inhibited by the "Defensive Upright Posture" than were males. This description is in agreement with that of Gandelmann & Svare (1974) and Gandelmann (1972) who record low latency and ferocity and also an absence of investigation and tail rattling.

Unlike males, female aggression is directed at both males and females. Rosensen & Asheroff (1975) tested isolated lactating females for aggression against various types of intruder and found that introduced lactating females were not attacked. Haug (1972,

1973) on the other hand, has reported the reverse, and demonstrated that virgin mice are attacked more if they have been caged with lactating females before testing. Another contrast with males occurs with the effect of the group size in which the mice are kept before testing. Males show most aggression if isolated, but females most if they are kept in groups of four (Blick, Gross & Ropartz, 1971).

Aggression and territoriality are closely related, and female territoriality is considered in a later section.

STRAIN COMPARISONS

As mentioned in the Introduction, the wide range of strains available in the mouse provides fertile material for genetical investigation. There have been a large number of comparisons between strains and quantitative differences in behavioural characters are general.

In simple strain comparisons, differences have been found in aggression (Scott, 1940; Calhoun, 1956; Levine, Diakow & Barsel, 1965; Van Oortmerssen, 1971; Eleftheriou, Bailey & Denenberg, 1974), escape behaviour (Fuller, 1967), sexual behaviour (Levine *et al.*, 1965; McGill, 1969), exploration (Fuller, 1967), nest building (Van Oortmerssen, 1971), learning (Denenberg, 1965), parental behaviour (Ressler, 1962) and activity (McClearn, 1960). One of the most interesting investigations is that of Van Oortmerssen (1971) who found that the behaviour of the inbred strains that he examined fell into two major groups, and on the basis of this, he advanced the theory that laboratory strains of mice are derived from two separate wild stocks with different behavioural adaptations.

Single gene differences have also been shown to involve behavioural effects (Denenberg, Sherman & Blumenfield, 1963; Van Abeelen, 1963; Henry & Essinger, 1967) and a number of investigators have examined F_1 hybrids and/or back-crosses, in addition to the parental strains (e.g. McGill & Tucker, 1964; Van Abeelen, 1966; McGill & Ransome, 1968; Eleftheriou *et al.*, 1974).

Heterosis has been detected on a number of occasions and Collins (1964) quotes Bruel's 1964 hypothesis that only quantitative polygenic traits which were biologically adaptive under past selective pressures would result in heterotic inheritance, and he uses this to suggest that facility in conditioned avoidance performance is adaptive. A parallel effect was found in our laboratory in an investigation of the behavioural characteristics of the CE and CBA inbred strains, and the reciprocal hybrids between them. The elements whose score in the hybrids significantly exceeded the parental mean were

"Circle", "Zigzag" and "Displacement Groom". These at first sight do not appear to be a particularly meaningful group, but as they occur as a response to the inhibition of aggression produced by the Defensive Upright Posture, then it may be suggested that it is the effectiveness of this element that has recently been subject to positive selection. This in itself would seem to be logical, as domestication has involved close grouping of mice at a level which would not be encountered in the wild.

This contention is supported by Noirot's (1965) finding referred to above, that aggressive females are not only less inhibited by the "Defensive Upright" but also show less tail rattle.

COMPARISONS BETWEEN WILD MICE AND LABORATORY STRAINS

This comparison would appear to be of particular interest; however, the results obtained have not shown any very striking differences. For example, in a series of experiments in which we compared the behaviour of six inbred strains and stocks of wild mice, it was found that the behavioural profile of the wild mice fell within the range of the inbred strains. Beilharz & Beilharz (1975) and Beilharz (1975) have also shown that the behaviour of wild and of laboratory mice is very similar, although as is usual all strains showed variations in threshold levels and in the frequency with which they display various elements. However, R. H. Smith (1972) showed differences that were consistent with a higher level of flight motivation in wild mice and Antalfi (1963) produced evidence which indicates that the laboratory mouse may well have lost its response to predators.

The differences that have been found, therefore, between these strains and wild mice have been exclusively quantitative. The investigation of mice from the Isle of May, however, revealed a possible qualitative difference. The samples of male mice were tested for aggression against laboratory mice which had been maintained in groups up to the time of the experiments. As will be shown later the wild males fell into two categories, one aggressive and the other not. Of the non-aggressive males, some showed extreme flight and retreated at all times from laboratory mice. However, others in this group adopted a form of the "Upright Posture" we had not seen previously. It was similar to an "Offensive Upright Posture" except that the hind legs were more extended so that they tended to stand on their toes. The mice were very active while in this posture, even bouncing forwards, and the forepaws were windmilled at the attacker. Additionally they frequently delivered a bite from this position.

The May population is known to differ from that of the mainland in terms of morphology, Clevedon-Brown & Twigg (1969) having demonstrated the presence of a skeletal variant. Whether this element is unique to May mice is not known. It is absent from the Midland samples of wild mice which have been examined, but too few detailed analyses have been carried out to be certain.

SOCIAL STRUCTURE

The social structure of the mouse differs under different conditions, and it will therefore be considered under three sub-headings, firstly the structure as it appears in the laboratory cage, secondly the structure under semi-natural conditions, i.e. in relatively large enclosures, and thirdly the social structure in the field.

Social Structure in Cages

The effect of housing conditions

Uhrich (1938) showed that if small groups of laboratory mice were caged together for a period, then a social hierarchy developed, in which one male was dominant, and all the others submissive. This type of despotic dominance has been the common finding of all other workers who have repeated this type of study. Stability of the rank order is variable, and Poole & Morgan (1973) demonstrated that this factor is related to the size of the group. They investigated groups of four, five, nine and 12, and showed that there were significantly more changes of dominance in the large groups.

The emergence of a hierarchy may depend on the degree of familiarity within the founding stock, as Poole & Morgan (1975) also showed that if groups were made up from litter mates then they tended to be amicable with little apparent hierarchical structure.

Dominance has been reported to give no priority of access to food; however, W. I. Smith & Ross (1951) showed that if a group of mice were given a vitamin deficient diet then the dominant animal outlived the subordinates. Social rank also has an influence on urination patterns (Desjardins, Maruniak & Bronson, 1973), the number of urine pools being much reduced in subordinates, and on activity (Fujimoto, 1953).

The group structure is probably maintained by individual recognition. Bowers & Alexander (1967) reported that mice can readily distinguish between two males on the basis of olfactory cues, and Kalkowski's (1967) studies indicated greater ability than this, as his mice were able to separate at least 18 mice in nine pairs on the basis of odour. He also showed that the discrimination was retained for 14 days.

The composition of the group affects the aggressiveness of its members, as for example the presence of females tends to make the males more aggressive when subsequently tested (Crawley, Schleitt & Contera, 1975; Goyens & Noirot, 1974).

The effect of isolation

Individual housing of male mice has frequently been observed to have the effect of increasing their tendency to be aggressive (Thiessen, 1963; Cairns & Nakelski, 1971; Krsiak & Janku, 1968; Valzelli, Vernasconi & Gomber, 1974), and the process has often been used to stimulate aggressiveness. However there has been some dispute as to the nature of the mechanisms which produce this effect.

There are two viewpoints: the first, exemplified by Brain and his co-workers who have measured a number of physiological parameters such as adrenal response and gonadal condition, states that the effects of isolation are not stressful and that isolation stimulates a "territorial" type of behaviour (Brain, Nowell & Wouters, 1971; Goldsmith, Brain & Benton, 1976; Benton *et al.*, 1978). On the other hand Essmann (1966) who studied activity differences between grouped and isolated mice, suggested that individual housing is stressful, as did Priestnall (1975), while Banerjee (1972) also indicated that isolation may be deleterious as (he states) it produces an aggravation of the normal senile and degenerative changes. The source of the disagreement may lie in the rather variable effects of isolation as Goldsmith *et al.* (1976) showed that aggression increased with the duration of isolation, reaching an asymptote at 56–58 days, whereas in a similar experiment (Fig. 5) it was found that the peak levels were reached in a much shorter time. It has also been shown (Krsiak & Borgosova, 1973) that isolation may produce timidity instead of aggression, more than half the animals in their experiment becoming less aggressive.

Ontogeny of behaviour

Another effect that has been found in the studies of isolation-induced aggression is that its effects may vary with the time in the life of the mouse at which it is applied. It has been clearly demonstrated that the rise in male aggression is closely associated with the pubertal surge in androgens (McKinney & Desjardins, 1973), but more complex maturational effects have been reported.

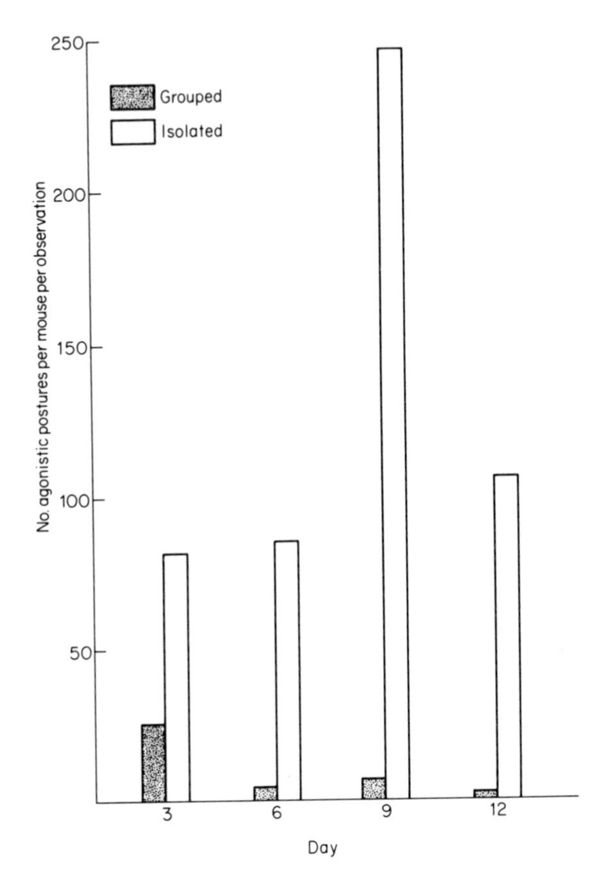

FIG. 5. Comparison of aggressive behaviour in group-housed mice (2 mice per cage) with that of mice isolated for 3, 6, 9 and 12 days respectively.

King & Gurney (1954) showed that isolation from weaning to 107 days was less effective than if there was a period of housing with a relative of either sex, and a possible mechanism for this effect is suggested by the findings of De Ghett (1975). In his experiments young mice exposed at an age of 21–30 days to the sight of adults fighting were themselves more aggressive when tested later in life. Dyer & Southwick (1974) produced evidence indicating that there may be short sensitive periods during development which affect aggressiveness. In the same way that males from groups of mixed sexes are more aggressive, males which have been brought up in litters containing females are more aggressive than those which have not (Namikas & Wehmer, 1978) and males derived from smaller litters are dominant over those from larger ones (Ryan & Wehmer, 1975).

Prenatal effects have also been detected. Kahn (1954) showed that males whose mothers had been subjected to defeat while pregnant were less aggressive than those of undefeated mothers and behavioural development was retarded in males whose mothers had been crowded during pregnancy (Keeley, 1962).

Social Structure in Enclosures

Territorial behaviour

The need for some compromise between the laboratory cage and pure field studies has led to the frequent use of a variety of enclosures. These have varied in size from what was virtually a large cage (1.32 m × 0.66 m) (Lloyd, 1975) to very considerable spaces, e.g. 26 m × 14.8 m (Lidicker, 1976). Enclosure experiments have been undertaken with the purpose of studying the effects of population density, and compared with cage experiments have frequently been carried out with wild mice.

The most striking effects of enlarging the area available to the mouse is the emergence of unambiguous territorial behaviour. Various authors have referred to aggression in cages as "territorial" and Davis (1958) regarded territorialism and social rank as the two poles of a behavioural continuum. However, the lack of a tendency for mice consistently to win encounters in their home cage (Mackintosh, 1965) and the emergence of additional behaviour patterns such as patrolling and lookout activity in large spaces (Crowcroft, 1966; Mackintosh, 1970) suggests that there may be a discontinuity. The characteristics of territorial behaviour were first described in detail by Crowcroft (1954, 1955a, b, 1966) and Crowcroft & Rowe (1957, 1961, 1963), and their findings stand with very little modification. Crowcroft used enclosures of several sizes from 1.3 m^2 to 6 m^2 and showed that in a population a minority of males defend territories which are stable and which, if the area is sufficient, may have zones of no-man's land between them. The remaining males are non-territorial and subordinate and restricted in their movements, often crowded together in a small nest box. Females mate with the territorial males and tend to remain with them, as do the offspring. This could lead to the establishment of what appears to be a group territory (Eibl-Eibesfeldt, 1958). Pregnant females assist in territorial defence. Reimer & Petras (1967) also showed very stable territories using a population cage consisting of nest boxes connected by runways. Aggression levels were high and most intruders were killed. Females were active in territorial defence and one held a territory for five weeks in the absence of males. Lloyd (1975) and Lidicker (1976) achieved substantially the same results, although in the former case, the territories were less stable.

Mackintosh (1970, 1973) describes a series of experiments in which the factors affecting territory formation were examined. In the enclosure used ($1.8\,m^2$) territories were established by first separating two groups of mice by means of a partition. This partition was removed after one week. The results obtained were similar to those of Crowcroft, except that there were no undisputed areas and no evidence of site attachment of females and juveniles. Mackintosh (1978) ascribed this difference to the relatively limited area available. However, the threshold of female participation in territorial defence apparently varies, as a recent series of experiments carried out by Tyack (in preparation) in the same situation has clearly demonstrated female territorial defence with or without the presence of males.

Other features which were demonstrated were that subordinate males showed a stratification into subdominant and at least two classes of subordinate (Evans & Mackintosh, 1976), and that, unexpectedly, boundaries were recognized by means of visual cues, even if there was a mismatch between these cues and the available olfactory information. Odours were used if the visual situation was neutral, and in a similar experiment Harrington (1976) reported that fighting between males took place in zones where there were discrepancies between olfactory and visual cues, although in his experiments vision appears less dominant.

In addition, Mackintosh showed that boundaries tended to form where there was a strong physical feature, and that dispersal was a role largely undertaken by subordinates which would establish their own territories if they encountered an unoccupied space. Adjacent males with established territories organized their activity phases and patrolling activity in such a way as to minimize contact, and subordinates reduced their overall activity to a minimum. In freely-growing populations, young males set up rank-ordered relationships and those which were dominant to their peers occasionally succeeded in establishing their own territories within the zones occupied by the older territorial males.

Population control

As mentioned above, many of the enclosure experiments which have been undertaken had the study of population control as a major objective. It has been recognized that the social structure of the mouse population is important in determining the population level (Southwick, 1955a), but the active mechanisms are numerous and appear to vary from population to population, or from one experimental situation to another. On some occasions there is a considerable drop of fertility (Crowcroft & Rowe, 1957; Lloyd, 1975; Lidicker, 1976), and Lloyd & Christian (1969) report that the reproductive status of females is related to their social status, one of the few reports of rank-ordered behaviour amongst females.

In other cases the major factor limiting population growth appears to be infant survival, either through poor nest structure at high densities (Brown, 1953; Southwick, 1955a, b) or the result of normal retrieving behaviour becoming too frequent at high population densities (Mackintosh, 1978).

The effects of aggression tend to be indirect and not always closely related to density, as an increase in available space can result in increase in conflict (Mackintosh, 1978), and Rowe, Taylor & Chudley (1964) showed that aggression in wild mouse populations in ricks was affected by the structure of the rick to a greater extent than it was by density. A carry-over of effect from an overcrowded generation to that which follows it might be expected, as Randt, Blizard & Friedman (1975) showed that male mice undernourished in infancy were more aggressive as adults.

Social Structures of Free-Living Populations

Anderson (1961) reviewed the relationship between density and population control and showed that the lower limits of density achieved by experimental populations just overlap with those which have been recorded in the field. As there is clearly a close link between social organization and population control the possibility also exists that natural populations may differ in their social behaviour, and although the way in which free-living populations of the house mouse organize their societies appears to be very similar to that shown in many enclosure experiments, direct evidence is lacking.

Established groups resist penetration by immigrants, as would be expected if territories were held (Anderson, 1964; Adamczyk & Ryskowski, 1965) and trapping results indicate the presence of site-attached males and females (Berry, 1970) together with a floating population of males. Similarly Mackintosh (1978) reported observation of a single encounter between two free-living mice which was identical to the type of border encounters seen in enclosures, even to the extent of occurring at a natural discontinuity in the environment.

It was mentioned previously that male mice from the Isle of May population showed a bimodal distribution of aggressive behaviour when tested against laboratory mice. Eight per cent were highly aggressive and it is possible that these were previous territory holders, whereas the remaining males showed relatively low levels of aggression and might therefore have corresponded to the itinerant groups. There is no direct evidence that this is so at present but weight and age characteristics support the idea.

The site-attached males, however, in one natural situation at least, are some 40 m apart and the means by which mice can maintain boundaries with this degree of spacing is not known. Presumably, as in some of Crowcroft's experiments, only a core area of the home range is defended, but how much, whether patrolling behaviour occurs, whether encounters between adjacent males are frequent, and what role females play in the defence of territories, remains to be discovered.

STRATEGIES

Currently much of the attention of students of animal behaviour is focused on the assessment of the costs and pay-offs of particular actions. All the habits of the mouse could of course be examined in this way, but discussion will be limited to a few examples.

Gilder & Slater (1978) found that females showed preference for sawdust which carried the odours of unrelated males of the same strain to sawdust smelling of brothers or of males of a different strain. If this preference was reflected in their choice of mating partners they would be adopting a strategy which would reduce inbreeding but would prevent large discrepancies in genetical make-up which might be deleterious. The result, however, does not quite match with that of Yanai & McClearn (1972) who found that when given a choice females selected mates of a different inbred strain.

A clearly adaptive strategy of parturient mice was recorded by Newton, Foshee & Newton (1966) who showed that if females in this condition are alternated between a familiar and unfamiliar situation, then significantly more pups are born in the familiar one.

A quite different category is that of the strategies developed by individual mice which are not common to all members of the same class. Crowcroft (1966) described the way in which subordinates would shake off a pursuer by turning quickly and our own observations reveal the tendency of some individuals to learn that a rapid turn just in front of a solid object is particularly effective. In the same way Lloyd & Christian (1967) described how some subordinates were unscarred as the result of adopting particular habits.

CONCLUSIONS

The house mouse is indisputably a highly successful animal able to exist in a wide variety of situations, from open grassland to the interior of carcases in cold stores, and it would be extremely satisfying if it were possible to point to the particular adaptations that enable it to do so. The gaps in our knowledge of the mouse, however, together with even greater ignorance of other rodent species which are not able to emulate it make this a difficult task. An animal that is primarily territorial would not appear to be a good candidate for a commensal existence as this way of life entails the exploitation of concentrated food sources and therefore the likelihood of high local population densities. However, as has been described, confined conditions lead to the replacement of territorial behaviour with a hierarchical system and evidence exists that suggests levels of aggression may fall as density increases. Opportunism is also revealed by the way in which visual cues are used to locate territory boundaries although they would be expected to be relatively unavailable in many situations. In a similar category is the ease with which modifications occur in individual behaviour which reduce the penalties of a subordinate role and the various methods by which the behaviour of females may reduce inbreeding and minimize the loss of young. The detailed analysis of social behaviour shows that although a true submissive posture such as is found in more social species may be absent, temporary inhibition of attack is achieved by the "Defensive Upright Posture" and it is possible that selection for the increased efficiency of this element has occurred under confined conditions. A major contribution to the success of the mouse is therefore probably a network of behavioural adaptations which allow it to exploit a variety of living situations by conferring a degree of social flexibility.

REFERENCES

Abeelen, J. H. F. Van (1963). Mouse mutants studied by means of ethological methods. *Genetics* **34**: 79–94.

Abeelen, J. H. F. Van (1966). Effects of genotype on mouse behaviour. *Anim. Behav.* **14**: 218–225.

Adamczyk, K. & Ryskowski, L. (1965). Settling of mice (*Mus musculus*) released in an uninhabited and inhabited place. *Bull. Acad. pol. Sci.* (Biol.) **13**: 631–637.

Anderson, P. K. (1961). Density, social structure and non-social environment in house mouse populations and the implication for the regulation of numbers. *Trans. N. Y. Acad. Sci.* (2) **23**: 447–451.

Anderson, P. K. (1964). Lethal alleles in *Mus musculus*: local distribution and evidence for isolation of demes. *Science, N. Y.*, **145**: 177–178.

Antalfi, S. (1963). Biological determination of the intensity of the alarm reaction in the house mouse. *J. comp. Physiol. Psychol.* **56**: 889–891.

Banerjee, V. (1972). Somatic, physiological and behavioural effects of prolonged isolation in male mice and behavioural response to treatment. *Physiol. Behav.* **9**: 63–67.

Barnett, S. A. & Cowan, P. E. (1976). Activity, exploration, curiosity and fear – an ethological study. *Interdiscipl. Sci. Rev.* **1**: 43–62.

Beeman, E. A. (1947). The effect of male hormone on aggressive behaviour in mice. *Physiol. Zool.* **20**: 373–405.

Beilharz, R. G. (1975). The aggressive response of male mice (*Mus musculus* L.) to a variety of stimulus animals. *Z. Tierpsychol.* **39**: 141–149.

Beilharz, R. G. & Beilharz, V. C. (1975). Observations on fighting behaviour of male mice. *Z. Tierpsychol.* **39**: 126–140.

Benton, D., Goldsmith, J. F., Brain, P. F. & Hucklebridge, F. (1978). Adrenal activity in isolated mice of different social status. *Physiol. Behav.* **20**: 459–464.

Berry, R. J. (1970). The natural history of the house mouse. *Fld Stud.* **3**: 219–262.

Blick, N., Gross, C. M. & Ropartz, P. (1971). Effect of density of grouping on aggression and reproductive function of female mice. *C. r. hebd. Séanc. Acad. Sci., Paris* **272**: 293–296.

Bowers, J. M. & Alexander, B. K. (1967). Mice: individual recognition by olfactory cues. *Science, N. Y.*, **158**: 1208–1210.

Brain, P. F., Nowell, N. W. & Wouters, A. (1971). Some relationships between adrenal function and the effectiveness of a period of isolation in inducing inter-male aggression in albino mice. *Physiol. & Behav.* **6**: 27–29.

Brown, R. Z. (1953). Social behavior, reproduction and population changes in the House mouse. *Ecol. Monogr.* **23**: 217–240.

Bruel, J. H. (1964). Inheritance of behavioral and physiological characters of mice and the problem of heterosis. *Am. Zool.* **4**: 125–138.

Cairns, R. B. & Nakelski, J. S. (1971). On fighting in mice: ontogenetic and experiential determinants. *J. comp. Physiol. Psychol.* **74**: 354–364.

Calhoun, J. B. (1956). A comparative study of the social behaviour of two inbred strains of house mouse. *Ecol. Monogr.* **26**: 81–103.

Clarke, L. H. & Schein, W., (1966). Activities associated with conflict behaviour in mice. *Anim. Behav.* **14**: 44–49.

Clevedon-Brown, J. & Twigg, G. I. (1969). Studies on the pelvis in British Muridae and Cricetidae (Rodentia). *J. Zool., Lond.* **158**: 81–132.

Collins, R. L. (1964). Inheritance of avoidance conditioning in mice. *Science, N. Y.* **143**: 1188–1190.

Crawley, J. N., Schleitt, W. M. & Contera, J. F. (1975). Does social environment decrease propensity to fight in male mice? *Behav. Biol.* **15**: 73–83.

Crowcroft, P. (1954). Mouse research in Suffolk. *Trans. Suffolk Nat. Soc.* **8**: 185–187.

Crowcroft, P. (1955a). Territoriality in wild house mice (*Mus musculus* L.). *J. Mammal.* **36**: 299–301.
Crowcroft, P. (1955b). Social organization in wild mouse colonies. *Br. J. Anim. Behav.* **3**: 36.
Crowcroft, P. (1966). *Mice all over.* London: Foulis.
Crowcroft, P. & Rowe, F. P. (1957). The growth of confined colonies of the wild house mouse (*Mus musculus*). *Proc. zool. Soc. Lond.* **129**: 359–370.
Crowcroft, P. & Rowe, F. P. (1961). The weights of wild house mice (*Mus musculus* L.) living in confined colonies. *Proc. zool. Soc. Lond.* **136**: 177–185.
Crowcroft, P. & Rowe, F. P. (1963). Social organization and territorial behaviour in the wild house mouse (*Mus musculus* L.). *Proc. zool. Soc. Lond.* **140**: 517–531.
Davis, D. E. (1958). The role of density on aggressive behaviour of house mice *Mus musculus. Anim. Behav.* **4**: 207–210.
Dawkins, M. & Dawkins, R. (1976). Hierarchical organization and postural facilitation: rules for grooming in flies. *Anim. Behav.* **24**: 739–755.
De Ghett, B. J. (1975). A factor influencing aggression in adult mice: witnessing aggression when young. *Behav. Biol.* **13**: 291–300.
Denenberg, V. H. (1965). Behavioural differences in two closely-related lines of mice. *J. Genet. Psychol.* **106**: 201–205.
Denenberg, V. H., Sherman, R. & Blumenfield, M. (1963). Behaviour differences between mutant and nonmutant mice. *J. comp. Physiol. Psychol.* **56**: 290–293.
Desjardins, C., Maruniak, J. A. & Bronson, F. H. (1973). Social rank in the house mouse: differentiation revealed by ultraviolet visualization of urinary marking patterns. *Science, N. Y.* **182**: 939–941.
Dews, P. B. (1953). The measurement of the influence of drugs on voluntary activity in mice. *Br. J. Pharmacol.* **8**: 46–48.
Dyer, D. P. & Southwick, C. H. (1974). A possible sensitive period for juvenile socialization in mice. *Behav. Biol.* **12**: 551–558.
Ebert, P. D. (1976). Agonistic behaviour in wild and inbred strains of *Mus musculus. Behav. Biol.* **18**: 291–294.
Eibl-Eibesfeldt, I. (1958). Das Verhalten der Nagerntiere. *Handb. Zool. Berl.* 8 lf12, 10(13): 1–88.
Eleftheriou, B. E., Bailey, D. W. & Denenberg, V. H. (1974). Genetic analysis of fighting behaviour in mice. *Physiol. & Behav.* **13**: 773–777.
Essmann, W. B. (1966). The development of activity differences in isolated and aggregated mice. *Anim. Behav.* **14**: 406–409.
Estep, D. W., Lanier, D. L. & Dewsbury, D. A. (1975). Copulatory behaviour and nest building behaviour of wild house mice (*Mus musculus*). *Anim. Learn. Behav.* 3: 329–336.
Evans, C. M. & Brain, P. R. (1978). Attempts to influence fighting and threat behaviour in adult isolated female CFW mice in standard opponent aggression tests using injected and subcutaneously implanted androgens. *Physiol. & Behav.* **14**: 551–556.
Evans, C. M. & Mackintosh, J. H. (1976). Endocrine correlates of territorial and subordinate behaviour in groups of male CFW mice under semi-natural conditions. *J. Endocr.* **71**: 91.
Fraser, D. G. & Barnett, S. A. (1975). Effects of pregnancy on parental and other activities of laboratory mice. *Horm. Behav.* **6**: 181–188.

Fujimoto, K. (1953). Diurnal activity of mice in relation to social order. *Seiro-Seitae* **5**: 97.

Fuller, J. L. (1967). Effects of the albino gene upon behaviour of mice. *Anim. Behav.* **15**: 467–470.

Gandelmann, R. (1972). Mice: post-partum aggression elicited by the presence of an intruder. *Horm. Behav.* **3**: 23–28.

Gandelmann, R. & Svare, B. (1974). Mice: pregnancy termination, lactation and aggression. *Horm. Behav.* **5**: 397–405.

Garcia-Segura, L. M. (1977). Effects of the sensory milieu on the exploratory behaviour of adult mice. *Arch. Neurobiol.* **40**: 307–312.

Gilder, P. M. & Slater, P. J. B. (1978). Interest of mice in conspecific male odours is influenced by degree of kinship. *Nature, Lond.* **274**: 364–365.

Goldsmith, J. F., Brain, P. F. & Benton, D. (1976). Effects of age at differential housing and the duration of individual housing/grouping on intermale fighting behaviour and adrenocortical activity in TO strain mice. *Aggress. Behav.* **2**: 307–323.

Goyens, J. & Noirot, E. (1974). Effects of cohabitation with females on aggressive behaviour in male mice. *Devl Psycho-biol.* **8**: 39–84.

Grant, E. C. & Mackintosh, J. H. (1963). A comparison of the social posture of some common laboratory rodents. *Behaviour* **21**: 246–259.

Gutmann, R., Lieblich, I. & Gross, R. (1975). Behavioural correlates of oestrous cycle stages in laboratory mice. *Behav. Biol.* **13**: 127–132.

Harrington, J. E. (1976). Recognition of territorial boundaries by olfactory cues in mice. *Z. Tierpsychol.* **41**: 295–306.

Haug, M. (1972). Aggression phenomenon related to the introduction of a virgin or strange female into a group of female mice. *C. r. hebd. Séanc. Acad. Sci. Paris* **275**: 2729–2732.

Haug, M. (1973). Demonstration of an odour linked to lactation, and stimulating aggressive behaviour in a group of female mice. *C. r. hebd. Séanc. Acad. Sci. Paris* **275**: 3457–3460.

Henry, K. R. & Essinger, K. (1967). Effects of the albino and dilute loci on mouse behaviour. *J. comp. Physiol. Psychol.* **63**: 320–323.

Hyde, J. S. & Sawyer, T. F. (1977). Oestrous cycle fluctuations in aggression of house mice. *Horm. Behav.* **9**: 290–295.

Jakob, J. & Claret, B. (1967). Experimental study of some parameters of habituation to a new situation in the mouse. *Therapie* **22**: 781–792.

Kahn, M. W. (1954). Infantile experience and mature aggressive behaviour in mice – some maternal influences. *J. genet. Psychol.* **84**: 65–77.

Kalkowski, W. (1967). Olfactory bases of social orientation in the white mouse. *Folia biol., Kraków* **15**: 69–87.

Kalkowski, W. (1968). Visual control of social environment in the white mouse. *Folia biol., Kraków* **16**: 215–233.

Keeley, K. (1962). Prenatal influence on behavior of offspring of crowded mice. *Science, N. Y.* **135**: 44–45.

King, J. A. & Gurney, N. L. (1954). Effect of early social experience on adult aggressive behaviour in C57BL/10 mice. *J. comp. Physiol. Psychol.* **47**: 326–330.

Krsiak, M. & Borgosova, M. (1973). Aggression and timidity induced in mice by isolation. *Activ. Nerv. (Praha)* Suppl. **15**: 21–22.

Krsiak, M. & Janku, I. (1968). The development of aggressive behaviour in mice by isolation. In *Aggressive behaviour* **1968**: 101–105. Garattini, S. & Sigg, E. B. (Eds). Amsterdam: Excerpta Med. Foundation.

Land, R. B. & McGill, T. E. (1967). The effects of the mating pattern of the mouse on the formation of corpora lutea. *J. Reprod. Fert.* **13**: 121–125.
Levine, L., Diakow, C. A. & Barsel, G. E. (1965). Inter-strain fighting in male mice. *Anim. Behav.* **13**: 52–58.
Lidicker, W. Z. (1976). Social behaviour and density regulation in the house mouse living in large enclosures. *J. Anim. Ecol.* **45**: 677–697.
Lloyd, J. A. (1975). Social structure and reproduction in two freely-growing populations of house mice (*Mus musculus* L.). *Anim. Behav.* **23**: 413–424
Lloyd, J. A. & Christian, J. J. (1967). Relationship of activity and aggression to density in two confined populations of house mouse (*Mus musculus*). *J. Mammal.* **48**: 262–269.
Lloyd, J. A. & Christian, J. J. (1969). Reproductive activity of individual females: three experimental freely-growing populations of house mouse (*Mus musculus* L.). *J. Mammal.* **50**: 49–59.
Mackintosh, J. H. (1965). The behaviour of small mammals. *J. Anim. Techn. Assoc.* **15**: 1–3.
Mackintosh, J. H. (1970). Territory formation by laboratory mice. *Anim. Behav.* **18**: 177–183.
Mackintosh, J. H. (1973). Factors affecting the recognition of territory boundaries by mice (*Mus musculus*). *Anim. Behav.* **21**: 464–470.
Mackintosh, J. H. (1978). The experimental analysis of overcrowding. In *Population control by social behaviour*: 157–180. Ebling, F. J. & Stoddart, D. M. (Eds). London: Institute of Biology.
Mackintosh, J. H., Chance, M. R. A. & Silverman, A. P. (1977). The contribution of ethological techniques to the study of drug effects. In *Handbook of psychopharmacology*: 3–35. Iverson, L. L., Iverson, S. D. & Snyder, S. H. (Eds). London: Plenum.
Manosevitz, M. (1967). Hoarding and inbred strains of mice. *J. comp. Physiol. Psychol.* **63**: 148–150.
McClearn, G. E. (1960). Strain differences in activity of mice. Influence of illumination. *J. comp. Physiol. Psychol.* **53**: 142–143.
McGill, T. E. (1962). Sexual behaviour in three inbred strains of mice. *Behaviour* **19**: 341–350.
McGill, T. E. (1963). Sexual behaviour of the mouse after long term and short term post-ejaculatory recovery periods. *J. genet. Psychol.* **103**: 53–57.
McGill, T. E. (1967). A double replication of a "small sample" study of the sexual behaviour of DBA/2J male mice. *Anat. Rec.* **157**: 151–153.
McGill, T. E. (1969). An enlarged study of genotype and recovery of sex drive in male mice. *Psychol. Sci.* **15**: 250–251.
McGill, T. E., Cowen, D. M. & Harrison, D. T. (1968). Copulatory plug does not produce luteal activity in mouse, *Mus musculus. J. Reprod. Fert.* **15**: 149–151.
McGill, T. E. & Ransome, T. W. (1968). Genotypic change affecting conclusions regarding the role of inheritance of elements of behaviour. *Anim. Behav.* **16**: 88–91.
McGill, T. E. & Tucker, G. R. (1964). Genotype and sex drive in intact and castrated male mice. *Science, N. Y.* **145**: 514–515.
McKinney, T. D. & Desjardins, C. (1973). Postnatal development of the testis and fighting behaviour and fertility in mice. *Biol. Reprod.* **9**: 279–294.
Namikas, J. & Wehmer, F. (1978). Gender, composition of litter affects behaviour of male mice. *Behav. Biol.* **23**: 219–224.

Newton, N., Foshee, D. & Newton, M. (1966). Parturient mice: effect of environment on labour. *Science, N. Y.* **151**: 1560–1561.

Noirot, E. (1958). Analyse du comportement dit maternel chez la souris. *Monogr. Fr. Psychol.* **1**.

Noirot, E. (1964a). Changes in responsiveness to young in the adult mouse: the effect of external stimuli. *J. comp. Physiol. Psychol.* **57**: 97–99.

Noirot, E. (1964b). Changes in responsiveness to young in the adult mouse: the effect of initial contact with strong stimulus. *Anim. Behav.* **12**: 442–445.

Noirot, E. (1965). Changes in responsiveness to young in the adult mouse: the effect of immediately preceding performances. *Behaviour* **24**: 318–325.

Noirot, E. (1969a). Changes in responsiveness to young in the adult mouse: priming. *Anim. Behav.* **17**: 542–546.

Noirot, E. (1969b). Serial order of maternal response in mice. *Anim. Behav.* **17**: 547–550.

Noirot, E. & Goyens, J. (1971). Changes in maternal behaviour during gestation in the mouse. *Horm. Behav.* **2**: 207–215.

Noirot, E., Goyens, J. & Buhot, M. C. (1976). Aggressive behaviour of pregnant mice toward males. *Horm. Behav.* **6**: 9–17.

Oortmerssen, G. A. Van (1971). Biological significance, genetics and evolutionary origin of variability in behaviour with and between inbred strains of mice. *Behaviour* **38**: 1–92.

Poirel, C. (1968). Variations temporelles du comportement d'exploration chez la souris. *C. r. Séanc. Biol.* **162**: 2312–2316.

Poole, T. E. & Morgan, H. D. R. (1973). Differences in aggressive behaviour between male mice (*Mus musculus* L.) in colonies of different sizes. *Anim. Behav.* **21**: 788–795.

Poole, T. E. & Morgan, H. D. R. (1975). Aggressive behaviour of male mice (*Mus musculus*) towards familiar and unfamiliar opponents. *Anim. Behav.* **23**: 470–479.

Priestnall, R. (1975). The effects of litter size and post-weaning isolation or grouping on adult emotionality in C3H mice. *Devl. Psychobiol.* **6**: 217–244.

Randt, C. T., Blizard, C. D. & Friedman, E. (1975). Early life undernutrition and aggression in two mouse strains. *Devl. Psychobiol.* **8**: 275–279.

Reimer, G. & Petras, M. L. (1967). Breeding structure of the house mouse (*Mus musculus*) in a population cage. *J. Mammal.* **48**: 88–99.

Ressler, R. H. (1962). Parental handling of two strains of mice reared by foster parents. *Science, N. Y.* **137**: 129–130.

Retzlaff, E. G. (1938). Studies in population physiology with albino mice. *Biologia gen.* **14**: 238–265.

Rosensen, L. M. & Asheroff, A. K. (1975). Maternal aggression in CD1 mice: influence of the hormonal condition of the intruder. *Behav. Biol.* **15**: 219–224.

Rowe, F. P., Smith, F. J. & Swinney, T. (1974). Field trials of Calciferol combined with Warfarin against wild house mice (*Mus musculus* L.). *J. Hyg., Camb.* **73**: 353–478.

Rowe, F. P., Taylor, E. J. & Chudley, A. H. J. (1964). The effect of crowding on the reproduction of the house mouse (*Mus musculus* L.) living in corn-ricks. *J. Anim. Ecol.* **33**: 477–483.

Ryan, V. & Wehmer, F. (1975). Effect of post natal litter size on adult aggression in the laboratory mouse. *Devl. Psychobiol.* **8**: 363–370.

St John, R. C. & Corning, A. (1973). Maternal aggression in mice. *Behav. Biol.* **9**: 635–639.

Schmidt, C. (1978). Olfactory threshold and its dependence on the sexual status of the female laboratory mouse. *Naturwissenschaften.* **65**: 601.

Scott, J. P. (1940). Heredity differences in social behavior. Fighting of males in 2 inbred strains of mice. *Anat. Rec.* **78** (Suppl.), 103 (Abstract).

Scott, J. P. (1945). Experimental modification of aggressive and defensive behaviour in the C57 inbred strain. *Genetics* **30**: 21 (Abstract).

Scott, J. W. & Pfaff, D. W. (1970). Behavioral and electrophysiological responses of female mice to male urine odours. *Physiol. Behav.* **5**: 407–411.

Slater, P. J. B. (1978). Data collection. In *Quantitative ethology*: 7–24. Colgen, P. W. (Ed.). London: Wiley.

Smith, R. H. (1972). Wildness and domestication in *Mus musculus* – behavioral analysis. *J. comp. Physiol. Psychol.* **79**: 22–29.

Smith, W. I. & Ross, S. (1951). Avitaminosis and social dominance in male C3H mice. *Physiol. Zool.* **24**: 238–241.

Southwick, C. H. (1955a). The population dynamics of confined house mice supplied with unlimited food. *Ecology* **36**: 212–225.

Southwick, C. H. (1955b). Regulatory mechanisms of house mouse populations: social behavior affecting litter survival. *Ecology* **36**: 627–634.

Southwick, C. H. (1958). Population characteristics of house mice living in English corn ricks: density relationships. *Proc. zool. Soc. Lond.* **131**: 163–175.

Tedeschi, R. E., Tedeschi, D. H., Mucha, A., Cook, L., Mattis, P. A. & Fellows, E. J. (1959). Effects of various centrally acting drugs on the fighting behaviour of mice. *J. Pharm. exp. Therap.* **125**: 28–34.

Thiessen, D. D. (1963). Varying sensitivity of C57 BL/CRY1 mice to grouping. *Science, N. Y.* **141**: 827–828.

Tolman, J. & King, J. A. (1956). Effects of testosterone propionate on aggression in male and female mice. *Br. J. Anim. Behav.* **4**: 147–149.

Tyack, B. A. (In preparation). *Territorial behaviour of female mice.*

Uhrich, J. (1938). Social hierarchy in white mice. *J. comp. Psychol.* **25**: 373–413.

Valzelli, L., Vernasconi, S. & Gomber, P. (1974). Effect of isolation on some behavioural characteristics in three strains of mice. *Biol. Psychiat.* **9**: 329–333.

Warne, M. C. (1947). A time analysis of certain aspects of the behaviour of small groups of caged mice. *J. comp. Physiol. Psychol.* **40**: 371–387.

Weipkema, P. R. (1968). Behaviour changes in CBA mice as a result of one gold thioglucose injection. *Behaviour* **32**: 179–210.

Weipkema, P. R. (1971). Behavioural factors in the regulation of food intake. *Proc. nutr. Soc.* **30**: 142–149.

Yanai, J. & McClearn, G. E. (1972). Assortative mating in mice. I. Female mating preference. *Behav. Genet.* **2**: 173–183.

Symp. zool. Soc. Lond. (1981) No. 47, 367–393

Senses and Communication

JANE C. SMITH

Rodent Research Branch, Tolworth Laboratory, Ministy of Agriculture, Fisheries and Food, Tolworth, Surbiton, Surrey, UK

SYNOPSIS

A review is made of current knowledge of mouse senses and communication, beginning with the role of the senses in mother–infant communication, how this role changes with the developing capabilities of the pups, and the effects of sensory environment on development in infant and juvenile mice. The factors involved in the recognition of juvenile mice are then described. Adult communication is broken down by sense: chemical, acoustic, and then the other senses. In each section a brief description of the reception and production of sensory signals is followed by a summary of evidence for communication. In conclusion, the constraints on communication in free-living mice are considered.

INTRODUCTION

The mouse is a mammal and apparently has the usual complement of mammalian senses: vision, hearing, smell, taste, touch, and heat and pain sensitivity. The mouse is also small and mainly nocturnal in habit. Because of its small size, the mouse has not been chosen as the type species for the investigation of mammalian senses as often as its relative the rat. Because the mouse is mainly nocturnal and until recently was thought to be rather non-vocal, it has been thought of as a primarily olfactory animal; and the wealth of knowledge about olfactory influences on behaviour and physiology in this species have made olfactory communication in mice a rewarding area for study.

In the last 20 years, however, it has also been found that mice and other murid and cricetid rodents, which are all small in size, use acoustic communication at frequencies above our hearing range, i.e. ultrasounds (reviewed by Sales & Pye, 1974). This has opened up an exciting new area of mouse communication. The remaining senses have received less study than they deserve, although they appear to have relatively minor roles in mouse communication.

MOTHER–INFANT COMMUNICATION AND SENSORY DEVELOPMENT

In mice, as in other mammals, the relationship of the young with their mother is vital for their successful physical and behavioural development. The initiation and maintenance of this relationship depends to a large extent on communication, and this may be by visual, auditory, olfactory, thermal or tactual means.

Baby mice are born in a relatively undeveloped state. They are naked, unable to eliminate waste products unaided, the eyes and ears are closed and they depend on the mother not only for food but also for warmth as they cannot maintain their own body temperatures. Since baby mice are also unable to crawl at this stage they rely on the mother coming to them to give them maternal care. As the young develop they become progressively more active and approach responses may be made by both mother and pups. During the last stages of development the pups readily approach the mother to nurse but she begins to reject them as they become independent.

Infant Cues and Maternal Behaviour

In mice, although maternal hormones appear to be responsible for the onset of maternal behaviour at parturition, there is strong evidence that stimuli from the pups play an important part in eliciting and/or maintaining maternal behaviour in lactating females either directly or through the hormonal system. Repeated exposure to pups will also induce "maternal" responses in virgin females and in males who may otherwise be aggressive towards pups to the extent of killing them (Noirot, 1964a, b). Stimuli from the pups are also thought to be responsible for protecting them from the post-partum aggression of the mother who often attacks other strange mice near the nest.

The cues from the pups which influence maternal behaviour are thought to be primarily olfactory and acoustic. The exact source of pup odours has not yet been established but may include birth fluids and pup gut effluvia. Olfactory cues appear to play a part in pup recognition by the mother and in her subsequent retrieving, nest building and suckling responses (Noirot, 1970). Gandelman, Zarrow & Denenberg (1971) have shown that olfactory bulbectomy will result in cannibalism of young by lactating or "maternal" virgin mice. Young (day 3) pups are much more likely to be killed than day 14 ones; nest building behaviour is also abolished by olfactory bulbectomy. Smotherman, Bell *et al.* (1974) found that olfactory cues from pups played a part in eliciting maternal approach responses to pups in a Y maze.

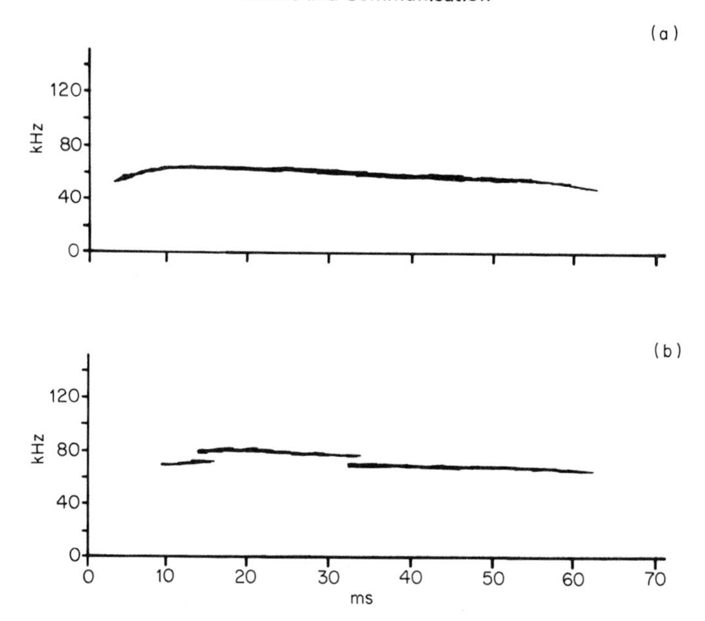

FIG. 1. (a) Traced sonagram of a call produced by a one-day-old mouse.
(b) Traced sonagram of a call produced by a seven-day-old mouse.

Acoustic cues from the pups are also important communication signals. Newborn mouse pups, like other murid rodent species, produce ultrasonic distress calls (Fig. 1a) in response to various adverse environmental conditions (Sales & Smith, 1978). A drop in ambient temperature below about 33°C, which may be experienced when the mother is absent from the nest for a while or if a pup is displaced from the nest, stimulates ultrasonic calling in pups which have not developed homoiothermy (Okon, 1970) (Fig. 2). Once homoiothermy is acquired, ultrasonic responses to cold disappear.

As well as the cold response, changes in the olfactory environment affect ultrasonic calling by mouse pups. Generally the home smell has a quieting effect on pups. Mild tactile stimulation provokes loud ultrasonic calls, particularly marked in newborn pups, but persisting in older pups after responses to cold have waned. The calls of older pups (Fig. 1b) differ in structure from those of the newborn (Sales & Smith, 1978); these may be related to the onset of adult social response patterns, rather than distress calls directed at the mother. Audible squeaks are only produced in response to strong tactile stimulation, and such calls may form a strong "contact desist" signal. The signal may only be used as a last resort as it would be more easily detected by predators than ultrasonic cries.

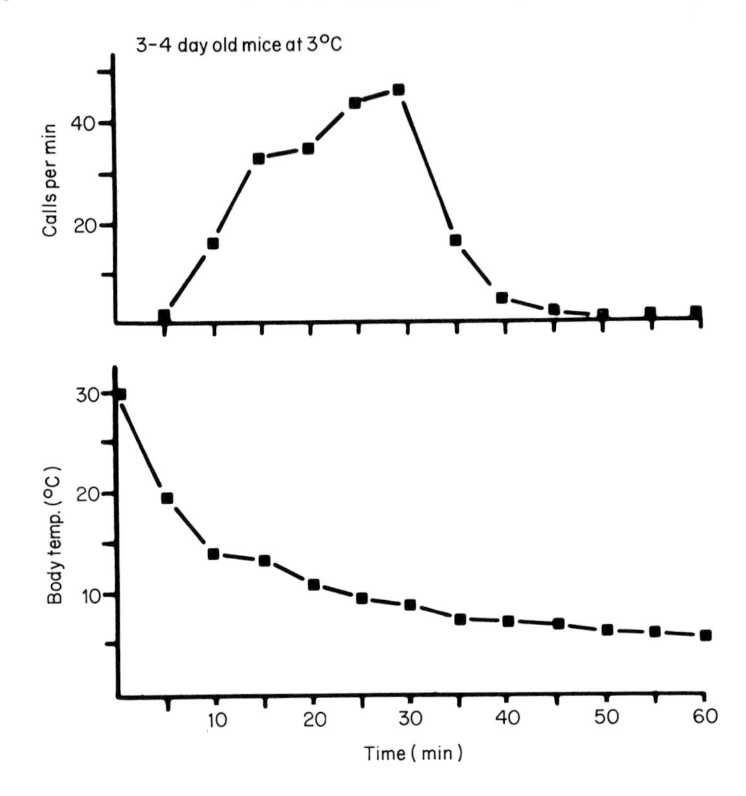

FIG. 2. Changes in ultrasonic calling and of body temperature in mice exposed for one hour at 3°C.

The ultrasonic calls of mouse pups can be shown to exert a powerful influence on maternal behaviour in the laboratory: tape-recordings of pup calls will elicit directed approach to the loudspeaker from lactating TO Swiss and wild mice (Smith, 1974), although C57/BL10 mice appear to be less responsive (Smotherman, Bell *et al.*, 1974). It is not known, however, over what range such a response would be possible in free-living mice, as ultrasounds are subject to severe attenuation with distance and in a cluttered environment (Smith, 1979).

There is circumstantial evidence that pup ultrasounds exert effects on other aspects of maternal behaviour than directing approach (Noirot, 1970, 1974). It has been suggested that rodent pup ultrasounds may play a part in mediating early experience or handling effects through their effect on maternal behaviour (Barnett & Walker, 1974; Bell *et al.*, 1974) while Smotherman, Wiener *et al.* (1977) have put forward the idea that, at least in rats, ultrasounds might mediate an elevated corticosterone response in mothers as a response to experimental treatment of their pups. Cues from rat pups also induce high and sustained levels of prolactin in their mothers (Leon, 1974) which are responsible not only for such effects as milk let-down but the production of a large amount of caecotrophe in the gut. This substance, when voded, has an attractive effect on the pups as a so-called "maternal pheromone" (p. 372). It is possible that similar effects exist in mice.

Although it is possible to demonstrate the effects of individual pup cues on maternal behaviour and physiology, it is likely that many aspects of maternal behaviour in mice are under multisensory control as it is also often possible to demonstrate normal maternal responses in the laboratory in the absence of one particular sensory cue. For instance, Busnel & Lehmann (1977) failed to show increased cannibalism of pups by their mothers in deaf mice (GFF *dn/dn*) compared with mice with normal hearing (GFF +/+), and mothers in the two groups were able to retrieve scattered young with similar latencies over short distances when other cues (visual and olfactory) were available. The fact that it is possible to cross foster mice and other species (e.g. Quadagno & Banks, 1970) supports the argument that there cannot be very tight stimulus control of maternal behaviour.

The importance of visual and tactual cues for mother-infant interactions has received less attention that that of auditory and olfactory stimuli. Experiments using tape-recorder play-back (Smith, 1974; Smotherman, Bell *et al.*, 1974) suggest that visual cues are not essential for eliciting maternal approach although they may normally be used, especially by deaf mothers, while tactual cues may operate at close range. For example, a mouse mother will show maternal care to pups from other litters if they are of fairly similar age to her own; the presence of body fur on a pup will elicit attack from a mother whose own pups are still naked, but the same pups shaved are apparently accepted. This response does not occur, however, when the mother's own pups are furry so presumably her sensory concept of "pup" changes with the development of her own litter (Sayler & Salmon, 1971).

Sensory Development and Infant Responses

Newborn mouse pups appear only to respond to thermal, tactual and probably olfactory and gustatory cues. These senses appear to be used in seeking out the mother's body and nipples in most altricial mammals (Rosenblatt, 1976). The vibrissae on the face of a baby mouse are present at birth, before the tactile guard hairs on the rest of the body which erupt about three days after birth. The olfactory system of the newborn rodent is anatomically relatively immature, and there is evidence that the development of the central olfactory structures is somewhat uneven (Alberts, 1976). Behavioural evidence for rats indicates that a number of airborne chemical stimuli can be perceived as early as day 2 (e.g. Schapiro & Salas, 1970). Again in rats, spontaneous gross electrical activity of the olfactory bulb cannot be detected in animals younger than three days and adult patterns of activity are not acquired until about day 12–15 (Salas, Guzman-Flores & Schapiro, 1969). Maternal odours produce changes in the electrical activity of the olfactory system by nine to 12 days; whereas food odours are not effective until day 21 (Salas, Schapiro & Guzman-Flores, 1970). It is thought that chemical sensitivity may be mediated by the trigeminal nerve in pups younger than three days. Although less work has been done on the mouse, it has been shown that pups with olfactory bulbectomies are slow to reach and attach to the nipples, and this results in growth retardation (Cooper & Cowley, 1976). Thermal, tactual and olfactory cues also trigger the ultrasonic distress response which is present on the first day.

Once the pups' motor abilities become sufficiently developed, they show pronounced aggregation behaviour, responding to thermal gradients (G. D. Sales, personal communication) and presumably also to olfactory and tactual cues from the nest and their sibs. This means that the pups tend to form a huddle with a dynamic structure in which the pups on the outside continually burrow their way back into the centre. This behavior is adaptive as huddled pups lose heat less quickly than isolated ones (Sales & Skinner, 1979).

It has also been shown that 18-day-old laboratory mouse pups are attracted to the odour of lactating females (Breen & Leshner, 1977). Part of the attractive odour appeared to come from the faeces, but greater attraction occurred when intact females were present behind a screen which allowed olfactory and, presumably, auditory cues to pass but no visual or tactual contact. This response is similar to that shown by laboratory rats which have been studied in greater detail. In rats, the "maternal pheromone" appears to be present in caecotrophe from the female, and the attractiveness of the female and the response of the pups both show a peak at 16 days post-partum (Leon, 1974). It seems likely that such a communication system could serve to keep the litter close to the nest site and the mother as they become mobile and the onus for maintaining mother–infant contact passes from mother to young.

There is evidence that the odour of the parents may affect the later odour-mediated social and sexual responses of mouse pups. Mainardi, Marsan & Pasquali (1965) reared mice with parents swabbed in perfume. The perfume-raised males showed no differences from control males in their later sexual preference for perfumed or normal female mice, whereas the perfume-raised females differed from control females by showing less aversion to perfumed males. Cross-fostering experiments (e.g. Quadagno & Banks, 1970) indicate changes in later social preference of male and female mice towards the species that reared them, again the greatest effect occurring in females. Sexual behaviour was less strongly affected.

The presence of adult males, or females other than the mother, in the rearing cage has been found to change the speed of sexual development, and possibly the overall growth and development of female mouse pups (e.g. Vandenbergh, 1967; Fullerton & Cowley, 1971). Female pups show accelerated development in the presence of males and retarded development in the presence of several females. The effect is partly odour-mediated as urine painted on the noses of pups has similar effects to the presence of whole animals (Cowley & Wise, 1972). Drickamer (1974), however, showed that an intact male was more effective than his urine alone because of the additional accelerating effects of tactile contact between the male and the pups.

The external ears of mouse pups unfold at four days of age. Anatomically, the development of the cochlea is mainly finished at day 8 (Mikaelian & Ruben, 1965) but growth is still taking place in the Organ of Corti up to day 18 (Ruben, 1967). Physiologically, cochlear potentials in a middle frequency range are present on day 8 and the auditory nerve starts functioning adequately after day 10. Electrical activity in the auditory nerve increases up to day 14 (Mikaelian & Ruben, 1965), probably as a result of continuing growth and maturation. Ehret (1976) has studied the development of absolute auditory thresholds in NMRI mice using unconditioned reflexes. Behavioural reactions to tones appear first on day 10 in the frequency range of greatest sensitivity in the adults (10–20 kHz). The low frequency peak at 15 kHz becomes evident on day 14; increases in sensitivities below 15 kHz continue to day 15, and above 15 kHz to day 18. The high frequency peak at 50 kHz does not reach adult levels until two to three months of age (Fig. 3).

This means that very young pups cannot use the cries of their mothers or of their sibs as directional cues, but I have found that 18-day-old pups respond to tape-recordings of the calls of their sibs by increasing their own call rate and approaching the loudspeaker. This communication channel could be used to keep a growing litter together.

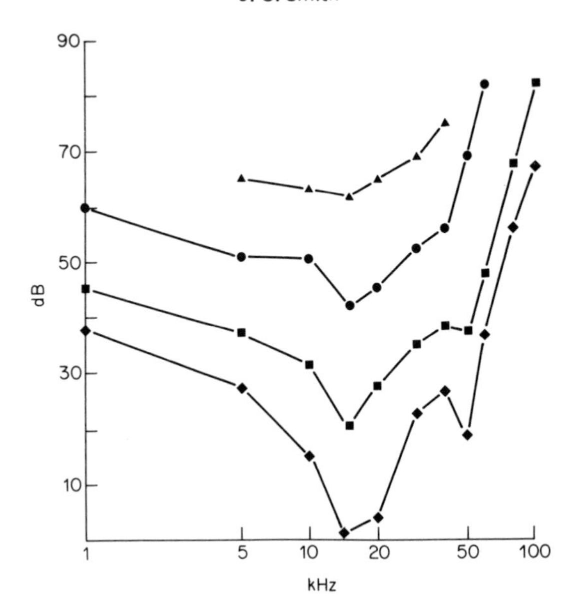

FIG. 3. Auditory threshold curves of young mice. ▲——▲ 11 days unconditioned stop; ●——● 14 days pinna reflex; ■——■ 17–19 days conditioning; ◆——◆ 2–3 months conditioning (redrawn after Ehret, 1976)

The structure of the adult retina is established by about the twelfth day after birth, just before the eyes open. The eye is not fully mature, however, as the hyaloid artery which supplies the embryonic lens is present until three weeks after birth and the internal organization of the rods is not complete until about four weeks of age (Sorsby *et al.*, 1954; Fuller & Wimer, 1966). Thus the two- to three-week-old mouse has two new sensory channels progressively opening to him, and this may be associated with an acceleration in the development of social responses during this period. However, the extent to which these senses are used by young mice remains to be established.

COMMUNICATION IN JUVENILES

Senses and Development

The previous section indicated that young mice do not complete their sensory development until after they would normally begin to lead independent lives. During the juvenile period, body growth and sexual development are also completed. The age of attainment of sexual maturity varies according to the social environment. The effect of explosure to males during the pre-weaning period on the sexual development of females has already been mentioned, and similar effects follow post-weaning exposure.

Vandenbergh (1967) found that placing a male with groups of immature females advanced vaginal opening and first oestrus compared with controls. Grouping of immature females, or exposure to female urine, delays the onset of puberty. Males behind a double mesh barrier, or bedding from male cages, also advance puberty, indicating an olfactory effect (Vandenbergh, 1969). The puberty-accelerating substance appears to be androgen-dependent since castrated males have no effect. Exposure of males of immature females releases LH and oestradiol in the same way as in mature females (Bronson, 1976).

The male odour occurs in urine, but the puberty-accelerating effect of the male is not entirely odour-mediated since tactile cues are also important. These can be supplied by a castrated male since a castrated male plus male urine is nearly as good as an intact male in promoting uterine growth, and far better than urine alone (Bronson, 1976).

The sexual development of juvenile male mice appears to be influenced by conspecific odours. Puberty occurs early in males housed individually with females (Vandenbergh, 1971), but exposure to females after males are 30 days of age apparently has no effect (Maruniak, Coquelin & Bronson, 1978). The LH elevation response of male mice to female urine develops at 24–36 days of age but falls off with repeated exposure, and exposure to female urine alone over long periods does not advance puberty. It therefore appears that, while female urine can advance puberty in males, the effect may not be by urinary stimulation of LH release (Maruniak *et al.*, 1978).

Recognition and Social Behaviour

It has been noticed during work on territorial behaviour of mice that whereas intruding adult males are attacked by the resident male, adult females and juveniles can cross territorial boundaries unmolested. Juvenile females appear to be protected from attack in the same way that adult females are, by an inhibitory factor in the urine. When urine from juvenile females is rubbed on the fur of male intruders, attack by resident males is reduced (Dixon & Mackintosh, 1976). Juvenile male mouse urine does not have this aggression-inhibiting property, however; it also lacks the androgen-dependent aggression-eliciting factor present in adult male urine. It appears that the lack of aggression shown by adult males to juvenile males may be merely because the latter do not promote aggression (Dixon & Mackintosh, 1976).

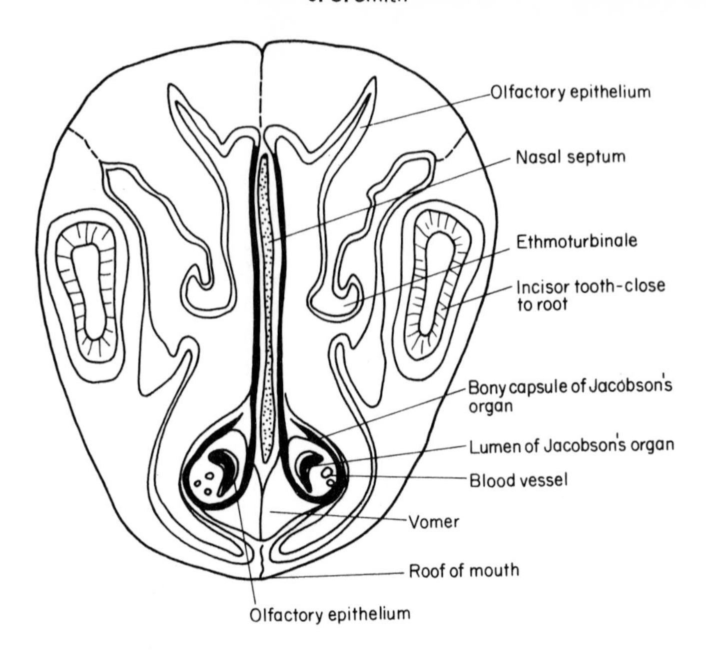

FIG. 4. Transverse section through the snout of an adult mouse (from Stoddart, 1976).

The role of other sensory cues in the social interactions of juveniles is unknown. Juveniles certainly present different visual cues from adults to other mice, and these may be used in recognition, especially at a distance. Acoustic communication is another possibility, but little work has been done on the acoustic behaviour of juvenile mice.

CHEMICAL COMMUNICATION IN ADULTS

Reception and Production

Adult mice have two main olfactory channels, the "primary" olfactory system and the vomero-nasal system. The primary olfactory system is composed of chemoreceptors in the olfactory mucosa, the first cranial nerve, and the main olfactory bulbs plus their central projections. The vomero-nasal system is composed of receptors in the mucosa of Jacobson's Organ (Fig. 4), the vomero-nasal nerve and the accessory olfactory bulbs which have central projections that appear not to overlap those of the primary olfactory system (Scalia & Winans, 1976). The trigeminal nerve which projects to the brainstem and spinal cord, and the terminal nerve from the nasal septum are also thought to contribute to olfactory sensation. Olfactory bulbectomy probably abolishes both channels, whereas zinc sulphate application to the snout is less likely to affect Jacobson's Organ than the primary olfactory system because the former is rather well enclosed. Differences between results obtained using these two methods may also be a result of either the central effects of bulbectomy or zinc sulphate poisoning.

Work with biologically significant odours on small rodents, including mice, indicates that both differences in the structure and chemical properties of the incoming olfactory stimuli and in the inhalation pattern which pulses olfactory information to the higher centres, are important in odour discrimination. The firing of units in the olfactory bulb in response to biologically significant odours has now been demonstrated for a number of rodents. The hypothalamus of both rat and mouse appears to show differentiation of conspecific scents according to functional groups, reflected in changes in overall firing rates of neurones. The hypothalamus appears to be involved in interactions between conspecific odours, hormones, and reproductive behaviour and not only may hypothalamic neurones discriminate natural odours but circulating hormones can apparently affect the activity of such neurones (Macrides, 1976).

The main sources of odour used in communication by mice appear to be the urine, preputial glands, coagulating glands and plantar glands. The odours are carried around on the bodies of the mice, particularly in the anogenital area, and they may also be deposited as scent marks. Marking takes the form of placing spots of urine around the edges of an enclosure or on conspicuous objects. The morphology of the male prepuce appears to be designed for easy marking, being relatively long with long brush-like hairs on the tip, and there is a reservoir inside the tip which holds urine. The ducts of the preputial glands open on the tip of the sheath (Bronson, 1976). Urine marking is much more frequent in males than in females. It appears to be promoted by changes in the environment or the presence of a strange mouse, and is under the influence of androgen levels and social rank. Marking is suppressed in subordinate animals (Bronson, 1976).

Odour, Reproductive Behaviour and Physiology

The effect of urinary odours on the reproductive system in mice have led to a classification into "primers" that affect neuroendocrine and endocrine activity and have no immediate effect on behaviour in the recipient, and "signallers" which have a more direct effect on the behaviour of the recipient.

It has been known for some time that urinary odours can act as primers which affect the reproductive state of female mice in a number of ways: inducing or inhibiting oestrus and ovulation; accelerating sexual maturity in young females and blocking implantation. These effects have been reviewed several times elsewhere (e.g. Bronson, 1971) and will be dealt with only briefly here.

The female mouse oestrous cycle is very much under olfactory control. Females caged together in small groups show suppressed oestrous cycling either by the induction of pseudopregnancy (the Lee-Boot effect: van der Lee & Boot, 1956); or, when grouped at greater densities, by the induction of prolonged anoestrus (Whitten, 1959). Reynolds & Keverne (1979) have recently shown that oestrus suppression is dependent on the integrity of the vomero-nasal organ, since grouped female mice in which vomero-nasals have been removed overcame oestrus suppression. Whitten (1956) found that the introduction of a stud male mouse immediately brought group-housed females into synchronous, short oestrous cycling (the Whitten effect). The male's urine was implicated by inducing the same effect with urine alone. Bladder urine is apparently as effective as voided urine so additions from the sex accessory glands appear not to be involved. The effectiveness of the urine is androgen-dependent, abolished by castration and reinstated by testosterone administration. As little as 0.1 ml male urine per day on the bedding is sufficient to induce oestrus (Bronson, 1974). The same odour may be responsible for the acceleration of sexual maturity in juvenile females (Vandenbergh, 1967, 1969) and probably works by stimulating the release of FSH and LH via the hypothalamic-pituitary-gonadal system.

The most curious olfactory priming effect is the Bruce effect. Removal of the stud male and introduction of a strange male will block the pregnancy of a recently inseminated female mouse by causing a failure of implantation and a return to the oestrous state (Bruce, 1959). Again male urine is implicated as the effect is abolished in females with olfactory bulbectomies and can be caused by exposure to strange male urine alone. The effect also implies that the female can discriminate between the odours of males as the odour of the male that inseminated her will not cause the Bruce effect. Recent evidence for voles, *Microtus agrestis*, indicates that the odour of the stud male is imprinted on the female memory within one hour of mating (Milligan, 1979). The odour of strange male urine apparently blocks normal corpus luteum formation (Dominic, 1970) and prolactin injections will override the effect (Dominic, 1969). The involvement of other reproductive hormones is not yet clear.

Apart from these priming effects male urine also acts as a signaller, having an attractive effect on female mice, increasing their activity, and hence the chance of male and female coming together. Intact male urine is preferred to that of castrates and the male's preputial gland has been implicated in this effect (Bronson, 1971). Similarly, female urine increases the activity of male mice and provokes the production of ultrasonic calls in socially experienced males. These calls are normally produced by the male in the initial stages of courtship behaviour (pp. 383–384). Male mouse urine elicits no ultrasonic calls and the urine of castrated subjects exerts an intermediate effect. At present it seems that the oestrous state of donor females does not affect the response. It is interesting in this respect that experienced male mice appear to use female vaginal tissue rather than urine to assess oestrous state. More mounts are made by males paired with dioestrous females smeared with oestrous female vaginal contents than by males paired with females smeared with oestrous female urine (Hayashi & Kimura, 1974).

The ultrasonic response of male mice to female urine differs from that to entire females in several important respects. Males vocalize to females from the first encounter and apparently without decrement, whereas the urine response only occurs after the males have had post-pubertal experience (not necessarily sexual) of females; the response improves with daily experience of females, to reach a maximum at three to eight days and subsequently wanes if experience with females is then withdrawn (Dizinno, Whitney & Nyby, 1978; Smith & Halls, unpublished). Dizinno and his colleagues therefore concluded that the urine response appears to be learned. Nyby and his co-workers (Nyby, Whitney *et al.*, 1978) found support for this idea by giving naïve males postpubertal experience with perfumed females and producing a similar ultrasonic response to the perfume. The effect was limited to postpubertal experience, as perfuming males' mothers did not elicit an ultrasonic response to the perfume in adult male mice. This work begs questions about the role of learning in odour-mediated effects on behaviour and physiology in mice.

It has also been found that female mouse urine can have a similar priming effect in males, as male urine has in females, i.e. eliciting the release of LH and gonadal steroids. However the LH-releasing power of female urine is independent of ovarian activity whereas the male urine effect is abolished by castration. Also the physiological response of males to female urine habituates rapidly (Macrides, Bartke & Dalterio, 1975; Maruniak & Bronson, 1976).

Odour, Individual Recognition, Social Status and Agonistic Behaviour

Odours play an important part in other aspects of mouse social behaviour. It appears that mice can recognize individuals on the basis of odour alone (Hahn & Simmel, 1968); indeed Kalkowski (1967) indicated that recognition was possible over distances as great as 17.5 cm. Ropartz (1977) has suggested that individual recognition could be based on plantar odour and a substance produced by the coagulating glands in grouped mice able to exchange tactile stimulation.

Bilateral olfactory bulbectomy decreases the social behaviour of mice whereas application of zinc sulphate to the nose does not (Ropartz, 1977). Possible reasons for this difference are discussed above.

The urine of dominant male mice differs from that of subordinates in quality (Rowe, peronal communicaiton) and in aversive potency. Jones & Nowell (1974a) found that male mice would tend to stay out of an area spotted with dominant male urine, preferring a clean area, whereas subordinate male urine had little effect on choice of area. In contrast, female mice spend more time in an area of dominant male urine than in a clean area, and subordinate male urine is less attractive to them (Jones & Nowell, 1974b). The aversion factor in dominant male urine for other males appears to be androgen-dependent, as urine from male castrates produces little aversion, and aversion is restored by treatment with testosterone (Jones & Nowell, 1973a). It has already been mentioned that dominant and subordinate mice have different urination patterns and these differences in behaviour and urine potency could be related to territorial demarcation (see below).

The role of the coagulating glands in the production of aversive odours by males is unclear. Ropartz (1977) found that coagulating gland odour from a group of mice produced increased adrenal size in isolated mice; and Jones & Nowell (1973b) found that coagulating gland secretion mixed with bladder urine resulted in aversion in non-dominant, group-housed, males by the choice of area method. However they also found that coagulating gland secretion rubbed on castrates reduced attack from trained fighter males (Jones & Nowell, 1973c).

The role of odour in signalling stress is also unclear. For example, Carr, Martorano & Krames (1970) observed that isolated males in a two-choice test spent less time sniffing a container vacated by an electrically shocked male than a container vacated by a control male. They related this to results in similar experiments in which isolated mice preferred victor's smell to defeated (and presumably therefore stressed) mouse smell, unless they were defeated mice offered the smell of the mouse which had just beaten them. How these results fit with Jones & Nowell's work on grouped mice, which apparently show aversion to dominant urine, is not clear because of differences in the experience of subjects and in the methods.

Strange male mice are attacked when introduced into the enclosure of another male (e.g. Mackintosh, 1970), whereas females and juveniles do not elicit the same response. This discrimination by resident males appears to be based on odours in the urine of the intruders. Male mice rubbed with strange male urine elicit high aggression from residents (Mackintosh & Grant, 1966; Mugford & Nowell, 1970) whereas male mice rubbed with female urine elicit less attack (Mugford & Nowell, 1970; Dixon & Mackintosh, 1971).

The aggression-eliciting potency of male urine is androgen-dependent; castrated males do not fight and urine from male castrates rubbed on an intruder's fur is no more aggression-eliciting than water (e.g. Mugford & Nowell, 1970). Furthermore, testosterone treatment restores aggression-eliciting effects to castrates (Lee & Brake, 1972).

Preputial gland secretion has been implicated in the aggression-eliciting odour of male mouse urine. Jones & Nowell (1973c) found that trained fighters would attack castrates treated with preputial gland secretion or with preputial gland secretion mixed with bladder urine, more readily than castrates treated with bladder urine alone. Similarly, Mugford (1973) showed that pairs of preputialectomized mice did not fight as much as intact mice; however, pairs of male mice placed in a cage recently vacated by preputialectomized mice showed increased fighting.

On the other hand female mice not only lack aggression-promoting odours but produce an aggression-inhibiting substance (Mugford, 1973; Dixon & Mackintosh, 1971, 1975). The urine-mediated reduction in aggression of males towards females appears not to be due merely to a competing increase in sexual behaviour, since non-social behaviour increases as well (Dixon & Mackintosh, 1971). Dixon & Mackintosh (1975) have also shown that whereas the aggression-inhibiting odour of females appears not to be affected by ovariectomy, the sexual attractant produced by females is abolished by such treatment. Evans *et al.* (1978) found that aggression-reducing cues present in female bladder urine are resistant to bacterial or oxidative breakdown and to change in pH. The cues are relatively non-volatile which may indicate that they are used as short-range easily localizable signals.

ACOUSTIC COMMUNICATION IN ADULTS

Reception and Production

Berlin (1963) studied hearing up to 40 kHz in laboratory mice. He found them to be most sensitive to 10–16 kHz. More recent work has used frequencies up to 100 kHz, and consistently found a bimodal hearing curve. For example Brown (1973a,b) recorded the neural responses to sounds at the cochlea and at the inferior colliculus (Fig. 5a) while Ehret (1974) used behavioural methods (Fig. 5b). Both workers found two peaks of sensitivity, one around 15 kHz and a second around 50 kHz. Similar hearing curves have been produced by Brown (1973a,b) for other species that use ultrasonic communication, and the position of the high frequency peak varies in a way that can be related to interspecific variation in pitch of the communication calls. These results suggest that ultrasonic communication is of importance to small mammals.

Sound production by mice and other small mammals has been studied by Roberts (1972, 1975a,b). He found that ultrasonic cries of rodents are mainly emitted through the mouth, although weak calls can be detected through the nose if the mouth is blocked (Roberts, 1975a). In pups, they are produced as one pulse at the onset of every exhalation, when pressure is high and air flow low (Roberts, 1972).

Sewell (1969) has suggested that a whistle mechanism might be involved in the production of rodent ultrasounds, as the large frequency steps seen in mouse calls often involve frequency changes in simple ratios of 2:1 or 3:2 similar to the effect of overblowing a recorder. Roberts (1975b) supported this idea by studying the effects of light gases on rodent vocalizations. In a helium–oxygen mixture, the audible squeaks of mice were affected in the same way as the human voice; the fundamental frequency and harmonic series produced by the vibration of the vocal cords were unchanged, whereas the resonances of the vocal tract were altered by the light gas, so that higher frequencies in the harmonic series were emphasized. For the ultrasounds, all frequencies were moved upward indicating that no vibrating part of the vocal tract is involved in the production of the calls. Rather, the vocal tract appears to act as some sort of whistle, although the exact structures involved are still unknown. As an extension of this idea, Roberts found that model bird-whistles could produce very similar ultrasonic calls to those of rodents.

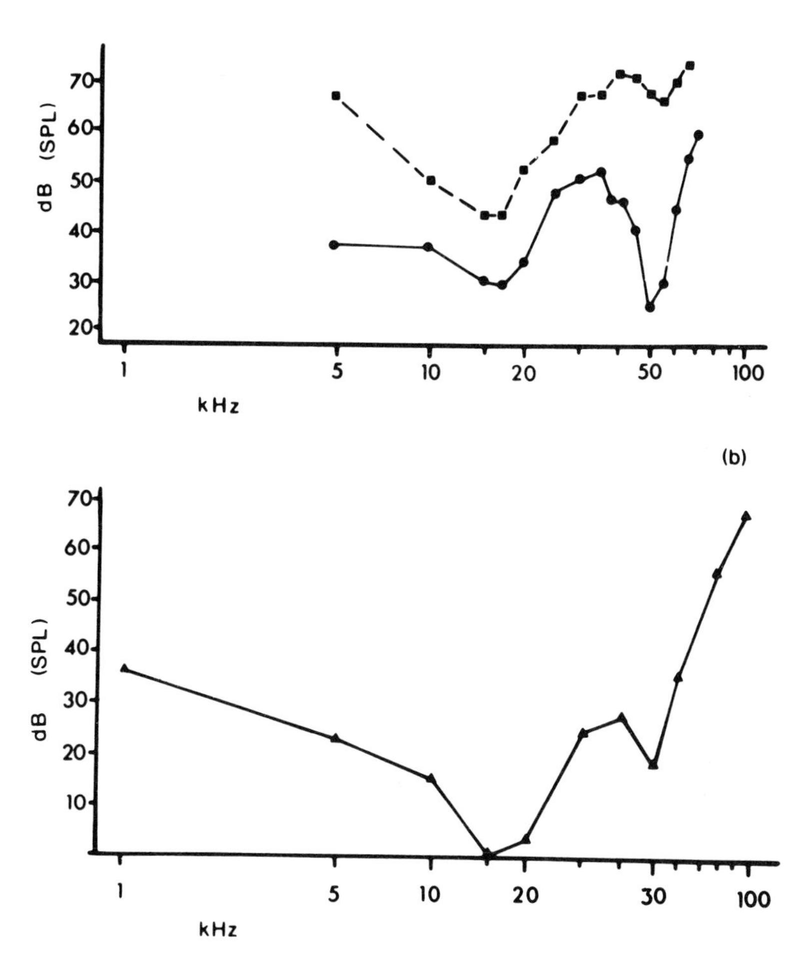

FIG. 5. Auditory threshold curves of adult mice. (a) ■— —■ cochlear microphonic response; ●——● inferior collicular response. (b) ▲——▲ conditioning. (From Smith & Sales, 1980.)

Communication

Compared with other myomorph rodents, particularly the rat, the mouse apparently restricts its acoustic communication mainly to mother–infant interactions and to heterosexual behaviour. Sales (1972a) was the first to detect ultrasounds during heterosexual behaviour in mice. The calls appeared to be correlated with the behaviour of the male, particularly approach, nosing, anogenital

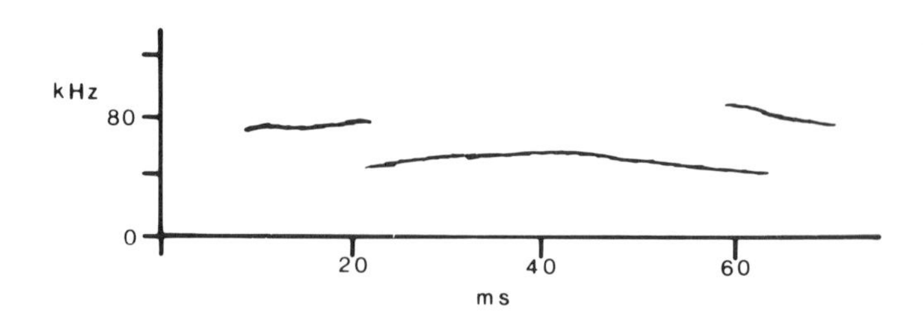

FIG. 6. Tracing of a sonagram of a call produced by a male mouse in a heterosexual encounter.

sniffing and mounting without intromission. Later stages of sexual behaviour were mostly silent. The calls are 50–300 ms long and in the range 50–110 kHz; they include frequency drifts, rapid frequency changes and frequency steps (Fig. 6). In heterosexual encounters most of the ultrasounds are produced by the male. No calls have been recorded when males are presented with a fresh female immediately after ejaculation (Sales, 1972a). Whitney, Stockton & Tilson (1971) detected no calls when females were paired with anaesthetized males, although a male–anaesthetized female pair did produce calls. Female mice do, however, produce loud audible squeals when rejecting the advances of a male. These often appear to silence a male temporarily (Smith & Halls, unpublished).

Females occasionally produce ultrasounds, in male–female encounters (Smith & Halls, unpublished), and also in female–female encounters (Sewell, 1969; Whitney *et al.,* 1971). A superficial survey indicates that they appear to be structurally rather similar to the calls of males (Smith & Halls, unpublished).

Dizinno & Whitney (1977) found that ultrasonic calling by mice was influenced by androgen. Castrates showed an increased latency to call to females, and this was reduced by testosterone propionate treatment; in contrast testosterone propionate treatment of females increased ultrasonic calling and mounting behaviour towards other females (Nyby, Dizzino & Whitney, 1977). Dominant males emit many more calls to normal females than do subordinates (Nyby, Dizinno & Whitney, 1976). Further work has shown that the response of male mice can be elicited by female urine alone. This is discussed above (pp. 380–381).

The function of ultrasonic communication in heterosexual encounters of mice remains to be determined. Sales (1972a) suggested that the calls may signal sexual motivation in the male and induce the female to take up the mating posture. The early suggestions of Whitney and his colleagues (1971) that the calls "resembled infant

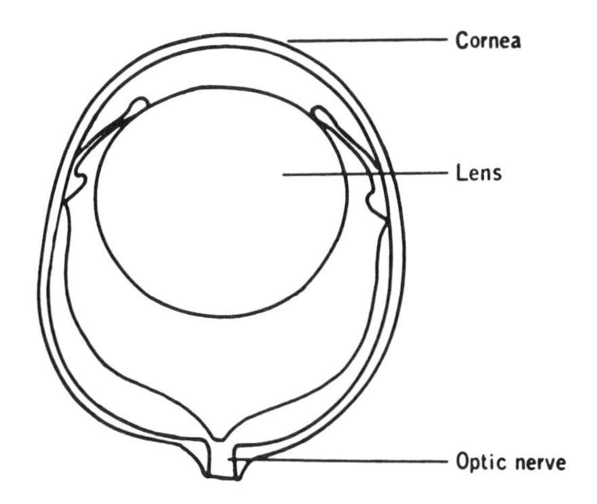

FIG. 7. Diagrammatic cross-section of the mouse eye (from Walls, 1942).

calls" and might represent a ritualized courtship display involving juvenile behaviour is unlikely, since the calls are quite easy to tell from those of young pups, even by a human being equipped with an ultrasound detector.

The role of acoustic communication in territorial and agonistic behaviour of mice may be small. Unlike rats, male mice do not produce ultrasounds when fighting in cages (Sales, 1972b), although loud audible squeals may be produced by the animal under attack. No sounds have been detected in observations of groups of mice in large enclosures (Smith, unpublished); in contrast, woodmice *Apodemus sylvaticus* are very vocal in such situations (Sales, 1972a). No work has been done on territorial interactions in mice. This would be particularly worthwhile where juveniles are involved.

OTHER SENSES AND COMMUNICATION IN MICE

The role of visual and tactile communication in mice is not fully known. Most data are negative. For example, blind mice fight, and all mice mate in complete darkness. However mice do have eyes and the lack of knowledge of how communication may operate in the wild may make generalizations from laboratory experiments misleading.

The mouse eye has a large, spherical lens (Fig. 7) and appears to be better designed for low-light vision than high visual acuity (Walls, 1942). There is disagreement as to whether the retina is all rods or has some cones. Using a brightness discrimination task Bonaventure

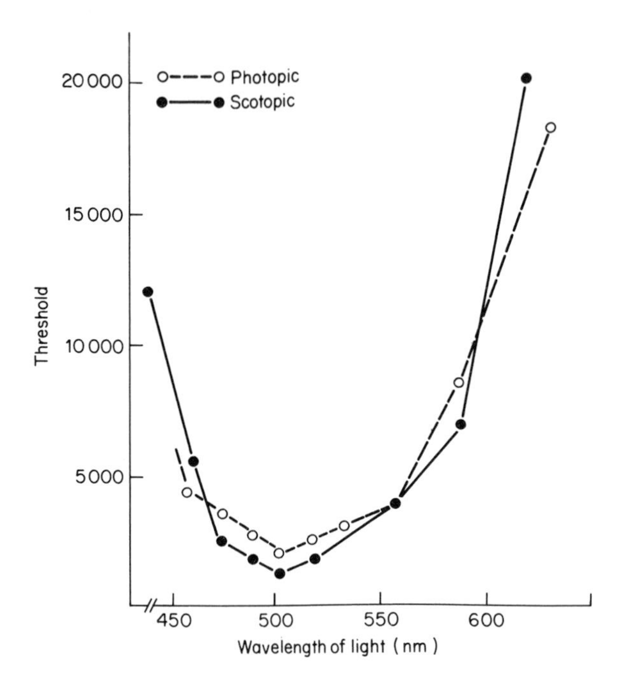

FIG. 8. Absolute thresholds in mouse photopic and scotopic vision (from Bonaventure, 1961).

(1961) found maximum sensitivity at around 500 nm, with very similar scotopic and photopic curves (Fig. 8). This indicates that cones cannot play a large part in mouse vision.

The small number of experiments on pattern discrimination by mice indicate that they are capable of quite difficult discriminations between shapes if they can approach stimuli closely. However, Kalkowski (1968) concluded that mice did not use visual cues in social communication. In his experiments, mice did not show different behaviour when exposed to conspecifics in glass boxes than if the boxes were empty. Mackintosh (1973) however, found that mice can use visual landmarks in artificial enclosures in assessing the position of territorial boundaries. When the only points of reference were 5 cm × 7 cm black rectangles fixed on each side of the enclosure 60 cm above the floor, and these were moved, the mice moved their territorial boundary in the same direction. In other experiments in which the whole enclosure was moved, the mice were found to be using visual information outside the enclosure, probably strip lights in the room.

It is well known that mice use characteristic body postures and movements in their social behaviour (e.g. Grant & Mackintosh, 1963). Although many of these are undoubtedly associated with the exchange of olfactory information, plenty of cues are also available for visual and tactile communication. For example, social and aggressive grooming as well as crawling over and under the other animal must involve tactile communication. Tactile communication is clearly involved in sexual behaviour, for example during mounting and intromission. The exact combinations of sensory channels used in conveying information through the body postures and movements characteristic of mouse behaviour have not been completely unravelled. Probably there is a certain amount of redundancy of information, since social exchanges between mice can still occur in the absence of one sensory channel. It is worth considering that the relative roles of various cues used in behaviour in the field might be rather different from those used under laboratory conditions.

CONCLUSION

Almost all the information about communication in mice has been gleaned from laboratory studies. This is to be expected, since it is more difficult to conduct research on the communication behaviour of mice in the wild. But as our knowledge of mouse communication under laboratory conditions grows, we must increasingly turn to the field for a true assessment of how systems are actually used by free-living mice.

Each sensory communication channel has its own particular properties and limitations under different environmental conditions. Olfactory cues are more persistent than any other type of communication signal. Cues on the body of a mouse require close contact between individuals, but those that a mouse deposits in its environment can communicate information while it is not there: it has been found that renewal of dominant mouse urine marks is only needed every two days for aversive potency to be continually effective (Jones & Nowell, 1977). The persistency of olfactory cues makes them potentially very useful for territorial demarcation, but persistency also means that a message cannot be so rapidly cancelled or changed. Olfactory communication over a distance depends on diffusion and air currents, and olfactory cues are not as easily localizable as visual and acoustic cues can be.

Acoustic communication, like visual communication, is more immediate than olfactory communication. Generally, such signals can be transmitted over greater distances and they are more easily localizable. However, these factors depend on the intensity, structure and attenuating properties of the environment in which they are used. High frequency sounds, such as mouse ultrasounds, are attenuated with distance more than lower frequency sounds, mainly by atmospheric attenuation, ground attenuation and scattering. These effects are not all linear so it is difficult to predict how far mouse sounds may travel without making measurements under conditions in which sounds are known to be produced in the wild. However, long-distance communication seems unlikely (Smith, 1979).

Visual communication requires light and must therefore be more restricted at night and in dark places, although mice appear to have eyes designed for low light conditions. Furthermore light is absorbed and scattered by objects in its path and animals would not therefore be able to see each other very far away in a cluttered environment. Here again we need to know about the field conditions before making predictions.

Tactile communication is essentially close range and must operate between individuals brought together by other means. Nevertheless, its importance in mother–infant relationships, sexual and social development is now becoming realized. The role of tactile communication in sexual and agonistic behaviour of adults should also not be underestimated.

The mouse has the full range of mammalian senses at its disposal. We are still a long way from a full understanding of how the mouse uses these senses in its social communication.

ACKNOWLEDGEMENTS

I am grateful to Professor J. D. Pye for the hospitality and facilities that have been made available to me at the Zoology Department, Queen Mary College, University of London, during the time that I have been out-stationed there.

REFERENCES

Alberts, J. R. (1976) Olfactory contributions to behavioral development in rodents. In *Mammalian olfaction, reproductive processes, and behavior* 4: 67–94. Doty, R. L. (Ed.). New York: Academic Press.

Barnett, S. A. & Walker, K. Z. (1974). Early stimulation parental behavior, and the temperature of infant mice. *Devl Psychobiol.* 7: 563–577.

Bell, R. W., Nitschke, W., Bell, N. J. & Zachman, T. A. (1974). Early experience,

ultrasonic vocalizations, and maternal responsiveness in rats. *Devl Psychobiol.* 7: 235–242.
Berlin, C. I. (1963). Hearing in mice via GSR audiometry. *J. Speech Hearing Res.* 6: 359–368.
Bonaventure, N. (1961). Sur la sensibilité spectrale de l'appareil visuel chez la souris. *C. r. Séanc. Soc. Biol.* 155: 908–921.
Breen, M. F. & Leshner, A. I. (1977). Maternal pheromone: a demonstration of its existence in the mouse (*Mus musculus*). *Physiol. Behav.* 18: 527–529.
Bronson, F. H. (1971). Rodent pheromones. *Biol. Reprod.* 4: 344–357.
Bronson, F. H. (1974). Pheromonal influences on reproductive activities in rodents. In *Frontiers of biology* 32 (*Pheromones* 18: 344–365). Birch, M. C. (Ed.). Amsterdam: North Holland.
Bronson, F. H. (1976). Urine marking in mice: causes and effects. In *Mammalian olfaction, reproductive processes, and behaviour.* 6: 119–141. Doty, R. L. (Ed.). New York: Academic Press.
Brown, A. M. (1973a). High frequency peaks in the cochlear microphonic response of rodents. *J. comp. Physiol. Psychol.* 83: 377–392.
Brown, A. M. (1973b). High levels of responsiveness from the inferior colliculus of rodents at ultrasonic frequencies. *J. comp. Physiol. Psychol.* 83: 393–406.
Bruce, H. M. (1959). An exteroreceptive block to pregnancy in the mouse. *Nature, Lond.* 184: 105.
Busnel, R. G. & Lehmann, A. (1977). Acoustic signals in mouse maternal behavior: retrieving and cannibalism. *Z. Tierpsychol.* 45: 21–24.
Carr, W. J., Martorano, R. D. & Krames, L. (1970). Responses of mice to odors associated with stress. *J. comp. Physiol. Psychol.* 71: 223–228.
Cooper, A. J. & Cowley, J. J. (1976). Mother–infant interaction in mice bulbectomized early in life. *Physiol. Behav.* 16: 453–459.
Cowley, J. J. & Wise, D. R. (1972). Some effects of mouse urine on neonatal growth and reproduction. *Anim. Behav.* 20: 499–506.
Dixon, A. K. & Mackintosh, J. H. (1971). Effects of female urine upon the social behaviour of adult male mice. *Anim. Behav.* 19: 138–140.
Dixon, A. K. & Mackintosh, J. H. (1975). The relationship between the physiological condition of female mice and the effects of their urine on the social behaviour of adult males. *Anim. Behav.* 23: 513–520.
Dixon, A. K. & Mackintosh, J. H. (1976). Olfactory mechanisms affording protection from attack to juvenile mice. *Z. Tierpsychol.* 41: 225–234.
Dizinno, G. & Whitney, G. (1977). Androgen influence on male mouse ultrasounds during courtship. *Horm. Behav.* 8: 188–192.
Dizinno, G., Whitney, G. & Nyby, J. (1978). Ultrasonic vocalizations by male mice (*Mus musculus*) in response to a female-produced pheromone: Effects of experience. *Behav. Biol.* 22: 104–113.
Dominic, C. J. (1969). Pheromonal mechanisms regulating mammalian reproduction. *Gen. comp. Endocr.* Suppl. 2: 260–267.
Dominic, C. J. (1970). Histological evidence for the failure of corpus luteum function in the olfactory block to pregnancy in mice. *J. anim. Morph. Physiol.* 17: 126–130.
Drickamer, L. C. (1974). Contact stimulation, androgenized females and accelerated maturation in female mice. *Behav. Biol.* 12: 101–111.
Ehret, G. (1974). Age-dependent hearing loss in normal hearing mice. *Naturwissenschaften* 11: 506.
Ehret, G. (1976). Development of absolute auditory thresholds in the house mouse (*Mus musculus*). *J. Am. Audio Soc.* 1: 179–184.

Evans, C. M., Mackintosh, J. H., Kennedy, J. F. & Robertson, S. M. (1978). Attempts to characterise and isolate aggression reducing olfactory signals from the urine of female mice *Mus musculus L. Physiol. & Behav.* **20**: 129–134.

Fuller, J. L. & Wimer, R. E. (1966). Neural, sensory, and motor functions. In *Biology of the laboratory mouse:* 609–629. Green, E. L. (Ed.). New York: McGraw-Hill.

Fullerton, C. E. & Cowley, J. J. (1971). The differential effect on the presence of adult male and female mice on the growth and development of the young. *J. genet. Psychol.* **119**: 89–98.

Gandelman, R., Zarrow, M. X. & Denenberg, V. H. (1971). Stimulus control of cannibalism and maternal behavior in anosmic mice. *Physiol. & Behav.* **7**: 583–586.

Grant, E. C. & Mackintosh, J. H. (1963). A comparison of the social postures of some common laboratory rodents. *Behaviour* **21**: 246–259.

Hahn, M. E. & Simmel, E. C. (1968). Individual recognition by natural concentrations of olfactory cues in mice. *Psychon. Sci.* **12**: 183–184.

Hayashi, S. & Kimura, T. (1974). Sex-attractant emitted by female mice. *Physiol. Behav.* **13**: 563–567.

Jones, R. B. & Nowell, N. W. (1973a). The effect of urine on the investigatory behaviour of male albino mice. *Physiol. & Behav.* **11**: 35–38.

Jones, R. B. & Nowell, N. W. (1973b). The coagulating glands as a source of aversive and aggression-inhibiting pheromone(s) in the male albino mouse. *Physiol. & Behav.* **11**: 455–462.

Jones, R. B. & Nowell, N. W. (1973c). Effects of preputial and coagulating gland secretions upon aggressive behaviour in male mice: a confirmation. *J. Endocr.* **59**: 203–204.

Jones, R. B. & Nowell, N. W. (1974a). The urinary aversive pheromone of mice: species, strain and grouping effects. *Anim. Behav.* **22**: 187–191.

Jones, R. B. & Nowell, N. W. (1974b). A comparison of the aversive and female attractant properties of urine from dominant and subordinate male mice. *Anim. Learn. Behav.* **2**: 141–144.

Jones, R. B. & Nowell, N. W. (1977). Aversive potency of male mouse urine: a temporal study. *Behav. Biol.* **19**: 523–526.

Kalkowski, W. (1967). Olfactory bases of social orientation in the white mouse. *Folia biol., Kraków* **15**: 69–87.

Kalkowski, W. (1968). Visual control of social environment in the white mouse. *Folia biol., Kraków* **16**: 215–233.

Lee, C. T. & Brake, S. C. (1972). Reaction of male fighters to male castrates treated with testosterone propionate or oil. *Psychon. Sci.* **27**: 287–288.

Leon, M. (1974). Maternal pheromone. *Physiol. & Behav.* **13**: 441–453.

Mackintosh, J. H. (1970). Territory formation by laboratory mice. *Anim. Behav.* **18**: 177–183.

Mackintosh, J. H. (1973). Factors affecting the recognition of territory boundaries by mice (*Mus musculus*). *Anim. Behav.* **21**: 464–470.

Mackintosh, J. H. & Grant, E. C. (1966). The effect of olfactory stimuli on the agonistic behaviour of laboratory mice. *Z. Tierpsychol.* **23**: 584–587.

Macrides, F. (1976). Olfactory influences on neuroendocrine function in mammals. In *Mammalian olfaction, reproductive processes, and behavior* **3**: 29–65. Doty, R. L. (Ed.). New York: Academic Press.

Macrides, F., Bartke, A. & Dalterio, S. (1975). Strange females increase plasma testosterone levels in male mice. *Science, N.Y.* **189**: 1104–1106.

Mainardi, D., Marsan, M. & Pasquali, A. (1965). Causation of sexual preferences of the house mouse. The behaviour of mice reared by parents whose odour was artificially altered. *Atti. Soc. ital. Sci. nat.* **104**: 325–338.

Maruniak, J. A. & Bronson, F. H. (1976). Gonadotropic responses of male mice to female urine. *Endocrinology* 99: 963–969.

Maruniak, J. A., Coquelin, A. & Bronson, F. H. (1978). The release of LH in male mice in response to female urinary odors: characteristics of the response of young males. *Biol. Reprod.* **18**: 251–255.

Mikaelian, D. & Ruben, R. J. (1965). Development of hearing in the normal CBA-J Mouse. *Acta Otolaryngol.* 59: 451–461.

Milligan, S. R. (1979). Pregnancy blockage and the memory of the stud male in the vole (*Microtus agrestis*). *J. Reprod. Fert.* 57: 223–225.

Mugford, R. A. (1973). Intermale fighting affected by home-cage odors of male and female mice. *J. comp. Physiol. Psychol.* **84**: 289–295.

Mugford, R. A. & Nowell, N. W. (1970). Pheromones and their effect on aggression in mice. *Nature, Lond.* **226**: 967–968.

Noirot, E. (1964a). Changes in the responsiveness to young in the adult mouse. I. The problematical effect of hormones. *Anim. Behav.* **12**: 52–58.

Noirot, E. (1964b). Changes in the responsiveness to young in the adult mouse. II. The effect of external stimuli. *J. comp. Physiol. Psychol.* 57: 97–99.

Noirot, E. (1970). Selective priming of maternal responses by auditory and olfactory cues from mouse pups. *Devl. Psychobiol.* **2**: 273–276.

Noirot, E. (1974). Nest-building by the virgin female mouse exposed to ultrasound from inaccessible pups. *Anim. Behav.* 22: 410–420.

Nyby, J., Dizinno, G. & Whitney, G. (1976). Social status and ultrasonic vocalizations of male mice. *Behav. Biol.* **18**: 285–289.

Nyby, J., Dizinno, G. & Whitney, G. (1977). Sexual dimorphism in ultrasonic vocalizations of mice (*Mus musculus*): Gonadal hormone regulation. *J. comp. Physiol. Psychol.* **91**: 1424–1431.

Nyby, J., Whitney, G., Schmitz, S. & Dizinno, G. (1978). Postpubertal experience establishes signal value of mammalian sex odor. *Behav. Biol.* 22: 545–552.

Okon, E. E. (1970). The effect of environmental temperature on the production of ultrasounds by isolated non-handled albino mouse pups. *J. Zool., Lond.* **162**: 71–83.

Quadagno, D. M. & Banks, E. M. (1970). The effect of reciprocal cross fostering on the behaviour of two species of rodents, *Mus musculus* and *Baiomys taylori ater*. *Anim. Behav.* **18**: 379–390.

Reynolds, J. & Keverne, E. B. (1979). The accessory olfactory system and its role in the pheromonally mediated suppression of oestrus in grouped mice. *J. Reprod. Fert.* 57: 31–35.

Roberts, L. H. (1972). Correlation of respiration and ultrasound production in rodents and bats. *J. Zool., Lond.* **168**: 439–449.

Roberts, L. H. (1975a). Evidence for the laryngeal source of ultrasonic and audible cries of rodents. *J. Zool., Lond.* 175: 243–257.

Roberts, L. H. (1975b). The rodent ultrasound production mechanism. *Ultrasonics* 13: 83–88.

Ropartz, P. (1977). Chemical signals in agonistic and social behaviour of rodents. In *Chemical signals in vertebrates:* 169–184. Müller Schwarze, D. & Mozell, M. M. (Eds). New York: Plenum.

Rosenblatt, J. S. (1976). Stages in the early behavioural development of altricial young of selected species of non-primate mammals. In *Growing points in ethology* **11**: 345–383. Bateson, P. & Hinde, R. A. (Eds). Cambridge: University Press.

Ruben, R. J. (1967). Development of the inner ear of the mouse: a radioautographic study of terminal mitoses. *Acta Otolaryngol.* (Suppl.) 20.

Salas, M., Guzman-Flores, C. & Schapiro, S. (1969). An ontogenetic study of olfactory bulb electrical activity in the rat. *Physiol. & Behav.* 4: 699–703.

Salas, M., Schapiro, S. & Guzman-Flores, C. (1970). Development of olfactory bulb discrimination between maternal and food odors. *Physiol. & Behav.* 5: 1261–1264.

Sales, G. D. (1972a). Ultrasound and mating behaviour in rodents with some observations on other behavioural situations. *J. Zool., Lond.* 168: 149–164.

Sales, G. D. (1972b). Ultrasound and aggressive behaviour in rats and other small mammals. *Anim. Behav.* 20: 88–100.

Sales, G. D. & Pye, J. D. (1974). *Ultrasonic communication by animals.* London: Chapman & Hall.

Sales, G. D. & Skinner, N. C. (1979). The effect of ambient temperature on body temperature and on ultrasonic behaviour in litters of albino laboratory mice deprived of their mothers. *J. Zool., Lond.* 187: 265–281.

Sales, G. D. & Smith, J. C. (1978). Comparative studies of the ultrasonic calls of infant murid rodents. *Devl Psychobiol.* 11: 595–619.

Sayler, A. & Salmon, M. (1971). An ethological analysis of communal nursing by the house mouse (*Mus musculus*). *Behaviour* 40: 61–85.

Scalia, F. & Winans, S. S. (1976). New perspectives on the morphology of the olfactory system: olfactory and vomero-nasal pathways in mammals. In *Mammalian olfaction, reproductive processes, and behavior* 2: 7–28. Doty, R. L. (Ed.). New York: Academic Press.

Schapiro, S. & Salas, M. (1970). Behavioral responses of infant rats to maternal odor. *Physiol. & Behav.* 5: 815–817.

Sewell, G. D. (1969). *Ultrasound in small mammals.* Ph.D. Thesis: London University.

Smith, J. C. (1974). *Sound communication in some myomorph rodents with special reference to adult-infant relationships.* Ph.D. Thesis: London University.

Smith, J. C. (1979). Factors affecting the transmission of rodent ultrasounds in natural environments. *Am. Zool.* 19: 432–442.

Smith, J. C. & Sales, G. D. (1980). Ultrasonic behavior and mother–infant interactions in rodents. In *Maternal influences and early behavior* 5: 105–133. Smotherman, W. P. & Bell, R. W. (Eds). New York: Spectrum.

Smotherman, W. P., Bell, R. W., Starzec, J., Elias, J. & Zachman, T. A. (1974). Maternal responses to infant vocalizations and olfactory cues in rats and mice. *Behav. Biol.* 12: 55–66.

Smotherman, W. P., Wiener, S. G., Mendoza, S. P. & Levine, S. (1977). Maternal pituitary-adrenal responsiveness as a function of differential treatment of rat pups. *Devl Psychobiol.* 10: 113–122.

Sorsby, A., Koller, P. C., Attfield, M., Davey, J. B. & Lucas, D. R. (1954). Retinal dystrophy in the mouse: histological and genetic aspects. *J. exp. Zool.* 125: 171–198.

Stoddart, D. M. (1976). *Mammalian odours and pheromones.* (Studies in Biology No. 73.) London: Edward Arnold.

Vandenbergh, J. G. (1967). Effect of the presence of a male on the sexual maturation of female mice. *Endocrinology* 81: 345–349.

Vandenbergh, J. G. (1969). Male odor accelerates female sexual maturation in mice. *Endocrinology* 84: 658–660.

Vandenbergh, J. G. (1971). The influence of the social environment on sexual maturation in male mice. *J. Reprod. Fert.* 24: 383–390.

van der Lee, S. & Boot, L. M. (1956). Spontaneous pseudopregnancy in mice. II. *Acta Physiol. Pharmacol. Neerl.* 5: 213–214.

Walls, G. L. (1942). *The vertebrate eye and its adaptive radiation.* (*Bull. Cranbrook Inst. Sci.* 19: 1–785). Michigan: Cranbrook.

Whitney, G., Stockton, M. D. & Tilson, E. F. (1971). Possible social functions of ultrasounds produced by adult mice (*Mus musculus*). *Am. Zool.* 11: 634.

Whitten, W. K. (1956). Modifications of the oestrous cycle of the mouse by external stimuli associated with the male. *J. Endocr.* 13: 399–404.

Whitten, W. K. (1959). Occurrence of anoestrus in mice caged in groups. *J. Endocr.* 18: 102–107.

Symp. zool. Soc. Lond. (1981) No. 47, 395–425

Population Dynamics of the House Mouse

R. J. BERRY

Department of Zoology, University College London, London, UK

SYNOPSIS

The numbers of any population are the result of a balance between births and deaths, immigration and emigration. The determinants of these are discussed for house mice living under different conditions: commensal and feral, tropical and Antarctic. Many of the regulating agents are still unknown, but:

(1) All four variables are subject to environmental modification. For example, breeding may be continuous or seasonal, depending on temperature and food supply; mortality may be "biological" (predation, parasitism, starvation) although more usually "physiological" (especially cold-dependent); and movement is affected by ecology, so that population churning ranges from virtual panmixis to highly viscous.

(2) All four variables have large inherited components. This has been best studied for mortality, and opens up new ways of studying population dynamics: death in an apparently uniform population of mice may result from the differential response of different genomes to environmental stresses.

In ecological terms, house mice are *r*-selected "weeds". They contrast in this way with (say) field mice, and consideration of their metabolism and population structure must recognize this biological basis. The most useful population studies will necessarily be long-term because every mouse population is genetically unique and extrapolation of the reactions of one population in the face of a particular environmental stress to other populations may be improper. Notwithstanding, mouse populations could be immensely valuable testing grounds of both ideas and reaction systems if proper collaboration of ecologists, geneticists, pathologists and physiologists could be achieved.

INTRODUCTION

The number (N) of individuals in any population is the difference between those recruited by birth (B) and immigration (I) and those disappearing through death (D) and emigration (E):

$$N = B - D + I - E.$$

All these values are very variable in house mice, and densities can fluctuate 10,000-fold. The most spectacular plagues have been recorded in Australia: between April and July 1917, 544 tons of mice (about 32 million individuals) were caught at Lascelles in Victoria; photographs exist of mounds of dead mice at this period – 126 000 collected in two nights and 500 000 in four nights (Cleland, 1918; Osborne, 1932). At such times the density of animals is about 875 per hectare (Newsome & Crowcroft, 1971). Less extreme outbreaks have been reported also from California and Russia (Hall, 1927; Piper, 1928; Fenyuk, 1934, 1941; Evans & Storer, 1944; Evans, 1949).

In every case where the causes of an explosive increase of numbers have been investigated, it has been found to result from a greater or more dense population of mice surviving an unfavourable period and able to reproduce at a high rate during a less stressed phase. The simplest situation is when a mild winter produces a lower mortality than usual, and leaves a large number of animals as parents for summer breeding. This has been well-documented by Pearson (1963) on populations in rough grassland in central California. He showed that densities of 500 to 700 mice per hectare followed unusually warm winters.

A similar correlation was found by Berry (1968) in a study of a closed population of mice on an 100 hectare island (Skokholm), three km off the Pembrokeshire (Dyfed) coast of Wales. Here a life table based on recruitment, mortality, and total number estimates over a period of five years showed a constant rate of increase of about six-fold during the breeding season (April to September) each year (homogeneity χ^2 of $P \simeq 90\%$), but vastly different death rates during the different winters ($\chi^2_5 = 43.9$, $P < 0.1\%$) (Berry & Jakobson, 1971). Peak numbers at the end of breeding in any year were thus entirely dependent upon the numbers surviving the previous winter. These in turn were apparently directly proportional to the coldness of the winter, most easily measured by the mean temperature in February, the coldest month (Table I).

A more complicated situation prevails in hot desert or semi-desert conditions in Australia. Here numbers are limited by the soil becoming too dry and hard for burrowing at the time when food becomes abundant. House mice survive in damp habitats where they can burrow at all seasons, but in which there is a chronic shortage of food (Newsome, 1969b). Increase in numbers takes place largely in emigrant populations established in temporally or seasonally food-rich habitats (Newsome, 1969a). In the normal hot summers of southern Australia, mice are wanderers in the huge wheat fields and rarely breed; they multiply only when substantial rain falls in the summer and softens the soil sufficiently for them to burrow. In central Australia where drought conditions may persist for several

TABLE I

Population numbers on Skokholm

Season	Estimate of numbers	Mean temperature in preceding February (°C)
Spring 1960	150–400	5.6
Spring 1961	500–1000	8.3
Spring 1962	150–400	6.1
Autumn 1962	1500–3000	
Spring 1963	100–300	1.5
Spring 1964	200–500	6.1
Autumn 1964	2000–4000	
Spring 1965	150–400	4.6
Autumn 1965	1000–3000	
Autumn 1966	2500–5000	6.6
Spring 1967	300–600	6.8
Autumn 1967	1000–2000	
Spring 1968	250–400	4.8
Autumn 1968	2000–3000	

After Berry & Jakobson, 1975b.

years and less than one mouse is caught for every 100 trap-nights, rains trigger rapid breeding so that the trapping success rate is raised to more than 50 mice per 100 trap-nights (Newsome & Corbett, 1975). The rain converts marginal into excellent habitat (Anderson, 1970). Newsome (1970) managed to produce plague densities of mice in a food-limited damp area (normally a marginal habitat) by providing unlimited food.

House mice have larger litters than most other murids, and are thus capable of rapidly increasing their numbers in any locality. However, their breeding seems to be easily halted by cold or food-shortage (Evans, 1949; Smith, 1954; Breakey, 1963; DeLong, 1967) and by disturbance through intra- or inter-specific interference (Strecker & Emlen, 1953; Southwick, 1955a; Crowcroft & Rowe, 1957, 1958; DeLong, 1966; Lidicker, 1966; Berry & Tricker, 1969). Bronson (1979) has reviewed the environmental factors which may affect breeding. He concluded that disease, exercise, and light intensity are not important under natural conditions, but that seven variables may at times be limiting: day–night regulation of diurnal

TABLE II

Characteristics of two common types of mouse population

Characteristic	Commensal & high density	Feral & low density
Food resources	Stable, plentiful	Seasonally and otherwise unstable, sparse
Reproduction	Usually non-seasonal	Usually seasonal
Population density	Up to 10 mice m^{-2}	Up to one mouse $(100m)^{-2}$
Home range	Less than 10 m^2	Up to 1000 m^2
Social organization	Territorial, with elements of hierarchical organization; some evidence for cooperation	Typically unstable, probably nearest neighbour dominant for resident males when and where possible, no cooperation
Potential for attaining plague numbers	Apparently none	High
Potential for dispersal	High	High

After Bronson, 1979.

TABLE III

Densities of some mouse populations

Habitat	Numbers per hectare	Reference
Plague (in Australia)	875	Newsome (1970)
Maximum in artificial enclosure	52 000	Lidicker (1976)
Chicken barn	70 000	Selander (1970a)
Open grassland	53	Stickel (1979)
Skokholm (maximum)	60	Berry & Jakobson (1975b)

rhythms; total caloric intake; specific nutrients in the diet; temperature variation; agonistic stimuli; specific tactile cues; and priming pheromones. The social environment affects gonadotropin release relatively directly, while pheromones and tactile clues act via sensory reception and brain mediation to control pituitary secretion of luteinizing hormone and prolactin, and dietary and temperature influences affect indirect pathways. Bronson comments that this rather loose type of ambient cueing is unusual, but ideally suited for the colonizing strategy that the mouse has evolved. For example, most rodents breed seasonally with their reproduction controlled by photoperiod; this control does not exist in house mice, and dispersing mice are thus able to maximize their rate of increase in a new habitat.

The reproductive ecology of mice leads in practice to low or high density populations (Table II). Crowding *per se* does not normally seem to inhibit reproduction (Table III). However, the normal response of mice faced with unfavourable conditions is to move in search of more favourable conditions (Justice, 1962; Rowe, Taylor & Chudley, 1964; Lidicker, 1965; Stickel, 1979). Mortality is discussed below (p. 404), but the most common causes seem to be physiological (particularly the effects of cold) rather than starvation, predation or cannibalism. Notwithstanding, food shortage is likely to cause periodic problems: when wild mice are given exercise wheels, they may achieve 60 000 revolutions in a single night (*c.* 20 km), which indicates a tremendous expenditure of energy (Schneider, 1946). In large cages, mice use about 2 kJ of energy for every gramme of body weight when indulging in normal activity (Brown, 1963). Mating may be depressed by as little as 48 hours of food deprivation (McClure, 1962, 1966). The importance of disease on vitality or as a cause of death is almost entirely unknown.

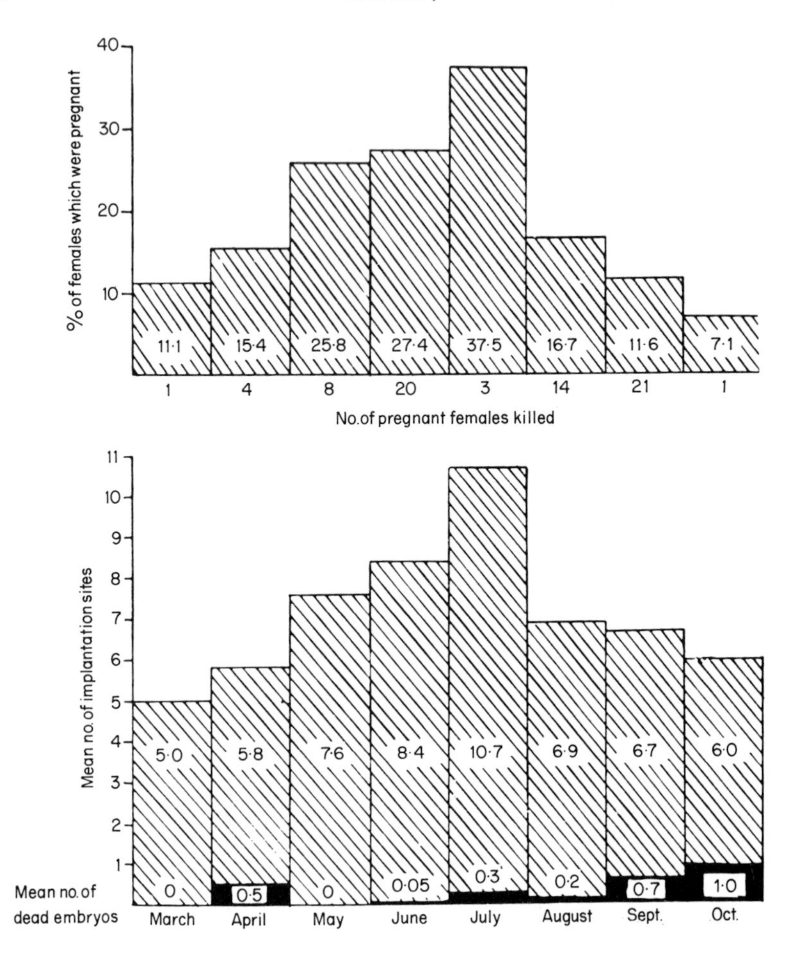

FIG. 1. Variation in breeding intensity and litter size in the Skokholm mouse population (after Batten & Berry, 1967, with additional data).

RECRUITMENT

Mice normally have their first litter at the age of six weeks to two months and thereafter breed about every four weeks. The number of young in a litter is most commonly six to eight, although this varies seasonally (early litters are smaller than later ones) and in different habitats (Laurie, 1946; Pelikán, 1974, this volume p. 215 Wallace, this volume p. 197). The mean litter size of a population containing a large number of young females (as towards the end of the breeding season in a seasonally breeding population) is reduced

TABLE IV

Correlation coefficients (r) between maternal size (head & body length) and litter size (number of implantations).

Source of mice	r
Welsh farm mice	0.53 ± 0.08
Skokholm	0.18 ± 0.18
Isle of May	0.02 ± 0.16
Foula	− 0.03 ± 0.27

After Batten & Berry, 1967.

in proportion to the number of mothers producing their first few litters (Batten & Berry, 1967; Berry, 1968) (Fig. 1). Ovulation rate increases with maternal size, but this correlation disappears in large island mice (Table IV), suggesting that litter size is an adaptive character in these populations (Lack, 1948; Batten & Berry, 1967; Roberts, this volume p. 249).

If a pair of mice produce six young in a litter and breed monthly with no mortality and equal numbers of males and females, their young having their first litter at six weeks, 2688 animals will be alive at the end of six months. If the first litter is assumed to occur at eight weeks, the population increase is reduced to 500-fold. Clearly there must be considerable mortality and/or female sterility in free-living populations. Most of the mortality seems to be post-natal: although a significant proportion of eggs fail to implant, most populations have a low incidence of post-implantation death (which means, among other things, that there does not seem to be a large number of recessive lethal alleles in mouse populations: Table V).

Perhaps the best studied example of population growth is that of Lidicker (1976) who set up two 385 m^2 outdoor enclosures in Sydney, Australia, where breeding could take place throughout the year (Pennycuik, 1972). In one, numbers increased from 24 (12 males and 12 females) to *c.* 2000 in nine months (reaching *c.* 1000 after six months, an 80-fold increase); in the other, escapes and predation (by *Rattus rattus*) reduced the increase to around 50-fold in the same period (12 to *c.* 900 after 11 months). Mean litter size was 7.5. Reproductive output decreased as the populations became dense, mainly by the failure of young mice to mature sexually. However, very much higher densities have been found in

TABLE V

Mean litter size and prenatal mortality

	Number of corpora lutea	Litter size (= no. live implants)	% prenatal loss
Wild mice			
Cereal ricks	9.2	6.4	31.1
Barn mice	7.5	4.9	34.4
Isle of May	10.1	6.4	36.2
Inbred strains			
DBA	8.9	6.7	23.9
C3H	10.1	6.9	31.2
CBA	9.3	6.0	35.7
CBA♂ × C3H♀	10.1	7.1	29.7
C3H♂ × CBA♀	9.7	6.9	28.4

From Batten & Berry, 1967.

chicken barns (Table III), showing that there is no sharp threshold to the effect of crowding on successful reproduction.

An obvious but important conclusion from Lidicker's (1976) study is that emigration and mortality are major factors when considering recruitment. In the completely natural Skokholm population when no emigration is possible, Berry & Jakobson (1971) calculated that more than a quarter of the young born fail to enter the adult population (i.e., that proportion of the population susceptible to trapping, q.v. Crowcroft & Jeffers, 1961) (Table VI). DeLong (1967) found a low rate of recruitment occurring almost throughout the year in mice living on grassland in the San Francisco Bay area, California. His population varied about three-fold in number annually as did ones living in grass and cereal fields in Maryland, studied by Stickel (1979). In the protected and benign environment of a wheat rick, Rowe, Taylor & Chudley (1963) showed that numbers increased 15-fold in nine months between the building and threshing of the rick, while Southern & Laurie (1946) record that numbers doubled every two months in ricks. Lidicker (1965) found a similar rate of increase in an indoor enclosure, where mortality was negligible.

One particular variety of recruitment failure in mouse populations needs mentioning, although it is probably a special case of the control of reproduction discussed in the previous section. House

TABLE VI

Pre-weaning mortality in the Skokholm mouse population

Period	Litter size	% females pregnant	Expected rate of increase per female	Observed rate of increase per female	% survival to weaning
March–April	5.6	20	2.24	1.34	59.8
May–June	8.2	30	4.92	3.44	69.9
July–August	7.6	45	6.84	5.55	81.1
September–October	6.7	35	4.69	3.67	78.3

From Berry & Jakobson, 1971.

mice seem to be peculiarly sensitive to interspecific competition. Their normal response to competition is to retreat and seek empty habitats (DeLong, 1966; Stickel, 1979). Local populations may become extinct either through movement away or (if emigration is prevented) through a failure to recruit enough young to replace animals dying. This latter has happened in at least two well-documented cases: with the only named British race of house mouse (*Mus musculus muralis*) on the Scottish island of St Kilda following the evacuation of the human population in 1930 and the invasion of the former village area by field mice (*Apodemus sylvaticus*) where previously only house mice had been found (Clarke, 1914; Harrisson & Moy-Thomas, 1933; Berry & Tricker, 1969); and on Brooks Island in San Francisco Bay following the colonization of the island by *Microtus californicus* (Lidicker, 1966). In both cases there can be little doubt that the failure of the populations was a consequence of lack of successful breeding rather than either starvation or fighting. (Mice are certainly predated by rats – Davis, 1979 – but they can co-exist with or out-compete both voles and *Peromyscus* in artificial situations: King, 1957; Caldwell, 1964; Gentry, 1966).

MORTALITY

Maximum numbers in a mouse population with excess food (water is rarely limiting: Fertig & Edmonds, 1969) and a non-varying physical environment seem to be mainly determined by suppression of successful breeding in females rather than cannibalism or other forms of violent death, although these certainly play a part in population regulation (Brown, 1953; Southwick, 1955a,b; Christian, 1956; Lloyd & Christian, 1969; Vandenbergh, 1971; Lidicker, 1976). In commensal habitats there is little variation in reproductive activity through the year, and absolute population size is thus controlled mainly by recruitment and emigration. However, few natural populations have the equable environment of commensal and laboratory animals, and both the causes and rate of mortality vary during the life of an individual mouse.

Probably the greatest hazard to a mouse in most situations is cold (Jakobson, 1978, this volume p. 325; Berry, Peters & Van Aarde, 1978). Small homoiotherms have to expend a very high proportion of their available energy in maintaining their body temperature, and a drop in ambient temperature may be fatal, particularly to an ageing animal (Berry, Jakobson & Triggs, 1973; Bellamy *et al.*, 1973). As we have seen, the survival strategy of house mice involves repeated prospecting and establishing of new colonies. This means spending a considerable time in unfavourable habitats, with a consequent risk to continued survival (Southwood, 1977). It is difficult to monitor food accessible to mice, but the available evidence suggests that there is usually sufficient food (seeds and insects) available to them (Lidicker, 1966; Berry, 1968; Berry, Jakobson & Triggs, 1973; Ford, this volume p. 260); except in plagues, no cases of overt starvation have been reported (Berry, 1968; Newsome & Crowcroft, 1971).

Few mice are taken by bird or mammal predators, with the exception once again of plagues when owls and foxes catch large numbers of individuals (Evans, 1949; Glue, 1967; Newsome & Crowcroft, 1971).

The only reports of epidemic disease in wild mice are from populations at high densities: pneumonia seems to have been an important cause of death in a mouse plague in Kern County, California (Piper, 1928), while large numbers of mice dead from disease were reported during two Russian outbreaks (Fenyuk, 1934, 1941). Pearson (1963) described "large haemorrhagic patches of unknown aetiology" in the lungs when one of the populations he studied was decreasing in number in the winter. DeLong (1967) found mice carrying "an enteric streptococcus in the spleen and liver" and apparently dying from septicaemia at a time when the population density decreased eight-fold in one month. Sherriff (quoted by Newsome & Crowcroft, 1971) has suggested that sick mice in one dense rick population might have been suffering from "a combination of an eperythryzoan infection coupled with murine hepatitis". At more usual densities, Anderson, Dunn & Beasley (1964) found young mice in an island population "fatally parasitized" by larvae of bot-flies (*Cuterebra* sp.), while Bellamy *et al.* (1973) recorded infertility and uterine oedema in a number of adult females infected with berry bugs (*Neotrombicula autumnalis*). As far as laboratory mice are concerned, Munro (1972) determined the causes of death in 12 of 29 which died out of a total group of 341 mated females. Six of them had colonic blockage (five with the nematode *Aspicularis tetraptera*), and five had liver necrosis.

Notwithstanding, there is no reason to dissent from the proposition that most mortality in nature is associated with low temperatures. Mice can adapt to cold both physiologically and genetically (Berry & Jakobson, 1975a; Barnett, Munro, Smart & Stoddart, 1975), but nevertheless seem to have difficulty in adjusting to a deterioration of their environment, however kind it may be initially. In this context it may be significant that cold adapted mice are

TABLE VII

Amount of allozymic variation as determined by electrophoresis

Location	No. of loci classified	% of loci polymorphic	% heterozygosity per locus	Reference
Enewetak Atoll	36	30	10.9	Berry, Sage, Lidicker & Jackson (1981)
Hawaii Island	36	50	16.6	Berry, Sage, Lidicker & Jackson (1981)
Lake Casitas, California	46	24	9.9	Rice & O'Brien (1980)
Bouquet Canyon, California	46	17	7.7	Rice & O'Brien (1980)
Central Peru	49	32	8.0	Berry & Peters (1981)
Southern England	22	27	4.7	Berry & Peters (1977)
Northern Scotland	63	30	8.5	Berry & Nash (unpublished)
British islands				
Skokholm	22	18	6.0	Berry & Peters (1977)
Isle of May	22	0	0	Berry & Peters (1977)
Orkney	22	21	5.0	Berry & Peters (1977)
Shetland	22	20	4.4	Berry & Peters (1977)
Jutland, Denmark	41	26	8.5	Selander, Hunt & Yang (1969)
South Atlantic islands				
Marion	24	36	7.7	Berry, Peters & Van Aarde (1978)
Macquarie	17	32	6.5	Berry & Peters (1975)
South Georgia	27	8	3.4	Berry, Bonner & Peters (1979)
Mus musculus brevirostris	40	30	11.0	Rice & O'Brien (1980)
"Swiss outbred"	46	17	7.1	Rice & O'Brien (1980)

more responsive to novelty in a cold environment than animals reared and kept in a warmer one (Barnett, Hocking & Wolfe, 1978). A study of the causes of death in mice living in an equable tropical environment would be of great interest (Tomich, Wilson & Lamoureux, 1968; Tomich, 1970; Berry & Jackson, 1979).

HETEROZYGOSITY AND POPULATION HETEROGENEITY

The use of electrophoresis since 1966 to identify the amount of allozymic variation in population samples has on the one hand radically changed our understanding of the factors maintaining inherited variation in populations, and on the other led to a realization of the amount of heterogeneity in virtually all populations of mice (Berry, 1979).

House mice are among the more variable mammalian species with a mean heterozygosity per locus in different populations ranging from 0% to 16.6% (Table VII). The overall average heterozygosity for mice is about 7% compared with 5.6% for all rodents and 3.6% for all mammals (Selander & Kaufman, 1973; Nevo, 1978). About twice as many alleles are detectable by heat denaturation techniques which distinguish between allelic products differing in thermostability, as by electrophoresis (Bonhomme & Selander, 1978).

In the context of this review, the important consequence of this inherited variety is the great number of different genetical types found within a population of apparently uniform mice. If all the variants revealed by electrophoresis were neutral in their effect, this would be irrelevant, but there is now a considerable body of evidence that allozymic variants may be adaptive (or at least label adaptive segments of chromosomes).

The first recognition of the effect of genetical heterogeneity came from sampling blood from animals caught on Skokholm in spring and autumn of two successive years – at the end of winter and at the end of the breeding season; the animals were released at the point of capture after bleeding, so there was no change in population density or composition (Berry & Murphy, 1970). Three of the six variable loci showed changes that could be attributed only to natural selection. The most spectacular (and subsequently regular) change was in genotypic proportions at the *Hbb* locus (Table VIII). Heterozygotes increased during the breeding season at the expense of the two homozygotes, but decreased again during the winter. This was a change in genotypic frequencies; the allelic frequencies changed little over the period. However, this repeated change took

TABLE VIII

Genetical changes in succeeding samples from the Skokholm population

Season	N	% excess of heterozygotes over binomial expectation *Hbb* (cliff-caught animals only)	*Pep-3*	
			%frequency of *Pep-3*[c]	% of heterozygotes
Spring 1968	99	−2.4	92.5	10.3
Autumn 1968	122	43.3	81.5	30.3
Spring 1969	66	27.0	80.7	17.5
Autumn 1969	111	72.8	66.3	42.3

From Berry & Murphy, 1970.

place only in the territorially organized part of the population around the cliffs; the mice of the interior of the island (which floods every winter, and has to be recolonized from the periphery every summer) tended to reflect the cliff changes, but in a less precise way. In crude terms, it appeared to be a good thing to be an *Hbb* heterozygote during the summer breeding season, but a bad thing during the winter when faced with harsh problems of survival.

In subsequent years only a single sample has been collected, in the autumn. The animals in such samples range from a few weeks to over a year old: the older ones are survivors of the previous winter and their immediate offspring, whereas the younger ones are summer-born animals. When the genetical constitution of the mouse cohort younger than three months of age is compared with that of the one older than three months, it is found that the younger (summer-born) animals contain a higher proportion of heterozygotes in every year except 1974 and 1976, when there were no significant differences between the two age-classes (Table IX).

These exceptions are important. The changes in heterozygote proportions point to different relative survivals of the *Hbb* genotypes under different conditions, implying the action of natural selection on the phenotypes produced by that locus. When no change takes place, this presumably means that there are no differences between phenotypes under the conditions being experienced. The winters of both 1973–1974 and 1975–1976 were particularly warm. Under commensal conditions Petras (1967a) found a *deficiency* of heterozygotes at the *Hbb* locus and Selander (1970a) different frequencies of the segregating alleles on different sides of the same barn, suggesting that here the *Hbb* chromosomal segment is behaving entirely neutrally. In contrast, Myers (1974a) found that different

TABLE IX

Differences in Hbb *heterozygous excess in samples caught on the cliffs on Skokholm in the autumn*

	N	Frequency of Hbb^s	% excess of heterozygotes		Diff. from February mean temp. (°C)
			Older[a]	Younger	
1968	89	0.438	34.8	75.0	−1.4
1969	111	0.464	67.8	100.0	−2.4
1971	117	0.271	−29.8	9.2	+1.4
1972	76	0.539	−11.0	31.0	+0.9
1974	62	0.621	40.2	31.6	+1.7
1975	92	0.426	−23.7	7.9	+1.1
1976	53	0.415	23.5	18.6	+1.4
1977	73	0.595	−16.7	8.3	−0.3

[a]More than three months
Berry, 1978a and unpublished.

Hbb genotypes affected reproduction or survival to different extents at different times of the year in populations living in abandoned fields in northern California, while Garnett & Falconer (1975) fixed an *Hbb* allele in six replicates by selecting for large size from a mixed stock (although allele frequencies at eight other loci were randomly distributed after 32 generations of selection). The two common alleles of *Hbb* affect the oxygen dissociation curve of mouse haemoglobin and through it the characteristics of the blood (Kano & Nishida, 1972; Berry & Jakobson, 1975a; Jakobson, 1978); this may be part of the explanation for the observed selection.

Berry & Peters (1977) scored electrophoretic variations at 22 loci in 27 population samples of mice. In their data, there were highly significantly different variances ($P \ll 0.001$) of allelic frequencies between different loci, whereas if all loci had been affected equally by inbreeding and no differential adaptive effects were operating, the variances should have been identical. All functional groups of enzymes except oxido-reductases showed heterogeneous variances. This test is not a definitive proof of selective action since it is affected by extremes of distributions, and many of the allelic frequencies in the data were close to zero (or one). Notwithstanding, it is an additional indication that the allozyme phenotypes affect the fitness of their carriers.

In speaking of allozyme phenotypes, it is pertinent to note that Bellamy *et al.* (1973) found that a range of metrical traits (organ

TABLE X

Size and relative amount of brown fat in some wild-caught mouse samples

Location	Latitude	Average temperatures (°C) Coldest month	Warmest month	Annual	*N*	Weight (g)	Lengths (mm) Head & body	Tail	Brown fat ($mg\,g^{-1}$)	Reference
South Georgia	54° S	− 2.5	4.3	1.0	89	18.6	89.5	73.7	10.95	Berry, Bonner & Peters (1979)
Macquarie	54° S	3.0	8.6	4.6	63	16.9	92.2	75.1	6.19	Berry & Peters (1975)
Marion	46° S	3.3	7.4	5.1	92	21.0	78.4	78.6	3.65	Berry, Peters & Van Aarde (1978)
Shetland	60° N	3.5	12.5	7.6	239	16.5	85.4	75.5	4.98	Berry & Jakobson (1975a)
Orkney	59° N	3.8	12.8	8.0	331	18.2	57.5	72.6	4.14	Berry & Jakobson (1975a)
Skokholm	52° N	5.7	15.9	10.4	117	18.2	92.6	83.6	3.36	Berry & Jakobson (1975a)
Southern England (Taunton)	51° N	16.5	5.0	10.5	104	13.4	86.6	74.3	4.90	Berry & Jakobson (1975a)
Central Peru	12° N	8.3	12.1	10.7	216	16.3	88.3	77.7	4.01	Berry & Peters (1981)
Hawaii Is. (NE Coast)	20° N	20.8	23.6	22.2	105	11.6	80.7	72.7	4.45	Berry & Jackson (1979)
Enewetak Atoll	11° N	27.2	28.5	27.9	121	9.2	75.9	67.4	2.16	Berry & Jackson (1979)

and body weights, ionic composition of bones, haematological characters, etc.) contributed to an efficient discrimination between animals with different alleles at the *Hbb* and *Es-2* loci, i.e. the electrophoretic phenotypes were functioning as markers for a switch involving a whole suite of characteristics, as well as indicating variation at one particular locus. Recently, examples of linkage disequilibria have begun to appear (Berry & Peters, 1981 and unpublished), giving additional evidence of the correlated action of certain chromosomal segments (cf. Selander, Hunt & Yang, 1969).

On Skokholm, genetical changes show the action of different selective forces during the two main phases of the annual cycle – the summer breeding and winter survival phases. These results are complemented by ones from two oceanic islands (Macquarie and Marion) lying almost on the Antarctic Convergence where temperate conditions give way to a sub-polar climate. The annual mean temperature on both islands is about 5°C, with a very small fluctuation between the warmest and coldest months (Table X). Despite the fact that the temperature only rarely reaches the level at which reproduction takes place in Britain or the USA, breeding occurs throughout the year. This means that there can be no question of genetical adjustment to different seasons. However, both mouse populations show significant genetical changes between the younger and older cohorts (Berry & Peters, 1975; Berry, Peters & Van Aarde, 1978): on Macquarie the frequency of Hbb^d declined from a frequency of 0.71 in animals aged three months and younger, to 0.54 in older animals; similar but less marked shifts took place at the *Hbb* and six other loci in the Marion population, the greatest change being at one of the malic enzyme loci (*Mod-1*), where one allele had a frequency of 0.38 in younger mice but only 0.23 in older ones. No deviations from the expected Hardy–Weinberg distributions were found. The obvious explanation for these changes is that the phenotypes produced by the genes in question (or the segment of chromosome linked to them) differentially affect survival in adult life, but have an opposite effect at some other stage – presumably during reproduction or embryonic life. Both populations have only been sampled on a single occasion and it is theoretically possible that all the loci in both the populations were being affected by directed selection to reduce or eliminate certain alleles, but the similar behaviour of both populations coupled with the evidence of selection on the *Hbb* segment makes it much more likely that the observed changes in adult life were reversed at some other stage of the life cycle.

Direct evidence of selection acting in different directions comes from a study of possibly the most environmentally stressed mouse population in the world, living on a remote south-facing (i.e. towards the South Pole) headland on the South Atlantic island of South Georgia where the annual mean temperature is less than 2°C and the monthly average below 0°C for four months each year. The mice are big animals (although not as large as those on some North Atlantic islands) and have much brown fat, showing their response to their cold environment. Only two out of 27 gene loci scored electrophoretically were segregating (3.4% heterozygotes/locus); these (*Es-6, Got-2*) are four cross-over units apart on chromosome 8, and were in strong linkage disequilibrium. However the most interesting fact is that genetical changes go in opposite directions in males and females: $Es\text{-}6^a$ increased from 16% in males of less than three months, to 35% in older animals; it declined in females from 36% to 19%. Parallel changes took place with *Got-2* alleles (Berry, Bonner & Peters, 1979). It will be fascinating to learn more about the activities of the *Es-2* – *Got-2* – *Es-6* segment of chromosome 8.

In contrast to the genetical adjustments in environmentally stressed sub-Antarctic island populations, and complementing the lack of adjustment at *Hbb* after a warm winter on Skokholm, a study of three mouse populations on equable but warm central Pacific islands showed no genetical heterogeneity between cohorts (Table X) (Berry, Sage *et al.,* 1981). Eighteen of 36 loci scored electrophoretically were segregating in at least one of the populations. There were no deviations from the expected Hardy–Weinberg distributions; no statistically significantly different allele frequencies between males and females; and only one significant change in frequency between younger and older cohorts out of 32 comparisons (at *Np-1* in the Medren sample: $\chi^2_1 = 4.84$, $P \simeq 0.03$).

An interesting feature of these tropical populations was the high mean heterozygosities in all of them (Table VII); the sample from Paauilo in the north-east of Hawaii Island, the "Big Island" of the State of Hawaii, has the highest heterozygosity reported so far in any house mouse population. There was no trend of heterozygosity with age in the tropical populations, although there was a slight reduction in the sub-Antarctic Marion population (Berry, Peters *et al.,* 1978). The significance of overall heterozygosity estimates are still debated, as to whether they have a biological meaning in themselves or whether they are nothing more than a convenient way of summarizing data (Smith, Manlove & Joule, 1978; Berry, 1978b), but there can be no argument that they give information about the need to take genetical heterogeneity into account when considering ecological variables. The birth, death, and movement variables which together determine population numbers (p. 395) are all subject to genetical variation.

POPULATION SUBDIVISION AND INDIVIDUAL MOVEMENT

In the preceding section, the mouse samples described were treated as if they were taken from a completely panmictic population. This is clearly an over-simplification. Indeed some authors have believed that mouse populations are so tightly organized into territorially-restricted social units that "it is quite conceivable the effective population size in natural populations is less than four" (DeFries & McClearn, 1972). If this was the case, allele frequencies would be highly affected by drift.

Three lines of evidence have been adduced to support this conclusion:

(1) Numerous laboratory studies have shown that mice adopt unambiguously territorial behaviour in large enclosures (Crowcroft, 1955; Crowcroft & Rowe, 1963; Mackintosh, this volume p. 355), which may be very stable. For example, Reimer & Petras (1967) comment about their own study, "Several incidences of female movement were observed. This may be a mechanism for gene exchange between breeding units. No males however, were ever found to enter successfully an established deme .... Migrant animals were bitten in 95% of the cases, and 91% of the males and 78% of the females were killed .... The genetically effective size of the random breeding units in the population cage (was) probably less than five if subordinate males do not contribute to the gene pool. If subordinate males contribute, the effective breeding size of the population could conceivably exceed ten. In any event the size of the breeding unit is small". Brown (1953), Oakeshott (1974) and Busser, Zweep & Van Oortmerssen (1974) have found that subordinate males may sometimes be reproductively successful.

(2) A number of field studies have shown a restricted home range of individuals in natural mouse populations (Table XI); an apparently random distribution of gene frequencies between neighbouring colonies (Anderson, 1964, 1970; Selander, 1970a, b); and a deficiency of heterozygotes in wild-caught population samples, indicating inbreeding in physically undivided populations (Petras, (1967a). Indeed one of the first ever population studies of house mouse "revealed an unexpected lack of heterozygotes... (indicating) that the mice were not mating at random, but break up into comparatively small mating units" (Philip, 1938). This conclusion

TABLE XI

Home ranges of mice in different habitats

Habitat	Range (m^2)	Reference
Chicken barn	1.9	Selander (1970a)
Oxford cellar	4.6–5.6	Southern (1954)
Open field (in presence of voles)	122.3	Quadagno (1968)
Open field (in absence of voles)	364.6	Quadagno (1968)

was based on breeding carried out on a mixed population of grey and yellow-bellied mice collected from a depth of 550 m in a coal mine in Ayrshire, Scotland, when they were living on the feed of pit ponies and on crusts thrown away by miners (Elton, 1936).

Anderson (1964) has studied populations living in small grain storage buildings ("granaries") on the Canadian prairies. Each building held a behaviourally defined unit of about ten weaned mice, of which four to seven were reproductively active, only one or two being males. Population size was controlled by excluding immigrants (and thus reproductively isolating the group) and exporting excess young. Granaries as little as one metre apart have genetically distinct populations which may persist for several generations. On the basis of approximately 18% heterozygous deficiency at two loci studied, Petras (1967a) estimated that the effective breeding size of the farm building populations he studied was between six and 80.

However, the classical example of a restricted house mouse population is the "hillock mouse" (*Mus musculus hortulanus*) (= *Mus hortulanus* of Marshall & Sage, this volume p. 18) of the Ukraine (Naumov, 1940). In the autumn of each year the adults of a pair gather a supply of weed seeds for the winter. At the beginning of the winter each "hillock" is occupied by a pair of adults and their last litter of the season. There is a high likelihood that the adults will die during the winter, resulting in the hillock and surrounding area being taken over by members of the same family, who will continue the family line by sib-matings on the same site.

(3) Populations containing a *t*-allele would be expected to have a high frequency of the allele, since these will spread until their segregation advantage (in male heterozygotes) is balanced by the frequency of lethal homozygotes. Most populations in the U.S.A. have been found to carry a *t*-allele (Dunn, Beasley & Tinker, 1960; see also Klein *et al.*, this volume p. 448). However, Anderson (1964)

found that different granaries on the same farm did not all carry the same allele – and some carried none – although they were spatially very close. A similar result has been reported by Petras (1967b) and Reimer & Petras (1968). If the segregation distortion is the only factor affecting gene frequency, *t*-alleles should be much commoner than they are – about 30% in most populations. Lewontin (Lewontin & Dunn, 1960; Lewontin, 1962) has simulated population units of various sizes and found that small breeding units (two males, six females) produce the frequencies found in nature, since with this size, alleles are often lost by chance (i.e. by genetic drift).

Experimental support for this result was obtained by Anderson, Dunn & Beasley (1964), who released six male mice carrying a *t*-allele on Great Gull Island (7 ha in extent) in Long Island Sound, Connecticut where the existing population had no *t*-alleles. The introduced allele became established in the immediate area of its introduction within the first two breeding seasons. However, it thereafter spread very slowly – much more slowly than would have occurred if the Great Gull population was behaving as a single breeding unit (Bennett *et al.*, 1967). The allele subsequently disappeared from the population (Myers, 1974b).

Convincing though these evidences have been to some people, they are sometimes interpreted to imply a tighter social organization than actually exists in most populations. Lidicker (1976) confirmed the existence of a strict and persisting territorial system in his outdoor enclosures by direct behavioural observation, documentation of home sites over long periods, attempts to introduce strange mice, and the distribution of coat colour morphs. Notwithstanding, a small amount of gene flow occurred between established social groups through the occasional successful immigration of individuals into adjacent groups (mostly females on the verge of sexual maturity), and more extensive genetical mixing took place by the formation of new social groups. An interesting finding was that 17% of emigrants were albinos, although only 5% of the enclosure animals had this genotype. Whilst accepting that abnormal animals are more likely to be excluded from their social group than a normal one, this finding demonstrates that colonizers and stay-at-homes may be genetically distinct.

On Skokholm Berry & Jakobson (1974) showed by release-recapture trapping that most adult animals were faithful to a locality for extended periods. However, more than a quarter of the animals on the island breed at a site other than the one at which they were born ("site" being defined as a trapping area, usually 50–100 m from

TABLE XII

Change in the residence of mice which survive the winter on Skokholm (% of total population)

	Between autumn and the following spring	Between spring and the following autumn
Males	33.6	25.7
Females	29.7	23.6

another) (Table XII). The maximum number of adult mice in a single group (i.e. within an area in which all the individuals have overlapping ranges) seems to be about six, but the composition of all groups constantly changes because of the relatively high mortality at all ages.

Confirmation of the genetical churning inferred from recording individual movements has been obtained from a study of rare alleles at allozymic loci (Berry & Jakobson, 1974; Berry, 1977). It is clear that such alleles are able to spread from the site of their original presumed mutation. There can be no doubt that individual territories form only temporary barriers to gene flow, even though in more commensal situations social structure may be much more rigid and population mixing correspondingly less.

Other studies which contradict the assumption of particularly small effective population sizes in mice could be quoted (e.g. Myers, 1974a; Stickel, 1979). Probably only those populations which have a plentiful and stable food supply possess a high degree of social stability; social plasticity would seem to be the more normal rule. However, it is of interest in the light of the theoretical work that has been invested in the genetical dynamics of the *t*-situation to note that in an Australian experiment involving the release of $t/+$ males into an enclosure population, *t* heterozygotes had a lower fitness than $+/+$ animals (contrary to one of the assumptions of Lewontin's model), and the introduced allele rapidly became extinct without any slowing of population growth (Pennycuik *et al.*, 1978).

The inevitable conclusion is that mice are as adaptable in their social organization as in their reproductive ecology, and do not lend themselves to the reductionism of model-building theoreticians. *Mus musculus* can clearly withstand a considerable degree of inbreeding (or it would have been impossible to establish virtually homozygous strains by continued brother–sister mating in the laboratory), but this does not mean that inbreeding is the normal fashion. The outstanding characteristic of mouse populations is opportunistic colonization. New groups are continually formed.

The variation carried by the founders of these new groups must be the basis of any adaptation they undergo subsequently – but this is a stochastic process unsusceptible to easy prediction (see next section).

UNIQUE POPULATIONS AND LONGITUDINAL STUDIES

The house mouse is a weed: quick to exploit opportunity, and able to withstand local adversity and extinction without harm to the species. It has to be able to breed rapidly, tolerate a wide variety of conditions, and adjust quickly to changes in its environment. Perhaps linked to these properties is the high genetical variation possessed by the species (Berry, 1979).

A consequence of the repeated formation of new populations by small numbers of founders is that every population is likely to be unique. The importance of subdivision in producing diversity was shown by an early study of Berry (1963), in which every population sample collected from 15 cereal ricks on a large (800 ha) farm at Odiham, Hampshire was significantly distinct from every other. However, when all the mice were pooled into a single group, the resulting population was entirely characteristic of southern English mice (Berry, 1964, 1977). Rick populations are short-lived, being founded every autumn by immigrants from surrounding fields and dispersing when the ricks are threshed, usually in early spring. Rowe, Taylor & Chudley (1963) have shown that about 100 animals may invade each rick. There is little movement of mice between ricks in the winter, so each harbours an effectively isolated population. The lack of time for progressive differentiation but the way that each isolate is effectively a distinct sample of a widespread population shows the power of the founder effect in promoting interpopulation diversity. Indeed, the mouse populations on different islands in the Faroe group which have been quoted as examples of particularly rapid evolutionary divergence (Huxley, 1942), are almost certainly the descendants of small groups of colonizers in the fairly recent past (Berry, Jakobson & Peters, 1978)–instances of "instant subspeciation" (Berry, 1967).

Virtually every collection of data involving different mouse populations shows apparently random divergence of each population (Fig. 2). These differences stem almost exclusively from the initial founding event ("intermittent drift" in the terminology of Waddington, 1957), although subsequent adaptation will produce further genetical change – which may involve divergence or

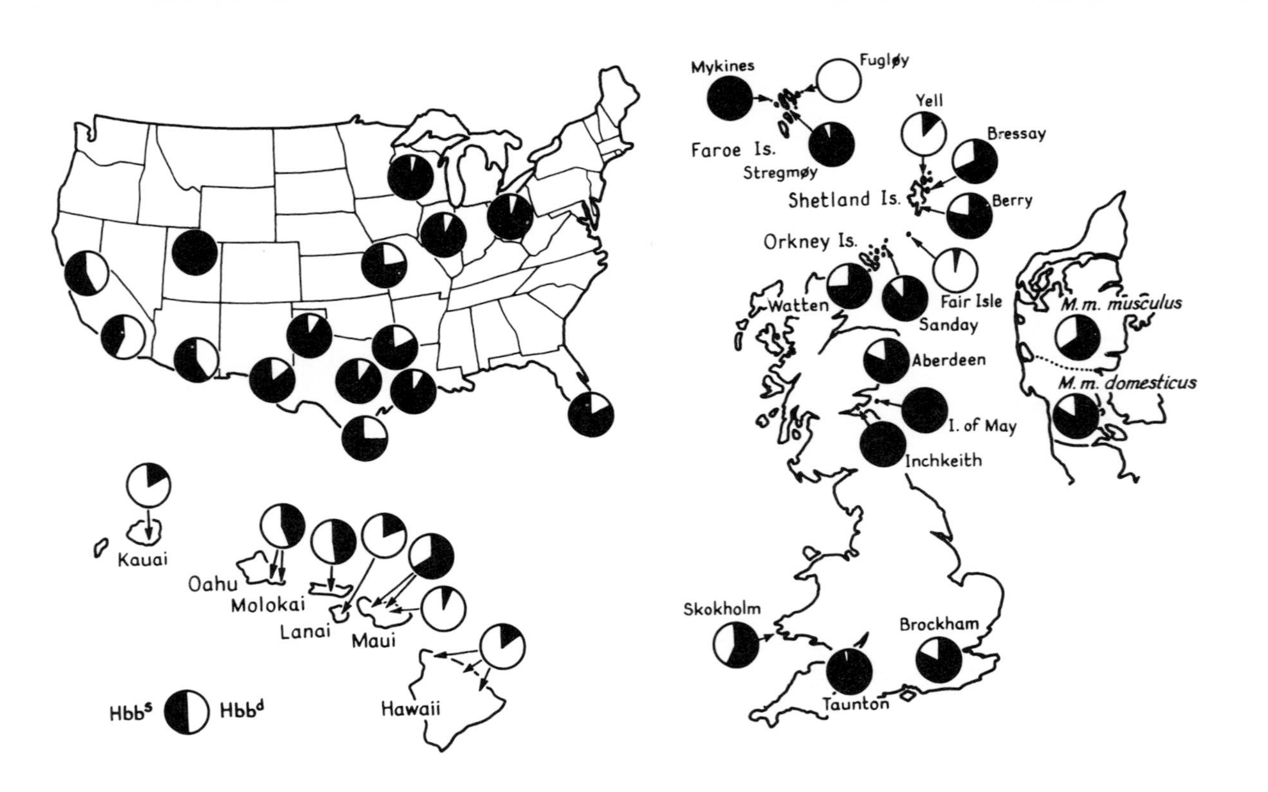

FIG. 2. Frequencies of the two common alleles at the *Hbb* locus, showing apparently random differentiation between populations (from various sources, q.v. Berry & Peters, 1977).

TABLE XIII

Effect of genotypes of Hbb *(DD, DS, SS) on survival (S) and reproduction (R) in mice on two grids in northern California*

Period	Grid one			Grid two		
	DD	DS	SS	DD	DS	SS
January–March		R	S	S		
April–July				S, R		
August–November		R		S, R		
December–April		S	R	R	S	

S and R indicate an effect 10% greater than that of the other genotypes. After Myers, 1974a.

convergence: Berry & Jakobson (1975b) showed that the Skokholm population was still diverging from its mainland ancestor after 80 years of isolation, while Berry & Peters (unpublished) found that descendants of CBA × C57BL F_1 animals released onto an uninhabited Shetland island showed rapid increases in the frequencies of the normally common allele at all loci monitored except *Hbb,* where the frequencies remained close to the 50% present in the introduced animals.

It is impossible to know if a single sample of wild-caught mice is genetically or physiologically adapted; only repeated sampling can show this (Berry & Jakobson, 1975a). Myers (1974a) trapped two 0.8 ha grids separated by 92 m in north Californian grassland, and monitored two alloyzmic loci. She found the frequencies of alleles at both loci were different between and within each grid, but that changes in frequencies associated with the differential reproduction or survival of different individuals took place at different times of the year (Table XIII).

All our knowledge of mouse populations leads to two general conclusions:

(1) Every population is genetically unique and highly adaptable.

(2) The adaptability involves genes affecting and potentially adjusting all the constituents of mortality, natality, and movement which control population numbers (p. 395).

Natural mouse populations contain an array of genomes which will be of great value in dissecting the factors involved in population dynamics once these have been separated into units (ideally determined by single genes). This is likely to be a much more revealing method of analysis of demographic traits than the classical biometric method. It may soon be practicable since we can now score unifactorial genetical characters enabling us to recognize "natural" differences between individuals in a population and to apportion the elements contributing to complex traits like birth or death rate. Furthermore there is a tremendous potential for investigating the effects of particular genes in different environments; Bulfield in this volume (p. 643) suggests that some wild-type alleles may be pathogenic in different environments (see also Coleman, 1978; Padua, Bulfield & Peters, 1978). Conversely the maintenance of histocompatibility polymorphisms can best be studied by using laboratory techniques of immunology and microbiology on wild-living populations, and perhaps by setting up natural populations of mice along the lines pioneered by Crowcroft, Lidicker, and their ilk. It is time to couple the expertise of laboratory and field workers in order to exploit the possibilities of wild-living house mouse populations as a source, model, and testing ground of both ecological concepts and genetical variants.

REFERENCES

Anderson, P. K. (1964). Lethal alleles in *Mus musculus*: local distribution and evidence for isolation of demes. *Science, N.Y.* **145**: 177–178.

Anderson, P. K. (1970). Ecological structure and gene flow in small mammals. *Symp. zool. Soc. Lond.* No. 26: 299–325.

Anderson, P. K., Dunn, L. C. & Beasley, A. B. (1964). Introduction of a lethal allele into a feral house mouse population. *Am. Nat.* **98**: 57–64.

Barnett, S. A., Hocking, W. E. & Wolfe, J. L. (1978). Effects of cold on activity and exploration by wild house mice in a residential maze. *J. comp. Physiol.* 123: 91–95.

Barnett, S. A., Munro, K. M. H., Smart, J. L. & Stoddart, R. C. (1975). House mice bred for many generations in two environments. *J. Zool., Lond.* 177: 153–169.

Batten, C. A. & Berry, R. J. (1967). Prenatal mortality in wild-caught house mice. *J. Anim. Ecol.* **36**: 453–463.

Bellamy, D., Berry, R. J., Jakobson, M. E., Lidicker, W. Z., Morgan, J. & Murphy, H. M. (1973). Ageing in an island population of the house mouse. *Age & Ageing* 2: 235–250.

Bennett, D., Bruck, R., Dunn, L. C., Klyde, B., Shutsky, F. & Smith, L. J. (1967). Persistence of an introduced lethal in a feral house mouse population. *Am. Nat.* **101**: 538–539.

Berry, R. J. (1963). Epigenetic polymorphism in wild populations of *Mus musculus*. *Genet. Res.* **4**: 193–220.

Berry, R. J. (1964). The evolution of an island population of the house mouse. *Evolution, Lancaster, Pa.* **18**: 468–483.

Berry, R. J. (1967). Genetical changes in mice and men. *Eugen. Rev.* **59**: 78–96.

Berry, R. J. (1968). The ecology of an island population of the house mouse. *J. Anim. Ecol.* **37**: 445–470.

Berry, R. J. (1977). The population genetics of the house mouse. *Sci. Progr., Oxf.* **64**: 341–370.
Berry, R. J. (1978a). Genetic variation in wild house mice: where natural selection and history meet. *Am. Scient.* **66**: 52–60.
Berry, R. J. (1978b). Genetical pressures and social organization in small mammal populations. In *Populations of small mammals under natural conditions:* 114–117. Snyder, D. A. (Ed.). Pittsburgh, Pennsylvania: University of Pittsburgh.
Berry, R. J. (1979). Genetical factors in animal population dynamics. In *Population dynamics:* 53–80. Anderson, R. M., Taylor, L. R. & Turner, B. D. (Eds). Oxford: Blackwell.
Berry, R. J., Bonner, W. N. & Peters, J. (1979). Natural selection in house mice from South Georgia (South Atlantic Ocean). *J. Zool., Lond.* **189**: 385–398.
Berry, R. J. & Jackson, W. B. (1979). House mice on Enewetak Atoll. *J. Mammal.* **60**: 222–225.
Berry, R. J. & Jakobson, M. E. (1971). Life and death in an island population of the house mouse. *Expl Geront.* **6**: 187–197.
Berry, R. J. & Jakobson, M. E. (1974). Vagility in an island population of the house mouse. *J. Zool., Lond.* **173**: 341–354.
Berry, R. J. & Jakobson, M. E. (1975a). Adaptation and adaptability in wild-living house mice. *J. Zool., Lond.* **176**: 391–402.
Berry, R. J. & Jakobson, M. E. (1975b). Ecological genetics of an island population of the house mouse. *J. Zool., Lond.* **175**: 523–540.
Berry, R. J., Jakobson, M. E. & Peters, J. (1978). The house mice of the Faroe Islands: a study in microdifferentiation. *J. Zool., Lond.* **185**: 73–92.
Berry, R. J., Jakobson, M. E. & Triggs, G. S. (1973). Survival in wild-living mice. *Mamm. Rev.* **3**: 46–57.
Berry, R. J. & Murphy, H. M. (1970). Biochemical genetics of an island population of the house mouse. *Proc. R. Soc. Lond.* (B) **176**: 87–103.
Berry, R. J. & Peters, J. (1975). Macquarie Island house mice: a genetical isolate on a sub-Antarctic island. *J. Zool., Lond.* **176**: 375–389.
Berry, R. J. & Peters, J. (1977). Heterogeneous heterozygosities in *Mus musculus* populations. *Proc. R. Soc.* (B) 197: 485–503.
Berry, R. J. & Peters, J. (1981). Allozymic variation in house mouse populations. In *Mammalian population genetics:* 242–253. Smith, M. H. & Joule, J. (Eds). Athens: University of Georgia.
Berry, R. J., Peters, J. & Van Aarde, R. J. (1978). Sub-Antarctic house mice: colonization, survival and selection. *J. Zool., Lond.* **184**: 127–141.
Berry, R. J., Sage, R. D., Lidicker, W. Z. & Jackson, W. B. (1981). Genetical variation in three Pacific house mouse populations. *J. Zool., Lond.* **193**: 391–404.
Berry, R. J. & Tricker, B. J. K. (1969). Competition and extinction: the mice of Foula, with notes on those of Fair Isle and St Kilda. *J. Zool., Lond.* **158**: 247–265.
Bonhomme, F. & Selander, R. K. (1978). Estimating total genetic diversity in the house mouse. *Biochem. Genet.* **16**: 287–297.
Breakey, D. R. (1963). The breeding season and age structure of feral house mouse populations near San Francisco Bay, California. *J. Mammal.* **44**: 153–168.
Bronson, F. H. (1979). The reproductive ecology of the house mouse. *Q. Rev. Biol.* **54**: 265–299.

Brown, R. Z. (1953). Social behavior, reproduction and population changes in the house mouse (*Mus musculus* L.). *Ecol. Monogr.* **23**: 217–240.

Brown, R. Z. (1963). Patterns of energy flow in populations of the house mouse. *Bull. ecol. Soc. Am.* **44**: 129.

Busser, J., Zweep, A. & van Oortmerssen, G. A. (1974). Variability in the aggressive behaviour of *Mus musculus domesticus,* its possible role in population structure. In *The genetics of behaviour:* 185–199. van Abeelen, G. H. F. (Ed.). Amsterdam: North-Holland.

Caldwell, L. D. (1964). An investigation of competition in natural populations of mice. *J. Mammal.* **45**: 12–30.

Christian, J. J. (1956). Adrenal and reproductive responses in mice from freely growing populations. *Ecology* **37**: 258–273.

Clarke, W. E. (1914). Notes on the mice of St. Kilda. *Scott. Nat.* **30**: 124–128.

Cleland, J. B. (1918). Rats and mice. *J. R. Soc. N.S.W.* **52**: 32–165.

Coleman, D. L. (1978). Diabetes and obesity – thrifty mutants? *Nutr. Rev.* **36**: 129–132.

Crowcroft, P. (1955). Territoriality in the wild house mouse (*Mus musculus* L.). *J. Mammal.* **36**: 299–301.

Crowcroft, P. & Jeffers, J. N. R. (1961). Variability in the behaviour of wild house mice (*Mus musculus* L.) towards live traps. *Proc. zool. Soc. Lond.* **137**: 573–582.

Crowcroft, P. & Rowe, F. P. (1957). The growth of confined colonies of the wild house mouse (*Mus musculus* L.). *Proc. zool. Soc. Lond.* **129**: 359–370.

Crowcroft, P. & Rowe, F. P. (1958). The growth of confined colonies of the wild house mouse: the effect of dispersal on female fecundity. *Proc. zool. Soc. Lond.* **131**: 357–365.

Crowcroft, P. & Rowe, F. P. (1963). Social organization and territorial behaviour in the wild house mouse (*Mus musculus* L.). *Proc. zool. Soc. Lond.* **140**: 517–531.

Davis, R. A. (1979). Unusual behaviour by *Rattus norvegicus. J. Zool., Lond.* **188**: 298.

DeFries, J. C. & McClearn, G. E. (1972). Behavioral genetics and the fine structure of mouse populations: a study in microevolution. In *Evolutionary biology* **5**: 279–291. Dobzhansky, Th., Hecht, M. K. & Steere, W. C. (Eds). New York: Appleton-Century-Crofts.

DeLong, K. T. (1966). Population ecology of feral house mice: interference by *Microtus. Ecology* **47**: 481–484.

DeLong, K. T. (1967). Population ecology of feral house mice. *Ecology* **48**: 611–634.

Dunn, L. C., Beasley, A. B. & Tinker, H. (1960). Polymorphisms in populations of wild house mice. *J. Mammal.* **41**: 220–229.

Elton, C. (1936). House mice (*Mus musculus*) in a coal mine in Ayrshire. *Ann. Mag. nat. Hist.* (10) **17**: 553–558.

Evans, F. C. (1949). A population study of house mice (*Mus musculus*) following a period of local abundance. *J. Mammal.* **30**: 351–363.

Evans, F. C. & Storer, T. I. (1944). Abundance of house mice at Davis, California in 1941–2. *J. Mammal.* **25**: 89–90.

Fenyuk, B. K. (1934). [The mass increase of mouse-like rodents in Stalingrad Region in the autumn of 1933.] *Vest. mikrobiol. Epidem. Parazit.* **13**: 235–247. [In Russian].

Fenyuk, B. K. (1941). [The mass increase of mouse-like rodents in the south-east of the R.S.F.S.R. in 1937. In *Rodents and rodent control* No. 1: 209–224.] Saratov: State Inst. Microbiol. and Epidemiol. for the south-east of the U.S.S.R. [In Russian].

Fertig, D. S. & Edmonds, V. W. (1969). The physiology of the house mouse. *Scient. Am.* 221: 103–110.

Garnett, I. & Falconer, D. S. (1975). Protein variation in strains of mice differing in body size. *Genet. Res.* 25: 45–57.

Gentry, J. B. (1966). Invasion of a one-year abandoned field by *Peromyscus polionotus* and *Mus musculus. J. Mammal.* 47: 431–439.

Glue, D. E. (1967). Prey taken by the barn-owl in England and Wales. *Bird Study* 14: 169–183.

Hall, E. R. (1927). An outbreak of house mice in Kern County, California. *Univ. Calif. Publs Zool.* 30: 189–203.

Harrisson, T. H. & Moy-Thomas, J. A. (1933). The mice of St. Kilda, with special reference to their prospects of extinction and present status. *J. Anim. Ecol.* 2: 109–115.

Huxley, J. S. (1942). *Evolution, the modern synthesis.* London: Allen & Unwin.

Jakobson, M.E. (1978). Winter acclimatization and survival of wild house mice. *J. Zool., Lond.* 185: 93–104.

Justice, K. E. (1962). *Ecological and genetical studies of evolutionary forces acting on desert populations of* Mus musculus. Tucson, Arizona: Arizona-Sonora Desert Museum Inc.

Kano, K. & Nishida, S. (1972). Genetic polymorphism of hemoglobin and its relation to some hematological parameters in several inbred strains of mice. *Tohuku J. agric. Res.* 23: 122–131.

King, J. A. (1957). Intra- and interspecific conflict of *Mus* and *Peromyscus. Ecology* 38: 355–357.

Lack, D. (1948). The significance of litter-size. *J. Anim. Ecol.* 17: 45–50.

Laurie, E. M. O. (1946). The reproduction of the house-mouse (*Mus musculus*) living in different environments. *Proc. R. Soc.* (B) 133: 248–281.

Lewontin, R. C. (1962). Interdeme selection controlling a polymorphism in the house mouse. *Am. Nat.* 96: 65–78.

Lewontin, R. C. & Dunn, L. C. (1960). The evolutionary dynamics of a polymorphism in the house mouse. *Genetics* 45: 705–722.

Lidicker, W. Z. (1965). Comparative study of density regulation in confined populations of four species of rodents. *Researches Popul. Ecol. Kyoto Univ.* 7: 57–72.

Lidicker, W. Z. (1966). Ecological observations on a feral house mouse population declining to extinction. *Ecol. Monogr.* 36: 27–50.

Lidicker, W. Z. (1976). Social behaviour and density regulation in house mice living in large enclosures. *J. Anim. Ecol.* 45: 677–697.

Lloyd, J. A. & Christian, J. J. (1969). Reproductive activity of individual females in three experimental freely growing populations of house mice (*Mus musculus*). *J. Mammal.* 50: 49–59.

McClure, T. J. (1962). Infertility in female rodents caused by temporary inanition at or about the time of implantation. *J. Reprod. Fert.* 4: 241.

McClure, T. J. (1966). Infertility in mice caused by fasting at about the time of mating. *J. Reprod. Fert.* 12: 243–248.

Munro, K. M. H. (1972). Causes of death in mated female mice. *Lab. Anim.* 6: 67–74.

Myers, J. (1974a). Genetic and social structure of feral house mouse populations on Grizzly Island, California. *Ecology* **55**: 747–759.

Myers, J. H. (1974b). The absence of *t* alleles in feral populations of house mice. *Evolution, Lancaster, Pa.* 27: 702–704.

Naumov, N. P. (1940). [The ecology of the hillock mouse, *Mus musculus hortulanus* Nordm.] *J. Inst. Evolut. Morph.* 3: 33–77 [In Russian].

Nevo, E. (1978). Genetic variation in natural populations: patterns and theory. *Theor. popul. Biol.* 13: 121–177.

Newsome, A. E. (1969a). A population study of house-mice temporarily inhabiting a South Australian wheatfield. *J. Anim. Ecol.* **38**: 341–359.

Newsome, A. E. (1969b). A population study of house-mice permanently inhabiting a reed-bed in South Australia. *J. Anim. Ecol.* **38**: 361–377.

Newsome, A. E. (1970). An experimental attempt to produce a mouse plague. *J. Anim. Ecol.* 39: 299–311.

Newsome, A. E. & Corbett, L. K. (1975). Outbreaks of rodents in semi-arid and arid Australia: causes, preventions, and evolutionary considerations. In *Rodents in desert environments*: 117–153. Prakash, I. & Ghosh, P. K. (Eds). The Hague: Junk.

Newsome, A. E. & Crowcroft, W. P. (1971). Outbreaks of house mice in South Australia in 1965. *CSIRO Wildl. Res.* **16**: 41–47.

Oakeshott, J. G. (1974). Social dominance, aggressiveness and mating success among male house mice (*Mus musculus*). *Oecologia* 15: 143–158.

Osborne, W. A. (1932). Mice plagues in Australia. *Nature, Lond. 129*: 755.

Padua, R. A., Bulfield, G. & Peters, J. (1978). Biochemical genetics of a new glucosephosphate isomerase allele (*Gpi-1*^{C}) from wild mice. *Biochem. Genet.* **16**: 127–143.

Pearson, O. P. (1963). History of two local outbreaks of feral house mice. *Ecology* **44**: 540–549.

Pelikán, J. (1974). On the reproduction of *Mus musculus* L. in Czechoslovakia. *Acta Sci. nat. acad. Sci. bohem. Brno* 8 (12): 3–42.

Pennycuik, P. R. (1972). Seasonal changes in reproductive productivity, growth rate, and food intake in mice exposed to different regimes of day length and environmental temperature. *Aust. J. biol. Sci.* 25: 627–635.

Pennycuik, P. R., Johnston, P. G., Lidicker, W. Z. & Westwood, N. H. (1978). Introduction of a male sterile allele (t^{w2}) into a population of house mice housed in a large outdoor enclosure. *Aust. J. Zool.* **26**: 69–81.

Petras, M. L. (1967a). Studies of natural populations of *Mus.* I. Biochemical populations and their bearing on breeding structure. *Evolution, Lancaster, Pa.* 21: 259–274.

Petras, M. L. (1967b). Studies of natural populations of *Mus.* II. Polymorphism at the *T* locus. *Evolution, Lancaster, Pa.* **21**: 466–478.

Philip, U. (1938). Mating system in wild populations of *Dermestes vulpines* and *Mus musculus. J. Genet.* **36**: 197–211.

Piper, S. E. (1928). The mouse infestation of Buena Vista Lake Basin, Kern County, California, September 1926 to February 1927. *Month. Bull. Calif. Dep. Agric.* 17: 538–560.

Quadagno, D. M. (1968). Home range size in feral house mice. *J. Mammal.* **49**: 149–151.

Reimer, J. D. & Petras, M. L. (1967). Breeding structure of the house mouse, *Mus musculus,* in a population cage. *J. Mammal.* **48**: 88–99.

Reimer, J. D. & Petras, M. L. (1968). Some aspects of commensal populations of *Mus musculus* in southwestern Ontario. *Can. Fld Nat.* **82**: 32–42.

Rice, M. C. & O'Brien, S. J. (1980). Genetic variance of laboratory outbred Swiss mice. *Nature, Lond.* **283**: 157–161.

Rowe, F. P., Taylor, E. J. & Chudley, A. H. J. (1963). The numbers and movements of house mice (*Mus musculus* L.) in the vicinity of four corn ricks. *J. Anim. Ecol.* **32**: 87–97.

Rowe, F. P., Taylor, E. J. & Chudley, A. H. J. (1964). The effect of crowding on the reproduction of the house mouse (*Mus musculus* L.) living in corn ricks. *J. Anim. Ecol.* **33**: 477–482.

Schneider, H. A. (1946). On breeding "wild" house mice in the laboratory. *Proc. Soc. exp. Biol. Med.* **63**: 161–165.

Selander, R. K. (1970a). Behavior and genetic variation in natural populations. *Am. Zool.* **10**: 53–66.

Selander, R. K. (1970b). Biochemical polymorphism in populations of the House mouse and Old-field mouse. *Symp. zool. Soc. Lond.* No. 26: 73–91.

Selander, R. K., Hunt, W. G. & Yang, S. Y. (1969). Protein polymorphism and genic heterozygosity in two European subspecies of the house mouse. *Evolution, Lancaster, Pa.* **23**: 379–390.

Selander, R. K. & Kaufman, D. W. (1973). Genic variability and strategies of adaptation in animals. *Proc. natn. Acad. Sci. U.S.A.* **70**: 1875–1877.

Smith, M. H., Manlove, M. N. & Joule, J. (1978). Spatial and temporal dynamics of the genetic organization of small mammal populations. In *Population of small mammals under natural conditions*: 99–113. Snyder, D. (Ed.). Pittsburgh, Pennsylvania: University of Pittsburgh.

Smith, W. W. (1954). Reproduction in the house mouse *Mus musculus* L. in Mississippi. *J. Mammal.* **35**: 509–515.

Southern, H. N. (Ed.). (1954). *Control of rats and mice.* 3. *House mice.* Oxford: University Press.

Southern, H. N. & Laurie, E. M. O. (1946). The house mouse (*Mus musculus*) in corn ricks. *J. Anim. Ecol.* **15**: 135–149.

Southwick, C. H. (1955a). Regulatory mechanisms of house mouse populations: social behaviour affecting litter survival. *Ecology* **36**: 627–634.

Southwick, C. H. (1955b). The population dynamics of confined house mice supplied with unlimited food. *Ecology* **36**: 212–225.

Southwood, T. R. E. (1977). Habitat, the templet for ecological strategies? *J. Anim. Ecol.* **46**: 337–365.

Stickel, L. C. (1979). Population ecology of house mice in unstable habitats. *J. Anim. Ecol.* **48**: 871–887.

Strecker, R. L. & Emlen, J. T. (1953). Regulatory mechanisms in house mouse populations: the effect of limited food supply on a confined population. *Ecology* **34**: 375–385.

Tomich, P. Q. (1970). Movement patterns of field rodents in Hawaii. *Pacif. Sci.* **24**: 195–234.

Tomich, P. Q., Wilson, N. & Lamoureux, C. H. (1968). Ecological factors on Manana Island, Hawaii. *Pacif. Sci.* **22**: 352–368.

Vandenbergh, J. G. (1971). The influence of the social environment on sexual maturation in male mice. *J. Reprod. Fert.* **24**: 383–390.

Waddington, C. H. (1957). *Strategy of the genes.* London: Allen & Unwin.

Symp. zool. Soc. Lond. (1981) No. 47, 427–437

Structure, Dynamics and Productivity of Mouse Populations: A Review of Studies Conducted at the Institute of Ecology, Polish Academy of Sciences

WIERA WALKOWA

Institute of Ecology, Polish Academy of Sciences, Dziekanów Leśny, nr Warsaw, Poland

SYNOPSIS

A review is presented of work on *Mus musculus* populations carried out in the Institute of Ecology of the Polish Academy of Sciences, which involved large numbers of confined white mice and of free-living wild house mice.

The original experiments were designed to examine population structure in relation to numbers in self-regulating populations. These led on to a study of the factors controlling the optimum production of population biomass in a small space with numbers kept constant. Population structure, production, bioenergetics and modifying ecological factors were all analysed.

BACKGROUND

This contribution reviews studies on *Mus musculus* carried out in the Institute of Ecology of the Polish Academy of Sciences. These have focused on population structure and its effect on population dynamics and productivity. Population in this context is defined very narrowly as a group of mice kept in a single cage. The experiments described were mostly conducted on white mouse populations bred for many years. In addition, some investigations were made under natural conditions – in a loft and other parts of a building with free access for mice to the outside.

The starting hypothesis was that a range of interactions in a population affect the life of individual animals and hence population size. The most readily observed interactions are fights, which lead to and demonstrate the dominance structure of the population. Therefore the first phase of the study attempted to identify the factors which affect the chance of a mouse becoming socially dominant. From a long series of experiments, Petrusewicz and his co-workers concluded (Petrusewicz, 1958a, 1959a,b; Petrusewicz & Wilska, 1959; Petrusewicz & Andrychowska, 1960):

(1) A male in his home cage is at a considerable advantage in a fight with a foreign male.

(2) Other members of the same population are a support to an animal in a fight with an intruder.

(3) The combined effects of home cage and members of the same population are greater than other factors which stimulate aggression, such as body weight, inherited or acquired aggressiveness, previous successful fights, etc.

These conclusions were developed further from observations of colonies each started with 20 males and 20 females in 6 m^2 pens (Andrzejewski, Petrusewicz & Walkowa, 1963; Walkowa, 1964). A rapid, albeit brief, increase in the number of fights follows the initial introduction of new individuals, and this increase is proportional to the number of animals introduced. However, the increase declines with successive introductions. In the first week after an introduction, the number of fights between resident and introduced males is higher than expected if these were merely random encounters, but this difference decreases with increasing numbers of foreign individuals. The frequency of fights between introduced males is no higher than expected on the basis of random encounters. Newcomers are gradually accepted by a resident population.

Introductions do not usually affect the established dominance structure of a population. Introduced males soon learn which fights they are likely to win in the same way as resident subordinates; a very few become dominants. In general, their mortality is higher than that of residents, and increases in proportion to the number of newcomers. In contrast, the survival of residents is not affected by introductions.

The effect on the social structure of removing dominant individuals was studied by Nowak (1971) in 66 populations of white mice. A dominant was removed from each population every week for three weeks. The first-removed animal was then returned to its cage (i.e. after a three-week absence). After the removal of the first two dominants, a dominance hierarchy was quickly re-established with the new dominant taking his position after two or three days. The original dominant released back into his population after three weeks only regained his former dominance in 37% of cases.

A variant of this experiment involved joining two separate populations by taking away a partition in their cage, and then successively removing the first three dominants that established themselves. It was found that they were usually replaced by an individual from the group to which the removed animal belonged before the populations were joined.

EXPERIMENTALLY INDUCED POPULATION GROWTH

Population growth was investigated by Petrusewicz (1958b, 1960, 1963) in two groups of experiments:

(1) Forty-seven populations which had been in existence for at least a year were transferred to another cage, which might be larger, smaller, or the same size as the original.

(2) One hundred and twenty-seven populations were manipulated by:

(a) Removing several individuals (both males and females) and then returning them a week later.

(b) Introducing a few virgin females for about a week at a time.

Such treatments significantly increased both birth rate and survival, and hence the probability of population growth. From this it was inferred that the introductions or removals disturbed the established structure of the populations (involving individual interactions, parental care, territorial behaviour, mating pattern, etc.), so that the animals behaved as if they were a newly-established population. i.e. an unorganized set of individuals, which always tends to grow. Hence it seems legitimate to speak of structure-dependent as well as density-dependent factors controlling population dynamics. Indeed Petrusewicz & Trojan (1963) showed that the smaller the area of a cage, the higher the density attainable. This result applies to other species besides house mice.

These ideas led on to work on wild mice living in the Field Station of the Institute of Ecology at Dziekanów Leśny, involving mark–release–recapture experiments to determine population dynamics, and hence the time individuals persisted in the population and the rate at which they disappeared, trappability, diurnal cycles of activity, etc. (Petrusewicz & Andrzejewski, 1962; Wierzbowska & Petrusewicz, 1963; Kaczmarzyk, 1964; Adamczyk & Petrusewicz, 1966). Although there were no physical barriers, populations differing in dynamics existed in the loft and the rest of the building, with very limited exchange of individuals. The loft mice were permanently supplied with additional food at four points. Except when the density was very low, particular mice were associated with each point.

In one experiment in the loft, Adamczyk & Ryszkowski (1965) removed all the resident mice and introduced 45 new wild ones, followed by another 20 two months later and a further 20 after another two months. The aim was to compare the animals released into the empty loft and those introduced when it was inhabited. They found:

(1) The survival rate was higher in the first releases (into the empty loft).

(2) The disappearance rate was stable for mice released in the inhabited loft, but variable for those in the empty loft.

(3) The reproductive rate was higher in the empty loft than in the occupied one.

All these facts suggest that interactions between individuals are of great importance to colonization processes, underling again the role of structure in population dynamics.

A further experiment involved the loft mice being replaced by 80 white and 52 wild mice in order to compare the two groups (Reimov, Adamczyk & Andrzejewski, 1968). The trappability of the white mice was higher than that of the wild mice and they tended to associate more with particular trapping points than wild ones. White females only mated with wild males on a few occasions. The authors suggested that there was competition between the two groups with the white animals being dominant.

PRODUCTIVITY

During the 1960s, the studies concentrated on productivity. In the laboratory it was relatively easy to determine the various parameters necessary to calculate population production, and the effect of varying different ecological factors on the productivity of white mouse populations was studied. The research was particularly concerned with developing methods for measuring such parameters, particularly ones which could be used to estimate production in other rodents under field conditions. For example, Ryszkowski, Walkowa & Wierzbowska (1967) devised a new method for comparing the age structure of populations at different times, and hence computing the average length of life.

Walkowa & Petrusewicz (1967) and Walkowa (1967) estimated biomass production in four white mouse colonies kept for several years. All populations were started with the same number of individuals and kept under identical conditions, but there were large differences in their productivity. The proportion of productivity attributable to reproduction and that to weight losses were fairly constant in time and between different populations. Individual growth was greatest in the densest populations. Biomass production was calculated by summing body weight increases and taking into account length of life, average number of individuals, mean weight, etc., and then transformed into energy units by measuring the ash content of the body and estimating age-related changes in the calorific value of the body (Myrcha & Walkowa, 1968). Newborn mice were characterized by a low energy content. However the calorific value of 1 g ash-free dry weight does not change during postnatal growth.

In addition, changes in the body weight of females during pregnancy and lactation was compared with the amount of food ingested and metabolizable energy produced (Myrcha, Ryszkowski & Walkowa, 1969). The total amount of metabolizable energy used by a pregnant female was 344 kilojoules, and 1487 more kilojoules were used during 26 days of lactation.

From these data, Myrcha (1975) estimated bioenergetic indices for singly kept mice. Measurements were made of food consumption, assimilation, faeces and urine, oxygen uptake, and carbon dioxide production. The same measurements were made on whole mouse populations raised in metabolism cages. Using the estimates for single animals, the expected instantaneous and cumulative energy budgets of the population could be computed and compared with the empirical measurement for the whole population. It was clear that the energy budget of the population was far more economical than that of the same number of individuals kept singly. Further experiments showed that the level of metabolic turnover is largely influenced by population size and density.

EXPLOITATION

Walkowa (1969, 1971) analysed the effect of exploitation by removing individuals at 12-week intervals from a set of 40 confined white mouse populations. The exploitation rate varied from zero to 67%. As it increased:

(1) Survival of young mice also increased.

(2) Numbers (and biomass) fell, body weight losses were reduced, and survival of 11–14-week-old mice declined.

Some variables, such as the average number of young per female and the number of adult females, fluctuated in proportion to the exploitation rate. Biomass production and total number of individuals were compensated for only when less than a third of the mice were removed. The main compensating factor was the survival rate of young mice.

A similar experiment was conducted on the wild mouse population in the loft (Adamczyk & Walkowa, 1971). This population was divided into two "subpopulations", one of which was exploited at a rate of 32% of the standing crop at monthly intervals, and the other not exploited at all. Biomass production was higher in the exploited subpopulation. Since no differences were observed in either the number of young born or the number of immigrants between the two parts of the loft, it was concluded that the compensation was due to a higher survival rate among the young and an increase in the time of residence of individuals in the loft.

STABLE NUMBERS

The experiments described so far were carried out on freely reproducing populations, i.e. their *numbers* were not manipulated – the exploitation experiments removed animals at various rates, but did not seek to regulate the numbers in any of the populations concerned.

The next series of experiments involved populations kept at a stable level by human culling. Their aim was to find the conditions under which optimum production could be achieved in a small space. The method used was to remove all young four to six weeks after birth, except for those needed to replace deaths in the parental generation.

Earlier experiments had shown that mortality among young is high at high population densities. Adamczyk & Walkowa (1974) estimated the number of mice born in 83 populations by three methods:

(1) Daily recording.

(2) Daily recording plus the estimated number of newborn mice that did not survive to be recorded.

(3) Number of pregnancies multiplied by average litter size.

The third method gave the most reliable estimates.

Adamczyk (1977) analysed the effect of numbers and density of adults on the survival, reproduction, growth, and other production parameters of 68 populations of white mice. In some experiments there was a direct relationship between the initial number of individuals in a population (which ranged from two to 220 mice) and population density (varying between three and 344 per m^2), while in others the density was kept constant by using cages of various sizes. Two groups of populations were distinguished: small (two to 20 mice) where population size and density both affected the parameters studied; and large (40–220 mice) where only changes in the number of mice were important (Table I).

Population size and sex ratio affect behaviour (Kozakiewicz, 1976). For example, the frequency of females moving nestlings from place to place is directly proportional to population size and number of females. Mortality among young animals is related to the

TABLE I

Direction of changes in population parameters as an effect of changes in population size and density

Index	Size of the basic stock			
	Small		Large	
Total production	↗	↘ (dashed)	↗	—
Biomass removed	↗	↘ (dashed)	—	—
Percentage of total production of population				
Biomass of newborn	—	—	↗	—
Biomass removed	—	↘ (dashed)	—	—
Dead biomass	—	↗ (dashed)	—	—
Total number newborn	—	↘ (dashed)	↗	—
Number of newborn per female	↗	↘ (dashed)	—	—
Cannibalism	↗	↗ (dashed)	↗	—
Survival of young mice	↘	↘ (dashed)	—	—

The arrows denote the effect of population size (continuous) and density (dashed). Horizontal dashes denote a lack of response.

frequency they are moved by females if there are less than 12 males in the population; when there are more males, mortality rate in this group rapidly approaches 100%.

ENDOCRINOLOGICAL FACTORS

Hormone levels affect the response of mice to demographic factors (Zaleska-Freljan, 1979a, b, in press a, b). No changes have been observed in hormone levels in populations of less than ten. However, corticosteroid secretion increases significantly in populations of 20 to 60, but no relation has been found between number and corticosteroid levels in populations of 100 animals.

Corticosteroid levels drop with individual age and are affected by the age structure of the population. The plasma level of these hormones in 27-week-old males in populations of 100 animals was found to be higher when all individuals were of the same age than when they were of different ages.

Not surprisingly, a relationship was found between the amount of corticosteroids in females and their physiological state. The lowest levels are observed in non-pregnant females. Levels increase during pregnancy. They are influenced also by population size.

Both males and females have a daily rhythm of plasma corticosteroid level. There are seasonal changes in females.

FOOD

Work has been carried out on the effect of food supply and quality, and in particular the factors that modify consumption in a population. Walkowa & Adamczyk (in preparation) have studied these problems in 99 populations differing in:

(1) Numbers (20, 60, and 100 mice, kept stable for each population).

(2) Average age at the beginning of the experiment (7, 10, 15, and 27 weeks).

(3) Rate of replacement by introduced animals (0.5%, 30%, and 50% of the total population per month).

(4) Physical complexity of the cage (i.e. the number of dividing walls, albeit incomplete so that individuals had access to the whole cage).

The first stage was to determine the amount of food removed (not only consumed) from the supply points, growth in biomass, and the proportion of food removed which was converted into biomass. The effect on these was most pronounced for population size – although a low rate of food removal or conversion, or biomass increase, was also produced by a high rate of population turnover (i.e. exchange of individuals) or a high average age in the population. High values of the indices were associated with cage complexity.

Food removal depended mainly on population size and cage complication. When the population was large, each mouse took less food but the whole population removed more because of the larger number of individuals. In more complicated cases both the whole population and each individual removed more food on average.

In general the response of the populations to food depended on their size. In populations of 20 animals, the response increased in proportion to other factors (age, turnover, cage complexity), but in larger populations the effect of total number was so great that some indices did not change even when other factors were varied.

REFERENCES

Adamczyk, K. (1977). Effect of size and density of population of laboratory mice on the parameters determining their production. *Acta theriol.* **22**: 459–484.

Adamczyk, K. & Petrusewicz, K. (1966). Dynamics, diversity and intrapopulation differentiation of a free-living population of house mouse. *Ekol. pol.* (A) **14**: 725–741.

Adamczyk, K. & Ryszkowski, L. (1965). Settling of mice (*Mus musculus* L.) released in an uninhabited and an inhabited place. *Bull. Acad. pol. Sci.* (Sci. biol.) **13**: 631–637.

Adamczyk, K. & Walkowa, W. (1971). Compensation of numbers and production in a *Mus musculus* population as a result of partial removal. *Annls Zool. Fenn.* **8**: 145–153.

Adamczyk, K. & Walkowa, W. (1974). Estimating the number of newborn animals in enclosed populations of laboratory mice. *Acta theriol.* **19**: 247–257.

Andrzejewski, R., Petrusewicz, K. & Walkowa, W. (1963). Absorption of newcomers by a population of white mice. *Ekol. pol.* (A) **11**: 223–241.

Kaczmarzyk, K. (1964). Alimentary activity of a free house mouse population (*Mus musculus*, L.). *Bull. Acad. pol. Sci.* (Sci. biol.) **12**: 201–205.

Kozakiewicz, A. (1976). The effect of population density and sex ratio on the mortality of juvenile laboratory mice in experimental populations. *Acta theriol.* **21**: 339–350.

Myrcha, A. (1975). Bioenergetics of an experimental population and individual laboratory mice. *Acta theriol.* **20**: 175–226.

Myrcha, A., Ryszkowski, L. & Walkowa, W. (1969). Bioenergetics of pregnancy and lactation in white mouse. *Acta theriol.* **14**: 161–166.

Myrcha, A. & Walkowa, W. (1968). Changes in the caloric value of the body during the postnatal development of white mice. *Acta theriol.* **13**: 391–400.

Nowak, Z. (1971). The effect of removing a dominant on the social organization of laboratory mice populations. *Acta theriol.* **16**: 61–71.

Petrusewicz, K. (1958a). Influence of the presence of their own population on the results of fights between male mice. *Bull. Acad. pol. Sci.* (Sci. biol.) **6**: 25–28.

Petrusewicz, K. (1958b). Investigation of experimentally induced population growth. *Ekol. pol.* (A) **5**: 281–309.

Petrusewicz, K. (1959a). Further investigation of the influence exerted by the presence of their home cages and own populations on the results of fights between male mice. *Bull. Acad. pol. Sci.* (Sci. biol.) 7: 319–322.

Petrusewicz, K. (1959b). Research on rapidity of the formation of interpopulation relations and sense of ownership of cages in mice. *Bull. Acad. pol. Sci.* (Sci. biol.) 7: 323–326.

Petrusewicz, K. (1960). An increase in mice population induced by disturbance of the ecological structure of the population. *Bull. Acad. pol. Sci.* (Sci. biol.) **8**: 301–304.

Petrusewicz, K. (1963). Population growth induced by disturbance in the ecological structure of the population. *Ekol. pol.* (A) **11**: 87–125.

Petrusewicz, K. & Andrychowska, R. (1960). Further investigation of the influence of the home cage on the result of fights between male mice. *Ekol. pol.* (A) **8**: 325–333.

Petrusewicz, K. & Andrzejewski, R. (1962). Natural history of a free-living population of house mice (*Mus musculus* Linnaeus) with particular reference to groupings within the population. *Ekol. pol.* (A) **10**: 85–122.

Petrusewicz, K. & Trojan, P. (1963). The influence of the size of the cage on the numbers and density of a self-ranging population of white mice. *Ekol. pol.* (A) **11**: 612–614.

Petrusewicz, K. & Wilska, T. (1959). Investigation of the influence of interpopulation relations on the result of fights between male mice. *Ekol. pol.* (A) 7: 357–390.

Reimov, R., Adamczyk, K. & Andrzejewski, R. (1968). Some indices of the behaviour of wild and laboratory house mice in a mixed population. *Acta theriol.* 13: 129–150.

Ryszkowski, L., Walkowa, W. & Wierzbowska, T. (1967). Estimation of average length of life of mice having variable survival rates. *Ekol. pol.* (A) **14**: 791–801.

Walkowa, W. (1964). Rate of absorption of newcomers by a confined white mouse population. *Ekol. pol.* (A) **12**: 325–335.

Walkowa, W. (1967). Production due to reproduction and to body growth in a confined mouse population. *Ekol. pol.* (A) **15**: 819–822.

Walkowa, W. (1969). Operation of compensation mechanisms in exploited populations of white mice. In *Energy flow through small mammal populations:* 247–253. Petrusewicz, K. & Ryszkowski, L. (Eds). Warsaw: Polish Sci. Publs.

Walkowa, W. (1971). The effect of exploitation on the productivity of laboratory mouse populations. *Acta theriol.* **16**: 295–328.

Walkowa, W. & Adamczyk, K. (In preparation). *Food utilization by a laboratory mouse population under different ecological conditions.*

Walkowa, W. & Petrusewicz, K. (1967). Net production of confined mouse populations. In *Secondary productivity of terrestrial ecosystems* 1: 335–347. Petrusewicz, K. (Ed.). Warsaw-Kraków: Polish Sci. Publ.

Wierzbowska, T. & Petrusewicz, K. (1963). Residency and rate of disappearance of two free-living populations of the house mouse (*Mus musculus* L.). *Ekol. pol.* (A) **11**: 557–574.

Zaleska-Freljan, K. I. (1979a). Diurnal changes in the level of blood plasma corticosteroids in laboratory mice in relation to sex and season. *Bull. Acad. pol. Sci.* (Sci. biol.) 7: 571–574.

Zaleska-Freljan, K. I. (1979b). Changes in the level of corticosteroids in blood plasma of male laboratory mice in relation to their number, age and type of population. *Bull. Acad. pol. Sci.* (Sci. biol.) 7: 575–580.

Zaleska-Freljan, K. I. (In press a). Characteristic of the changes of the corticosteroid level in blood plasma of mice in relation to their physiological state and the size of population. *Bull. Acad. pol. Sci.* (Sci. biol.)

Zaleska-Freljan, K. I. (In press b). The changes of the corticosteroid level in the blood plasma of the laboratory mice in relation to size of group, isolation and age of mice. *Bull. Acad. pol. Sci.* (Sci. biol).

Symp. zool. Soc. Lond. (1981) No. 47, 439–453

Population Immunogenetics of Murine *H–2* and *t* Systems

J. KLEIN, D. GÖTZE, J. H. NADEAU and E. K. WAKELAND

Abteilung Immungenetik, Max-Planck Institut für Biologie, Corrensstrasse 42, 7400 Tübingen, West Germany

SYNOPSIS

The results of H-2 typing of 445 wild mice from the United States, Europe, and Northern Africa are summarized. The mice were typed for 34 class I antigenic determinants (18 *K*-locus controlled and 16 *D*-locus controlled), and 13 class II determinants (10 *A*-locus controlled and 3 *E*-locus controlled). When the data from the individual populations are pooled and the global population of *Mus musculus* considered, and when artefactually high frequencies are disregarded, the average gene frequency of the two class I loci can be calculated to be about 0.025. All class I alleles thus far detected have this same low frequency and even this is probably an over-estimate. We predict that the world-wide population of *Mus musculus* contains at least 100 class I alleles.

The average phenotype frequencies of private determinants controlled by the class II *A* locus ranged from 8% to 29%; those of the *E*-locus controlled determinants ranged from 40% to 80%. Hence the class II loci, in particular the *E*-locus, are probably less polymorphic than the class I loci, although more reliable data will be necessary to support this conclusion. More than 90% of the wild mice tested are heterozygous at the *K*, *D*, and *A* loci; about 50% are heterozygous at the *E* locus. Each local population contains only about a dozen or so identifiable alleles at the class I loci. The frequencies of individual antigenic determinants in each local population fall on a negative exponential curve. The composition of alleles in each population is so characteristic that *H-2* typing can be used for tracing the origin of wild mice. The strong linkage disequilibrium between the *H-2* and *t* complexes observed in laboratory stocks, has been confirmed by typing wild mice.

A POPULATION GENETICIST'S DREAM

If a population geneticist were asked to describe an ideal genetical system, he might come up with something like this. Firstly, the products of the system would have to be easily detectable, expressed in codominant fashion, and present in many different tissues. Secondly, the system would have to be highly polymorphic so that genetical variants would be present in every population tested. The polymorphism, however, would have to be limited so that variants characterizing individual populations could become fixed. Thirdly, the system would have to be multigenic so that parameters such as linkage relations or gene interaction could be studied. Fourthly, the function of the loci composing the system would have to be partially known so that this system would be amenable to selection studies. The fastidious population geneticist might also insist upon a fifth feature, namely on this system's association with a recombination suppressor capable of maintaining strong linkage disequilibrium that could be used to trace the origin of different populations.

We believe that the *H-2* complex of the mouse is such a system. Virtually all molecules controlled by the *H-2* loci can be assayed serologically, using methods that are among the fastest and the most sensitive in gene-product detection (Klein, 1975, 1979). The *H-2* antigens are codominantly expressed and are present in virtually all somatic tissues. The *H-2* loci are the most polymorphic known in the mouse, so much so that a population in which these loci do not vary is yet to be found. The *H-2* complex consists of at least 10 loci, all clustered within a short chromosomal region. Although the function of these loci is not yet fully understood, enough is known to allow one to design meaningful selection experiments. The *H-2* complex is linked to the *t* system which, among other things, drastically reduces crossing-over frequency in the *t* – *H-2* interval (Hammerberg & Klein, 1975a).

Bearing this knowledge in mind, one may wonder why not all mouse population geneticists are studying the *H-2* system. The answer to this question lies in the fact that any study of *H-2* population genetics must be preceded by a long preparatory phase in which one has to produce the necessary typing reagents and characterize the antigens present in wild mouse populations. In this laboratory, we started to study the *H-2* system of wild mice some 10 years ago (Klein, 1970, 1971, 1972) and only now are we in a position to cash in on this investment (Zaleska-Rutczynska & Klein, 1977; Duncan, Wakeland & Klein, 1979a, b; Wakeland & Klein, 1979; Götze *et al.,* 1980). In this communication we shall summarize, after a brief description of the *H-2* complex, our most recent findings concerning *H-2* population genetics, and then discuss some of the implications of these findings.

THE *H-2* COMPLEX

The *H-2* complex, the major histocompatibility complex (MHC) of the mouse, is a cluster of closely linked loci controlling molecules involved in the recognition of foreign substances by lymphocytes (Klein, 1979). According to a currently popular hypothesis, a subpopulation of lymphocytes, the so-called T cells, recognize foreign substances (antigens) only when these are presented to the cells together with the individual's own H-2 molecules. There are two main groups of loci in the *H-2* complex: class I loci controlling molecules that participate in antigen recognition by cytolytic T cells (i.e., T cells capable of killing appropriate target cells), and class II loci involved in antigen recognition by regulatory T cells (i.e., T cells that enhance or suppress immune response of other lymphocytes). The chromosomal region occupied by class I and class II loci also contains other loci, whose relationship to the MHC proper is not clear. The most prominent among these accompanying loci are those coding for certain complement components (complement being a system of enzymes involved in the destruction of certain bacteria and other cells sensitized by specific, complement-fixing antibodies). It is convenient to group the *H-2*-associated, complement-encoding loci under the heading of class III loci; various other loci intimately associated with the *H-2* complex can be referred to as class IV loci.

In this communication, we shall concern ourselves only with the class I and class II loci. There are two class I loci in the *H-2* complex, *K* and *D*, located at opposite ends of the *H-2* map (Fig. 1). Of the

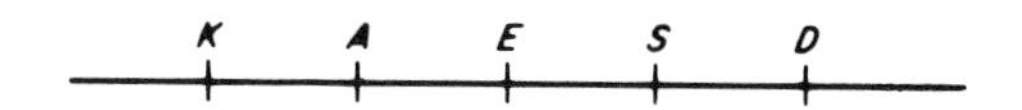

FIG. 1. A simplified map of the *H-2* complex. *K*, *D* = class I loci; *A*, *E* = class II loci; *S* = class III (complement-encoding) locus.

five known class II loci (*A*, *B*, *J*, *E*, and *C*), we shall deal here only with two: *A* and *E* (the others are difficult to detect serologically). The molecules controlled by the *K*, *D*, *A*, and *E* loci are anchored in the cell membrane and exposed on the cell surface. Two individuals often differ in the biochemical composition of their H-2 molecules and this difference is recognized by the recipient's immune system when H-2 molecules of one individual are introduced into the other individual. The recipient responds to such an introduction by producing antibodies to the foreign H-2 antigenic determinants, and it is these antibodies that we use as typing reagents to distinguish between H-2 molecules of different mice. The individual determinants recognized by the antibodies are referred to by numbers with one series of numbers used for class I (K and D) determinants, and another series of numbers for class II (A and E) determinants. A particular combination of class I and class II alleles carried by a single chromosome is referred to as an *H-2* haplotype.

The *H-2* complex is located in chromosome 17, some 17 map units from the centromere and some 15 map units from the *t* (T) complex. The properties of the *t* complex are discussed by Dr Mary F. Lyon (this volume p. 455); here it suffices to mention that there are at least seven *t* complementation groups and that a genetical element in the *t* complex reduces the crossing-over frequency between *T* and *H-2* from 15% to less than 1%.

PHENOTYPE AND GENE FREQUENCIES

Two H-2 molecules controlled by distantly related alleles differ in some 50 of the approximately 400 amino acids composing the H-2 polypeptide chain. There is no known eukaryotic locus besides the *H-2* complex, the allelic products of which differ by so many amino acid substitutions. The biochemical disparity between H-2 molecules is reflected in their serological complexity. The surface of each allelic H-2 molecule can be compared to a distinctive landscape in which the individual features are recognized by different antibodies. In serological terminology, one says that each H-2 molecule carries several antigenic determinants. Some of the features (determinants) are repeated in many different landscapes, others are much more limited in their appearance. The former determinants are referred to as public and the latter as private. The difference between the two is that sharing of private determinants between two molecules usually implies close relatedness of these molecules, whereas sharing of public determinants has no such significance. However, private determinants are not allele-specific: it is known that mutations leading to single amino acid substitutions and thus to the generation of new alleles often do not change private determinants in a serologically demonstrable way (for a review see Klein, 1978). To identify individual alleles serologically it would be best to type mice for all private and as many public determinants as possible; different alleles would then be recognized by their particular, unique constellations of determinants. Such an approach was not feasible at the time we initiated the studies described here. We therefore chose the second best approach, based on typing mice for their private antigenic determinants. The limitations of this approach, however, should be kept in mind when translating phenotypic into genetic frequencies.

Another important point is that, because of the specificity of the typing reagents, the only molecules detected are those carrying determinants defined by one of the reagents in the battery. A molecule carrying a private determinant for which no reagent is available will appear on typing as a "blank". Although individual blanks may represent different molecules (alleles), they will appear as an undifferentiated group. Hence, the greater the frequency of blanks the lower the information about a given population.

After these explanatory notes, we can now examine the data on antigenic frequencies gathered by our laboratory thus far (Tables I–III). The data consist of two different samplings carried out in

TABLE I

Frequencies of K-*locus antigenic determinants in two samples of wild mice.*

Antigenic determinants	*Phenotypic frequency (%)*[a] USA	Europe	Average	Average allele frequency (%)
15	0.0	3.4	2.4	1.26
16	1.6	13.7	10.3	5.41
17	3.2	2.0	2.4	1.26
19	4.0	7.8	6.7	3.51
20	3.2	0.3	1.1	0.58
21	4.8	7.1	6.4	3.36
23	2.4	5.9	4.9	2.57
26	2.4	6.5	5.3	2.78
31	11.2	14.6	13.4	7.03
33	1.6	7.6	5.9	3.04
103	13.6	4.0	6.7	3.51
108	4.8	8.7	7.6	3.99
109	4.0	28.4	21.5	11.28
113	3.2	2.5	2.7	1.42
115	15.2	3.5	6.8	3.75
116	6.4	1.2	2.7	1.42
123	8.0	2.4	4.0	2.10
124	8.0	6.5	7.0	3.67
Blank				38.01

[a]Sample size: USA 125 mice, Europe 320 mice.
Data from Duncan *et al.* (1979a); Götze *et al.* (1980).

North America and in Europe (together with a few localities in North Africa).

The total sample consisted of 445 wild mice from two continents (125 mice from the United States and 320 mice from Europe). The sample was tested for the presence of 34 class I antigenic determinants (18 *K*-locus determinants and 16 *D*-locus determinants) and 13 class II determinants (10 A and 3 E determinants).

TABLE II

Frequencies of D-*locus antigenic determinants in two samples of wild mice.*

Antigenic determinants	Phenotypic frequency (%)[a] USA	Europe	Average	Average allele frequency (%)
2	8.8	2.5	4.3	2.26
4	1.6	15.9	11.6	6.09
9	8.8	1.0	3.6	1.89
18	0.0	0.6	0.5	0.24
22	nt[b]	0.0	0.0	0.00
30	3.2	6.0	5.2	2.73
32	1.6	5.4	4.2	2.26
106	3.2	4.4	4.0	2.11
107	nt	25.9	25.9	13.60
110	3.2	0.3	1.1	0.59
111	4.0	4.1	4.1	2.11
112	0.8	1.0	0.9	0.47
114	1.6	7.3	5.6	2.97
117	nt	13.7	13.7	7.19
118	22.4	13.4	15.9	8.34
122	3.2	13.1	10.3	5.40
Blank				41.90

[a] Sample size: USA 125 mice, Europe 320 mice.
[b] nt, not tested.
Data from Duncan *et al.* (1979a); Götze *et al.* (1980).

Let us consider the class I determinants first. All the determinants but one (H-2.22) were found in the total sample of mice; in the sample from the United States all but two determinants (H-2.15 and H-2.18) were found. There was no striking difference in frequency between K and D determinants. Among the K determinants, determinant H-2.20 occurred at the lowest (1.1%) and determinant H-2.109 at the highest (21.5%) frequency. The frequencies of most of the K determinants were between 2% and 7%. Among the D determinants detected in the population the least frequent was H-2.18 (0.5%) and the most frequent was H-2.107 (25.9%); the average frequency of most of the determinants was between 4% and 6%. Thus, in both groups, the frequencies of the determinants were fairly uniform, except for a few that were twice (or more) as frequent as the rest of the determinants. The more frequent determinants were H-2K.16 (10.3%), H-2K.31 (13.4%), H-2K.109 (21.5%), H-2D.4 (11.6%), H-2D.107 (25.9%), H-2D.117 (13.7%0, H-2D.118 (15.9%), and H-2D.122 (10.3%). In most of these cases, the high frequency is very likely artefactual in the sense that each single determinant frequency is in fact a combined frequency of at least two determinants. Reagents detecting H-2 determinants 16, 109, 107, 117, 118, and

TABLE III

Frequencies of class II antigenic determinants in two samples of wild mice.

Antigenic determinant	Phenotypic frequency (%)[a] USA	Europe
A-locus determinants		
1	36	nt[b]
2	12	25
4	4	17
11	7	8
12	18	37
14	nt	9
19	16	nt
20	nt	29
24	31	nt
31	34	nt
E-locus determinants		
7	82	nt
21	nt	38
32	58	61

[a] Sample size: USA 125 mice, Europe 320 mice.
[b] nt, not tested.
Data from Duncan *et al.* (1979a); Götze *et al.* (1980).

122 were produced in strain combinations differing at both *K* and *D* loci, and they very likely contain at least two different antibodies. (At the time of the reagent production, strains were not available that would have allowed the specific removal of one of these antibodies and thus the production of a monospecific antiserum.) Furthermore, anti-H-2.107, 109, 117, 118, and 122 sera were originally thought to be detecting private antigens, but in fact turned out to be directed against public antigens. These reagents, therefore, each identify a group of alleles, and the members in each group may not even be closely related. Reagents detecting determinants H-2.4 and H-2.31 appear to be monospecific, however, and the high frequencies of these determinants are almost certainly the result of a sampling bias (these two determinants were among the most frequent ones in the two largest samples in our collection – one from Denmark and the other from Egypt). After elimination of the above determinants, the frequencies of other H-2 determinants do not exceed 7% and usually are lower than this value.

The interpretation of the class II antigen frequencies is complicated by two main factors. The first is the technical difficulty in typing. Since the determinants are expressed only in a subpopulation of lymphocytes constituting a maximum of 50% to 60% of cells, positive reactions, especially in *H-2* heterozygotes carrying a single antigen dose, are sometimes difficult to distinguish from false positives due to a high background. The second factor is the reactivity of reagents suitable for class II antigen typing. We have been able to type for 13 class II determinants; of these, however, only five were of the private category – the rest were public determinants. Also, the quality of the reagents was inferior to that of the class I reagents. Because of these complicating factors, the data obtained concerning class II determinants must be considered as highly tentative.

The most striking finding of this preliminary analysis is that class II determinants are more frequent than class I determinants. The frequencies of the private class II determinants (i.e. determinants 2, 4, 11, 14, and 20, all encoded by the *A* locus) range from 8% to 29%; those of the public determinants are even higher. Whether this observation means that there are fewer class II than class I alleles or whether it merely indicates complexity of the class II determinants remains to be decided. The second striking finding is that the frequencies of the *E*-encoded determinants are unusually high (relative to class I loci determinants): they range from 40% to 80%. It may be that thus far only the most public of the *E*-encoded determinants have been detected; alternatively there may not be any *E*-encoded private determinants.

Heterozygosity of the *H-2* Loci

If antigenic markers were available for all the alleles present in a given wild mouse population, one could determine the frequency of *H-2* heterozygotes by gene counting. However, since we are still far from this ideal situation, this simple method cannot be used. The presence of only one private antigenic determinant controlled by a given locus in a given mouse could mean either that the mouse is homozygous at this locus or that the second allele is serologically silent. Thus, at this time, the only reliable way of heterozygosity determination is by progeny testing. Barring the presence of *H-2* linked segregation distorters, absence of segregation in the progeny of a cross between a wild and an inbred mouse indicates that the wild parent is an *H-2* homozygote.

We have progeny-tested thus far over 60 informative wild males from localities in the United States and in Europe. Since no significant difference in *H-2* heterozygosity between the two samples could be observed, the data were pooled (Table IV). The data indicate that *K*, *D* and *A* homozygotes are rare: some 90% or more of

TABLE IV

Frequency of H-2 *heterozygotes among wild mice.*

Locus	No. heterozygotes / total tested	Frequency of herterozygotes
K	61/64	0.95
D	48/54	0.88
A	10/11	0.90
E	9/19	0.47

Data from Duncan *et al.* (1979b); Nadean *et al.* (1981).

the animals are heterozygotes at these three loci. At the class II *E* locus, homozygous and heterozygous animals are about equally represented.

Gene Frequencies

There are several problems associated with the translation of phenotypic into gene frequencies. Two of them we have already mentioned: the fact that private determinants are, at best, markers for groups of related alleles rather than for individual alleles, and also that some of our typing reagents might simultaneously detect products of two loci. Another problem is the sampling error: since there are significant differences in determinant frequencies among individual populations, the global frequencies would only be reliable if – ideally – all the populations were sampled, and the sample size was proportional to the size of the individual populations. We have, obviously, not sampled all the populations, and we have only rough indications of correspondence between the sample and the population size from the "trappability" of mice at a given locality. All three factors tend to inflate the gene frequencies so that the actual gene frequencies are probably lower than those indicated by the phenotypic frequencies.

When one considers the crude data, one obtains the class I gene frequencies given in Tables I and II. Ignoring the few frequent determinants, one can estimate the average gene frequency to be about 2.5% to 3%. This figure suggests that there are, in the worldwide population of *Mus musculus*, some 40 to 50 *equally frequent* alleles. Such a situation is most unusual in two respects: in the high number of alleles, and the fact that all these alleles occur at equal frequencies. However, if this figure is an over-estimate, as we believe it is, then the actual number of alleles could be at least twice as high, which would mean over 100 alleles present in the world-wide population. The high frequency of blanks tends to support the latter estimate (unless, of course, there is a highly frequent allele among the blanks; if so, it would be a strange coincidence that we have not detected it thus far).

The class II data are too premature to allow any firm conclusions. There must be, however, at least 20 to 30 *A*-locus alleles, and at least five *E*-locus alleles, though these values, again, may be considerable underestimates.

Geographical Differentiation

Figure 2 shows the frequencies of class I determinants at widely separated localities in Europe and North Africa. The numbers of determinants detected at each locality at appreciable frequencies vary surprisingly little: from eight to 15 for the *K*-encoded determinants and from seven to 12 for the *D*-encoded determinants. The composition of these determinants, however, varies considerably from locality to locality (at each locality, however, the gradation of the determinant frequencies from the most common to the least frequent is usually exponential). In fact, the frequency of the most common K and the most common D determinants is an excellent marker that unambiguously identifies a given locality. In the near future, we should be able to refine these markers to the degree that we will be able to tell where in the world a small sample of mice sent to us came from. This observation brings us rather close to the fulfilment of even the wildest dream of any population geneticist.

H-2—t Disequilibrium

The presence of several loci in the *H-2* cluster provides an opportunity to search for associations of the individual loci at the population level. We are presently evaluating our data for just such associations. However, one association leading to a strong linkage disequilibrium is already clearly established, namely that between the *H-2* and the *t* complexes. Some time ago, we observed that in laboratory mouse stocks, certain *t* alleles were always associated with certain *H-2* haplotypes (Hammerberg & Klein, 1975b; Hammerberg *et al.*, 1976). We have now extended this observation to *t* alleles extracted directly from wild mouse populations. Table V lists 23 typed t chromosomes. With one exception, similar lethal *t* alleles (i.e., alleles belonging to the same complementation group) are associated with the same *H-2* haplotype. The one exception (t^6) may be a laboratory recombinant. This striking linkage disequilibrium

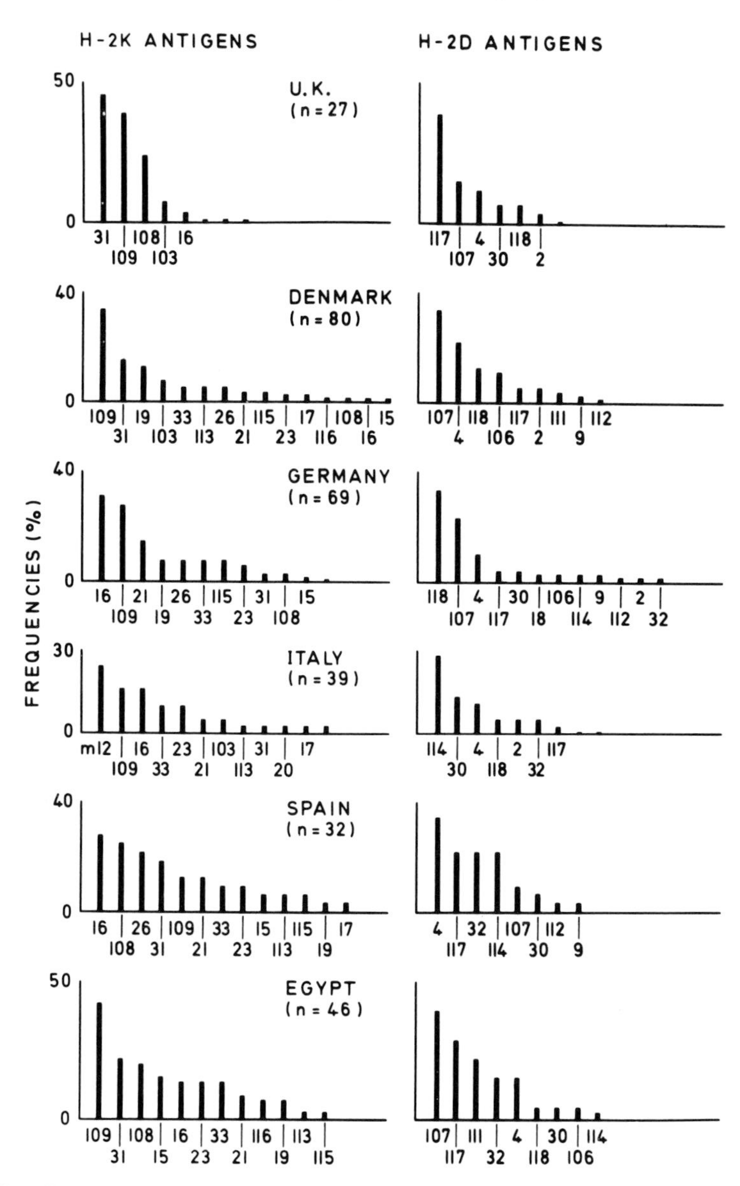

FIG. 2. Frequencies of class I antigenic determinants at different localities in Europe and North Africa.

between *t* and *H-2* is undoubtedly maintained because of the recombination suppression by the *t* complex, but the fact that it is so universal must be taken into account when considering the evolutionary origin of the *t* complex.

TABLE V

Correlation between t *assignment and* H-2 *classification of* t*-bearing chromosomes.*

t chromosome	Origin	t group	*H-2* haplotype
t^{12}	Lab. stock?	t^{12}	$w28$
t^{w32}	Montana, USA		$w28$
t^{0}	France	t^{0}	$w29$
t^{1}	France		$w29$
t^{6}	Lab. stock		$w30$
t^{w1}	New York, USA	t^{w1}	$w30$
t^{w12}	California, USA		$w30$
t^{w71}	Denmark		$w30$
t^{w5}	New York, USA	t^{w5}	$w31$
t^{w75}	East Germany		$w31$
t^{w93}	New York, USA		$w31$
t^{w94}	New York, USA		$w31$
t^{w95}	New York, USA		$w31$
t^{w97}	New York, USA		$w31$
t^{w105}	New York, USA		$w31$
t^{Tu1}	Texas, USA		$w31$
t^{w73}	Denmark	t^{w73}	$w32$

Based on Hammerberg & Klein (1975b), and on E. Hsu, D. Bennett, K. Arzt & J. Klein (unpublished data).

INTERPRETATION

The picture emerging from our *H-2* studies of wild mouse populations is that in the world population, there are a large number of alleles (at least of class I loci), of which only a dozen or so occur in each local population (the latter number will undoubtedly increase when all the blanks are replaced by identifiable alleles). In the world-wide population, all *H-2* alleles thus far tested occur with equally low frequencies of less than 2.5%. In each local population, however, the frequencies of individual antigenic determinants fall on a negative exponential curve. The composition of alleles in each local population is characteristic of that population.

How could this arrangement have arisen? The interpretation that we favour at this time is this. Each locality was originally colonized by a few mice bringing in a small pool of *H-2* alleles. At some localities, this colonization occurred a long time ago, in others it is of more recent date. At each locality, the founder population expanded and during this expansion the present-day profile of *H-2* alleles became fixed. A strong force modelling this profile was probably selection exercised by the local assortment of pathogens. As mentioned earlier, the main function of the *H-2* complex appears to be the marking of self for the recognition of non-self. The fault in this system of recognition is that in certain combinations of self and non-self the latter for some reason fails to be identified as foreign, and the individual is thus unresponsive to such combinations. The pathogens probably exploit this weakness to their own advantage. The *H-2* polymorphism is most likely a way of overcoming this imperfection at the population level: although some individuals in a given population may be defenceless against a given pathogen because they carry the wrong *H-2* alleles, others, with different alleles, are normally responsive and thus survive the attack of the pathogen. Which alleles increase in frequency in a given population may, therefore, depend on the predominant pathogens in this population. Since different environments may be infested with different pathogens (or genetical variants thereof), different *H-2* alleles will be favoured in different populations.

The question that remains is how stable are the allelic profiles of the individual populations? The mutation rates of at least some of the *H-2* genes appear to be unusually high (Klein, 1978), and mutations must have been the prime force in generating the extensive *H-2* polymorphism of the world-wide population in the first place. Someone might, therefore, argue that mutations will tend to change the profiles rapidly. However, the argument may be false, as drift and selection will work against such changes, and the profiles may remain stable for as long as the spectrum of pathogens in a given environment does not change.

An important marker for tracing the pattern of colonization will be the *t* alleles. The limited number of *t* complementation groups and the persistence of linkage disequilibrium between *t* and *H-2* in each group suggest that the generation of *t* complexes is a rare event that might have occurred only a few times in the history of *Mus musculus*. If so, then by sampling individual populations for *t* alleles, one should be able to establish a genealogical relationship among these populations and arrive at an understanding of how the present-day organization arose historically. Thus far, *t* alleles have been extensively sampled only in North American populations, which, however – because of their recent origin – may be the most uniform and least informative. Sampling of European populations should prove to be much more rewarding in this respect.

ACKNOWLEDGEMENTS

We thank Ms D. Smith, Ms Eva Illgen, and Mr S. Forster for technical assistance, Ms R. Franklin for editorial help, Ms L. Yakes and Ms K. Bartels for typing the manuscript. The following persons have supplied us with wild mice:

Professor R. J. Berry, University College London, University of London, London, UK;

Dr. F. Bonhomme, Laboratoire d'évolution des vertébrés, Paléontologie et Génétique, Faculté des Sciences, Montpellier, France;

Dr I. K. Egorov, Division of Immunology, Duke University Medical Center, Durham, North Carolina, USA;

Dr A. Gropp, Institut für Pathologie, Medizinische Hochschule Lübeck, Lübeck, Federal Republic of Germany;

Dr J. P. Hjorth, Institute of Ecology and Genetics, University of Århus, Århus, Denmark;

Dr H. Hoogstraal, United States Naval Medical Research Unit No. 3, Cairo, Egypt; and

Dr J. Vives, Hospital Clinico y Provincial, Servicio de Immunologia, Barcelona, Spain.

This study was supported by NIH grant AI 14736 (to J. K.) and grants from the VW Foundation (to J. K.), and the Deutsche Forschungsgemeinschaft, Forschergruppe Leukämie, Tübingen (to D. G.). E. K. Wakeland is the recipient of the US Public Health Service Research Fellowship 2F32 A105531–02; J. Nadeau is the recipient of a Fellowship of the Max Planck-Gesellschaft, Müchen, FRG.

REFERENCES

Duncan, W., Wakeland, E. K. & Klein, J. (1979a). Histocompatibility-2 system in wild mice. VIII. Frequencies of H-2 and Ia antigens in wild mice from Texas. *Immunogenetics* 9: 261–272.

Duncan, W., Wakeland, E. K. & Klein, J. (1979b). Heterozygosity of the *H-2* loci in wild mice. *Nature, Lond.* **281**: 603–605.

Götze, D., Nadeau, J., Wakeland, E. K., Berry, R. J., Bonhomme, F., Egorov, I. K., Gropp, A., Hjorth, J. P., Hoogstraal, H., Vives, J. & Klein, J. (1980). Histocompatibility-2 system in wild mice. X. Frequencies of *H-2* and *Ia* antigens in wild mice from Europe and Africa. *J. Immunol.* **124**: 2675–2681.

Hammerberg, C. & Klein, J. (1975a). Linkage relationships of markers on chromosome 17 of the house mouse. *Genet. Res.* **26**: 203–211.

Hammerberg, C. & Klein, J. (1975b). Linkage disequilibrium between *H-2* and *t* complexes in chromosome 17 of the mouse. *Nature, Lond.* **258**: 296–299.

Hammerberg, C., Klein, J., Arzt, K. & Bennett, D. (1976). Histocompatibility-2 system in wild mice. II. *H-2* haplotypes of *t*-bearing mice. *Transplantation* 21: 199–212.

Klein, J. (1970). Histocompatibility-2 *H-2* polymorphism in wild mice. *Science, N. Y.* 168: 1362–1364.

Klein, J. (1971). Private and public antigens of the mouse *H-2* system. *Nature, Lond.* 229: 635–637.

Klein, J. (1972). Histocompatibility-2 system in wild mice. I. Identification of five new *H-2* chromosomes. *Transplantation* 13: 221–299.

Klein, J. (1975). *The biology of the mouse histocompatibility complex.* Berlin: Springer-Verlag.

Klein, J. (1978). *H-2* mutations: Their genetics and effect on immune functions. *Adv. Immunol.* 26: 55–146.

Klein, J. (1979). The major histocompatibility complex of the mouse. *Science, N. Y.* 203: 516–521.

Nadeau, J., Wakeland, E. K., Götze, D. & Klein, J. (1981). Genetic differentiation of *H-2* antigens in European and African populations of the house mouse, *Mus musculus. Genet. Res.* 37: 17–31.

Wakeland, E. K. & Klein, J. (1979). Histocompatibility-2 system in wild mice. VII. Serological analysis of 29 wild-derived *H-2* haplotypes with antisera to inbred I-region antigens. *Immunogenetics* 8: 27–39.

Zaleska-Rutczynska, Z. & Klein, J. (1977). Histocompatibility-2 system in wild mice. V. Serological analysis of sixteen B10.W congenic lines. *J. Immunol.* 119: 1903–1911.

Symp. zool. Soc. Lond. (1981) No. 47, 455–477

The *t*– Complex and the Genetical Control of Development

M. F. LYON

M. R. C. Radiobiology Unit, Harwell, Didcot, UK

SYNOPSIS

The *t*-haplotypes of the mouse are naturally occurring genetic factors, located on chromosome 17, and found in a low proportion of wild mice. They produce a characteristic syndrome of harmful effects including embryonic lethality when homozygous, male sterility in some genotypes, distortion of segregation ratios in fertile males and interaction with the gene *T* to produce taillessness and suppression of recombination in a segment of chromosome 17. In the laboratory it is possible to find "mutant" forms of *t*-haplotypes in which the various parts of the syndrome are separated. Study of these "mutants" has shown that the *t*-haplotypes are not single-gene changes but represent a region of altered chromatin occupying at least one G-band and with factors for different parts of the syndrome located at particular points. It is suggested that the alteration may involve intercalary DNA. The *T*-complex is of general interest to developmental genetics as an example of an effect on development of DNA other than that involved in coding for structural genes. Furthermore, *t*-haplotypes involve alterations in so-called differentiation antigens on spermatogenic cells and embryos, and may prove useful models for the study of the role of these antigens in development.

INTRODUCTION

In addition to its importance in relation to population genetics, the *t*-complex of the mouse has aroused great interest in developmental genetics. Both its proximity on chromosome 17 to the major histocompatibility complex, *H-2,* and the presence within a relatively short region of chromosome of a number of genes affecting early development, have given grounds for investigation and speculation (Snell, 1968; Gluecksohn-Waelsch & Erickson, 1970; Artzt & Bennett, 1975). Two main questions to be considered in relation to the developmental genetics of the *t*-complex are firstly, what type of change in the genetic material gives rise to the unusual properties of the *t*-haplotypes, and secondly, what is the role in development of their normal homologues?

GENETICAL NATURE OF CHANGES IN *t*-HAPLOTYPES

This question has been addressed by studies of the properties of mutant *t*-haplotypes which arise spontaneously when naturally occurring *t*-haplotypes are maintained in the laboratory. The majority of the mutants prove to have arisen by crossing-over, and correlation of the apparent position of the crossover with the alteration in genetic properties has enabled the location of various factors governing particular traits.

Properties of Naturally Occurring *t*-haplotypes

Naturally occurring *t*-haplotypes found in wild populations have a characteristic set of properties (Bennett, 1975; Klein, 1975; Klein & Hammerberg, 1977; Sherman & Wudl, 1977; Gluecksohn-Waelsch & Erickson, 1970). The character by means of which they were originally detected is that of interaction with the mutant gene for brachyury, *T*. This gene when heterozygous with wild-type (*T*/+) results in a short tail, but heterozygotes with *t* (*T*/*t*) are completely tailless, although +/*t* and *t*/*t* (in cases where such a genotype is viable) have normal tails.

When homozygous, naturally occurring *t*-haplotypes are lethal or semi-lethal during embryogeny, and fall into six or more lethal complementation groups and one semi-lethal group (Bennett, 1975; Sherman & Wudl, 1977) (Table I). Haplotypes of different groups complement each other at least partially in terms of viability but alleles within a group show no complementation.

A third property is male sterility. Males carrying two complementing lethals, t^x/t^y, are sterile, as also are those homozygous for a semi-lethal. Males heterozygous for a single *t*-haplotype (+/*t* or *T*/*t*) are fertile, but transmit the haplotype to their offspring with an abnormal frequency, usually much greater than the expected 50%.

Yet a further distinct property is that of cross-over suppression. Natural *t*-haplotypes suppress crossing-over strongly in both sexes in the segment of chromosome 17 extending from the brachyury (*T*) locus, approximately to the *H-2* complex (Hammerberg & Klein, 1975a). Distally the suppression is known to end at or just beyond *H-2,* but lack of suitable markers proximal to *T* has so far precluded accurate mapping of the proximal end of the suppressed region.

A corollary of the cross-over suppression to *H-2* is that each naturally occurring *t* has its own characteristic *H-2* haplotype (Table I). In almost all cases members of the same lethal complementation group share an *H-2* type. An unforeseen point is

TABLE I

Complementation groups of t*-haplotypes*

Complementation group	Haplotypes	*H-2* haplotype	Lethal stage
t^0	t^0, t^1	$w5$	Egg cylinder
	t^6	$tw1$	Ectodermal differentiation fails
t^{12}	t^{12}, t^{w32}	$t12$	Morula. Metabolic and ultrastructural abnormalities
t^{w1}	t^{w1}, t^{w12}, t^{w71} (t^{w20}, t^{w21}, t^{w72})	$tw1$	Day 9 onwards Neural tube abnormal
	t^{w75}	$tw5$	
t^{w5}	t^{w5}, t^{w75} (t^{w74} and many others)	$tw5$	Egg cylinder Embryonic ectoderm fails
t^{w73}	t^{w73} (t^{w74})	$tw1$	Blastocyst. Trophectoderm abnormal
t^{w2}	t^{w2}, t^{w8}	$w5$	Semi lethal
t^4	t^4, t^9, t^{w18}	–	Day 8–9. Primitive streak stage. Mesoderm formation fails
T	–	–	Day 10. Primitive streak and notochord abnormal

Data from Bennett (1975) and Artzt *et al.* (1979) for lethality and Hammerberg *et al.* (1976) for *H-2*. The haplotypes enclosed in brackets were not tested for *H-2*.

that these *H-2* types appear to be unique to *t*-haplotypes and are different from those found in other wild mice (Hammerberg & Klein, 1975b; Hammerberg *et al.*, 1976).

Detection of Mutants

t-haplotypes are normally maintained in the laboratory in balanced lethal stocks with brachyury (*T*) and the marker gene tufted, *tf.* Since both *T*/*T* and *t*/*t* are lethal prenatally, and since crossing-over between *T* and *tf* is suppressed, all the liveborn offspring of such a cross are typically tailless non-tufted like the parents. Occasionally, however, a normal-tailed animal may be found, or a tailless animal homozygous for tufted (Fig. 1). These animals are the "mutants" from which information has been derived. Further "mutants" have come from linkage tests in which *Ttf*/*t* + animals have been crossed to + *tf*/+ *tf* or a similar genotype. Although crossing-over is strongly

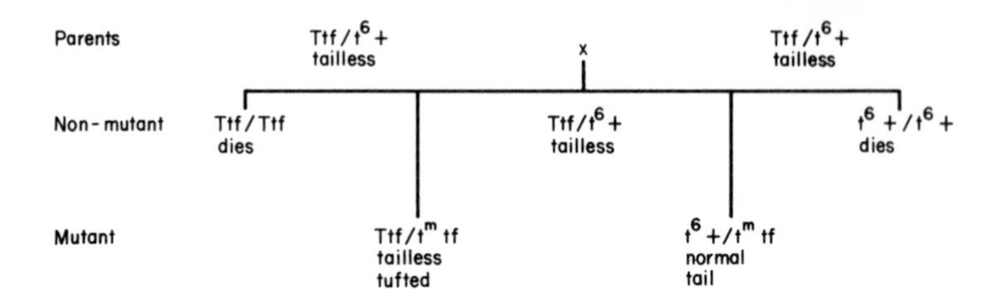

FIG. 1. Detection of mutants in a balanced lethal tailless line.

suppressed, rare cross-overs do occur and these prove to have alterations in some of the properties of the *t*-haplotype.

The most detailed work has involved the haplotype t^6, but broadly similar results have been obtained with other haplotypes. Although a natural rather than mutant haplotype, t^6 is unusual in that it was found in a laboratory stock (Carter & Phillips, 1950). It belongs to the t^0 complementation group, but differs from other members of this group in carrying the *H-2* haplotype $H\text{-}2^{tw1}$ (Table I). Over 40 mutants derived directly or indirectly from t^6 have now been studied (Lyon & Meredith, 1964a, b c; Lyon & Mason, 1977; Lyon, Jarvis *et al.*, 1979; Lyon, Evans *et al.*, 1979; Bechtol & Lyon, 1978). The mutation rate is approximately 5×10^{-3}, rather higher than the rate of $1\text{–}2 \times 10^{-3}$ found for other haplotypes (Lyon & Bechtol, 1977).

Genetical Deduction from Properties of Mutants

Tail-interaction, lethality and sterility

In almost all cases animals detected as "mutants" (e.g. normal tailed animals in a balanced lethal tailless stock) prove to have arisen by crossing-over in the region between *T* and *tf*, and animals detected as cross-overs proved to have "mutant" properties.

The normal-tailed mutants in balanced lethal stocks in all except one case carried a *t*-haplotype which had lost the embryonic lethality, so that these mutant *t*'s were viable when homozygous or when heterozygous with t^6. Loss of lethality was always accompanied by crossing-over between *T* and *tf*, and this suggested that a factor for lethality was located near to the locus of *tf* and was lost whenever the $+^{tf}$ allele (carried by the parental t^6 chromosome) was lost (Fig. 2). This was confirmed by the finding in linkage tests of the complementary type of cross-over (i.e. carrying the $T+^{tf}$ chromosome). Animals with such a chromosome proved to carry a mutant

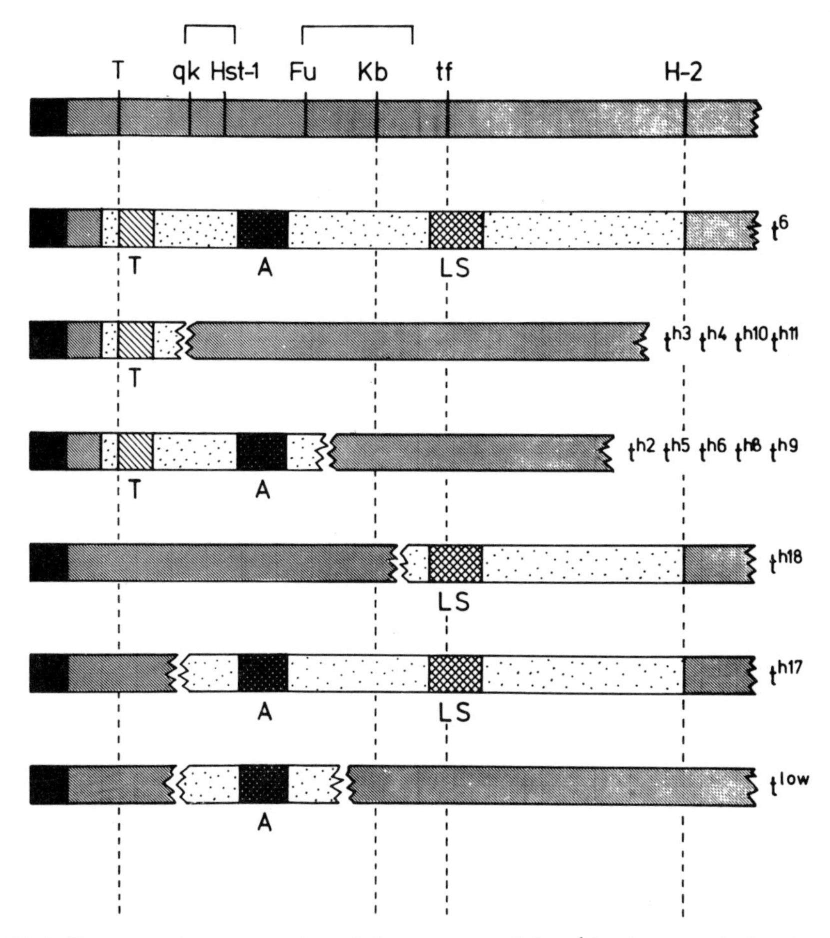

FIG. 2. Diagrammatic representation of the structure of the t^6-haplotype and of various mutants derived from it. The corresponding normal segment of chromosome 17 is shown at the top. Centromere, black; loci in normal chromosome 17 are brachyury, T; quaking, qk; hybrid sterility, Hst-1; fused, Fu; knobbly, Kb; tufted, tf; and H-2. The postulated T-, A- and LS-factors are hatched and the intervening altered chromatin is stippled. (Reprinted with permission from *Nature.*)

t-haplotype which retained the lethal factor, resulting in lethality either when homozygous or when in combination with t^6, but which had lost the factor for interaction with *T* to produce taillessness. When in combination with *T* (e.g. $Ttf/t^{h18}+$) such haplotypes (t^{h17} and t^{h18} in Fig. 2) resulted in a short-tailed animal just as in *T*/+ heterozygotes. The interpretation was that there was a factor for interaction with *T*, the T-factor or T-int factor, located at or very close to the locus of *T*, and a factor for lethality close to the locus of *tf* (Fig. 2).

All the viable mutants permitted fertility in males, i.e. they had lost the factor for male sterility, whereas the two lethal distal-end mutants t^{h17} and t^{h18} both caused male sterility when in combination with the complementing lethal t^{w5} (Lyon & Mason, 1977). It thus seems that a factor for male sterility lies close to that for lethality. For simplicity, the two factors are depicted as one, the LS-factor, in Fig. 2, but they may later prove to be separable.

Segregation distortion

Mutant viable haplotypes derived from t^6 are of two main types; those with a normal ratio (t^{h3}, etc. in Fig. 2) and those with a low transmission ratio from males (t^{h2}, etc. in Fig. 2). This suggests that a factor for low transmission ratio might be located between *T* and *tf*, so that the position of the cross-over which gave rise to a viable haplotype would determine whether it carried this low ratio factor or not. This was confirmed by the finding of a further type of mutant, which arose as cross-overs from the haplotype t^{h17}, and which had the property of causing low transmission of the chromosome 17 on which they lay, but which were not lethal and did not interact with *T*. The first of these mutants was considered by Dunn & Bennett (1968), who found it not to be part of the *t*-complex, and was designated as a distinct locus, termed *Low*.

The grounds for this conclusion were that *Low*'s interactions with other *t*-haplotypes were different from those of other mutant haplotypes. Lyon & Mason (1977), however, showed that other mutants derived from t^6 did interact in a way similar to *Low* and re-interpreted *Low* as a *t*-haplotype termed t^{low}. Since then at least three further mutants of the t^{low} type have been found, all derived by crossing-over in animals of the genotype t^{h2}/t^{h17} (two reported by Lyon, Evans *et al.*, 1979, and one unpublished). The repeated finding of such mutants in such circumstances lends further support to the interpretation that they are *t*-haplotypes.

These t^{low}'s show crossing-over with both *T* and *tf* (Dunn & Bennett, 1971) and are regarded as constituting a middle piece of *t*-chromatin carrying a factor for low transmission ratio, the A-factor, but not the T- or LS-factors (Fig. 2). If an animal carries t^{low} and another *t* with either high or low ratio (e.g. t^{low}/t^{h2}), the two chromosomes are transmitted equally. Similarly, when an animal is doubly heterozygous for other t^6 mutants (or t^6 itself) which both possess the A-factor (e.g. t^{h2}/t^{h17}, t^{h2}/t^6), the transmission rates are equalized. It thus seems that, for an abnormal ratio to occur, the A-factor must be present and heterozygous (Lyon & Mason, 1977). However, the A-factor alone gives a low ratio. The high transmission ratio of natural haplotypes seems to depend on other parts of the *t*-chromatin, since loss of either the proximal end (in t^{h17} and t^{h18}) or the distal end (in the various viable mutants, t^{h2} etc.) causes a loss of the high transmission. It is suggested that the spaces between the T-, A- and LS-factors are occupied by abnormal *t*-chromatin and that the high ratio of natural haplotypes is due to interaction of the A-factor with all parts of the *t*-chromatin (Lyon & Mason, 1977; Lyon, Evans *et al.*, 1979).

This interpretation relates to the haplotype t^6. It is probable that the structures of other natural haplotypes (t^{12}, t^{w5}, t^{w1} etc.) are broadly similar but that there are also some differences. Mutant viables arising from any natural haplotype can be divided into those with a normal or with a low transmission ratio (Bennett, Dunn & Artzt, 1976), as in the case of t^6, suggesting the presence of an A-factor. However, double heterozygosis for mutant haplotypes with a low and high ratio may not equalize the ratios, but rather may cause exacerbated disturbance of the ratio (Bennett & Dunn, 1971), and reduced fertility. This suggests the presence of additional as yet unidentified factors affecting both transmission ratio and fertility. Furthermore, transmission ratio is affected by the "background genotype" of both the male (Bennett, 1978) and of the female to which he is mated (Braden, 1972; Demin & Safronova, 1979).

It is not known whether this "background" effect involves factors on chromosome 17.

Suppression of crossing-over

Like the segregation distortion, the recombination suppressing effect of t^6 seems to depend on all parts of the *t*-chromatin. Mutant haplotypes which have arisen by crossing-over between *T* and *tf*, whether they consist of proximal ends (e.g. t^{h2}) or distal ends (e.g. t^{h17}, t^{h18}) permit recombination between *T* and *tf*, although in some cases this recombination is slightly suppressed. However, Bennett, Artzt *et al.* (1979) have shown that the viable t^{38} (derived from t^0) (Fig. 3) retains a strong cross-over suppressing effect over the shorter segment *T–qk*. This suggests that *t*-chromatin suppresses recombination over its own length and that a short mutant haplotype thus shows an effect only over a short length. Lyon, Evans *et al.* (1979) tested this by studying recombination in the *T–tf* segment in animals heterozygous for two complementary and overlapping mutant haplotypes, t^{h2} and t^{h17} (Fig. 2). There was a strong suppression of recombination in such animals despite the fact that either t^{h2} or t^{h17} separately permit crossing-over between *T* and *tf*. Similarly, Bennett, Artzt *et al.* (1979) found suppressed

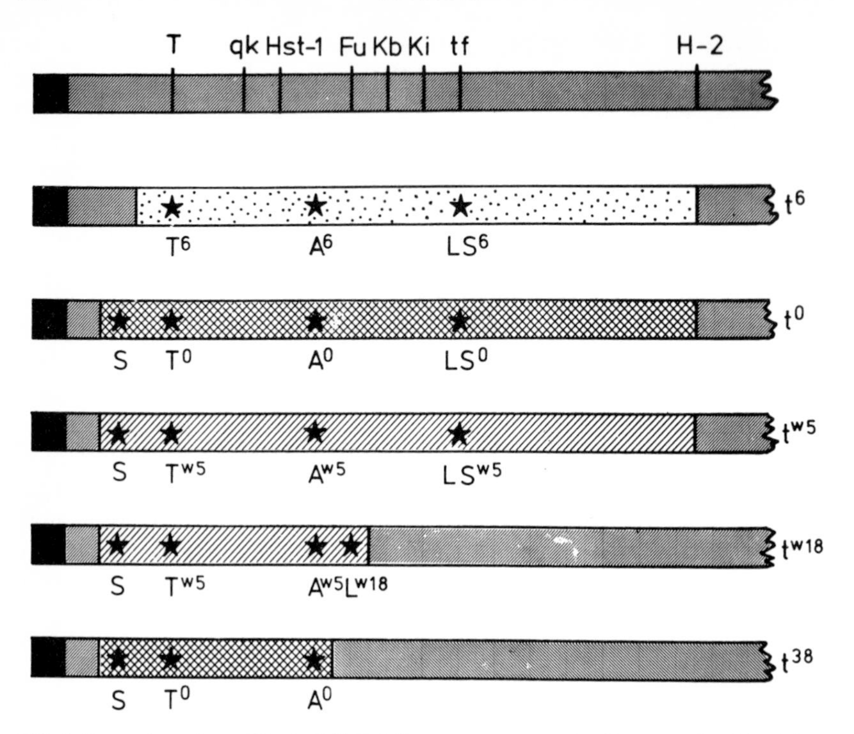

FIG. 3. Examples of possible "allelic" differences among naturally occurring t-haplotypes, t^6, t^0 and t^{w5}, together with two mutants, t^{w18}, which arose from t^{w5} (Bennett, 1975) and t^{38} which arose from t^0 (Bennett, Dunn & Artzt, 1976). t^0 and t^{w5} are believed to have a sterility factor, S, not present in t^6. The sequences in intercalary DNA in t^0 and t^{w5} may differ from that in t^6 (hatched), and the T-, A- and LS-factors also may show "allelic" differences.

recombination between t^{38} and t^{low}, which again must be overlapping haplotypes, since t^{38} has a low transmission ratio and thus must, like t^{low}, carry the A-factor (Fig. 3). Conversely, earlier results had shown that crossing-over could occur between *T* and *tf* in double heterozygotes for two non-overlapping haplotypes (e.g. t^{h2} and t^{h18}) (Lyon & Meredith, 1964b; Lyon, Evans *et al.*, 1979).

The interpretation is that any stretch of t-chromatin, whether a proximal (t^{h2}), medial (t^{low}), or distal (t^{h17}) fragment, suppresses crossing-over strongly in its own length. Bennett, Artzt *et al.* (1979) showed that in a segment adjacent to t-chromatin (i.e. between *qk* and *tf* or t^{low} and *tf* in t^{38} heterozygotes) crossing-over was enhanced. Thus, the recombination between two markers, such as *T* and *tf*, in a segment only partially covered by t-chromatin, does not give a true measure of the length of normal chromatin remaining.

Cross-over suppression could result either from normal chiasma formation with elimination of products of crossing-over or from reduced chiasma formation. Chiasma counts in animals with chromosome 17 marked by translocations have shown, for four different naturally occurring *t*-haplotypes, t^6, t^{12}, t^{w5} and t^{w32}, that in fact chiasma formation is reduced (Forejt, 1972; Forejt & Gregorova, 1977; Lyon, Evans *et al.*, 1979).

It is of interest to know whether chiasma formation is also suppressed when *t*-chromatin is present on both homologous chromosomes, rather than heterozygous with wild type. The strong cross-over suppression in animals carrying complementary overlapping haplotypes (t^{h2}/t^{h17} or t^{38}/t^{low}) suggests that chiasmata do not occur freely in the overlap region, but since the length of overlap is not known this point is not certain. Linkage tests have been made, using animals heterozygous for $t^{h7}+/t^{h2}tf$. t^{h7} is a haplotype derived from t^6, and believed to differ from it only by an altered T-factor (Fig. 2) (Lyon & Meredith, 1964b). It suppresses crossing-over strongly between *T* and *tf*. As t^{h2} carries the T- and A-factors, and as the A-factor from the work of Dunn & Bennett (1971) on the location of t^{low} must be distal to the locus of *qk*, $t^{h7}+/t^{h2}tf$ must have *t*-chromatin on both homologues at least over the distance *T–qk*, or 3 cM (Fig. 2). Thus, if crossing-over occurred freely in regions homozygous for *t*-chromatin one would expect at least 3% recombination between the altered T-int factor and *tf* in $t^{h7}+/t^{h2}tf$ heterozygotes. In fact Lyon & Jarvis (1980) found only two recombinants in over 200 informative offspring, or $< 1\%$. It therefore appears that reduced chiasma formation is an intrinsic property of *t*-chromatin, rather than being due to some incompatibility between *t*- and normal chromatin.

Hypothesis of change in intercalary DNA

The information gained from mutants concerning the genetical structure of *t*-haplotypes makes it clear that they cannot be considered as changes at one or a few genetical loci. They must be regarded as a change involving a sizable segment of chromosome 17. The cross-over suppression extends at least from the *T* to the *H-2* locus, a genetic distance of about 15 cM.

The cytogenetic position of the chromosome region involved has been mapped approximately by correlating genetic and cytogenetic breakpoints in translocations involving chromosome 17. The region from *T* to *H-2*, and hence the *t*-complex, must occupy at least the greater part of band 17B, and may extend into the adjacent bands A and C (Lyon, Evans *et al.*, 1979). Whatever the alteration may be in *t*-chromatin it is not of a gross visible nature, as band 17B shows no visible difference in size or stainability in *t*-haplotypes. Nor is there any evidence of chromosomal inversions. Thus, *t*-chromatin probably does not involve structural changes in highly repetitious DNA of the type located in discrete blocks. On the other hand the

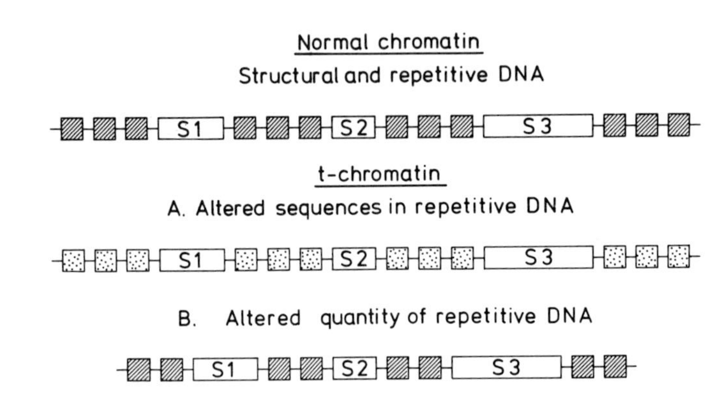

FIG. 4 . Suggested altered DNA in *t*-chromatin. Structural sequences, depicted in S1, S2, S3 are unaltered. Intervening intercalary repetitious DNA (hatched) is altered either in sequence (stippled) or number of repeats.

change is apparently not at the level of changes in structural gene sequences. There are many known genes in the chromosomal segment between *T* and *H-2,* including the loci of quaking, *qk,* knobbly, *Kb*, tufted, *tf* and glyoxylase-1 (*Glo-1*)(Fig. 5), and *t*-chromatin appears to carry wild-type alleles of these genes, so that normal sequences must be present.

Gluecksohn-Waelsch & Erickson (1970) were the first to suggest that *t*-haplotypes might involve an alteration in repetitious DNA. Lyon, Evans *et al.* (1979) elaborated this by suggesting that the change was in the moderately repetitious DNA intercalated among the structural sequences (Fig. 4). These intercalary sequences are believed to be involved in some way in the control of gene expression and Stern & Hotta (1977) have suggested that they are also concerned in meiotic crossing-over. Lyon, Evans *et al.* (1979) suggested that in *t*-haplotypes the intercalary sequences were modified in such a way that both the meiotic and the developmental functions were impaired. The function of any particular structural gene within the altered region might or might not be affected according to its dependence on the exact type of regulation brought about by the intercalary DNA.

The exact manner in which gene expression would be affected is a matter for speculation. Obviously effects on the timing or rate of

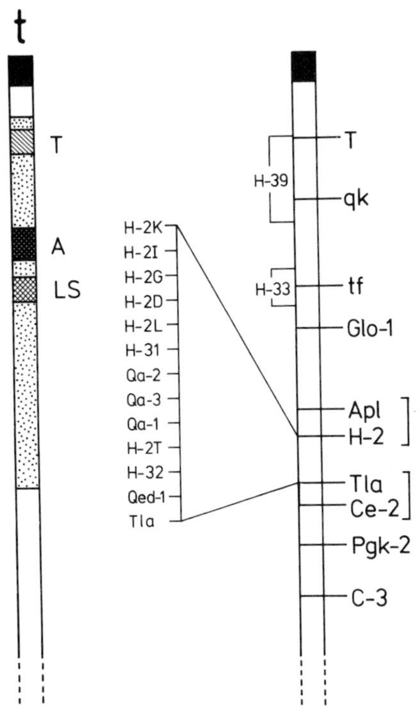

FIG. 5. Antigenic loci in the segment of chromosome 17 involved in the *t*-haplotypes. Normal chromosome 17 to the right, with marker loci to the right and antigenic loci to the left. Data from Davisson & Roderick (1979) with position of *Apl* altered according to Eicher (personal communication).

transcription should be considered. Another possibility is an effect on gene splicing.

The effect observed in the gene product clearly might be a change in quantity of a protein or activity of an enzyme. Possible evidence of such an effect is provided by the work of Shur & Bennett (1979) who showed that galactosyl transferase activity of sperm was greatly increased in mice carrying *t*-haplotypes, and attributed this to decreased activity of an inhibitor.

However, it is also possible that a change in molecular weight or charge might be observed, if the gene whose function was affected was a "processing gene". Paigen (1978) has classified genes affecting the control of enzyme activity into structural, processing, regulatory, and temporal genes (see below, pp. 468–470). Processing genes are regarded as those involved in such matters as attachment to membranes, localization in organelles, or the addition or removal of residues such as glycosides. It is this last function which can lead to a change in molecular charge.

This possibility of an altered charge is relevant, because of the work on the protein called p63/6.9 by Silver, Artzt & Bennett (1979). This protein of testicular cells was found to have an altered charge in mice carrying *t*-haplotypes. By study of mutants, the gene concerned was shown to be located in the proximal part of the *t*-region. The allele of brachyury known as hairpin tail, T^{hp}, is presumed (on other evidence) to involve a deletion covering the region from *T* to *qk* (Bennett, 1975). This deletion resulted in absence of the p63 protein, i.e. whereas *T*/*t* animals had both normal (p63b) and *t*-like (p63a) protein, $T^{hp}/+$ had only normal and T^{hp}/t had only p63a protein. This seems to indicate that a gene near the locus of brachyury codes for the p63 protein, and that a factor near the proximal end of *t*-haplotypes causes a change in the charge on the protein. This could be explained if *t*-chromatin caused an altered activity of a processing gene located near the structural gene and cis-acting. Paigen (1978) has shown that both regulatory and temporal genes may be located adjacent to a structural gene and cis-acting, but he gives no such example for processing genes. Thus, whether this is an adequate explanation for the alteration in p63 protein caused by *t*-haplotypes is not yet clear.

Differences among natural *t*-haplotypes

Since the work on details of genetical structure has largely involved the haplotype t^6 it is important to consider to what extent this haplotype is representative of others.

As mentioned above (p. 461) Bennett, Dunn & Artzt (1976) have described many mutants arising in balanced lethal stocks of haplotypes of the complementation groups t^0, t^{12}, t^{w5}, t^{w1}. As with t^6, loss of lethality was always accompanied by crossing-over in the segment between *T* and *tf*, and the viable haplotypes so produced were of two main types, with low or with normal transmission ratios. This suggests that the broad outline of three factors, T-, A- and LS-, affecting tail-interaction, abnormal ratio and lethality, applies also to these other haplotypes. There is a difference, however, concerning male sterility. t^6 appears to have only one factor, the LS-factor, involved in male sterility, so that viable mutants have almost full fertility. Other haplotypes appear to have more than one sterility factor, and males doubly heterozygous for a mutant viable and a lethal have markedly reduced fertility (Bennett & Dunn, 1971; Dunn & Bennett, 1969). Erickson, Lewis & Slusser (1978) have shown that a locus concerned with male sterility is located within the segment covered by the presumed deletion involved in the brachyury allele T^{Orl} (Bennett, 1975). Since this deletion covers the segment from *T* to *qk*, this sterility locus must be proximal to *tf* and therefore different from the one identified in t^6 (Fig. 3).

A further difference concerns the lethality. Although the longest known and best studied haplotypes, belonging to the complementation groups t^0, t^{12}, t^{w1} and t^{w5}, appear to have lethal factors located near to each other and to the *tf* locus, and thus near the position of the LS-factor of t^6, some recently discovered haplotypes show differences. The haplotypes t^{w74} and t^{w75} fail to complement lethals of two groups (t^{w74}: t^{w5} and t^{w73}; t^{w75}: t^{w1} and t^{w5}) (Artzt, Babiarz & Bennett, 1979). Furthermore, the haplotype t^{w73} carries a lethal at a locus proximal to *tf* and near the locus of *T* in the segment covered by the deletion in the brachyury allele T^{hp}, but not by deletion T^{Orl} (Babiarz, 1979).

Detailed study of haplotypes from apparently new complementation groups, such as t^{Pa} (J. L. Guenet, personal communication) and t^{Lub1} (Winking, 1979; Guenet & Winking, 1979) may provide further information on the extent of variation in genetic structure among natural *t*-haplotypes, and in turn on the type of alteration in the DNA.

If *t*-haplotypes involve a change in repetitious DNA, then this might be either quantitative in number of repeats, or qualitative in the nature of the repeated sequence, or a combination of these (Fig. 4). Clearly, any quantitative change must be sufficiently small that it would not cause a visible alteration in size of the relevant G-bands of chromosome 17. Different naturally occurring *t*-haplotypes, belonging to different lethal complementation groups, could show "allelic" variation in the repeated sequence, and this could lead to the observed differences in properties.

Origin of natural *t*-haplotypes

Views on the nature of the genetical differences among *t*-haplotypes are interdependent with views on their possible mode of origin.

The mechanism by which a change in repetitive DNA over a considerable chromosomal length could arise is obscure. One possibility is that the change arose stepwise. A short length of altered DNA might arise by mutation, and then become elongated by unequal crossing-over until it reached a structure which resulted in high transmission ratio, when it would be maintained in the population.

Unequal crossing-over has been postulated to explain the altered T-int factor in t^{h7} (Lyon & Meredith, 1964b) and there is some evidence from studies of the p63 protein that it may have occurred in other cases (Silver & Kintanar, 1979).

Another possibility is that, since the sequences are repeated, there may be some mechanism, not yet understood, by which a change in all repeats can arise in a single event, e.g. by a mutation in a "proof-reading" system. One could then envisage that several natural haplotypes had arisen by separate mutational events. The different mutational events would give rise to specific alterations in the sequence and hence the *t*-haplotypes generated would show "allelic" differences. Silver *et al.* (1979) on the other hand, espouse the view that, whatever event may have originally given rise to *t*-chromatin, it occurred only once, and that the present differences observed among natural *t*-haplotypes are the result of evolutionary divergence. This view is put forward on the basis that the p63a protein is found in all *t*-haplotypes so far tested. According to this concept the lethality of natural *t*'s is due to mutant genes which have arisen during evolution and become "entrapped" by the recombination suppression and high transmission ratio of the *t*-haplotype. This seems not to explain why so many of the lethals appear to involve the same or neighbouring functional units, nor the apparent association of particular *H-2* haplotypes with complementation groups.

Further work will be needed before the question of the origin of *t*-haplotypes can be settled definitively.

RELEVANCE OF THE *t*-COMPLEX TO NORMAL DEVELOPMENT

In view of the number and variety of the early lethals involved in the various haplotypes of the *t*-complex it is valuable to consider what information may be derived from them concerning the genetical control of early development, and the particular role, if any, of this segment of chromosome 17.

Genetical Control of Early Development

Genes causing lethality in early development presumably do so through malfunction of a protein essential for some aspect of development. Such proteins obviously include enzymes, and also molecules acting as inducers, as for example when differentiation of ectodermal structures is brought about by inducers from the underlying mesenchyme. They also include proteins concerned with the correct functioning of organelles such as mitochondria, Golgi apparatus, endoplasmic reticulum, lysosomes and the cell surface.

The abnormality may consist of an incorrect structure of the protein, rendering it inactive. Other possibilities, however, include incorrect timing or rate of production of the protein, incorrect response to inducing agents such as hormones, or inappropriate attachment to intracellular membranes. Paigen (1978) has classified genes affecting enzyme function into various types, as follows:

(*a*) *Structural genes.* These code for the amino-acid sequence of the enzyme.

(*b*) *Processing genes.* These genes are concerned in modification of the protein at some stage after transcription. The functions of such genes may include the addition or modification of prosthetic groups. An example is the sialylation of acid phosphatase and mannosidase, in some tissues but not others, by a gene located on chromosome 17 close to the *H-2* complex, and denoted *Apl* or *Map-2* (Fig. 5). Another function of processing genes is the control of the rate of degradation of an enzyme, and an example of such a gene is again found on chromosome 17, in the *Ce-2* locus which controls the rate of degradation of catalase, and which again is located near *H-2* (Fig. 5). Yet another processing role is the cleavage of a protein either into two or more functional subunits, or so as to remove or attach a sequence concerned in attachment to membranes. All the examples of processing genes given by Paigen (1978) are located at a distance from the structural gene the product of which they modify, and are not cis-acting.

(*c*) *Regulatory genes.* Regulatory genes include those controlling the rate of synthesis of the gene product, and also those involved in response to inducing agents, such as hormones. These genes may either be remote from the loci they regulate, or adjacent to them and cis-acting.

(*d*) *Temporal genes.* These genes are concerned in the developmental programming of the production of a protein at various stages of life. As with the regulatory genes, Paigen found that temporal genes might be either remote, or proximate and cis-acting to the locus they controlled.

An example of a cis-acting temporal gene on chromosome 17 is the gene found by Boubelík *et al.* (1975) to control the time of expression of *H-2* on erythrocytes.

Although Paigen made this classification for enzymes, presumably some similar but not necessarily identical system would be appropriate also for other proteins. It should be noted that, as Paigen points out, complexities arise when the classification is considered in detail. A gene which regulates a second locus may do so by coding for a protein which takes part in the control of this locus; and it is therefore a structural locus for the regulatory protein.

An example is the *Tfm* locus, which is a structural locus for an androgen receptor protein (Ohno, 1979), and thereby acts as a regulatory locus for other androgen inducible loci. Thus, regulatory loci which are also structural genes may themselves also have regulatory and temporal sequences controlling them.

Gene Expression in *t*-chromatin

It has been suggested above that the abnormality in *t*-chromatin which gives rise to the lethality and effects on male fertility lies in the control of the gene expression, rather than in the structural genes. In terms of Paigen's (1978) classification it is thus in proximate cis-acting regulatory, temporal or processing genes. The effect of the altered sequences in *t*-chromatin is that the function of some of the structural genes located in the *T–H-2* segment of chromosome 17 is altered in timing, or rate of transcription, or some form of processing.

The differences among *t*-haplotypes, in lethal complementation groups, or in degree of male sterility, could be due to "allelic" differences in the altered repetitious sequences, leading to differences in timing, or level of synthesis of the critical protein.

It has been mentioned above that there are a number of known structural genes in the region of chromosome 17 occupied by the *t*-haplotypes which are apparently not affected by the altered chromatin. It is of interest to consider those genes which are affected, and attempt to determine whether they have some common property. Thus, we must consider in more detail the factors involved in T-interaction, lethality, sterility and abnormal segregation.

Embryonic lethality

Details of the embryological defects in *t*-lethals have been extensively reviewed by others (Bennett, 1975; McLaren, 1976; Sherman & Wudl, 1977). The major points will be briefly summarized here.

(a) Each lethal complementation group acts at a specific stage of development, ranging from the morula (t^{12}) to the foetal stage (t^{w1}) (Table I).

(b) Some lethals (t^0, t^{w5}) appear to act as cell lethals, and others (t^4, T) are not generalised cell lethals but affect a specific cell type.

(c) It should be borne in mind that the various lethals are *not* alternatives at a single locus. On the contrary, the loci of the T, t^4 and combined group of t^0, t^{12}, t^{w1} and t^{w5} lethals can be shown to recombine; t^{w1} and t^{w5} can be present together (in t^{w75}); and t^{w73} carries a lethality not at the same locus as t^0, t^{12}, t^{w1} and t^{w75}.

Thus, it would be fallacious to attempt an explanation in terms of a single polypeptide. There may in fact be no unitary explanation for the various lethals.

(d) Two main theories concerning the action of the *t*-lethals have been put forward (reviewed by Sherman & Wudl, 1977). That of Bennett and her colleagues explains the lethality in terms of abnormalities of the cell surface, concentrating on the importance of surface antigens in cell–cell interactions, and also possibly cell–environment interactions. Hillman (1975) on the other hand believes the abnormality to be one of cellular metabolism. As Sherman & Wudl (1977) point out, these two theories are not mutually incompatible, since abnormalities of the cell surface could lead to defective intracellular metabolism, or conversely, altered intracellular metabolism could lead to changes in the carbohydrate moieties of cell surface glycoproteins.

Male sterility and distortion of transmission ratio

There appear to be at least two types of male sterility caused by *t*-haplotypes. In males homozygous for a semi-lethal haplotype, spermatogenesis is abnormal resulting in azoospermia (Dooher & Bennett, 1974).

The sterility of males heterozygous for two lethals t^x/t^y, however, appears to result mainly from defects of sperm function rather than sperm formation. These males possess numerous sperm, although there may be some defects in spermiogenesis (Olds, 1971a; Dooher & Bennett, 1977). Olds (1971b) found that sperm from sterile males were defective in ability to fertilize eggs, and this was confirmed by McGrath & Hillman (1977) who showed by means of tests with *in vitro* fertilization that sperm from t^6/t^{w32} males failed to penetrate eggs.

Sperm penetration appears also to be involved in the distortion of transmission ratio. Various studies have indicated that the abnormality which gives rise to segregation distortion must occur after the time of entry of sperm into the female reproductive tract. Hammerberg & Klein (1975c) by using a translocation as a marker for chromosome 17 showed that at the time of male meiosis the segregation ratio was normal, and there is no evidence of elimination of any class of spermatids during spermiogenesis (Hillman & Nadijcka, 1978). Conversely, the facts that transmission ratio is affected by delayed mating (Braden, 1972), depends on the genotype of the female used (Demin & Safronova, 1979), and even on the *t*-complex genotype of the egg fertilized (Bateman, 1960), show that events occurring in the female tract are important. Olds-Clarke & Carey

(1978) showed that sperm from *T/t* males penetrated eggs more rapidly than normal sperm *in vitro* and suggested that the more rapidly penetrating sperm were the *t*-bearing ones, thus accounting for the high transmission.

Again, there are opposing theories to account for these phenomena. Ginsberg & Hillman (1974) showed that the metabolic function of *t*-sperm is altered, and that sperm from *T/t* males have increased oxygen uptake, whereas Bennett and colleagues explain the phenomena in terms of cell surface antigens.

Cell surface antigens and the *t*-complex

The suggestion that the abnormalities brought about by *t*-haplotypes might result from changes in cell surface antigens was first put forward by Gluecksohn-Waelsch & Erickson (1970). Bennett, Goldberg *et al.* (1972) detected an antigen on sperm cells associated with the brachyury gene, *T*, and later Yanagisawa *et al.* (1974) found further antigens associated with various *t*-haplotypes. Artzt & Bennett (1977) analysed the *t*-haplotype specificities further, and compared mutant viable haplotypes with the naturally occurring lethals from which they arose. The mutants had lost some of the specificities of the parent haplotype but had retained others, thus showing that more than one locus was involved. Some specificities may have been associated with the lethality.

The antigens are thought to be expressed on sperm and embryonic cells. Bennett's group have further suggested that the F9 antigen, found on undifferentiated teratocarcinoma cells, is a product of the wild-type allele of t^{12} (Bennett, 1975; Marticorena, Artzt & Bennett, 1978).

More recently, Shur & Bennett (1979) studied glycosyltransferases in testicular cells of mice with *t*-haplotypes. Sperm from *T/t* mice showed greatly increased galactosyltransferase activity, which appeared to be due to decreased activity of an inhibitor present in wild-type sperm. Studies of mutants showed that, as with the antigenic specificities, the effect appeared to be due to more than one locus.

Thus, this work on galactosyltransferase is consistent with our views on the genetical structure of *t*-haplotypes in two ways: (i) more than one locus is involved; (ii) the effect appears to be a regulatory one controlling level of the enzyme.

Shur & Roth (1975) reviewed the role of cell surface glycosyltransferases and produced evidence for their involvement in cell–cell interactions in embryogenesis, in sperm–egg interaction at fertilization, and in cell–environment interactions. Shur, Oettgen & Bennett (1979) produced evidence that galactosyltransferase, together with F9 antigen, was in some way involved with blastocyst formation in the mouse embryo, and hence that abnormal galactosyltransferase activity might be connected with the lethal embryonic effects of *t*-haplotypes. Clearly the increased galactosyltransferase activity of *t*-sperm (Shur & Bennett, 1979) might also be associated with the more rapid penetration of eggs found by Olds-Clarke & Carey (1978). As discussed above, however, the relation of the cell surface changes to the metabolic changes found by Hillman's group remains to be elucidated.

We have mentioned above that, apart from the *t*-complex, regulatory genes are already known in the relevant segment of chromosome 17, including a temporal gene for *H-2* expression (Boubelík *et al.*, 1975), and a processing gene affecting sialylation. Furthermore, again apart from the *t*-complex, numerous genes in this segment are known to affect cell surface antigens, either as histocompatibility antigens or as differentiation antigens (Fig. 5).

This raises the possibility that the structural genes whose regulation is affected by the altered intercalary DNA postulated to be present in *t*-haplotypes do indeed have a property in common. They may all be genes connected in some way with the expression of cell surface antigens, possibly through the action of glycosyltransferases. Thus, this would be a chromosome segment which has, whether by gene duplication or other means, come to be rich in genes of a particular type. There may be some analogy with mouse chromosome 8 which carries a large number of loci, separated by appreciable map distances, for esterases.

CONCLUSION

Clearly, much remains to be elucidated concerning the *t*-complex and its relation to early development, particularly the interrelation between the cell surface and the metabolic changes. Whatever facts may emerge, however, it seems likely that the *t*-complex will continue to prove a valuable model, both for study of the role of intercalary repetitious DNA in development and for a better understanding of the role of the cell surface in differentiation.

ACKNOWLEDGEMENTS

I am grateful to Mrs S. E. Jarvis for assistance and diagrams.

REFERENCES

Artzt, K., Babiarz, B. & Bennett, D. (1979). A *t*-haplotype (t^{w75}) overlapping two complementation groups. *Genet. Res.* 33: 279–285.

Artzt, K. & Bennett, D. (1975). Analogies between embryonic (*T/t*) antigens and adult major histocompatibility (*H-2*) antigens. *Nature, Lond.* **256**: 545–547.

Artzt, K. & Bennett, D. (1977). Serological analysis of sperm of antigenically cross-reacting *T/t*-haplotypes and their recombinants. *Immunogenetics* 5: 97–107.

Babiarz, B. (1979). Another lethal mutation in a *t*-haplotype? *Mouse News Letter* **61**: 61.

Bateman, N. (1960). Selective fertilization at the T-locus of the mouse. *Genet. Res.* **1**: 226–238.

Bechtol, K. B. & Lyon, M. F. (1978). *H-2* typing of mutants of the t^{6} haplotype in the mouse. *Immunogenetics* **6**: 571–583.

Bennett, D. (1975). The T-locus of the mouse. *Cell* **6**: 441–454.

Bennett, D. (1978). Fluctuations in transmission ratio. *Mouse News Letter,* **59**: 60–61.

Bennett, D., Artzt, K., Cookingham, J. & Calo, C. (1979). Recombinational analysis of the viable *t*-haplotype t^{38}. *Genet. Res.* 33: 269–277.

Bennett, D. & Dunn, L. C. (1971). Transmission ratio distorting genes on chromosome IX and their interactions. In *Proc. Symp. Immunogenetics of the H-2 System:* 99–103. Lengerova, A. & Vojtiskova, M. (Eds). Basel: Karger.

Bennett, D., Dunn, L. C. & Artzt, K. (1976). Genetic change in mutations at the T/t-locus in the mouse. *Genetics* **83**: 361–372.

Bennett, D., Goldberg, E., Dunn, L. C. & Boyse, E. A. (1972). Serological detection of a cell surface antigen specified by the T (Brachyury) mutant gene in the house mouse. *Proc. natn. Acad. Sci. U.S.A.* **69**: 2076–2080.

Boubelík, M., Lengerová, A., Bailey, D. W. & Matousek, V. (1975). A model for genetic analysis of programmed gene expression as reflected in the development of membrane antigens. *Devl. Biol.* 47: 206–214.

Braden, A. W. H. (1972). T-locus in mice: segregation distortion and sterility in the male. In *The genetics of the spermatozoon*: 289–305. Beatty, R. A. & Gluecksohn-Waelsch, S. (Eds). Edinburgh & New York: The Librarian, Dept. of Genetics, University of Edinburgh.

Carter, T. C. & Phillips, R. J. S. (1950). Three recurrences of mutants in the house mouse. *J. Hered.* **41**: 252.

Davisson, M. & Roderick, T. H. (1979). Linkage map of the mouse. *Mouse News Letter* **61**: 19.

Demin, Yu. S. & Safronova, L. D. (1979). Effect of the female genotype on non-Mendelian segregation in progeny of male carriers of *t*-haplotypes in the house mouse (*Mus musculus*). *Dokl. Biol. Sci.* 243: 582–584.

Dooher, G. B. & Bennett, D. (1974). Abnormal microtubular systems in mouse spermatids associated with a mutant gene at the T-locus. *J. Embryol. exp. Morph.* 32: 749–761.

Dooher, G. B. & Bennett, D. (1977). Spermiogenesis and spermatozoa in sterile mice carrying different lethal *T/t* locus haplotypes: a transmission and scanning electron microscropic study. *Biol. Reprod.* 17: 269–288.

Dunn, L. C. & Bennett, D. (1968). A new case of transmission ratio distortion in the house mouse. *Proc. natn. Acad. Sci. U.S.A.* **61**: 570–573.

Dunn, L. C. & Bennett, D. (1969). Studies of effects of *t*-alleles in the house mouse on spermatozoa. II. Quasi-sterility caused by different combinations of alleles. *J. Reprod. Fert.* **20**: 239–246.

Dunn, L. C. & Bennett, D. (1971). Further studies of a mutation (Low) which distorts transmission ratios in the house mouse. *Genetics* **67**: 543–558.

Erickson, R. P., Lewis, S. E. & Slusser, K. S. (1978). Deletion mapping of the *t*-complex of chromosome 17 of the mouse. *Nature, Lond.* **274**: 163–164.

Forejt, J. (1972). Chiasmata and crossing-over in the male mouse (*Mus musculus*): suppression of recombination and chiasma frequencies in the ninth linkage group. *Folia Biol., Praha* **18**: 161–170.

Forejt, J. & Gregorova, S. (1977). Meiotic studies of translocations causing male sterility in the mouse. I. Autosomal reciprocal translocations. *Cytogenet. Cell Genet.* **19**: 159–179.

Ginsberg, L. & Hillman, N. (1974). Meiotic drive in t^n-bearing mouse spermatozoa: a relationship between aerobic respiration and transmission frequency. *J. Reprod. Fert.* **38**: 157–163.

Gluecksohn-Waelsch, S. & Erickson, R. P. (1970). The *T*-locus of the mouse: implications for mechanisms of development. *Current Topics Devl. Biol.* **5**: 281–316.

Guenet, J. L. & Winking, H. (1979). Results of complementation tests with t^{Lub1}. *Mouse News Letter* **60**: 72.

Hammerberg, C. & Klein, J. (1975a). Linkage relationships of markers on chromosome 17 of the house mouse. *Genet. Res.* **26**: 203–211.

Hammerberg, C. & Klein, J. (1975b). Linkage disequilibrium between *H-2* and *t*-complexes in chromosome 17 of the mouse. *Nature, Lond.* **258**: 296–299.

Hammerberg, C. & Klein, J. (1975c). Evidence for postmeiotic effect of *t* factors causing segregation distortion in mouse. *Nature, Lond.* **253**: 137–138.

Hammerberg, C., Klein, J., Artzt, K. & Bennett, D. (1976). Histocompatibility-2 system in wild mice. II. *H-2* haplotypes of *t*-bearing mice. *Transplantation* **21**: 199–212.

Hillman, N. (1975). Studies of the T-locus. In *The early development of mammals*: 182–206. Balls, M. & Wild, A. E. (Eds). London: Cambridge University Press.

Hillman, N. & Nadijcka, M. (1978). A comparative study of spermiogenesis in wild-type and *T*: *t*-bearing mice. *J. Embryol. exp. Morph.* **44**: 243–261.

Klein, J. (1975). *Biology of the mouse histocompatibility-2 complex.* Berlin: Springer Verlag.

Klein, J. & Hammerberg, C. (1977). The control of differentiation by the *T* complex. *Immunol. Rev.* **33**: 70–104.

Lyon, M. F. & Bechtol, K. B. (1977). Derivation of mutant *t*-haplotypes of the mouse by presumed duplication or deletion. *Genet. Res.* **30**: 63–76.

Lyon, M. F. & Jarvis, S. E. (1980). Recombination in double heterozygotes for two *t*-haplotypes. *Mouse News Letter* **62**: 48.

Lyon, M. F., Evans, E. P., Jarvis, S. E. & Sayers, I. (1979). *t*-haplotypes of the mouse may involve a change in intercalary DNA. *Nature, Lond.* **279**: 38–42.

Lyon, M. F., Jarvis, S. E., Sayers, I. & Johnson, D. R. (1979). Complementation reactions of a lethal mouse *t*-haplotype believed to include a deletion. *Genet. Res.* 33: 153–161.

Lyon, M. F. & Mason, I. (1977). Information on the nature of *t*-haplotypes from the interaction of mutant haplotypes in male fertility and segregation ratio. *Genet. Res.* 29: 255–266.

Lyon, M. F. & Meredith, R. (1964a). Investigations of the nature of *t*-alleles in the mouse. I. Genetic analysis of a series of mutants derived from a lethal allele. *Heredity* 19: 301–312.

Lyon, M. F. & Meredith, R. (1964b). Investigations of the nature of *t*-alleles in the mouse. II. Genetic analysis of an unusual mutant allele and its derivatives. *Heredity* 19: 313–325.

Lyon, M. F. & Meredith, R. (1964c). Investigations of the nature of *t*-alleles in the mouse. III. Short tests of some further mutant alleles. *Heredity* 19: 327–330.

McGrath, J. & Hillman, N. (1977). The inability of spermatozoa from sterile (t^6/t^{w32}) mice to effect *in vitro* fertilization. *J. Cell Biol.* 75: 170a.

McLaren, A. (1976). Genetics of the early mouse embryo. *A. Rev. Genet.* 10: 361–388.

Marticorena, P., Artzt, K. & Bennett, D. (1978). Relationship of F9 antigen and genes of the *T/t* complex. *Immunogenetics* 7: 337–347.

Ohno, S. (1979). *Major sex determining genes.* (Monographs on Endocrinology 11.) Berlin: Springer-Verlag.

Olds, P. J. (1971a). Effect of the T locus on sperm ultrastructure in the house mouse. *J. Anat.* 109: 31–37.

Olds, P. J. (1971b). Effect of the T locus on fertilization in the house mouse. *J. exp. Zool.* 177: 417–434.

Olds-Clarke, P. & Carey, J. E. (1978). Rate of egg penetration *in vitro* accelerated by *T/t* locus in the mouse. *J. exp. Zool.* 206: 323–332.

Paigen, K. (1978). Genetic control of enzyme activity. In *Origins of inbred mice*: 255–278. Morse, H. C. (Ed.) New York: Academic Press.

Sherman, M. I. & Wudl, L. R. (1977). T-complex mutations and their effects. In *Concepts in mammalian embryogenesis:* 136–234. Sherman, M. I. (Ed.). Cambridge, Mass: MIT Press.

Shur, B. D. & Bennett, D. (1979). A specific defect in galactosyltransferase regulation on sperm bearing mutant alleles of the *T/t* locus. *Devl Biol.* 71: 243–259.

Shur, B. D., Oettgen, P. & Bennett, D. (1979). UDP galactose inhibits blastocyst formation in the mouse. Implications for the mode of action of *T/t*-complex mutations. *Devl Biol.* 73: 178–181.

Shur, B. D. & Roth, S. (1975). Cell surface glycosyltransferases. *Biochim. Biophys. Acta* No. 415: 473–512.

Silver, L. Artzt, K. & Bennett, D. (1979). A major testicular cell protein specified by a mouse *T/t*-complex gene. *Cell* 17: 275–284.

Silver, L. & Kintanar, A. (1979). The p63 gene. *Mouse News Letter* 61: 63.

Snell, G. D. (1968). The *H-2* locus of the mouse: observations and speculations concerning its comparative genetics and its polymorphism. *Folia biol. Praha* 14: 335–358.

Stern, H. & Hotta, Y. (1977). Biochemistry of meiosis. *Phil. Trans. R. Soc.* (B) 277: 277–294.

Winking, H. (1979). Characteristics of t^{Lub1} on the Rb(4.17)13Lub metacentric. *Mouse News Letter* **60**: 56.
Yanagisawa, K., Bennett, D., Boyse, E. A., Dunn, L. C. & Dimeo, A. (1974). Serological identification of sperm antigens specified by lethal *t*-alleles in the mouse. *Immunogenetics* **1**: 57–67.

Symp. zool. Soc. Lond. (1981) No. 47, 479–516

Enzyme and Protein Polymorphism

JOSEPHINE PETERS

MRC Radiobiology Unit, Harwell, Didcot, UK

SYNOPSIS

Since the start of studies of biochemical genetics in the mouse about 30 years ago 115 biochemical variants have been described and more are continually being found. Both inbred strains and mice derived from wild-living populations have been valuable for the identification of genetically determined biochemical variation. Nearly 80% of the enzyme and protein variation in the mouse has been found by electrophoretic methods. Not only has variation been described for structural genes (those that code for the amino acid sequence of the polypeptide chain of an enzyme or protein) but also for regulatory genes (which control rate of protein synthesis) and for processing genes (which modify enzyme and protein structure after primary synthesis). In addition, the structure of many enzymes and proteins is defined by more than one gene locus. Overall, the extent of polymorphism and degree of heterozygosity for mice from wild-living populations agree well with estimates of these parameters in other mammals; but there is still controversy regarding the reasons for the maintenance of polymorphisms. In the forseeable future it should be possible to examine the DNA directly; this seems likely to be the ultimate step in the search for variation.

INTRODUCTION

Visible variation in mice has been known for thousands of years; a word for "spotted" mice appears in the earliest Chinese dictionary compiled in 1100 B.C. and the Japanese developed their own fancy stocks incorporating such alleles as non-agouti, chocolate, waltzing and spotting (Berry, 1977). In modern mouse genetics one of the first problems to be tackled was the inheritance of coat colour by C. C. Little who started working on this problem in 1907. Although genetically determined visible variation has been known for a very long time, the realization that cryptic variation exists in the forms of quantitative and qualitative variation in the amounts and structure of proteins has happened relatively recently. The first example of quantitative variation was found by Khanolkar & Chitre in 1942 when they reported that the level of serum esterase in the strain

C57BL was half that found in the strains C3H and A, but unfortunately no breeding studies were carried out. The biochemical genetics of the mouse can truly be said to start when Law, Morrow & Greenspan (1952) showed that a single gene determines high or low activity of β-glucuronidase in different inbred strains, and this gene was later shown to be the structural gene for the enzyme (Paigen, 1961). Variants have now been described at 26 loci by quantitative methods.

The first example of qualitative variation was reported by Ranney & Gluecksohn-Waelsch (1955) when they discovered genetically controlled differences in the electrophoretic mobility of mouse haemoglobins. The demonstration of qualitative variation in protein structure is closely linked with the development of electrophoretic techniques. The combination of these with histochemical detection methods for specific enzymes and proteins (Hunter & Markert, 1957) was a landmark in the search for qualitative variation. By 1971 22 electrophoretically detectable enzyme and protein variants had been identified. This figure has been steadily increasing since then and at the present time 90 variants of this sort are known. Thus, nearly 80% of the enzyme and protein variation in the mouse has been found using electrophoretic methods (Table I). These

TABLE I

Enzyme and protein variation

Enzyme	No. of variant genes detected by quantitative methods	No. of variant genes detected by qualitative methods	Total
Oxidoreductases	[a]6	[a]23	28
Transferases	6	11	17
Hydrolases	10	34	44
Lyases	3	3	6
Isomerases	0	2	2
Total	25	73	97
Proteins	1	17	18
Total	26	90	115

[a]Variation at aldehyde oxidase (*Aox-1*) detected by both quantitative and qualitative methods.

figures do not include the considerable amount of variation in protein structure demonstrated by immunological techniques, such as variation of immunoglobulin proteins, or of proteins coded by genes of the *H-2* complex. The enzyme and protein variants in mice are listed in the Appendix.

The wealth of enzyme and protein variation has been exploited by investigators wishing to use these biochemical markers as research tools, for instance in work on speciation, mutation, regulation of gene expression during development, in gene mapping studies, and especially in work on protein subunit structure and in the ascertainment of the number of genes responsible for a particular enzyme activity. However, the very abundance of biochemical variation is of interest for its own sake and has prompted many investigations regarding its nature, extent and maintenance.

SOURCES OF ENZYME AND PROTEIN VARIATION

Mice originating in one of two very different ways are the reservoirs of enzyme and protein variation. Firstly there are the inbred strains and laboratory stocks of mice and derivations of these, and secondly natural populations living in the wild (Table II).

TABLE II

Sources of enzyme and protein variation

Inbred strains and laboratory stocks	1. Long established strains 2. Strains of recent origin e.g. MOR/Cv Mol-O Mol-A 3. Recombinant inbred lines 4. Congenic lines 5. Irradiated laboratory stocks
Wild-caught mice	1. *Mus musculus* (Europe, North America, South America, Antarctic) 2. *Mus musculus castaneus* (Thailand) 3. *Mus musculus molossinus* (Japan) 4. *Mus spretus*

Inbred Strains

There are now over 300 inbred strains of mice (Staats, 1979) but many of the early inbred strains originated in the north-east United States from a small number of stocks. During recent years new inbred strains have been developed from animals taken from the wild from divergent sources. For example the MOR/Cv strain which carries a variant allele of malate dehydrogenase, *Mor-1*, was developed from feral mice from Ohio, and the Mol-A and Mol-O strains which carry variants of chymotrypsin, *Prt-2*, and trypsin, *Prt-3*, are inbred strains derived from Japanese feral mice (Chapman, Paigen *et al.*, 1979). Although different strains differ markedly from each other owing to their original traits and subsequent selection, it is apparent that they contain only a proportion of the variation present in wild-living populations. For example, of the 115 enzyme and protein variants found in mice, 21 have been found only in wild-caught mice and not in inbred strains. It is not surprising that inbred strains carry only a proportion of the variation that exists since each inbred strain represents only one genotype, and every individual from a natural population has its own unique genotype.

Recombinant Inbred (RI) Lines

Over the years inbred strains have been genetically manipulated to establish other strains such as congenic, coisogenic and recombinant inbred strains. The latter have been especially useful in murine biochemical genetics. They are produced by inbreeding the F_2 generation of a cross between two unlike progenitor inbred strains, so that segregation and recombination from the F_1 generation occurs until the RI lines approach homozygosity (Bailey, 1971). RI lines have been derived in this way from 13 different crosses and as many as 31 RI lines have been obtained from one original cross (Taylor, 1979). Swank & Bailey (1973) put forward the advantages of using RI lines in biochemical genetics. RI lines are useful for ascertaining the number of genes responsible for determining the expression of a biochemical trait. This is especially true in instances where quantitative assay methods are used to detect variation. If all the RI lines have phenotypes resembling one or other of the progenitor strains then probably the trait is determined by a single gene. If new phenotypes are found, these represent the occurrence of new genetical combinations and more than one gene must be involved in determining the trait. RI lines can also be used in linkage testing, so that the distribution of alleles at a new locus can be compared with the distribution of alleles at other known loci. Womack, Lynes & Taylor (1975) used data from RI lines as well as from a conventional back-cross to show that leucine arylaminopeptidase (*Lap-1*) is linked to dilute (*d*) and supernatant malic enzyme (*Mod-1*) on chromosome 9. RI lines are of great value for experiments where it is necessary to follow a process through time, such as enzyme expression during development or induction of enzyme activity, and where an individual animal can only be tested once because testing requires that the animal be killed. Once an RI line is established there will be a source of animals of identical genotype as when an inbred strain is used.

Congenic Lines

These are inbred strains genetically identical to an already established strain except for a small chromosomal segment. One of the uses of these has been in linkage studies. For example the strains C57BL/6 and C3H carry different alleles at the *Gus-s* gene, the structural gene for β-glucuronidase, and the *Gus-t* gene which regulates the expression of β-glucuronidase during development. There is a C57BL/6 congenic line where chromosome 5 is mainly derived from C57BL/6, but is also composed of a small piece of chromosome 5 from C3H. This small segment carries both the *Gus-s* and *Gus-t* genes and the congenic line has *Gus-s* and *Gus-t* alleles characteristic of C3H, but not C57BL/6, indicating that these genes are linked (Paigen, Swank *et al.,* 1975).

Variants Arising from Use of Mutagens

In a few instances biochemical variants have been induced by irradiation of laboratory stocks: at the *Cs-1* locus, the structural locus for catalase (Feinstein *et al.*, 1966); at the *spf* locus coding for ornithine carbamoyl transferase (DeMars *et al.*, 1976); and at the *Hba* and *Hbb* loci coding for α- and β-globin respectively (Russell, Russell *et al.*, 1976).

Wild-caught Mice

At about a quarter of the loci examined, variation has only been found in mice recently derived from the wild or taken directly from feral populations. Also for ten genes which exhibit allelic variation between strains, further alleles have been found in wild-caught mice which are not found in inbred strains (Table III). The mice derived from the wild came from a wide variety of sources ranging from the Japanese and Thai subspecies, *Mus musculus molossinus* and *Mus musculus castaneus*, to mice from Europe, the Americas and the Antarctic. Mice from natural populations have not only been very valuable in the search for new enzyme and protein variants and for finding additional alleles at established loci, but also in finding genetical combinations not present among existing laboratory stocks.

TABLE III

Enzyme and protein variation in inbred strains and wild-caught mice

	No. of loci where variant present in inbred strains	[a, b]No. of loci where variant found in wild living mice
Oxidoreductases	22	6
Transferases	13	4
Hydrolases	35	8
Lyases	5	1
Isomerases	2	0
Total	77	19
Proteins	16	2
Total	93	21

[a]The loci are: *Adh-1, Sdh-1, Ldh-2, Pgd, Hao-2, Np-1, Got-1, Pk-1, Pgk-1, Es-6, Es-7, Es-8, Es-14, Apk, Ags, Pep-2, Ada, His, Alb-1, Erp-1.*
[b]Loci where variants found in inbred strains but where further alleles are found in wild derived mice: *Gdc-1, Mor-1, Idh-1, Gpt-1, Es-1, Es-2, Es-3, Es-10, Car-1, Gpi-1.*

POLYMORPHISM

For enzymes and proteins this term describes situations in which members of a population can be sharply classified into several distinct phenotypes in terms of particular characteristics of an enzyme or protein. The definition of a polymorphic locus is arbitrary but in general a locus may be considered polymorphic if the frequency of the most commonly occurring allele is less than 0.99.

Usually the term polymorphism is only applicable to wild-living populations or laboratory stocks which are not maintained by inbreeding, and cannot be applied to inbred strains. One exception is the HRS/J strain which has been maintained by forced heterozygosity at the hairless locus, *hr*, by crossing males homozygous for the *hr* allele with heterozygous females for at least 60 generations (Cumming, Walton *et al.*, 1979). Not only is there segregation at the *hr* locus but the HRS/J strain is also heterozygous for alleles at two closely linked genes, esterase-10, *Es-10*, and formamidase-5, *For-5*. (Womack, Davisson *et al.*, 1977; Cumming, Walton *et al.*, 1979). The *hr* allele segregates with the *Es-10*^{b} and *For-5*^{d} alleles and the normal allele with the *Es-10*^{a} and *For-5*^{b} alleles. In all inbred strains examined so far except BUB/BnJ the *Es-10*^{b} and *For-5*^{d} alleles are associated, as are the *Es-10*^{a} and *For-5*^{d} alleles. BUB/BnJ has the *For-5*^{b} allele and the rare *Es-10*^{c} allele. It has been suggested that this allelic association is evidence of very close linkage between the *hr*, *Es-10* and *For-5* loci, but another possibility is that it is maintained by linkage disequilibrium.

Inbred strains have been very valuable in the elucidation of the biochemical and genetical nature of enzyme and protein variation and for this reason have been included in this review of polymorphism.

THE DETECTION OF VARIATION

Electrophoresis

By far the most widely used and successful method for the detection of enzyme and protein variation has been one-dimensional electrophoresis followed by the identification of gene products by specific staining procedures. Using these techniques two major types of protein variant have been found. The first class includes variants which lead to synthesis of a protein with altered electrophoretic mobility. In general, variations of this sort are due to alterations in the net charge on the enzyme or protein being considered, although on certain separation media, such as starch or acrylamide, alterations due to differences in molecular size may also be detected, since the gel matrix itself imposes a molecular sieving effect. Although relatively few of the enzyme and protein variants have been purified and sequenced it appears that the technique is capable of detecting single amino acid changes, which it is assumed have arisen from single base charges. The best evidence for this comes from work carried out on enzyme and protein variants in man. The best known example is the variant HbS haemoglobin in man which differs in one amino acid from the normal HbA haemoglobin.

Theoretically about 30% of all possible base substitutions will lead to the replacement of one amino acid by another of different charge (Harris, 1974). Of course it is not possible to detect base substitutions which give rise to synonymous codons by electrophoresis, and those which lead to the substitution of one amino acid by another of similar charge are probably not easily detected. A further problem is that an unknown proportion of those base substitutions which result in proteins altered in charge also give rise to inactive enzymes which would not be detected by activity staining.

The second type of enzyme and protein variation which can be detected by electrophoretic techniques are those variants that cause loss of enzyme or protein activity. Electrophoretic techniques are ill suited to observing this type of variation since even the loss of 50%

of enzyme activity due to the synthesis of a product which is completely inactive enzymically are often difficult to detect. In general, when a polymorphism is found, a "null" allele may be observed by apparent increased homozygosity and the appearance of rare phenotypes which exhibit no enzyme activity. This is true for the enzyme esterase-2 in the mouse where two common alleles exist in wild populations, the "null" $Es\text{-}2^a$ allele and the $Es\text{-}2^b$ allele which has enzyme activity. The heterozygote $Es\text{-}2^a/Es\text{-}2^b$ has an appearance similar to the $Es\text{-}2^b/Es\text{-}2^b$ homozygote (Petras, 1963; Peters & Nash, 1977).

Another way of detecting the presence of "null" alleles or alleles with very greatly reduced enzyme activity is by observing the production of functional hybrid isozymes. Sometimes an allele which is reduced in activity or even inactive in the homomeric state is capable of forming functional heteromeric molecules when hybridized to the normal polypeptide. If the low activity protein differs in charge from the normal protein, heterozygotes for low activity or "null" alleles may be detected by the presence of heteromeric isozymes. Thus in mice heterozygous for the low activity glucose phosphate isomerase allele, $Gpi\text{-}1^c$ and one of the full activity alleles $Gpi\text{-}1^a$ or $Gpi\text{-}1^b$, only two isozymes are present and the homodimeric isozyme formed from the low activity polypeptide is not observed. In heterozygotes for the full activity alleles the expected three-banded electrophoretic pattern is seen (Fig. 1) (Padua, Bulfield & Peters, 1978).

It is apparent that electrophoretic techniques by themselves

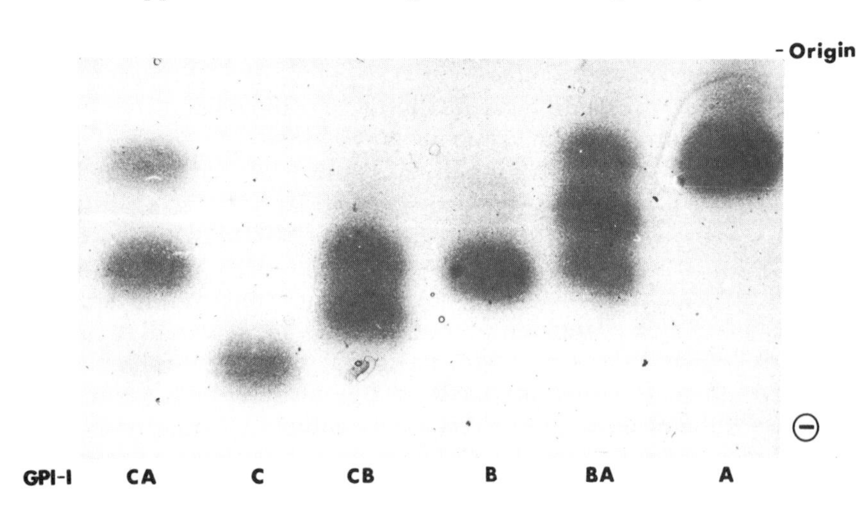

FIG. 1. Glucose phosphate isomerase isozyme patterns in haemolysates. (Reproduced with permission from Padua, Bulfield & Peters, 1978.)

underestimate the number of alleles that actually occur, but there are two further reasons for underestimation of the number of alleles One is that the discriminative power of the system varies from enzyme to enzyme, and probably inadequacy of technique explains the failure to find variants which actually exist. The second reason is that it is assumed for practical purposes that electrophoretically indistinguishable variants are attributable to the same allele. This of course is not necessarily so since two quite different amino acid substitutions can produce the same change in electrophoretic mobility. Just because different inbred strains and wild mice have the same electrophoretic form of an enzyme or protein does not mean they carry the same allele.

Recently some attempts have been made to solve this problem. One approach has been to study thermostability properties of enzymes and proteins as well as electrophoretic mobility. Bonhomme & Selander (1978) studied the products of 14 structural loci in 40 strains of *Mus musculus* and found four new variants, all of which were esterases, which differed in thermostability characteristics but could not be detected by electrophoresis alone. Thermostability variants which are not detectable electrophoretically have also been found for glycerol-3-phosphate dehydrogenase (*Gdc-1*) (Kozak & Erdelsky, 1975), pyruvate kinase-1 (*Pk-1*) (Bulfield, Moore & Peters, 1978), and β-glucuronidase (*Gus-s*) (Paigen, Swank *et al.*, 1975).

The most complete estimation of the number of alleles at a locus has been carried out on *Drosophila* where studies of the xanthine dehydrogenase locus, *Xdh*, in *Drosophila pseudoobscura* and *D. persimilis*, have shown that a considerable increase in the number of detectable alleles is obtained by using a wide variety of electrophoretic conditions and also tests of heat sensitivity (Singh, Lewontin & Felton, 1976; Coyne, 1976). Similar findings have been made for the esterase-5 locus, *Est-5* in *Drosophila pseudoobscura* (Coyne, Felton & Lewontin, 1978) but not for malic dehydrogenase, *Mdh*, or for octanol dehydrogenase *Odh* (Coyne & Felton, 1977). Interestingly, a large increase in allelic diversity has only been found for loci which were already known to be highly polymorphic; but in one instance (the highly polymorphic alcohol dehydrogenase locus *Adh-6*) genetical diversity was not increased to nearly the same extent as it was for the *Xdh* and *Est-5* loci (Coyne & Felton, 1977). So the phenomenon of finding greatly increased variation using a variety of techniques does not appear to be general, but is more likely for loci which are known to be highly polymorphic.

Isoelectric Focusing

Isoelectric focusing has greater resolving power than electrophoresis and has recently been used to demonstrate the polymorphism of the haemoglobin α-chain locus (*Hba*) (Whitney, Copland *et al.*, 1979). Prior to this the different alleles or haplotypes at the *Hba* locus could only be detected by looking for differences in haemoglobin solubility in buffered salt solutions of high molarity, or by the presence or absence of a particular tryptic peptide (Russell, Blake & McFarland, 1972). The application of isoelectric focusing to the separation of *Hba* haplotypes is interesting for several reasons (Table IV). Firstly, two new haplotypes (Hba^f and Hba^g) were resolved

TABLE IV

Haemoglobin alpha-chain haplotypes

Hba type	α polypeptide chains	Primary structure differences		
		α-25	α-62	α-68
a	1	Gly	Val	Asn
b	2, 3	Gly	Val	Ser
		Gly	Val	Thr
c	1, 4	Gly	Val	Asn
		Val	Ile	Ser
d	1, 2	Gly	Val	Asn
		Gly	Val	Ser
e	4	Val	Ile	Ser
f	5	?	?	?
g	1, 5	Gly	Val	Asn
		?	?	?

among mice that had previously been considered on the basis of haemoglobin solubility to be Hba^a, so this new technique has greater resolving power than previous methods. Secondly the amino acid sequence of four of the α-globin chains was known and all four differ from one another in non-charged amino acids only (Hilse & Popp, 1968). This type of substitution is probably difficult to detect by standard electrophoretic methods, which have been unsuccessful in resolving different *Hba* haplotypes. It seems that the neutral amino acid substitutions in the α-globin chains produce subtle changes in the net charge of the tetrameric haemoglobins which can be detected by the high resolving power of isoelectric focusing. If this result is generally applicable then isoelectric focusing should be a very powerful technique for resolving polymorphisms which involve neutral–neutral amino acid mutations.

Quantitative Methods

Many gene mutations are expressed by an alteration in the activity of a specific enzyme which is best detected by quantitative methods. Alterations in the level of enzyme activity may be caused by modifications to the catalytic properties of the enzyme, its stability, or its rate of synthesis. Enzyme activity may also be changed as a consequence of mutations which lead to the alteration of levels of activators or inhibitors. Of the 26 enzyme and protein variants known in the mouse which are observed by quantitative methods, nine have been found to be due to genetically determined alterations in thermostability of the enzyme. In these instances the mutation is assumed to have occurred in the "structural" gene, the gene that codes for the amino acid sequence of the enzyme. Alterations in the rates of protein synthesis may also be due to a mutation in the structural gene but can be caused as well by a mutation occurring in a "regulatory" gene which controls the rate of synthesis of the structural gene product.

REGULATORY GENES

The most detailed work on regulatory genes in mice has been carried out on the enzyme β-glucuronidase (Paigen, Swank *et al.*, 1975; Swank *et al.*, 1978), but regulatory genes have also been identified for α-galactosidase (Lusis & Paigen, 1975), β-galactosidase (Paigen, Meisler *et al.*, 1976; Berger & Paigen, 1979) and aryl sulphatase (Daniel, 1976 a, b).

Paigen, Swank *et al.* (1975) have described two types of regulatory gene: those that affect enzyme activity levels at a particular time during development (temporal genes) and those whose effect on enzyme activity is independent of time (regulatory genes). For example, closely linked to the structural gene (*Gus-s*) for β-glucuronidase on chromosome 5 are two genes which affect activity levels of the enzyme: β-glucuronidase regulator, *Gus-r,* which controls the responsiveness of the enzyme in kidney to induction by androgens and β-glucuronidase temporal (*Gus-t*) which controls the rate of synthesis of β-glucuronidase during development. Both *Gus-r* and *Gus-t* have been shown to control the rate of enzyme synthesis (Paigen, Swank *et al.,* 1975). In the absence of any certain recombinants between *Gus-s*, *Gus-r*, and *Gus-t* (Swank *et al.,* 1978) it is not known if *Gus-r* and *Gus-t* are separate genes from *Gus-s* or if they are all one and the same gene. If they are the same gene then this means in this case that rate of enzyme synthesis is controlled by the structural gene itself.

Other examples show that temporal and regulatory genes can be separate entities from their respective structural genes. There is no strain concordance between alleles of the β-galactosidase structural gene (*Bgl-e*) and a linked temporal gene *Bgl-t* (Breen, Lusis & Paigen, 1977), and the temporal gene for α-galactosidase, *Tag*, is autosomal whereas the structural gene, *Ags*, is X-linked (Lusis & Paigen, 1975; Lusis & West, 1978). Another regulatory gene *Bgl-s,* which controls levels of β-galactosidase and whose effect is independent of development, is also genetically distinct from the temporal gene *Bgl-t* and the structural gene *Bgl-e* (Breen *et al.,* 1977).

The manner of action of the *Gus-r* gene appears to be *cis* so that in heterozygotes that carry a different *Gus-r* allene, each chromosome is regulated according to the *Gus-r* allele it carries and acts independently of the other (Swank *et al.,* 1978). In contrast Meredith & Ganschow (1978) concluded that *Gus-t* control is exerted *trans. Cis* action implies that the regulatory gene does not code for a diffusible product; otherwise the products would be expected to mix and both chromosomes should respond equivalently. *Cis* action has also been shown for the *Bgl-s* locus which controls the rate of synthesis of β-galactosidase in mouse tissue (Berger & Paigen, 1979).

PROCESSING GENES

In general it is assumed that electrophoretic variants result from mutations that have occurred in the structural gene coding for the amino acid sequence of the enzyme or protein being examined. However, the enzyme or protein may be modified after primary synthesis by conjugation with glycosyl, phosphoryl or other groups. Modification of protein structure may be due to the action of a separate gene, termed a processing gene (Paigen, Swank *et al.*, 1975). For example, electrophoretic studies of lysosomal acid phosphatase from mouse liver show that two zones of enzyme activity (AP-1 and AP-2) are present in all strains examined except one (SM/J), in which only the more anodally migrating form, AP-2, is found. However, if liver samples are treated with sialidase prior to electrophoresis the AP-2 zone disappears in all strains examined, including SM/J, and the activity of the AP-1 zone is increased and then AP-1 is found in SM/J. It is postulated that sialic acid residues are bound to acid phosphatase and that these are removed by the action of sialidase. The variation observed in SM/J does not appear to be due to a mutation in the acid phosphatase structural gene but is due to variation of a processing gene which affects the degree of sialylation of the enzyme (Lalley & Shows, 1977). This processing gene is located on chromosome 17 (Womack & Eicher, 1977). Interestingly, Dizik & Elliott (1978) reported a variation in SM/J of another liver lysosomal enzyme, α-mannosidase, which also appears to be due to the result of differences in sialylation. This gene *Map-2* has also been mapped on chromosome 17 and no recombination has been found so far between *Map-2* and *Apl* (Dizik & Elliott, 1978). It appears that *Map-2* and *Apl* may be one and the same gene. Other examples of processing genes are *Map-1*, another gene affecting α-mannosidase (Dizik & Elliott, 1977); *Ce-1* and *Ce-2* affecting catalase (Ganschow & Schimke, 1970; Rechcigl & Heston, 1967); *Eg* which acts on β-glucuronidase (Paigen, Swank *et al.*, 1975). All the examples so far except *Apl* show dominant-recessive inheritance.

MULTIPLE LOCI

In many instances two or more separate gene loci are involved in defining the structure of a protein and electrophoretic techniques have been very useful in demonstrating this. Table V lists those where genetically determined variants have been found at more than one locus and where the variation results in an alteration in electrophoretic mobility.

The most striking example of multiple genes determining enzyme structure are esterase genes. Twelve loci have been described which code for carboxyl esterase: *Es-1*, *Es-2*, *Es-3*, *Es-5*, *Es-6*, *Es-7*, *Es-8*, *Es-9*, *Es-10*, *Es-11*, *Es-12*, *Es-14* (Taylor, 1977; Peters & Nash, 1978; Bonhomme, Britton-Davidian & Thaler, 1979). For no other enzyme have so many loci been described.

In some cases there is differential expression of the products of each structural locus in different tissues. For example two structural loci code for the enzyme phosphoglycerate kinase (*Pgk-1*, *Pgk-2*). *Pgk-1* is expressed in many tissues but *Pgk-2* is only expressed in the testis (VandeBerg, Cooper & Close, 1976). Similarly there is a testis specific form of lactate dehydrogenase but two other genes, *Ldh-1* and *Ldh-2*, also code for this enzyme and these are differentially expressed in different tissues. In mouse kidney both these loci are expressed to about the same extent, but the *Ldh-1* locus is preferentially expressed in the liver.

For some enzymes gene expression is associated with different cellular structures. For example, two genes *Mod-1* and *Mod-2* code

TABLE V

Some enzymes whose structure is determined by more than one gene

Oxidoreductases	Alcohol dehydrogenase Lactate dehydrogenase Malic enzyme α-hydroxyacid oxidase
Transferases	Aspartate amino transferase Phosphoglycerate kinase Phosphoglucomutase
Hydrolases	Carboxyl esterase Alkaline phosphatase Amylase Peptidase
Lyases	Carbonic anhydrase

for malic enzyme but *Mod-1* codes for the cytoplasmic form and *Mod-2* for the mitochondrial form of the enzyme (Shows, Chapman & Ruddle, 1970). Malic enzyme is composed of four subunits but heteromeric isozymes due to the interactions of the subunits produced by the separate *Mod-1* and *Mod-2* loci are not observed, presumably because of the different subcellular location of the gene products. On the other hand, lactate dehydrogenase, which is also a tetramer, does form heteromeric isozymes between the products of the *Ldh-1* and *Ldh-2* loci, presumably because the products of both loci are found in the cytoplasm.

In some isntances there is close linkage of the genes coding for a specific enzyme. For example *Car-1* and *Car-2* coding for carbonic anhydrase are closely linked on chromosome 3 (Eicher, Stern *et al.*, 1976); and seven esterase genes (*Es-1*, *Es-2*, *Es-5*, *Es-6*, *Es-7*, *Es-9*, and *Es-11*) are linked on chromosome 8 (Peters & Nash, 1978). In these cases it is postulated that the genes have arisen from random duplication of an ancestral gene.

THE EXTENT OF VARIATION

Theoretically a very large number of alleles of a single gene can occur. Harris (1975) considered the β-globin gene in man and calculated that 287 alleles could exist differing in the substitution of a single base leading to the substitution of one amino acid by another of different charge. This means that 287 different electrophoretic variants could exist. Surprisingly over 60 different haemoglobin variants have been identified which appear to have this type of substitution in the β-globin polypeptide chain, inplying that a great many alleles of the β-globin gene must occur. Moreover the β-globin gene does not appear to be unusual in this respect, for over 80 different allelic variants of the enzyme glucose-6-phosphate dehydrogenase have been reported (Harris, 1975). Many of these alleles are very rare, but for those enzymes and proteins which are polymorphic several variant alleles occur at high frequencies in populations.

Obviously it is of some interest to investigate the extent of allelic diversity in populations and since the first surveys on man and *Drosophila pseudoobscura* (Harris, 1966; Lewontin & Hubby, 1966) many studies have been carried out on a wide variety of species. Overall the data for house mouse populations (Berry & Peters, 1981; Ruddle, Roderick, Shows, Weigl, Chipman & Anderson, 1969; Wheeler & Selander, 1972) show that about a quarter of the loci surveyed are polymorphic (polymorphic loci were defined as those at which the most common allele does not exceed a frequency of either 0.95 or 0.99) and any individual animal is likely to be heterozygous at about 6% of its loci. These figures are of the same order as those reported for other rodents (Selander & Kaufman, 1973) and man (Harris & Hopkinson, 1972). All these data relate to electrophoretically detectable variation.

One of the cornerstones in population surveys of electrophoretically detectable variation is that a random selection of enzymes and proteins are studied and that these are the products of structural genes. It has been suggested by several workers that different enzymes may have different variabilities depending on their function (Johnson, 1974; Gillespie & Kojima, 1968), subunit structure (Harris, Hopkinson & Edwards, 1977), or size (Ward, 1977). However, in many population studies there has been an emphasis towards surveying certain types of enzyme, in particular nonspecific hydrolases and dehydrogenases, mainly because methods are readily available for detecting these types of enzyme. In general cytoplasmic enzymes are studied, rather than those associated with cell organelles, and furthermore most of the proteins studied are enzymes. Recently Leigh Brown & Langley (1979) have used a different criterion for the selection of enzymes and proteins used to survey allelic variation in populations. They chose proteins which occurred in greatest abundance in *Drosophila melanogaster*. Interestingly they found much lower levels of heterozygosity (4%) after screening the protein products of 54 loci than by conventional gel electrophoresis where the figure is 14%. It remains to be seen if this is true for other organisms.

THE SIGNIFICANCE OF VARIATION

The widespread occurrence of polymorphism and heterozygosity has provoked much discussion about the biological significance of these phenomena. There are two opposing views: one is that the occurrence of most polymorphisms is due to neutral or near neutral mutations and random genetic drift, and the other is that polymorphisms are maintained by balancing selection.

For over a decade geneticists have addressed themselves to these issues but the problems can hardly be said to have been solved. It is probably true to say, though, that the complexities of the problems have been more closely identified.

The contribution that work on the house mouse has made to this debate has been to measure allelic variation in different populations and to try and interpret the results in the light of developing knowledge of the ecology and physiology of the animal (for review see Berry, 1978).

In a survey of genetic variation in mouse populations from 29 localities in Britain, the nearby Shetland, Orkney and Faroe islands, three islands in the Antarctic Ocean, as well as Central Peru (Berry, Bonner & Peters, 1979; Berry & Peters, 1977, 1981) different populations had mean heterozygosities ranging from 0% for two Scottish islands and the Faroese island of Fugløy to 11.4% for one of the British mainland samples from Caithness in the North of Scotland (see Berry, this volume, p. 406). Allele frequencies differed greatly from place to place. Overall mice from mainland populations have a higher average heterozygosity, 8.6% for the British mainland and 8.0% for Peru, but lower than those on islands (4.9% or 6.3% if three invariant populations are omitted), but the difference is not very striking. Findings such as these make it difficult to attach great importance to either overall heterozygosity or the frequencies of particular alleles. Mean heterozygosity was weakly correlated with population size but not at all with latitude which contrasts with the findings of Selander, Smith *et al.* (1971) who found a correlation between heterozygosity and latitude for *Peromyscus polionotus*.

There have been many attempts to correlate mean heterozygosity with environmental heterogeneity. For example Bryant (1974) found that 83% of the allelic variation in a number of studies of house mouse could be accounted for by temperature and rainfall data. The most important climatic variable was the coefficient of variation of mean monthly rainfall over the year but this is of little importance in the ecology of mice in most environments (see Berry

& Peters, 1977), quite apart from the fact that most of the data used came from mice living in buildings and protected from the weather.

Environmental homogeneity has been put forward as a reason for reduced levels of heterozygosity in the pocket gopher *Thomomys talpoides* (Nevo *et al.*, 1974), but this explanation is confounded, for another pocket gopher, *Thomomys bottae*, has a relatively high level of heterozygosity (Patton & Yang, 1977). It has also been suggested that the low levels of heterozygosity found in some island populations of rodents reflect reduced environmental heterogeneity compared to mainland situations. Schnell & Selander (1981), have suggested that drift, including founder effect, may be responsible for reduced heterozygosity in island populations. In all these studies there is a profound difficulty in separating the effects of selection from those of drift and founder effect.

The action of founder effect has been invoked by Berry (1977, 1978) and Berry & Peters (1977) as a partial explanation of the wide range of heterozygosity between different house mouse populations. It was found that the mean genetic distance between samples in three successive years from a closed population was 0.03, whereas the mean genetic distance between populations from three neighbouring mainland farms was 0.09. It was assumed that the difference between these two figures represents the sort of difference that can be produced by founder effect alone.

The best evidence that selective forces maintain polymorphism in the house mouse have come from studies of allelic variation at the *Hbb* locus which codes for the β-globin polypeptide chain, where allele (or genotype) frequencies have been shown to change with season and social status (Berry & Murphy, 1970; Myers, 1974) and with age (Berry & Peters, 1975; Berry, Peters & Van Aarde, 1978). It has also been shown to respond to selection for size under laboratory conditions (Garnett & Falconer, 1975). Whether or not selection is acting directly on the *Hbb* locus or on a closely linked locus is not known.

If selection is not acting on the *Hbb* locus itself but on a closely linked locus then this implies the instance of a linkage disequilibrium. Recombination is known to be suppressed on chromosome 17 between the *T* and *H-2* segments (Dunn, Bennett & Beasley, 1962; Forejt, 1972; Womack & Roderick, 1974) and recently two other examples have been found in the mouse, one on chromosome 7 between *Gpi-1* and *Hbb* (Berry & Peters, 1981) and the other on chromosome 8 between *Es-6* and *Got-2* (Berry, Bonner & Peters, 1979).

Linkage disequilibria may arise and be maintained by selection if there is an epistatic interaction between loci or they may be due to chance in which case they will be temporary. It is not known if the linkage disequilibria between *Gpi-1* and *Hbb* and between *Es-6* and *Got-2* are maintained by selection or if they are transitory, since each has only been observed in a single population sample. However, if selective forces are maintaining the linkage disequilibria these may fluctuate in both time and space.

FUTURE PERSPECTIVES

Mouse genetics started with studies of the inheritance of visible variants. More recently genetically determined variation in enzyme and protein structure has been investigated. Most of the enzyme variants described can be presumed to have arisen from single amino acid substitutions which are attributable to single base changes in DNA. Ultimately it will be of interest to look at the DNA itself. Recently it has become possible to do this because of the rapid development of recombinant DNA technology and the increasing availability of DNA probes.

Using these techniques Jeffreys & Flavell (1977) analysed the genes for rabbit β-globin, and made a very remarkable finding. This was that the β-globin mRNA was not a faithful copy of the base sequence of the gene. Whereas in the β-globin mRNA the sequence of bases specifying the protein is continuous, in the DNA the base sequence coding for β-globin is not. It appears that the sequence in the gene is interrupted by intervening sequences which are not transcribed. This has also been found for the mouse β-globin gene (Tilghman *et al.*, 1978).

In man four genes coding for the β-globin polypeptides are linked on the short arm of chromosome 11. Two of these, the ${}^{G}\gamma$ and ${}^{A}\gamma$ code for the ${}^{G}\gamma$ and ${}^{A}\gamma$ polypeptide chains which are found in foetal haemoglobin and the other two, the β- and δ-globin genes, code for the β- and δ-globin polypeptide chain of adult haemoglobin. The base sequence of all four globin genes is interrupted by intervening sequences (Flavell *et al.*, 1978; Little *et al.*, 1979).

Recently Jeffreys (1979) has started to look for DNA sequence variants of the four globin genes in man. He has analysed globin DNA from 60 normal individuals using eight different restriction endonucleases. Three different variant restriction enzyme patterns of globin DNA were found. One individual was heterozygous for the presence of an additional cleavage site for the restriction endonuclease Pst1 within the δ-globin gene intervening sequence. A polymorphism was found for the presence or absence of a restriction endonuclease cleavage site (Hind III) in the intervening sequence of the $^{A}\gamma$-globin gene. Individuals homozygous for the presence and for the absence of this cleavage site were observed as well as heterozygotes. The Hind III cleavage site polymorphism was also found to occur in the intervening sequence of the $^{G}\gamma$-globin gene.

It is interesting that all the variants detected so far are found within the intervening base sequences of the globin genes. Since these sequences only account for about 8% of the sequences examined and since the screening procedure would have detected variation within the coding sequences as well as in the intervening sequences, it appears unlikely that the three variants observed occurred in the intervening sequences by chance. It has been found that whereas the coding sequence of rabbit and mouse β-globin genes are well conserved the intervening sequences of these two animals diverge substantially from each other (van den Berg *et al.*, 1978).

The existence of variation in DNA sequence is apparently not peculiar to β-globin genes, for Lai *et al.* (1979) have found two cleavage site polymorphisms in the chicken ovalbumin gene, and both of these polymorphisms occur in intervening sequences.

Detailed analysis of mouse DNA will be possible as soon as the necessary gene probes become available and then maybe the mouse genome will yield some more of its secrets; an exciting prospect for the future!

APPENDIX

Enzyme Variation

EC Number	Enzyme	Names of variant genes affecting enzyme	Gene symbol	Method of detection	[c]References
Oxido-reductases					
1.1.1.1	Alcohol dehydrogenase	Alcohol dehydrogenase-1	*Adh-1*	Electrophoresis	145
		Alcohol dehydrogenase-3	*Adh-3e*	Electrophoresis	71
		Alcohol dehydrogenase-3 temporal	*Adh-3t*	Electrophoresis	74
1.1.1.8	Glycerol-3-phosphate dehydrogenase	NAD alpha-glycerol phosphate dehydrogenase	*Gdc-1*	Electrophoresis and thermostability	89
1.1.1.14	Sorbitol dehydrogenase	Sorbitol dehydrogenase-1	*Sdh-1*	Electrophoresis	76
1.1.1.27	Lactate dehydrogenase	Lactate dehydrogenase-1	*Ldh-1*	Electrophoresis	155
		Lactate dehydrogenase-2	*Ldh-2*	Electrophoresis	120, 15
		Lactate dehydrogenase regulator	*Ldr-1*	Electrophoresis	58, 81, 150
1.1.1.37	Malate dehydrogenase	Mitochondrial malate dehydrogenase	*Mor-1*	Electrophoresis	149, 183
1.1.1.40	Malic enzyme	Malic enzyme, supernatant	*Mod-1*	Electrophoresis	65, 149
		Malic enzyme, mitochondrial	*Mod-2*	Electrophoresis	149
1.1.1.42	Isocitrate dehydrogenase	Isocitrate dehydrogenase-1	*Idh-1*	Electrophoresis	65

APPENDIX (*continued*)

EC Number	Enzyme	Names of variant genes affecting enzyme	Gene symbol	Method of detection	References
1.1.1.44	Phosphogluconate dehydrogenase (decarboxylating)	6-phosphogluconate dehydrogenase	*Pgd*	Electrophoresis	20
1.1.1.47	Glucose dehydrogenase	Glucose phosphate dehydrogenase	*Gpd-1*	Electrophoresis	81, 134, 138
1.1.1.49	Glucose-6-phosphate dehydrogenase	G6PD regulator-1	*Gdr-1*	Quantitative assay	80
		G6PD regulator-2	*Gdr-2*	Quantitative assay	80
1.1.3.1	L-α-Hydroxyacid oxidase	Alpha-hydroxyacid oxidase-1	*Hao-1*	Electrophoresis	71
		Alpha-hydroxyacid oxidase-2	*Hao-2*	Electrophoresis	71, 72
1.2.1.3	Aldehyde dehydrogenase	Aldehyde dehydrogenase-1	*Ahd-1*	Electrophoresis	73
1.2.3.1	Aldehyde oxidase	Aldehyde oxidase	*Aox-1*	Quantitative assay and electrophoresis	75, 169
		Aldehyde oxidase-2	*Aox-2*	Electrophoresis	75
1.4.3.3	D-Amino acid oxidase	D-amino-acid oxidase	*Dao*	Electrophoresis	70
1.6.4.2	Glutathione reductase	Glutathione reductase	*Gr-1*	Electrophoresis	51, 113
1.11.1.6	Catalase	Catalase-1	*Cs-1*	Quantitative assay	40, 50, 68
		Liver catalase degradation	*Ce-1*	Quantitative assay	54, 66, 133
		Kidney catalase	*Ce-2*	Electrophoresis	69

APPENDIX *(continued)*

EC Number	Enzyme	Names of variant genes affecting enzyme	Gene symbol	Method of detection	References
1.14.18.1	Monophenol monooxygenase	Albino	*c*	Quantitative assay	27
1.15.1.1	Superoxide dismutase	[a]Superoxide dismutase soluble	*Sod-1*	Electrophoresis	148
Transferases					
2.1.3.3	Ornithine carbomoyl transferase	Sparse-fur	*spf*	Quantitative assay	39
2.4.2.1	Purine nucleoside phosphorylase	Nucleoside phosphorylase-1	*Np-1*	Electrophoresis	181
		Nucleoside phosphorylase-2	*Np-2*	Electrophoresis	14
2.6.1.1	Aspartate aminotransferase	Glutamate oxalo acetate transaminase-1	*Got-1*	Electrophoresis	23, 46
		Glutamate oxalo acetate transaminase-2	*Got-2*	Electrophoresis	38
2.6.1.2	Alanine aminotransferase	Glutamic pyruvic transminase	*Gpt-1*	Electrophoresis	25, 49
2.6.1.38	Histidine aminotransferase	Histidine amino transferase-1	*Hat-1*	Quantitative assay and thermostability	17
2.7.1.2	Glucokinase	Glucokinase activity	*Gk*	Quantitative assay	29
2.7.1.6	Galactokinase	Galactokinase	*Glk*	Quantitative assay	105
2.7.1.38	Phosphorylase kinase	Phosphorylase kinase	*Phk*	Quantitative assay and thermostability	59, 78, 101
2.7.1.40	Pyruvate kinase	Pyruvate kinase-1	*Pk-1*	Quantitative assay and thermostability	18
		[b]Pyruvate kinase-2	*Pk-2*	Electrophoresis	109

EC Number	Enzyme	Names of variant genes affecting enzyme	Gene symbol	Method of detection	References
		[b] Pyruvate kinase-3	*Pk-3*	Electrophoresis	110
2.7.2.3	Phosphoglycerate kinase	Phosphoglycerate kinase-1	*Pgk-1*	Electrophoresis	114
		Phosphoglycerate kinase	*Pgk-2*	Electrophoresis and thermostability	163, 44
2.7.5.1	Phosphoglucomutase	Phosphoglucomutase-1	*Pgm-1*	Electrophoresis	151, 104
		Phosphoglucomutase-2	*Pgm-2*	Electrophoresis	24, 151
Hydrolases					
3.1.1.1	Carboxyl esterase	Serum esterase-1	*Es-1*	Electrophoresis	131, 135, 136
		Serum esterase-2	*Es-2*	Electrophoresis	125, 139
		Kidney esterase-3	*Es-3*	Electrophoresis	130, 135
		Esterase-5	*Es-5*	Electrophoresis	127
		Esterase-6	*Es-6*	Electrophoresis	128
		Esterase-7	*Es-7*	Electrophoresis	21
		Esterase-8	*Es-8*	Electrophoresis	21
		Esterase-9	*Es-9*	Electrophoresis	144
		Esterase-10	*Es-10*	Electrophoresis	121, 181
		Esterase-11	*Es-11*	Electrophoresis	122
		Esterase-12	*Es-12*	Electrophoresis	160
		Esterase-13	*Es-13*	Electrophoresis	186
		Esterase-14	*Es-14*	Electrophoresis	11
3.1.3.1	Alkaline phosphatase	Alkaline phosphatase-1	*Akp-1*	Electrophoresis	178

APPENDIX *(continued)*

EC Number	Enzyme	Names of variant genes affecting enzyme	Gene symbol	Method of detection	References
		Alkaline phosphatase-2	*Akp-2*	Electrophoresis	178
		Alkaline phosphatase-3	*Akp-3*	Electrophoresis	178
		Alkaline phosphatase-4	*Akp-4*	Electrophoresis	178
3.1.3.2	Acid phosphatase	Acid phosphatase-kidney	*Apk*	Electrophoresis	180
		Acid phosphatase-liver	*Apl*	Electrophoresis	92
3.1.6.1	Aryl sulphatase	Aryl sulphatase B	*As-1*	Quantitative assay and thermostability	35, 36
		Aryl sulphatase B regulation-1	*Asr-1*	Quantitative assay	36
3.2.1.1	α-Amylase	Salivary amylase	*Amy-1*	Electrophoresis	86
		Pancreatic amylase	*Amy-2*	Electrophoresis	86
3.2.1.20	α-Glucosidase	Alpha-glucosidase processing	*Aglp*	Electrophoresis	124
3.2.1.22	α-Galactosidase	Alpha-galactosidase	*Ags*	Thermostability	99
		Temporal alpha-galactosidase	*Tag*	Quantitative assay	97
3.2.1.23	β-Galactosidase	Beta-galactosidase electrophoresis	*Bgl-e*	Electrophoresis	13
		Beta-D-galactosidase activity	*Bgl-s*	Quantitative assay	13, 118
		Beta-galactosidase temporal	*Bgl-t*	Quantitative assay	117
3.2.1.24	α-Mannosidase	Alpha-mannosidase processing-1	*Map-1*	Electrophoresis	41
		Alpha-mannosidase processing-2	*Map-2*	Electrophoresis	42

APPENDIX *(continued)*

EC Number	Enzyme	Names of variant genes affecting enzyme	Gene symbol	Method of detection	References
3.2.1.31	β-Glucuronidase	Beta-glucuronidase	*Gus-s*	Electrophoresis and thermostability	91, 118
		Beta-glucuronidase regulation	*Gus-r*	Quantitative assay	118, 159
		Beta-glucuronidase temporal	*Gus-t*	Quantitative assay	118, 159
3.4.11.1	Leucine amino peptidase	Leucyl aryl amino peptidase-1	*Lap-1*	Electrophoresis	184
3.4.11 or 13	Peptidase	[a]Peptidase-2	*Pep-2*	Electrophoresis	8, 94
		Peptidase-3	*Pep-3*	Electrophoresis	24, 94
3.4.21	Tamase	Tosyl arginine methylesterase-1	*Tam-1*	Electrophoresis	154
3.4.21.1	Chymotrypsin	Pancreatic proteinase-2	*Prt-2*	Electrophoresis	167, 168
3.4.21.4	Trypsin	Pancreatic proteinase-1	*Prt-1*	Electrophoresis	167
		Pancreatic proteinase-3	*Prt-3*	Electrophoresis	167
3.5.1.9	Formamidase	Formamidase-1	*For-1*	Quantitative assay and thermostability	33
		Formamidase-5	*For-5*	Thermostability	34
3.5.4.4	Adenosine deaminase	[a]Adenosine deaminase	*Ada*	Electrophoresis	8
Lyases					
4.1.1.22	Histidine decarboxylase	Histidine decarboxylase (kidney)	*Hdc*	Quantitative assay and thermostability	106

APPENDIX (*continued*)

EC Number	Enzyme	Names of variant genes affecting enzyme	Gene symbol	Method of detection	References
4.2.1.1	Carbonic anhydrase	Carbonic anhydrase-1	*Car-1*	Electrophoresis	47
		Carbonic anhydrase-2	*Car-2*	Electrophoresis	47
4.2.1.24	Amino laevulinate dehydratase	Delta-amino laevulinate dehydratase	*Lv*	Quantitative assay	28
4.3.1.3	Histidase	Histidinaemic	*His*	Quantitative assay	85
4.4.1.5	Glyoxalase 1	Glyoxylase-1	*Glo-1*	Electrophoresis	102
Isomerases					
5.3.1.8	Mannose phosphate isomerase	Mannose phosphate isomerase-1	*Mpi-1*	Electrophoresis	112
5.3.1.9	Glucose phosphate isomerase	Glucose phosphate isomerase-1	*Gpi-1*	Electrophoresis	19, 37, 115

[a]No breeding data available.
[b]*Pk-2* and *Pk-3* may be the same locus.
[c]Numbers refer to the list of references.

Protein	Names of variant genes affecting protein	Gene symbol	Method of detection	[a]References
Albumin	Serum albumin variant	*Alb-1*	Electrophoresis	126
	Albumin conformation factor-1	*Acf-1*	Electrophoresis	176
Egasyn	Endoplasmic β-glucuronidase	*Eg*	Quantitative assay	98
Erythrocytic protein	Erythrocytic protein-1	*Erp-1*	Electrophoresis	88
Haemoglobin, α-chain	Haemoglobin alpha-chain	*Hba*	Solubility, chromatography, isoelectric focusing	141, 171
	Alpha thalassemia	Hba^{*th}	Reaction with sodium chloride	172, 173
Haemoglobin, β-chain	Haemoglobin beta-chain	*Hbb*	Electrophoresis	141
Haemoglobin, embryonic x-chain	Haemoglobin X	*Hbx*	Electrophoresis	172
Haemoglobin, embryonic y-chain	Haemoglobin Y-chain	*Hby*	Electrophoresis	57, 157
Major liver protein	Major liver protein-1	*Lvp-1*	Electrophoresis	175
Major urinary protein	Major urinary protein	*Mup-1*	Electrophoresis	77
Plasma protein	Plasma protein	*Plp*	Electrophoresis	174
Prealbumin protein	Prealbumin component-1	*Pre-1*	Electrophoresis	177, 179
	Prealbumin component-4	*Pre-4*	Electrophoresis	177, 179
Serum protein	Serum protein-1	*Sep-1*	Electrophoresis	48

APPENDIX (*continued*)

Protein	Names of variant genes affecting protein	Gene symbol	Method of detection	References
Seminal vesicle protein	Seminal vesicle protein-1	*Svp-1*	Electrophoresis	129
	Seminal vesicle protein-2	*Svp-2*	Electrophoresis	107
Transferrin	Transferrin	*Trf*	Electrophoresis	26, 152

[a]Numbers refer to the list of references.

REFERENCES

1. Bailey, D. W. (1971). Recombinant-inbred strains: an aid to finding identity, linkage and function of histocompatibility and other genes. *Transplantation* **11**: 325–327.
2. Berger, F. G. & Paigen, K. (1979). *Cis*-active control of mouse β-galactosidase biosynthesis by a systemic regulatory locus. *Nature, Lond.* **282**: 314–316.
3. Berry, R. J. (1977). The population genetics of the house mouse. *Sci. Progr., Oxf.* **64**: 337–366.
4. Berry, R. J. (1978). Genetic variation in wild house mice: where natural selection and history meet. *Am. Scient.* **66**: 52–60.
5. Berry, R. J., Bonner, W. N. & Peters, J. (1979). Natural selection in house mice from South Georgia (South Atlantic Ocean). *J. Zool., Lond.* **189**: 385–398.
6. Berry, R. J. & Murphy, H. M. (1970). The biochemical genetics of an island population of the house mouse. *Proc. R. Soc.* (B.) **176**: 87–103.
7. Berry, R. J. & Peters, J. (1975). Macquarie Island house mice: a genetical isolate on a sub-Antarctic island. *J. Zool., Lond.* **176**: 375–389.
8. Berry, R. J. & Peters, J. (1977). Heterogeneous heterozygosities in *Mus musculus* populations. *Proc. R. Soc.* (B.) **197**: 485–503.
9. Berry, R. J. & Peters, J. (1981). Allozymic variation in house mouse populations. In *Mammalian population genetics*: 242–253. Smith, M. H. & Joule, J. (Eds). Athens: University of Georgia Press.
10. Berry, R. J., Peters, J. & Van Aarde, R. J. (1978). Sub-Antarctic house mice: colonization, survival and selection. *J. Zool., Lond.* **184**: 127–141.
11. Bonhomme, F., Britton-Davidian, J. & Thaler, L. (1979). *Mouse News Letter* **60**: 63.
12. Bonhomme, F. & Selander, R. K. (1978). Estimating total genic diversity in the house mouse. *Biochem. Genet.* **16**: 287–297.
13. Breen, G. A. M., Lusis, A. J. & Paigen, K. (1977). Linkage of genetic determinants for mouse β-galactosidase electrophoresis and activity. *Genetics* **85**: 73–84.
14. Bremner, T. A., Premkumar-Reddy, E., Nayar, K. & Kouri, R. E. (1978). Nucleoside phosphorylase 2 (*Np-2*) of mice. *Biochem. Genet* **16**: 1143–1151.
15. Britton, J. & Thaler, L. (1978). Evidence for the presence of two sympatric species of mice (genus *Mus* L.) in southern France based on biochemical genetics. *Biochem. Genet.* **16**: 213–225.
16. Bryant, E. H. (1974). On the adaptive significance of enzyme polymorphisms in relation to environmental variability. *Am. Nat.* **108**: 1–19.
17. Bulfield, G. (1978). Genetic variation in the activity of the histidine catabolic enzymes between inbred strains of mice: a structural locus for a cytosol histidine aminotransferase isozyme (*Hat-1*). *Biochem. Genet.* **16**: 1233–1241.
18. Bulfield, G., Moore, K. & Peters, J. (1978). *Mouse News Letter* **59**: 29.
19. Carter, N. D. & Parr, C. W. (1967). Isoenzymes of phosphoglucose isomerase in mice. *Nature, Lond.* **216**: 511.
20. Chapman, V. M. (1975). 6-Phosphogluconate dehydrogenase (PGD) genetics in the mouse: linkage with metabolically related enzyme loci. *Biochem. Genet.* **13**: 849–856.

21. Chapman, V. M., Nichols, E. A. & Ruddle, F. H. (1974). Esterase-8 (*Es-8*): characterization, polymorphism, and linkage of an erythrocyte esterase locus on chromosome 7 of *Mus musculus. Biochem. Genet.* **11**: 347–358.
22. Chapman, V. M., Paigen, K., Syracusa, L. & Womack, J. E. (1979). Biochemical variation: mouse. In *Inbred and genetically defined strains of laboratory animals. 1. Mouse and rat*: 77–95. Altman, P. L. & Katz, D. D. (Eds). Bethesda: Fed. Am. Socs. exp. Biol.
23. Chapman, V. M. & Ruddle, F. H. (1972). Glutamate oxaloacetate transaminase (GOT) genetics in the mouse: polymorphism of GOT-1. *Genetics* **70**: 229–305.
24. Chapman, V. M., Ruddle, F. H. & Roderick, T. H. (1971). Linkage of isozyme loci in the mouse: phosphoglucomutase-2 (*Pgm-2*), mitochondrial NADP malate dehydrogenase (*Mod-2*), and dipeptidase-1 (*Dip-1*). *Biochem. Genet.* **5**: 101–110.
25. Chen, S-H., Donahue, R. P. & Scott, C. R. (1973). The genetics of glutamic-pyruvic transaminase in mice: inheritance, electrophoretic phenotypes, and postnatal changes. *Biochem. Genet.* **10**: 23–28.
26. Cohen, B. L. (1960). Genetics of plasma transferrins in the mouse. *Genet. Res.* **1**: 431–438.
27. Coleman, D. L. (1962). Effect of genic substitution on the incorporation of tyrosine into the melanin of mouse skin. *Archs Biochem. Biophys.* **96**: 562–568.
28. Coleman, D. L. (1971). Linkage of genes controlling the rate of synthesis and structure of aminolevulinate dehydratase. *Science, Wash.* **173**: 1245–1246.
29. Coleman, D. L. (1977). Genetic control of glucokinase activity in mice. *Biochem. Genet.* **15**: 297–305.
30. Coyne, J. A. (1976). Lack of genic similarity between two sibling species of *Drosophila* as revealed by varied techniques. *Genetics* **84**: 593–607.
31. Coyne, J. A. & Felton, A. A. (1977). Genic heterogeneity at two alcohol dehydrogenase loci in *Drosophila pseudoobscura* and *Drosophila persimilis. Genetics* **87**: 285–304.
32. Coyne, J. A., Felton, A. A. & Lewontin, R. C. (1978). Extent of genetic variation at a highly polymorphic esterase locus in *Drosophila pseudoobscura. Proc. natn Acad. Sci. U.S.A.* **75**: 5090–5093.
33. Cumming, R. B., Gaertner, R. H., Walton, M. F. & O'Donnell, S. C. (1977). A structural gene for formamidase. *Mouse News Letter* **56**: 53–54.
34. Cumming, R. B., Walton, M. F., Fuscoe, J. C., Taylor, B. A., Womack, J. E. & Gaertner, F. H. (1979). Genetics of formamidase-5 (brain formamidase) in the mouse: localization of the structural gene on chromosome 14. *Biochem. Genet.* **17**: 415–431.
35. Daniel, W. L. (1976a). Genetics of murine liver and kidney arylsulfatase B. *Genetics* **82**: 477–491.
36. Daniel, W. L. (1976b). Genetic control of heat sensitivity and activity level of murine arylsulfatase B. *Biochem. Genet.* **14**: 1003–1018.
37. DeLorenzo, R. J. & Ruddle, F. H. (1969). Genetic control of two electrophoretic variants of glucosephosphate isomerase in the mouse (*Mus musculus*). *Biochem. Genet.* **3**: 151–162.
38. DeLorenzo, R. J. & Ruddle, F. H. (1970). Glutamate oxalate transaminase (GOT) genetics in *Mus musculus*: linkage, polymorphism and phenotypes of the *Got-2* and *Got-1* loci. *Biochem. Genet.* **4**: 259–273.

39. DeMars, R., LeVan, S. L., Trend, B. L. & Russell, L. B. (1976). Abnormal ornithine carbamoyltransferase in mice having the sparse-fur mutation. *Proc. natn. Acad. Sci. U. S. A.* **73**: 1693–1697.
40. Dickerman, R. C., Feinstein, R. N. & Grahn, D. (1968). Position of the acatalasemia gene in linkage group V of the mouse. *J. Hered.* **59**: 177–178.
41. Dizik, M. & Elliott, R. W. (1977). A gene apparently determining the extent of sialylation of lysosomal α-mannosidase in mouse liver. *Biochem. Genet.* **15**: 31–46.
42. Dizik, M. & Elliott, R. W. (1978). A second gene affecting the sialylation of lysosomal α-mannosidase in mouse liver. *Biochem. Genet.* **16**: 247–260.
43. Dunn, L. C., Bennett, D. & Beasley, A. B. (1962). Mutation and recombination in the vicinity of a complex gene. *Genetics* **47**: 285–303.
44. Eicher, E. M., Cherry, M. & Flaherty, L. (1978). Autosomal phosphoglycerate kinase linked to mouse major histocompatibility complex. *Molec. Gen. Genet.* **158**: 225–228.
45. Eicher, E. M. & Reynolds, S. (1978). *Mouse News Letter* **58**: 48.
46. Eicher, E., Reynolds, S. & Southard, J. L. (1977). *Mouse News Letter* **56**: 42.
47. Eicher, E. M., Stern, R. H., Womack, J. E., Davisson, M. T., Roderick, T. H. & Reynolds, S. C. (1976). Evolution of mammalian carbonic anhydrase loci by tandem duplication: close linkage of *Car-1* and *Car-2* to the centromere region of chromosome 3 of the mouse. *Biochem. Genet.* **14**: 651–660.
48. Eicher, E. M., Taylor, B. A., Leighton, S. C. & Womack, J. E. (1980). A serum protein polymorphism determinant on chromosome 9 of *Mus musculus. Molec. Gen. Genet.* **177**: 571–576.
49. Eicher, E. M. & Womack, J. E. (1977). Chromosomal location of soluble glutamic-pyruvic transaminase-1 (*Gpt-1*) in the mouse. *Biochem. Genet.* **15**: 1–8.
50. Feinstein, R. N., Howard, J. B., Braun, J. T. & Seaholm, J. E. (1966). Acatalasemic and hypocatalasemic mouse mutants. *Genetics* **53**: 923–933.
51. Firth, S., Peters, J. & Bulfield, G. (1979). Activity and electrophoretic mobility of glutathione reductase allozymes in different tissues of the mouse. *Biochem. Genet.* **17**: 229–232.
52. Flavell, R. A., Kooter, J. M., DeBoer, E., Little, P. F. R. & Williamson, R. (1978). Analysis of the β-δ-globin gene loci in normal and Hb Lepore DNA: direct determination of gene linkage and intergene distance. *Cell* **15**: 25–41.
53. Forejt, J. (1972). Chiasmata and crossing-over in the male mouse (*Mus musculus*): suppression of recombination and chiasma frequencies in the ninth linkage group. *Folia biol., Praha* **18**: 161–170.
54. Ganschow, R. E. & Schimke, R. T. (1970). Murine catalase phenotypes. *Biochem. Genet.* **4**: 157–167.
55. Garnett, I. & Falconer, D. S. (1975). Protein variation in strains of mice differing in body size. *Genet. Res.* **25**: 45–57.
56. Gillespie, J. H. & Kojima, K. (1968). The degree of polymorphism in enzymes involved in energy production compared to that in non-specific enzymes in two *Drosophila ananassae* populations. *Proc. natn. Acad. Sci. U.S.A.* **61**: 582–585.

57. Gilman, J. G. & Smithies, O. (1968). Fetal hemoglobin variants in mice. *Science, N. Y.* **160**: 885–886.
58. Glass, R. D. & Doyle, D. (1972). Genetic control of lactate dehydrogenase expression in mammalian tissues. *Science, N. Y.* **176**: 180–181.
59. Grass, S. R., Longshore, M. A. & Pangburn, S. (1975). The phosphorylase kinase deficiency (*Phk*) locus in the mouse: evidence that the mutant allele codes for an enzyme with an abnormal structure. *Biochem. Genet.* **13**: 567–584.
60. Harris, H. (1966). Enzyme polymorphisms in man. *Proc. R. Soc.* (B.) **164**: 298–310.
61. Harris, H. (1974). Common and rare alleles. *Sci. Progr., Oxf.* **61**: 495–514.
62. Harris, H. (1975). *The principles of human biochemical genetics.* 2nd edn. Amsterdam: North-Holland.
63. Harris, H. & Hopkinson, D. A. (1972). Average heterozygosity per locus in man: an estimate based on the incidence of enzyme polymorphisms. *Ann. hum. Genet.* **36**: 9–20.
64. Harris, H., Hopkinson, D. A. & Edwards, Y. H. (1977). Polymorphism and the subunit structure of enzymes: a contribution to the neutralist-selectionist controversy. *Proc. natn. Acad. Sci. U.S.A.* **74**: 698–701.
65. Henderson, N. S. (1965). Isozymes of isocitrate dehydrogenase: subunit structure and intracellular location. *J. exp. Zool.* **158**: 263–273.
66. Heston, W. E., Hoffman, H. A. & Rechcigl, M. Jr. (1965). Genetic analysis of liver catalase activity in two substrains of C57BL mice. *Genet. Res.* **6**: 387–397.
67. Hilse, K. & Popp, R. A. (1968). Gene duplication as the basis for amino acid ambiguity in the alpha-chain polypeptides of mouse hemoglobins. *Proc. natn. Acad. Sci. U.S.A.* **61**: 930–936.
68. Hoffman, H. A. & Grieshaber, C. K. (1974). Genetic studies of murine catalase: liver and erythrocyte catalase controlled by independent loci. *J. Hered.* **65**: 277–279.
69. Hoffman, H. A. & Grieshaber, C. K. (1976). Genetic studies of murine catalase: regulation of multiple molecular forms of kidney catalase. *Biochem. Genet.* **14**: 59–66.
70. Holmes, R. A. (1976). Genetics of peroxisomal enzymes in the mouse: non-linkage of D-amino acid oxidase locus (*Dao*) to catalase (*Cs*) and L-α-hydroxyacid oxidase (*Hao-1*) loci on chromosome 2. *Biochem. Genet.* **14**: 981–987.
71. Holmes, R. S. (1977). The genetics of α-hydroxyacid oxidase and alcohol dehydrogenase in the mouse: evidence for multiple gene loci and linkage between *Hao-2* and *Adh-3*. *Genetics* **87**: 709–716.
72. Holmes, R. S. (1978a). Genetics of hydroxyacid oxidase isozymes in the mouse: localisation of *Hao-2* on linkage Group XVI. *Heredity* **41**: 403–406.
73. Holmes, R. S. (1978b). Genetics and ontogeny of aldehyde dehydrogenase isozymes in the mouse: localization of *Adh-1* encoding the mitochondrial isozyme on chromosome 4. *Biochem. Genet.* **16**: 1207–1218.
74. Holmes, R. S. (1979a). Genetics and ontogeny of alcohol dehydrogenase isozymes in the mouse: evidence for a *cis*-acting regulator gene (*Adt-1*) controlling C_2 isozyme expression in reproductive tissues and close linkage of *Adh-3* and *Adt-1* on chromosome 3. *Biochem. Genet.* **17**: 461–472.
75. Holmes, R. S. (1979b). Genetics, ontogeny and testosterone inducibility

of aldehyde oxidase isozymes in the mouse: evidence for two genetic loci (*Aox-1* and *Aox-2*) closely linked on chromosome 1. *Biochem. Genet.* **17**: 517–527.

76. Holmes, R. S., Jones, J. T. & Peters, J. (1978). Genetic variation, cellular distribution and ontogeny of sorbitol dehydrogenase and alcohol dehydrogenase isozymes in male reproductive tissues of the mouse. *J. exp. Zool.* **206**: 279–288.
77. Hudson, D. M., Finlayson, J. S. & Potter, M. (1967). Linkage of one component of the major urinary protein complex of mice to the brown coat color locus. *Genet. Res.* **10**: 195–198.
78. Huijing, F., Eicher, E. M. & Coleman, D. L. (1973). Location of phosphorylase kinase (*Phk*) in the mouse X chromosome. *Biochem. Genet.* **9**: 193–196.
79. Hunter, R. L. & Markert, C. L. (1957). Histochemical demonstration of enzymes separated by zone electrophoresis in starch gels. *Science, N. Y.* **125**: 1294–1295.
80. Hutton, J. J. (1971). Genetic regulation of glucose 6-phosphate dehydrogenase activity in the mouse. *Biochem. Genet.* **5**: 315–331.
81. Hutton, J. J. & Roderick, T. H. (1970). Linkage analyses using biochemical variants in mice. III. Linkage relationships of eleven biochemical markers. *Biochem. Genet.* **4**: 339–350.
82. Jeffreys, A. J. (1979). DNA sequence variants in the $^{G}\gamma$-, $^{A}\gamma$-, δ- and β-globin genes of man. *Cell* **18**: 1–10.
83. Jeffreys, A. J. & Flavell, R. A. (1977). The rabbit β-globin gene contains a large insert in the coding sequence. *Cell* **12**: 1097–1108.
84. Johnson, G. B. (1974). Enzymes, polymorphism and metabolism. *Science, Wash.* **184**: 28–37.
85. Kacser, H., Bulfield, G. & Wallace, M. E. (1973). Histidinaemic mutant in the mouse. *Nature, Lond.* **244**: 77–79.
86. Kaplan, R. D., Chapman, V. & Ruddle, F. H. (1973). Electrophoretic variation of α-amylase in two inbred strains of *Mus musculus*. *J. Hered.* **64**: 155–157.
87. Khanolkar, V. R. & Chitre, R. G. (1942). Studies in esterase (butyric) activity. *Cancer Res.* **2**: 567–570.
88. Kirby, G. C. (1974). The genetics of an electrophoretic variant of an erythrocytic protein in the house mouse (*Mus musculus*). *Anim. Blood Groups biochem. Genet.* **5**: 153–157.
89. Kozak, L. P. & Erdelsky, K. J. (1975). The genetics and development regulation of L-glycerol 3-phosphate dehydrogenase. *J. cell. Physiol.* **85**: 437–447.
90. Lai, E. C., Woo, S. L., Dagaiczyk, A. & O'Malley, B. W. (1979). The ovalbumin gene: alleles created by mutations in the intervening sequences of the natural gene. *Cell* **16**: 201–211.
91. Lalley, P. A. & Shows, T. B. (1974). Lysosomal and microsomal glucuronidase: genetic variant alters electrophoretic mobility of both hydrolases. *Science, Wash.* **185**: 442–444.
92. Lalley, P. A. & Shows, T. B. (1977). Lysosomal acid phosphatase deficiency: liver specific variant in the mouse. *Genetics* **87**: 305–317.
93. Law, L. W., Morrow, A. G. & Greenspan, E. M. (1952). Inheritance of low liver glucuronidase activity in the mouse. *J. natn. Cancer Inst.* **12**: 909–916.

93a. Leigh Brown, A. J. & Langley, C. H. (1979). Reevaluation of genic heterozygosity in natural populations of *Drosophila melanogaster* by two-dimensional electrophoresis. *Proc. natn. Acad. Sci. U.S.A.* **76**: 2381–2384.

94. Lewis, W. H. P. & Truslove, G. M. (1969). Electrophoretic heterogeneity of mouse erythrocyte peptidases. *Biochem. Genet.* **3**: 493–498.

95. Lewontin, R. C. & Hubby, J. L. (1966). A molecular approach to the study of genic heterozygosity in natural populations. II. Amount of variation and degree of heterozygosity in natural populations of *Drosophila pseudoobscura*. *Genetics* **54**: 595–609.

96. Little, P. F. R., Flavell, R. A., Kooter, J. M., Annison, G. & Williamson, R. (1979). Structure of the human foetal globin gene locus. *Nature, Lond.* **278**: 227–231.

97. Lusis, A. J. & Paigen, K. (1975). Genetic determination of the α-galactosidase developmental program in mice. *Cell* **6**: 371–378.

98. Lusis, A. J., Tomino, S. & Paigen, K. (1977). Inheritance in mice of the membrane anchor protein egasyn: the *Eg* locus determines egasyn levels. *Biochem. Genet.* **15**: 115–122.

99. Lusis, A. J. & West, J. D. (1976). X-linked inheritance of a structural gene for α-galactosidase in *Mus musculus*. *Biochem. Genet.* **14**: 849–855.

100. Lusis, A. J. & West, J. D. (1978). X-linked and autosomal genes controlling mouse α-galactosidase expression. *Genetics* **88**: 327–342.

101. Lyon, J. B. Jr. (1970). The X-chromosome and the enzymes controlling muscle glycogen: phosphorylase kinase. *Biochem. Genet.* **4**: 169–185.

102. Meo, T., Douglas, T. & Rijnbeek, A-M. (1977). Glyoxalase I polymorphism in the mouse: a new genetic marker linked to H-2. *Science, Wash* **198**: 311–313.

103. Meredith, S. A. & Ganschow, R. F. (1978). Apparent *trans* control of murine β-glucuronidase synthesis by a temporal gene element. *Genetics* **90**: 725–734.

104. Miner, G. D. & Wolfe, H. G. (1972). Phosphoglucomutase isozymes of red cells in *Mus musculus*: additional polymorphisms at *Pgm-1* compared by two electrophoretic systems. *Biochem. Genet.* **7**: 247-252.

105. Mishkin, J. D., Taylor, B. A. & Mellman, W.J. (1976). *Glk*: a locus controlling galactokinase activity in the mouse. *Biochem. Genet.* **14**: 635–640.

106. Mortimer, A. & Bulfield, G. (1978). *Mouse News Letter* **59**: 30.

107. Moutier, R., Toyama, K. & Charrier, M. F. (1971). Contrôle génétique des protéines de la sécrétion des glandes seminales chez la souris et le rat. *Exp. Anim.* **4**: 7–18.

108. Myers, J. H. (1974). Genetic and social structure of feral house mouse populations on Grizzly Island, California. *Ecology* **55**: 747–59.

109. Nash, H. R., Peters, J. & Bulfield, G. (1978). *Mouse News Letter* **59**: 31.

110. Nash, H. R., Peters, J., Eicher, E. M. & Bulfield, G. (In preparation). *Polymorphism and linkage of pyruvate kinase in* Mus musculus.

111. Nevo, E., Kim, Y. J., Shaw, C. R. & Thaelar, C. S. (1974). Genetic variation, selection and speciation in *Thomomys talpoides* pocket gophers. *Evolution* **28**: 1–23.

112. Nichols, E. A., Chapman, V. M. & Ruddle, F. H. (1973). Polymorphism and linkage for mannosephosphate isomerase in *Mus musculus*. *Biochem. Genet.* **8**: 47–53.

113. Nichols, E. A. & Ruddle, F. H. (1975). Polymorphism and linkage of

glutathione reductase in *Mus musculus. Biochem. Genet.* **13**: 323–329.

114. Nielson, J. T. & Chapman, V. M. (1977). Electrophoretic variation for X-chromosome-linked phosphoglycerate kinase (*Pgk-1*) in the mouse. *Genetics* **87**: 319–325.
115. Padua, R. A., Bulfield, G. & Peters, J. (1978). Biochemical genetics of a new glucosephosphate isomerase allele ($Gpi\text{-}1^c$) from wild mice. *Biochem. Genet.* **16**: 127–143
116. Paigen, K. (1961). The effect of mutation on the intracellular location of β-glucuronidase. *Expl Cell. Res.* **25**: 286–301.
117. Paigen, K., Meisler, M., Felton, J. & Chapman, V. M. (1976). Genetic determination of the β-galactosidase developmental program in mouse liver. *Cell* **9**: 533–539.
118. Paigen, K., Swank, R. T., Tomino, S. & Ganschow, R. E. (1975). The molecular genetics of mammalian glucuronidase. *J. cell. Physiol.* **85**: 379–392.
119. Patton, J. L. & Yang, S. Y. (1977). Genetic variation in *Thomomys bottae* pocket gophers: macrogeographic patterns. *Evolution* **31**: 697–720.
120. Peters, J. (1978). *Mouse News Letter* **59**: 32.
121. Peters, J. & Nash, H. R. (1976). Polymorphism of esterase-10 in *Mus musculus. Biochem. Genet.* **14**: 119–125.
122. Peters, J. & Nash, H. R. (1977). Polymorphism of esterase-11 in *Mus musculus*, a further esterase locus on chromosome 8. *Biochem. Genet.* **15**: 217–226.
123. Peters, J. & Nash, H. R. (1978). Esterases of *Mus musculus*: substrate and inhibition characteristics, new isozymes and homologies with man. *Biochem. Genet.* **16**: 553–569.
124. Peters, J. & Swallow, D. M. (1979). α-glucosidase. *Mouse News Letter* **60**: 46.
125. Petras, M. L. (1963). Genetic control of a serum esterase component in *Mus musculus. Proc. natn. Acad. Sci. U.S.A.* **50**: 112–116.
126. Petras, M. L. (1972). An inherited albumin variant in the house mouse, *Mus musculus. Biochem. Genet.* **7**: 273–277.
127. Petras, M. L. & Biddle, F. G. (1967). Serum esterases in the house mouse, *Mus musculus. Can. J. Genet. Cytol.* **9**: 704–710.
128. Petras, M. L. & Sinclair, P. (1969). Another esterase variant in the kidney of the house mouse, *Mus musculus. Can. J. Genet. Cytol.* **11**: 97–102.
129. Platz, R. D. & Wolfe, H. G. (1969). Mouse seminal vesicle proteins: the inheritance of electrophoretic variants. *J. Hered.* **60**: 187–192.
130. Popp, R. A. (1966). Inheritance of an erythrocyte and kidney esterase in the mouse. *J. Hered.* **57**: 197–201.
131. Popp, R. A. & Popp, D. M. (1962). Inheritance of serum esterases having different electrophoretic patterns among inbred strains of mice. *J. Hered.* **53**: 111–114.
132. Ranney, H. M. & Gluecksohn-Waelsch, S. (1955). Filter-paper electrophoresis of mouse haemoglobin: a preliminary note. *Ann. hum. Genet.* **19**: 269–272.
133. Rechcigl, M. Jr. & Heston, W. E. (1967). Genetic regulation of enzyme activity in mammalian system by the alteration of the rates of enzyme degradation. *Biochem. Biophys. Res. Commun.* **27**: 119–124.
134. Roderick, T. H., Ruddle, F. H., Chapman, V. M. & Shows, T. B. (1971).

Biochemical polymorphisms in feral and inbred mice (*Mus musculus*). *Biochem. Genet.* **5**: 457–466.

135. Ruddle, F. H. & Roderick, T. H. (1966). The genetic control of two types of esterases in inbred strains of the mouse. *Genetics* **54**: 191–202.
136. Ruddle, F. H. & Roderick, T. H. (1968). Allelically determined isozyme polymorphisms in laboratory populations of mice. *Ann. N.Y. Acad. Sci.* **15**: 531–539.
137. Ruddle, F. H., Roderick, T. H., Shows, T. B., Weigl, P. G., Chipman, R. K. & Anderson, P. K. (1969). Measurement of genetic heterogeneity by means of enzyme polymorphisms in wild populations of the mouse. *J. Hered.* **60**: 321–322.
138. Ruddle, F. H., Shows, T. B. & Roderick, T. H. (1968). Autosomal control of an electrophoretic variant of glucose-6-phosphate dehydrogenase in the mouse (*Mus musculus*). *Genetics* **58**: 599–606.
139. Ruddle, F. H., Shows, T. B. & Roderick, T. H. (1969). Esterase genetics in *Mus musculus*: expression, linkage, and polymorphism of locus *Es-2*. *Genetics* **62**: 393–399.
140. Russell, E. S., Blake, S. L. & McFarland, E. C. (1972). Characterization and strain distribution of four alleles at the hemoglobin α-chain structural locus in the mouse. *Biochem. Genet.* **7**: 313–330.
141. Russell, E. S. & McFarland, E. C. (1974). Genetics of mouse hemoglobins. *Ann. N.Y. Acad. Sci.* **241**: 25–38.
142. Russell, L. B., Russell, W. L., Popp, R. A., Vaughan, C. & Jacobson, K. B. (1976). Radiation-induced mutations at mouse hemoglobin loci. *Proc. natn. Acad. Sci. U.S.A.* **73**: 2843–2846.
143. Schnell, G. D. & Selander, R. K. (1981). Environmental and morphological correlates of genetic variation in mammals. In *Mammalian population genteics*: 60–99. Smith, M. H. & Joule, J. (Eds). Athens: University of Georgia Press.
144. Schollen, J., Bender, K. & von Deimling, O. (1975). Esterase. XXI. *Es-9*, a possibly new polymorphic esterase in *Mus musculus* genetically linked to *Es-2*. *Biochem. Genet.* **13**: 369–377.
145. Selander, R. K., Hunt, W. G. & Yang, S. Y. (1969). Protein polymorphism and genic heterozygosity in two European subspecies of the house mouse. *Evolution* **23**: 379–390.
146. Selander, R. K. & Kaufman, D. W. (1973). Genic variability and strategies of adaptation in animals. *Proc. natn. Acad. Sci. U.S.A.* **70**: 1875–1877.
147. Selander, R. K., Smith, M. H., Yang, S. Y., Johnson, W. E. & Gentry, J. B. (1971). Biochemical polymorphism and systematics in the genus *Peromyscus*. 1. Variation in the old-field mouse (*Peromyscus polionotus*). (Studies in Genetics VI.) *Univ. Texas Publs* No. 7103: 49–90.
148. Selander, R. K. & Yang, S. Y. (1969). Protein polymorphism and genic heterozygosity in a wild population of the house mouse (*Mus musculus*). *Genetics* **63**: 653–667.
149. Shows, T. B., Chapman, V. M. & Ruddle, F. H. (1970). Mitochondrial malate dehydrogenase and malic enzyme: Mendelian inherited electrophoretic variants in the mouse. *Biochem. Genet.* **4**: 707–718.
150. Shows, T. B. & Ruddle, F. H. (1968). Function of the lactate dehydrogenase B gene in mouse erythrocytes: evidence for control by a regulatory gene. *Proc. natn. Acad. Sci. U.S.A.* **61**: 574–581.

151. Shows, T. B., Ruddle, F. H. & Roderick, T. H. (1969). Phosphoglucomutase electrophoretic variants in the mouse. *Biochem. Genet.* 3: 25–35.
152. Shreffler, D. C. (1960). Genetic control of serum transferrin type in mice. *Proc. natn. Acad. Sci. U.S.A.* **46**: 1378–1384.
153. Singh, R. S., Lewontin, R. C. & Felton, A. A. (1976). Genetic heterogeneity within electrophoretic "alleles" of xanthine dehydrogenase in *Drosophila pseudoobscura. Genetics* **84**: 609–629.
154. Skow, L. C. (1978). Genetic variation at a locus (*Tam-1*) for submaxillary gland protease in the mouse and its location on chromosome 7. *Genetics* **90**: 713–724.
155. Soares, E. R. (1977). *LDH. Mouse News Letter* **57**: 33.
156. Staats, J. (1979). Inbred strains: mouse. In *Inbred and genetically defined strains of laboratory animals. 1. Mouse and rat:* 21–29. Altman, P. L. & Katz, D. D. (Eds). Bethesda: Fedn Am. Socs exp. Biol.
157. Stern, R. H., Russell, E. S. & Taylor, B. A. (1976). Strain distribution and linkage relationship of a mouse embryonic hemoglobin variant. *Biochem. Genet.* **14**: 373–381.
158. Swank, R. T. & Bailey, D. W. (1973). Recombinant inbred lines: value in the genetic analysis of biochemical variants. *Science, Wash.* **181**: 1249–1252.
159. Swank, R. T., Paigen, K., Davey, R., Chapman, V., Labarca, C., Watson, G., Ganschow, R., Brandt, E. J. & Novak, E. (1978). Genetic regulation of mammalian glucuronidase. *Recent Progr. Horm. Res.* **34**: 401–436.
160. Taylor, B. A. (1977). New esterase variant (*Es-12*). *Mouse News Letter* **56**: 41.
161. Taylor, B. A. (1979). Recombinant inbred lines: mouse. In *Inbred and genetically defined strains of laboratory animals. 1. Mouse and rat*: 37. Altman, P. L. & Katz, D. D. (Eds). Bethesda: Fedn. Am. Socs exp. Biol.
162. Tilghman, S. M., Tiemeier, D. C., Seidman, J. G., Peterlin, B. M., Sullivan, M., Maizel, J. V. & Leder, P. (1978). Intervening sequence of DNA identified in the structural portion of a mouse β-globin gene. *Proc. natn. Acad. Sci. U.S.A.* **75**: 725–729.
163. VandeBerg, J. L., Cooper, D. W. & Close, P. J. (1973). Mammalian testis phosphoglycerate kinase. *Nature New Biol.* **243**: 48–50.
164. VandeBerg, J. L., Cooper, D. W. & Close, P. J. (1976). Testis specific phosphoglycerate kinase B in mouse. *J. exp. Zool.* **198**: 231–240.
165. van den Berg, J., van Ooyen, A., Mantei, N., Schamböck, A., Grosveld, G., Flavell, R. A. & Weissmann, C. (1978). Comparison of cloned rabbit and mouse β-globin genes showing strong evolutionary divergence of two homologous pairs of introns. *Nature, Lond.* **276**: 37–44.
166. Ward, R. D. (1977). Relationship between enzyme heterozygosity and quaternary structure. *Biochem. Genet.* **15**: 123–135.
167. Watanabe, T., Ogasawara, N. & Goto, H. (1976a). Genetic study of pancreatic proteinase in mice (*Mus musculus*): genetic variants of trypsin and chymotrypsin. *Biochem. Genet.* **14**: 697–707.
168. Watanabe, T., Ogasawara, N. & Goto, H. (1976b). Genetic study of pancreatic proteinase in mice: linkage of the *Prt-2* locus on chromosome 8. *Biochem. Genet.* **14**: 999–1002.
169. Watson, J. G., Higgins, T. J. Collins, P. B. & Chaykin, S. (1972). The mouse liver aldehyde oxidase locus (*Aox*). *Biochem. Genet.* **6**: 195–204.

170. Wheeler, L. L. & Selander, R. K. (1972). Genetic variation in populations of the house mouse, *Mus musculus*, in the Hawaiian Islands. (Studies in Genetics VII.) *Univ. Texas Publs* No. 7213: 269–296.
171. Whitney, J. B. III, Copland, G. T., Skow, L. C. & Russell, E. S. (1979). Resolution of products of the duplicated hemoglobin α-chain loci by isoelectric focusing. *Proc. natn. Acad. Sci. U.S.A.* **76**: 867–871.
172. Whitney, J. B. & Russell, E. S. (1978a). Alpha thalassemia ($Hba^{th\text{-}J}$). *Mouse News Letter* **58**: 47–48.
173. Whitney, J. B. & Russell, E. S. (1978b). Alpha thalassemia ($Hba^{th\text{-}J}$). *Mouse News Letter* **59**: 26.
174. Whitney, J. B., Taylor, B. A. & Cherry, M. (1978). Plasma protein (Plp). *Mouse News Letter* **58**: 49.
175. Wilcox, F. H. (1972). Genetic variation of a major liver protein in the mouse. *J. Hered.* **63**: 60–62.
176. Wilcox, F. H. (1973). Genetic differences in alteration of albumin in the house mouse, *Mus musculus*. *Biochem. Genet.* **10**: 69–78.
177. Wilcox, F. H. (1975). Genetic variation in plasma prealbumin of the house mouse. *J. Hered.* **66**: 19–22.
178. Wilcox, F. H. (1978). New mutations and biochemical variants. *Mouse News Letter* **59**: 23.
179. Wilcox, F. H. & Shreffler, D. C. (1976). Terminology for prealbumin loci in the house mouse. *J. Hered.* **67**: 113.
180. Womack, J. E. & Auerbach, S. B. (1978). An acid phosphatase locus expressed in mouse kidney (*Apk*) and its genetic location on chromosome 10. *Biochem. Genet.* **16**: 239–245.
181. Womack, J. E., Davisson, M. T., Eicher, E. M. & Kendall, D. A. (1977). Mapping of nucleoside phosphorylase (*Np-1*) and esterase-10 (*Es-10*) on mouse chromosome 14. *Biochem. Genet.* **15**: 347–355.
182. Womack, J. E. & Eicher, E. M. (1977). Liver-specific lysosomal acid phosphatase deficiency (*Apl*) on mouse chromosome 17. *Molec. gen. Genet.* **155**: 315–317.
183. Womack, J. E., Hawes, N. L., Soares, E. R. & Roderick, T. H. (1975). Mitochondrial malate dehydrogenase (*Mor-1*) in the mouse: linkage to chromosome 5 markers. *Biochem. Genet.* **13**: 519–525.
184. Womack, J. E., Lynes, M. A. & Taylor, B. A. (1975). Genetic variation of intestinal leucine arylaminopeptidase (*Lap-1*) in the mouse and its location on chromosome 9. *Biochem. Genet.* **13**: 511–518.
185. Womack, J. E. & Roderick, T. H. (1974). T-alleles in the mouse are probably not inversions. *J. Hered.* **65**: 308–310.
186. Womack, J. E., Taylor, B. A. & Barton, J. E. (1978). Esterase 13, a new mouse esterase locus with recessive expression and its genetic location on chromosome 9. *Biochem. Genet.* **16**: 1107–1112.

Symp. zool. Soc. Lond. (1981) No. 47, 517–546

Mouse Pharmacogenetics

IAN E. LUSH

Department of Genetics and Biometry, University College London, UK

SYNOPSIS

This review deals with a number of genes which affect either the responses of mice to drugs or the metabolism of drugs by mice. Genetical variation in response to hormones is not included. New genetical tools such as recombinant inbred strains and congenic strains are of great value in pharmacogenetics. Differences between substrains are also potentially useful but can be a pitfall for the unwary. Genetical variation affecting behavioural responses to scopolamine, ethanol, amphetamine and morphine is critically discussed. Variation affecting physiological responses to morphine, mescaline and oxotremorine is reviewed. A short introduction to the subject of drug metabolism is followed by some examples of variation of this system. Variation of drug-metabolizing enzymes can affect the carcinogenicity of certain carcinogens, some of which are pollutants of the environment.

INTRODUCTION

Pharmacogenetics deals with genetically-determined variation in the way a drug acts on an animal, or in the way a drug is metabolized by an animal. A drug is defined as "any substance used in the composition of medicine: a substance used to stupefy or poison or for self-indulgence" (Geddie, 1952). Perhaps this definition should be widened slightly to include harmless foreign substances used in experiments. Some authors understand the above definition to include insulin and other hormones but the present author considers that the effects of these substances are best dealt with under the heading of physiological genetics (Kalow, 1962; Meir, 1963; see Shire, this volume). For some reason (perhaps expense) most workers on experimental pharmacogenetics have used drugs which affect the central nervous system and produce alterations in behaviour which can be measured. This work has recently been thoroughly and critically reviewed by Broadhurst (1978). The present paper will therefore concentrate on those examples of mouse pharmacogenetics in which there has been some attempt to identify and measure the effects of individual genes. Some other examples will also be given which illustrate ideas and methods which are relevant to pharmacogenetics as a whole.

Pharmacogenetics suffers from a certain diffuseness of subject matter. Examples of pharmacogenetic variation are often discovered by accident (particularly in Man) and the subject as a whole has not yet acquired a logical structure. A large number of papers merely record the fact that different strains differ significantly in their reaction to a certain drug. This is a good start, but it is only a first step across the threshold of genetics. It is also necessary that both the results and conclusions obtained with behavioural tests should be confirmed by other workers. This may seem a rather uncharitable remark, but unfortunately results obtained with behavioural tests, particularly of the more sophisticated variety, can be strongly influenced by experimental techniques. According to Iversen & Iversen (1975) "there is no area in behavioural pharmacology more fraught with experimental traps than the study of pharmacologically-induced changes in learning and memory. Only the most carefully designed experiments avoid the pitfalls". If worker B cannot confirm the results of worker A the reason might also be that, although they used the same techniques and strains, they happened to employ different substrains. It is becoming increasingly clear that substrain differences are too common and too large to be ignored (Morse, 1978). Some workers measure the effect of a drug in two or more strains and then attempt to correlate it with some other measurement in the same strains, for example the level of a neurotransmitter in a part of the brain. However, even if there are good reasons for expecting these two measurements to be correlated, the laws of chance still apply and a reasonable number of strains must be tested. Two strains are not enough!

It sometimes happens that when several inbred strains are surveyed with a particular drug the results fall into two or three classes, indicating the presence of one gene of major effect with two or three alleles. This opens the way to the conventional procedure of crosses, back-crosses, etc. to confirm the existence of the major gene. When two substrains of the same strain fall into different classes a mutation from one allele to another must have occurred after the substrains were separated from the original common progenitor strain. Even when strains cannot be easily classified it may be possible to obtain evidence of major genes by the use of recombinant inbred (RI) strains. The theory and use of RI strains are now well documented (Bailey, 1971; Swank & Bailey, 1973; Taylor, 1978; Oliverio, 1979) and they will be increasingly exploited by mouse geneticists interested in pharmacological, physiological and behavioural characteristics.

Congenic strains are also powerful tools for the same purposes, although they are at present mainly used by immunogeneticists.

One fundamental difficulty encountered in mouse pharmacogenetics is that when a drug is introduced into an animal the degree to which the animal reacts to the drug is influenced by two quite different factors. The first factor is the sensitivity of the tissue, or receptor, which is acted on by the drug. The second factor is the rate at which the drug is chemically metabolized, usually by the cytochrome P-450, or "microsomal", system of enzymes found mainly in the liver. Both these factors can vary genetically. It is often difficult to know which factor is varying in any given example of pharmacogenetic variation. If one can remove the target tissue or receptor from the body and measure its sensitivity *in vitro,* then clearly the metabolic factor is eliminated from the measurement. Similarly if one can measure the concentration of the drug *in vivo* and obtain a time curve of its elimination from the target tissue, then the sensitivity factor is excluded. A less direct method is as follows. Suppose that a drug acts on a particular receptor and is metabolized mainly by hydroxylation. Suppose also that there is another drug which acts on the same receptor but is metabolized mainly by conjugation. Any genetical variation which can be demonstrated equally well with either drug is more likely to be due to variation of the receptor than to variation of the two different pathways of metabolism. This reasoning is weakened by the discovery that some examples of inherited variation of drug metabolism do indeed affect several different metabolic pathways as described below (p. 538). Nevertheless this method can still be useful as a first step, particularly with drugs which are metabolized by several different pathways.

EXPLORATORY ACTIVITY AND SCOPOLAMINE

Oliverio, Eleftheriou & Bailey (1973a) presented data which has led to the inclusion of two new genes in the *Mouse News Letter* gene list: *Exa* (exploratory activity) on chromosome 4, and *Sco* (scopolamine modification of exploratory activity) on chromosome 17. The evidence for both genes deserves some critical examination.

The characteristic exploratory activities of mice (sex not stated) of strains BALB/cBy and C57BL/6By, and of the seven CXB RI strains derived from them, were measured in a toggle-floor box after an injection of saline. The toggle-floor box is a simple two-compartment apparatus which records automatically the number of times the mouse crosses from one compartment to the other. During the ten-minute test period the C57BL/6 micre had a mean of 27.1 crossings while the BALB/c mice had a mean of only 13.4 crossings. The authors tested a B6.C congenic strain (HW26) in which a portion of chromosome 4 of BALB/c origin was present on a C57BL/6 background. This congenic strain had a mean exploratory activity similar to BALB/c, and this result was given as evidence that the difference between C57BL/6 and BALB/c is due to a single gene on chromosome 4. The RI strains appeared to fall into two classes such that D, I and J resembled BALB/c and E, G, H and K resembled C57BL/6. The authors state that this strain distribution pattern (SDP) matches that for a histocompatibility gene which is present on chromosome 4. There are five histocompatibility genes on chromosome 4. Their SDPs are given in Table I, from which it can be seen that none of them has the above SDP. This does not support the assignment of *Exa* to chromosome 4. Oliverio *et al.* (1973a) found that both reciprocal types of F_1 mice had mean exploratory activities within the high range, which suggested that the C57BL/6 allele was dominant. They went on to test 30 progeny from an $F_1 \times$ BALB/c back-cross and their data show evidence of equal segregation into the two parental phenotypes. Unfortunately the back-cross progeny were not typed for any other gene on chromosome 4 and so again the assignment of *Exa* to chromosome 4 was not confirmed. One must conclude that the evidence for a single gene on chromosome 4 which determines the difference in exploratory activity between C57BL/6 and BALB/c is still incomplete.

Scopolamine (hyoscine) is an antagonist of the muscarinic actions of acetylcholine. After peripheral injection it reaches the CNS and produces a variety of behavioural effects including an impairment of learning (Iversen & Iversen, 1975). Oliverio *et al.* (1973a) injected mice from C57BL/6, BALB/c and the CXB RI strains with graded doses of scopolamine, i.e. 2, 4 and 6 mg kg^{-1}. In general they found that strains which had a low exploratory activity after saline (see above) became increasingly active after increasing dosages of scopolamine. On the other hand those strains which had a high exploratory activity after saline became increasingly inactive after increasing dosages of scopolamine. RI strains D and H both deviated from this general pattern. They both became more active after 2 mg kg^{-1} scopolamine but higher dosages produced no further significant increase in activity. The authors considered this plateau effect to derive from the BALB/c progenitor strain. They suggest that there is a gene, *Sco*, which has two alleles. One allele, Sco^a, determines the plateau effect found in RI strains D and H. The other allele, Sco^b, determines the dose-related response characteristic of the other

TABLE I

The strain distribution patterns (SDP) in the CXB RI lines of each of the five histocompatibility genes[a] *on chromosome 4, and of genes* Exa *and* Eam.

Gene	SDP						
	D	E	G	H	I	J	K
Exa	C	B	B	B	C	C	B
Eam	B	B	C	C	C	C	C
H-15	C	B	C	C	B	C	B
H-16	C	B	C	C	C	B	B
H-18	B	B	C	B	C	B	B
H-20	B	B	C	B	C	B	B
H-21	B	B	C	C	C	C	B

[a] Data from D. W. Bailey (personal communication).

strains. Since the *Sco*^{a} allele is supposed to have come from the BALB/c progenitor strain the absence of the plateau effect in BALB/c mice calls for an explanation. The authors suggest that "the lack of a plateauing effect in the BALB/c strain. . . can only be explained by yet other modifier gene(s) in BALB/c background". Clearly this makes nonsense of the classification since the hypothetical modifier gene(s) are also likely to be present in some of the RI strains. It so happens that the gene *H-2* has an SDP in the CXB RI strains such that the allele of BALB/c origin is found in the RI strains D and H. This led the authors to suggest that *Sco* is on chromosome 17, as is *H-2*. They tested the scopolamine response of two B6.C congenic strains (HW19 and HW41) which both contain the *H-2*^{d} allele of the BALB/c progenitor. They state that their results showed that "one strain was BALB/cBy-like;and the other was not". It is not clear what this means since neither of the B6.C congenic strains showed a plateau effect, indeed HW41 was only tested at one dosage (4 mg kg^{-1}). Clearly these results cannot be construed as support for the existence of the *Sco* gene.

AVOIDANCE LEARNING AND CHLORPROMAZINE

Different strains of mice differ in the rate at which they learn avoidance responses in a shuttle box. Oliverio, Eleftheriou & Bailey (1973b) showed that strain BALB/cBy mice learn relatively quickly to associate a visual stimulus with subsequent electric shocks and they then take appropriate action to avoid the shocks. C57BL/6By mice were found to be slower learners. The authors used the CXB RI strains and two congenic strains and showed clearly that the above difference in learning ability is determined by one gene, which they called *Aal* (active avoidance learning).

In a subsequent paper Castellano *et al.* (1974) showed that an injection of chlorpromazine given to a trained mouse 15 minutes before a shuttle box test caused a decrease in its level of avoidance. The authors again used the CXB RI strains and tested each strain at four doses of chlorpromazine (1.0, 1.5, 2.0 and 4.0 mg kg^{-1}). The decrease in avoidance performance was found to be dose dependent in all strains, although different strains showed different degrees of dose dependence. The differences in dose dependence were interpreted in terms of two loci which act either on low doses (locus 1) or high doses (locus 2, *Cpz*) of chlorpromazine. The whole scheme is very hypothetical and cannot be accepted without further evidence. (Readers of the paper should note that in Table 4 the SDPs in the first two rows should be interchanged, and chromosome 3 should be substituted for chromosome 9 on the final page).

ETHANOL

Two methods have been widely used to measure the effect of alcohol on mice. The first method is simply to inject a mouse with a dose of alcohol sufficiently large to render it incapable of standing on its feet. It is then laid on its back and the time which elapses until it regains its righting reflex is recorded as its "sleep time". The second method is to allow each mouse a choice between drinking tap water and drinking a dilute solution of alcohol. Some strains of mice prefer to drink water and other strains prefer the alcohol. Using the first method several authors have shown that after an intraperitoneal injection of about 4 g alcohol kg^{-1}, BALB/cJ mice sleep about twice as long as C57BL/6J mice (Damjanovich & MacInnes, 1973; Randall & Lester, 1974). The strain difference is thought to be due to the more rapid metabolism of alcohol in C57BL/6J mice. However, Moisset (1978) has recently reported that "A mutation appears to have occurred in BALB/cJ mice, changing them from highly sensitive to resistant to ethanol. Sleep time after 4 g kg^{-1} of ethanol is 35.91 ± 5.59 minutes as compared to previous reports that vary from 66 to 110 minutes. This possible mutation may have occurred quite recently and it is possible that BALB/cJ mice are still segregating for this gene. BALB/cBy mice, on the other hand, are very sensitive to ethanol. Their sleep time is 90.68 ± 7.52 minutes. This value resembles that of previous authors using BALB/cJ".

Using the alcohol–water choice method, Poley (1972) found

evidence of another possible mutation, this time in a subline of C57BL/6. He tested C57BL/10J, C57BL/6J and C57BL/6A. The last named is a subline which was separated from C57BL/6J in 1956 and is now kept at Alberta. Poley's data show very clearly that whereas C57BL/10J and C57BL/6J prefer alcohol to water, C57BL/6A mice prefer water to alcohol. The probable explanation of the difference is that a mutation occurred in C57BL/6A after its separation from C57BL/6J and the mutant allele is now fixed in the new strain. On the basis of the above two brief reports it seems probable that strain differences in alcohol tolerance and preference may be due to a few genes of large effect. Moisset (1978) comments that both sensitive and tolerant BALB/c strains show a preference for water given an alcohol–water choice, thus indicating a dissociation between tolerance of injected alcohol and preference for drinking alcohol.

Oliverio & Eleftheriou (1976) tested the effect of injected alcohol on motor activity. They used the same toggle-box technique that they had previously used in their work on the effect of scopolamine. Once again they worked with C57BL/6By, BALB/cBy and the CXB RI strains. Mice were injected with 0.5, 1.0, 1.5 or 2.0 mg alcohol kg^{-1} ten minutes before the activity test. The larger the dose of alcohol, the greater was the reduction in activity which followed it. Their data showed that the effect of alcohol in reducing motor activity was more marked in C57BL/6 mice than in BALB/c mice. Two of the seven CXB RI strains (D and E) resembled C57BL/6 in this respect and the other five resembled BALB/c. This variation was explained in terms of a new gene *Eam* (ethanol activity modifier). The Eam^h allele present in C57BL/6 was said to be recessive to the Eam^c allele of the BALB/c strain, and the two alleles were allocated the SDP given above. So far so good. Unfortunately the further evidence adduced to support these conclusions does not do so. The authors state that the above SDP is "closely matched" by four histocompatibility genes on chromosome 4, but Table I shows that in fact none of the histocompatibility genes on chromosome 4 match this SDP. The authors state that a B6.C congenic strain containing the *H-16* allele from BALB/c resembled BALB/c in its reaction to alcohol, while the B6.C congenic strains containing *H-15* or *H-18* alleles from BALB/c resembled C57BL/6, but no data are offered to support these statements. Their conclusion that "the gene exerting a major effect on alcohol-induced decrement of basal activity is linked to, or possibly identical with, the gene *H-16*" cannot be accepted on present evidence. Finally, the authors give data intended to show that 34 back-cross offspring fell into two activity groups after alcohol treatment, as would be expected if a single gene were segregating.

The back-cross data do not in fact appear to be bimodal. There is therefore insufficient evidence to establish the existence of the gene *Eam*.

MORPHINE

Morphine has several measurable effects on mice, e.g. analgesia, hypothermia or hyperthermia (depending on the dose: Glick, 1975) and changes in locomotor activity. Tolerance can be assessed from the decreasing responses to repeated doses of morphine. Physical dependence can be demonstrated by the characteristic jumping behaviour of morphine-dependent mice when they are injected with naloxone, an opiate receptor antagonist. The naloxone ED50 decreases with an increase in morphine dependence (Way, Loh & Shen, 1969; Fernandes, Kluwe & Coper, 1977).

Inbred strains differ considerably in their responses to morphine and the differences are probably genetically determined. However some responses to morphine are also very susceptible to environmental influences. This means that small changes in experimental design or procedure can produce apparently contradictory results. For example, how should one measure the effect of morphine on the locomotor activity of C57BL/6 mice? Eidelberg *et al.* (1975) found that a dose of $10\,mg\,kg^{-1}$ reduced their activity, and higher doses up to $80\,mg\,kg^{-1}$ merely restored their activity to control levels. However Brase, Loh & Way (1977), using an essentially similar experimental design, found that a dose of $20\,mg\,kg^{-1}$ increased their activity four-fold, and higher doses maintained it at that level. There must have been some environmental factor which was absent from the former experiment but present in the latter experiment, and this factor interacted with the morphine-treated mice so as to excite them. Oliverio & Castellano (1974) have found that when morphine-treated C57BL/6 mice are placed in a very familiar environment they become inactive. If the environment is less familiar their activity is the same as that of control (saline-treated) mice. However, if there is some element of novelty in the environment, as for example a change in lighting or an unfamiliar smell, then the morphine-treated mice show greatly increased locomotor activity.

In spite of experimental difficulties such as the above there is good evidence that strains DBA/2J and A/J are exceptionally resistant to the effect of morphine on locomotor activity (Shuster, Webster & Yu, 1975; Frigeni *et al.*, 1978; Brase *et al.*, 1977; Oliverio & Castellano, 1974). It is not yet clear whether they are also resistant to the analgesic effects of morphine (Collins & Whitney, 1978). The answer may depend on whether the test used is a tail flick test, a hot plate test or an abdominal constriction test. As Jacob & Barthelemy (1967) have remarked, "les diverses réactions utilisées an algésimétrie expérimentale ne sont pas toujours équivalentes". Notwithstanding, Hayashi & Takemori (1971) considered that in the case of morphine the three analgesic assays are probably measuring similar analgesic-receptor interactions. Attempts to use the CXB RI lines to unravel the genetics of sensitivity to morphine have produced contradictory results, probably because of the inadequacy of the techniques (Shuster, Webster, Yu & Eleftheriou, 1975; Oliverio, Castellano & Eleftheriou, 1975; Baran *et al.*, 1975).

A much more promising approach is to concentrate on the properties of the opiate receptors rather than the behavioural reactions of the whole mouse. Kosterlitz and his collaborators (Henderson & Hughes, 1976; Waterfield *et al.*, 1978) have published extensive data on the reactions of the isolated *vas deferens* to morphine, normorphine and endogenous opioid peptides. The motor innervation of the mouse *vas deferens* is adrenergic and morphine acts by inhibiting the release of noradrenaline from the motor nerve terminals. This can be demonstrated in electrically-stimulated *vasa deferentia* by noting the reduction in noradrenaline release, or the reduction in smooth muscle contractions, as morphine or normorphine are applied to the preparation. In some respects TO strain males are much more sensitive than C57BL/6 ones. For example, Fig. 1 shows the effect of morphine on the contractions of the TO *vas deferens*. With the C57BL/6 *vas deferens* a mean concentration of $30.88 + 9.47\ \mu\text{M}$ morphine was required to produce a 50% inhibition. This difference was not apparent when morphine was replaced by an enkephalin which reacts with a different type of receptor in the *vas deferens*. Waterfield *et al.* (1978) conclude that "the difference between the strains is in the sensitivity of their μ-receptors, mainly interacting with opioid alkaloid morphine and its surrogates". This system seems ripe for a more extensive genetical investigation.

d-AMPHETAMINE AND ADRENALINE

Amphetamine is a drug which is a stimulant and an anorexic in Man. It acts peripherally by displacing noradrenaline from adrenergic nerve endings. It therefore has sympathomimetic properties, although amphetamine itself has little or no direct effect on adrenergic receptors. Amphetamine easily penetrates into the brain where it

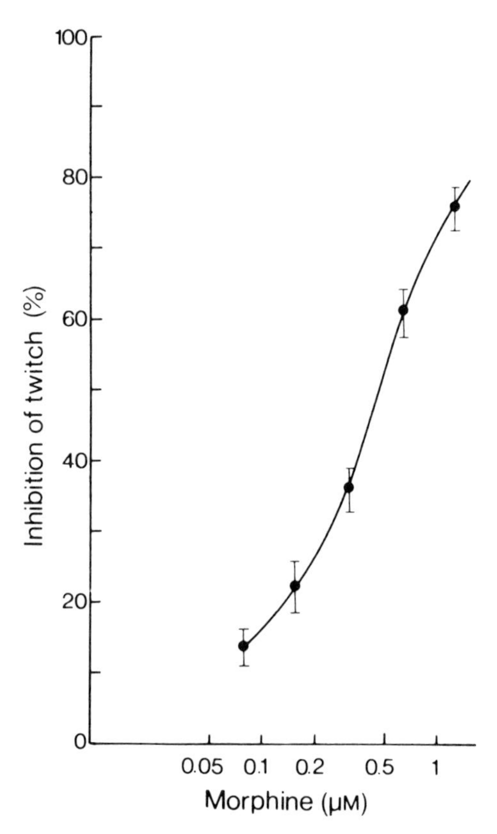

FIG. 1. Effect of morphine on contractions of the mouse *vas deferens* stimulated continuously at 0.1 Hz. Mean results from six experiments. Vertical lines show s.e. mean. Ordinate scale: inhibition of twitch as a percentage of the initial response. Abscissa scale: morphine concentration (From Henderson & Hughes, 1976.)

releases both noradrenaline and dopamine from neurones containing these neurotransmitters. In mice amphetamine causes changes in body temperature and locomotor activity; it also induces aggressiveness and a number of stereotyped actions such as gnawing and grooming (Thomas & Handley, 1978 a, b; Handley & Thomas, 1978).

Using a simple open-field technique, Moisset (1977) has noted an interesting difference between strains C57BL/10J, C57BL/6J and C57BL/6By in their locomotor response to amphetamine. Males of all three strains had the same initial (control) activity, but after injection with $5\,mg\,kg^{-1}$ of amphetamine the C57BL/10J and C57BL/6J mice both doubled their mean activity, whereas the C57BL/6By mice showed no change. C57BL/6By was separated from C57BL/6J in 1951, and it seems that we may have here another example of a mutation becoming established in a new inbred substrain after its separation from the original strain. Since amphetamine is metabolized in mammals via at least four different pathways (Cho & Wright, 1978) it is likely that the mutation has altered the sensitivity of an amphetamine receptor rather than the rate at which amphetamine is metabolized.

The effect of amphetamine on body temperature depends on the dose and the strain used. Results from two different laboratories (Scott, Lee & Ho, 1971; Caccia *et al.,* 1973; Jori & Rutczynski, 1978) are essentially in agreement and can be summarized as follows. BALB/c mice are made hypothermic by very low doses but become hyperthermic after doses higher than about 3.0 mg kg^{-1}. On the other hand C3H and C57BL/6 mice cannot be made significantly hyperthermic by any dose, and react to most doses except the highest by becoming hypothermic. NMRI resembles BALB/c in becoming hyperthermic after higher doses but has not yet been tested in the lower dose range. It will be interesting to see if other strains can be classified in the above way with respect to their susceptibility to amphetamine-induced hyperthermia. The isolation or aggregation of mice in their cages is known to be an important factor influencing survival after very large doses of amphetamine, although Wolf & Bunce (1973) have shown that body temperature is not affected by this factor.

Terpstra & Raaijmakers (1976) have described a case of variation in response to adrenaline. They found one batch of "Swiss Albino" mice to differ from previous batches in that no elevation of blood eosinophil count followed an injection of 0.3 mg/kg of adrenaline. Normally this dose of adrenaline causes a two-fold increase in blood eosinophil count. Their results suggest that a loss of β-adrenergic (receptor?) responsiveness had occurred, but no genetical work on this interesting anomaly has been published.

EMOTIONAL DEFAECATION AND MESCALINE

Mescaline (3, 4, 5-trimethoxyphenylethylamine) is one of a series of substituted phenylethylamines which are hallucinogenic in man (Nieforth, 1971; Shulgin, 1973). It occurs naturally in the cactus *Lophophora williamsii* (*Anhalonium lewinii*) where it is used by certain North American Indians to induce a ritual ecstatic state. Corne & Pickering (1967) found that in mice mescaline provokes a type of head-twitch which is also seen after treatment with 5-hydroxytryptophan. This is in line with the current idea that phenylalkylamine hallucinogens act by mimicking, releasing or potentiating serotonin centrally (Kier & Glennon, 1978). Other reported effects of mescaline in mice include changes in motor activity (Shah & Himwich, 1971; Cooper & Walters, 1972) and inhibition of emotional defaecation (Lush, 1975).

When a mouse is placed in an unfamiliar environment it may respond by defaecating. This response, which is sometimes considered to be a measure of emotionality, varies considerably from strain to strain. Lush (1975) measured the emotional defaecation of male mice from seven strains, A2G, C57BR/cd, C3H/He, CBA/Cam, F/St, ICFW and Schneider. Mice from each strain were tested after a saline injection and, on a later occasion, after a mescaline injection. The data are displayed in Fig. 2, which illustrates the curious result that the degree of inhibition of emotional defaecation by mescaline increases as the amount of defaecation in the saline control experiment increases. No explanation has been found for this relationship, but this type of experiment in which a behavioural or physiological variable is found to be correlated with the effect of a drug may prove to be a means of gaining insight into the biochemical basis of normal variation.

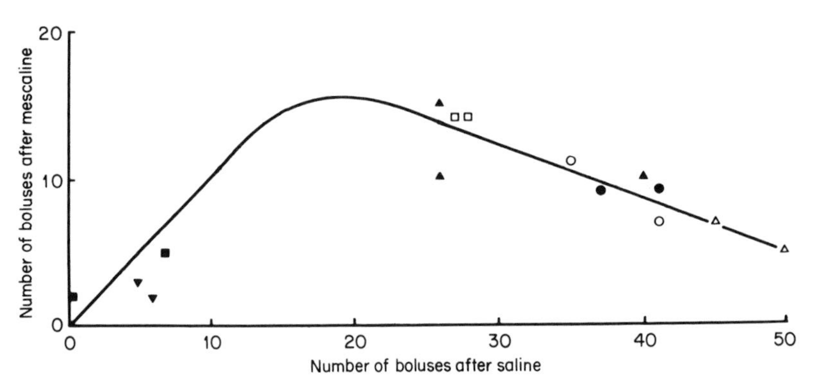

FIG. 2. The relationship between emotional defaecation after mescaline injection and after saline injection in males from seven strains of mice. Each symbol represents the total number of faecal boluses produced by six males tested individually. Two or three lots of six males from each strain were tested. The strains were ■ Schneider; ▼ ICFW; □ F/St; ▲ CBA; ○ C3H; ● C57BR; △ A2G. The line drawn through the points shows that the inhibitory effect of mescaline is greatest in the high-defaecating strains. (Data from Lush, 1975).

HYPOTHERMIA AND NEUROTRANSMITTER AGONISTS

Hypothermia in the mouse can be produced by a variety of drugs ranging from chlorpromazine to Δ^9-tetrahydrocannabinol (Abel,

1973; Haavik, 1977). Agonists of dopamine, serotonin and acetylcholine (muscarinic) also produce hypothermia in the mouse and the degree of hypothermia produced by each of these three types of drug varies widely between strains of mice (Lapin, 1975; Lush, unpublished). For example, BALB/c mice are more sensitive than C57BL/6 to the hypothermic effect of the muscarinic agonist oxotremorine which acts directly on the temperature-controlling centre of the hypothalamus. By the use of the CXB RI strains preliminary evidence has been obtained that this strain difference is largely determined by one gene, *Ots* (Lush & Andrews, 1978a). Whether the difference in hypothermia is a result of a difference in muscarinic receptor sensitivity or oxotremorine metabolism is not yet known. The CXB RI strains can also be classified with respect to their hypothermic responses to the dopamine agonists apomorphine and piribedil (Lush, unpublished); however, the situation is complicated by the fact that the dopaminergic pathway acts on the hypothalamic temperature centre indirectly through a serotonergic link (Cox & Lee, 1979), which is itself genetically variable in sensitivity. Preliminary results indicate that a gene influencing sensitivity to the serotonin agonist quipazine is situated on chromosome 17 (Lush, unpublished).

TASTE SENSITIVITY TO SUCROSE OCTAACETATE, PHENYLTHIOCARBAMIDE AND SACCHARIN

A striking example of inherited variation of taste sensitivity in mice has been reported by Warren & Lewis (1970) using a sugar derivative, sucrose octaacetate, which tastes intensely bitter to man. Mice were housed individually and could choose between drinking either water or a solution of sucrose octaacetate (SOA). The daily consumption of each liquid was measured to the nearest 0.05 ml and the concentration of SOA was increased every two days. The acceptability of SOA was assessed by expressing the SOA consumed as a percentage of the total fluid intake during each two-day period. With this simple but effective experimental design the authors found that, over a wide range of concentrations, mice of strains C57L/N, C57BL/6N and C3Hf/HeN drank equal amounts of SOA and water, presumably because they could taste no difference between them. However, mice of strain CFW/N showed an increasing aversion to drinking SOA when the concentration rose above 1.0 μM as shown in Fig. 3. Appropriate crosses showed that the difference is determined by a single autosomal gene with the allele determining tasting ability

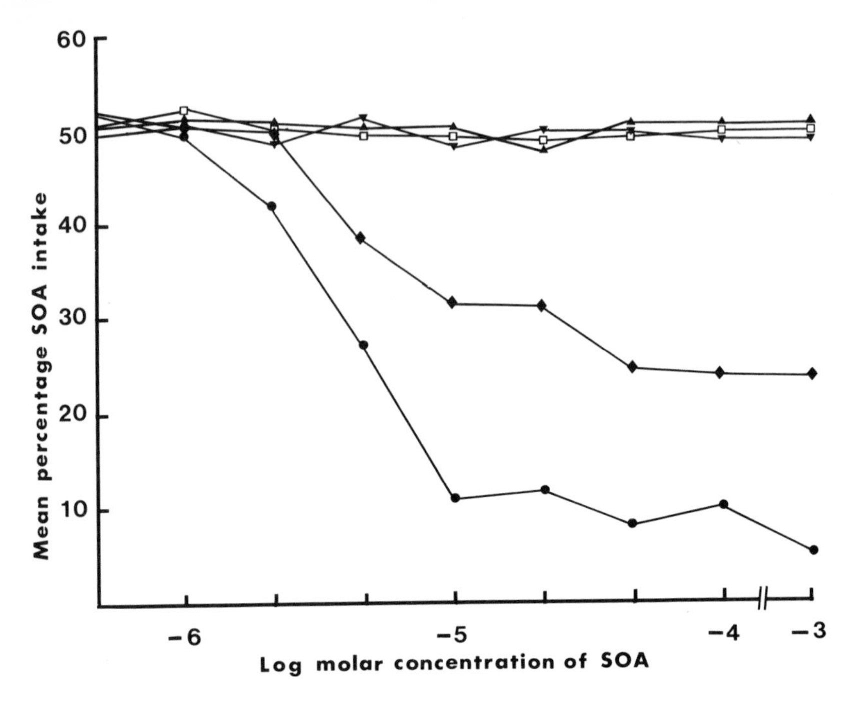

FIG. 3. The effect of increasing concentration of sucrose octaacetate (SOA) on the amount of SOA solution consumed by mice from different strains. ▲ C57L; □ C57BL/6; ▼ C3Hf/He; ● CFW; ♦ Wild house mice. (From Warren & Lewis, 1970.)

(aversion) being dominant. Ten wild house mice showed an intermediate mean phenotype but the authors state that "the difference between the CFW and the wild-type can be traced to non-tasters among the wild mice; those that rejected SOA seemed to be as sensitive as the CFW animals, avoiding it at the same concentrations". In other words the wild mice were polymorphic for this character. Warren & Lewis (1970) did not coin a symbol for the new gene. The symbol *Soa* would seem to be appropriate, with Soa^a (aversion) and Soa^b (blind) for the alleles present in CFW and the other strains respectively.

Klein & DeFries (1970) showed that strains of mice vary in their ability to taste phenylthiocarbamide (PTC). Using the same experimental technique as described above, they classified BALB/cIby as a "taster" strain and A/Ibg, C3H/Ibg, DBA/2Ibg and C57BL/Ibg as "non-taster" strains. The difference in taste sensitivity for PTC was not so clear-cut as it was for SOA, nevertheless they were able to analyse their back-cross and F_2 data to show that segregation at a single locus could account for their results, the "taster" allele being dominant.

Mice prefer to drink a sodium saccharin solution rather than water, although at concentrations above about 5% this preference becomes an aversion (Pelz, Whitney & Smith, 1973; Fuller, 1974). Fuller showed that C57BL/6J mice have a significantly greater degree of preference for saccharin than do DBA/2J mice. He considered that DBA/2J mice seem to have a lower incentive to drink saccharin rather than an impaired ability to taste it. F_2 and back-cross progeny gave evidence of segregation of this character. As with PTC tasting, there was some overlap between the phenotypic classes but Fuller's statistical analysis presents a reasonable case for an interpretation in terms of one gene, *Sac,* with a dominant allele Sac^b (C57BL/6-like) and a recessive allele Sac^d (DBA/2-like). Fuller (1974) concludes that "the two-locus model is not excluded with the size of the sample available, but it is distinctly a second choice".

CADMIUM-INDUCED TESTICULAR DAMAGE

In males from some strains of mice a single injection of cadmium chloride causes necrosis of the testes, while males from other strains are resistant to this effect (Gunn, Gould & Anderson, 1965). Resistance to cadmium-induced testicular damage is conferred by a recessive gene, *cdm,* on chromosome 3 (Taylor, Heiniger & Meir, 1973; Taylor, 1976a). The above authors made the very reasonable suggestion that "the primary effect of the *cdm* gene might be a quantitative or qualitative alteration of the cadmium-binding proteins of one or more tissue". However the most recent results (Godowicz & Pawlus, 1979) seem to show that the testes of resistant mice are enclosed by a thicker *tunica albuginea* (fibrous coat) than is found in non-resistant mice.

DRUG METABOLISM

It was stated in the Introduction that there are two important variables in pharmacogenetics. The first variable is the sensitivity or accessibility of the target tissue or receptor. The second variable is the rate at which a drug is metabolized. Most of the examples discussed so far are probably due to variation of the first kind. Some of the examples to be discussed below involve variation in drug metabolism. They are therefore prefaced with a very brief introduction to this topic.

Drugs are metabolized by mammals in a bewildering variety of ways, but the usual result is that they become more easily excretable from the body (Parke, 1968). Some drugs are metabolized by enzymes present in the cytosol or mitochondria of liver cells or in the plasma. For example, atropine is hydrolysed in the rabbit by a plasma esterase which shows genetically-determined variation in activity (Ecobichon & Comeau, 1974). Isoniazid and several other drugs are acetylated by an *N*-acetyl-transferase present in the liver of some species, and this enzyme is also genetically variable in the rabbit (Weber *et al.*, 1976) and Man (Propping, 1978). Most drugs are metabolized by a system of enzymes attached to the endoplasmic reticulum, mainly in the liver but also to some extent in other tissues. When the cells are homogenized and centrifuged the enzymes are found in the microsomal fraction which contains fragmented endoplasmic reticulum. The actions of these microsomal enzymes on drugs are mainly oxidations, reductions and hydrolyses. The oxidations are catalysed by haemoprotein enzymes which require NADPH and O_2 and are collectively referred to as cytochrome P-450 because of the characteristic Soret absorption maximum of their reduced form when complexed with carbon monoxide. The oxidations include a range of reactions such as hydroxylations, dealkylations and deaminations, all of which may be ascribed to one common oxidative mechanism (Parke, 1968). The synthesis of microsomal enzymes can be temporarily stimulated, or induced, by injection or feeding an animal with certain drugs such as phenobarbitone, 3-methylcholanthrene or ethanol.

Cytochrome P-450 is in fact a mixture of at least six enzymes each of which must be able to accept a wide range of substrates since the number of different drugs which are oxidized by the system is enormous. The nomenclature of these enzymes is at present in a rather confused state, but they can be divided into three main classes on the basis of their inducibility and spectral characteristics as follows:

(1) Cytochrome P-450 (in the strict sense), which is inducible by phenobarbitone but not inducible by polycyclic aromatic hydrocarbons such as 3-methylcholanthrene.

(2) Cytochrome P_1-450, which is inducible by polycyclic hydrocarbons but not inducible by phenobarbitone.

(3) Cytochrome P-448, which differs from cytochrome P_1-450 by an approximately 2 nm shift to the blue in the Soret peak of the reduced CO-complex during induction by polycyclic hydrocarbons. (Nebert, Atlas *et al.*, 1978).

Further subdivisions can be made by the use of techniques such as differential inhibitors and gel electrophoresis, but there is at present no generally agreed classification or nomenclature. Some authors use the term cytochrome P-448 to mean all forms which are inducible by polycyclic hydrocarbons.

Some drugs are inactive until they are oxidized by cytochrome P-450 to a pharmacologically active metabolite. For example tremorine is inactive but it is oxidized to oxotremorine which is a powerful muscarinic agent, as mentioned on p. 000.

Tremorine → Oxotremorine

Other drugs are active in their injected form but are oxidized to inactive metabolites. For example hexobarbitone is metabolically inactivated by hydroxylation of the 3-carbon of its cyclohexanyl side chain.

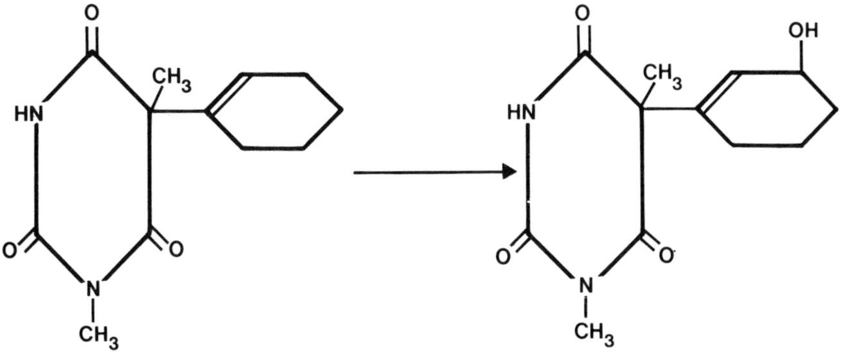

Hexobarbitone → 3-Hydroxyhexobarbitone

Some drugs are excreted in the oxidized form but most undergo a further metabolic step by being conjugated with glucuronic acid, sulphate, glucose, glutathione or taurine. The synthesis of these glucuronides, sulphates, etc. is catalysed by enzymes which are not part of the cytochrome P-450 system although one of them, UDP-glucuronyltransferase, is found in the microsomal fraction.

HEXOBARBITONE AND PENTOBARBITONE

The length of the sleeping time which follows an injection of a short-acting barbiturate has been much used as a convenient measure of the activity of the cytochrome P-450 enzymes, particularly in studies concerned with the induction of these enzymes. However several authors (Jay, 1955; Vesell, 1968) have shown that there are large strain differences in the hexobarbitone sleeping times of normal (uninduced) mice. These differences are due to differences in the rate of metabolic hydroxylation of hexobarbitone and not to variation in brain sensitivity to the drug. Lush & Lovell (1978) compared the normal sleeping times of 15 strains with respect to both hexobarbitone and pentobarbitone and found that they are very highly correlated (Fig. 4). The genetical basis of the strain differences is under investigation (Lovell, 1978) and the provisional conclusions

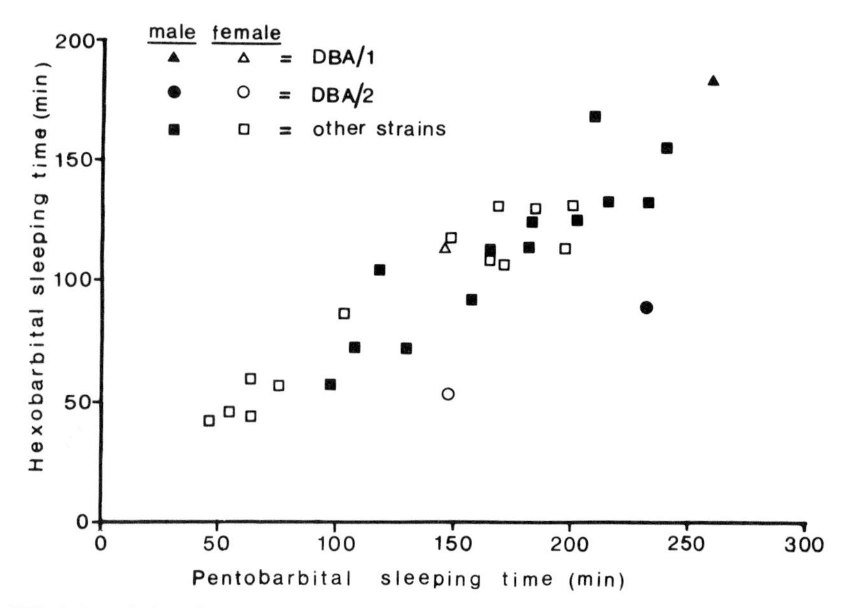

FIG. 4. Correlation between mean pentobarbital and hexobarbital sleeping times of 15 strains (Data from Lush & Lovell, 1978.)

are that for pentobarbitone sleeping time at least two genes are involved, with predominantly additive gene action and no evidence of epistasis.

Mitoma *et al.* (1967) found that within and between five strains of rats hexobarbitone sleeping time was inversely correlated not only with liver microsomal hexobarbitone hydroxylase activity but also with three other microsomal enzyme activities; acetanilide hydroxylase, *O*-nitroanisole demethylase and aminopyrine demethylase. Lush (1976) also found that strain differences in hexobarbitol sleeping time in mice were correlated with zoxazolamine paralysis time (which is an inverse measure of microsomal zoxazolamine hydroxylase activity) and with differences in survival time on a warfarin-containing diet. The biochemical and genetical reasons for these correlations are not known. It may be that there are several genes of minor effect which influence the structure of the endoplasmic reticulum and which therefore influence the activities of all enzymes which are attached to the endoplasmic reticulum.

COUMARIN AND WARFARIN

Wood & Conney (1974) showed that mouse liver microsomal enzymes can hydroxylate coumarin at the 7-position to form the highly fluorescent product umbelliferone, and that strain DBA/2 has a higher liver cytochrome P-450 coumarin 7-hydroxylase activity than other strains. Wood & Taylor (1979; Wood, 1979) found that one gene, *Coh,* is responsible for the strain difference and mapped it on chromosome 7. The same authors surveyed 16 strains and showed that strains with high and low activity also differ in the degree to which the enzyme is inhibited *in vitro* by aniline and metyrapone. This is evidence that the difference in activity is the consequence of a qualitative difference in the enzyme. Lush & Andrews (1978b) surveyed 19 strains and considered that they fell into three groups; a high-activity group (DBA/1 and DBA/2), a medium-activity group (CBA, NZB, NZW and 129/Rr) and a low-activity group (the other strains). The same strains fell into the same groups when they were tested for their ability to *O*-deethylate 7-ethoxycoumarin. It therefore appears that there are three alleles of the *Coh* gene and that the microsomal enzyme determined by this gene can both hydroxylate and deethylate the appropriate substrate at the 7-position. The data can also be explained in terms of two closely linked genes, *Coh-1* and *Coh-2* (Lush & Andrews, 1978b).

Warfarin [3-(d-acetonylbenzyl)-4-hydroxycoumarin] is a derivative of coumarin which is used as a rodenticide because of its ability to inhibit the synthesis of several blood clotting factors. Mice which ingest food containing warfarin usually die within a few days as a result of internal haemorrhages. Laboratory strains vary in their susceptibility to warfarin. In a survey of females from 15 strains Lush (1976) showed that on a diet containing 0.05% racemic warfarin the strain mean survival times ranged from four days to over two weeks. Although mice can 7-hydroxylate the warfarin molecule, those strains with medium or high ability to 7-hydroxylate coumarin did not survive longer than strains with low hydroxylating ability, thus indicating that this pathway is not important in the detoxification of warfarin in the mouse (Lush & Arnold, 1975). An unexpected finding (Lush, 1976) was that strain mean survival time correlated positively with strain mean hexobarbitone sleeping time ($r = 0.68; P < 0.005$).

The widespread use of warfarin to control mouse infestations has had the predictable result that highly warfarin-resistant mice have appeared in some wild populations (Rowe & Redfern, 1968). Wallace & MacSwiney (1976) have taken advantage of this unintended exercise in natural selection and have shown that in one population near Cambridge the resistance is due to a gene, *War,* which is closely linked to the gene frizzy (*fr*) on chromosome 7. They used four marker genes and proposed the following linear order: *War*–*fr*–*sh-1*–c^{ch}–*p*. Homozygotes for the gene *War* survive indefinitely on a warfarin-containing diet which kills almost all laboratory mice in a matter of days. *War* is dominant in females but incompletely so in males, where it is influenced by modifier genes. In the rat (*Rattus norvegicus*) a gene (*Rw*) conferring warfarin resistance has also been found in some wild populations (Bishop & Hartley, 1976). It is linked to albino (*c*) and pink-eyed dilution (*p*) in the order *Rw*–*c*–*p* (Greaves & Ayres, 1977). There is therefore good reason to conclude that the genes for warfarin resistance in these two species are homologous and probably produce resistance by the same mechanism. The nature of this mechanism is not yet completely clear. Resistant rats have a higher dietary requirement for vitamin K than do sensitive rats (Hermondson, Suttie & Link, 1969). Vitamin K is metabolized to an epoxide which is recycled by reduction to vitamin K. The enzyme vitamin K epoxide reductase is inhibited by warfarin, and this leads to an accumulation of the epoxide which is a competitive inhibitor of the role of vitamin K in the synthesis of certain blood clotting factors (Townsend, Odam & Page, 1975). Vitamin K epoxide reductase is less inhibited by warfarin in resistant rats than it is in normal rats (Bell & Caldwell, 1973). The genetical difference may therefore be in the nature of the reductase enzyme, which in resistant rats is less efficient but also less susceptible to inhibition by warfarin. In wild rats the *Rw* resistant allele is apparently maintained as a balanced polymorphism in certain areas where warfarin is used (Greaves *et al.,* 1977).

ARYL HYDROCARBON HYDROXYLASE INDUCTION

Since its independent discovery in two laboratories (Nebert, Goujon *et al.,* 1972; Thomas, Kouri & Hutton, 1972), this locus has been the subject of a flow of publications that shows no sign of abating (Thorgeirsson & Nebert, 1977; Nebert, Atlas *et al.,* 1978; Nebert & Jensen, 1979).

Mouse liver contains a cytochrome P_1-450 enzyme aryl hydrocarbon hydroxylase (AHH) which can hydroxylate benz(a)pyrene to form the fluorescent product 3-hydroxybenze(a)pyrene and probably other phenols having similar wavelengths of fluorescence excitation and emission. Strains of mice differ in the degree to which their AHH levels can be induced by pre-treatment with 3-methylcholanthrene. In some strains, for example DBA/2J and AKR/J, induction is slight or insignificant. In other strains, for example C3H/HeJ and C57BL/6J, induction is considerable and may be as high as five or six-fold. Crosses using C57BL/6J and DBA/2 have shown that a single autosomal gene, named *Ah*, is responsible for the difference in inducibility between these two strains. The C57BL/6J allele for inducibility Ah^b is dominant over the DBA/2J allele determining non-inducibility, Ah^d. In crosses using C3H/HeJ and DBA/2J the heterozygotes show intermediate inducibility (codominance), while in crosses using C57BL/6N and AKR/N, the AKR/N allele for non-inducibility is dominant over the C57BL/6N allele for inducibility. The genetical basis of all the above differences is not yet clear but Nebert, Atlas *et al.* (1978) at present favour an explanation involving at least two loci (*Ah-1* and *Ah-2*) each with three alleles. Neither of these loci has yet been mapped or allocated to a particular chromosome. The level of basal AHH activity present in uninduced mice of different strains is not correlated with the inducibility of AHH. This fact and others support the conclusion that the AHH induced by 3-methylcholanthrene is synthesized *de novo* and is a different form of the enzyme from that present in uninduced mice.

Nebert and his colleagues have concentrated their attention on the difference between C57BL/6J and DBA/2J and have shown that the inducibility of several other microsomal enzyme activities is also associated with the Ah^b allele. These are listed in Table II together with some microsomal enzyme activities whose inducibility is not affected by the *Ah* locus. The *Ah* gene affects the inducibility of AHH and the other associated microsomal enzymes in several tissues, e.g. kidney, intestine, lung and skin, although the genetical difference in inducibility is not so well defined as it is in liver. One result of great practical interest to emerge from this work is that the AHH-inducible phenotype enhances the carcinogenicity of polycyclic hydrocarbons such as 3-methylcholanthrene, benz(a)pyrene, 7,12-dimethylbenzanthracene (DMBA), and 2-acetylaminofluorene, particularly in skin and lung. Presumably the application of such chemicals to inducible mice leads to a local induction of AHH which then increases their conversion to metabolites which are known to be carcinogenic (Kouri, 1976; Wang *et al.*, 1976). The results with some

TABLE II

Mouse microsomal enzyme activities whose inducibility is (A) affected by the Ah *gene, or (B) not affected by the* Ah *gene*[a]

A	B
Aryl hydrocarbon hydroxylase (AHH)	Aminopyrine *N*-demethylase
p-Nitroanisole *O*-demethylase	*d*-benzphetamine *N*-demethylase
7-Ethoxycoumarin *O*-deethylase	Diphenylhydantoin hydroxylase
Zoxazolamine 6-hydroxylase	Hexobarbital hydroxylase
Dimethylaminoazobenzene *N*-demethylase	Pentobarbital hydroxylase
2-Acetylaminofluorene *N*-hydroxylase	Aniline hydroxylase
Phenacetin *O*-deethylase	Benzenesulphonamide hydroxylase
Biphenyl 2-hydroxylase	Chlorcyclizine *N*-demethylase
Biphenyl 4-hydroxylase	Ethylmorphine *N*-demethylase
Acetanilide 4-hydroxylase	Testosterone 7α-, 16α-, and 6β-hydroxylases
Naphthalene monooxygenase	
Ethoxyresorufin *O*-deethylase	
Acetaminophen *N*-hydroxylase	
p-chloroacetanilide *N*-hydroxylase	
UDP glucuronyltransferase	

[a]Data from Thorgeirsson & Nebert (1977) and Nebert, Atlas *et al.* (1978).

strains are not entirely concordant with this generalization, but it should be remembered that the carcinogenic intermediates are themselves further metabolized by other enzymes of the microsomal system and that genetical variation of these enzymes would be expected to modify any simple relationship between *Ah* phenotype and carcinogenicity. Schmid, Elmer & Tarnowski (1969) first showed that topical application of DMBA to the skin of some strains of mice caused inflammation, ulceration and sometimes papilloma formation. With other strains little or no reaction was visible. The same authors showed that the strain difference in responsiveness was due to a single gene which they called *In* (see also Taylor, 1971). Thomas, Hutton & Taylor (1973) then showed that there is an excellent correlation between responsiveness to DMBA and inducibility by 3-methylcholanthrene, and they concluded that *In* and *Ah* are probably identical. Thus Schmid *et al.* (1969) first identified the *Ah* gene, although under a different name.

The mechanism by which the inducibility of so many different microsomal enzyme activities is controlled by one gene, or perhaps two, is at present unclear. It is in fact possible to induce the AHH of DBA/2 mice by using the very potent inducer, 2,3,7,8-tetrachlorodibenzo-*p*-dioxin (TCDD) although the induction is still not as effective as it is in C57BL/6 mice. It has been suggested that the *Ah* gene may be the structural gene of a cytosol receptor molecule which combines with polycyclic hydrocarbons and in some way mediates the enzyme induction which they cause. The Ah^d allele product is probably a less efficient form of the receptor molecule than that determined by the Ah^b allele. It therefore seems that the *Ah* gene variation should be classified as an example of receptor variation, even though it is expressed in terms of the inducibility of microsomal drug-metabolizing enzymes.

LUNG TUMOURS AND URETHANE

Urethane (ethyl carbamate) was formerly used as an hypnotic and anaesthetic but it is now known to be mutagenic, carcinogenic and teratogenic (Bateman, 1976). In mice, urethane causes lung tumours and it can also enhance the carcinogenicity of other carcinogens. Bloom & Falconer (1964) injected mice with urethane at three and at nine weeks of age, and then examined their lungs for visible tumours at 23 weeks of age. They used six inbred strains, A/Fa, C57BL/Fa, CBA/Fa, RIII/Fa, JU/Fa and KL, and also F_1 offspring from all the 15 possible crosses between these six strains. Strain C57BL developed fewest tumours and this characteristic was recessive in crosses with all the other strains. When (C57BL × A)F_1 mice were back-crossed to C57BL the progeny segregated equally into two classes, thus indicating the segregation of two alleles of a single gene. Bloom & Falconer (1964) named the recessive C57BL allele *ptr* (pulmonary tumour resistance) and the A strain allele *Ptr* and showed that this gene accounted for about three quarters of the total difference in urethane susceptibility between the two strains. Their results greatly extend the earlier work of Cowen (1950) who also found evidence of a single gene difference in urethane susceptibility between strains C57BL and A. Surprisingly, no further work has been published on the *Ptr* gene. Opinions differ as to whether urethane is itself carcinogenic or whether it is metabolized to a carcinogenic metabolite such as *N*-hydroxyurethane (Mirvish, 1968; Williams & Nery, 1971; Yamamoto, Weisburger & Weisburger, 1971). Perhaps the *Ptr* gene could be used to throw some light on this matter.

CONCLUSIONS

This review has been largely restricted to differences which have been analysed in terms of Mendelian genes. In some other published examples of strain differences, authors have attempted genetical analyses without identifying any individual genes of major effect. For example Taylor (1976b) measured the lethality of a large dose of isoniazid (4 mg per mouse, injected subcutaneously) in ten inbred strains and four sets of RI strains. The data suggested the existence of up to four major genes affecting susceptibility to isoniazid, but in order to identify them individually a more precise and discriminating criterion than lethality will probably be necessary. Many such strain differences in reaction are reported in the literature, but in most cases the genetical determinants have not been identified. For example Hill *et al.* (1975; see also Vesell *et al.*, 1976) have shown that strain C57BL/6J is about four times as resistant to the lethal effects of injected chloroform as strain DBA/2J.

The mouse is an excellent subject for pharmacogenetic research, especially now that new RI strains and congenic strains are being developed. I have tried to systematize the literature in terms of variation of either drug receptors or drug metabolism. In both of these fields biochemists and pharmacologists are intensely active at the present time. It is hoped that geneticists will soon be able to make a significant contribution.

ACKNOWLEDGEMENTS

I am grateful to the Nuffield Foundation for giving me the opportunity to develop my interest in pharmacogenetics, and to Miss Felicity Peadon for help in obtaining the literature for this review.

REFERENCES

Abel, E. L. (1973). Comparative effects of Δ^9-THC on thermoregulation. In *Marijuana:* 120–141. MacLoulan, R. (Ed.). New York: Academic Press.

Bailey, D. W. (1971). Recombinant-inbred strains. An aid to finding identity, linkage, and function of histocompatibility and other genes. *Transplantation* **11**: 325–327.

Baran, A., Shuster, L., Eleftheriou, B. E. & Bailey, D. W. (1975). Opiate receptors in mice: genetic differences. *Life Sci.* **17**: 633–640.

Bateman, A. J. (1976). The mutagenic action of urethane. *Mutation Res.* **39**: 75–96.

Bell, R. G. & Caldwell, P. T. (1973). Mechanism of Warfarin resistance. Warfarin and the metabolism of vitamin K. *Biochemistry* **12**: 1759–1762.

Bishop, J. A. & Hartley, D. J. (1976). The size and age structure of rural populations of *Rattus norvegicus* containing individuals resistant to the anticoagulant poison warfarin. *J. Anim. Ecol.* **45**: 623–646.

Bloom, J. L. & Falconer, D. S. (1964). A gene with major effect on susceptibility to induced lung tumors in mice. *J. natn. Cancer Inst.* **33**: 607–617.

Brase, D. A., Loh, H. H. & Way, E. L. (1977). Comparison of the effects of morphine on locomotor activity, analgesia and primary and protracted physical dependence in six mouse strains. *J. Pharmac. exp. Ther.* **201**: 368–374.

Broadhurst, P. L. (1978). *Drugs and the inheritance of behavior. A survey of comparative psychopharmacogenetics.* New York: Plenum Press.

Caccia, S., Cecchetti, G., Garattini, S. & Jori, A. (1973). Interaction of (+)-amphetamine with cerebral dopaminergic neurones in two strains of mice, that show different temperature responses to this drug. *Br. J. Pharmac.* **49**: 400–406.

Castellano, C., Eleftheriou, B. E., Bailey, D. W. & Oliverio, A. (1974). Chlorpromazine and avoidance: a genetic analysis. *Psychopharmacologia* **34**: 309–316.

Cho, A. K. & Wright, J. (1978). Pathways of metabolism of amphetamine and related compounds. *Life Sci.* **22**: 363–372.

Collins, R. L. & Whitney, G. (1978). Genotype and test experience determine responsiveness to morphine. *Psychopharmacology* **56**: 57–60.

Cooper, P. D. & Walters, G. C. (1972). Stereochemical requirements for the mescaline receptor. *Nature, Lond.* **238**: 96–98.

Corne, S. J. & Pickering, R. W. (1967). A possible correlation between drug-induced hallucinations in man and a behavioural response in mice. *Psychopharmacologia* **11**: 65–78.

Cowen, P. N. (1950). Strain differences in mice to the carcinogenic action of urethane and its non-carcinogenicity in chicks and guinea-pigs. *Br. J. Cancer* **4**: 245–253.

Cox, B. & Lee, T. F. (1979). Possible involvement of 5-hydroxytryptamine in dopamine-receptor mediated hypothermia in the rat. *J. Pharm. Pharmac.* **31**: 352–354.

Damjanovich, R. P. & MacInnes, J. W. (1973). Factors involved in ethanol narcosis: analysis in mice of three inbred strains. *Life Sci.* **13**: 55–65.

Ecobichon, D. J. & Comeau, A. M. (1974). Genetic polymorphism of plasma carboxylesterases in the rabbit: correlation with pharmacologic and toxicologic effects. *Toxicol. App. Pharmac.* **27**: 28–40.

Eidelberg, E., Erspammer, R., Kreinick, C. J. & Harris, J. (1975). Genetically determined differences in the effects of morphine on mice. *Eur. J. Pharmac.* **32**: 329–336.

Fernandes, M., Kluwe, S. & Coper, H. (1977). Quantitative assessment of tolerance to and dependence on morphine in mice. *Naunyn-Schmiedebergs Arch. Pharmac.* **297**: 53–60.

Frigeni, V., Bruno, F., Carenzi, A., Recagni, G. & Santini, V. (1978). Analgesia and motor activity elicited by morphine and enkephalins in two inbred strains of mice. *J. Pharm. Pharmac.* **30**: 310–311.

Fuller, J. L. (1974). Single-locus control of saccharin preference in mice. *J. Hered.* **65**: 33–36.

Geddie, W. (Ed.). (1952). *Chambers twentieth century dictionary.* Edinburgh: Chambers.

Glick, S. D. (1975). Hyperthermic and hypothermic effects of morphine in mice: interactions with apomorphine and pilocarpine and changes in sensitivity after caudate nucleus lesions. *Arch. int. Pharmacodyn.* **213**: 264–271.

Godowicz, B. & Pawlus, M. (1979). Inheritance of sensitivity to cadmium-induced testicular damage in mice. *Folia Histochem. Cytochem.* **17**: 267–274.

Greaves, J. H. & Ayres, P. B. (1977). Unifactorial inheritance of warfarin resistance in *Rattus norvegicus* from Denmark. *Genet. Res.* **29**: 215–222.

Greaves, J. H., Redfern, R., Ayres, P. B. & Gill, J. E. (1977). Warfarin resistance: a balanced polymorphism in the Norway rat. *Genet. Res.* **30**: 257–263.

Gunn, S. A., Gould, T. C. & Anderson, W. A. D. (1965). Strain differences in susceptibility of mice and rats to cadmium-induced testicular damage. *J. Reprod. Fert.* **10**: 273–275.

Haavik, C. O. (1977). Profound hypothermia in mammals treated with tetrahydrocannabinols, morphine, or chlorpromazine. *Fedn Proc. Fedn Am. Socs exp. Biol.* No. **36**: 2595–2598.

Handley, S. L. & Thomas, K. V. (1978). On the mechanism of amphetamine-induced behavioural changes in the mouse. II. Effects of agents stimulating noradrenergic receptors. *Drug Res.* **23**: 834–837.

Hayashi, G. & Takemori, A. E. (1971). The type of analgesic-receptor interaction involved in certain analgesic assays. *Eur. J. Pharmac.* **16**: 63–66.

Henderson, G. & Hughes, J. (1976). The effects of morphine on the release of noradrenaline from the mouse *vas deferens*. *Br. J. Pharmac.* **57**: 551–557.

Hermondson, M. A. Suttie, J. W. & Link, K. P. (1969). Warfarin metabolism and vitamin K requirement in the Warfarin-resistant rat. *Am. J. Physiol.* **217**: 1316–1319.

Hill, R. N., Clemens, T. L., Lin, D. K. & Vesell, E. S. (1975). Genetic control of chloroform toxicity in mice. *Science, N.Y.* **190**: 159–161.

Iversen, S. D. & Iversen, L. L. (1975). *Behavioural pharmacology*. Oxford: University Press.

Jacob, J. & Barthelemy, C. (1967). Réactivé nociceptive et sensibilité à la morphine de souris de diverses souches. *Thérapie* **22**: 1435–1448.

Jay, G. E. 1955). Variation in response of various mouse strains to hexobarbital (Evipal). *Proc. Soc. exp. Biol. Med.* **90**: 378–380.

Jori, A. & Rutczynski, M. (1978). A genetic analysis of the hyperthermic response to d-amphetamine in two inbred strains of mice. *Psychopharmacology* **59**: 199–203.

Kalow, W. (1962). *Pharmacogenetics. Heredity and the response to drugs.* Philadelphia: W. B. Saunders.

Kier, L. B. & Glennon, R. A. (1978). Psychotomimetic phenalkylamines as serotonin agonists—sar analysis. *Life Sci.* **22**: 1589–1594.

Klein, T. W. & DeFries, J. C. (1970). Similar polymorphism of taste sensitivity to PTC in mice and man. *Nature, Lond.* **225**: 555–557.

Kouri, R. E. (1976). Relationship between levels of aryl hydrocarbon hydroxylase activity and susceptibility to 3-methylcholanthrene and benzo(a)pyrene-induced cancers in inbred strains of mice. In *Carcinogenesis: a comprehensive survey* **1**: 139–151. Freudenthal, R. F. & Jones, P. W. (Eds). New York: Raven Press.

Lapin, I. P. (1975). Effects of apomorphine in mice of different strains. In *Psychopharmacogenetics:* 19–32. Eleftheriou, B. E. (Ed). New York: Plenum Press.

Lovell, D. (1978). The inheritance of barbiturate sleeping time in mice. *Heredity* **41**: 418.

Lush, I. E. (1975). A comparison of the effect of mescaline on activity and emotional defaecation in seven strains of mice. *Br. J. Pharmac.* **55**: 133–139.

Lush, I. E. (1976). A survey of the response of different strains of mice to substrates metabolised by microsomal oxidation; hexobarbitone, zoxazolamine and Warfarin. *Chem. Biol. Interactions* **12**: 363–373.

Lush, I. E. & Andrews, K. M. (1978a). Genetical differences in sensitivity to tremorine and oxotremorine in mice. *Eur. J. Pharmac.* **49**: 95–103.

Lush, I. E. & Andrews, K. M. (1978b). Genetic variation between mice in their metabolism of coumarin and its derivatives. *Genet. Res.* **31**: 177–186.

Lush, I. E. & Arnold, C. J. (1975). High coumarin 7-hydroxylase activity does not protect mice against Warfarin. *Heredity* **35**: 279–381.

Lush, I. E. & Lovell, D. (1978). A correlation between hexobarbitone and phenobarbitone sleeping times in different inbred strains of mice. *Gen. Pharmac.* **9**: 167–170.

Meir, H. (1963). *Experimental pharmacogenetics. Physiopathology of heredity and pharmacological responses.* New York: Academic Press.

Mitoma, C., Neubauer, S. E., Badger, N. L. & Sorich, T. J. (1967). Hepatic microsomal activities in rats with long and short sleeping times after hexobarbital: a comparison. *Proc. Soc. exp. Biol. Med.* **125**: 284–288.

Mirvish, S. S. (1968). The carcinogenic action and metabolism of urethan and N-hydroxyurethan. *Adv. Cancer Res.* **11**: 1–42.

Moisset, B. (1977). Genetic analysis of the behavioural response to d-amphetamine in mice. *Psychopharmacology* **53**: 263–267.

Moisset, B. (1978). Subline differences in behavioural responses to pharmacological agents. In *Origins of inbred mice:* 483–484. Morse, H. C. (Ed.). New York: Academic Press.

Morse, H. C. (Ed.). (1978). *Origins of inbred mice.* New York: Academic Press.

Nebert, D. W., Atlas, S. A., Guenthner, T. M. & Kouri, R. B. (1978). The *Ah* locus: genetic regulation of the enzymes which metabolize polycyclic hydrocarbons and the risk for cancer. In *Polycyclic hydrocarbons and cancer: environment chemistry, molecular and cell biology:* Gelboin, H. V. & Ts'o, P. O. P. (Eds). New York: Academic Press.

Nebert, D. W., Goujon, F. M. & Gielen, J. E. (1972). Aryl hydrocarbon hydroxylase induction by polycyclic hydrocarbons: simple autosomal dominant trait in the mouse. *Nature, Lond.* **236**: 107–110.

Nebert, D. W. & Jensen, N. M. (1979). The *Ah* locus: genetic regulation of the metabolism of carcinogens, drugs and other environmental chemicals by cytochrome P-450-mediated monooxygenases. *C. R. C. Critical Rev. Biochem.* **6**: 401–437.

Nieforth, K. A. (1971). Psychotomimetic phenethylamines. *J. Pharmac. Sci.* **60**: 655–665.

Oliverio, A. (1979). Uses of recombinant inbred lines. In *Quantitative genetic variation:* 197–218. Thompson, J. N. & Thoday, J. M. (Eds). New York: Academic Press.

Oliverio, A. & Castellano, C. (1974). Experience modifies morphine-induced behavioural excitation of mice. *Nature, Lond.* **252**: 229–230.

Oliverio, A., Castellano, C. & Eleftheriou, B. E. (1975). Morphine sensitivity and tolerance: a genetic investigation in the mouse. *Psychopharmacologia* **42**: 219–224.

Oliverio, A. & Eleftheriou, B. E. (1976). Motor activity and alcohol: genetic analysis in the mouse. *Physiol. & Behav.* **16**: 577–581.

Oliverio, A., Eleftheriou, B. E. & Bailey, D. W. (1973a). Exploratory activity: genetic analysis of its modification by scopolamine and amphetamine. *Physiol. & Behav.* **10**: 893–899.

Oliverio, A., Eleftheriou, B. E. & Bailey, D. W. (1973b). A gene influencing activity avoidance performance in mice. *Physiol. & Behav.* **11**: 497–501.

Parke, D. V. (1968). *The biochemistry of foreign compounds.* Oxford: Pergamon Press.

Pelz, W. E., Whitney, G. & Smith, J. C. (1973). Genetic influences on saccharin preference of mice. *Physiol. & Behav.* **10**: 263–265.

Poley, W. (1972). Alcohol-preferring and alcohol-avoiding C57BL mice. *Behav. Genet.* **2**: 245–248.

Propping, P. (1978). Pharmacogenetics. *Rev. Physiol, Biochem. Pharmacol.* **83**: 123–173.

Randall, C. L. & Lester, D. (1974). Differential effects of ethanol and pentobarbital on sleep time in C57BL and BALB mice. *J. Pharm. exp. Ther.* **188**: 27–33.

Rowe, F. P. & Redfern, R. (1968). The effect of Warfarin on plasma clotting time in wild house mice (*Mus musculus* L.). *J. Hyg., Camb.* **66**: 159–174.

Schmid, F. A., Elmer, I. & Tarnowski, G. S. (1969). Genetic determination of differential inflammatory reactivity and subcutaneous tumor susceptibility of AKR/J and C57BL/6J mice to 7,12-dimethylbenz(d)anthracene. *Cancer Res.* **29**: 1585–1589.

Scott, J. P., Lee, C. T. & Ho, J. E. (1971). Effects of fighting, genotype and amphetamine sulphate on body temperature of mice. *J. comp. Physiol. Psychol.* **76**: 349–352.

Shah, N. S. & Himwich, H. E. (1971). Study with mescaline-8-C^{14} in mice: effect of amine oxidase inhibitors on metabolism. *Neuropharmacology* **10**: 547–556.

Shulgin, A. T. (1973). Mescaline: the chemistry and pharmacology of its analogues. *Lloydia* **36**: 46–58.

Shuster, L., Webster, G. W. & Yu, G. (1975). Increased running responses to morphine in morphine pre-treated mice. *J. Pharm. exp. Ther.* **192**: 64–72.

Shuster, L., Webster, G. W., Yu, G. & Eleftheriou, B. E. (1975). A genetic analysis of the response to morphine in mice: analgesia & running. *Psychopharmacologia* **42**: 249–254.

Swank, R. T. & Bailey, D. W. (1973). Recombinant inbred lines: value in the genetic analysis of biochemical variants. *Science, N.Y.* **181**: 1249–1251.

Taylor, B. A. (1971). Strain distribution and linkage tests of 7,12-dimethylbenzathracene (DMBA) inflammatory response in mice. *Life Sci.* **10**: 1127–1134.

Taylor, B. A. (1976a). Linkage of the cadmium resistance locus to loci on mouse chromosome 12. *J. Hered.* **67**: 389–390.

Taylor, B. A. (1976b). Genetic analysis of susceptibility to isoniazid-induced seizures. *Genetics* **83**: 373–377.

Taylor, B. A. (1978). Recombinant inbred strains: use in gene mapping. In *Origins of inbred mice:* 423–438. Morse, H. C. (Ed.). New York: Academic Press.

Taylor, B. A., Heiniger, H. J. & Meir, H. (1973). Genetic analysis of resistance to cadmium-induced testicular damage in mice. *Proc. Soc. exp. Biol. Med.* **143**: 629–633.

Terpstra, G. K. & Raaijmakers, J. A. M. (1976). Loss of an adrenergic effect. *Eur. J. Pharmac.* **38**: 373–376.

Thomas, K. V. & Handley, S. L. (1978a). On the mechanism of amphetamine-induced behavioural changes in the mouse. I. An observational analysis of dexamphetamine. *Drug Research* **28**: 827–833.

Thomas, K. V. & Handley, S. L. (1978b). On the mechanism of amphetamine-induced behavioural changes in the mouse. III. Effects of apomorphine and Fla63. *Drug Research* **28**: 993–997.

Thomas, P. E., Hutton, J. J. & Taylor, B. A. (1973). Genetic relationship between aryl hydrocarbon hydroxylase inducibility and chemical carcinogen induced skin ulceration in mice. *Genetics* **74**: 655–659

Thomas, P. E., Kouri, R. E. & Hutton, J. J. (1972). The genetics of aryl hydrocarbon hydroxylase induction in mice: a single gene difference between C57BL/6J and DBA/2J. *Biochem. Genet.* **6**: 157–168.

Thorgeirsson, S. S. & Nebert, D. W. (1977). The *Ah* locus and the metabolism of chemical carcinogens and other foreign compounds. *Adv. Cancer Res.* **25**: 149–193.

Townsend, M. G., Odam, E. M. & Page, J. M. (1975). Studies of the microsomal drug metabolism system in Warfarin-resistant and -susceptible rats. *Biochem. Pharmac.* **24**, 729–735.

Vesell, E. S. (1968). Factors altering the responsiveness of mice to hexobarbital. *Pharmacology* **1**: 81–97.

Vesell, E. S., Lang, C. M., White, W. J., Passanati, G. T., Hill, R. N., Clemens, T. L., Lin, D. K. & Johnson, W. D. (1976). Environmental and genetic factors affecting the response of laboratory animals to drugs. *Fedn Proc. Fedn Am. Socs exp. Biol.* **35**: 1125–1132.

Wallace, M. E. & MacSwiney, F. J. (1976). A major gene controlling Warfarin-resistance in the house mouse. *J. Hyg., Camb.* **76**: 173–181.

Wang, I. Y., Rasmussen, R. E., Petrakis, N. L. & Wang, A. C. (1976). Enzyme induction and the difference in the metabolite patterns of benzo(a)pyrene produced by various strains of mice. In *Carcinogenesis; a comprehensive survey* **I**: 77–89. Freudenthal, R. F. & Jones, P. W. (Eds). New York: Raven Press.

Warren, R. P. & Lewis, R. C. (1970). Taste polymorphism in mice involving a bitter sugar derivative. *Nature, Lond.* **227**: 77–78.

Waterfield, A. A., Lord, J. A. H., Hughes, J. & Kosterlitz, H. W. (1978). Differences in the inhibitory effects of normorphine and opioid peptides on the responses of the *vasa deferentia* of two strains of mice. *Eur. J. Pharmac.* **47**: 249–250.

Way, E., Loh, H. H. & Shen, F. H. (1969). Simultaneous quantitative assessment of morphine tolerance and physical dependence. *J. Pharmac. exp. Ther.* **167**: 1–8.

Weber, W. W., Miceli, J. N., Hearse, D. J. & Drummond, G. S. (1976). N-acetylation of drugs. Pharmacogenetics studies in rabbits selected for their acetylator characteristics. *Drug Metab. Disp.* **4**: 94–101.

Williams, K. & Nery, R. (1971). Aspects of the mechanism of urethane carcinogenesis. *Xenobiotica* **1**: 545–550.

Wolf, H. H. & Bunce, M. E. (1973). Hyperthermia and the amphetamine aggregation phenomenon: absence of a causal relation. *J. Pharm. Pharmac.* **25**: 425–427.

Wood, A. W. (1979). Genetic regulation of coumarin hydroxylase activity in mice. Biochemical characterization of the enzyme from two inbred strains and their F_1 hybrid. *J. biol. Chem.* **254**: 5641–5646.

Wood, A. W. & Conney, A. H. (1974). Genetic variation in coumarin hydroxylase activity in the mouse (*Mus musculus*). *Science, N.Y.* **185**: 612–613.

Wood, A. W. & Taylor, B. A. (1979). Genetic regulation of coumarin hydroxylase activity in mice. Evidence for single locus control on chromosome 7. *J. biol. Chem.* **254**: 5647–5651.

Yamamoto, R. S., Weisburger, J. H. & Weisburger, E. K. (1971). Controlling factors in urethane carcinogenesis in mice: effect of enzyme inducers and metabolic inhibitors. *Cancer Res.* **31**: 483–486.

Symp. zool. Soc. Lond. (1981) No. 47, 547–574

Genes and Hormones in Mice

JOHN G. M. SHIRE

Department of Biology, University of Essex, Colchester, UK

SYNOPSIS

Genetical variants are known that affect the differentiation of endocrine organs; the synthesis, transport and degradation of hormones; and the action of hormones on their target organs. The ascertainment of mutations that cause pathological defects in endocrine systems is biased away from those affecting critical homeostatic systems. A wealth of variation within the range of normality has been found by comparing inbred strains. Such variants may be present in coadapted combinations. The few studies that there are suggest that the endocrine systems of wild mice are also subject to genetical variation.

Environmental factors interact with genetical variables. The endocrine systems of mice of different genotypes respond differently to factors such as light, diet and pathogens. Other mice form an important part of an individual's environment. Some of the hormonal consequences of situations involving more than one mouse are described. These include crowding, pregnancy, parent–offspring interactions before weaning and social encounters of adults. The existence of lasting maternal effects on the endocrine phenotype of offspring is discussed.

It is concluded that laboratory investigations of the genetics of endocrine systems can already give some insight into the way that homeostatic mechanisms function in natural, polymorphic populations of mice.

INTRODUCTION

Mice are made up of very large numbers of cells of many diverse kinds. Hormones control and regulate these cells so that each mouse functions as an integrated whole. Each endocrine system consists of one or more endocrine units (Fig. 1), each made up of an endocrine tissue and its target organs, and each responsive to control stimuli. Every component part is subject to genetical variation. The following section describes the sources of such variation being investigated in the house mouse and gives examples of some of the variants known. More extensive lists can be found elsewhere of variation affecting the endocrine systems of mice in general (Chai & Dickie, 1966; Shire, 1976) and in particular: anterior pituitary (Bartke, 1979), posterior

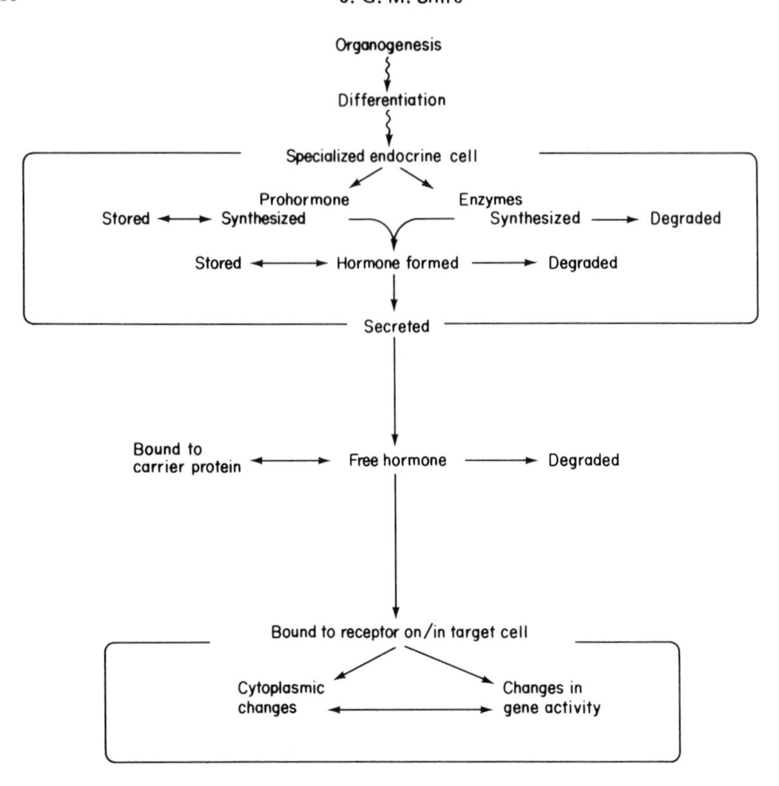

FIG. 1. A generalized scheme of hormone synthesis and action. Reprinted from *Genetic Variation in Hormone Systems,* Vol. I (1979), Shire, J. G. M. (Ed.) with permission of the publishers, CRC Press, Boca Raton, Florida.

pituitary (A. D. Stewart, 1979), adrenal glands (Dunn, 1970; Shire, 1974, 1979a), thyroid (Shire, 1979b), catecholamines (Barchas *et al.*, 1974) and carbohydrate metabolism (Stauffacher *et al.*, 1971). The subsequent sections discuss whether the variants occur at random in laboratory populations and describe the extent of the genotype–environment interactions revealed by studies in controlled environments. The final section considers the importance and difficulty of extrapolating laboratory findings on the biology of homeostatic mechanisms to natural, polymorphic populations.

SOURCES OF VARIATION

Mutations

Over the years mutations with pathological effects have occurred in many mouse colonies. Some of these mutations have been shown to affect endocrine systems. Two recent lists (Beamer, 1979; Shire, 1979c) implicate 21 and 35 loci. Undoubtedly there are more loci with effects on the endocrine system waiting to be recognized, even amongst loci already named.

Mutations affecting most of the components shown in Fig. 1 are known, though only a few for each individual endocrine unit.

Testicular feminization, *Tfm*, is the most striking of the mutations that affect the responsiveness of target organs (Lyon & Hawkes, 1970; Bullock, 1979). In *Tfm*/Y mice androgen-responsive tissues such as kidney, brain and submandibular gland contain only traces of the cytoplasmic receptor proteins that bind testosterone and consequently these tissues do not respond to the administration of testosterone. The seminal vesicles, epididymis and Wolffian duct derivatives, which in normal males are most responsive to dihydrotestosterone, fail to develop (Goldstein & Wilson, 1972).

Mice having the X-linked hypophosphataemia mutation (*Hyp*) show disturbances of calcium and phosphorus metabolism and have skeletal abnormalities (Eicher *et al.*, 1976). The mice are not cured either by vitamin D or its natural metabolites, but the genetical lesion can be overcome by treatment with the synthetic analogue 1-α-hydroxy vitamin D (Beamer, Wilson & DeLuca, in press). This situation may be analogous to that found for the induction of arylhydrocarbon hydroxylase by dioxin. This compound induces hydroxylase activity in DBA mice even though such mice do not respond to methylcholanthrene and benzanthracene, which are the normal inducers of this enzyme (Poland *et al.*, 1974).

Mutations affecting target-organ responsiveness have been found in mouse cells in culture. These include variants affecting corticosteroid receptors (Yamamoto *et al.*, 1976) and receptors for those hormones that act by activating adenylcyclase (Coffino *et al.*, 1976).

Mutations affecting the production of hormones are well exemplified by those that are known to affect pituitary function. Both pituitary dwarf mutations, *dw* and *df*, affect the differentiation of acidophil cells of the pituitary and thus cause a deficiency of both growth hormone and prolactin (Bartke, 1964, 1979). The lesion caused by the little mutation, *lit*, produces a primary deficiency of growth hormone alone (Beamer & Eicher, 1976; Beamer, in press). The levels of gonadotrophic hormones are increased four-fold in W^x/W^v mice relative to their litter mates because of the failure of feedback inhibition in the mutant mice (Murphy & Beamer, 1973). Gonadotrophin production is negligible in hypogonadal, *hpg*, homozygotes because of the failure of the hypothalamus to stimulate the pituitary with the appropriate releasing hormone (Cattanach *et al.*, 1977). Hypothalamic defects may be involved in lethargic mice (*lh*) whose adrenals are overactive when the mice are young (Dung & Swigart, 1971) and petite (*pet*) mice, which grow and reproduce normally after treatment with thyroxine (Southard & Eicher, 1977; W. G. Beamer, personal communcation).

A mutation segregating in a stock of animals may reveal the presence of genetical variation that would otherwise remain hidden (Rendel, 1979). The phenotypes of both the diabetes (*db*) and obese-hyperglycemic (*ob*) mutations were found to be much more severe on the C57BL/KsJ background than on the related C57BL/6J background (Hummel, Coleman & Lane, 1972; Coleman & Hummel, 1973). This indicates that the two strains differ at a locus or loci that act as modifiers of both these mutations. These modifier genes may cause quantitative, or "polygenically-determined" (Thompson, 1975), differences in carbohydrate metabolism between the two kinds of C57 mice.

Laboratory Stocks

Much genetical variation, producing phenotypes all of which are viable and reasonably fertile under laboratory conditions, has been discovered by comparing different laboratory stocks. Marked quantitative differences have been found between inbred strains for many endocrine parameters. The variation in size of endocrine organs is considerable, as can be seen for the relative weight of the testis in Fig. 2. The factors causing the relatively small testes of C57BL/1O mice are autosomal (Shire & Bartke, 1972), and some of them may be in the *H-2* region (Iványi, Gregorova & Micková, 1972). In contrast, factors on the Y chromosome are responsible for much of the difference in testis size between CBA mice and those of other inbred strains (Hayward & Shire, 1974). Adrenal weight varies threefold over strains in both males and females, and as much as five-fold in young adult females when selection lines are included (Fig. 3). Some genetical factors affect males and females differently, causing the points in the figure to be dispersed. The pituitary dwarf mutation, *dw*, has a marked effect on the adrenal weight of females but none on those of males (Shire & Hambly, 1973).

Table I shows which components of three steroid hormone systems are known to differ between strains and gives some of the references to those differences. In 17 cases such strain differences have been shown to be caused by variation at a single locus (Shire, 1979c). Eight of these loci have been mapped, often by the use of

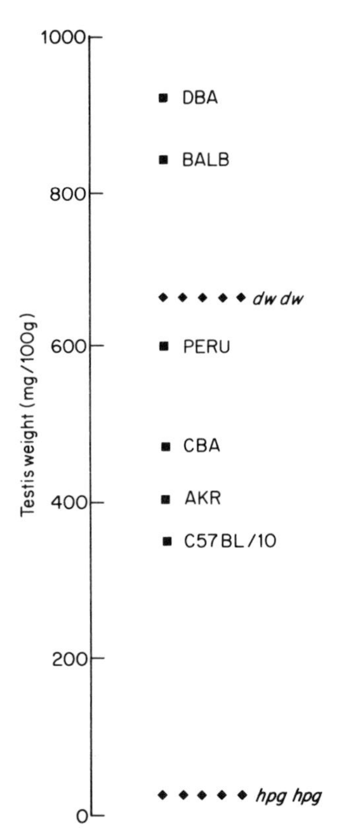

FIG. 2. Relative testis weight, in mg per 100 g body weight, for some mutants and inbred strains. Data from Cattanach *et al.*, 1977; Shire & Bartke, 1972; Shire & Hambly, 1973.

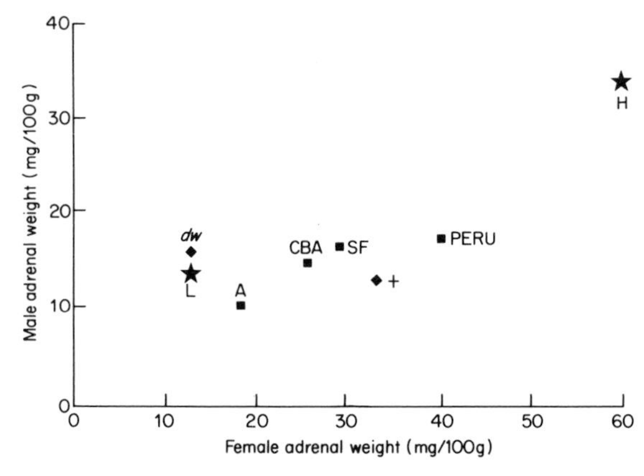

FIG. 3. Relative adrenal weight, in mg per 100 g body weight, for male and female mice aged eight weeks. The points refer to four strains, a pair of lines (H & L) at generation 32 of selection for adrenal weight derived from these strains, and homozygotes for pituitary dwarfism (*dw*) and their normal sibs (+).

TABLE I

Strain differences in steroid hormone systems

	Androgens	Oestrogens	Corticosteroids
Size of endocrine tissue	Fig. 2; Curé, Valatx & Delobel (1971).	Taylor & Waltman (1940). Mos, Vriend & Poley (1974).	Fig. 3; Shire (1974).
Precursor stores	Bartke & Shire (1972) Bartke, Weir *et al.* (1974).	—	Doering, Kessler & Clayton (1970). Doering, Shire *et al.* (1973).
Biosynthetic pathway	—	—	Badr & Spickett (1965). Hawkins *et al.* (1975).
Steroid output	Bartke, Roberson & Dalterio (1977).	—	Doering, Shire *et al.* (1972).
Plasma levels	Bartke (1974).	—	Levine & Treiman (1964). Eleftheriou & Bailey (1972). Shire (1979a).
Steroid degradation	Ford, Lee & Engel (1979).	Pelkonen, Boobis & Nebert (1978).	Lindberg *et al.* (1972) Shire (1979a).
Steroid receptors	—	Sato *et al.* (1979).	Goldman *et al.* (1977). Butley, Erickson & Pratt (1978).
Target-organ responses	Bartke & Shire (1972). Stylianopolou & Clayton (1976). Wilson, Erdos, Dunn & Wilson (1977). Gregorova *et al.* (1977).	Claringbold & Biggers (1955). Westburg, Bern & Barnawell (1957). Stylianopoulou & Clayton (1976).	Wragg & Speirs (1952). Eleftheriou (1974).

TABLE I (*continued*)

Strain differences in steroid hormone systems

	Androgens	Oestrogens	Corticosteroids
Behavioural responses	McGill (1970). McGill & Manning (1976). Vale, Ray & Vale (1974).	Vale, Ray & Vale (1973). Gorzalka & Whalen (1974).	Levine & Levin (1970).

recombinant inbred strains (Oliverio, 1979). Ten strain differences that affect other endocrine systems have also been shown to be monogenically inherited, and two of the loci mapped. Differences found by comparing closely-related sublines, such as those in the turnover of the enzymes that synthesize catecholamines (Ciaranello, Lipsky & Axelrod, 1974) and that in the incidence of true hermaphroditism (Beamer, Whitten & Eicher, 1978), should also have a relatively simple genetical basis.

Heterogeneous stocks of mice have been successfully selected for a number of endocrine parameters. These include granularity of the renal juxtaglomerular apparatus, which secretes renin (Rapp, 1965, 1967), adrenal weight (Badr & Spickett, 1971), and several aspects of gonadal function. Mice have been selected for early and late sexual maturation (Bartke, Weir *et al.*, 1974), for oestrogen responsiveness (Biggers & Claringbold, 1955), for large and small testes (Islam, Hill & Land, 1976) and for ovulation rate (Land & Falconer, 1969). In these last two selection experiments differences in testis size were positively correlated with differences in ovulation rate, suggesting that selection in either sex was picking out genes affecting gonadal function in both sexes (Land, 1973; Islam *et al.*, 1976). The effect of the pituitary dwarfism mutation on body weight was greater when it was crossed into a line of mice selected for large size than when it was crossed into the corresponding low line. The selected lines were also found to differ in their growth responses to exogenous bovine growth hormone, the low line being least responsive (Pidduck & Falconer, 1978).

Wild Mice

Very few studies have been made of genetical variation affecting endocrine function in wild mice or in laboratory stocks derived from feral mice. Mice of the SF/Cam strain, inbred from wild mice captured near San Francisco, have adrenals in which the X-zone is relatively well developed (Shire & Spickett, 1968) and the glomerular zone is difficult to distinguish (Shire & Spickett, 1967). The zona glomerulosa is also poorly developed in mice of the Peru-Atteck stock but the locus involved, *ezg*, is not the same as the one in SF/Cam mice (Shire & Spickett, 1967). The zona glomerulosa is generally considered to synthesize aldosterone, which is an important mineralocorticoid hormone. The regulation of electrolyte metabolism in Peru mice is unaffected by the administration of either aldosterone (J. Stewart, 1975) or an aldosterone antagonist (J. Stewart, 1969). Less tritiated aldosterone was bound to the nuclei of kidney cells from Peru mice than was bound to nuclei from the kidneys of CBA mice (J. Stewart, 1975). It is surprising, therefore, that Peru mice synthesize aldosterone and have plasma levels of this hormone that are similar to those found in CBA mice (J. Stewart *et al.*, 1972). They also regulate the level of this apparently inactive compound in response to variations in the sodium content of the diet (Papaioannou & Fraser, 1974). The zona glomerulosa can be made to disappear in CBA mice by administering large doses of corticotrophin (ACTH). Distinct glomerular cells appeared in the adrenals of Peru mice in which ACTH levels had been reduced by treatment with dexamethasone (Shire & Stewart, 1972). The *ezg* phenotype, which may be an indicator of high levels of ACTH, is associated in the Peru mice with adrenal glands in which both the zona fasciculata and the X-zone are large. The correlation between adrenal size and *ezg* phenotype broke down when segregating generations were studied, implying that Peru and CBA mice differ at several loci affecting adrenal function (Shire & Stewart, 1972). Levine & Treiman (1969) found the rise in plasma corticosterone following stress to be much higher in feral house mice that had been trapped in California than in mice of the laboratory strain, C57BL/10J, with the most pronounced stress response. Mice born to wild Canadian house mice and then fostered onto laboratory mice had much larger adrenals in both males (41 mg $(100\text{ g})^{-1}$) and females (70 mg $(100\text{ g})^{-1}$) than did mice of a Swiss strain that had been comparably housed. When superfused with a solution containing ACTH, adrenals from the male wild mice produced corticosterone at a significantly greater rate, per mg adrenal, than adrenals from Swiss males (Seabloom & Seabloom, 1975). The comparison of the females is complicated by the possibility of differences in the amount and developmental state of the X-zone.

Genetical variation has been shown to affect other endocrine systems in wild mice. Peru mice have relatively low levels of thyroxine in plasma but a relatively high basal metabolic rate (A. D. Stewart, Batty & Harkiss, 1977). Peru mice also have 8-lysine vasopressin as their antidiuretic hormone, in contrast to the 8-arginine vasopressin found in laboratory strains (A. D. Stewart, 1971, 1979). Differences have been found between CBA and Peru mice in their responses to these hormones (see p. 557). The offspring of wild Texan mice maintained in a standard animal house compared with the offspring of CF-1 laboratory mice show plasma levels of the gonadotrophins FSH and LH about two-and-one-half times higher than laboratory mice. Reproduction was markedly suppressed by high light intensities in the wild mice but not in the laboratory mice (Bronson, 1979).

Apart from a successful study of differences in the inducibility of renal glucuronidase by testosterone (Chapman, 1978) few comparisons have been made between mice of different subspecies. Measurements of selected endocrine parameters on the F_1 progeny of wild *Mus musculus domesticus, M.m. musculus* and *M.m. molossinus* crossed to standard laboratory strains would be interesting, as would studies of differences in the severity of the phenotype of major mutants transferred into these subspecies. The effects of substituting individual chromosomes from *Mus poschiavinus* for the corresponding chromosomes of laboratory *Mus musculus* have not yet been looked at but could prove rewarding.

Uses of Endocrine Variation

Variants in endocrine function, whether originating in the wild or in the laboratory, can be used in several ways in experimental endocrinology. Such uses have been reviewed elsewhere (Thoday, 1967; Shire, 1974, 1976, 1979d) and include the production of animal models of human disease (Bartke, 1979), and optimal experimental material for bioassay, or when the absence of or insensitivity to a particular hormone is required. Knowledge of the forms that genetical variation actually takes can give insight into the way that the endocrine system functions, while studies on post-segregational generations can elucidate the true nature of the relations between variables that appear to be correlated when genetically uniform stocks are compared.

IS THE VARIATION RANDOM?

Sources of Bias

Mutants that are viable but have externally visible pleiotropic effects, such as dwarfing or skeletal changes or obesity, are most likely to be noticed. Some mutations resulting in sterility will be detected. Many mutations that are homologous to known human endocrine defects, such as congenital goitre and adrenogenital virilism, are likely to cause rather unspecific pre-weaning lethality and so are difficult to detect and identify. Differences between inbred strains will be restricted to those that do not reduce viability and fertility too much, at least within the relatively untaxing environment of an animal house.

The genetical variants known are bound to represent a biased sample of those that exist because of the influence of scientific fashion in concentrating attention on particular systems or chromosomal regions. Several loci with endocrine effects have been located near *H-2*. We are not yet in a position to determine whether this represents gene clustering or is merely a consequence of the availability of strains congenic for the *H-2* region. The interests of individual workers in particular systems will lead to an uneveness of knowledge about endocrine systems as a whole. Similarly stocks of mice differ greatly in their availability and popularity. This has resulted in a great deal of knowledge about a few strains and the neglect of other interesting strains, such as the *H* line of Furtado Dias (1959) which had a high frequency of testicular and mammary tumours in *males*.

Coadapted Combinations

Arginine vasopressin is more potent as an antidiuretic hormone than lysine vasopressin in most mammals. Figure 4 shows that the two hormones were equally potent in *in vivo* assays in Peru mice and that arginine vasopressin was significantly more potent in assays in CBA mice (A. D. Stewart, 1973, 1979). Thus the renal target organs of Peru mice appear adapted to the kind of vasopressin produced by their pituitaries as do CBA mice and pigs, which also produce lysine vasopressin. Mice of the A strain have smaller adrenals than do CBA

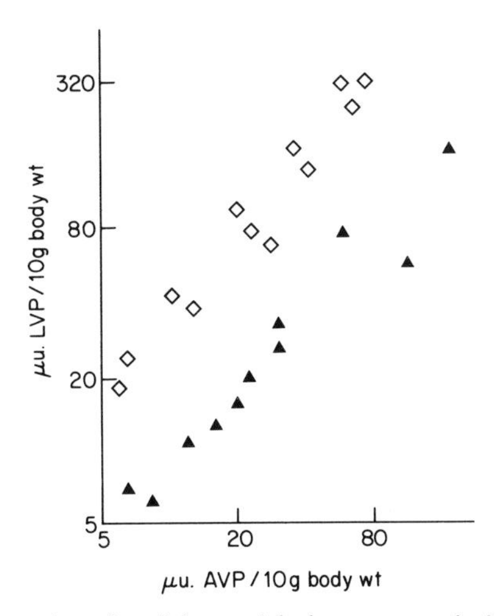

FIG. 4. The relative potencies of arginine and lysine vasopressin in Peru (▲) and CBA (◇) mice. Each point represents the doses of the two hormones that produced the same antidiuretic effect. Redrawn from A. D. Stewart (1973) with permission of the author and publisher.

mice. This difference in the number of adrenocortical cells is compensated for by a higher biosynthetic activity of these cells, leading to similar outputs of hormone in both genotypes (Spickett, Shire & Stewart, 1967; Shire, 1979a). The existence of coadapted systems may only be revealed when they are upset in out-crosses. Carbohydrate metabolism is regulated effectively in both C3HF and I strain mice, but the genes involved must differ as the F_1 hybrids between these two strains develop diabetes mellitus (Jones, 1964; Stauffacher *et al.*, 1971).

Although AKR and DBA mice differ strikingly in gonadal size (Fig. 1), the adrenocortical physiologies of these two strains are practically indistinguishable. Both strains show only a small adrenocortical response to stress (Levine & Treiman, 1964). The corticosterone output *in vitro* by adrenals from both strains is relatively low, as are the cholesterol ester stores in adult mice (Doering, Kessler & Clayton, 1970). In both strains these stores of hormone precursors rise after castration and can be made to fall by injections of testosterone (Arnesen, 1974; Doering, Shire *et al.*, 1972; Shire, 1979a). Crosses between the strains showed, by the recovery of significant numbers of mice with stores of hormone precursor (Fig. 5), that the genetical basis of the apparently similar phenotypes must differ between the strains (Doering, Shire *et al.*, 1973).

DBA mice differ from C57BL mice in the output of corticosterone by adrenal cells and adrenal stores of cholesterol esters (Doering, Shire *et al.*, 1972; Shire, 1979a). A difference between the strains in the rate of catabolism of corticosterone counteracts the differences in hormone production and results in similar circulating levels of

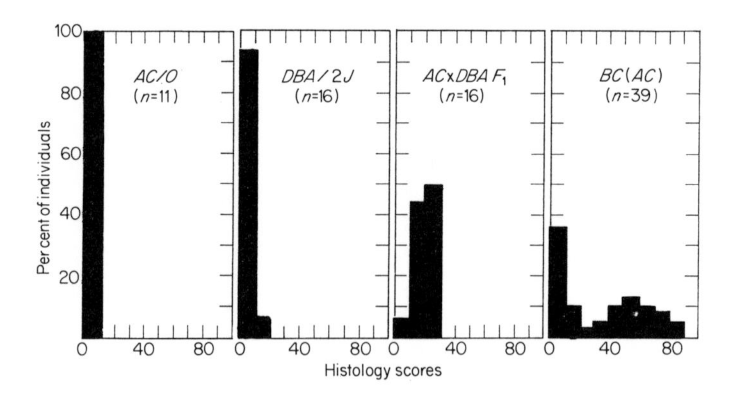

FIG. 5. The extent of sudanophilia, a measure of cholesterol ester stores, in the adrenals of mice of the DBA and AC strains, their F_1, and the back-cross to AC. Reprinted from Doering, Shire *et al.* (1973) with permission.

corticosterone in both strains under resting conditions (Levine & Treiman, 1964; Lindberg *et al.*, 1972). Exposure to stress, however, results in a much larger and more prolonged rise in corticosteroids in C57BL mice than in DBA mice (Levine & Treiman, 1964; Shire, 1979a). Although the passive avoidance behaviour of DBA mice appears to be unaffected by exogenous corticosteroids (Levine & Levin, 1970), the induction by such hormones of tyrosineamino-transferase in liver (Monroe, 1968), phenylethanolamine-*N*-methyl transferase in adrenal medulla (Ciaranello, Dornbusch & Barchas, 1972) and cleft palate in foetuses (Salomon & Pratt, 1979) is greater in DBA mice than in C57BL mice. The small adrenal reponse to stress, however, results in a low spontaneous rate of foetal cleft palate (Salomon & Pratt, 1979), a fall in the levels of tyrosine-aminotransferase during starvation (Blake, 1970) and little disruption of spatial discrimination by ether stress (Elias & Simmerman, 1971). A low spontaneous rate of cleft palate is also found in C57BL mice, despite their marked adrenal response to stress, at least in part because of the relatively low level of corticosteroid receptors found in foetal facial mesenchyme (Goldman *et al.*, 1977; Salomon & Pratt, 1979). Similarly, exogenous corticosteroids produce only a low frequency of palatal malformations in C57BL mice. Glucocorticoid receptors are also present at relatively low levels in the liver of C57BL mice (Butley, Erickson & Pratt, 1978) and tyrosineamino-transferase is relatively poorly induced by exogenous corticosteroids (Monroe, 1968). The stress of starvation is, however, sufficient in C57BL mice to increase tyrosineaminotransferase levels in the face of falling body weight (Blake, 1970). Similarly the response of C57BL mice to the stress of surgery was sufficiently large to increase their survival after exposure to 650 rad of γ-rays, while a similar stress did not increase the survival of DBA mice significantly (Pachiarz & Teague, 1975). Both strains have adrenocortical systems that regulate metabolism effectively under laboratory conditions, but in quantitatively different ways. The causes of mortality differ between the strains and may not be unrelated to adrenocortical function during life. C57BL mice suffer from autoimmune diseases (Pantelouris, 1974), which are associated with hypofunction of the immune system, perhaps as a result of immunosuppression by corticosteroids. DBA mice, and also AKR mice, die of neoplastic diseases in which a contributory factor may be a relative deficiency of anti-mitotic corticosteroids.

The fitness of breeding pairs must be sufficient for an inbred strain to survive. The reproductive phenotype of C57BL males, particularly of the 6 and 10 sublines, is such that C57BL strains would be in danger of extinction unless the genes causing the male phenotype were compensated for by loci elswhere in the genome. C57BL males are poor breeders in out-crosses and delay the onset of cleavage in all of their offspring (Shire & Whitten, 1980a). C57BL mice have small testes (Fig. 1; Shire & Bartke, 1972) and produce a high proportion of morphologically abnormal spermatozoa (Krzanowska, 1970). The levels of androgens in plasma (Bartke, 1974) and luteinizing hormone in both plasma and pituitary are low (Bartke, Roberson & Dalterio, 1977; Selmanoff, Goldman & Ginsberg, 1977; Sustarsic & Wolfe, 1979). The pituitaries of C57BL males contain relatively few gonadotrophs (Håkansson & Lundin, 1975) and are relatively unresponsive to luteinizing hormone releasing hormone from the hypothalamus (Sustarsic & Wolfe, 1979). The absence of pituitary or adrenal hyperplasia following neonatal castration in C57BL males (Woolley, 1950) also points to altered hypothalamic function. As a consequence of low androgen output by the testes several androgen target organs, including the kidneys and submandibular glands (Bartke & Shire, 1972), show little sexual dimorphism in C57BL mice. C57BL males normally produce negligible amounts of the pregnancy-blocking pheromone but can be made into effective producers of the pheromone by injections of testosterone (Hoppe, 1975). Exogenous testosterone increases β-glucuronidase levels in the kidneys of C57BL mice, but to a lesser extent than in other strains. The locus responsible, *Gur*, is on chromosome 5 (Paigen *et al.,* 1975). The renin activity of the submandibular gland is readily inducible by testosterone in most strains of mice, but not in C57BL mice (Wilson, Erdos, Dunn & Wilson, 1977). The locus responsible for this difference, *Rnr*, is on chromosome 1 (Wilson, Erdos, Wilson & Taylor, 1978). The adrenal stores of cholesterol ester are not depleted by testosterone in C57BL males (Stylianopolou & Clayton, 1976). The brain is also relatively unresponsive to androgen in C57BL males; aggression (Legrand, 1970) and sex drive are low and the mice require a long period for recovery between bouts of sexual activity (McGill, 1970; McGill & Manning, 1976). The low frequency of mating may compensate in part for the low rate of production of spermatozoa. The seminal vesicles of C57BL mice, perhaps because of their importance for the success of reproduction in the male, are not reduced in size (Bartke & Shire, 1972; Sustarsic & Wolfe, 1976). They are more responsive to testosterone than are the seminal vesicles of DBA mice (Bartke, 1974) perhaps because of the synergistic effects of prolactin (Barkey *et al.,* 1979) which is present at relatively high levels in the plasma of C57BL males (Sinha & Baxter, 1979; Sinha, Baxter & Vanderlaan, 1979).

In contrast to C57BL males, C57BL females breed well in outcrosses. Their eggs are fertilized rapidly (Braden, 1958; Krzanowska, 1972; Nicol & McLaren, 1974) and begin cleavage soon afterwards (McLaren & Bowman, 1973; Shire & Whitten, 1980b). Their pregnancies are not disturbed by the presence of strange males (Chapman & Whitten, 1969). The pituitaries, hypothalamus (Gardner & Strong, 1940; Gorzalka & Whalen, 1974), vagina (Claringbold & Biggers, 1955) and adrenals (Stylianopolou & Clayton, 1976) of C57BL females are very responsive to oestrogen. Thus the poor reproductive success of C57BL males appears to be compensated for, within the inbred lines, by the breeding performance of the females. The way in which the C57BL genotype came into being could be studied by comparing sublines separated at different times. The testes are larger in sublines, such as C57BL/Ka and C57BL/Tb, that diverged early from the C57BL/6 and 10 mainline.

INTERACTIONS WITH THE ENVIRONMENT

Physical Factors

House mice have successfully adapted to life in cold stores at −10°C (Laurie, 1946; Barnett, 1965) and to life at environmental temperatures above 34°C (Pennycuik, 1966, 1967, 1969). The adaptation of these populations must have involved considerable selection for genetical variants resulting in appropriate changes in thyroid function and the metabolism of carbohydrate and fat. Light is another factor that is known to affect the endocrine systems of mice, and to affect different genotypes differentially. High light intensity reduced the reproductive success of wild mice from Texas, but not that of CF-1 laboratory mice (Bronson, 1979). Variation in light intensity affected the weights of endocrine tissues differently in C57BL and BALB mice (Mos, Vriend & Poley, 1974). Immature BALB females ovulate in response to a bright flash of light, if appropriately primed with follicle-stimulating hormone, whilst immature C57BL females do not (Eleftheriou & Kristal, 1974). The gene responsible for this difference between these two strains is on chromosome 4.

Diet

The different responses of the activity of liver enzymes to starvation in DBA and C57BL mice have already been noted (p. 559; Blake,

1970). Variation in the fat content of the diet also affects mice of these two strains differently, even when the amount of food is not restricted. Corticosterone catabolism is increased in C57BL mice on a fat-free diet but is unchanged in DBA mice (Shire, 1980). A diet high in fat increased the susceptibility of C57BL mice to cleft-palate induced by cortisone, perhaps by reducing the rate of breakdown of this steroid (Miller, 1977). A fatty diet had different effects on both skeletal growth and adipose tissue in DBA and C57BL mice. Only the C57BL mice became very fat, and only they suffered from osteoarthritis (Silberberg & Silberberg, 1950). Nayudu & Moog (1967) demonstrated strain differences in the level of alkaline phosphatase, an enzyme induced by corticosteroids, in the intestine.

Alcohol acts as a stressor for DBA mice, producing a greater rise in plasma corticosterone in these mice than it does in C57BL mice (Kakihana, Noble & Butte, 1968), which prefer to drink 10% ethanol rather than water. Some plants, such as brassicas, contain goitrogens which interfere with the function of the thyroid gland. In the laboratory, C57BL mice are highly resistant to a range of goitrogenic chemicals (Gorbman, 1947; Shire, 1979b). Wild mice with such a genotype might be able to eat a wider range of foodstuffs than could other mice, leading to an advantage in times of food shortage. The pathways and rates of metabolism of xenobiotic chemicals, such as drugs and carcinogens, frequently depend on the hormonal status of the mice (Pelkonen, Boobis & Nebert, 1978; Lush, this volume p. 517). Thus their effects may differ from mouse to mouse, according to their genetically determined endocrine constitution.

Pathogens

When mice of certain genotypes become infected with encephalocardiomyelitis virus their pancreatic islets are destroyed and they develop diabetes mellitus. The islet cells of other mice, including C57BL, are resistant to the damaging effects of this virus (Yoon *et al.*, 1976). Large doses of dexamethasone, a synthetic corticosteroid, induce C-type virus particles in the pancreas of C57BL mice, but not in those of BALB or C3H mice (Boiocchi, della Torre & della Porta, 1975). C3H and C57BL mice differ in the plasma levels and turnover of prolactin. Part of the difference between the strains is caused by the presence of mammary tumour viruses in the C3H mice, but the rest must be due to other differences between the strains because they persisted when animals free of virus were compared (Sinha, Salocks *et al.*, 1977; Sinha & Baxter, 1979). Differences in the metabolism of growth hormone and prolactin, and in their target organs the mammary glands, are associated with susceptibility to mammary tumours caused by milk-borne virus (Nandi & Bern, 1960; Yanai & Nagasawa, 1968; Bartke, 1979).

Differences in the immune system also result in differences in infection by micro-organisms. Infected animals will show endocrine changes not found in the resistant animals. Conversely individuals already under physiological stress will be more susceptible than mice whose homeostatic systems are not as severely taxed (Christian & Davis, 1964). Similarly genetical differences in susceptibility to protozoan and nematode parasites (Ali-Khan, 1974; Wakelin, 1975; Morrison & Roelants, 1978) will lead to differences in endocrine phenotype.

Social Factors

Other mice form an important part of the environment of every mouse. Many social encounters have endocrine consequences which may be beneficial or harmful to one or both of the participants. Crowding mice together stresses them and reduces their reproductive success (Christian & Davis, 1964). Increased population density affects mice of different genotypes in different ways, some strains showing marked responses in the adrenal X-zone (Shire, 1968) and in adrenal and testicular weight (Thiessen, 1964), whilst mice of other strains showed little apparent change. Even the change from a mouse being alone to a situation where it has a companion has different effects on different genotypes (Simmel & Walker, 1972). Aggressive encounters can lead to stress and tissue damage (Levin, Vandenbergh & Cole, 1974), and are in part controlled by the individuals' endocrine and neuroendocrine status (McGill, 1970; Vale, Ray & Vale, 1973; Ciaranello, Lipsky *et al.*, 1974).

Reproduction involves interactions between male and female and between parent and offspring. In these situations pheromones are important. Mice differ greatly in their ability to produce and respond to pheromones (Chapman & Whitten, 1969; Zarrow, Christenson & Eleftheriou, 1971; Hoppe, 1975). The responses of some genotypes are abolished if the animals are even mildly stressed (Bruce, Land & Falconer, 1968). Whilst most female mice ovulate spontaneously, ovulation can be induced in BALB/cWt and BALB/cBy females within hours of pairing with a male (Champlin *et al.*, 1980). Studies on recombinant-inbred strains showed that this characteristic of BALB/c mice could be separated from the induction of ovulation by flashes of light also found in this strain (p. 561).

Lactation is both strongly dependent on maternal hormones and of great importance for the growth of the next generation. There are notable differences between strains in the lactational performance of females (Nagasawa, Kanzawa & Kuretani, 1967) and in its hormonal control (Nandi & Bern, 1960; Sinha, Baxter *et al.,* 1979). Before birth maternal—foetal interactions are affected by genetical variation expressed in the mother and by variation expressed in the foetus. Maternal metabolism modulates the influence of maternal hormones on the foetus, and also the influence of foetal hormones, such as placental lactogen, on their target organs in the mother. Conversely foetal genotype determines the uptake and fate of the maternal hormones that can cross the placenta. The relations of inherited foetal and maternal variation to differences in the metabolism of corticosteroids are discussed by Nguyen-Trong-Tuan, Rekdal & Burton (1971), Wong & Burton (1971) and Salomon & Pratt (1979). Prenatal exposure to corticosteroids, catecholamines or stress has been shown to produce permanent changes in the behaviour of the offspring (Keeley, 1962; Lieberman, 1963; Belyaev, Schuler & Borodin, 1977). The magnitude, and in some cases the direction, of the changes depend on the strain of mouse (Thompson & Olian, 1961; Simeonsson & Meier, 1970). Similarly postnatal effects, acting through maternal handling (Ressler, 1962) and ultrasonic communication (Bell, Nitschke & Zachman, 1972), can modify the endocrine phenotype of the offspring in ways that depend on the maternal genotype.

The endocrine consequences of interactions between individuals are more complicated in out-crosses and natural populations than within inbred strains, for the genotypes, and consequently the phenotypes, of the interacting individuals may differ. The same individual may interact, on different occasions, with mice of several different genotypes. Such interactions have been found in studies on the effects of handling on dominance relationships (Porter, 1972), in studies of the effects of *H-2* haplotypes on mating behaviour (Yamazaki *et al.,* 1978), and in studies of the pheromones involved in aggression (Jones & Nowell, 1974) and reproduction (see above). In both cases C57BL males did not trigger the responses that other males did. Differences in histocompatibility type between mother and offspring can affect the size of the placenta (Beer & Billingham, 1976), an important site of hormone synthesis. Some of the complexities that arise when mother and foetus are of different genotypes are discussed, in the context of susceptibility to cleft palate induced by corticosteroids, by Francis (1973) and Biddle & Fraser (1977).

Maternal influences affecting the endocrine phenotype of genetically distinct foetuses can sometimes be detected when differences are found between females from reciprocal F_1 crosses between inbred strains. Adrenal function appears to be depressed in the offspring of females from strains, such as C57BL, with active adrenocortical systems and to be stimulated in the offspring of females from strains, such as DBA, that have relatively hypofunctional adrenals. Characters showing evidence of this effect include adrenal weight (Meckler & Collins, 1965), X-zone involution (Daughaday, 1941), and plasma corticosterone levels after stress (Treiman, Fulker & Levine, 1970). The phenotypes of males from reciprocal F_1's and reciprocal back-crosses between DBA and C57BL suggest a similar maternal influence, partially counteracting the effect of the nuclear genome, on corticosteroid production by adrenal cells in culture (P. R. W. Wood, personal communcation; Wood, 1977). Imprinting of young animals plays a part in the determination of behaviour and endocrine function, but the extent to which genetical variation in both parents and offspring modulates the process has yet to be determined.

CONCLUSIONS

The preceding sections have shown the extent of genetical variation affecting endocrine function, have demonstrated the importance of interactions between genetical and environmental variables, and have suggested that the inheritance of physiological characters may have a non-nuclear, social component. Neither the genetical constitution nor the environment of any natural population is constant. The size of natural populations can also change dramatically, either seasonally or cyclically, as on Skokholm, or catastrophically, as in the mouse plagues that occur occasionally in Australia. The individual alleles at particular loci carried by individual wild mice will need to be identified. For endocrine characters this will have to be done either by progeny testing with well-defined laboratory stocks or by detecting clear-cut differences in proteins or in the sequence of nucleic acids.

ACKNOWLEDGEMENTS

I should like to thank Peter R. W. Wood for interesting discussions about coadapted genes and maternal effects and Mrs G. Morrison for typing the manuscript.

REFERENCES

Ali-Khan, Z. (1974). Parasite biomass and antibody response in three strains of inbred mice against graded doses of *Echinococcus multilocularis*. *J. Parasit.* **60**: 231–235.

Arnesen, K. (1974). Adrenocortical lipid depletion and leukemia in mice. *Acta path. microbiol. scand.* (A) Supplement **248**: 15–19.

Badr, F. M. & Spickett, S. G. (1965). Genetic variation in the biosynthesis of corticosteroids in *Mus musculus*. *Nature, London* **205**: 1088–1090.

Badr, R. M. & Spickett, S. G. (1971). Genetic variation in adrenal weight in young adult mice. *J. Endocr.* **49**: 105–111.

Barchas, J. D., Ciaranello, R. D., Dominic, J. A., Deguchi, T., Orenberg, O., Renson, J. & Kessler, S. (1974). Genetic aspects of monoamine mechanisms. *Adv. biochem. Psychopharm.* **12**: 195–204.

Barkey, R. J., Shari, J., Amit, T. & Barzilai, D. (1979). Characterization of the specific binding of prolactin to binding sites in the seminal vesicle of the rat. *J. Endocr.* **80**: 181–189.

Barnett, S. A. (1965). Adaptation of mice to cold. *Biol. Rev.* **40**: 5–51.

Bartke, A. (1964). Histology of the anterior hypophysis, thyroid and gonads of two types of dwarf mice. *Anat. Rec.* **149**: 225–236.

Bartke, A. (1974). Increased sensitivity of seminal vesicles to testosterone in a mouse strain with low plasma testosterone levels. *J. Endocr.* **60**: 145–148.

Bartke, A. (1979). Genetic models in the study of anterior pituitary hormones. In *Genetic variation in hormone systems* **1**: 113–126. Shire, J. G. M. (Ed.). Boca Raton, Florida: CRC Press.

Bartke, A., Roberson, C. & Dalterio, S. (1977). Concentration of gonadotrophins in the plasma and testicular responsiveness to gonadotrophic stimulation in androgen deficient C57BL/10J mice. *J. Endocr.* **75**: 441–442.

Bartke, A. & Shire, J. G. M. (1972). Differences between mouse strains in testicular cholesterol levels and androgen target organs. *J. Endocr.* **55**: 173–184.

Bartke, A., Weir, J. P., Mathison, P., Roberson, C. & Dalterio, S. (1974). Testicular function in mouse strains with different age of sexual maturation. *J. Hered.* **65**: 204–208.

Beamer, W. G. (1979). Mutant genes with endocrine effects: mouse. In *Biological handbooks: Inbred and genetically defined strains of animals. Part 1: Mouse and rat:* 101–102. Altman, P. L. & Katz, D. D. (Eds). Bethesda, Maryland: Fedn Am. Socs exp. Biol. Med.

Beamer, W. G. (In press). Pituitary profile of the ateliotic dwarf mouse, little. *Endocrinology*.

Beamer, W. G. & Eicher, E. M. (1976). Stimulation of growth in the little mouse. *J. Endocr.* **71**: 37–45.

Beamer, W. G., Whitten, W. K. & Eicher, E. M. (1978). Spontaneous sex mosaicism in BALB/c Wt. mice. In *Genetic mosaics and chimeras in mammals:* 195–208. Russell, L. B. (Ed.). New York: Plenum.

Beamer, W. G., Wilson, M. C. & De Luca, H. F. (In press). Stimulation of phosphate transport in genetically hypophosphatemic mice by 1-α hydroxyvitamin D_3. *Endocrinology*.

Beer, A. E. & Billingham, R. E. (1976). *The immunobiology of mammalian reproduction.* Englewood Cliffs, N.J.: Prentice-Hall.

Bell, R. W., Nitschke, W. & Zachman, T. A. (1972). Ultrasounds in three inbred strains of young mice. *Behav. Biol.* 7: 805–814.

Belyaev, D. K., Schuler, L. & Borodin, P. M. (1977). [Problems of stress genetics. III. Differential effects of stress on the fertility of mice of different genotypes.] *Genetika* 13: 52–60. [In Russian].

Biddle, F. G. & Fraser, F. C. (1977). Cortisone-induced cleft palate in the mouse. A search for the genetic control of the embryonic response trait. *Genetics* 85: 289–302.

Biggers, J. D. & Claringbold, P. J. (1955). Selection for local (intravaginal) action of oestrogens. *J. Endocr.* 12: 1–8.

Blake, R. L. (1970). Regulation of liver tyrosine aminotransferase activity in inbred strains and mutant mice. I. Strain variance in fasting enzyme levels. *Int. J. Biochem.* 1: 361–370.

Boiocchi, M., della Torre, G. & della Porta, G. (1975). Genetic control of endogenous C-type virus production in pancreatic acinar cells of C57BL/He and C57BL/6J mice. *Proc. natn. Acad. Sci. U.S.A.* 72: 1892–1894.

Braden, A. W. H. (1958). Variation between strains of mice in phenomena associated with sperm penetration and fertilization. *J. Genet.* 56: 37–47.

Bronson, F. H. (1979). Light intensity and reproduction in wild and domestic house mice. *Biol. Reprod.* 21: 235–239.

Bruce, H. M., Land, R. B. & Falconer, D. S. (1968). Inhibition of pregnancy-block by handling. *J. Reprod. Fert.* 15: 289–294.

Bullock, L. P. (1979). Genetic variations in sexual differentiation and sex steroid action. In *Genetic variation in hormone systems* 1: 69–88. Shire, J. G. M. (Ed.). Boca Raton, Florida: CRC Press.

Butley, M. S., Erickson, R. P. & Pratt, W. B. (1978). Hepatic glucocorticoid receptors and the H-2 locus. *Nature, Lond.* 275: 136–138.

Cattanach, B. M., Iddon, C. A., Charlton, H. M., Chiappa, S. A. & Fink, G. (1977). Gonadotrophin-releasing hormone deficiency in a mutant mouse with hypogonadism. *Nature, Lond.* 269: 338–340.

Chai, C. K. & Dickie, M. M. (1966). Endocrine variations. In *The biology of the laboratory mouse:* 387–403. 2nd edn. Green, E. L. (Ed.). New York: McGraw-Hill.

Champlin, A. K., Beamer, W. G., Carter, S. C., Shire, J. G. M. & Whitten, W. K. (1980). Genetic and social modifications of mating patterns of mice. *Biol. Reprod.* 22: 164–172.

Chapman, V. M. (1978). Biochemical polymorphisms of wild mice. In *Origins of inbred mice:* 555–568. Morse, H. C. (Ed.). New York: Academic Press.

Chapman, V. M. & Whitten, W. K. (1969). The occurrence and inheritance of pregnancy block in inbred mice. *Genetics* 61: 59.

Christian, J. J. & Davis, D. E. (1964). Endocrines, behaviour and population. *Science, N.Y.* 146: 1550–1560.

Ciaranello, R. D., Dornbusch, J. N. & Barchas, J. D. (1972). Regulation of adrenal phenylethanolamine-*N*-methyltransferase activity in three inbred mouse strains. *Molec. Pharmacol.* 8: 511–520.

Ciaranello, R. D., Lipsky, A. & Axelrod, J. (1974). Association between fighting behaviour and catecholamine biosynthetic enzyme activity in two inbred mouse sublines. *Proc. natn. Acad. Sci. U.S.A.* 71: 3006–3008.

Claringbold, P. J. & Biggers, J. D. (1955). The response of inbred mice to oestrogens. *J. Endocr.* 12: 9–14.

Coffino, P., Bourne, H. R., Friedrich, U., Hochman, J., Insel, P. A., Lemains, I., Melmon, K. & Tomkins, G. M. (1976). Molecular mechanisms of cyclic AMP action: a genetic approach. *Rec. Progr. Horm. Res.* **32**: 669–699.

Coleman, D. L. & Hummel, K. P. (1973). The influence of genetic background on the expression of the obese gene in the mouse. *Diabetologia* **9**: 287–293.

Curé, M., Valatx, J. -L. & Delobel, B. (1971). Étude comparative des glandes endocrines chez diverses souches de souris consanguines: données pondérales. *C.r. hebd. Séanc. Soc. Biol.* **165**: 1619–1623.

Daughaday, W. (1941). A comparison of the X-zone of the adrenal cortex in two inbred strains of mice. *Cancer Res.* **1**: 883–885.

Doering, C. H., Kessler, S. & Clayton, R. B. (1970). Genetic variation of cholesterol ester content in mouse adrenals. *Science, Wash.* **170**: 1220–1222.

Doering, C. H., Shire, J. G. M., Kessler, S. & Clayton, R. B. (1972). Cholesterol ester concentration and corticosterone production in adrenals of the C57BL/10 and DBA/2 strains in relation to adrenal lipid depletion. *Endocrinology* **90**: 93–101.

Doering, C. H., Shire, J. G. M., Kessler, S. & Clayton, R. B. (1973). Genetic and biochemical studies of the adrenal lipid depletion phenotype in mice. *Biochem. Genet.* **8**: 101–111.

Dung, H. C. & Swigart, R. H. (1971). Experimental studies of "lethargic" mice. *Texas Rep. Biol. Med.* **29**: 273–288.

Dunn, T. B. (1970). Normal and pathologic anatomy of the adrenal gland of the mouse, including neoplasms. *J. natn. Cancer Inst.* **44**: 1323–1389.

Eicher, E. M., Southard, J. L., Scriver, C. R. & Glorieux, F. H. (1976). Hypophosphatemia: mouse model for human familial hypophosphatemic (vitamin-D resistant) rickets. *Proc. natn. Acad. Sci., U.S.A.* **73**: 4667–4671.

Eleftheriou, B. E. (1974). Genetic analysis of hypothalamic retention of corticosterone in two inbred strains of mice. *Brain Res.* **69**: 77–81.

Eleftheriou, B. E. & Bailey, D. W. (1972). Genetic analysis of plasma corticosterone levels in two inbred strains of mice. *J. Endocr.* **55**: 415–420.

Eleftheriou, B. E. & Kristal, M. B. (1974). A gene controlling bell- and photically-induced ovulation in mice. *J. Reprod. Fert.* **38**: 41–47.

Elias, M. E. & Simmerman, S. J. (1971). Proactive and retroactive effects of ether on spatial discrimination learning in inbred mouse strains DBA/2J and C57BL/6J. *Psychon. Sci.* **22**: 299–301.

Ford, H. C., Lee, E. & Engel, L. L. (1979). Circannual variation and genetic regulation of hepatic testosterone hydroxylase activities in inbred strains of mice. *Endocrinology* **104**: 857–861.

Francis, B. M. (1973). Influence of sex-linked genes on embryonic sensitivity to cortisone in three strains of mice. *Teratology* **7**: 119–126.

Furtado Dias, M. T. (1959). Lésions de surrénale chez les souris de la legnée H. *Archos port. Sci. biol.* **12**: 107–121.

Gardner, W. V. & Strong, L. C. (1940). Estrogens in female mice of ten strains varying in susceptibility to spontaneous neoplasms. *Yale J. Biol. Med.* **12**: 543–553.

Goldman, A. S., Katsumata, M., Yaffe, S. J. & Glasser, D. L. (1977). Palatal cytosol cortisol-binding protein associated with cleft-palate susceptibility and H-2 genotype. *Nature, Lond.* **265**: 643–645.

Goldstein, J. L. & Wilson, J. D. (1972). Studies on the pathogenesis of pseudohermaphroditism in the mouse with testicular feminization. *J. clin. Invest.* **51**: 1647–1658.

Gorbman, A. (1947). Thyroidal and vascular changes in mice following chronic treatment with goitrogens and carcinogens. *Cancer Res.* 7: 746–766.

Gorzalka, B. B. & Whalen, R. E. (1974). Accumulation of estradiol in brain, uterus and pituitary: strain, species, suborder and order comparisons. *Brain Behav. Evol.* 9: 376–392.

Gregorova, S., Ivanyi, P., Simonova, D. & Mickova, M. (1977). *H-2* associated differences in androgen-influenced organ weights of A and C57BL/10 mouse strains and their crosses. *Immunogenetics* **4**: 301–313.

Håkansson, E. M. & Lundin, L. G. (1975). Genetic variation in number of pituitary PAS-purple cells in the house mouse. *J. Hered.* **66**: 144–146.

Hawkins, E. F., Young, P. N., Hawkins, A. M. C. & Bern, H. A. (1975). Adrenocortical function: corticosterone levels in BALB/c and C3H mice under various conditions. *J. exp. Zool.* **194**: 479–494.

Hayward, P. & Shire, J. G. M. (1974). Y chromosome effect on adult testis size. *Nature, Lond.* **250**: 499–500.

Hoppe, P. C. (1975). Genetic and endocrine studies of the pregnancy-blocking pheromone of mice. *J. Reprod. Fert.* **45**: 109–115.

Hummel, K. P., Coleman, D. L. & Lane, P. W. (1972). Influence of genetic background on expression of mutations at the diabetes locus in the mouse. I. C57BL/ksJ and C57BL/6J strains. *Biochem. Genet.* 7: 1–13.

Islam, A. B. M., Hill, W. G. & Land, R. B. (1976). Ovulation rates of lines of mice selected for testis weight. *Genet. Res.* 27: 23–32.

Iványi, P., Gregorova, S. & Micková, M. (1972). Genetic differences in thymus, lymph node, testes and vesicular gland weights among inbred mouse strains. Association with the major histocompatibility (*H-2*) system. *Folia biol., Praha* **18**: 81–97.

Jones, E. (1964). Structure and metabolism of pancreatic islets. *Proc. Int. Wenner-Gren Symp.* **3**: 189–191.

Jones, R. B. & Nowell, N. W. (1974). The urinary aversive pheromone of mice: species, strain and grouping effects. *Anim. Behav.* **22**: 187–191.

Kakihana, R., Noble, E. P. & Butte, J. C. (1968). Corticosterone response to ethanol in inbred strains of mice. *Nature, Lond.* **218**: 360–361.

Keeley, K. (1962). Prenatal influence on behavior of offspring of crowded mice. *Science, N.Y.* **135**: 44–45.

Krzanowska, H. (1970). Relation between fertilization rate and penetration of eggs by supplementary spermatozoa in different mouse strains and crosses. *J. Reprod. Fert.* **22**: 199–204.

Krzanowska, H. (1972). Rapidity of removal *in vitro* of the cumulus oophorus and the zona pellucida in different strains of mice. *J. Reprod. Fert.* **31**: 7–14.

Land, R. B. (1973). The expression of female sex-limited characters in the male. *Nature, Lond.* **241**: 208–209.

Land, R. B. & Falconer, D. S. (1969). Genetic studies of ovulation rate in the mouse. *Genet. Res.* 13: 25–46.

Laurie, E. M. O. (1946). The reproduction of the house mouse (*Mus musculus*) living in different environments. *Proc. R. Soc. Lond.* (B) 133: 248–281.

Legrand, R. (1970). Successful aggression as a reinforcer of runaway behavior of mice. *Psychon. Sci.* **20**: 303–305.

Levin, B. H., Vandenbergh, J. G. & Cole, J. L. (1974). Aggression, social pressure and asymptote in laboratory mouse populations. *Psychol. Rep.* **34**: 239–244.

Levine, S. & Levin, R. (1970). Pituitary-adrenal influences on passive avoidance in two inbred strains of mice. *Horm. Behav.* **1**: 105–110.

Levine, S. & Treiman, D. M. (1964). Differential plasma corticosterone response to stress in four inbred strains of mice. *Endocrinology* **75**: 142–144.

Levine, S. & Treiman, D. (1969). Determinants of individual differences in the steroid response to stress. In *Physiology and pathology of adaptation mechanisms:* 171–184. Bajusz, E. (Ed.). Oxford: Pergamon Press.

Lieberman, M. W. (1963). Early developmental stress and later behavior. *Science, N.Y.* **141**: 824–825.

Lindberg, M., Shire, J. G. M., Doering, C. H., Kessler, S. & Clayton, R. B. (1972). Reductive metabolism of corticosterone in mice: differences in NADPH requirement of liver homogenates of males of two inbred strains. *Endocrinology* **90**: 81–92.

Lyon, M. F. & Hawkes, S. G. (1970). X-linked gene for testicular feminization in the mouse. *Nature, Lond.* **227**: 1217–1219.

McGill, T. E. (1970). Genetic analysis of male sexual behavior. In *Contributions to behavior-genetic analysis: The mouse as prototype:* 57–88. Lindzey, G. & Thiessen, D. D. (Eds). New York: Appleton-Century-Crofts.

McGill, T. E. & Manning, A. (1976). Genotype and retention of the ejaculatory reflex in castrated male mice. *Anim. Behav.* **24**: 507–518.

McLaren, A. & Bowman, P. (1973). Genetic effects on the timing of early development in the mouse. *J. Embryol. exp. Morph.* **30**: 491–498.

Meckler, R. J. & Collins, R. L. (1965). Histology and weight of the mouse adrenal: a diallel genetic study. *J. Endocr.* **31**: 95–103.

Miller, K. K. (1977). Commercial dietary influences on the frequency of cortisone-induced cleft-palate in C57BL/6J mice. *Teratology* **15**: 249–252.

Monroe, C. B. (1968). Induction of tryptophan oxygenase and tyrosine aminotransferase in mice. *Am. J. Physiol.* **214**: 1410–1414.

Morrison, W. I. & Roelants, G. E. (1978). Susceptibility of inbred strains of mice to *Trypanosoma congolense*. *Clin. exp. Immunol.* **32**: 25–40.

Mos, L., Vriend, J. & Poley, W. (1974). Effects of light environment on emotionality and the endocrine system of inred mice. *Physiol. Behav.* **12**: 981–989.

Murphy, E. D. & Beamer, W. G. (1973). Plasma gonadotropin levels during early stages of ovarian tumorigenesis in mice of the W^x/W^v genotype. *Cancer Res.* **33**: 721–723.

Nagasawa, H., Kanzawa, F. & Kuretani, K. (1967). Lactation performance of the high and low mammary tumor strains of mice. *Gann* **58**: 331–336.

Nandi, S. & Bern, H. A. (1960). Relation between mammary-gland responses to lactogenic hormone combinations and tumor susceptibility in various strains of mice. *J. natn. Cancer Inst.* **24**: 907–931.

Nayudu, P. R. V. & Moog, F. (1967). The genetic control of alkaline phosphatase activity in the duodenum of the mouse. *Biochem. Genet.* **1**: 155–170.

Nguyen-Trong-Tuan, Rekdal, D. J. & Burton, A. F. (1971). The uptake and metabolism of ^{3}H-corticosterone and fluorimetrically determined corticosterone in fetuses of several mouse strains. *Biol. Neonate* **18**: 78–84.

Nicol, A. & McLaren, A. (1974). An effect of the female genotype on sperm transport in mice. *J. Reprod. Fert.* **39**: 421–424.

Oliverio, A. (1979). Uses of recombinant inbred lines. In *Quantitative genetic variation:* 197–218. Thompson, J. N. & Thoday, J. M. (Eds). New York: Academic Press.

Pachiarz, J. A. & Teague, P. O. (1975). Different responses of two mouse strains to 650 rads and protection by surgical stress. *Proc. Soc. exp. Biol. Med.* **148**: 1095–1100.

Paigen, K., Swank, R. T., Tomino, S. & Ganschow, R. E. (1975). The molecular genetics of mammalian glucuronidase. *J. cell. Physiol.* **85**: 379–392.

Pantelouris, E. M. (1974). Common parameters in genetic athymia and senescence. *Exp. Geront.* **9**: 161–167.

Papaioannou, V. E. & Fraser, R. (1974). Plasma aldosterone concentration in sodium deprived mice of two strains. *J. steroid Biochem.* **5**: 191–192.

Pelkonen, O., Boobis, A. R. & Nebert, D. W. (1978). Genetic differences in binding of reactive carcinogenic metabolites to DNA. In *Carcinogenesis* **3**: 383–400. Jones, P. W. & Freudenthal, R. I. (Eds). New York: Raven Press.

Pennycuik, P. R. (1966). Differences between mice gestating and lactating at 21° and at 34–36° with particular reference to the mammary gland. *Aust. J. exp. Biol. Med. Sci.* **44**: 419–438.

Pennycuik, P. R. (1967). A comparison of the effects of a variety of factors on the metabolic rate of the mouse. *Aust. J. exp. Biol. Med. Sci.* **45**: 331–346.

Pennycuik, P. R. (1969). Reproductive performance and body weights of mice maintained for 12 generations at 34°C. *Aust. J. biol. Sci.* **22**: 667–675.

Pidduck, H. G. & Falconer, D. S. (1978). Growth hormone function in strains of mice selected for large and small size. *Genet. Res.* **32**: 195–206.

Poland, A. P., Glover, E., Robinson, J. R. & Nebert, D. W. (1974). Genetic expression of aryl hydrocarbon hydroxylase activity: induction by p-dioxin. *J. biol. Chem.* **249**: 5599–5606.

Porter, R. H. (1972). Infantile handling differentially affects inter-strain dominance interactions in mice. *Behav. Biol.* **7**: 415–420.

Rapp, J. P. (1965). Alteration of juxtaglomerular index by selective inbreeding. *Endocrinology* **76**: 486–490.

Rapp, J. P. (1967). Effects of DCA and Na intake on mice bred for high and low juxtaglomerular indices. *Am. J. Physiol.* **212**: 1135–1146.

Rendel, J. M. (1979). Canalisation and selection. In *Quantitative genetic variation:* 139–156. Thompson, J. N. & Thoday, J. M. (Eds). New York: Academic Press.

Ressler, R. (1962). Parental handling in two strains reared by fostered parents. *Science, N.Y.* **137**: 129–130.

Salomon, D. S. & Pratt, R. M. (1979). Involvement of glucocorticoids in development of the secondary palate. *Differentiation* **13**: 141–154.

Sato, B., Spomer, W., Huseby, R. A. & Samuels, L. T. (1979). The testicular estrogen receptor system in two strains of mice differing in susceptibility to estrogen-induced Leydig cell tumors. *Endocrinology* **104**: 822–831.

Seabloom, R. W. & Seabloom, N. R. (1975). Response to ACTH by superfused adrenals of wild and domestic house mice. *Life Sci.* **15**: 73–82.

Selmanoff, M. K., Goldman, B. D. & Ginsberg, B. E. (1977). Developmental changes in serum luteinizing hormone, follicle stimulating hormone and androgen levels in males of two inbred mouse strains. *Endocrinology* **100**: 122–127.

Shire, J. G. M. (1968). Genes, hormones and behavioural variation. In *Genetic and environmental influences on behaviour:* 194–205. Thoday, J. M. & Parkes, A. S. (Eds). Edinburgh: Oliver & Boyd.

Shire, J. G. M. (1974). Endocrine genetics of the adrenal gland. *J. Endocr.* **62**: 173–207.

Shire, J. G. M. (1976). The forms, uses and significance of genetic variation in endocrine systems. *Biol. Rev.* **51**: 105–141.

Shire, J. G. M. (1979a). Corticosteroids and adrenocortical function in animals. In *Genetic variation in hormone systems* **1**: 43–67. Shire, J. G. M. (Ed.). Boca Raton, Florida: CRC Press.

Shire, J. G. M. (1979b). The thyroid gland and thyroid hormones. In *Genetic variation in hormone systems* **2**: 1–20. Shire, J. G. M. (Ed.). Boca Raton, Florida: CRC Press.

Shire, J. G. M. (1979c). Mutations affecting the hormone systems of rodents. In *Genetic variation in hormone systems* **2**: 151–156. Shire, J. G. M. (Ed.). Boca Raton, Florida: CRC Press.

Shire, J. G. M. (1979d). Uses and consequences of genetic variation in hormone systems. In *Genetic variation in hormone systems* **1**: 1–10. Shire, J. G. M. (Ed.). Boca Raton, Florida: CRC Press.

Shire, J. G. M. (1980). Corticosterone metabolism by mouse liver: interactions between genotype and diet. *Horm. Metab. Res.* **12**: 117–119.

Shire, J. G. M. & Bartke, A. (1972). Strain differences in testicular weight and spermatogenesis with special reference to C57BL/10J and DBA/2J mice. *J. Endocr.* **55**: 163–171.

Shire, J. G. M. & Hambly, E. A. (1973). The adrenal glands of mice with hereditary pituitary dwarfism. *Acta path. microbiol. scand.* (A) **81**: 225–228.

Shire, J. G. M. & Spickett, S. G. (1967). Genetic variation in adrenal structure: qualitative differences in the zona glomerulosa. *J. Endocr.* **39**: 277–284.

Shire, J. G. M. & Spickett, S. G. (1968). Genetic variation in adrenal structure: strain differences in quantitative characters. *J. Endocr.* **40**: 215–229.

Shire, J. G. M. & Stewart, J. (1972). The zona glomerulosa and corticotrophin: a genetic study in mice. *J. Endocr.* **55**: 185–193.

Shire, J. G. M. & Whitten, W. K. (1980a). Genetic variation in the timing of first cleavage: effect of paternal genotype. *Biol. Reprod.* **22**: 363–368.

Shire, J. G. M. & Whitten, W. K. (1980b). Genetic variation in the timing of first cleavage in mice: effect of maternal genotype. *Biol. Reprod.* **22**: 369–376.

Silberberg, R. & Silberberg, M. (1950). Growth and articular changes in slowly and rapidly developing mice fed a high-fat diet. *Growth* **14**: 213–230.

Simeonsson, R. J. & Meier, G. W. (1970). Strain-specific effects of chronic prenatal irradiation in mice. *Devl Psychobiol.* **3**: 197–206.

Simmell, E. C. & Walker, D. A. (1972). The effects of a companion on exploratory behavior in two strains of mice. *Behav. Genet.* **2**: 249–254.

Sinha, Y. N. & Baxter, S. R. (1979). Metabolism of prolactin in mice with high incidence of mammary tumours: evidence for greater conversion into a non-immunoassayable form. *J. Endocr.* **81**: 299–314.

Sinha, Y. N., Baxter, S. R. & Vanderlaan, W. P. (1979). Metabolic clearance rate of prolactin during various physiological states in mice with high and low incidences of mammary tumors. *Endocrinology* **105**: 680–684.

Sinha, Y. N., Salocks, C. B., Vanderlaan, W. P. & Vlahakis, G. (1977). Evidence for an influence of mammary tumour virus on prolactin secretion in the mouse. *J. Endocr.* **74**: 383–392.

Southard, R. L. & Eicher, E. M. (1977). Petite. *Mouse News Letter* **54**: 40.

Spickett, S. G., Shire, J. G. M. & Stewart, J. (1967). Genetic variation in adrenal and renal structure and function. *Mem. Soc. Endocr.* **15**: 271–288.

Stauffacher, W., Orci, L., Cameron, D. P., Burr, I. M. & Renold, A. E. (1971). Spontaneous hyperglycemia and/or obesity in laboratory rodents: an example of the possible usefulness of animal disease models with both genetic and environmental components. *Rec. Progr. Horm. Res.* **27**: 41–95.

Stewart, A. D. (1971). Genetic variation in the neurohypophysical hormones of the mouse, *Mus musculus. J. Endocr.* **51**: 191–201.

Stewart, A. D. (1973). Sensitivity of mice to 8-arginine and 8-lysine vasopressins as antidiuretic hormones. *J. Endocr.* **59**: 195–196.

Stewart, A. D. (1979). Genetic variation in the endocrine system of the neurohypophysis. In *Genetic variation in hormone systems* **1**: 144–176. Shire, J. G. M. (Ed.). Boca Raton, Florida: CRC Press.

Stewart, A. D., Batty, J. & Harkiss, G. D. (1977). Genetic variation in plasma thyroxine levels and minimal metabolic rates of the mouse, *Mus musculus. Genet. Res.* **31**: 303–306.

Stewart, J. (1969). Diuretic responses to electrolyte loads in four strains of mice. *Comp. Biochem. Physiol.* **30**: 977–987.

Stewart, J. (1975). Genetic studies on the mechanism of action of aldosterone in mice. *Endocrinology* **96**: 711–717.

Stewart, J., Fraser, R., Papaioannou, V. & Tait, A. (1972). Aldosterone production and the zona glomerulosa: a genetic study. *Endocrinology* **90**: 968–972.

Stylianopolou, F. & Clayton, R. B. (1976). Strain-dependent gonadal effects upon adrenal cholesterol ester concentration and composition in C57BL/10J and DBA/2J mice. *Endocrinology* **99**: 1631–1637.

Sustarsic, D. L. & Wolfe, H. G. (1976). Differences in male reproductive physiology between C3H/HeWe and C57BL/6We inbred strains of mice. *Genetics* **83**: s74–s75.

Sustarsic, D. L. & Wolfe, H. G. (1979). A genetic study of luteinizing hormone levels and induced luteinizing hormone release in male mice. *J. Hered.* **70**: 226–230.

Taylor, H. C. & Waltman, C. A. (1940). Hyperplasias of the mammary gland in the human being and in the mouse. *Arch. Surg.* **40**: 733–820.

Thiessen, D. D. (1964). Population density, mouse genotype and endocrine function in behavior. *J. comp. Physiol.* **57**: 412–416.

Thoday, J. M. (1967). Uses of genetics in physiological studies. *Mem. Soc. Endocr.* **15**: 297–311.

Thompson, J. N. (1975). Quantitative variation and gene number. *Nature, Lond.* **258**: 665–668.

Thompson, W. R. & Olian, S. (1961). Some effects on offspring behavior of maternal adrenalin injection during pregnancy in three inbred mouse strains. *Psychol. Rep.* **8**: 87–90.

Treiman, D. M., Fulker, D. W. & Levine, S. (1970). Interaction of genotype and environment as determinants of corticosteroid response to stress. *Devl Psychobiol.* **3**: 131–140.

Vale, J. R., Ray, D. & Vale, C. A. (1973). Interaction of genotype and exogenous neonatal estrogen: aggression in female mice. *Physiol. & Behav.* **10**: 181–183.

Vale, J. R., Ray, D. & Vale, C. A. (1974). Neonatal androgen treatment and sexual behavior in males of three inbred strains of mice. *Devl Psychobiol.* **7**: 483–488.

Wakelin, D. (1975). Genetic control of immune responses to parasites: immunity to *Trichuris muris* in inbred and random-bred strains of mice. *Parasitology* **71**: 51–60.

Westberg, J. A., Bern, H. A. & Barnawell, E. B. (1957). Strain differences in the response of the mouse adrenal to oestrogen. *Acta endocr., Kbh* **25**: 70–82.

Wilson, C. M., Erdos, E. G., Dunn, J. F. & Wilson, J. D. (1977). Genetic control of renin activity in the submaxillary gland of the mouse. *Proc. natn. Acad. Sci. U.S.A.* **74**: 1185–1189.

Wilson, C. M., Erdos, E. G., Wilson, J. D. & Taylor, B. A. (1978). Location on chromosome 1 of *Rnr*, a gene that regulates renin in the submaxillary gland of the mouse. *Proc. natn. Acad. Sci. U.S.A.* **75**: 5623–5626.

Wong, M. D. & Burton, A. F. (1971). Inhibition by corticosteroids of glucose incorporation into fetuses of several strains of mouse. *Biol. Neonate* **18**: 146–152.

Wood, P. R. W. (1977). Responsiveness of adrenal cells to ACTH. *Mouse News Letter* **57**: 13.

Woolley, G. W. (1950). Experimental endocrine tumors with special reference to the adrenal cortex. *Rec. Progr. Horm. Res.* **5**: 383–405.

Wragg, L. E. & Speirs, R. S. (1952). Strain and sex differences in response of inbred mice to adrenal cortical hormones. *Proc. Soc. exp. Biol. Med.* **80**: 680–684.

Yamamoto, K. R., Gehring, U., Stampfer, M. R. & Sibley, C. H. (1976). Genetic approaches to steroid hormone action. *Rec. Progr. Horm. Res.* **32**: 3–32.

Yamazaki, K., Yamaguchi, M., Andrews, P. W., Peake, B. & Boyse, E. A. (1978). Mating preference of F_2 segregants of crosses between MHC-congenic mouse strains. *Immunogenetics* **6**: 253–259.

Yanai, R. & Nagasawa, H. (1968). Age-, strain- and sex-differences in the anterior pituitary growth-hormone content of mice. *Endocr. japon.* **15**: 395–402.

Yoon, J. W., Lesniak, M. A., Fussganger, R. & Notkins, A. L. (1976). Genetic differences in susceptibility of pancreatic β cells to virus-induced diabetes mellitus. *Nature, Lond.* **264**: 178–179.

Zarrow, M. X., Christenson, C. M. & Eleftheriou, B. E. (1971). Strain differences in the ovulatory response of immature mice to PMS and to the pheromonal facilitation of PMS-induced ovulation. *Biol. Reprod.* **4**: 52–56.

Symp. zool. Soc. Lond. (1981) No. 47, 575–589

Wild House Mouse Biology and Control

F. P. ROWE

Rodent Pests Department, Tolworth Laboratory, Ministry of Agriculture, Fisheries and Food, Tolworth, Surrey, UK

SYNOPSIS

The house mouse (*Mus musculus*) is a remarkably adaptable and successful rodent, world-wide in distribution. It lives in close association with man, commonly occupying dwellings and food stores in urban and rural areas. In some regions of the world it also thrives in the field. The house mouse is a significant pest of mankind, inflicting economic losses by destroying and contaminating stored foodstuffs and growing crops and causing other incidental damage. It can also transmit diseases to man and is therefore of public health concern. Biological and behavioural traits of the wild house mouse that are of particular significance in relation to its control are considered together with present-day control measures.

INTRODUCTION

The benefit derived from the use of the laboratory mouse as a research tool is offset by the various problems posed to mankind as the result of the continued success of the wild animal. *Mus musculus* is one of three rodents, the other two being the Norway rat, *Rattus norvegicus,* and the roof or ship rat, *R. rattus*, that have become particularly well-adapted to a commensal existence while still maintaining natural field-living populations. Although the house mouse has been a recognized pest of man since earliest times, active research into its control was not undertaken until the 1939–45 war. The systematic investigations conducted by Southern (1954) and his co-workers in the Bureau of Animal Population, Oxford, were begun when it was found that house mice were causing serious damage to stockpiled foodstuffs in Britain and that rat control techniques were largely ineffective against them. The pioneer Oxford work not only helped to establish basic control principles and practices, but also stimulated further study of wild house mouse biology and behaviour. In the laboratory, observations on penned colonies of mice provided deeper insight into a number of intraspecific behavioural traits, notably social, aggressive and territorial behaviour, and of factors influencing the growth and dispersal of populations. The life-history, physiology and dynamics of island and mainland populations of feral house mice also received attention, the ecological studies pursued by Berry (1968) and his colleagues (Berry, Jakobson & Triggs, 1973) and by Newsome (1969a, b) being of particular significance in this respect. House mouse control was not of direct concern in most of these later studies but, nevertheless, the findings have been of importance to the applied biologist working in this field.

This chapter reviews salient biological and behavioural characteristics of the house mouse that have contributed to its success as a pest. Control methods in current use and their further progress are also considered.

DISTRIBUTION

The house mouse probably originated as a wild species in the steppe zone of the southern Palaearctic and, helped by the development of agriculture and methods of transportation, spread to most parts of the world. It is still extending its range in some remote areas, having reached South Seymour Island in the Galapagos Group (Eibl-Eibesfeldt, 1955) and Guadalcanal in the Solomon Islands (Rowe, 1967) in fairly recent times. At the present time, *M. musculus* is probably more widely distributed than any other mammal apart from man; it exists in temperate, tropical, sub-Antarctic, steppe and semi-desert regions, occupying such varied field habitats as arable land, the banks of rice fields, sugar cane plantations, refuse tips, hedgerows, salt marshes and coal mines and it also commonly lives indoors inhabiting diverse types of building, in both urban and rural areas.

The successful establishment of permanent outdoor populations of the house mouse in regions where it has been introduced appears to have been dependent on the existing mammal fauna as well as on the prevailing climate and vegetation. For example, *M. musculus* has successfully colonized Skokholm Island (Berry, 1968) and the Faroes (Berry, Jakobson & Peters, 1978) off Britain and Gough Island in the South Atlantic where it encountered little or no competition. Again, where competition proved to be weak, as in parts of Australia, the house mouse has replaced some indigenous species over much of their former ranges. *M. musculus* is less abundant than other small rodents of the genus *Apodemus* in mainland Europe, however, and feral house mice in North America have also been found to be at a disadvantage in competition with *Peromyscus* spp. (Caldwell, 1964), which occupy a similar ecological niche to *Apodemus,* and with *Microtus* spp. (Lidicker, 1966). The house mouse has also failed to become well established in parts of Africa and South America in competition with local species.

ECONOMIC IMPORTANCE

Precise information on the damage, direct and indirect, inflicted by house mouse populations is extremely difficult to obtain and few critical assessments of losses have been made. Crop damage can reach spectacular proportions when populations mass on arable land (Newsome & Crowcroft, 1971); in most countries however, the house mouse is of more concern as a pest of food stores, being the most prevalent rodent in many habitats of this type. Southern (1954) concluded that the accumulated losses to foodstuffs and the incidental destruction arising from widespread indoor infestations of mice probably far outweighed those attributable to feral populations, and this assessment would seem equally valid today.

The house mouse can be a troublesome pest of domestic premises, notably in dilapidated buildings and in situations where poor hygiene is combined with neglect, but the damage it produces is often more irritating than severe. Control problems in residential areas tend to be exacerbated by a degree of public tolerance towards this species and often an unawareness of its presence, at least until a population has become firmly established and already spread. Some appreciation of the scale of infestation that can exist in a large urban area can be gathered from the findings of a rodent survey conducted in London in 1972: of the 8534 premises surveyed at random, 8.7% were found to be mouse infested (Rennison & Shenker, 1976).

More serious losses to foodstuffs attributable to house mice occur in bulk food stores, food processing plants, shops, mills, bakeries and restaurants in urban areas, and granaries, animal feed stores, cribs and silos on farmland. The mouse consumes only about 4 g of food daily but estimates of losses based on this intake alone are unrealistic because of its wasteful feeding habits. Southern (1954), for example, found that 10% or more of the grain in corn ricks was fragmented by mice and rendered unsuitable for milling purposes. Ricks are rarely built nowadays but *M. musculus* is opportunistic and it has rapidly taken advantage of other changes in farming practice. Livestock has become more intensively husbanded and reared indoors and the house mouse is a common and often serious pest of poultry houses, piggeries, calf pens and dairy units.

House mice are responsible for much of the rodent filth found to contaminate foodstuffs. The mouse voids 50 or more droppings daily and these can prove difficult to remove at economical cost; the droppings, moreover, invariably contain hairs, the detection of which can cause the rejection of processed foods. The loss in reputation that can follow the marketing of contaminated foodstuffs is often of more concern to food manufacturers and retailers than the amount of food actually eaten by mice (Goldenberg & Rand, 1971).

Other indirect but significant economic losses can occur as the result of mouse infestation. Food spillage, involving recollection and repair costs, is common when mice attack containers in their search for nesting material. Most modern materials used to insulate walls and to line ceilings are easily shredded by mice and their damage, in poultry houses for example, can reach significant proportions. House mice also succeed in penetrating and nesting in poorly guarded electrical installations on occasion, and their tendency to gnaw wiring in that event creates fire hazards.

PUBLIC HEALTH IMPORTANCE

The house mouse is much less dangerous than the rat in the dissemination of such diseases as rat-bite fever, murine typhus and tularaemia (Cameron, 1949), and it is not thought to play a significant role in the transmission of plague (*Yersinia pestis*). The occurrence of leptospirosis (Weil's disease) in man is also largely attributable to rats although *M. musculus* was said to be a major carrier of the disease in Hawaii (Minette, 1964). The disease that is probably most commonly transmitted to humans via the mouse is salmonellosis (infectious food poisoning) while livestock in mouse-infested farm buildings are at similar risk (although see Blackwell, this volume p. 591).

Firm assessment of the importance of the house mouse in relation to public health is not possible because of the surprising lack of detailed study of the diseases borne by wild mice and their transmission. However, the reports of human infection that have occurred following close contact with mice add support to economic considerations for the implementation of effective control measures.

BIOLOGY AND BEHAVIOUR

The particular difficulties that are encountered in controlling *M. musculus* populations can be most readily appreciated by considering pertinent aspects of the biology and behaviour of the species.

Important factors that have influenced the success of the house mouse in addition to its ability to exploit variable types of habitat are reproductive potential, acutely developed senses, omnivorous and erratic feeding habits, limited indoor range, and social organization.

Reproduction

Under optimum environmental conditions the house mouse breeds throughout the year. Fertility is largely unaffected by low temperatures provided that nutritious food and ample nesting material are available; thus, Laurie (1946) found that mice thrived in meat stores kept at $-10°C$. The same author also examined mice drawn from urban premises, flour stores and corn ricks in Britain but observed no seasonal differences in reproductive performance while more recent work (Rowe & Swinney, 1977) has shown that house mice occupying farm buildings also reproduce at a fairly even rate throughout the year. W. W. Smith (1954) however, in a study of urban and rural mice in North America, found that pregnancy rates were highest in the spring and autumn months.

Breeding females readily compensate for lack of nesting facilities by sharing a communal nest and raising their young together. In confined colony studies, reproductive performance has been found to be impaired when overcrowding becomes acute, populations ceasing to increase either as the result of a reduction in female fertility (Crowcroft & Rowe, 1957) or because of abnormally high embryonic resorption and poor survival of young (Southwick, 1955). However, there is little evidence that naturally occurring house mouse populations are much regulated by the operation of such mechanisms. Free-living mice are open to the additional effects of dispersal and predation while farm building populations have been found to be essentially unstable, their size and composition fluctuating constantly (Petrusewicz & Andrzejewski, 1962; Rowe & Swinney, 1977).

Information on the reproductive ecology of feral mouse populations is less well documented. In temperate climates, breeding activity tends to be seasonal. Berry (1968) found that the main breeding season of Skokholm Island mice was April to September, insignificant reproduction occurring at other times of the year. In California, Breakey (1963) observed a delay in the onset of maturity in females as the breeding season advanced, late-born individuals not coming into breeding condition until the following spring.

Senses

The early work of Waugh (1910) indicated that the mouse has poor visual acuity, the main functions of the eyes being the detection of movement at close quarters and the appreciation of different light intensities. Hopkins (1953) has since stated, however, that house mice can identify objects at least 15 m distant. Mice have an acute sense of hearing, responding to any sudden noise and being particularly sensitive to sounds of high frequency. Smell is important in the location of foods and taste in discerning their palatability. Olfactory cues also play a part in recognition between related and unrelated mice and of territories (Mackintosh & Grant, 1966; Bowers & Alexander, 1968; Archer, 1968). Nestlings respond immediately to touch and this sense is also helpful in movements away from the nest. Characteristically, the house mouse explores its home range intensively, its movements tending to become repetitive through repeated sequences of muscular movements, the kinaesthetic sense described by Southern (1954). Thus, the mouse rapidly becomes acquainted with changes in its surroundings and is able to take cover immediately.

Food and Water Requirements

Cereals comprise much of the diet of house mice infesting food stores, but fat and protein foods of all kinds are consumed. Mixed food stores are particularly attractive habitats but mice living in premises holding a single, relatively poor diet, such as flour, have been found to supplement their intake by seeking out insect and other food items. Cereals, grass and weed seeds are consumed by mice occupying arable, grass and waste land, but insect larvae are sometimes eaten in high proportion and the roots and stems of various plants are also known foods (Whitaker, 1966).

Water requirements are low, particularly for mice living on foods high in moisture content. A supply of drinking water is of more importance to animals subsisting on very dry foods in hot environments (Southern, 1954) and its deficiency helps to lower reproductive performance. Feral mice may occupy arid regions where there is little or no open water for long periods; dew is probably an important source of water for them. *M. musculus* is adapted physiologically to reduce water loss (Fertig & Edmonds, 1969).

Feeding Activity and Movements

M. musculus is mainly nocturnal but day-time feeding is common, particularly when food is in short supply or populations are crowded and little disturbed. The mouse is a light and intermittent feeder, the two main feeding periods at dusk and near dawn being interspersed with shorter feeding bursts, but individuals vary considerably in their behaviour (Southern, 1954; Crowcroft, 1959). House mice are drawn to new feeding points, including poison baits, more quickly than are rats, but the diversion is often transitory unless an attractive bait is used or the baiting points are re-sited periodically (Crowcroft, 1959).

The movements of indoor-living mice are most restricted in habitats where food and cover coincide and they may then amount to a few metres only (Young, Strecker & Emlen, 1950). Feral mouse movements tend to be less localized, particularly when food is scarce or there is a sudden change in environmental conditions. Rowe, Taylor & Chudley (1963) found that mice living on arable land remained in the vicinity of the fields at harvest time, seeking natural crevices and feeding on spilled grain, but movement into hedgerows and to a lesser extent into farm buildings occurred when the fields were ploughed and rolled in the autumn. Minor movements in the reverse direction in spring, from buildings to fields, were traced by Macleod (1959). In South Australia, Newsome (1969b) found that a reed-bed served as an important off-season refuge for mice in an arable district, nearby wheatfields being invaded in the spring. Mice have also been found to reside in isolated patches of wasteland outside farm buildings, the individuals living in these marginal habitats being well-situated to rapidly re-colonize premises cleared of infestation (Rowe & Swinney, 1977).

Aggressive and Social Behaviour

Indoor populations of house mice appear to be socially organized through territorial behaviour in a way similar to that analysed in laboratory situations (Crowcroft, 1955; Anderson, 1961; Crowcroft & Rowe, 1963; Rowe & Redfern, 1969). Fighting has been observed between the members of different populations occupying distinct parts of the same building (Brown, 1953). Again, mice infesting buildings holding bagged foodstuffs tend to become rapidly dispersed throughout, isolated stacks providing ideal conditions for the establishment of separate territories. There is little movement of individuals between stacks once the latter have all become occupied by mice (Rowe, 1973) and similar restricted exchange of animals has been recorded in the case of mouse populations inhabiting different but closely adjacent farm buildings (Rowe & Swinney, 1977). Aggressive behaviour apparently serves to aid the dispersal of mice, resulting in the early and full exploitation of available habitat and hence the spread of infestation.

CONTROL

Operational Methods

No serious attempts appear to have been made to control feral populations of *M. musculus,* although some consideration has been given to forecasting their development in arable areas periodically affected by outbreaks. It is likely that such field operations would be extremely difficult to conduct although the findings of Newsome (1969b) suggest that the off-season treatment of mice inhabiting marginal habitats could significantly reduce the subsequent build-up of populations in nearby arable areas.

The methods most commonly adopted to prevent the establishment of and to control indoor populations of house mice in urban and rural areas involve the protection of buildings, environmental improvement, poisoning and trapping. Effective control, in the long term, is often dependent on the operation of more than one of these methods, applied in combination or in sequence.

Protection of Buildings

The need to deny mice entry into buildings as a first line of defence is continually stressed by authorities concerned with rodent control. The basic principles underlying the protection of food stores and other buildings from invasion by rodents are well known (Jenson, 1979), the proofing of large food stores, in particular, being best done in close consultation with architects and builders. The complete exclusion of mice from buildings is difficult since accidental introductions can occur and adult animals can penetrate apertures only 10 mm wide. Even so, the effort and expense involved in inspecting incoming goods and in proofing is justified when valuable foodstuffs and commercial reputation need to be safeguarded. Rather simple protective measures, necessitating little additional cost, can prove worthwhile. For example, many farm buildings present almost insurmountable external proofing problems but often minor improvements to the internal structure can do much to reduce the size and distribution of mouse populations inside them. At the simplest level of a village grain store, Taylor (1972) found that stores made of sun-baked mud were rodent-proof.

Environmental Improvement

The effectiveness of proofing measures can be quickly eroded unless the fabric of a building is inspected regularly. The removal of plant cover, waste food and harbourage in the vicinity of a building and the maintenance of a high standard of hygiene inside, can help deter establishment and population increases of mice. The need to detect the presence of mice at an early stage is important and, for inspection purposes, stacked foodstuffs, for example, should be stored 0.5 m or more from walls and the stacks built not more than 10 m wide. Bulk storage on farms of cereals rather than in bags is advantageous as this reduces the amount of cover and nesting material available to mice; this benefit is soon lost, however, when loose grain is retained inside straw bales or other barriers providing favourable mouse harbourage.

Application of Rodenticides

Most control operations against infestations of house mice involve the application of rodenticides, either in solid bait, in dust, or in water. It is likely that the use of poisonous compounds will remain the principal means of control in the foreseeable future. Those in current use are either of the single dose, acutely toxic type such as zinc phosphide and the more recently introduced alphachloralose (Cornwell & Bull, 1967), or of the multiple dose chronic type, mainly anticoagulant. Each type has its merits and demerits. In the case of acute poisons, inadequate control can occur as the result of some individuals feeding insufficiently on bait and consuming a sub-lethal dose of poison. Improved success in an acute poison baiting treatment against mice can be achieved by the laying of plain bait for two or three days before the poison is included but this "pre-baiting" technique (Southern, 1954) is not always adopted for economic or other reasons.

The development of the anticoagulant rodenticides some 30 years ago marked a major advance in rodent control. These compounds have a common action, disrupting the mechanism controlling clotting of the blood and causing fatal haemorrhages to develop. Their use demands no prior pre-baiting and vitamin K_1 is strongly antidotal. Even so, the baiting method adopted in controlling house mice with anticoagulants differs little from that involving acute poisons. The limited movement of mice in most indoor habitats and their unpredictable feeding habits require closely spaced baiting points and bait that is highly attractive in comparison with other available foods. The use of anticoagulants in mouse control received a serious set-back when populations resistant to warfarin and other existing anticoagulants began to occur (Rowe & Redfern, 1965). In recent years however, three new rodenticides, calciferol (Rowe, Smith & Swinney, 1974) and the anticoagulants difenacoum (Hadler & Shadbolt, 1975) and brodifacoum (Redfern, Gill & Hadler, 1976) have been found to kill mice resistant to other anticoagulants.

Dust preparations, usually incorporating an anticoagulant or Y-BHC, are also used in the treatment of house mouse infestations. Mice travelling through strategically sited dust patches become contaminated with dust and ingest the poison during grooming. The effectiveness of this control method is therefore independent of any idiosyncrasies in feeding behaviour. For efficiency relatively high concentrations of poison are included in formulated dusts and safety problems can limit their use. Anticoagulant-treated water baits are less frequently used; this poisoning technique is effective against mice in dry environments, but quicker control is achieved by placing a poisonous dust around drinking points (Rowe & Chudley, 1963).

Fumigation

Fumigation is a quick and efficient method of controlling mice living in stockpiled foods (Thompson, 1959). Methyl bromide is most commonly employed in this manner and, with the correct dosage level and exposure period, it is possible to treat insect and mouse infestations simultaneously. However, fumigation is an expensive control method, requiring skilled labour, and fumigation of the free space or some other additional control measure is required in buildings where mice infest the fabric.

Trapping

Snap-traps are still commonly employed in house mouse control but they have only a limited value in dealing with populations that are dense or widespread. They are most effective when infestation is light, to control, for example, individual mice entering houses, stores or factories. Traps are also useful in helping to remove the survivors of poison treatments and in situations where the use of rodenticides is undesirable. The effectiveness of trapping campaigns is largely determined by the sensitivity of the traps, their siting and abundance. Even in carefully planned and intensive campaigns difficulty can be experienced in completely eliminating even a small mouse population, for trapping is selective, the heavier animals being at greatest risk while some individuals may tend to avoid traps. For the best success, frequent trapping sessions, conducted over a few days, are preferable to continuous campaigns.

Biological Control

Control methods based on the use of bacterial cultures and of reproduction inhibitors have been practised in the past or considered in relation to house mouse control, but their future development or use is problematical. The use of bait containing cultures of the bacterium *Salmonella* in rodent control was first made about 75 years ago. Such preparations were not found to be completely effective against mice, however, and the risk of side-effects on human populations has led to the banning of their use in most countries. Several chemosterilant compounds, notably synthetic oestrogens, cause long-lasting infertility in rats but house mice have not received the same attention. It seems improbable, in any event, that the application of a reproduction inhibitor would be considered a method of first choice against indoor populations of mice because there is invariable need for a more immediate form of control.

FUTURE DEVELOPMENTS

The basic work of Southern (1954) emphasized the need for detailed information on all aspects of house mouse biology, behaviour and ecology in the design and implementation of effective and long-lasting control measures. Our knowledge of this ubiquitous rodent has been greatly extended in more recent years and there has been steady advance too in the development of control techniques. Outstanding problems still remain, however, and further research is required both in relation to the improvement of existing techniques and in the development of alternative control methods.

Over the years, considerable attention has been focused on rodenticides, their efficacy, best mode of application and related problems. The occurrence of commensal rat and mouse populations resistant to anticoagulant poisons stimulated genetical and biochemical studies in order to elucidate the mode of inheritance and the physiological mechanisms involved. It also led to a more rational and fruitful search than hitherto for alternative rodenticides although resistance problems will almost inevitably recur. Less advance has been made in understanding the physiological mechanisms involved in the development of "poison bait shyness". The occurrence of this phenomenon in rodents, the outcome of the ingestion of a sub-lethal dose of a quick-acting poison and subsequent bait refusal, was clearly demonstrated in experimental studies conducted by Rzoska (1953). Attempts have been made to increase the acceptance of such poisons in bait and to delay the onset of poisoning symptoms by the use of microencapsulation techniques (Greaves *et al.*, 1968; Cornwell, 1970). Improvement in efficacy was found to be marginal but continued research in this field is well justified.

All experience of pest research on rodents suggests that new control techniques are more likely to stem from sustained behavioural studies than from less well-founded investigations. There has been, for example, an exhaustive and mainly abortive search for compounds showing suitable rodent repellent or attractant properties, for inclusion in packaging or other materials and bait respectively. The investigation of ultrasonic and electromagnetic wave-producing devices for repellency purposes, has indicated they are largely worthless (Greaves & Rowe, 1969; R. J. Smith, 1979). Deserving greater attention and of more potential practical value is investigation into the responses of rodents to specific chemical social signals (pheromones) and to ultrasonic calls used in communication. J. C. Smith (1976) has shown, for example, that the activity of adult mice can be much influenced by their response to the calls of infants. Taking advantage of other behavioural traits of the mouse such as investigative and grooming activity, Jenkins & Gibson (1979) have recently reported on a non-baiting method of poisoning mice, low in hazard.

While behavioural and associated research studies will almost certainly lead to the introduction of more successful control measures against *M. musculus* in the future, it is evident that more significant contributions to the reduction of infestation problems will have to be found elsewhere. The success of the house mouse as a commensal pest is to a large extent the result of human failings, difficulties in practical control being aggravated by such factors as low standards of hygiene, the acceptance of infestation, inferior, ill-designed and poorly maintained buildings and, at the operational level, the failure to plan and conduct control programmes efficiently. Fundamental changes in these areas are needed to bring about a major and long-term reduction of this rodent problem.

REFERENCES

Anderson, P. K. (1961). Density, social structure, and nonsocial environment in house mouse populations and the implications for regulation of numbers. *Trans. N. Y. Acad. Sci.* 23: 447–451.

Archer, J. (1968). The effect of strange male odour on aggressive behavior in male mice. *J. Mammal.* **49**: 572–575.
Berry, R. J. (1968). The ecology of an island population of the house mouse. *J. Anim. Ecol.* 37: 445–470.
Berry, R. J., Jakobson, M. E. & Peters, J. (1978). The house mice of the Faroe islands: a study in micro-differentiation. *J. Zool., Lond.* **185**: 73–92.
Berry, R. J., Jakobson, M. E. & Triggs, G. S. (1973). Survival in wild living mice. *Mammal Rev.* 3: 47–57.
Bowers, J. M. & Alexander, B. K. (1968). Mice: individual recognition by olfactory cues. *Science, N. Y.* **158**: 1208–1210.
Breakey, D. R. (1963). The breeding season and age structure of feral house mouse populations near San Francisco Bay, California. *J. Mammal.* **44**: 153–168.
Brown, R. Z. (1953). Social behavior reproduction and population changes in the house mouse (*Mus musculus* L.). *Ecol. Monogr.* 23: 217–240.
Caldwell, L. D. (1964). An investigation of competition in natural populations of mice. *J. Mammal.* 45: 12–30.
Cameron, T. M. W. (1949). Diseases carried by house mice. *Pest Control* 17(9): 9–11.
Cornwell, P. B. (1970). Studies in microencapsulation of rodenticides. *Int. Pest Control* 12(4): 35–42.
Cornwell, P. B. & Bull, J. O. (1967). Alphakil, a new rodenticide for mouse control. *Pest Control* 35(8). 31–32.
Crowcroft, P. (1955). Territoriality in wild house mice *Mus musculus* L. *J. Mammal.* 36: 299–301.
Crowcroft, P. (1959). Spatial distribution of feeding activity in the wild house mouse (*Mus musculus* L.). *Ann. appl. Biol.* 47: 150–155.
Crowcroft, P. & Rowe, F. P. (1957). The growth of confined colonies of the wild house mouse (*Mus musculus* L.). *Proc. zool. Soc. Lond.* **129**: 359–370.
Crowcroft, P. & Rowe, F. P. (1963). Social organization and territorial behaviour in the wild house mouse (*Mus musculus* L.) *Proc. zool. Soc. Lond.* **140**: 517–531.
Eibl-Eibesfeldt, I. (1955). Über das Massenauftreten des Hausmaus auf Süd – Seymour, Galapagos. *Saugetierk. Mitt.* 3: 175–176.
Fertig, D. S. & Edmonds, V. W. (1969). The physiology of the house mouse. *Scient. Am.* 221: 103–110.
Goldenberg, N. & Rand, C. (1971). Rodents and the food industry: an in-depth analysis for a large British food handler. *Pest Control* 39(11): 24–26, 28–29, 45, 47.
Greaves, J. H. & Rowe, F. P. (1969). Responses of confined rodent populations to an ultrasound generator. *J. Wildl. Mgmt* 33: 407–417.
Greaves, J. H., Rowe, F. P., Redfern, R. & Ayres, P. (1968). Microencapsulation of rodenticides. *Nature, Lond.* 219: 402–403.
Hadler, M. R. & Shadbolt, R. S. (1975). Novel 4-hydroxycoumarin anticoagulants active against resistant rats. *Nature, Lond.* 253: 275–277.
Hopkins, M. (1953). Distance perception in *Mus musculus. J. Mammal.* 34: 393.
Jenkins, D. L. & Gibson, J. A. (1979). New ways of mouse control. *Food Manuf.* 54(8): 33 & 35.
Jenson, A. G. (1979). *Proofing of buildings against rats, mice and other pests.* London: HMSO. (Replacement for Technical Bulletin 12: 1–33).

Laurie, E. M. O. (1946). The reproduction of the house mouse (*Mus musculus*) living in different environments. *Proc. R. Soc.* (**B**)**133**: 248–281.

Lidicker, W. Z. (1966). Ecological observations on a feral house mouse population declining to extinction. *Ecol. Monogr.* **36**: 27–50.

Mackintosh, J. H. & Grant, E. C. (1966). The effect of olfactory stimuli on the agonistic behaviour of laboratory mice. *Z. Tierpsychol.* **23**: 584–587.

Macleod, C. E. (1959). The population dynamics of unconfined populations of the house mouse (*Mus musculus*) in Minnesota. *Diss. Abstr.* **20**: 1492.

Minette, H. P. (1964). Leptospirosis in rodents and mongooses on the island of Hawaii. *Am. J. trop. Med. Hyg.* **13**: 826–832.

Newsome, A. E. (1969a). A population study of house mice temporarily inhabiting a South Australian wheatfield. *J. Anim. Ecol.* **38**: 341–359.

Newsome, A. E. (1969b). A population study of house mice permanently inhabiting a reed bed in South Australia. *J. Anim. Ecol.* **38**: 361–377.

Newsome, A. E. & Crowcroft, P. (1971). Outbreaks of house-mice in south Australia in 1965. *CSIRO Wildl. Res.* **16**: 41–47.

Petrusewicz, K. & Andrzejewski, R. (1962). Natural history of a free-living population of house mice (*Mus musculus* L.) with particular references to groupings within the population. *Ekol. pol.* (**A**)**10**: 85–112.

Redfern, R., Gill, J. E. & Hadler, M. R. (1976). Laboratory evaluation of WBA 8119 for use against warfarin resistant rats and mice. *J. Hyg., Camb.* **77**: 491–426.

Rennison, B. D. & Shenker, A. M. (1976). Rodent infestation in some London boroughs in 1972. *Environ. Health* **84**: 9–10, 12–13.

Rowe, F. P. (1967). Notes on rats in the Solomon and Gilbert Islands. *J. Mammal.* **48**: 649–650.

Rowe, F. P. (1973). Aspects of mouse behaviour related to control. *Mammal Rev.* **3**: 58–63.

Rowe, F. P. & Chudley, A. H. J. (1963). Combined use of rodenticidal dust and poison solution against house mice infesting a food store. *J. Hyg., Camb.* **61**: 169–174.

Rowe, F. P. & Redfern, R. (1965). Toxicity tests on suspected Warfarin resistant house mice (*Mus musculus* L.) *J. Hyg., Camb.* **63**: 417–425.

Rowe, F. P. & Redfern, R. (1969). Aggressive behaviour in related and unrelated wild house mice (*Mus musculus* L.). *Ann. appl. Biol.* **64**: 425–431.

Rowe, F. P., Smith, F. J. & Swinney, T. (1974). Field trials of calciferol combined with Warfarin against wild house mice (*Mus musculus* L.) *J. Hyg., Camb.* **73**: 353–360.

Rowe, F. P. & Swinney, T. (1977). Population dynamics of small rodents in farm buildings and on arable land. *EPPO Bull.* **7**: 431–437.

Rowe, F. P., Taylor, E. J. & Chudley, A. H. J. (1963). The numbers and movements of house mice (*Mus musculus* L.) in the vicinity of four corn ricks. *J. Anim. Ecol.* **32**: 87–97.

Rzoska, J. (1953). Bait shyness, a study in rat behaviour. *Br. J. Anim. Behav.* **1**: 128–135.

Smith, J. C. (1976). Responses of adult mice to models of infant calls. *J. comp. Physiol. Psychol.* **90**: 1105–1115.

Smith, R. J. (1979). Rodent repellers attract EPA strictures. *Science, N. Y.* **204**: 484–486.

Smith, W. W. (1954). Reproduction in the house mouse, *Mus musculus* L. in Mississippi. *J. Mammal.* **35**: 509–515.

Southern, H. N. (1954). *Control of rats and mice.* **3**. Oxford: Clarendon Press.
Southwick, C. H. (1955). Regulatory mechanisms of house mouse populations: social behaviour affecting litter survival. *Ecology* **36**: 627–634.
Taylor, K. D. (1972). The rodent problem. *Outl. Agric.* **7**: 60–67.
Thompson, R. H. (1959). Fumigation against mice. *Pest Technol.* **2**: 7–11.
Waugh, K. T. (1910). The role of vision in the mental life of the mouse. *J. Neurol.* **20**: 549–599.
Whitaker, J. O. (1966). Food of *Mus musculus, Peromyscus maniculatus bairdi* and *Peromyscus leucopus* in Vigo County, Indiana. *J. Mammal.* **47**: 473–486.
Young, H., Strecker, R. L. & Emlen, J. T. (1950). Localization of activity in two indoor populations of house mice *Mus musculus. J. Mammal.* **31**: 403–410.

Symp. zool. Soc. Lond. (1981) No. 47, 591–616

The Role of the House Mouse in Disease and Zoonoses

JENEFER M. BLACKWELL

Ross Institute, London School of Hygiene and Tropical Medicine, Keppel Street, London, UK

SYNOPSIS

A review of the literature reveals the relative unimportance of *Mus musculus* in zoonoses of human infectious disease. Only four situations in which mice have been implicated as natural reservoir hosts of infrequent, and hence incidental, human disease have been found. Three other situations occur where the agents of infection are found in both man and mice but do not appear to be directly transmitted between the two.

The real importance of mice in human infectious disease is found in their role as experimental hosts to infection. Prior to the development of modern sero-diagnostic techniques, clinical diagnosis of a disease could rely upon the ability of the infective agent to produce disease or a characteristic reaction in the experimental host. In recent times, however, the importance of mice as experimental hosts has been in the study of innate and acquired immunity to disease, and in drug trials.

No attempt has been made to review all of the literature on experimental infections in mice but examples of rickettsial, bacterial, fungal, protozoan, helminth and arthropod infections (or infestations) are examined. In so doing, emphasis is placed on some of the more interesting recent advances in the study of genetic control of innate and acquired responses to infection. In concluding, some speculative questions concerning the possible mechanisms of innate resistance are discussed.

INTRODUCTION

Although literally meaning "disease of animals" the term zoonoses is today employed, except in the Russian literature, for "those diseases and infections (the agents of) which are naturally transmitted between (other) vertebrate animals and man" (World Health Organisation, 1959, 1979). Euzoonoses occur when the infection (or infestation) is common in man and the reservoir host, and parazoonoses when man is an infrequent and hence incidental host (Faust &

Russell, 1964). Only four parazoonoses in which mice have been implicated as natural reservoir hosts were found in the literature, and no euzoonoses above the level of virus infections (which are the subject of another chapter in this volume). Three other situations occur where the agents of infection are found in both man and mice but do not appear to be directly transmitted between the two. This may be due in part to lack of a suitable contact between them. It is likely, however, that biochemical taxonomic procedures may yet distinguish between apparently similar infectious agents of man and mouse.

Wild *Mus musculus* thus appear to be quite unimportant as natural rodent reservoir hosts of human disease. The subject of natural infections of wild house mice which are not agents of human disease will not be examined.

The real importance of mice in human infectious disease lies in their role as experimental hosts, a subject which would in itself require an entire volume to deal with satisfactorily. In the time of the early masters such as Pasteur and Koch, the proof that an organism was capable of causing disease rested, in many cases, on its ability to produce disease or a characteristic reaction in a lower vertebrate. A very strict protocol was adhered to when this procedure was subsequently developed into a test for clinical diagnosis. The practice reached its peak in the tropics, for example, where in excess of 30 000 mice might be used in one area in one year in the diagnosis of yellow fever virus. Bacterial diseases formerly diagnosed by inoculation of infected material from the patient (e.g. fluid exudates, tissue homogenates, or serum) included pneumococcus, anthrax, and rat-bite fevers (Whitby & Hynes, 1951). Since the development of modern culture methods and serodiagnostic tests, however, the importance of mice has been as experimental models of disease in the study of innate and acquired immunity and in drug trials. This paper will be limited to an examination and appraisal of some of the more interesting recent advances in the study of genetically controlled resistance to infection.

NATURAL INFECTIONS

Parazoonoses

Rickettsial pox

Rickettsial pox is a self-limiting, acute febrile illness in man caused by *Rickettsia akari* and accidentally transmitted from mice to man by a bite from the sucking mite *Allodermanyssus sanguineus*

(Woodward & Jackson, 1965). It was first observed in New York City in 1946 (Sussman, 1946; Shankman, 1946) and later in New England, Philadelphia, and Cleveland. For several years about 180 cases were reported annually in New York City alone but by the early 1960s the incidence had declined, with only a few reported cases each year (Lackman, 1963). The identity of *R. akari* isolated from humans, mice and mites has been demonstrated (Huebner, Jellison & Pomerantz, 1946; Huebner, Stamps & Armstrong, 1946; Huebner, Jellison & Armstrong, 1947) and the role of mice in the epidemiology of the disease outlined (Greenberg, Pelliteri & Jellison, 1947). The disease is now optionally reportable, hence few cases have been recorded in recent years by the United States Public Health Service. Nevertheless, because of the widespread distribution of the vector mite (Pratt, 1963), it seems likely that the disease would be found throughout the United States if an effort was made to diagnose and report cases.

Soviet workers in 1949–50 described a similar benign, febrile illness and proved its rickettsial nature by isolating the organism from the blood of patients (see Zdrodovskii & Golinevich, 1960). Epidemiologically the disease, designated "vesicular rickettsiosis", follows the same urban pattern as in the United States. Incidence in recent years has declined, probably owing to rodent control (World Health Organisation, 1963).

Experimental infection of *R. akari* in inbred mice indicates genetical control of susceptibility in the host (see p. 595).

Haverhill rat-bite fever

Rat-bite fever due to *Actinobacillus muris* is an irregularly relapsing, febrile, septicaemic disease characterized by metastatic arthritis and morbilliform and petechial cutaneous eruptions (Wilson & Miles, 1964). The bacteria occur in the nasopharynx of the rat and are normally transmitted to humans following a rat-bite. Outbreaks of Haverhill fever studied by Parker & Hudson (1926) and Place & Sutton (1934) were both, however, traced to the consumption of raw milk. It is thought possible that excretion of the bacteria in the urine of infected mice could account for the contamination of the milk.

The natural disease "infective arthritis of mice", produced in laboratory mice by *Actinobacillus muris,* has been described by Levaditi, Selbi & Schoen (1932) and Mackie, van Rooyen & Gilroy (1933), and in wild mice by S. Williams (1941).

Salmonellosis

Mice have often been regarded as possible vectors of human and animal salmonellosis because of their indiscriminate habits in feeding and defaecation. In the only two documented cases of mice transmitting infection to man through contamination of food (Brown & Parker, 1957), evidence existed for the mice having acquired the infection from living cultures of *Salmonella enteritidis* var. *danysz* in rodenticide baits. A general survey of wild mice in the city (Manchester) in which the cases were reported produced only one infected with *S. enteritidis* var. *danysz* out of 256 examined (0.4%). In this case it seemed likely that the infection had again been acquired from bait laid in a nearby factory. Similarly, of 364 house mice examined from estates in Berkshire, Oxfordshire and Surrey (Jones & Twigg, 1976), *S. typhimurium* was isolated from one and *S. dublin* from seven. *S. dublin* had been acquired from experimentally infected cattle. Natural infection of house mice is so uncommon in comparison with other reservoir hosts that mice were not even considered as possible reservoir hosts in the Zoonoses Order 1975 (Bennett, 1976; McCoy, 1976; Osborne, 1976). However, Shimi, Keyhani & Hedayati (1979) isolated salmonellae from the faeces of 17 out of 170 (10%) wild mice examined in Iran. The serotypes isolated were all similar to those frequently isolated from man and other animals in Iran. From this circumstantial evidence Shimi *et al.* (1979) concluded that the house mouse population may play an important role in human and animal salmonellosis. Direct evidence is still, however, required.

Genetical control of susceptibility to salmonella infection in mice is examined on p. 597.

The rat tapeworm

The rat tapeworm, *Hymenolepis diminuta,* is a common parasite in rats and mice as well as a variety of "native" rodents. Humans are occasionally infected by accidental ingestion of the ectoparasites of the murine host or from ingesting insects infesting pre-cooked cereals (Faust & Russell, 1964). The relative importance of mice as the source of human infection is not well documented.

Infections Common to Mouse and Man

Murine typhus

The aetiologic agent of murine typhus, *Rickettsia mooseri,* occurs as a natural infection in rats and mice and causes an acute febrile illness in man when accidentally transmitted by the rat flea *Xenopsylla cheopis* (Snyder, 1965).

Rat-bite fever, *Spirillum minus*

Numerous strains of the causative agent of the second disease known as rat-bite fever, *Spirillum minus*, have been isolated from mice, rats, and human patients (Schockaert, 1928). All apparently constitute a single species although mice have never been implicated in transmission of the disease to man. Bacteriological diagnosis of the disease is made by subcutaneous or intraperitoneal inoculation of blood taken at the height of a febrile paroxysm, or of serum expressed from the local lesion, into mice and guinea pigs (Brown & Nunemaker, 1942).

The dwarf tapeworm

The dwarf tapeworm, *Hymenolepis nana*, is the only human tapeworm typically requiring no intermediate host (Faust & Russell, 1964). Experimental infection of young children with eggs from murine strains suggests that, under exceptional circumstances, they may be susceptible to infection from murine sources. Man is probably, however, the only common source for human infection.

EXPERIMENTAL INFECTIONS

Rickettsial infections

Marked differences in susceptibility to infection with *Rickettsia tsutsugamushi*, the agent of human scrub typhus, have been observed in inbred strains (Table I) by Kekcheeva & Kokorin (1976). A more extensive study by Groves & Osterman (1978) demonstrated that resistance was dominant and under the control of a single autosomal gene which has subsequently been designated *Ric* and mapped to chromosome five (Rosenstreich, O'Brien, Groves & Taylor, 1980). Studies of peritoneal exudate cells from intraperitoneally infected mice (Catanzaro *et al.*, 1976; Kokorin, Chyong-din-Kyet *et al.*, 1976) suggest that rickettsiacidal macrophages with increased activities of hydrolytic and oxidizing–reducing enzymes suppress the infection at the site of inoculation in resistant strains. In highly susceptible strains fatal infection is accompanied by death of macrophages and necrotization of peritoneal exudate cells.

Inbred mouse strains also show marked differences in susceptibility to *R. akari* (Rybkina, Kabanova & Chyong-din-Kyet, 1976; Chyong-din-Kyet, 1977), the agent of human rickettsial pox (see p. 592). For the two strains examined (see Table I) the susceptibilities were exactly opposite to *R. tsutsugamushi* infection. *In vitro* infection

TABLE I

Strain differences in resistance to rickettsial and bacterial infections

Infective agent	Susceptible strains	Resistant strains	Mechanism	References
Rickettsiae				
R. tsutsugamushi	A/HeJ, C3H/A, C3H/HeDub, C3H/HeJ, C3H/HeN, C3H/St, CBA/J, DBA/1J, DBA/2J, SJL/J	AKR/J, BALB/cDub, BALB/cJ, C57BL/J, C57BL/6, CBA(Russian), SWR/J	Innate, single gene chr. 5, r dominant, rickettsiacidal PE macrophages	Catanzaro *et al.* (1976) Kekcheeva & Kokorin, (1976) Kokorin, Chyong-din-Kyet *et al.* (1976) Groves & Osterman (1978)
R. akari	C57BL/6	DBA/2	Innate, rickettsiacidal spleen macrophages	Rybkina *et al.* (1976) Chyong-din-Kyet (1977) Kokorin, Kabanova *et al.* (1978)
Bacteria				
Salmonella[a] *typhimurium* i.p.	BALB/cJ, C57BL/6J, C3H/HeJ	A/J		Robson & Vas (1972)
S. typhimurium s.c.	BALB/c, B10.D2/n, B10.M, C57BL, C57BL/10J, DBA/1	A/J, CBA, C3H, C57L, DBA/2, I/St	Innate, single gene (*Ity*), r dominant, chr. 1	Plant & Glynn (1974, 1976, 1979)
Mycobacterium[b] *bovis* BCG	CBA/JG, C3H/HeCr, C3H.SW	B10.BR, C57BL/6, C57BL/Ks, C57BR/cdJ	Acquired, responder strains: enlarged spleen & lung	Allen, Moore & Stevens (1977)
M. lepraemurium[c]	BALB/c, C3H	C57BL	Acquired, difference in DH kinetics, dose dependent, lymphocyte infiltration in r mice	Closs & Haugen (1974) Closs (1975) Alexander & Curtis (1979)
Listeria monocytogenes	A/J, BALB/cJ, CBA/H, C3H/HeJ, C3H.OH/Sf, C3H, OL/Sf, DBA/1J, DBA/2J, LP.RIII, WB/Re, 129/J	B10.A/SgSn, B10.D2/Sn, B6.C-*H-2*[d]By, B6.PL(74NS)/Cy, C57BL/6J, C57BL/Sn, C57BL/10ScSn, NZB/WEHI, SJL/WEHI[c]	Innate, single gene, r ?inc. dominant, nonimmune macrophages kill or inhibit bacteria, early CMI in r mice	Cheers & MacKenzie (1978) Cheers *et al.* (1978) Skamene, Kongshavn & Sachs (1979)
Pseudomonas[d] *aeruginosa*	A/J, A/WySn, A.BY	C3H/HeJ, C3H/HeSn, C3H.SW/Sn	Innate, not serum factor	Pennington & Williams (1979)

[a] DBA/2J intermediate in susceptibility.
[b] Susceptible = non-responder, resistant = responder.
[c] AKR, WL, & ?BALB/c intermediate in susceptibility (Closs & Haugen, 1974).
[d] 10 strains intermediate in susceptibility: BALB/c, B10.A/SgSn, B10.A(2R)/SgSn, B10.A(5R)/SgSn, B10.BR, B10.D2/nSn, B10.D2/oSn, C57BL36J, C57BL/10Sn, DBA/2J.

of primary cultures of spleen macrophages derived from susceptible C57BL/6 and resistant DBA/2 mice (Kokorin, Kabanova *et al.*, 1978) resulted in intensive multiplication of rickettsiae in susceptible macrophages which eventually died and released rickettsiae into the culture medium, whereas rickettsiae were eliminated from resistant macrophages three days after being infected. A rickettsiacidal mechanism is therefore operating in macrophages from the strain of mouse which fails to control *R. tsutsugamushi* infection. Since *R. tsutsugamushi* resistance involves killing the rickettsia in macrophages at the site of inoculation, it would be interesting to know whether primary cultures of spleen macrophages from resistant mice could eliminate rickettsiae.

Bacterial Infections

Salmonella

Robson & Vas (1972) observed three patterns of survival in inbred mice infected intraperitoneally with *S. typhimurium* (Table I): susceptible, intermediate and resistant. With subcutaneous inoculation, Plant & Glynn (1974, 1976) demonstrated that inbred strains fall into sharply defined susceptible and resistant groups (Table I). Resistant mice were more readily immunized with salmonella antigen (Robson & Vas, 1972) and develop better salmonella-specific cell-mediated immunity (Plant & Glynn, 1974). Differences in results obtained with the different routes of inoculation were explained on the basis of effective increase of the inoculum by rapid bacterial growth in the poorly protected peritoneal cavity. Resistance was shown to be under single gene control with resistance dominant. The gene has been named *Ity* and located close to *ln* on chromosome one (Plant & Glynn, 1979).

Mycobacterium

Acquired resistance to infection with *M. lepraemurium* differs among inbred strains of mice (Table I). Granulomas in the livers of susceptible mice examined 15–16 weeks after intraperitoneal inoculation of $4–16 \times 10^7$ bacilli contain large numbers of bacteria with few or no surrounding lymphocytes (Closs & Haugen, 1974). Granulomas in resistant mice contain fewer bacilli and regularly show infiltration of lymphocytes. This difference in host response is less evident after a larger ($\times$ 10) initial inoculum of bacilli.

After foot-pad inoculation bacilli multiply in both susceptible and resistant mice during the first four weeks (Closs, 1975). Inhibition of multiplication of bacilli in resistant mice is preceded by a local swelling of the foot pad, suggesting the onset of an immune reaction. In susceptible mice this response is not detected and bacilli continue to multiply. Foot-pad reactions to mycobacterial antigen (MLM lepromin) in the uninoculated foot of susceptible and resistant mice (Alexander & Curtis, 1979) peak between six and ten days after inoculation. This is followed by a period of low reactivity before the development, in the third week, of a stable foot-pad reaction. No difference between susceptible and resistant mice in the size of the reaction at 24 hours is observed, but they differ markedly in their kinetics. Susceptible mice give a Jones–Mote–type of response while resistant mice give a tuberculin-type reaction. Further, with high and low doses of lepromin the prolonged foot-pad reaction of resistant mice is accompanied by mononuclear cell infiltration (Curtis & Turk, 1979). Susceptible mice show only mononuclear cell infiltration with the high dose. Dissemination of organisms from the infected foot-pad is minimal in resistant mice. In susceptible mice, dissemination to the draining lymph node and the liver occurs by five months after infection.

Inbred strains of mice also differ in their ability to kill *M. bovis* BCG (Table I). Responder strains show a dose-dependent enlargement of the spleen and lung four weeks after intravenous inoculation; non-responder strains do not (Allen, Moore & Stevens, 1977).

Listeria

The independent researches of Cheers & McKenzie (1978) and Skamene, Kongshavn & Sachs (1979) have recently demonstrated that innate resistance to *Listeria monocytogenes* in inbred strains of mice is under single gene control. Although acquired cell-mediated immunity develops earlier in resistant strains, resistance appears to be primarily related to the innate ability of nonimmune macrophages to kill or inhibit the growth of the organism during the first 24–28 hours after infection (Cheers *et al.*, 1978).

Adoptive transfer of cell-mediated immunity to naïve animals is restricted by *H-2* (Zinkernagel, 1974), specifically by the I region (Zinkernagel *et al.*, 1977).

Pseudomonas

Natural resistance to intraperitoneal infection with *P. aeruginosa* varies in inbred strains (Table I; Pennington & Williams, 1979). Killing activity of serum against the bacteria is not greater in resistant mice.

Fungal Infections

Reproducible strain-dependent differences in susceptibility to infection with *Histoplasma capsulatum* (Chick & Roberts, 1974) and with *Blastomyces dermatitidis* (Morozumi & Stevens, 1979) have been demonstrated (Table II). Administration of antilymphocyte serum markedly increases susceptibility of mice resistant to *H. capsulatum* (Patton *et al.*, 1976). The genetic basis for these strain differences has not been determined.

Other murine models of acquired immunoregulatory responses to human fungal pathogens utilizing inbred strains include: *Candida albicans* (Cutler, 1976; Domer & Moser, 1978; Giger, Domer & McQuitty, 1978; Giger, Domer, Moser & McQuitty, 1978); *Cryptococcus neoformans* (Hay & Reiss, 1978; Cauley & Murphy, 1979); *Coccidioides immitis* (Beaman, Pappagianis & Benjamini, 1977, 1979); and *Histoplasma capsulatum* (Howard & Otto, 1977; Williams, Graybill & Drutz, 1978; Artz & Bullock, 1979). Robson & Vas (1972) noted that A/J mice were susceptible while C57BL/6J mice were resistant to *C. neoformans.* Beyond that, genetical regulation of these responses has not been examined.

Protozoan Infections

Leishmania

Innate susceptibility to *L. donovani* infection in mice (measured over two to four weeks) is under the control of a single autosomal gene (*Lsh*) segregating for incompletely dominant resistant (*r*) and recessive susceptible (*s*) alleles (Bradley, 1974, 1977). Strain distribution of susceptible and resistant mice (Table III) compares precisely with *S. typhimurium* (Table I). *Lsh* maps away from the known histocompatibility loci to a position between the centromere and *Id-1* on chromosome 1 (Bradley *et al.*, 1979). Resistance is unaltered by thymectomy (Bradley & Kirkley, 1972) or serum factors (unpublished data). Autoradiographic studies demonstrate much lower parasite proliferation rates in resistant than in susceptible mice (Bradley, 1979). Some differential microbicidal activity of macrophages cannot, however, be excluded.

Amongst homozygous recessive Lsh^s strains of mice two patterns of recovery are observed when the course of infection is followed over a longer term (Bradley & Kirkley, 1977). In some strains a dramatic fall in parasite numbers with histological liver damage occurs while other strains maintain immense parasite loads for up to two years involving mononuclear phagocytes throughout the body. Recent

TABLE II

Strain differences in resistance to fungal infections in mice

Infective agent	Susceptible strains	Intermediate strains	Resistant strains	References
Histoplasma capsulatum	C57BL/10Sn	A/HeJ, DBA/2J	A/J, SJL/J, SWR/J	Chick & Roberts (1974)
Blastomyces dermatitidis	C3H/HeJ	A/HeJ, BALB/c/St, C57BL/10J, C3H.SW, DBA/2J, SJL/J	DBA/1J	Morozumi & Stevens (1979)

TABLE III

Strain differences in resistance to Leishmania *infections*

Infective agents	Susceptible strains	Intermediate strains	Resistant strains	Mechanisms	References
L. donovani Early phase	BALB/c, B10.A, B10.BR, B10.D2/n, B10.LP-a, C57BL, C57BL/10ScSn, SE, DBA/1, ICFW, NMRI, NZW		A, AKR, A2G, CBA/Ca, CBA/H-T6, C3H/He-*mg*, C57BR/cd, C57L, DBA/2, F/ST, NZB, 129/RrJ	Innate, single gene (*Lsh*), r dominant, chr. 1, low parasite proliferation in r mice	Bradley (1974, 1977, 1979) Bradley *et al.* (1979)
Late phase[a] (to 130 days)	BALB/c, B10.D2/a, B10.G, B10.M	BALB, C57BL/10ScSn	B10.S, B10.RIII	Acquired, *H-2* linked, chr.17, cure recessive & CM non-cure ?suppression	Blackwell, Freeman & Bradley (1980)
Late phase[a] (to 35 days)	B10.129(10M)	C57BL/10Sn	B10, LP-*H-3*[b]	Acquired, *Ir-2* linked dominant early cure with DH; H-11 linked non-cure	DeTolla *et al.* (1980)
L. tropica[b]		A2G, B10.D2, C57BL, DBA/2, NMRI	CBA/LAC, C3H/He, C57BR	Acquired, dose dependent CM	Preston, Behbehani & Dumonde (1978)
	BALB/c	DBA/1, DBA/2	A, ASW, C57BL, C57BL/6 C57BL/10, CBA	Extreme non-healing is single gene suppressor, F1 intermediate, dose independent	Howard, Hale & Chan-Liew (in press) Howard, Hale & Liew (1980)
L. mexicana[b]	BALB/c		C57BL/6		Perez, Arrendondo & Gonzalez (1978)

[a] Susceptible = non-cure, intermediate = cure, resistant = early cure
[b] Susceptible = extreme non-healing, intermediate = non-healing, resistant = self-healing at 10^5 or lower

studies (Blackwell, Freeman & Bradley, 1978, 1980) suggest that this difference in long-term response (up to 130 days) is largely controlled by a gene(s) situated in the K end of *H-2*. *Rld-1* (recovery from *L. donovani*) has been proposed as the provisional designation for this locus, while the two basic phenotype patterns are referred to as "cure" and "non-cure". "Cure" behaves as a recessive trait in $H\text{-}2^{b/d}$ mice. Liver parasite burdens decrease earlier in some "cure" mice ($H\text{-}2^{s}$, $H\text{-}2^{r}$) than in others ($H\text{-}2^{b}$). DeTolla *et al.* (1980) have provided evidence that non-*H-2* linked genes may also play a role in acquired resistance to *L. donovani* (Table III). On a B10($H\text{-}2^{b}$) genetical background strains carrying the alternative $Ir\text{-}2^{b}$ allele mimic our early "cure" haplotypes, while strains carrying the alternative $H\text{-}11^{b}$ allele may mimic our "non-cure" haplotypes. Infections were only followed for 35 days by DeTolla *et al.*

Preston, Behbehani & Dumonde (1978) found 12 mouse strains differed in their ability to heal cutaneous infection with 10^5 promastigotes (a low inoculum) of *L. tropica* (Table III). The patterns of response indicated a spectrum of host resistance to primary infection which ran parallel with acquisition or retention of delayed hypersensitivity to leishmanial antigen. Resistant strains healed their lesions and developed delayed hypersensitivity while susceptible strains were not able to heal their lesions. Susceptible mice could further be divided into two groups, those which failed to develop substantial delayed hypersensitivity and those in which a desensitization of delayed hypersensitivity occurred as the infection progressed. Resistant mice failed to heal after higher inocula of promastigotes.

Slightly different strain distribution of healing and non-healing mice (Table III) in *L. tropica* infection has been observed by Howard, Hale & Chan-Liew (in press). In addition to the normal non-healing response, BALB/c mice exhibit extreme dose-independent susceptibility to infection. The lesion continues to grow, the infection visceralizes and mice begin to die 12 weeks after infection. Early delayed hypersensitivity disappears later in infection. Immunological studies (Howard, Hale & Liew, 1980) indicate that a suppressor mechanism operates in extreme non-cure mice. Analysis of (BALB/c × C57BL) F_1, F_2 and back-cross progeny indicates that this mechanism is under single gene control (Howard, Hale & Chan-Liew, in press). F_1 mice display an intermediate response in which the size of lesion reaches a plateau after an initial increase. Some F_1 mice die later in infection.

This extreme susceptibility to cutaneous leishmaniasis in BALB/c mice has been confirmed for *L. mexicana* (Perez, Arrendondo & Gonzalez, 1978). Like Howard, these workers also found C57BL/6 mice capable of healing lesions (Table III). Preston *et al.* (1978) found C57BL mice to be nonhealing.

Trypanosoma

Clarkson (1976) demonstrated a marked rise in serum IgM concentration in inbred strains which survive longest after infection with *T. brucei.* Differential parasitaemias were not observed. With the NIM6 antigenic clone of *T. brucei* Clayton (1978) found that, in most cases, there was an inverse relation between survival time and parasite load during the first peak of parasitaemia. For the same strains examined (Table IV) the IgM response of Clarkson correlated precisely with the susceptible, intermediate and resistant strains of Clayton. Resistance appeared to be dominant. The two strains most resistant and most susceptible to *T. brucei* were also resistant and susceptible, respectively, to *T. cruzi* infection (Cunningham, Kuhn & Rowland, 1978). Biozzi high and low antibody-responding lines also differed in their response to *T. cruzi* infection (Kierszenbaum & Howard, 1976). Low responder mice were significantly more susceptible to infection in terms of shorter survival times, increased mortality rate, higher parasitaemias and lower LD_{50}. Mice of this line also developed much lower titres of anti-*T. cruzi* antibody.

Plasmodium

Strain distribution for variation in survival times and parasitaemias of inbred mice infected with *P. bergei* assessed by different workers (Greenberg & Kendrick, 1957; Most *et al.*, 1966; R. Vasquez, personal communication) are shown in Table IV. The picture is confusing with little agreement between workers even when using apparently similar criteria for measuring susceptibility. Greenberg & Kendrick (1957) suggest that the degree of parasitaemia and the extent of mortality after one week are "two partially dependent variables apparently under separate genetic control". There are thus four major categories into which strains of mice fall: low parasitaemia–low tolerance (i.e. high mortality); low parasitaemia–high tolerance; high parasitaemia–low tolerance; high parasitaemia–high tolerance (Wakelin, 1978). Inheritance of resistance between high and low parasitaemia strains (Greenberg & Kendrick, 1958, 1959) appears to be polygenically controlled.

Babesia

Susceptibility of four inbred strains of mice (Table IV) to infection with *B. microti* has been examined (Ruebush & Hanson, 1979). Strains vary early in infection but all eventually eliminate the

TABLE IV

Strain differences in resistance to other protozoan infections

Infective agent	Susceptible strains	Intermediate strains	Resistant strains	Mechanisms	References
Trypanosoma					
T. brucei[a]	C3H/mg	AKR, BALB/c, CBA, NZB	C57BL	IgM low in S, high in R	Clarkson (1976)
T. brucei	C3H/He	BALB/c, CBA/H, DBA/2	C57BL/6, (CBA/HxC57BL/6)	?resistance dominant	Clayton (1978)
T. cruzi	Biozzi Ab/L	–	Biozzi Ab/H	Deficient Ab response in S	Kierszenbaum & Howard (1976)
T. cruzi	C3H/He	–	C57BL/6		Cunningham *et al.* (1978)
Plasmodium					
P. berghei[b]	DBA/2JN	A/LN, BALB/cANN, C57BL/6JN, C3H/Hen	STR/N		Greenberg & Kendrick (1957)
P. berghei[c]	A/LN, C3H/Hen, DBA/2JN,	BALB/cANN, STR/N	C57BL/6JN	Polygenic control	Greenberg & Kendrick (1957)
P. berghei[c]	A/J, ST/6J, C57L	A/HeJ, ?AKR/J	DBA/2J, DBA/1J, C58/J		Most *et al.* (1966)
P. berghei[b]	AKR, CBA/J, C3H/He, C57BL	BALB/c, DBA/2	CBA/Ca, CBA/HTG, (CBA/CaxDBA/2)		R. Vasquez (pers. comm.)
Babesia					
B. microti	C3H/HeDub	BALB/c, CBA/J	C57BL/6J		Ruebush & Hanson (1979)

[a] Susceptibility based on IgM responses.
[b] Susceptibility based on survival time.
[c] Susceptibility based on parasitaemia.

parasite. No genetical analysis has been carried out to determine whether this difference in innate susceptibility is under single gene control.

Toxoplasma

Susceptibility of one outbred and six inbred strains of mice to infection with trophozoites of *T. gondii* at two different doses has been examined (Araujo *et al.*, 1976). Striking strain differences in susceptibility and changes of susceptibility with dosage were observed and suggested to be under genetical control.

Entamoeba

Neal & Harris (1977) reported C3H/mg and CBA/Ca mice to be susceptible to *E. histolytica* by intracardial inoculation. To investigate possible genetical and immunological differences among mice Gold & Kagan (1978) inoculated *E. histolytica* into the right ventral lobe of the liver of laparotomised ICR outbred mice and BALB/c × B10.D2, A/J, AKR, CBA, C57BL, DBA/2J and NMRI inbred strains, and a strain of hairless mice. Only eight of 209 (3.8%) mice examined had lesions as well as live amoebae in the liver and no single strain showed greater susceptibility.

Helminth Infections

Strain differences in susceptibility to infection with the cestode worms *Cysticercus fasciolaris* (Dow & Jarrett, 1960), *Echinococcus granulosus* (Lubinsky & Desser, 1963; Lubinsky, 1964; Ali-Khan, 1974), *Hydatigena taeniaeformis* (Gagarin *et al.*, 1972) and *Opisthorchis felineus* (Zelentsov, 1974) have been observed. BCG-induced resistance to infection with the trematode worm *Schistosoma mansoni* also varies between strains (Civil & Mahmoud, 1978) and is apparently under the control of genes not linked to *H-2*. Although hybrids between susceptible and resistant strains have been examined in some cases (Lubinsky & Desser, 1963; Lubinsky, 1964), the precise nature of the genetical control of these differences has not been thoroughly investigated.

Higher resistance of hybrids between strains susceptible and resistant to infection with the nematode worm *Nematospiroides dubius* (Si-Kwang Liu, 1966) suggests polygenic control of susceptibility. Protection against infection with *N. dubius* after immunization also varies between strains (Cypess & Zidian, 1975; Lueker & Hepler, 1975; Manger, 1976). No significant difference in susceptibility to infection with *Aspicularis tetraptera* was observed between outbred CF1, and inbred C57BR, C57BL and DBA/2 mice (Dunn & Brown, 1963). C3H mice, however, showed a reduction in worm burden. Genetical control of susceptibility to infection with *Ascaris suum* in mice has been analysed in terms of the two phases, natural or innate susceptibility to first infection and acquired resistance to second infection (Mitchell *et al.*, 1976). Susceptibility to first infection was shown to be under the polygenic control of genes not linked to *H-2.* T cell deficiency does not affect the innate response. Attempts to dissect out the mechanisms of acquired resistance were complicated by failure to show adoptive transfer of acquired resistance to naive recipients. Wakelin (1975) has suggested that the dominant ability to terminate an infection with *Trichuris muris* is immunologically mediated and controlled by only a few genes. Susceptibility of mice to *Trichinella spiralis* infection and protection against reinfection following a single immunizing infection is also shown to vary amongst mouse strains (Wakelin & Lloyd, 1976). More extensive examination of inbred strains of mice by Tanner (1978) failed, however, to demonstrate any significant differences in levels of infection with *T. spiralis.*

Ectoparasites

Clifford *et al.* (1967) examined susceptibility to infestation with the louse, *Polyplax serrata,* in amputee mice from several inbred strains. The greatest number of lice developed on C57BL/6JN mice and the smallest number on CFN mice. Those strains which supported the largest number of lice had the greatest mortality. Cross-breeding of resistant and susceptible strains suggested a genetical mechanism, with sex of the resistant parent influencing the reaction of the offspring. Coat colour of the mice appeared to influence louse infestation, but this was inconclusive.

CONCLUSIONS

As a result of lack of evidence for any major role of *Mus musculus* in zoonoses of human infectious disease, this paper has dealt primarily with experimental mouse models of disease; in particular, with genetical control of susceptibility and resistance to infection. In concluding, some very specific questions concerning genetical analysis of innate and acquired resistance and possible mechanisms involved in the innate response will be examined. For a more general discussion on genetical control of resistance and mechanisms involved in both innate and acquired responses, readers are referred to the recent excellent review by Wakelin (1978).

Firstly, the difficulties inherent in analysing genetical control of acquired responses without first examining the innate response should be emphasized. Many workers claim, for example, to have eliminated the possibility of *H-2* involvement in strain differences in susceptibility to infection purely on the basis that strains with the same *H-2* haplotype show different acquired responses. With these criteria it would not have been possible to demonstrate *H-2* linked acquired resistance to *L. donovani* infection since Lsh^r mice may carry the same *H-2* haplotype as "non-cure" Lsh^s mice. The value of *H-2* congenic resistant strains on the Lsh^s background must be emphasized. The possibility that *H-2* might also influence acquired resistance of Lsh^r mice has not been investigated because the levels of parasitaemia maintained in Lsh^r mice fall close to or below the threshold level for finding one parasite on an impression smear (see Blackwell *et al.*, 1980).

The question arises as to whether *H-2* influences the long-term acquired responses of *M. lepraemurium* and *L. tropica,* for which the same strains show acquired resistance (C57BL) and long-term susceptibility (BALB/c), in the same way as we have observed for *L. donovani.* In the case of *M. lepraemurium* this seems unlikely since C3H ($H\text{-}2^k$) mice are also susceptible to infection over long periods. $H\text{-}2^k$ Lsh^s mice usually recover from infection with *L. donovani* (Blackwell *et al.*, 1978). The same series of congenic resistant mice on BALB and B10 genetic backgrounds as was used in *L. donovani* studies by Blackwell *et al.* (1980) have now been examined for long-term response to *L. tropica* (Howard, Hale & Chan-Liew, in press). The extreme non-healing suppressor mechanism of BALB mice appears to be in the non-*H-2* background with only a minor *H-2* linked regulatory influence on the later stages of infection. Results obtained for B10 mice are inconclusive, varying according to the dose of parasite administered. Mice infected with 2×10^5 *L. tropica* show a similar rank order for haplotypes showing a mild residual disease and early arrest of lesion growth as found for "non-cure", "cure" and "early cure" haplotype mice infected with *L. donovani.* At a higher dose (2×10^7 *L. tropica*), the rank order for non-healing strains differs. These experiments are currently being repeated (J. G. Howard, personal communication).

A warning should also be given of the possible effect on experimental infections of natural infections with *Eperythrozoon coccoides.* This parasite normally remains undetected in the blood until after splenectomy (Thurston, 1955) and can dramatically alter the course of infection with another organism (Peters, 1965; Ott & Stauber, 1967).

The next question concerns the possible identity of genes controlling different infective agents. It has long been supposed, for example, that the genes controlling innate susceptibility to *S. typhimurium* and *L. donovani* (*Ity* and *Lsh*) might be identical (Bradley, 1974). Recent collaborative work by A. O'Brien and B. Taylor (personal communication) has suggested, however, that one of the recombinant inbred strains examined by Bradley *et al.* (1979) in mapping the *Lsh* gene differs in its susceptibility to *S. typhimurium.* The possibility that this represents a rare recombinant between two very closely linked genes is now being explored. Even very detailed formal genetical analysis and mapping of genes may therefore mask separate gene control. The possibility that several genes controlling macrophage function may be closely linked is, however, of considerable interest.

From the present examination of the innate resistance of some mice to intracellular macrophage interactions it is obvious that macrophages from the same strain of mouse can kill or control multiplication of one infective agent while another may multiply uncontrollably leading eventually to death of the cell. Two questions arise: (a) are different subpopulations of macrophages involved in the control of different infective agents; and (b) do the different genes represent control at different points of the same microbicidal biochemical pathways? The second question will be examined first.

Although the observations of Bradley (1979) suggest a mechanism involving inhibition of parasite proliferation in mice innately resistant to *L. donovani,* differential macrophage microbicidal activity in resistant and susceptible mice cannot be discounted. Optimum conditions for enhanced microbicidal activity in the resistant mouse (e.g. low intravacuolar pH) might, in itself, reduce the parasite proliferation rate. Differential microbicidal activity by macrophages from strains susceptible and resistant to rickettsial infection has been demonstrated (see p. 597). Evidence is not available to distinguish between the two possible mechanisms of resistance in mice infected with *Listeria* or *Salmonella.*

Of the possible antimicrobial systems of mononuclear phagocytes (Klebanoff & Hamon, 1975), the myeloperoxidase (MPO)-mediated hydrogen peroxide-dependent system seems the most likely mechanism to be involved in the innate resistance mechanism. Release of hydrogen peroxide from mouse peritoneal macrophages *in vitro* has been shown to depend upon activation and triggering (Nathan & Root, 1977). In an *in vivo* situation this triggering may well be accomplished by the very presence of the microbe. Sensitivity of most of the microbes eliciting innate resistance mechanisms to MPO-mediated hydrogen peroxide antimicrobial activity has been clearly demonstrated (Avila *et al.*, 1976; Nathan *et al.*, 1979). Whether the microbe is exposed to the antimicrobial system intracellularly depends, however, upon its behaviour following phagocytosis. For example, lysosomes fuse with phagosomes containing viable *M. lepraemurium* but not with phagosomes containing live *M. tuberculosis* (Draper & D'Arcy Hart, 1975). *T. cruzi* is able to escape from the original phagosome and live in direct contact with the cytoplasm (Kress *et al.*, 1975). Phagosome–lysosome fusion following phagocytosis has been demonstrated for *L. donovani* in hamster peritoneal macrophages (Chang & Dwyer, 1976, 1978) and for *L. mexicana* in mouse peritoneal macrophages (Alexander, 1974; Lewis, 1975; Alexander & Vickerman, 1975). For *L. donovani* the stage is set for analysis of the processes involved in the resistance mechanism. The critical experiments analysing the relevant biochemical pathways in resistant and susceptible macrophages infected with the microbe have yet to be performed. For *Listeria* and *Salmonella* the question of phago–lysosome fusion has not, to the author's knowledge, been examined.

If it is demonstrated that the same biochemical processes are involved in microbicidal activity against different microbes, resistance of the same strain of mouse to one microbe and susceptibility to another has still to be explained. Functional heterogeneity of macrophage sub-populations has been clearly demonstrated in relation to immunological processes (Walker, 1971, 1974). The concept in relation to innate resistance mechanisms is supported by our unpublished observations that liver and spleen macrophages express the action of the *Lsh* gene *in vitro* while peritoneal exudate cells do not. Further examination of this concept in relation to other intracellular macrophage parasites will be of considerable interest.

ACKNOWLEDGEMENTS

I am indebted to my colleagues at the Ross Institute, in particular Jacqueline Channon and Orysia Ulczak, for their generous help in the preparation of this review.

REFERENCES

Alexander, J. (1974). Fate of *L. mexicana* amastigotes in cultured mouse macrophages. *Trans. R. Soc. trop. Med. Hyg.* **68**: 6.

Alexander, J. & Curtis, J. (1979). Development of delayed hypersensitivity responses in *Mycobacterium lepraemurium* infections in resistant and susceptible strains of mice. *Immunology* **36**: 563–567.

Alexander, J. & Vickerman, K. (1975). Fusion of host cell secondary lysosomes with the parasitophorous vaculoes of *Leishmania mexicana* infected macrophages. *J. Protozool.* **22**: 502–508.

Ali-Khan, Z. (1974). Host-parasite relationship echinococcosis. I. Parasite biomass and antibody response in three strains of inbred mice against graded doses of *Echinococcus multilocularis* cysts. *J. Parasit.* **60**: 231–235.

Allen, E. M., Moore, V. L. & Stevens, J. O. (1977). Strain variation in BCG-induced chronic pulmonary inflammation in mice. I. Basic model and possible genetic control by non-*H-2* genes. *J. Immunol.* **119**: 343–347.

Araujo, F. G., Williams, D. M., Grumet, F. C. & Remington, J. S. (1976). Strain-dependent differences in murine susceptibility to Toxoplasma. *Infect. Immun.* **13**: 1528–1530.

Artz, R. P. & Bullock, N. E. (1979). Immunoregulatory responses in experimental disseminated histoplasmosis: depression of T cell-dependent and T-effector responses by activation of splenic suppressor cells. *Infect. Immun.* **23**: 893–902.

Avila, J. L., Convit, J., Pinardi, M. E. & Jacques, P. J. (1976). Loss of infectivity of mycobacterial and protozoal exoplasmic parasites after exposure *in vitro* to the polyenzymic cocktail "PIGO". *Biochem. Soc. Trans.* **4**: 680–681.

Beaman, L., Pappagianis, D. & Benjamini, E. (1977). The significance of T cells in resistance to experimental murine coccidioidomycosis. *Infect. Immun.* **17**: 580–585.

Beaman, L., Pappagianis, D. & Benjamini, E. (1979). Mechanisms of resistance to infection with *Coccidioides immitis* in mice. *Infect. Immun.* **23**: 681–685.

Bennett, G. H. (1976). The Zoonoses Order 1975. *J. R. Soc. Health* **96**: 19–20.

Blackwell, J. M., Freeman, J. C. & Bradley, D. J. (1978). Private communication. *Mouse News Letter* **59**: 56–57.

Blackwell, J., Freeman, J. & Bradley, D. J. (1980). Influence of *H-2* complex on acquired resistance to *Leishmania donovani* infection in mice. *Nature, Lond.* **283**: 72–74.

Bradley, D. J. (1974). Genetic control of natural resistance to *Leishmania donovani* in mice. *Nature, Lond.* **250**: 353.

Bradley, D. J. (1977). Regulation of *Leishmania* populations within the host. II. Genetic control of acute susceptibility of mice to *Leishmania donovani* infection. *Clin. exp. Immunol.* **30**: 130–140.

Bradley, D. J. (1979). Regulation of *Leishmania* populations within the host. IV. Parasite and host cell kinetics studied by radioisotope labelling. *Acta trop.* **36**: 171–179.

Bradley, D. J. & Kirkley, J. (1972). Variation in susceptibility of mouse strains to *Leishmania donovani* infection. *Trans. R. Soc. trop. Med. Hyg.* **66**: 527–528.

Bradley, D. J. & Kirkley, J. (1977). Regulation of *Leishmania* populations within the host. I. The variable course of *Leishmania donovani* infections in mice. *Clin. exp. Immunol.* **30**: 119–129.

Bradley, D. J., Taylor, B. A., Blackwell, J., Evans, E. P. & Freeman, J. (1979).

Regulation of *Leishmania* populations within the host. III. Mapping of the locus controlling susceptibility to visceral leishmaniasis in the mouse. *Clin. exp. Immunol.* **37**: 7–14.

Brown, C.M. & Parker, M.T. (1957). Salmonella infections in rodents in Manchester. *Lancet* 273: 1277–1279.

Brown, T.M. & Nunemaker, J.C. (1942). Rat-bite fever: a review of American cases with reevaluation of etiology; report of cases. *Bull. Johns Hopkins Hosp.* **70**: 201–327.

Catanzaro, P. J., Shirai, A., Hildebrandt, P.K. & Osterman, J.V. (1976). Host defences in experimental scrub typhus: histopathological correlates. *Infect. Immun.* **13**: 861–875.

Cauley, L.K. & Murphy, J.W. (1979). Response of congenitally athymic (nude) and phenotypically normal mice to *Cryptococcus neoformans* infection. *Infect. Immun.* **23**: 644–651.

Chang, K.P. & Dwyer, D.M. (1976). Multiplication of a human parasite (*Leishmania donovani*) in phagolysosomes of hamster macrophages *in vitro*. *Science, Wash.* **193**: 678–681.

Chang, K.P. & Dwyer, D.M. (1978) *Leishmania donovani*. Hamster macrophage interactions *in vitro:* cell entry, intracellular survival, and multiplication of amastigotes. *J. exp. Med.* **147**: 515–530.

Cheers, C. & McKenzie, I.F.C. (1978). Resistance and susceptibility of mice to bacterial infection: genetics of listeriosis. *Infect. Immun.* **19**: 755–762.

Cheers, C., McKenzie, I.F.C., Pavlov, H., Waid, C. & York, J. (1978). Resistance and susceptibility of mice to bacterial infection: course of listeriosis in resistant or susceptible mice. *Infect. Immun.* **19**: 763–770.

Chick, E.W. & Roberts, G.D. (1974). The varying susceptibility of different genetic strains of laboratory mice to *Histoplasma capsulatum*. *Mycopath. Mycol. Appl.* **52**: 251–253.

Chyong-din-Kyet (1977). [Differences in the sensitivity of various mouse lines to the causative agent of rickettsial pox.] *Zh. Mikrobiol.* **1977**: 78–81. [In Russian].

Civil, R.H. & Mahmoud, A.A.F. (1978). Genetic differences in BCG-induced resistance to *Schistosoma mansoni* are not controlled by genes within the major histocompatibility complex of the mouse. *J. Immunol.* **120**: 1070–1072.

Clarkson, M.J. (1976). IgM in *Trypanosoma brucei* infection of different strains of mice. *Parasitology* **73**: viii.

Clayton, C.E. (1978). *Trypanosoma brucei:* influence of host strain and parasite antigenic type on infections in mice. *Expl Parasit.* **44**: 202–208.

Clifford, C.M., Bell, J.F., Moore, G.J. & Raymond, G. (1967). Effects of limb disability on lousiness in mice. IV. Evidence of genetic factors in susceptibility to *Polyplax serrata*. *Expl Parasit.* **20**: 56–67.

Closs, O. (1975). Experimental murine leprosy. Growth of *Mycobacterium lepraemurium* in C3H and C57BL mice after foot-pad inoculation. *Infect. Immun.* 12: 480–489.

Closs, O. & Haugen, O.A. (1974). Experimental murine leprosy. II. Further evidence for varying susceptibility of outbred mice and evaluation of response of five inbred mouse strains to infection with *Mycobacterium lepraemurium*. *Acta path. microbiol. scand.* **(A) 82**: 459–474.

Cunningham, D.S., Kuhn, R.E. & Rowland, E.C. (1978). Suppression of humoral response during *Trypanosoma cruzi* infections in mice. *Infect. Immun.* **22**: 155–160.

Curtis, J. & Turk, J. L. (1979). Mitsuda-type lepromin reactions as a measure of host resistance in *Mycobacterium lepraemurium* infection. *Infect. Immun.* 24: 492–500.

Cutler, J. E, (1976). Acute systemic candidiasis in normal and congenitally thymic-deficient (nude) mice. *J. Reticuloendothel. Soc.* **19**: 121–124.

Cypess, R. H. & Zidian, J. L. (1975). *Heligmosomoides polygyrus* (= *Nematospiroides dubius*): the development of self-cure and/or protection in several strains of mice. *J. Parasit.* **61**: 819–824.

DeTolla, L. J., Semprevivo, L. H., Palczuk, N. C. & Passmore, H. C. (1980). Genetic control of acquired resistance to visceral leishmaniasis in mice. *Immunogenetics* **10**: 353–361.

Domer, J. E. & Moser, S. A. (1978). Experimental murine candidiasis: cell-mediated immunity after cutaneous challenge. *Infect. Immun.* **20**: 88–98.

Dow, C. & Jarrett, W. F. H. (1960). Age, strain and sex differences in susceptibility to *Cysticercus fasciolaris* in the mouse. *Expl Parasit.* **10**: 72–74.

Draper, P. & D'Arcy Hart, P. (1975). Phagosomes, lysosomes and mycobacteria: cellular and microbial aspects. In *Mononuclear phagocytes in immunity, infection and pathology:* 575–594. Van Furth, R. (Ed.). Oxford: Blackwell Scientific.

Dunn, M. C. & Brown, H. W. (1963). Comparison of the susceptibility of CF1, C57 Brown, C3H, C57 Black and DBA 2 strains of mice to *Aspiculuris tetraptera. J. Parasit.* **49** (2): 32–33.

Faust, E. C. & Russell, P. F. (1964). *Craig and Faust's clinical parasitology.* 7th edn. Philadelphia: Lea & Febiger.

Gagarin, V. G., Dushkin, V. A., Solonenko, I. G. & Zelentsov, A. G. (1972). [Infection with *Hydatigena taeniaeformis* strobilocerci in mice of inbred and undetermined strains.] *Byull. vses. Inst. Gel.* **8**: 17–19. [In Russian].

Giger, D. K., Domer, J. E. & McQuitty, J. T. (1978). Experimental murine candidiasis: pathological and immune responses to cutaneous inoculation with *Candida albicans. Infect. Immun.* **19**: 499–509.

Giger, D. K., Domer, J. E., Moser, S. A. & McQuitty, J. T. (1978). Experimental murine candidiasis: pathological and immune responses in T-lymphocyte-depleted mice. *Infect. Immun.* **21**: 729–737.

Gold, D. & Kagan, I. G. (1978). Susceptibility of various strains of mice to *Entamoeba histolytica. J. Parasit.* **64**: 937–938.

Greenberg, J. & Kendrick, L. P. (1957). Parasitaemia and survival in inbred strains of mice infected with *Plasmodium berghei. J. Parasit.* **43**: 413–419.

Greenberg, J. & Kendrick, L. P. (1958). Parasitaemia and survival in mice infected with *Plasmodium berghei.* Hybrids between Swiss (high parasitaemia) and STR (low parasitaemia) mice. *J. Parasit.* **44**: 492–498.

Greenberg, J. & Kendrick, L. P. (1959). Resistance to malaria in hybrids between Swiss and certain other strains of mice. *J. Parasit.* **45**: 263–267.

Greenberg, M., Pelliteri, O. & Jellison, W. L. (1947). Rickettsial pox. A newly recognised rickettsial disease. III. Epidemiology. *Am. J. publ. Health* **37**: 860–868.

Groves, M. G. & Osterman, J. V. (1978). Host defences in experimental scrub typhus: genetics of natural resistance to infection. *Infect. Immun.* **19**: 583–588.

Hay, R. J. & Reiss, E. (1978). Delayed-type hypersensitivity responses in infected mice elicited by cytoplasmic fractions of *Cryptococcus neoformans. Infect. Immun.* **22**: 72–79.

Howard, D. H. & Otto, V. (1977). Experiments in lymphocyte-mediated cellular immunity in murine histoplasmosis. *Infect. Immun.* **16**: 226–231.
Howard, J. G., Hale, Christine & Chan-Liew, W. Ling. (In press). Immunological regulation of experimental cutaneous leishmaniasis. I. Immunogenetic aspects of susceptibility to *Leishmania tropica* in mice. *Parasite Immunol.*
Howard, J. G., Hale, Christine & Liew, F. Y. (1980). Immunological regulation of experimental cutaneous leishmaniasis. III. The nature and significance of specific suppression of cell mediated immunity in mice highly susceptible to *Leishmania tropica. J. exp. Med.* **152**: 594–607.
Huebner, R. J., Jellison, W. L. & Armstrong, C. (1947). Rickettsial pox. A newly recognised rickettsial disease. V. Recovery of *R. akari* from a house mouse (*Mus musculus*). *Public Health Rep.* **62**: 777–780.
Huebner, R. J., Jellison, W. L. & Pomerantz, C. (1946). Rickettsial pox. A newly recognised rickettsial disease. IV. Isolation of a rickettsia apparently identical with the causative agent of rickettsial pox from *Allodermanyssus sanguineus*, a rodent mite. *Public Health Rep.* **61**: 1677–1682.
Huebner, R. J., Stamps, P. & Armstrong, C. (1946). Rickettsial pox. A newly recognised rickettsial disease. I. Isolation of the etiological agent. *Public Health Rep.* **61**: 1605–1614.
Jones, P. W. & Twigg, G. I. (1976). Salmonellosis in wild mammals. *J. Hyg., Camb.* **77**: 51–54.
Kekcheeva, N. G. & Kokorin, I. N. (1976). Different allotypic susceptibility of mice to *Rickettsia tsutsugamushi. Acta virol.* **20**: 142–146.
Kierszenbaum, F. & Howard, J. G. (1976). Mechanisms of resistance against experimental *Trypanosoma cruzi* infection: the importance of antibodies and antibody-forming capacity in the Biozzi high and low responder mice. *J. Immunol.* **116**: 1208–1211.
Klebanoff, S. J. & Hamon, C. B. (1975). Antimicrobial systems of mononuclear phagocytes. In *Mononuclear phagocytes in immunity, infection and pathology:* 507–531. Van Furth, R. (Ed.). Oxford: Blackwell Scientific.
Kokorin, I. N., Chyong-din-Kyet, Kekcheeva, N. G. & Miskarova, E. D. (1976). Cytological investigation of *Rickettsia tsutsugamushi* infection of mice with different allotypic susceptibility to the agent. *Acta virol.* **20**: 147–151.
Kokorin, I. N., Kabanova, E. A., Chyong-din-Kyet, Miskarova, E. D. & Abrosimova, G. E. (1978). Differences in the susceptibility of mouse lines to the rickettsial pox agent. *Acta virol.* **22**: 497–501.
Kress, Y., Bloom, B. R., Wittner, M., Rowen, A. & Tanowitz, H. (1975). Resistance of *Trypanosoma cruzi* to killing by macrophages. *Nature, Lond.* **257**: 394–396.
Lackman, D. B. (1963). A review of information on rickettsial pox in the United States. *Clin. Pediat.* **2**: 296–301.
Levaditi, C., Selbie, R. F. & Schoen, R. (1932). Le rhumatisme infectieux spontané de la souris provoqué par le *Streptobacillus moniliformis. Annls Inst. Pasteur* **48**: 308–343.
Lewis, D. H. (1975). *In vitro* studies on the host/parasite interactions between *L.m. mexicana* and peritoneal macrophages taken from normal and sensitized mice. *J. Protozool.* **22**: 53A.
Lubinsky, G. (1964). Growth of the vegetatively propagated strain of larval *Echinococcus multilocularis* in some strains of Jackson mice and in their hybrids. *Can. J. Zool.* **42**: 1099–1103.
Lubinsky, G. & Desser, S. (1963). Growth of the vegetatively propagated strain

of larval *Echinococcus multilocularis* in C57BL/J, B6AF$_1$, and A/J mice. *Can. J. Zool.* **41**: 1213–1216.

Lueker, D. C. & Hepler, D. I. (1975). Differences in immunity to *Nematospiroides dubius* in inbred and outbred mice. *J. Parasit.* **61**: 158–159.

Mackie, T. J., Rooyen, C. E. van & Gilroy, E. (1933). An epizootic disease occurring in a breeding stock of mice: bacteriological and experimental observations. *Br. J. exp. Path.* **14**: 132–136.

Manger, B. R. (1976). Studies on the resistance of mice to *Nematospiroides dubius* Baylis 1962. *Parasitology* **73**: xiii.

McCoy, J. H. (1976). Salmonella infections of man derived from animals. *Jl. R. Soc. Health* **96**: 25–30.

Mitchell, G. F., Hogarth-Scott, R. S., Edwards, R. D., Lewers, H. M., Cousins, G. & Moore, T. (1976). Studies on immune responses to parasite antigens in mice. I. *Ascaris suum* larvae numbers and antiphosphorylcholine responses in infected mice of various strains and in hypothymic *nu/nu* mice. *Int. Archs Allergy appl. Immunol.* **52**: 64–78.

Morozumi, P. A. & Stevens, D. A. (1979). Susceptibility differences in inbred strains of mice to pulmonary blastomycosis. Abstract No. F63. *A. Meet. Am. Soc. Microbiol.* 1979.

Most, H., Nussensweig, R. S., Vanderberg, J., Herman, R. & Yoeli, M. (1966). Susceptibility of genetically standardised (JAX) mouse strains to sporozoite- and blood-induced *Plasmodium berghei* infections. *Milit. Med.* **131**: 915–918.

Nathan, C. F., Nogueira, N., Juangbhanich, C., Ellis, J. & Cohn, Z. (1979). Activation of macrophages *in vivo* and *in vitro*. Correlation between hydrogen peroxide release and killing of *Trypanosoma cruzi*. *Jl. exp. Med.* **149**: 1056–1068.

Nathan, C. F. & Root, R. K. (1977). Hydrogen peroxide release from mouse peritoneal macrophages. Dependence on sequential activation and triggering. *J. exp. Med.* **146**: 1648–1662.

Neal, R. A. & Harris, W. G. (1977). Attempts to infect inbred strains of rats and mice with *Entamoeba histolytica*. *Protozoology* **3**: 197–199.

Osborne, A. D. (1976). Salmonella infections in animals and birds. *J. R. Soc. Health* **96**: 30–33.

Ott, K. J. & Stauber, L. A. (1967). *Eperythrozoon coccoides:* influence on course of infection of *Plasmodium chabaudi* in mouse. *Science, Wash.* **155**: 1546–1548.

Parker, F. & Hudson, N. P. (1926). Etiology of Haverhill fever (erythema arthriticum epidemicum). *Am. J. Path.* **2**: 357–379.

Patton, R. M., Riggs, A. R., Compton, S. G. & Chick, E. W. (1976). Histoplasmosis in pure bred mice: influence of genetic susceptibilty and immune depression on treatment. *Mycopathology* **60**: 39–43.

Pennington, J. E. & Williams, R. M. (1979). Influence of genetic factors on natural resistance of mice to *Pseudomonas aeruginosa*. *J. inf. Dis.* **139**: 396–400.

Perez, H., Arrendondo, B. & Gonzalez, M. (1978). Comparative study of American cutaneous leishmaniasis and diffuse cutaneous leishmaniasis in two strains of inbred mice. *Infect. Immun.* **22**: 301–307.

Peters, W. (1965). Competitive relationship between *Eperythrozoon coccoides* and *Plasmodium berghei* in the mouse. *Expl Parasit.* **16**: 158–166.

Place, E. H. & Sutton, L. E. (1934). Erythema arthriticum epidemicum (Haverhill fever). *Arch. int. Med.* **54**: 659–684.

Plant, J. & Glynn, A. A. (1974). Natural resistance to *Salmonella* infection, delayed hypersensitivity and Ir genes in different strains of mice. *Nature, Lond.* **248**: 345–347.

Plant, J. & Glynn, A. A. (1976). Genetics of resistance to infection with *Salmonella typhimurium* in mice. *J. inf. Dis.* **133**: 72–78.

Plant, J. & Glynn, A. A. (1979). Locating salmonella resistance gene on mouse chromosome 1. *Clin. exp. Immunol.* **37**: 1–6.

Pratt, H. D. (1963). Mites of public health importance and their control. *Publ. Public Health Service* No. 772.

Preston, P., Behbehani, K. & Dumonde, D. C. (1978). Experimental cutaneous leishmaniasis. VI. Anergy and allergy in the cellular immune response during non-healing infection in different strains of mice. *J. clin. lab. Immunol.* **1**: 207–219.

Robson, H. G. & Vas, S. I. (1972). Resistance of inbred mice to *Salmonella typhimurium. J. inf. Dis.* **126**: 378–386.

Rosenstreich, D. L., O'Brien, A. D., Groves, M. G. & Taylor, B. A. (1980). Genetic control of natural resistance to infection in mice. In *The molecular basis of microbial pathogenicity:* 101–114. Smith, H., Skehel, J. J. & Turner, M. J. (Eds). Dahlem Konferenzen 1980. Weinheim: Verlag Chemie GmbH.

Ruebush, N. J. & Hanson, W. L. (1979). Susceptibility of five strains of mice to *Babesia microti* of human origin. *J. Parasit.* **65**: 430–433.

Rybkina, N. N., Kabanova, E. A. & Chyong-din-Kyet (1976). Interlinear susceptibility of mice to the causative agent of vesiculous rickettsiosis. *Folia microbiol., Praha* **21**: 501.

Schockaert, J. (1928). Sur l'unicité des souches de *Spirillum minus. C. r. hebd. Séanc. Soc. Biol.* **98**: 595–597.

Shankman, B. (1946). Report of an outbreak of endemic febrile illness, not yet identified, occurring in New York City. *N.Y. Med. J.* **46**: 2156–2159.

Shimi, A., Keyhani, M. & Hedayati, K. (1979). Studies on salmonellosis in the house mouse, *Mus musculus. Lab. Animals* **13**: 33–34.

Si-Kwang Liu (1966). Genetic influence on resistance of mice to *Nematospiroides dubius. Expl Parasit.* **18**: 311–319.

Skamene, E., Kongshavn, P. A. L. & Sachs, D. H. (1979). Resistance to *Listeria monocytogenes* in mice: genetic control by genes that are not linked to the *H-2* complex. *J. inf. Dis.* **139**: 228–231.

Snyder, J. C. (1965). Typhus fever rickettsiae. In *Viral and rickettsial infections of man*: 1059–1094. 4th Edn. Horsfall, F. L. & Tamm, I. (Eds). London: Pitman.

Sussman, L. N. (1946). Kew Garden's spotted fever. *N.Y. Med.* **2**: 27–28.

Tanner, C. E. (1978). The susceptibility to *Trichinella spiralis* of inbred lines of mice differing at the *H-2* histocompatibility locus. *J. Parasit.* **64**: 956–957.

Thurston, J. P. (1955). Observations on the course of *Eperythrozoon coccoides* infections in mice, and the sensitivity of the parasite to external agents. *Parasitology* **45**: 141–151.

Wakelin, D. (1975). Genetic control of immune responses to parasites: immunity to *Trichuris muris* in inbred and random-bred strains of mice. *Parasitology* **71**: 51–60.

Wakelin, D. (1978). Genetic control of susceptibilty and resistance to parasitic infection. *Adv. Parasit.* **16**: 219–308.

Wakelin, D. & Lloyd, M. (1976). Immunity to primary and challenge infections of *Trichinella spiralis* in mice: a re-examination of conventional parameters. *Parasitology* 72: 173–182.

Walker, W. S. (1971). Macrophage functional heterogeneity in the *in vitro* induced immune response. *Nature New Biol.* 229: 211–212.

Walker, W. S. (1974). Functional heterogeneity of macrophages: subclasses of peritoneal macrophages with different antigen-binding activities and immune complex receptors. *Immunology* 26: 1025–1037.

Whitby, L. & Hynes, M. (1951). *Medical bacteriology.* 5th edn. London: J. & A. Churchill Ltd.

Williams, D. M., Graybill, J. R. & Drutz, D. J. (1978). *Histoplasma capsulatum* infection in nude mice. *Infect. Immun.* 21: 973–977.

Williams, S. (1941). Outbreak of infection due to *Streptobacillus moniliformis* among wild mice. *Med. J. Austra.* 1: 357–359.

Wilson, G. S. & Miles, A. A. (1964). *Topley and Wilson's principles of bacteriology and immunity.* 5th edn. London: Edward Arnold.

Woodward, T. E. & Jackson, E. B. (1965). Spotted fever rickettsiae. In *Viral and rickettsial infections of man:* 1095–1129. 4th edn. Horsfall, F. L. & Tamm, I. (Eds). London: Pitman.

World Health Organisation. (1959). *Zoonoses.* Second report of the joint WHO/FAO Expert Committee. *Tech. rep. Ser. World hlth Org.* No. 169.

World Health Organisation. (1963). *Meeting of the scientific group on rickettsial diseases of man.* Report to the Director-General, MHO/PA/144.63, Geneva.

World Health Organisation. (1979). *Parasitic zoonoses.* Report of a WHO Expert Committee with the participation of FAO. *Techn. rep. Ser. World hlth Org.* No. 637.

Zdrodovskii, P. F. & Golinevich, E. M. (1960). *The rickettsial diseases.* New York: Pergamon Press.

Zelentsov, A. G. (1974). [Susceptibility of inbred mice to helminths. III. The development of *Opisthorcis felineus* in mice of strains A/He, CBA/Lac, CC57W/MJ, C57BL/6J, DBA/2J and SWR/J.] *Medskaya Parasit.* 43: 95–98. [In Russian].

Zinkernagel, R. M. (1974). Restriction by *H-2* gene complex of transfer of cell-mediated immunity to *Listeria monocytogenes. Nature, Lond.* 251: 230–233.

Zinkernagel, R. M., Althage, A., Adler, B., Blanden, R. V., Davidson, W. F., Kees, U., Dunlop, M. B. C. & Shreffler, D. C. (1977). *H-2* restriction of cell-mediated immunity to an intracellular bacterium. Effector T cells are specific for *Listeria* antigen in association with *H-2I* region-coded self-markers. *J. exp. Med.* 145: 1353–1367.

Symp. zool. Soc. Lond. (1981) No. 47, 617–625

Mutant Mice as Models for Human Genetical Deafness

M. S. DEOL

Department of Genetics and Biometry, University College London, London, UK

SYNOPSIS

Normal hearing is an extremely complex phenomenon dependent on the integration of a large number of factors. It can therefore be affected by a large number of genetical loci, and numerous hereditary disorders of hearing have been described in man. In most cases very little is known about them apart from their mode of transmission, whether they are congenital or have a late onset, and whether they are uncomplicated or occur in association with some other abnormality. The reasons for this lack of knowledge are mainly technical because the inner ear in man is exceptionally difficult to study. Fortunately genetical disorders of hearing are also very common in the mouse, and as the anatomical, physiological and biochemical aspects of the inner ear are virtually identical in man and the mouse, we can use mouse mutants for the elucidation of human disorders. The question is whether mouse mutants provide good models. In some cases the similarities with man are so great that the conditions in the two species can be regarded as homologous. In other cases they may not be exactly homologous, but have sufficient features in common for the mouse to provide reliable working models for man. Even in those cases where there may be no indubitable similarity the mouse mutants could yield valuable clues to the understanding of the normal function of the inner ear, and so indirectly illuminate causes of deafness in man.

INTRODUCTION

Mutant animals are frequently found to reproduce human disorders, but this does not mean that they can always be reliably used as models for human conditions. This is particularly true if the condition is a malformation, for different genes may lead to the same malformation by different developmental pathways. Extrapolation from animals to man is especially hazardous if the disorder is a "negative" trait, that is, if it consists in the absence of a normal feature rather than in the presence of a different form of it. One such trait is deafness.

As normal hearing is an extremely complex phenomenon, absence of hearing may be caused by innumerable factors, and scores, probably hundreds, of gene loci must be involved. The development of the inner ear from a simple disc of cells into a labyrinthine arrangement of ducts and sacs with all their attendant sensory areas has to proceed in a certain way; its multitude of physiological and biochemical functions must be very finely integrated, and its elaborate neural pathways must be precisely established if normal hearing is to develop. Any fault anywhere could lead to deafness (Fig. 1). Nevertheless, mutant mice with defective hearing can

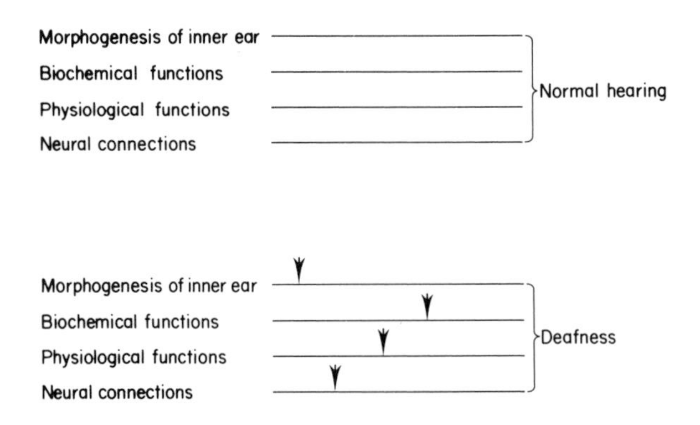

FIG. 1. Possible origins of deafness. Any fault in any of the pathways could lead to loss of hearing.

reasonably be treated as valuable models for investigations into human deafness. There are several reasons for this. Not only are the form and organization of the inner ear almost identical in both mouse and man, but even the dimensions of the organ are not very different in spite of the disparity in body size. (Indeed, the inner ear is probably the least variable organ in the mammalian world in this respect.) As far as is known, the metabolic, physiological and neural aspects are also very similar. Furthermore, because of the extreme morphological complexity of the inner ear it is possible to say with greater confidence than usual whether the disorders compared are likely to be truly homologous. Then there are syndromes in both mouse and man which combine deafness with the same abnormalities in other tissues and organs. Finally, for technical reasons – such as the inaccessibility of the inner ear and the need for material preserved in the living state – disorders of the inner ear are virtually impossible to study in man, and one must perforce resort to alternative material. Mouse mutants are by far the best source of it.

Deafness is of two kinds, conductive and non-conductive. Conductive deafness results from some abnormality of the middle ear or the otic capsule such that sound fails to be transmitted to the inner ear, although this organ may be perfectly normal. Non-conductive deafness, also known as sensory or neural or sensory-neural deafness, results – or is assumed to result because of the absence of any noticeable defect of conduction – from some abnormality of the cochlea, the eighth nerve or the auditory pathways.

CONDUCTIVE DEAFNESS

This type of deafness is quite common in man, there being several forms of it. A clear inherited basis has been established only for certain syndromes; the other types can only be described as familial. The most important among these, otosclerosis, does not appear to have any known counterpart in the mouse, but the next most important, otitis media or suppurative inflammation of the middle ear, has. Interestingly, in the mouse too it is a familial condition, certain inbred strains having a much stronger tendency to develop it than others.

NON-CONDUCTIVE DEAFNESS

A large number of disorders of this type have been described in man (Fraser, 1964, 1976; Ruben & Rozycki, 1971; Deol, 1973a; McKusick, 1978), and many of them seem to have counterparts in the mouse (Deol, 1968). But whereas in the mouse the genetical identity of these disorders has been established by means of test crosses, in man it is obviously very difficult to say whether any two cases with similar clinical pictures are caused by the same gene or different ones. Nor can much reliance be placed on the pathology of the disorder, because in most cases this is totally obscure and in others the samples are very small and the material poorly preserved (Altmann, 1953; Ormerod, 1960). As a result, identification of various types of non-conductive genetical deafness in man is dependent on other factors, such as its mode of inheritance, whether it is congenital or has a late onset, and whether it is uncomplicated or complicated by association with abnormalities of other tissues and organs.

Uncomplicated Deafness

This is the commonest type of genetical deafness reported in man. With few exceptions the disorder appears to be congenital in all cases. The vast majority of pedigrees show clear autosomal recessive inheritance. On the basis of progeny from marriages between deaf persons it has been estimated that at least six loci are involved (Stevenson & Cheeseman, 1956). However, it is generally agreed that this is a very conservative estimate; another analysis of the same data indicated that close to 100 loci might be concerned (Chung, Robison & Morton, 1959). Considering the ways in which congenital deafness can arise, this number is quite possible, but in the absence of corroborative evidence of some other kind this estimate must remain conjecture.

Stevenson & Cheeseman (1956) thought that all cases of uncomplicated congenital deafness encountered during their exhaustive survey of Northern Ireland could be explained on the basis of autosomal recessive inheritance, but it is now generally believed that autosomal dominant genes with incomplete penetrance must also be concerned. In any event, Everberg's (1960) discoveries in Denmark show that such genes are not uncommon. He found numerous cases of unilateral deafness, accompanied by malformations of the inner ear which could be identified radiographically, most of which could best be explained on the basis of autosomal dominant genes with incomplete penetrance.

Families with sex-linked recessive genes causing uncomplicated deafness have been reported from many parts of the world. It is difficult to say how many loci are involved, but they are thought to account for a significant proportion of affected males. Fraser (1965) has estimated that 6.2% of genetically deaf males in the UK may fall into this class.

Uncomplicated deafness with late onset is much less common, probably because once the development of the inner ear is complete and its neural connections established without error, far fewer things are likely to go wrong. Only two good pedigrees of this type have been reported, one showing autosomal dominant and the other sex-linked inheritance.

The pathology of these disorders is very poorly understood. It can only be discussed in general terms, applicable to genetical and non-genetical deafness alike. Congenital deafness, it seems, is nearly always associated with some malformation of the inner ear, and can therefore be attributed to some error in the morphogenesis of the organ. Ormerod (1960) has classified these malformations into five categories in descending order of severity, beginning with cases in which the whole of the inner ear is so grossly abnormal that no part of it may bear any resemblance to any part of the normal organ. Deafness with late onset, on the other hand, appears to be associated with degenerative changes in the cochlea, especially in the organ of Corti, rather than with any gross malformation.

Which mouse mutants should be regarded as providing situations comparable to human uncomplicated congenital deafness is somewhat difficult to decide. If the term "uncomplicated" is to be taken in the strictest sense and restricted to those mutants in which deafness is the sole known abnormality, then only two mutants fall into this category. One of these, *deaf* (Deol, 1956a), has become extinct. The other, *deafness* (Deol & Kocher, 1958), is still in existence. However, if we may apply this term to mutants in which apart from the inner ear no tissue or organ is known to be affected, then a large number of mutants fall into this class. The difficulty lies in the fact that in this latter group of mutants the behaviour is not normal. The animals are hyperactive, make jerking movements of the head more or less continuously and have a strong tendency to run in circles. There is reason to believe that the cause of this anomalous behaviour does not lie in the inner ear but in the central nervous system (Deol, 1966). The question is whether they can still be regarded as being comparable to the human conditions described above.

There are two reasons for giving an affirmative answer. Firstly, we know that the early stages of the morphogenesis of the inner ear are dependent on the inductive influence of the neural tube, and in all those human disorders in which the inner ear has gross malformations the neural tube would be under strong suspicion of being abnormal, and it is unlikely that these abnormalities would not leave a legacy in the brain. This view is supported by the fact that mental retardation, apparently independent of the effects of deafness, is common in such cases. If the assumption is correct then the mouse disorders are essentially similar to the human ones. The reason for the difference with regard to behaviour may lie in some neurological peculiarity of the rodent brain (this type of behaviour appears to be confined to rodents). It is also possible that this neurological feature is not peculiar to rodents but is present in man as well, but cannot affect behaviour in this case because of its subservience to higher centres in the brain. The second reason for regarding these mutants as belonging to the uncomplicated class is that the gene *deaf*, which does not affect behaviour, is an allele of *waltzer*, which does affect it (Kocher, 1960a). If two mutations at the same locus can give rise to deafness with or without the behaviour anomaly, then this anomaly is of no fundamental importance in the classification of mutants, and we can regard all mutants without known abnormalities of any other tissue or organ as being of the uncomplicated type.

There are about 15 such mutants (Deol, 1968). As in man, the inheritance is recessive in the vast majority of cases, but no sex-linked gene is known. These mutants can roughly be divided into two classes, morphogenetic and degenerative, depending on whether the morphogenesis of the inner ear went awry somewhere or whether the morphogenesis was normal but degenerative changes set in some time before the onset of hearing. Almost every kind of pathological picture encountered in human disorders can be found in one or other of these mutants, and the conclusion seems to be unavoidable that we are dealing with conditions that are not just superficially similar but are comparable at a deeper level.

As to deafness with late onset, it appears also to be much less common in the mouse. There is only one good example. Animals of the inbred strain C57BL (at least some substrains such as C57BL/Gr) begin to lose their hearing from about two months onwards and may become totally deaf during the next three months (Kocher, 1960b). It is not clear how many genes are involved. Loss of hearing is accompanied by degenerative changes in the cochlea, especially in the organ of Corti and the spiral ganglion. The mutant *shaker-1* may also be regarded as belonging to this category. In this case deafness is clearly not congenital, but it sets in early, the period of hearing lasting only a few weeks at the most, and progresses very fast. Pathologically, it is of the degenerative type (Deol, 1956b; Mikaelian & Ruben, 1964). As suggested above, genetical deafness with late onset may be rare because once the inner ear has differentiated normally fewer things can go wrong with it, but it is quite possible that this rarity is merely a reflection of the difficulties attending the study of deafness with late onset. As deafness of this type develops gradually over a long period, and as it can result from a multiplicity of causes, it is hard to detect and its genetical basis harder to establish. On general grounds, it would seem that such deafness is commoner than is realized, and a thorough search in large mouse colonies may prove rewarding.

Complicated Deafness

Deafness associated with abnormalities of structures other than the inner ear is common in man. Again, it may be congenital or have a late onset. The associated abnormalities cover a wide range, and only the most important syndromes will be mentioned here.

Congenital deafness occurs together with retinitis pigmentosa in Usher's syndrome. It is associated with hypothyroidism in Pendred's syndrome – quite possibly there are two genetically distinct varieties of this. In Lange–Nielsen's syndrome it is accompanied by an abnormality of the heart, which results in a marked prolongation of the QT interval in electrocardiograms. Then there is a group of four syndromes – the best known among them being Waardenburg's syndrome – in which it is associated with abnormalities of pigmentation. These abnormalities are always of the white spotting kind (hypopigmentation). The loss of pigment may be localized or total. When it is total the condition is not to be confused with albinism, but is to be regarded as being one very large spot. The reason is that whereas in albinos melanocytes are present but cannot form pigment as a result of a biochemical block, in white spots they cannot be identified at all. Another difference from albinism is that in hypopigmentation some pigment is always present in the retina. Loss of hearing is also variable. In most cases there is some residual hearing. The two sides may be affected to a different extent. The cochlea shows Scheibe type abnormalities. The inheritance is autosomal dominant in two cases, sex-linked recessive in the third, and uncertain in the fourth.

Deafness with late onset is accompanied by abnormalities of the kidneys in at least four different syndromes, the best known among which is Alport's syndrome. It occurs in association with abnormalities of the eyes in Norrie's disease and Flynn–Aird's syndrome, and with that of a number of structures of ectodermal origin in anhidrotic ectodermal dysplasia. It is also known to be associated with intestinal diverticula, perforated ulcers on the feet, and a number of other disorders.

There are a large number of mouse mutants in which congenital deafness occurs as a part of a syndrome, but with certain exceptions it is difficult to say whether they can be regarded as corresponding to any particular human condition. The exceptions are the mutants in which loss of hearing is accompanied by white spotting or hypopigmentation. The correspondence here is striking, extending even to fine detail, and it is difficult not to be convinced that we are dealing with homologous disorders (Deol, 1970). The extent of hypopigmentation is equally variable, ranging from a small white spot to complete absence of pigment from all parts of the body with the exception of the retina. The loss of hearing is also of the same type, there being normally some residual hearing, and the two ears may be affected to differing extents. The pathology of the cochlea is remarkably similar to that observed in the human disorders. Altogether, about a dozen loci can produce white spotting, and although the inner ear has been examined in only six of them, there is no reason to believe that it is normal in the others. Many of these loci have synergistic effects with regard to both hypopigmentation and hearing. The association of the two abnormalities is believed to be based on the origin in the neural crest of the melanocytes as well as a part of the acoustic ganglion.

Although no gene is known in the mouse that can lead to deafness associated with hypothyroidism, the syndrome can be experimentally produced in the inbred strain C57BL. Mice of this strain can breed and rear their young in the complete absence of thyroxine. If a female is given 0.1% propylthiouracil in the drinking water from the beginning of the pregnancy and the treatment is continued with the offspring after weaning, then they remain hypothyroidic and never develop normal hearing (Deol, 1973b). It appears that thyroxine is essential for the normal development of the tectorial membrane (Deol, 1976).

Complicated deafness with late onset is not known in the mouse.

CONCLUDING REMARKS

It is impossible to make categorical statements regarding the homology of apparently similar conditions in species that are not closely related. However, in favourable circumstances, when certain criteria are met, it becomes difficult to resist the conclusion that we are dealing with essentially the same disorder in both cases. A few mouse mutants, for instance those in which deafness is accompanied by hypopigmentation, meet the necessary criteria, and can be regarded as providing exact counterparts of human disorders. Then there are some others which may not provide exact counterparts, but which can be treated as valuable working models for certain types of human deafness. Mutants with morphogenetic defects of the inner ear fall into this category. Finally, there are many more mutants which may or may not have any feature that is also found in some human disorder, but which can nevertheless be useful for investigations into human deafness. One major problem in research on human deafness is that many aspects of the functioning of the inner ear are still obscure. Since disorders often provide important clues to normal mechanisms, these mutants could help in the elucidation of the functioning of the inner ear in man, and so indirectly to the understanding of the causes of deafness.

REFERENCES

Altmann, F. (1953). Malformations, anomalies and vestigial structures of the inner ear. *Arch. Otolaryngol. Chicago* 57: 591–602.

Chung, C. S., Robison, O. W. & Morton, N. E. (1959). A note on deaf mutism. *Ann. hum. Genet.* 23: 357–366.

Deol, M. S. (1956a). A gene for uncomplicated deafness in the mouse. *J. Embryol. exp. Morph.* 4: 190–195.

Deol, M. S. (1956b). The anatomy and development of the mutants pirouette, shaker-1 and waltzer in the mouse. *Proc. R. Soc.* (B) **145**: 206–213.

Deol, M. S. (1966). The probable mode of gene action in the circling mutants of the mouse. *Genet. Res.* 7: 363–371.

Deol, M. S. (1968). Inherited diseases of the inner ear in man in the light of studies on the mouse. *J. med. Genet.* 5: 137–158.

Deol, M. S. (1970). The relationship between abnormalities of pigmentation and of the inner ear. *Proc. R. Soc.* (B) **175**: 201–217.

Deol, M. S. (1973a). The ear. In *Clinical genetics*: 339–355. 2nd edn. Sorsby, A. (Ed.). London: Butterworths.

Deol, M. S. (1973b). An experimental approach to the understanding and treatment of hereditary syndromes with congenital deafness and hypothyroidism. *J. med. Genet.* **10**: 235–242.

Deol, M. S. (1976). The role of thyroxine in the differentiation of the organ of Corti. *Acta Otolaryngol.* **81**: 429–435.

Deol, M. S. & Kocher, W. (1958). A new gene for deafness in the mouse. *Heredity* 12: 463–466.

Everberg, G. (1960). Unilateral anacusis. Clinical, radiological and genetic investigations. *Acta Otol.* (Suppl.) **158**: 366–374.

Fraser, G. R. (1964). Profound childhood deafness. *J. med. Genet.* **1**: 118–151.

Fraser, G. R. (1965). Sex-linked recessive congenital deafness and the excess of males in profound childhood deafness. *Ann. hum. Genet.* **29**: 171–196.

Fraser, G. R. (1976). *The causes of profound deafness in childhood.* London: Baillière Tindall.

Kocher, W. (1960a). Untersuchungen zur Genetik und Pathologie der Entwicklung von 8 Labyrinthmutanten (deaf-waltzer-shaker-Mutanten) der Maus (*Mus musculus*). *Z. VererbLehre.* **91**: 114–140.

Kocher, W. (1960b). Untersuchungen zur Genetik und Pathologie der Entwicklung spät einsetzender hereditärer Taubheit bei der Maus (*Mus musculus*). *Arch. Ohr.-, Nas.- Kehlk.-Heilk.* **177**: 108–145.

McKusick, V. A. (1978). *Mendelian inheritance in man.* 5th edn. Baltimore & London: Johns Hopkins University Press.

Mikaelian, D. O. & Ruben, R. J. (1964). Hearing degeneration in shaker-1 mice. Correlation of physiological observations with behavioral responses and with cochlear anatomy. *Arch. Otolaryngol. Chicago* **80**: 418–430.

Ormerod, F. C. (1960). The pathology of congenital deafness. *J. Laryngol.* **74**: 919–950.

Ruben, R. J. & Rozycki, D. L. (1971). Clinical aspects of genetic deafness. *Ann. Otolaryngol., St Louis* **80**: 255–263.

Stevenson, A. C. & Cheeseman, E. A. (1956). Hereditary deaf mutism with particular reference to Northern Ireland. *Ann. hum. Genet.* **20**: 177–231.

Symp. zool. Soc. Lond. (1981) No. 47, 627–641

Expression of Murine Leukemia Viruses in Inbred Strains of Mice

JANICE D. LONGSTRETH and HERBERT C. MORSE III

Laboratory of Microbial Immunity, National Institutes of Allergy and Infectious Diseases, National Institutes of Health, Bethesda, Maryland, USA

SYNOPSIS

Although murine leukemia viruses (MuLV) exist as integrated components of mouse cellular DNA in every inbred strain, the expression of these viruses is not constitutive but under the control of both host and viral genes. By mechanisms which are still not well understood, complex interactions between host and viral gene products can result in complete or incomplete virus expresssion. Both complete and partial expression MuLV can be induced by a wide variety of stimuli resulting in both intriguing and complexing observations. In particular, incomplete expression, e.g. the appearance of molecules serologically related to the MuLV gp70 on the surfaces of cells not producing virus, may be important to activation and differentiation. Alternatively such expression may serve only to complicate studies of other cell surface antigens. MuLV, by their endogenous nature, provide an ever present resource and/or problem for researchers who use inbred strains of mice. An awareness of MuLV and their potential expression can often separate clarity from confusion.

INTRODUCTION

Oncology, and particularly the area of murine leukemogenesis, owes much to the early mammalian geneticists who began developing inbred mouse strains to facilitate their studies of the inheritance of a variety of traits (Morse, in press). The breeding of inbred mice with high incidences of leukemia (Furth, 1978) was a major turning point leading to the recognition that the genetic constitution of a mouse could be a critical factor in the development of leukemia. This, accompanied by Gross' discovery that a virus was the agent producing leukemia in the AKR strain (Gross, 1951), sparked an enormous amount of research which led to the realization that the genomes of these viruses are carried as apparently normal chromosomal genes in the mouse. This in turn stimulated a vast amount of research on the expression of these viral genes in both oncogenesis and other, widely disparate, events. The observations from such studies include:

(1) the expression of C-type virus (see below) antigens during embryogenesis (Huebner *et al.*, 1970);

(2) changes in the expression of viral antigens during lymphocyte differentiation (Stockert, Old & Boyse, 1971; Morse, Chused, Sharrow & Hartley, 1979); and

(3) the cross reactions of "specific" antisera due to the reaction of contaminating antiviral antibodies with viral proteins expressed on the cell surface (Klein, 1975).

As Rowe points out in his Harvey Lecture (1978) "Regardless of whether C-type viruses have this much biologic importance or are just good press agents, they presently constitute a major problem in biological research, even apart from their importance as models for the possible etiology of cancer."

CLASSIFICATION AND CHARACTERIZATION

The first classification of murine tumor viruses was based on electron microscopic studies of cell free tumor extracts (Bernhard & Guerin, 1958). Three types of particles, A, B and C, could be distinguished in these studies, but type C particles predominated in spontaneous leukemias. In this way, mouse leukemia viruses (MuLV) came to be characterized as C-type viruses.

Subsequent studies demonstrated that two classes of type-C MuLV could be distinguished on the basis of the cell surface antigens which they induced. Exogenous viruses such as those isolated by Friend (1957), Rauscher (1962) and Moloney (1960) from long passaged tumors, and which have never been found in spontaneous leukemias, induced a cell surface determinant called FMR (short for Friend, Moloney, Rauscher) (Steeves, 1961), whereas endogenous viruses (those found in spontaneous leukemias) such as that isolated by Gross (1951) induced the G (Gross) cell surface antigen (Old, Boyse & Stockert, 1964).

Since then the development of many new assays for infectious virus, virally coded proteins and viral genetic information, has led to the recognition that MuLV are highly polymorphic. As a result, the current taxonomy of MuLV is quite complex and, because of the ever developing methodology, in a constant state of flux. An appreciation of the basic structure of MuLV and of the general organization of their genomes is fundamental to understanding the distinctions made among the various classes and subclasses of this family.

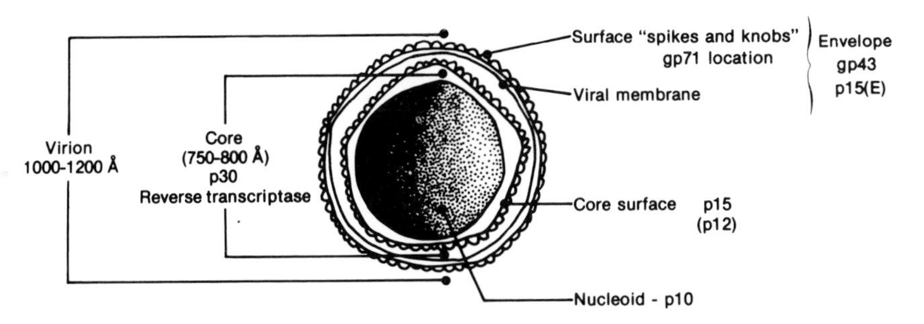

FIG. 1. An MuLV particle: anatomy of and speculations as to the location of the major proteins.

The major proteins of these viruses and their location in a virus particle are shown in Fig. 1. The structural elements of these viruses include a central core surrounded by an envelope acquired by the virus as it buds from the host cell. The core is composed of a nucleoid containing the RNA and an associated protein, p10, and an outer coat of core proteins designated p30, p15 (C), p12, and polymerase (reverse transcriptase). The envelope contains the major glycoprotein, gp70, which is linked by disulphide bonds to p15(E). Gp43, another envelope component, is generally found in low amounts. Evidence indicates that it is a less glycosylated form of gp70 (Schafer *et al.*, 1977).

The physical organization of the MuLV genome appears to be similar for all classes of MuLV. There are thought to be three structural genes involved (Baltimore, 1975): the *env* (*env*elope) gene codes for a glycosylated 90 000 dalton precursor which is cleaved into gp70 and p15(E); the *pol* (*pol*ymerase) gene codes for reverse transcriptase; and the *gag* (*g*roup specific *a*nti*g*en) gene codes for a 70 000 dalton precursor which is cleaved to give p30, p15(C), p12 and p10. The arrangement of these genes is thought to be 5′-*gag* (p15-p12-p30-p10)-*pol*-*env*-3′ (Barbacid, Stephenson & Aaronson, 1976).

The polymorphism of MuLV reflects differences in the fine structure of these common genetical and structural elements which will some day be detailed by nucleotide sequencing. At present, the definition of these differences is dependent primarily on biochemical techniques and on biological assays of infectious virus which are less refined than sequencing.

The various classes of MuLV (Table I) are generally classified by their ability to infect different types of cells. Ecotropic MuLV are infectious for mouse cells (but not xenogeneic cells) and can be

TABLE I

Classification of murine leukemia viruses. Family: Retroviridae

Classes		
1. Ecotropic		
N-tropic (AKR-Gross)	↘	
(B-tropic)		
NB-tropic (FMR)		
	→ Recombination	4. Dualtropic
2. Xenotropic		
α (Type III, not inducible)	↗	
β (Type II, inducible)		
3. Amphotropic		

subdivided into two major host-range subclasses based on their ability to replicate in cells of mice with a particular *Fv-1* genotype (see below). The subclass of ecotropic MuLV termed N-tropic replicates preferentially in *Fv-1*^{n} mouse strains whereas the B-tropic viruses replicate more efficiently in *Fv-1*^{b} mice (Pincus, Rowe & Lilly, 1971). These subclasses have been further defined by other techniques including competition radioimmunoassays of their p30 and p12 proteins (Benade, Ihle & DeCleve, 1978; Gautsch *et al.*, 1978), and oligonucleotide mapping (Rommelaere, Davis-Keller & Hopkins, 1979).

A second class of MuLV has been termed xenotropic (Levy & Pincus, 1970) on the basis of its ability to establish exogenous infection in xenogeneic but not mouse cells. At present there are no firm biological criteria available to distinguish subclasses of xenotropic MuLV. However, competition radioimmunoassays of purified viral proteins (Stephenson *et al.*, 1975), peptide maps of gp70 molecules (Elder, Jensen *et al.*, 1977), and nucleic acid hybridization studies of xenotropic MuLV RNA's (Callahan, Lieber & Todaro, 1975) all indicate that xenotropic viruses are probably at least as polymorphic as ecotropic MuLV.

A third class of MuLV demonstrates tropism for both mouse and xenogeneic cells and is termed amphotropic MuLV (Rasheed, Gardner & Chan, 1976; Hartley & Rowe, 1976). To date, infectious virus of this class has only been obtained from wild mice.

Finally, a fourth type of MuLV has been detected in both preleukemic and leukemic tissues of several mouse strains. These MuLV have a dual host range pattern similar to amphotropic MuLV but differ from them in interference and serological specificity as well as a unique ability to induce foci of morphological alternation in mink lung cell cultures. This latter characteristic led to their designation as dualtropic MCF (mink cell-focus-inducing) MuLV (Hartley *et al.*, 1977). By criteria of interference, neutralization (Hartley *et al.*, 1977), polypeptide maps of their gp70 molecules (Elder, Gautsch *et al.*, 1977) and oligonucleotide mapping (Rommelaere, Faller & Hopkins, 1978) these viruses have been shown to be genetical recombinants between ecotropic and xenotrophic MuLV in the *env* portion of the genome. Differences among MCF isolates from animals of a single strain (Cloyd, Hartley & Rowe, 1979) strongly suggest that recombinants are generated independently within the lifetime of a mouse and do not reflect activation of an inherited genome. Other studies indicate that recombination may occur at sites other than the *env* region of the viral genome. For example, it was shown that two isolates of B-tropic MuLV from C57BL/6 mice have p12 proteins serologically indistinguishable from the xenotropic MuLV of C57BL/6, yet possess p30 proteins indistinguishable from the C57BL/6 *N*-tropic MuLV, suggesting that recombination within the *gag* region has occurred (Benade *et al.*, 1978).

INTEGRATED VIRAL GENOMES AND EXPRESSION OF INFECTIOUS VIRUS

Testing of inbred strains for infectious MuLV of the different classes described above revealed several different patterns of virus expression. For both ecotropic and xenotropic MuLV there are three patterns (Table II). In high virus strains, moderate to high titres of MuLV can be detected in all mice from birth. Low virus strains yield virus in only a fraction of individuals, and then at considerably lower titres and at a later age than the high virus strains. Finally, there are virus-negative strains from which ecotropic and/or xenotropic MuLV have never been recovered at any age tested. As noted previously, no inbred strain spontaneously produces amphotropic MuLV.

The most compelling lines of evidence which suggest that these virus phenotypes are determined by chromosomally integrated viral genomes come from classical genetic studies of matings between high and low virus mice and from nucleic acid hybridization studies of inbred strains. Analyses of segregating crosses between mice of the high ecotropic MuLV phenotype, AKR (Rowe, Hartley & Bremner, 1972) or the high xenotropic MuLV phenotype, NZB (Datta & Schwartz, 1977; Chused & Morse, 1978), with low virus or virus negative mice of Swiss origin, showed that high expression of virus is inherited as a dominant trait and that both of these strains carry two

TABLE II

Spontaneous expression of MuLV in inbred strains of mice

High ecotropic MuLV producers		*High xenotropic MuLV producers*	
AKR	C3H/Fg	NZB	
C58		F/St	
PL		BXD-14	
F/St		C58	
BXH-2			
Low ecotropic MuLV producers		*Low xenotropic MuLV producers*	
BALB/c		BALB/c	AKR
C57BL/6		C57BL/6	
DBA/2		DBA/2	
SEA		C57L	
Ecotropic MuLV negative		*Xenotropic MuLV negative*	
129		129	
C57L		SWR	
NFS			
NZB			

unlinked loci which produce the high virus phenotype in descendants carrying either locus. Additional important aspects of these virus-inducing loci include the observations that:

(1) high ecotropic MuLV strains may have from one (strain PL) to as many as four (strain C58) loci governing the high virus phenotype (Rowe & Hartley, 1978);

(2) the integration sites for MuLV genomes are not allelic in different mouse strains (Table III); and

(3) the number of loci in a given strain can increase if there is reintegration of inherited viral information in germ cells of mice with the high ecotropic MuLV phenotype (Rowe, Hartley, & Kozak, in press; Chattopadhyay, Lowy *et al.*, 1974a, b). It should also be noted that not all viral loci behave in a dominant fashion in crosses with virus negative strains. Recent studies of the F/St and BXD-14 xenotropic viruses show that the high virus phenotype of these strains segregates as a recessive trait in crosses with several strains of mice (Morse, Wolford & Hartley, unpublished data).

Nucleic acid hybridization studies have yielded further evidence for integration of viral genes into mouse cellular DNA. Data obtained from binding studies using c-DNA prepared from different classes of MuLV and mouse embryo DNA, can provide four types of information:

(1) the proportion of the whole viral genome present in the cell DNA;

TABLE III

Integrated genomes of murine leukemia viruses

Virus	Strain	Locus	Chromosomal Location	References
Ecotropic	AKR	*Akv-1*	7	Rowe, Hartley & Bremner (1972)
	AKR	*Akv-2*	16	Kozak & Rowe (pers. comm.)
	BALB/c	*Cv-1*	5	Kozak & Rowe (pers. comm.)
	C57BL/10		8	Kozak (pers. comm.)
	B10.BR		11	Kozak (pers. comm.)
	C3H/Fg	*Fgv-1*	7	Rowe (1978)
Xenotropic	C57BL/6	*Bxv-1*	1	Kozak & Rowe (1978)
	BALB/c	*Bxv-1*	1	Kozak & Rowe (1978)
	AKR	*Bxv-1*	1	Kozak & Rowe (pers. comm.)
	C57L	*Bxv-1*	1	Kozak & Rowe (pers. comm.)
	C57Br	*Bxv-1*	1	Kozak & Rowe (pers. comm.)
	NZB	*Nzv-1*	Not 1	Kozak, Morse, Hartley & Chused (unpubl.)
	NZB	*Nzv-2*	Unknown	Kozak, Morse, Hartley & Chused (unpubl.)

(2) the number of copies of viral sequences in the cell genome;

(3) if there is more than one type of sequence, the numbers of each sequence type can be determined; and

(4) the relatedness of the viral probe to the integrated viral sequences can be assessed.

With the use of ecotropic MuLV c-DNA, it was shown that strains of mice with high and low ecotropic MuLV phenotypes had, respectively, three to four, or one to two, complete copies of ecotropic MuLV genome, whereas ecotropic virus negative strains had no complete copies. In addition it was found that all mouse strains have between 10 and 50 sequences which do not contain the entire ecotropic genome (Chattopadhyay, Lowy *et al.*, 1974a, b). These sequences could be related to xenotropic MuLV genomes or could be incomplete or defective genome sequences.

By comparison, studies with a xenotropic MuLV probe demonstrated that all mouse strains have multiple complete copies of the xenotropic genome though the numbers of such copies in high, low and negative virus phenotype strains is yet to be determined (Rowe & Hartley, 1978).

Hybridization studies from different laboratories have given conflicting data as to whether inbred strains or even wild mice carry complete sequences coding for amphotropic MuLV. One group reported that all mice carry from three to six complete sequences in their DNA (Chattopadhyay, Hartley *et al.*, 1978). By comparison, a second group found that no complete sequences were present in any mouse (Barbacid, Robbins & Aaronson, 1979), suggesting that the amphotropic virus of wild mice is acquired exogenously. More recent work (U. Rapp, personal communication) is consistent with the view that partial if not complete amphotropic MuLV genomes are present in at least one inbred strain, NFS. These investigators showed that several MCF viruses obtained after injection of this strain with ecotropic MuLV contained from 5–20% amphotropic MuLV-specific sequences.

There is currently no evidence on whether sequences coding for MCF or other recombinant MuLV are present in mouse cellular DNA or not; MCF MuLV specific probes are not yet available and current techniques are too insensitive to accurately assess the limited genomic differences between N- and B-tropic viruses obtained from a single strain.

PARTIAL GENOME EXPRESSION

Reports from a number of laboratories indicate that antigens serologically related to the viral proteins p30 (Taylor, Meier & Huebner, 1973; McClintock, Ihle & Joseph, 1977) and gp70 (Stockert, Old & Boyse, 1971; Lerner *et al.*, 1976; McClintock *et al.*, 1977; Morse, Chused, Boehm-Truitt *et al.*, 1979) can be detected in the serum and tissues of mice which produce no infectious MuLV. There are major difficulties in such studies, however, due to the potential inadequacies of the assays used to detect infectious MuLV or their proteins. The full extent of the problems can be appreciated by recalling that NZB mice were long known to express gs antigen (p30) in their spleens yet no infectious virus was detected until assays for xenotropic MuLV were developed (Levy & Pincus, 1970). Similar problems might plague all the reports listed above but in particular the expression of gs antigen (p30) in B10.D2(58N) mice (Taylor *et al.*, 1973), a strain of mouse bred to be congenic to C57BL/10 (B10) at the *H-1* locus. Initially it was thought that this strain differed from B10 by a small region of the seventh chromosome which contained the DBA/2 alleles of *H-1* and *Hbb* but it was then found to differ from B10 by an additional gene which segregated independently of both *H-1* and

Hbb. This gene governs the expression of a p30-related antigen in spleen, in the absence of infectious virus (Taylor *et al.*, 1973). This work was done prior to the recognition of xenotropic MuLV, so, given the extensive homology and serological cross-reactivity between ecotropic and xenotropic p30 proteins (Gautsch *et al.*, 1978), an alternative explanation was that B10.D2(58N) mice express xenotropic MuLV whereas B10 mice do not. Subsequent work has indicated, however, that both B10.D2(58N) and B10 mice produce low but equivalent amounts of xenotropic MuLV in spleen (Morse, Graff, Chused & Hartley, unpublished observations), and so the problem turns into one of discordant partial viral genome expression between two virus-producing strains. The situation is further complicated by an additional observation made by the same authors, that the B10.D2(58N) strain differs from B10 by the expression of a second antigen, this one serologically related to xenotropic MuLV gp70. This cell surface gp70, termed XenCSA (Morse, Chused, Boehm-Truitt *et al.*, 1979), has been shown not to correlate well with infectious virus production in NZB lymphocytes. NZB thymocytes express higher levels of XenCSA than spleen T cells, yet produce less infectious xenotropic MuLV (Morse, Chused, Sharrow *et al.*, 1979).

There are several possible explanations for the kind of discordant expression of viral antigens and infectious virus seen in B10.D2(58N) and NZB mice. The simplest one is that an infectious virus is being produced and the assay systems presently in use cannot detect it, or possibly that a defective particle is produced which is not infective but can code for the synthesis of the viral proteins gp70 and p30. In either case, the explanation of the B10, B10.D2(58N) difference would be that viral sequences were transferred in the process of making the congenic strain. A precedent for such an occurrence is seen in strain B10.AKM which is congenic to B10 at the *H-2* locus (Datta *et al.*, 1978). These mice apparently acquired an ecotropic viral genome from their high virus parent AKR.M. Several cycles of infection of an embryo followed by integration into the germ-line DNA produced a strain which is the B-tropic virological equivalent of AKR. An alternative explanation is that a mutation occurred in the B10.D2(58N) strain such that viral sequences present in it (and its B10 parent) could be transcribed. Again, there is precedent for such an occurrence for both. Stockert and her co-workers (Stockert, Old & Boyse, 1971) and our laboratory (Morse, 1978) have found what appear to be mutations resulting in the expression of antigens serologically related to xenotropic gp70. The concomitant expression of two virally related proteins seen in the B10.D2(58N) strain suggests, however, that if mutation is responsible, it most probably occurred in a regulator gene.

A third possible explanation for the apparent discordant expression of viral proteins, is that these proteins serve some function in the cell other than the assembly of virus (Lerner *et al.*, 1976; Stockert, Old & Boyse, 1971). This point of view is based on the unusual properties shown by certain gp70 molecules. Some act as alloantigens (Stockert, Old & Boyse, 1971; Morse, Chused, Boehm-Truitt *et al.*, 1979) while others are found associated with secretory and glandular epithelia (Lerner *et al.*, 1976). Expression of these gp70's has been demonstrated repeatedly in 129 mice, which, while they carry the genetic information for MuLV, have never been shown to produce infectious virus of any class. If these findings do not reflect the lack of assay systems capable of detecting a "strain 129 virus", then they raise the question of whether this represents partial transcription of complete virus sequences or complete transcription of partial viral sequences. Indeed a further question must be raised: is it fair to call these viral sequences? In part, this is a problem in semantics, but it raises interesting questions related to "molecular mimicry" (Sprent, 1959). If MuLV have adopted a cellular gene to make their major envelope protein this might give them a selective advantage against the host immune system for they would be recognized as "self".

DISCUSSION AND SUMMARY

The existence of MuLV genomes integrated into the cellular DNA of every inbred mouse strain, as well as their possible expression at any given time, has both positive and negative aspects. On the positive side:

(1) it seems highly likely that the study of the induction of virus expression can provide insight into the mechanisms underlying the expression of other cellular genes;

(2) study of the ways the cell regulates this expression may add to our general knowledge of mammalian gene regulation;

(3) an understanding of the way the host immune response regulates expression and spread of these viruses may give a clue to the host regulation of other events which may be inimical to survival, e.g. expression of autoreactivity.

(4) it seems likely that the study of the viral antigens expressed on the cell surface, in particular the gp70's, may indicate how various insults or changes in differentiation state may affect the expression of surface proteins.

On the negative side, the expression of MuLV or their proteins under a variety of stimuli, e.g. blastogenesis and graft-versus-host disease (Hirsch *et al.*, 1972) as well as in the course of normal growth of the host, can make interpretation of experimental data extremely difficult. In chemical carcinogenesis studies, one has to question if an observed effect is due to the metabolic activity of a chemical or its activation of latent leukemogenic viruses. A second problem arises from the observations that most mice have antibody reactive with the cell surface antigens induced by MCF-MuLV (Stockert, De Leo *et al.*, 1979; Longstreth, Sharrow *et al.*, in preparation). Such antibody as well as antibody to ecotropic and xenotropic MuLV associated cell surface antigens (Longstreth, Ihle & Hanna, 1976; Longstreth, Sharrow *et al.*, in preparation) may considerably confuse the interpretation of experiments using alloantisera, particularly since the expression of viral-like gp70 on normal lymphoid cell surfaces is quite common (Stockert, Old & Boyse, 1971; Morse, Chused, Boehm-Truitt *et al.*, 1979). Cell mediated lympholysis (CML) of tumor cells grown *in vitro* may also be complicated by the expression of MuLV. Several laboratories have noted striking increases in the expression of surface gp70 during *in vitro* cultivation of tumor cells (Civin & Wunderlich, personal communication; Longstreth & Kripke, unpublished observations), and others have noted that purified viral gp70 can induce cytotoxicity in certain *in vitro* CML assays (Lee & Ihle, 1977).

In conclusion, it seems likely that studies of MuLV and the problems they pose will provide insights into a number of interesting areas of mammalian biology. The most exciting future research probably lies in the area of the surface expression of gp70 in the absence of virus expression. Whether these molecules are cellular or viral is a moot point, but what function or functions they serve in the cell (or the virus) is likely to be a very interesting question.

ACKNOWLEDGEMENTS

The authors gratefully acknowledge the assistance of Ms E. Brown and Ms C. Bradley in the studies of natural immunity to MuLV. We deeply appreciate the excellent work of Mr O. Childers and Ms V. Shaw in the preparation of the manuscript and we wish to thank Dr P. R. McClintock for his helpful editorial suggestions.

REFERENCES

Baltimore, D. (1975). Tumor viruses: 1974. *Cold Spr. Harb. Symp. quant. Biol.* 39: 1187–1207.

Barbacid, M., Robbins, K. C. & Aaronson, S. A. (1979). Wild mouse RNA tumor viruses. A nongenetically transmitted virus group closely related to exogenous leukemia viruses of laboratory mouse strains. *J. exp. Med.* 149: 254–266.

Barbacid, M., Stephenson, J. R. & Aaronson, S. (1976). *Gag* gene of mammalian type-C RNA tumor viruses. *Nature, Lond.* 262: 564–566.

Benade, L. E., Ihle, J. H. & DeCleve, A. (1978). Serological characterization of B-tropic viruses of C57BL mice: possible origin by recombination of endogenous N-tropic and xenotropic viruses. *Proc. natn. Acad. Sci. U.S.A.* 75: 4553–4557.

Bernhard, W. & Guerin, M. (1958). Présence de particules d'aspect virusal dans les tissus tumoraux de souris atteintes de leucémie spontanée. *C.r. hebd. séanc. Acad. Sci., Paris* 247: 1802–1805.

Callahan, R., Leiber, M. M. & Todaro, G. J. (1975). Nucleic acid homology of murine xenotropic type-C viruses. *J. Virol.* 15: 1378–1385.

Chattopadhyay, S. K., Hartley, J. W., Lander, M. R., Kramer, B. S. & Rowe, W. P. (1978). Biochemical characterization of the amphotropic group of murine leukemia viruses. *J. Virol.* 26: 29–39.

Chattopadhyay, S. K., Lowy, D. R., Teich, N. M., Levine, A. S. & Rowe, W. P. (1974a). Evidence that the AKR murine-leukemia-virus genome is complete in DNA of the high-virus AKR mouse and incomplete in the DNA of the "Virus-Negative" NIH Mouse. *Proc. natn. Acad. Sci. U.S.A.* 71: 167–171.

Chattopadhyay, S. K., Lowy, D. R., Teich, N. M., Levine, A. S. & Rowe, W. P. (1974b). Qualitative and quantitative studies of AKR type murine leukemia virus sequences in mouse DNA. *Cold Spr. Harb. Symp. quant. Biol.* 39: 1085–1101.

Chused, T. M. & Morse, H. C. III (1978). Expression of XenCSA, a cell surface antigen related to the major glycoprotein (gp70) of xenotropic murine leukemia virus by lymphocytes of inbred mouse strains. In *Origins of inbred mice*: 297–319. Morse, H. C. III (Ed.). New York: Academic Press.

Cloyd, M. W., Hartley, J. W. & Rowe, W. P. (1979). Cell surface antigens associated with recombinant mink cell focus inducing murine leukemia viruses. *J. exp. Med.* 149: 702–712.

Datta, S. K. & Schwartz, R. S. (1977). Mendelian segregation of loci controlling xenotropic virus production in NZB crosses. *Virology* 83: 449–452.

Datta, S. K., Tsichlis, P., Schwartz, R. S., Chattopadhyay, S. K. & Melief, C. J. M. (1978). Genetic differences unrelated to *H-2* in *H-2* congenic mice. *Immunogenetics* 7: 359–365.

Elder, J. H., Gautsch, J. W., Jensen, F. C., Lerner, R. A., Hartley, J. W., & Rowe, W. P. (1977). Biochemical evidence that MCF murine leukemia viruses are envelope (*env*) gene recombinants. *Proc. natn. Acad. Sci. U.S.A.* 74: 4676–4680.

Elder, J. H., Jensen, F. C., Bryant, M. L. & Lerner, R. A. (1977). Polymorphism of the major envelope glycoprotein (gp70) of murine C-type viruses: virion associated and differentiation antigens coded for by a multigene family. *Nature, Lond.* 267: 23–30.

Friend, C. (1957). Cell free transmission in adult Swiss mice of a disease having the character of a leukemia. *J. exp. Med.* 105: 307–311.

Furth, J. (1978). The creation of the AKR strain, whose DNA contains the genome of a leukemia virus. In *Origins of inbred mice*: 69–97. Morse, H. C. III (Ed.). New York: Academic Press.

Gautsch, J. W., Elder, J. H., Schindler, J., Jensen, F. C. & Lerner, R. A. (1978).

Structural markers on core protein p30 of murine leukemia virus: functional correlation with *Fv-1* tropism. *Proc. natn. Acad. Sci. U.S.A.* 75: 4170–4174.

Gross, L. (1951). Spontaneous leukemia developing in C3H mice following inoculation in infancy with Ak leukemic extracts, or Ak embryos. *Proc. Soc. exp. Biol. Med.* 76: 27–30.

Hartley, J. W. & Rowe, W. P. (1976). Naturally occurring murine leukemia virus in wild mice: characterization of a new "amphotropic" class. *J. Virol.* 19: 19–25.

Hartley, J. W., Wolford, N. K., Old, L. J. & Rowe, W. P. (1977). A new class of murine leukemia virus associated with development of spontaneous lymphomas. *Proc. natn. Acad. Sci. U.S.A.* 74: 789–792.

Hirsch, M. S., Phillips, S. M., Solnik, C., Black, P. H., Schwartz, R. S. & Carpenter, C. B. (1972). Activation of leukemia viruses by graft-versus-host and mixed lymphocyte reactions *in vitro*. *Proc. natn. Acad. Sci. U.S.A.* 69: 1069–1072.

Huebner, R. J., Sarma, P. S., Kelloff, G. J., Gilden, R. V., Meier, H., Myers, D. D. & Peters, R. L. (1970). Immunological tolerance to RNA tumor virus genome expression: significance of tolerance and prenatal expression in embryogenesis and tumorgenesis. *Ann. N.Y. Acad. Sci.* 181: 246–258.

Klein, P. A. (1975). Anomalous reactions of mouse alloantisera with cultured tumor cells. I. Demonstration of widespread occurrence with reference typing sera. *J. Immunol.* 115: 1254–1260.

Kozak, C. A. & Rowe, W. P. (1978). Genetic mapping of xenotropic murine leukemia virus inducing loci in two mouse strains. *Science, Wash.* 199: 1448–1449.

Lee, J. C. & Ihle, J. N. (1977). Characterization of the blastogenic and cytotoxic responses of normal mice to ecotropic C-type viral gp 71. *J. Immunol.* 118: 928–934.

Lerner, R. A., Wilson, C. B., Del Villano, B. C., McConahey, P. J. & Dixon, F. J. (1976). Endogenous oncornaviral gene expression in adult and fetal mice: quantitative, histologic and physiologic studies of the major viral glycoprotein, gp70. *J. exp. Med.* 143: 151–164.

Levy, J. A. & Pincus, T. (1970). Demonstration of biological activity of a murine leukemia virus of New Zealand black mice. *Science, N.Y.* 170: 326–329.

Longstreth, J. D., Ihle, J. N. & Hanna, M. G., Jr. (1976). Autogenous immunity to endogenous RNA tumor virus: virion and virus induced cell surface antigens. *Ann. N.Y. Acad. Sci.* 276: 343–353.

Longstreth, J. D., Sharrow, S. O., Morse, H. C., III & Chused, T. M. (In preparation). *Natural antibody to xenotropic murine leukemia virus associated cell surface gp70.*

McClintock, P. R., Ihle, J. N. & Joseph, D. J. (1977). Expression of AKR MuLV gp71-like and BALB(x) gp71-like antigens in normal mouse tissues in the absence of overt virus-expression. *J. exp. Med.* 146: 422–434.

Moloney, J. B. (1960). Biological studies on a lymphoid leukemia virus extracted from sarcoma S.37.I. Origin and introductory investigation. *J. natn. Cancer Inst.* 24: 933–945.

Morse, H. C. III (1978). Differences among sublines of inbred mouse strains. In *Origins of inbred mice:* 441–444. Morse, H. C. III. (Ed.). New York: Academic Press.

Morse, H. C. III (In press). The laboratory mouse – an historical perspective. In *The mouse in biomedical research.* Foster, H. & Small, J. D. (Eds New York: Academic Press.

Morse, H. C. III, Chused, T. M., Boehm-Truitt, M., Mathieson, B. J., Sharrow, S. O. & Hartley, J. W. (1979). XenCSA: cell surface antigens related to the major envelope glycoproteins (gp70) of xenotropic murine leukemia viruses. *J. Immunol.* **122**: 443–454.

Morse, H. C. III, Chused, T. M., Sharrow, S. O. & Hartley, J. W. (1979). Variations in the expresssion of xenotropic murine leukemia virus genomes in lymphoid tissues of NZB mice. *J. Immunol.* **122**: 2345–2348.

Old, L. J., Boyse, E. A. & Stockert, E. (1964). Typing of mouse leukemias by serological methods. *Nature, Lond.* **201**: 777–788.

Pincus, T., Rowe, W. P. & Lilly, F. (1971). A major genetic locus affecting resistance to infection with murine leukemia viruses. II. Apparent identity to a major locus described for resistance to Friend murine leukemia virus. *J. exp. Med.* **133**: 1234–1241.

Rasheed, S., Gardner, M. B. & Chan, E. (1976). Amphotropic host range of naturally occurring wild mouse leukemia viruses. *J. Virol.* **19**: 13–18.

Rauscher, F. J. (1962). A virus-induced disease of mice characterized by erythrocytopoiesis and lymphoid leukemia. *J. natn. Cancer Inst.* **29**: 515–521.

Rommelaere, J., Davis-Keller, H. & Hopkins, H. (1979). RNA sequencing provides evidence for allelism of determinants of the N-, B- or NB-tropism of murine leukemia viruses. *Cell* **16**: 43–50.

Rommelaere, J., Faller, D. V. & Hopkins, N. (1978). Characterization and mapping of RNase T1 resistant oligonucleotides derived from the genomes of AKv and MCF murine leukemia viruses. *Proc. natn. Acad. Sci. U.S.A.* **75**: 495–499.

Rowe, W. P. (1978). Leukemia virus genomes in the chromosomal DNA of the mouse. *Harvey Lectures* **71**: 173–192.

Rowe, W. P. & Hartley, J. W. (1978). Chromosomal location of C-type virus genome in the mouse. In *Origins of inbred mice:* 289–295. Morse, H. C. III (Ed.). New York: Academic Press.

Rowe, W. P., Hartley, J. W. & Bremner, T. (1972). Genetic mapping of a murine virus inducing locus of AKR mice. *Science, Wash.* **178**: 860–862.

Rowe, W. P., Hartley, J. W. & Kozak, C. A. (In press). Murine leukemic viruses as chromosomal genes of the mouse. In *A century of mammalian genetics and cancer, 1929–2029; A view from midpassage.*

Schafer, W., Fishinger, P. J., Collins, J. J. & Bolognesi, D. P. (1977). Role of carbohydrate in biological functions of Friend murine leukemia virus gp70. *J. Virol.* **21**: 35–43.

Sprent, J. F. S. (1959). In *The evolution of living organisms:* 149–165. Melbourne: University Press.

Steeves, R. A. (1961), Cellular antigen of Friend virus induced leukemias. *Cancer Res.* **28**: 338–352.

Stephenson, J. R., Reynolds, R. K., Tronic, S. R. & Aaronson, S. A. (1975). Distribution of three classes of endogenous type-C RNA viruses among inbred strains of mice. *Virology* **67**: 404–414.

Stockert, E., Old, L. J. & Boyse, E. A. (1971). The G_{IX} system. A cell surface allo-antigen associated with murine leukemia virus: implications regarding chromosomal integration of the viral genome. *J. exp. Med.* **133**: 1334–1349.

Stockert, E., De Leo, A. B., O'Donnell, P. V., Obata, Y. & Old, L. J. (1979). $G_{(AKSL2)}$: A new cell surface antigen of the mouse related to the dual tropic mink cell focus-inducing class of murine leukemia virus detected by naturally occurring antibody. *J. exp. Med.* **149**: 200–214.

Taylor, B. A., Meier, H. & Huebner, R. J. (1973). Genetic control of the group-specific antigen of murine leukemia virus. *Nature New Biol., Lond.* **241**: 184–188.

Symp. zool. Soc. Lond. (1981) No. 47, 643–665

Inborn Errors of Metabolism in the Mouse

GRAHAME BULFIELD

Department of Genetics, The University, Leicester, UK

SYNOPSIS

Although over 600 mutants have been reported in the mouse, until recently none was known to be equivalent to any of the 147 inborn errors of metabolism in man. Attempts had been made with normal mice (and other laboratory animals) to produce animal models of human syndromes by experimental procedures such as drug treatment. Models produced in this way are usually inadequate for investigations of biochemical and clinical problems associated with the human syndromes.

Over the last few years, however, several extremely useful mutant mouse homologues of human inborn errors have been discovered. In some cases discovery was by chance and in others the result of a deliberate screening programme; some of these programmes are described in detail. Once a mutant is discovered it is important to ensure that it is biochemically homologous to the equivalent human condition and examples of the techniques involved are described.

These mouse mutants have become important tools in several areas of research. They have been used to investigate the relationship between structural and regulatory genes in the control of gene expression. In experimental therapy they are becoming valuable in obtaining information from experiments that might be difficult to perform on human patients. Finally they have introduced new concepts and ideas into the investigation of several areas of trace element metabolism and renal transport mechanisms.

INTRODUCTION

There are 147 human inherited enzyme deficiencies listed in McKusick's (1975) catalogue. Many of these syndromes are difficult or impossible to treat with present therapeutic methods, especially as they often occur as geographically isolated cases. Laboratory animals with homologous lesions would therefore be important in establishing rational approaches to therapy. It is, however, essential to have an animal that is cheap to maintain, breeds rapidly and has a substantial body of genetic knowledge associated with it. The house mouse satisfies all these criteria.

This was recognized nearly 30 years ago by J. B. S. Haldane on whose suggestion the urine of inbred strains of mice was screened for amino acid levels by paper chromatography (Harris & Searle, 1953). Unfortunately this technique only revealed substantial amounts of taurine, although present ion-exchange procedures reveal over 100 ninhydrin positive substances in mouse urine (Kacser & Bulfield, unpublished observations). Despite only finding taurine, Harris & Searle (1953) did find a ten-fold difference between the C57BL and the A or CBA inbred strains (a difference not apparent in the plasma). They ascribed this to a renal effect, a fact exploited in renal clearance studies many years later (Scriver, Chesney & McInnes, 1976; see below).

Over the last 20 years a large number of animal models has been discovered in the mouse and other species. A review by Cornelius (1969) lists 93 in farm animals, 98 in dogs and cats and 116 in laboratory rodents, mainly the mouse. This was updated in 1975 by a further 890 items (Bustad, Hegreberg & Padgett, 1975) and can be expanded by the abstracts (ILAR, 1971–1976) and three-monthly listings (ILAR News, 1979) of the Institute of Laboratory Animal Resources. The area is also covered by two series of articles on individual models (Jones, Hackel & Migaki, 1972–1977; Schwartz, 1978) and recent books (Morse, 1978; Festing, 1979; Altman & Katz, 1979; Hommes, 1979). The need to screen for specific animal models and to conserve and exploit them has also been stressed (Cornelius, 1969, 1978; Prichard, 1978).

Despite this apparent wealth of material, few of these models are homologous to specific inborn errors of metabolism in man (although some have been transiently induced by nutritional means: Gerritsen & Siegel, 1972; Stephens *et al.,* 1979), nor are they readily available in laboratory stocks (Bulfield, 1977). It cannot be stressed too strongly that an animal model, to be of value, must be an inherited deficiency of the same enzyme as that deficient in the human syndrome. It is not satisfactory to have a model which has a similar phenotype to the human condition, such as the *diabetes* and *dystrophia-muscularis* mutants in the mouse. If this criterion is adhered to, most of the criticisms of the use of animal models disappear, and a remarkable identity between man and mouse becomes apparent.

In this review I shall first evaluate the methods that have been used to screen for suitable mutants in the mouse and then discuss the present and potential value of these mutants in introducing new information and concepts into mammalian biochemical genetics. I shall not discuss animal models of haemorrhagic disease as they have recently been reviewed (Brinkhous, 1978).

SCREENING FOR BIOCHEMICAL MUTANTS IN THE MOUSE

Inbred Strains

For many years inbred strains of mice have been the main source of material for biologists because of the demand for genetically standardized and identical animals. The availability of these strains makes it understandable that they are initially used in screening for genetical variation in any parameter (Morse, 1978; Festing, 1979; Altman & Katz, 1979). Although there are over 200 of these strains we must be aware that several limitations exist.

Two hundred strains represent only 200 different genotypes and furthermore many of the strains are related. Despite this they have proved useful in supplying variants of the histocompatibility systems (Klein, 1975) and electrophoretic variants of enzymes (Roderick *et al.*, 1971). Most variation in enzyme levels between inbred strains has, however, proved to be in the order of two- to three-fold (Paigen, 1970, 1978; Bulfield, Moore & Kacser, 1978), whereas most inborn errors in man show a reduction to about 5%. Indeed, the strong selection used in maintaining inbred strains would be likely to remove even a mildly deleterious phenotype. The exceptions to this general rule are the type VIII glycogen storage disease (phosphorylase kinase deficiency) found in the I strain (J. B. Lyon, Porter & Robertson, 1967) and the alcohol preference associated with aldehyde dehydrogenase deficiency in the C57BL/6J strain (Sheppard, Albersheim & McClearn, 1970).

Phenotype

Some phenotypes of spontaneously occurring mouse mutants have suggested that they might be caused by particular enzyme deficiencies. These are, of course, the result of chance observations and not a rational screening programme.

For example, in a crossing experiment to produce a new inbred strain it was observed that the urine of some mice stained the sawdust bright yellow. This was found to be due to hyperprolinuria associated with a deficiency of the liver proline oxidase complex (hyperprolinaemia type I: Blake & Russell, 1972; Blake, 1972).

A further example was the result of a chance discussion at a research seminar which led to the discovery that the bladderstones in the X-linked *sparse-fur* mutant were mainly composed of orotic acid accumulated as a result of a deficiency of ornithine carbamoyl transferase (De Mars *et al.*, 1976). Similarly when a spontaneous

X-linked mutant was found which had a shortened trunk and hind limbs, comparison with human X-linked disorders led to the discovery that the animals were hypophosphataemic (Eicher, Southard *et al.*, 1976).

Deliberate screening of over 50 known (Searle, 1979) behavioural/neurological mutants for aminoacidopathies revealed histidinaemia amongst animals of a balance-defective strain of mice (Kacser, Bulfield & Wallace, 1973; Bulfield & Kacser, 1974, 1975). A similar deliberate screening of four anaemic mutants for enzymopathies of erythrocyte glucose metabolism, however, failed to show any specific enzyme deficiencies (Hutton & Bernstein, 1973). It is probable that many other mutants have been examined for their similarity to human disorders and the lack of identity has gone unpublished.

Mutation

There is only one case in the literature of the exploitation of spontaneous mutation rate and inbreeding to produce an inborn error. This was the deliberate search amongst 15 462 mice for a phenotype resembling cystic fibrosis in man. Three mice (all related) were found with small body size, yellow fatty faeces and *exocrine pancreatic insufficiency* (Pivetta & Green, 1973).

Deliberate use of mutagens to create new variation has been little used in mammals. This is because of the difficulty of increasing mutation rate to a greater frequency than 10^{-5} to 10^{-4} per gamete per generation either with irradiation or with chemical mutagens.

The most extensive search amongst mutagenized mice was for deficiency of blood catalase (Feinstein, Seaholm *et al.*, 1964; Feinstein, Howard *et al.*, 1966). A total of 12 306 offspring of irradiated males (600 R, fractionated dose) were screened semi-quantitatively for blood catalase activity. Forty-two presumptive heterozygotes were found and bred; five segregated as true heterozygotes for a single gene; one of these heterozygotes produced acatalasaemic offspring.

The scale of this screening programme is daunting to most research workers and there is a further problem. Irradiating mice produces chromosomal aberrations, rearrangements and deletions. Therefore homozygotes of "alleles" produced in this way are often lethal or, if viable, have multiple gene deficiencies and deleted genes which produce no protein cross-reacting material. With most inborn errors in man the enzyme deficiency is associated with normal levels of cross-reacting material (Sutton & Wagner, 1975). This problem is exemplified by the radiation-induced albino-lethal deletion mutants and the haemoglobin deficiency mutants of the mouse.

The albino-lethal deletions were discovered during a specific locus mutation rate experiment and have pleiotropic effects. They were originally found to be deficient in glucose-6-phosphatase (Gluecksohn-Waelsch & Cori, 1970) and recently abnormalities of the endoplasmic reticulum have been found associated with deficiencies in the activity of tyrosine aminotransferase, serine dehydratase (Gluecksohn-Waelsch *et al.*, 1974), cytochrome P-450, bilirubin glucuronyltransferase (Thaler, Erickson & Pelger, 1976), plasma protein synthesis (Garland *et al.*, 1976), and mitochondrial malic enzyme (Eicher, Lewis *et al.*, 1978). The mutants can be best explained as having extensive and variable deletions of chromosome 7 as they fall into several complementation groups (Gluecksohn-Waelsch *et al.*, 1974); indeed the structural locus for mitochondrial malic enzyme (*Mod*-2), which is deleted in some cases, is known to be one centimorgan distal to the *albino* locus (Eicher, Lewis *et al.*, 1978). Interesting though these mutants are in investigating the organization of the mammalian chromosome (Gluecksohn-Waelsch, 1979), they are of little use for studying inherited metabolic disease.

The radiation-induced haemoglobin mutants are potentially of more value as they were selected by screening for lack of gene product at the α and β globin loci and are therefore more likely to be point mutations. In this experiment SEC strain males were irradiated and mated to 101 strain females; the two strains have allelic differences at both the *Hba* and *Hbb* loci, thus facilitating recognition of mutants among the offspring. Of 8621 animals screened three had a similar α-chain abnormality and two had different β-chain abnormalities (L. B. Russell *et al.*, 1976). The α-chain deficiencies cause an α-thalassaemia similar to human α-thalassaemia (Popp *et al.*, 1979); one of the β locus variants is a tandem duplication while the other is probably the result of non-disjunction (L. B. Russell *et al.*, 1976; Popp *et al.*, 1979). It is not yet known whether it will be possible to exploit these mutants as models of human haemoglobinopathies.

Because of the problems of deletions and chromosomal abnormalities associated with radiation-induced mutations, it is interesting to speculate on the potential use of chemical mutagens which operate through more acceptable mechanisms. These mutagens would still probably only increase the mutation rate up to ten-fold and so far they have not been used for this purpose. It is, however, possible that useful mutants might arise in recently initiated screening programmes for environmental mutagens. These involve determining changes in the electrophoretic mobility and the specific activity of enzymes (especially of glycolysis) as a measure of mutation rate

(Wolff, 1977; Soares & Malling, 1977) and should produce some inherited enzyme deficiencies more acceptable as models of human inborn errors.

Feral Mice

It has recently become recognised that feral mice are an extensive source of genetical variation (Altman & Katz, 1979). This is especially so in European populations where the proximity of farms prevents isolation of populations and the resulting reduction in the frequency of deleterious genes associated with inbreeding. It is, however, always a wise precaution to take samples from as wide a variety of sites as possible.

The pattern of speciation of feral mice in Europe is also fortunate. There are several reproductively isolated populations, some allopatric (as in Jutland; Hunt & Selander, 1973) and some sympatric (as in Southern France and Spain; Britton & Thaler, 1978; Sage, 1978; Altman & Katz, 1979). These separate populations accumulate differences in nature but they interbreed with laboratory animals. The extent of this phenomenon has not yet been fully described or its immense value fully appreciated or exploited.

The disadvantage of feral mice is that it is unreasonable to suppose homozygotes for enzyme deficiencies will be found, owing both to their expected low frequency and to their expected lethality in nature. It is therefore necessary to screen for heterozygotes, which limits the use of wild mice to enzymes that can be directly and easily assayed (see below).

The frequency of heterozygotes is not unmanageably low even when we assume the homozygous condition to be lethal. For example, with a mutation rate of 10^{-6} the heterozygotes should have a frequency of approximately 2×10^{-3}; whereas with a mutation rate of 10^{-5} the heterozygotes should have a frequency of approximately 6×10^{-3} (Falconer, 1960). Where the homozygote is not a complete genetical lethal, the gene frequency and hence heterozygote frequency will be higher. Thus if it is possible to screen for the deficiency of many enzymes in one animal at the same time, the chance of finding a heterozygote is proportionately increased.

Recently several inherited enzyme deficiencies have been found by direct screening of feral mouse populations. Glucosephosphate isomerase deficiency (Padua, Bulfield & Peters, 1978) was discovered in a population of mice in Somerset and pyruvate kinase deficiency in one in Edinburgh (Bulfield & Moore, 1978). Histidinaemia was found in a strain of mice originating in wild-caught mice from Peru (Kacser, Bulfield & Wallace, 1973).

Biochemical Techniques

The area of metabolism in which screening is performed depends on the ease of establishing a screening system and the intended use of the mutants discovered. In the case of catalase deficiency (Feinstein, 1970) it was the desire to study the role of hydrogen peroxide in carcinogenesis; *exocrine pancreatic insufficiency* (Pivetta & Green, 1973) was discovered during a search for a model of cystic fibrosis; with the variants of β-glucuronidase activity and levels (Paigen *et al.*, 1975) it was because the enzyme was hormone inducible and distributed between two sub-cellular compartments, making it valuable for studies on the regulation of gene expression.

Mice have been screened systematically for variants in two areas of metabolism: amino acid metabolism (Kacser, Bulfield & Wallace, 1973) and erythrocyte glucose metabolism (Bulfield & Moore, 1978). Many human inherited metabolic diseases are caused by deficiencies of enzymes in these pathways (McKusick, 1975), and it has been possible to establish automated screening procedures for both areas. For amino acid metabolism automated column chromatography is employed to detect the large number of ninhydrin positive substances present in animal tissue. As metabolite pools are being investigated rather than direct gene products, it is necessary to use potentially homozygous animals. For erythrocyte glucose metabolism semi-automated assays through NAD(H) or NADP(H) linked reactions (Bulfield & Moore, 1974) permit direct determination of the activity of 17 enzymes and are therefore more suitable for screening wild mice for potential heterozygotes.

CHARACTERISTICS OF THE MOUSE MUTANTS

Structural Genes

When an enzyme deficiency has been discovered, one major problem is to decide whether it is due to a mutation at the structural locus or at a regulatory locus. This could have practical implications as it may be possible to alleviate the symptoms of a regulatory mutation by short-circuiting the mechanism using knowledge of the molecular events involved.

The classical method of determining whether a mutation is at the structural locus is by raising antibodies to a purified enzyme and assaying the mutant for immunologically cross-reacting material. This has, for instance, been successfully performed with catalase deficiency (Feinstein, 1970), although several problems may be associated with the technique. Purification of the enzyme can be extremely time-consuming and contamination of the final preparation with even a small amount of other highly immunogenic materials can give spurious results. The method also assumes that a mutation at a structural locus will affect the activity and not the immunogenic properties of the enzyme. An example of this problem is demonstrated by *histidinaemia* (Kacser, Bulfield & Wallace, 1973) where 5% of normal enzyme activity is associated with 5% cross-reacting material (S. A. Arfin, personal communication). This small amount of cross-reaction could be due either to a mutant histidase or to a complete absence of histidase and the presence of another enzyme with partial activity towards histidine and partial cross-reaction with the antibody. The technique can fail therefore to distinguish whether the mutation is at the structural locus or at a regulatory locus.

These problems are further exemplified by *phosphorylase kinase deficiency.* The I strain of mice was reported by Lyon (J. B. Lyon *et al.,* 1967; J. B. Lyon, 1970) to be totally deficient in skeletal muscle phosphorylase kinase which causes three-fold higher levels of muscle glycogen (glycogen storage disease type VIII). The enzyme was only partially deficient in brain, heart and kidney. Despite showing such severe muscle enzyme deficiency the animals were able to break down tissue glycogen after adrenalin stimulation. This paradox seemed to be resolved when, with a different extraction technique, small (4%) amounts of the enzyme were found in a centrifugable particulate fraction of skeletal muscle (Huijing, 1970). Later work, however, indicates that this particulate activity is due to the production of AMP and the activation of phosphorylase b rather than phosphorylase kinase (Cohen & Cohen, 1973), although it has been demonstrated that I strain mice can phosphorylate troponin B at half the rate of normal mice (Gross & Mayer, 1974).

This confusion over the presence, or absence, of residual enzyme activity in I strain mice is further complicated by immunological studies. Initially it was demonstrated that I strain mice had normal amounts of phosphorylase kinase material cross-reacting to purified rabbit muscle phosphorylase kinase antisera raised in hens (Cohen & Cohen, 1973) and goats, but not guinea-pigs (Gross, Longshore & Pangburn, 1975). When mouse phosphorylase kinase was itself purified and antibodies raised in hens the antisera failed to detect any of the four subunits of the enzyme in I strain animals

(Cohen, Burchell & Cohen, 1976). On the assumption that the α and α' subunits are coded for by different genes, these authors suggest that the mutation in I mice is pleiotropic and must therefore be in a regulatory gene. At present however there is no conclusive evidence that this is not a structural gene mutation.

As immunological techniques require purified enzyme and their results can be equivocal other techniques have been used to probe for changes in enzyme structure. These include determining the electrophoretic mobility (or isoelectric point), heat stability, Km's, Ki's and pH optima of the enzyme in crude homogenates (Paigen, 1970). Various combinations of these methods have been used to assign several partial enzyme deficiencies to their corresponding structural loci including: glucosephosphate isomerase deficiency (Padua, Bulfield & Peters, 1978); catalase deficiency (Feinstein, 1970); ornithine carbamoyl transferase deficiency (De Mars *et al.,* 1976). In this last syndrome the deficiency of OCTase is caused by an X-linked gene (*spf*) and results in a *sparse-fur* phenotype. The mutant enzyme has an altered pH profile with about 10% of normal activity at pH 7.2 but with twice the normal activity at pH 10.0. Another allele at this locus (spf^{ash}) causes 5% of normal enzyme activity throughout the pH range 6.0 to 10.0. Females heterozygous for the *spf* gene vary in the proportion of abnormal enzyme in accordance with the random X-inactivation hypothesis (M. F. Lyon, 1961, 1971).

None of the methods described above, when considered on their own, can provide an unequivocal conclusion in assigning variation to a structural locus. Apparent changes in the structure of an enzyme may be due to the mutation affecting its post-translation modification. This would be variation at a regulatory rather than the structural locus. The problem is exemplified by neuraminidase deficiency in SM/J mice which affects the electrophoretic pattern of at least three lysosomal enzymes (Womack & Potier, 1979).

Regulatory Genes

Several regulatory genes affecting enzyme levels have been discovered in laboratory mice. These mutants have a special use in our understanding of the control of mammalian gene expression (Paigen, 1970, 1978; Altman & Katz, 1979). The more important and unusual phenomena associated with them will be briefly discussed.

Some enzymes are synthesized from the same structural locus in all or several tissues. Their levels in different tissues and organelles can be changed by the intracellular environment or their turnover. Intracellular environment can be modified by regulatory genes as in the case of deficiency of microsomal β-glucuronidase (e.g. Swank & Paigen, 1973; Paigen *et al.*, 1975; Lusis & Paigen, 1977). Other examples include alterations in the levels of lactic dehydrogenase isozymes in erythrocytes (*Ldr*-1: Glass & Doyle, 1972) and increased catalase degradation (*Ce*: Ganschow & Schimke, 1969). A decreased enzyme stability can produce varying levels in different tissues: glucosephosphate isomerase deficiency (Padua *et al.*, 1978) and catalase deficiency (Feinstein, 1970) both have low erythrocyte levels with near normal levels in other tissues. In both cases the enzyme is coded for by one structural locus but the enzymic instability only affects its erythrocyte levels where there is no active protein synthesis. Genes affecting the structure of organelles, such as the effect *beige* and others on the lysosomes, can alter enzyme levels (Swank *et al.*, 1978).

Regulatory mutations can also alter the levels of enzyme by changing their rate of synthesis. These are known for β-galactosidase (*Bgs*: Breen, Lusis & Paigen, 1977) and δ-amino levulinate dehydratase (Coleman, 1971) and map at or near the structural locus where they operate in cis-configuration. Another cis-operating regulatory gene (*Gur*; Swank, Paigen & Ganschow, 1973; Paigen *et al.*, 1975; Swank *et al.*, 1978) alters the response of the β-glucuronidase gene to testosterone induction.

A different phenotype with an altered response to testosterone induction is *testicular feminisation* (*Tfm*: M. F. Lyon & Hawkes, 1970). This X-linked gene is homologous to a human syndrome (McKusick, 1975). Affected animals are deficient in a cytosol androgen receptor complex (Gehring, Tomkins & Ohno, 1971; Bullock & Bardin, 1974) resulting in XY animals having female secondary sexual characteristics despite possessing testes, albeit small ones. Enzymes inducible (Dofuku, Tettenborn & Ohno, 1971) or repressible (Bulfield & Nahum, 1978) by testosterone have female-like levels and are unresponsive to injections of the hormone.

Finally, the level of an enzyme can be affected by the concentration of co-factors, activators, or inhibitors. Examples are the effect of erythrocyte NADP on glucose-6-phosphate dehydrogenase activity (Erickson, 1974) and the effect of erythrocyte NAD levels on glyceraldehydephosphate dehydrogenase activity (Bulfield & Trent, 1981).

There are therefore a large variety of ways that regulatory genes can affect enzyme levels and undoubtedly further mechanisms will be discovered.

Therapy

A homologue of human histidinaemia occurs in the mouse. This is caused by a deficiency (< 5%) of the enzyme histidase (histidine ammonia lyase). High levels of histidine in pregnant mice cause some of their offspring to be balance defective (Kacser, Bulfield & Wallace, 1973; Bulfield & Kacser, 1974, 1975; Kacser, Bulfield & Wright, 1979; Kacser, Mya Mya & Bulfield, 1979). A maternal effect has also been observed with human histidinaemia (I. C. T. Lyon, Gardner & Veale, 1974; Harper, 1975) and human phenylketonuria (Denniston, 1963).

The maternal effect in histidinaemic mice has been extensively studied. Histological examination of this (*in utero*) effect demonstrates that the otoliths, semicircular canals and cochlea of the inner ear are extensively affected. Some of these effects are present in a mild form in behaviourally normal sibs of affected animals (Kacser *et al.*, 1977; Kacser, Bulfield & Wright, 1979; Kacser, Mya Mya & Bulfield, 1979).

With the use of dietary treatments it has been possible to investigate the developmental timing of the lesion and modulate its frequency and severity (Kacser *et al.*, 1977; Kacser, Mya Mya & Bulfield, 1979). When heterozygous females are fed on 8% histidine supplemented diet in various stages of pregnancy, their plasma histidine levels are elevated to those of histidinaemic females. Heterozygous animals in the second trimester of the 21-day gestation placed on this diet produced some balance defective offspring (14 out of 130 in 21 litters), whereas no abnormalities were found when heterozygotes were fed the diet in the first or third week of pregnancy. This critical (second week) period corresponds to that for *in utero* damage of the ear by X-irradiation (Rugh, 1964).

Knowledge of this critical period of *in utero* damage made it possible to study the effects of diets containing various levels of histidine on the number and severity of balance defective offspring produced by histidinaemic mothers. Homozygous animals placed on a low histidine diet (20% of normal levels) in the second week of pregnancy produced 3% affected offspring compared with 26% on a normal diet. The diet was again ineffective during the first and third weeks of pregnancy. Treatment of pregnant histidinaemic females with a 2% supplemented diet produced the same proportion of affected offspring as the controls (Kacser *et al.*, 1977; Kacser, Mya Mya & Bulfield, 1979).

It was also discovered that the genetical background has an effect on the penetrance and expression of the balance defective phenotype. Females from a Peru stock (in which the mutation arose and was initially maintained) produced 37% affected animals whereas those of hybrid Peru/C57BL background produced only 16% affected animals. This indicates that genes besides the one causing histidinaemia affect the production of the balance defective phenotype.

Prolinaemia (Blake & Russell, 1972) and phosphorylase kinase deficiency (J. B. Lyon *et al.*, 1967) are benign syndromes presenting no clinical phenotype. In the case of prolinaemia it is however possible by supplemented diet to increase tissue levels of proline to the levels of histidine in histidinaemia. This, however, does not produce any adverse effects on the offspring *in utero* (Kacser *et al.*, 1977; Bulfield, unpublished observations).

Animals with ornithine carbamoyl transferase deficiency (De Mars *et al.*, 1976) have a hairless phenotype (*sparse-fur*) with wrinkled skin, have urinary bladder stones (of orotic acid), are very small, and often do not survive to weaning. These mutants do survive longer if placed on a low protein diet but the phenomenon has not yet been extensively examined (De Mars *et al.*, 1976). As with histidinaemia, the genetical background has an effect; animals survive longer on the C57BL than on the SCFCP background.

A novel therapeutic method has been tried with G_{M1} gangliosidosis in the cat (Holmes & O'Brien, 1978a, b). This mutant has 7% of normal levels of β-galactosidase activity using G_{M1}-ganglioside as the substrate, although its activity towards other substrates is not as much reduced. The mutant enzyme has an increased thermolability, higher isoelectric point, lower molecular weight and altered antigenic specificity (Holmes & O'Brien, 1978a). Therefore this G_{M1} gangliosidosis appears to be a classical lysosomal storage disease. Reynolds, Baker & Reynolds (1978) have attempted enzyme replacement therapy with fibroblasts from the mutants by incubating them for six hours in the present of liposomes containing partially purified feline β-galactosidase. Within 48 hours a substantial amount of enzyme activity was recoverable from the fibroblasts and their previous inability to metabolise ^{14}C-galactose labelled glycopeptides was cured. Both these effects persisted for the length of the experiment, 10 days. This work indicates that liposomes carry the entrapped β-galactosidase into fibroblast lysosomes where it remains active for a considerable time. If it is possible to extend this experiment to live cats this animal model will provide the most valuable yet discovered. The complex area of enzyme replacement therapy in lysosomal storage diseases is one of the most difficult in clinical genetics (Bergsma, 1973).

Over 15 inherited anaemias have been recognized in the mouse (Bernstein, in Altman & Katz, 1979). In man there are 11 inherited deficiences of erythrocyte glucose metabolism associated with haemolytic anaemia (Valentine, 1968; McKusick, 1975). Five of the mouse anaemias have been examined for erythrocyte enzyme deficiency in this area of metabolism but none has been found (Hutton & Bernstein, 1973; Bulfield, *van* mutant, unpublished observations).

It may also be possible to detect enzyme deficiency anaemias, by screening directly for variation in the levels of enzymes (as discussed in detail above). This approach has been attempted with limited success using inbred strains (Bulfield & Moore, 1974; Bulfield *et al.*, 1978), but with greater success using wild mice (Padua *et al.*, 1978; Bulfield & Moore, 1978).

To date no mutant with severe haemolytic anaemia has been found. Several mutants have been found with substantial enzyme deficiency and mild haemolytic symptoms, which will be useful for therapeutic experiments in this continuing programme (Bulfield & Moore, 1978).

Thalassaemia-like mutations have been induced at the haemoglobin loci using radiation treatment (as discussed above, L. B. Russell *et al.*, 1976). Three mutants are deficient in α-chains and two in β-chains, although so far only their chemical nature and effect on erythrocyte structure have been examined and no therapeutic experiments have been attempted (Popp *et al.*, 1979).

Disorders of Trace Element Metabolism

Recent work with mice has revealed that six phenotypes have abnormal trace element metabolism. Two of these syndromes were initially recognized as coat colour genes; two by the deleterious effect of the mother's milk on the offspring and two as neonatal anaemias. All these syndromes involve disorders in metal metabolism or transport.

The most extensively studied disorder is caused by the X-linked *mottled* mutations (Hunt, 1974). There are seven alleles at this locus (Searle, 1979) affecting coat colour and viability; they range from *blotchy* (Mo^{blo}) which dilutes the coat of males a little and is viable and fertile, to *mottled* (*Mo*) which is lethal *in utero*. *Brindled* mice (Mo^{br}/Y) which die at about 14 days, have been examined in detail (Hunt, 1974).

Brindled mice around 8–13 days appear to have widespread disorders in brain amino acid and amine metabolism (Hunt & Johnson, 1972a, b). They have reduced noradrenalin, increased brain tyrosine (Hunt & Johnson, 1972a) and a decreased *in vivo* flux (to 13%) from dopamine to noradrenaline (Hunt & Johnson, 1972b). This low flux to noradrenaline is probably the result of low *in vivo* activity of dopamine β-hydroxylase and might explain the paralysis and neurological symptoms present in these animals prior to death (Hunt & Johnson, 1972b).

In vitro assay of dopamine beta hydroxylase in *brindled* brains gave elevated activities. The assay did however contain enough cupric ions to activate the enzyme. This difference between the *in vivo* and *in vitro* measurement of enzyme activity led Hunt (1974) to measure copper levels in the mutant. He found that the copper content of brain, liver and ceruloplasmin levels were reduced but intestinal levels were increased two-fold. The results indicate a primary defect in copper transport reminiscent of the human X-linked copper deficient – Menkes kinky hair disease (Hunt, 1974).

It has been possible to overcome the lethality and pigment deficiency of *brindled* mice (Mo^{br}/Y) by injections of copper chloride. The low copper levels of brain and liver remain unchanged after this treatment but the elevated kidney levels (three- to five-fold) are increased a further two-fold. The synthesis of brain noradrenaline does, however, return to normal and the animals survive, albeit with a small (one-third of normal) body weight. The viable *blotchy* mutant (Mo^{blo}/Y) was mildly affected in all the parameters measured (Hunt, 1976; Evans & Reis, 1978; Starcher *et al.*, 1978).

The deficiency of copper in the liver and the accumulation in the kidney of *brindled* mice is associated with four protein fractions of about 14 000 molecular weight (Hunt & Port, 1979). The kidney and liver proteins appear identical making it unlikely that they are the *Mo*-gene products (Port & Hunt, 1979). By 21–30 days of age the gut copper content of *viable-brindled* mice has returned to normal and liver levels in normal mice decline to those of mutant mice. The syndrome may therefore be related to the inability of neonatal mutant mice to release copper from the kidney and the gut (Hunt & Port, 1979; Port & Hunt, 1979). This indicates that successful treatment of Menkes disease in humans (where a variety of copper treatments have failed) will have to bypass the gut and perhaps overcome a lack of cellular uptake (Hunt, 1976).

The *pallid* mutation in the mouse causes a reduction in pigmentation and affects otolith formation in the inner ear. This leads to behavioural abnormalities such as head tossing (M. F. Lyon, 1954). As manganese deficiency causes a phenocopy of these symptoms, Erway, Fraser & Hurley (1971) supplemented the diet of *pallid* mice with manganese and prevented the otolith defect. It was later shown that transport of manganese, dopa and tryptophan was slower through the tissues of *pallid* mice (Cotzias *et al.*, 1972).

Two syndromes were discovered because they were associated with a severe maternal effect: *lethal milk* and *toxic milk*. In both cases trace element metabolism is involved.

In the *lethal milk* syndrome, mortality, morbidity and hair loss of suckling pups are due to a reduction in the level of zinc in the milk (Piletz & Ganschow, 1978). These symptoms can be largely overcome by supplementing the diet of lactating dams with zinc. Older animals show dermatitis, bronzing and loss of the hair, ataxia, and otolith abnormalities. This damage occurs *in utero* as it is not affected by prolonged zinc supplementation or fostering (Erway, Ganschow & Piletz, 1979). It has been suggested that the adult syndrome involves manganese, zinc and copper metabolism and this is being investigated (L. Erway, personal communcation).

The *toxic milk* syndrome displays a similar phenotype to *lethal milk*. Offspring of affected females have severe tremors and ataxia and die at 10 to 14 days; fostering alleviates the symptoms (Rauch, 1977). Mutant females produce four times as much tin as normal females and have reduced copper levels in their milk. Consequently at five days of age their pups have 10- to 15-fold more tin in their bodies than normal animals. Neonates also show a severe reduction in the copper levels of their liver and other tissues. Conversely adult animals have increased tissue copper levels and a blond pelage. This excessive storage of copper suggests a syndrome comparable to Wilson's disease (McKusick, 1975) in man and this possibility is being investigated (H. Rauch, personal communication).

There are two inherited anaemias in the mouse associated with erythrocyte iron deficiency. *Microcytic anaemia* presents as a generalised impairment of cellular iron uptake (E. S. Russell *et al.*, 1970; Edwards & Hoke, 1975a) possibly caused by a defect in red cell transferring receptor sites (Edwards & Hoke, 1975a). Similarly *sex-linked anaemia* (Bannerman & Copper, 1966) is also a defect in iron uptake although it can be temporarily reversed by a low iron diet (Sorbie & Valberg, 1971; Edwards & Hoke, 1975b; Valberg, Sorbie & Ludwig, 1977). The nature and site of the primary lesion in these two mutants remain to be elucidated.

Transport Disorders

There are three mouse syndromes which appear to affect the renal tubular transport of taurine, proline and phosphate respectively (Scriver, Chesney & McInnes, 1976).

The difference in urine taurine levels between the C57BL (high levels) and A (low levels) strains was quantified by Harris & Searle (1953) using paper chromatography. Plasma levels of taurine were identical in the strains, implying that the urine differences may be attributable to alterations in renal tubular reabsorption (Harris & Searle, 1953). Further investigations (Scriver, Chesney *et al.*, 1976) showed that although the intracellular taurine concentration in the outer cortex of the kidney of the two strains is similar *in vivo*, the net tubular reabsorption of the filtered load was 97% in C57BL and 84% in A. Steady-state uptake of taurine by thin outer cortex slices, however, is greater in C57BL than A, but this is not the result of altered efflux from the basilar membrane slice. These *in vivo* and *in vitro* data have been interpreted (Scriver, Chesney *et al.*, 1976) as indicating that the uptake of taurine is blocked in the luminal membrane of the proximal tubule. This is the first topological assignment of a hereditary transport defect to a specific nephron membrane site.

Prolinaemia in the mouse has been discussed above (Blake & Russell, 1972; Blake, 1972). The elevated proline levels of this syndrome are unequally distributed between the tissues and the urine. Whereas plasma proline levels in prolinaemic animals are eight-fold higher than those of normal mice and kidney cortex levels are four-fold higher, the urine proline levels may be increased more than 50-fold (Scriver, McInnes & Mohyuddin, 1975). Proline oxidase activity is reduced to 1% of normal oxidation in the kidney as well as in other tissues. Therefore the excessive prolinuria of these animals is a simple consequence of the increase in intracellular proline concentrations in the kidney, whilst the membrane transport systems are unimpaired (Scriver, McInnes *et al.*, 1975). These results indicate that intracellular metabolism of a substrate is independent of membrane transport but can still affect its rate of tubular absorption. This phenomenon in the mouse led to reanalysis of the situation with human hyperprolinaemia, indicating that the low kidney proline oxidase levels can also affect proline reabsorption in man (Scriver, Chesney *et al.*, 1976).

The third mouse syndrome with altered renal transport is *hypophosphataemia* (Eicher, Southard *et al.*, 1976). This mutant has a small body size with diminished bone ash and symptoms resembling rickets. It can be partially cured by the provision of phosphate-supplemented drinking water (Eicher, Southard *et al.*, 1976). Renal tubular reclamation of orthophosphate is reduced in these animals, although the uptake of phosphate by cortical and medullary slices and the intracellular concentration of phosphate appear normal. It has been suggested that the mutation permits excessive luminal efflux of cytoplasmic phosphate ion, and this accounts for the negative reabsorption characteristic of the syndrome (Eicher, Southard *et al.*, 1976; Scriver, Chesney *et al.*, 1976). An intrinsic defect has been found in brush border membrane vesicles isolated from hypophosphataemic mice. This is the first report of the location of gene product affecting transport to a specific position in renal epithelial cells (Tenenhouse & Scriver, 1978; Tenenhouse *et al.*, 1978). There is also a reduction in parathyroid hormone stimulation of adenyl cyclase in the proximal tubules and calcitonin stimulation of the enzyme in distal tubules, although these effects are probably secondary to the defect in the brush border (Brunette *et al.*, 1979).

A disorder of intestinal phosphate transport has also been found in hypophosphataemic mice (O'Doherty, De Luca & Eicher, 1976) but any identity between the two transport lesions has not yet been investigated. Finally, the mouse syndrome appears, so far, to be identical to the classic human vitamin D resistant rickets, and therefore provides an excellent animal model.

CONCLUSIONS

The usefulness of animal models of human inborn errors of metabolism depends on the same enzyme, or other gene product, being deficient as in the human syndrome. Most of the criticisms of their use disappear if this criterion is adhered to.

Many mouse mutants have been discovered by chance, and although there are some screening programmes in operation, these are not on a large scale. An homologous animal model would be of immense value in investigating novel therapeutic methods of intractable syndromes such as lysosomal storage diseases. Larger and more co-ordinated screening programmes must therefore be established, especially as the extensive genetical variation in feral mice populations has only recently begun to be exploited.

Research work on existing mouse syndromes has produced a surprisingly large amount of information on new phenomena and led to new concepts that otherwise would not have been revealed. This is especially true of the control of gene expression, experimental therapy, trace element metabolism and renal transport mechanisms. There are large areas of biochemistry and physiology, especially the study of *in vivo* regulation, that could be significantly advanced with further use of appropriate mutants.

ACKNOWLEDGEMENTS

I should like to thank S. A. Martin for her comments on the manuscript of this review. The work from the author's laboratory is supported by an MRC project grant.

REFERENCES

Altman, P. L. & Katz., D. (Eds). (1979). *Inbred and genetically defined strains of laboratory animals.* 1. *Mouse and rat.* Bethesda: Fed. Am. Soc. Exp. Biol.

Bannerman, R. M. & Copper, R. G. (1966). Sex-linked anemia: a hypochromic anemia of mice. *Science, N.Y.* **151**: 581–583.

Bergsma, D. (Ed.) (1973). *Enzyme therapy in genetic disease.* Birth Defects. (Original Article Series, 9 (2.). (National Foundation – March of Dimes.) Baltimore: Williams and Wilkins.

Blake, R. L. (1972). Animal model for hyperprolinaemia: deficiency of mouse proline oxidase activity. *Biochem. J.* **129**: 987–989.

Blake, R. L. & Russell, E. S. (1972). Hyperprolinaemia and prolinuria in a new inbred strain of mice, PRO/Re. *Science, N.Y.* **176**: 809–811.

Breen, G. A. M., Lusis, A. J. & Paigen, K. (1977). Linkage of genetic determinants for mouse β-galactosidase electrophoresis and activity. *Genetics* **85**: 73–84.

Brinkhous, K. (1978). Animal models: importance in research on haemorrhage and thrombosis. *Adv. exp. Med.* **102**: 123–133.

Britton, J. & Thaler, L. (1978). Evidence for the presence of two sympatric species of mice (Genus *Mus* L.) in southern France based on biochemical genetics. *Biochem. Genet.* **16**: 213–225.

Brunette, M. G., Chabardes, D., Imbert-Teboul, M., Clique, A., Motegut, M. & Morel, F. (1979). Hormone-sensitive adenylate cyclase along the nephron of genetically hypophosphataemic mice. *Kidney Int.* **15**: 357–369.

Bulfield, G. (1977). Nutrition and animal models of inherited metabolic disease. *Proc. Nutr. Soc.* **36**: 61–67.

Bulfield, G. & Kacser, H. (1974). Histidinaemia in man and mouse. *Archs Dis. Child* **49**: 545–552.

Bulfield, G. & Kacser, H. (1975). Histamine and histidine levels in the brain of the histidinaemic mutant mouse. *J. Neurochem.* **24**: 403–405.

Bulfield, G. & Moore, E. A. (1974). Semi-automated assays for enzymopathies of carbohydrate metabolism in liver and erythrocytes using a reaction rate analyser. *Clin. Chim. Acta* **53**: 265–271.

Bulfield, G. & Moore, K. (1978). Screening and study of inherited enzyme deficiencies of erythrocyte glycolysis and associated pathways in wild mice. *Hereditas* **89**: 140.

Bulfield, G., Moore, E. A. & Kacser, H. (1978). Genetic variation in activity of the enzymes of glycolysis and gluconeogenesis between inbred strains of mice. *Genetics* **89**: 551–556.

Bulfield, G. & Nahum, A. (1978). Effect of the mouse mutants testicular feminization and sex reversal on hormone-mediated induction and repression of enzymes. *Biochem. Genet.* **16**: 743–750.

Bulfield, G. & Trent, J. (1981). Genetic variation in erythrocyte NAD levels in the mouse and its effect on glyceraldehydephosphate dehydrogenase activity and stability. *Biochem. Genet.* **19**: 87–93.

Bullock, L. P. & Bardin, C. W. (1974). Androgen receptors in mouse kidney: a study of male and female and androgen-insensitive (*tfm*/Y) mice. *Endocrinology* **94**: 746–756.

Bustad, L. K., Hegreberg, G. A. & Padgett, G. A. (1975). *Naturally occurring animal models of human disease.* Washington D. C.: Institute of Laboratory Animal Resources, National Academy of Sciences.

Cohen, P. T. W., Burchell, A. & Cohen, P. (1976). The molecular basis of skeletal muscle phosphorylase kinase deficiency. *Eur. J. Biochem.* **66**: 347–356.

Cohen, P. T. W. & Cohen, P. (1973). Skeletal muscle phosphorylase kinase deficiency: detection of a protein lacking any activity in ICR/IAn mice. *FEBS Lett.* **29**: 113–116.

Coleman, D. L. (1971). Linkage of genes controlling the rate of synthesis and structure of aminoevulinate dehydratase. *Science, N.Y.* **173**: 1245–1246.

Cornelius, C. E. (1969). Animal models – a neglected medical resource. *N. Engl. J. Med.* **281**: 934–944.

Cornelius, C. E. (1978). Animal models: whose responsibility. *J. Lab. Clin. Med.* **91**: 187–190.

Cotzias, G. C., Tang, L. C., Miller, S. T. & Sladic-Simic, D. (1972). A mutation influencing the transportation of manganese, L-dopamine and L-tryptophan. *Science, N.Y.* **176**: 410–412.

De Mars, R., Le Van, S. L., Trend, B. L. & Russell, L. B. (1976). Abnormal ornithine carbamoyltransferase in mice having the sparse-fur mutation. *Proc. natn. Acad. Sci. U.S.A.* **73**: 1693–1697.

Denniston, J. C. (1963). Children of mothers with phenylketonuria. *J. Pediat.* **63**: 461–464.

Dofuku, R., Tettenborn, U. & Ohno, S. (1971). Testosterone- "regulon" in the mouse kidney. *Nature, New Biol.* **232**: 5–7.

Edwards, J. A. & Hoke, J. E. (1975a). Red cell iron uptake in hereditary microcytic anaemia. *Blood* **46**: 381–388.

Edwards, J. A. & Hoke, J. E. (1975b). Effect of dietary iron manipulation and phenobarbitone treatment on *in vivo* intestinal absorption of iron in mice with sex-linked anemia. *Am. J. Clin. Nutr.* **28**: 140–149.

Eicher, E., Lewis, S. E. Turchin, H. A. & Gluecksohn-Waelsch, S. (1978). Absence of mitochondrial malic enzyme in mice carrying two complementing lethal albino alleles. *Genet. Res.* **32**: 1–7.

Eicher, E., Southard, J. L., Scriver, C. R. & Glorieux, F. H. (1976). Hypophosphatemia: mouse model for human familiar hypophosphatemic (vitamin-D resistant) rickets. *Proc. natn. Acad. Sci. U.S.A.* **73**: 4667–4671.

Erikson, R. P. (1974). Erythrocyte nicotinamide-adenine dinucleotide phosphate levels and the genetic regulation of erythrocyte glucose-6-phosphate dehydrogenase activity in the inbred mouse. *Biochem. Genet.* **11**: 33–42.

Erway, L. C., Fraser, A. S. & Hurley, L. S. (1971). Prevention of congenital otolith defect in pallid mutant mice by manganese supplementation. *Genetics* **67**: 97–108.

Erway, L. C., Ganschow, R. E. & Piletz, J. E. (1979). Personal communication. *Mouse News Letter* **60**: 43.

Evans, G. W. & Reis, B. L. (1978). Impaired copper homeostasis in neonatal male and adult female brindled (Mo^{br}) mice. *J. Nutr.* **108**: 554–560.

Falconer, D. S. (1960). *Introduction to quantitative genetics.* Edinburgh and London: Oliver and Boyd.
Feinstein, R. N. (1970). Acatalasemia in the mouse and other species. *Biochem. Genet.* 4: 135–155.
Feinstein, R. N. Howard, J. B., Braun, J. T. & Seaholm, J. E. (1966). Acatalasemic and hypocatalasemic mouse mutants. *Genetics* 53: 923–933.
Feinstein, R. N., Seaholm, J. E., Howard, J. B. & Russell, W. L. (1964). Acatalasemic mice. *Proc. natn. Acad. Sci. U.S.A.* 52: 661–662.
Festing, M. F. W. (1979). *Inbred strains in biomedical research.* London: Macmillan.
Ganschow, R. E. & Schimke, R. T. (1969). Independent genetic control of the catalytic activity and the rate of degradation of catalase in mice. *J. biol. Chem.* 244: 4649–4658.
Garland, R. C., Satrustegui, J., Gluecksohn-Waelsch, S. & Cori, C. F. (1976). Deficiency in plasma protein synthesis caused by X-ray-induced lethal albino animals in the mouse. *Proc. natn. Acad. Sci. U.S.A.* 73: 3376–3380.
Gehring, U., Tomkins, G. M. & Ohno, S. (1971). Effect of the androgen-insensitive mutation on a cytoplasmic receptor for dihydrotestosterone. *Nature, New Biol.* 232: 106–107.
Gerritsen, T. & Siegel, F. L. (1972). The use of animal models for the study of amino acidopathies. (Proc. 3rd int. Congr. Neuro-Genetics and Neuro-Ophthalmology.) *Monogr. human Genet.* 6: 22–36.
Glass, R. D. & Doyle, D. (1972). Genetic control of lactic dehydrogenase expression in mammalian tissues. *Science, N.Y.* 176: 180–181.
Gluecksohn-Waelsch, S. (1979). Genetic control of morphogenetic and biochemical differentiation: lethal albino deletions in the mouse. *Cell* 16: 225–237.
Gluecksohn-Waelsch, S. & Cori, C. F. (1970). Glucose-6-phosphatase deficiency: mechanisms of genetic control and biochemistry. *Biochem. Genet.* 4: 195–201.
Gluecksohn-Waelsch, S., Schiffman, M. B., Thorndike, J. & Cori, C. F. (1974). Complementation studies of lethal alleles in the mouse causing deficiencies of glucose-6-phosphatase, tyrosine aminotransferase and serine dehydratase. *Proc. natn. Acad. Sci. U.S.A.* 71: 825–829.
Gross, S. R., Longshore, M. A. & Pangburn, S. (1975). The phosphorylase kinase deficiency (*Phk*) in the mouse: evidence that the mutant allele codes for an enzyme with an abnormal structure. *Biochem. Genet.* 13: 567–584.
Gross, S. R. & Mayer, S. E. (1974). Characterization of the phosphorylase b to a converting activity in skeletal muscle extracts of mice with the phosphorylase b kinase deficiency mutation. *J. biol. Chem.* 249: 6710–7618.
Harper, P. S. (1975). Interactions of fetal and maternal genotype in human mendelian disease. *Heredity* 34: 289.
Harris, H. & Searle, A. G. (1953). Urinary amino-acids in mice of different genotypes. *Ann. Eugen.* 17: 165–167.
Holmes, E. W. & O'Brien, J. S. (1978a). Feline G_{M1} gangliosidosis: characterization of the residual acid β-galactosidase. *Am. J. hum. Genet.* 30: 505–515.
Holmes, E. W. & O'Brien, J. S. (1978b). Hepatic storage of oligosaccharides and glycolipids in a cat affected with G_{M1} gangliosidosis. *Biochem. J.* 175: 945–953.

Hommes, F. A. (Ed.) (1979). *Models for the study of inborn errors of metabolism.* Amsterdam: North Holland Biomedical Press.
Huijing, F. (1970). Phosphorylase kinase deficiency in mice. *FEBS Lett.* **10**: 328–332.
Hunt, D. M. (1974). Primary defect in copper transport underlies mottled mutants in the mouse. *Nature, Lond.* **249**: 952–854.
Hunt, D. M. (1976). A study of copper treatment and tissue copper levels in the murine congenital copper deficiency mottled. *Life Sci.* **19**: 1913–1920.
Hunt, D. M. & Johnson, D. R. (1972a). Aromatic amino acid metabolism in brindled (Mo^{br}) and viable-brindled (Mo^{vbr}), two alleles at the mottled locus in the mouse. *Biochem. Genet.* **6**: 31–40.
Hunt, D. M. & Johnson, D. R. (1972b). An inherited deficiency in noradrenaline biosynthesis in the brindled mouse. *J. Neurochem.* **19**: 2811–2819.
Hunt, D. M. & Port, A. E. (1979). Trace element binding in the copper deficient mottled mutants in the mouse. *Life Sci.* **24**: 1453–1466.
Hunt, W. G. & Selander, R. K. (1973). Biochemical genetics of hybridisation in European house mice. *Heredity* **31**: 11–33.
Hutton, J. J. & Bernstein, S. E. (1973). Metabolic properties of erythrocytes of normal and genetically anemic mice. *Biochem. Genet.* **10**: 297–307.
I.L.A.R. (1971–1976). *Selected abstracts on animal models for biomedical research,* I–IV. Washington D.C.: Institute of Laboratory Animal Resources, Natn. Acad. Sci.
I.L.A.R. News (1979). Washington D.C.: Institute of Laboratory Animal Resources, Natn. Acad. Sci.
Jones, T. C. Hackel, D. B. & Migaki, G. (1972–1977). *Handbook. Animal models of human disease.* First to Sixth Fascicles. Registry of Comparative Pathology. Washington D.C.: Armed Forces Institute of Pathology.
Kacser, H., Bulfield, G. & Wallace, M. E. (1973). Histidinaemic mutant in the mouse. *Nature, Lond.* **244**: 77–79.
Kacser, H., Bulfield, G. & Wright, A. (1979). The biochemistry and genetics of histidinaemia in mice. In *Models for the study of inborn errors of metabolism:* 33–42. Hommes, F. A. (Ed.). Amsterdam: North Holland.
Kacser, H., Mya Mya, K., Duncker, M., Wright, A. F., Bulfield, G., McLaran, A. & Lyon, M. F. (1977). Maternal histidine metabolism and its effect on foetal development in the mouse. *Nature, Lond.* **265**: 262–266.
Kacser, H., Mya Mya, K. & Bulfield, G., (1979). Endogenous teratogenesis in maternal histidinaemia. In *Models for the study of inborn errors of metabolism.* Hommes, F. A. (Ed.). Amsterdam: North Holland.
Klein, J. (1975). *Biology of the mouse histocompatability-2 complex: principles of immunogenetics applied to a single system.* Berlin: Springer-Verlag.
Lusis, A. J. & Paigen, K. (1977). Relationships between levels of membrane-bound glucuronidase and the associated protein egasyn in mouse tissues. *J. Cell Biol.* **73**: 728–735.
Lyon, I. C. T., Gardner, R. J. M. & Veale, A. M. O. (1974). Human maternal histidinaemia. *Archs Dis. Child* **49**: 581–583.
Lyon, J. B. (1970). The X-chromosome and enzymes controlling muscle glycogen: phosphorylase kinase. *Biochem. Genet.* **4**: 169–185.
Lyon, J. B., Porter, J. & Robertson, M. (1967). Phosphorylase b kinase inheritance in mice. *Science, N.Y.* **155**: 1550–1551.
Lyon, M. F. (1954). Stage of action of the litter-size effect on absence of otoliths in the mouse. *Z. Ind. Abst. Verebl.* **86**: 289–292.
Lyon, M. F. (1961). Gene action in the X-chromosome of the mouse (*Mus musculus* L.). *Nature, Lond.* **190**: 372–373.

Lyon, M. F. (1971). Possible mechanisms of X-chromosome inactivation. *Nature, New Biology* **232**: 229–232.

Lyon, M. F. & Hawkes, S. G. (1970). X-linked gene for testicular feminization in the mouse. *Nature, Lond.* **227**: 1217–1218.

McKusick, V. A. (1975). *Mendelian inheritance in man.* 4th ed. Baltimore: Johns Hopkins.

Morse, H. C. (Ed.) (1978). *Origins of inbred mice.* New York: Academic Press.

O'Doherty, P. J. A., De Luca, H. F. & Eicher, E. M. (1976). Intestinal calcium and phosphate transport in genetic hypophosphatemic mice. *Biochem. Biophys. Res. Comm.* **71**: 617–621.

Padua, R. A., Bulfield, G. & Peters, J. (1978). Biochemical genetics of a new glucosephosphate isomerase (*Gpi-1c*) from wild mice. *Biochem. Genet.* **16**: 127–143.

Paigen, K. (1970). The genetics of enzyme realisation. In *Enzyme synthesis and degradation in mammalian systems:* Rechigl, M. (Ed.). Basel: Karger.

Paigen, K. (1978). Genetic control of enzyme activity. In *Origins of inbred mice:* 255–280. Morse, H. C. (Ed.). New York: Academic Press.

Paigen, K., Swank, R. T., Tomino, S. & Ganschow, R. E. (1975). The molecular genetics of mammalian glucuronidase. *J. cell. Physiol.* **85**: 379–392.

Piletz, J. E. & Ganschow, R. E. (1978). Zinc deficiency in murine milk underlies expression of the *lethal milk* (*lm*) mutation. *Science, N.Y.* **199**: 182–183.

Pivetta, O. H. & Green, E. L. (1973). Exocrine pancreatic insufficiency: a new recessive mutation in the mouse. *J. Hered.* **64**: 301–302.

Port, A. E. & Hunt, D. M. (1979). A study of copper binding proteins in liver and kidney tissues of neonatal normal and mottled mutant mice. *Biochem. J.* **183**: 721–730.

Popp, R. A., Stratton, L. P., Hawley, D. K. & Effron, K. (1979). Hemoglobin of mice with radiation-induced mutations at the haemoglobin loci. *J. mol. Biol.* **127**: 141–148.

Prichard, R. W. (1978). The need for new animal models – a philosophic approach. *J. Am. Vet. med. Assoc.* **173**: 1208–1209.

Rauch, H. (1977). Personal communication. *Mouse News Letter* **56**: 48.

Reynolds, G. D. Baker, B. J. & Reynolds, R. H. (1978). Enzyme replacement using liposome carriers in feline G_{M1} gangliosidosis fibroblasts. *Nature, Lond.* **275**: 754–755.

Roderick, T. H., Ruddle, R. H., Chapman, V. M. & Shows, T. B. (1971). Biochemical polymorphism in feral and inbred mice (*Mus musculus*). *Biochem. Genet.* **5**: 457–466.

Rugh, R. (1964). Why radiobiology? *Radiology* **82**: 917–920.

Russell, E. S., Nash, D. J., Bernstein, S. E., Kent, E. L., McFarland, E. C., Matthews, S. M. & Norwood, M. S. (1970). Characterization and genetic studies of microcytic anaemia in the house mouse. *Blood* **35**: 838–850.

Russell, L. B., Russell, W. L., Popp, R. A., Vaughan, C. & Jacobson, K. B. (1976). Radiation-induced mutations at mouse haemoglobin loci. *Proc. natn. Acad. Sci. U.S.A.* **73**: 2843–2846.

Sage, R. D. (1978). Genetic heterogeneity of Spanish house mice (*Mus musculus* complex) In *Origins of inbred mice:* 519–553. Morse, H. C. (Ed.). New York: Academic Press.

Schwartz, A. (1978). Animal models of human disease, preface for the series. *Yale J. Biol. Med.* **51**: 191–192.

Scriver, C. R., Chesney, R. W. & McInnes, R. R. (1976). Genetic aspects of renal

tubular transport: diversity and topology of carriers. *Kidney Int.* **9**: 147–171.
Scriver, C. R., McInnes, R. R. & Mohyuddin, F. (1975). Role of epithelial architecture and intracellular metabolism in proline uptake and trans-tubular reclamation in PRO/Re mouse kidney *Proc. natn. Acad. Sci., U.S.A.* **72**: 1431–1435.
Searle, A. (1979). Editor. *Mouse News Letter* **61**.
Sheppard, J. R., Albersheim, P. & McClearn, G. (1970). Aldehyde dehydrogenase and ethanol preference in mice. *J. biol. Chem.* **245**: 2876–2882.
Soares, E. R. & Malling, H. V. (1977). Personal communication. *Mouse News Letter* **57**: 33.
Sorbie, J. & Valberg, L. S. (1971). Reversibility of the defect in intestinal iron transport in mice with sex-linked anaemia (Abstract). *Clin. Res.* **19**: 780.
Starcher, B., Madaras, J. A., Fisk, D., Perry, E. F. & Hill, C. H. (1978). Abnormal cellular copper metabolism in the blotchy mouse. *J. Nutr.* **108**: 1229–1233.
Stephens, M. C., Bernatsky, A., Legler, G. & Kaufer, J. N. (1979). The Gaucher mouse: additional biochemical alterations. *J. Neurochem.* **32**: 969–972.
Sutton, H. E. & Wagner, R. P. (1975). Mutation and enzyme function in humans. *A. Rev. Genet.* **9**: 187–212.
Swank, R. T. & Paigen, K. (1973). Biochemical and genetic evidence for a macro molecular β-glucuronidase complex in microsomal membranes. *J. molec. Biol.* **77**: 371–389.
Swank, R. T., Paigen, K., Davey, R., Chapman, V., Labarca, C., Watson, G., Ganschow, R., Brandt, E. J. & Novak, E. (1978). Genetic regulation of mammalian glucuronidase. In *Recent progress in hormone research*: 401–436. Greep, R. O. (Ed.). New York and London: Academic Press.
Swank, R. T., Paigen, K. & Ganschow, R. E. (1973). Genetic control of glucuronidase induction in mice. *J. molec. Biol.* **81**: 225–243.
Tenenhouse, H. S. & Scriver, C. R. (1978). The defect in transcellular transport of phosphate in the nephron is located in brush-border membranes in X-linked hypophosphataemia (*Hyp* mouse model). *Can. J. Biochem.* **56**: 640–646.
Tenenhouse, H. S., Scriver, C. R., McInnes, R. R. & Glorieux, F. H. (1978). Renal handling of phosphate *in vivo* and *in vitro* by the X-linked hypophosphataemic male mouse: evidence for a defect in the brush border membrane. *Kidney Int.* **14**: 236–244.
Thaler, M. M., Erickson, R. P. & Pelger, A. (1976). Genetically determined abnormalities of microsomal enzymes in liver of mutant newborn mice. *Biochem. Biophys. Res. Comm.* **72**: 1244–1250.
Valberg, L. S., Sorbie, J. & Ludwig, J. (1977). Mucosal iron-binding proteins in mice with X-linked anaemia. *Br. J. Haemat.* **35**: 321–330.
Valentine, W. N. (1968). Hereditary hemolytic anaemias associated with specific erythrocyte enzymopathies. *Calif. Med.* **108**: 280–294.
Wolff, G. L. (1977). Personal communication. *Mouse News Letter* **57**: 32.
Womack, J. & Potier, M. (1979). Personal communication. *Mouse News Letter* **61**: 64.

Author Index

Numbers in italics refer to pages in the References at the end of each article

C

E

H

I

J

M

N

O

P

S

Subject Index

B

C

Q

R

S